The Almanac of Science and Technology

What's New and What's Known

The Almanac of Science and Technology

What's New and What's Known

Prepared by
World Information Systems
Edited by
Richard Golob and Eric Brus
World Information Systems

Harcourt Brace Jovanovich, Publishers
Boston San Diego New York

Printed in the United States of America

Library of Congress Cataloging-in-Publication Data
The Almanac of Science and Technology: What's New and What's Known
Edited by Richard Golob and Eric Brus — 1st ed.
p. cm.
Includes index.
ISBN 0-15-105050-3 (alk. paper) — ISBN 0-15-600049-0 (pbk. : alk. paper)
1. Science. 2. Technology. I. Golob, Richard. II. Brus, Eric.
Q158.5.A47 1990
500—dc20
89-34650
CIP

To my mother and father, who opened the door.

You knew that I would finish this voyage! Your love and encouragement helped guide me through the storms and rough waters.

—Richard S. Golob

Contents

Chapter Opening Figure Captions

Astronomy: **Supernova 1987A.** The first nearby stellar explosion, or supernova, in over 300 years was detected on 24 February 1987 in the Large Magellanic Cloud, a satellite galaxy of the Milky Way. In a very short time, the brightness of the exploding star increased many billion-fold — and the star's brightness rivaled that of the entire galaxy. This wide-angle photograph was taken on 25 February by a telescope at the National Optical Astronomy Observatories facility in Cerro Tololo, Chile. *Courtesy: National Optical Astronomy Observatories.*

Biology: **Father of Modern Evolutionary Theory.** This portrait of Charles Darwin was completed by artist Robert Collier in 1883, the year after the English naturalist died at the age of 73. Darwin is best known for his theory of evolution by natural selection, described in his landmark 1859 book *On the Origin of Species*. Today, about 130 years later, fundamentalist religious groups continue to dispute Darwin's evolutionary theory, despite its widespread acceptance by scientists. *Courtesy: National Portrait Gallery, London.*

Brain and Behavior: **Seat of Intelligence.** Although it weighs only about 3 pounds, the human brain contains about 10 billion to 20 billion nerve cells, or neurons, that regulate essential body functions, organize and interpret information from the senses, and transform all the data received into the set of perceptions, ideas, and emotions that make each person unique. *Courtesy: M. Christine de Lacoste, University of Texas Health Science Center.*

Chemistry: **Individual Atoms in a Rhodium Catalyst.** An instrument called a field ion microscope was used to study catalytic reactions that occur at the surface of a sharply pointed needle made of rhodium. The concentric rings in the image are the edges of layers composed of individual atoms of the element. Rhodium is used in automobile catalytic converters to catalyze the reaction of carbon monoxide with oxygen to form carbon dioxide. *Courtesy: Gary Kellogg, Sandia National Laboratories.*

Computers: **From ENIAC to the Laptop.** The ENIAC computer (top), developed in 1946, occupied an entire room, with its 18,000 vacuum tubes and massive banks of switches. In contrast, the Zenith TurbosPort 386 (bottom), a state-of-the-art laptop computer in 1989, weighs a scant 15 pounds. *Courtesy: The Bettmann Archive, Inc. (top), and Zenith Data Systems Corp. (bottom).*

Earth Sciences: **World's Major Plates.** The Earth's crust is divided into segments called plates that are slowly moved by currents deep within the planet. Most earthquakes occur at plate boundaries: mid-ocean ridges (double lines), where plates are spreading apart and new crust is created; subduction zones (barbed lines), where one plate slides beneath another; and transform faults (single lines), where plates slide past one another. *Courtesy: USGS.*

Environment: **Plastic Pollution.** Floating plastic debris is becoming an increasingly serious threat to marine wildlife — in the open ocean as well as in coastal waters. Gulls and other waterfowl often choke on plastic pellets; marine turtles ingest plastic bags, which block their intestines; and seals become entangled in fishing nets. This Hawaiian Monk Seal, unable to dislodge a plastic cylinder from his muzzle, starved to death. *Courtesy: Center for Marine Conservation. Photo: Dorris Alcorn, National Marine Fisheries Service.*

Medicine: **Human Immunodeficiency Virus.** The human immunodeficiency virus (HIV), which causes AIDS, typically infects white blood cells called lymphocytes. The virus replicates inside lymphocytes, eventually destroying the host cells. The electron micrograph on the left shows a lymphocyte that has been almost entirely destroyed but continues to host numerous viral particles (the small, dark objects). The electron micrograph series on the right, taken at a much higher magnification, shows a virus particle forming under a lymphocyte cell membrane (top), breaking away from the lymphocyte (middle), and finally becoming a free extracellular viral particle (bottom) containing the nucleus-like structure that distinguishes HIV from related viruses that do not cause AIDS. *Courtesy: S. Zaki Salahuddin (photo on the left) and Matthew Gonda (series of photos on the right), National Cancer Institute.*

Physics: **Particle–Antiparticle Pair.** An electron and its oppositely charged antiparticle, a positron, spiral to the left and to the right, respectively, in this helium bubble chamber photograph. The particle pair was created by the decay of a gamma ray, about halfway up the picture field; the gamma ray itself leaves no track in the bubble chamber. Each particle follows a spiral path as it loses energy through collisions with helium atoms and is deflected in the bubble chamber's magnetic field. *Courtesy: Brookhaven National Laboratories.*

Acknowledgments

This book has been a very long time in the making. Since science and technology are constantly changing, we have been focusing on a moving target, and with each new discovery or development, we have revised the text to make it current. We are grateful to all those who patiently helped with each incarnation of the project and especially to those who stayed with the book until its completion.

First and foremost, we thank Dorothy and Meyer Golob for their support, encouragement, and enthusiasm throughout the project. Although they both passed away before we completed the book, their spirit sustained and guided us to the finish.

Nan White deserves special commendation for her work as photo researcher, chief fact-gatherer, and proofreader. While sacrificing many evenings and weekends for the book, Nan diligently pursued the details, always challenging us to rethink our interpretation of the facts.

Steve Bennett, who joined our team in the last phase of the project, gave us the final burst of inspiration to turn the manuscript into printed pages. Steve coordinated the final revisions and served as our production liaison with Harcourt Brace Jovanovich. Thanks for helping us across the finish line, Steve.

We are also indebted to Lisa Jasak for devoting many, many hours to this project beyond the call of duty. Lisa deciphered the comments of our authors and reviewers, producing accurate and clean copy for each article in the book. Remarkably, she helped us complete the project without missing a beat in her duties as office manager at World Information Systems.

We also thank Ward and Margaret Fearnside who were instrumental in making this book a reality. Their constant encouragement and generous support throughout this project helped us through the critical times and to the finish. Life is good, Ward. In addition, we would like to thank the following people whose faith in us and the project made it possible for us to persevere and reach our goal: Pearl and Leonard Barron, Neil Beneck, Richard Berner, Ronald Bernstein, Lynne and Daniel Gelfman, Bernard Goldhirsh, Jonathan Guttmacher, BeeBee Horowitz Kendall, James Koehlinger, Mark Shwayder, Neil Westreich, Benjamin Williams, and Grant and Jonathan Winthrop. A special thank-you goes to Jim Kobak who provided support and thoughtful advice throughout this project. The patience and encouragement of John Lesanto also helped make it possible for us to complete the project.

Throughout the voyage, we have appreciated the insights and assistance of our attorney, Robert Licht, and our literary agent, Julian Bach. Although Robert Johnson, who preceded Robert Licht as our attorney, passed away before we completed the book, his generosity with time in the early stages and his confidence in us made it possible to begin this book. We also appreciate the longtime friendship and help of Helen Haralampu, who saw this project through its numerous revisions.

In addition, we would like to express our gratitude to several people for their efforts at various stages of the project: Liz Parish, Jon Queijo, Pamela Novak, and David Frutkoff for many long hours of photo research and fact gathering; Anne Lougee for her early help in managing the project; Kay Campbell, Marcy Ketcham, Inga Parsons, Claudia Roeber, Peter Smith, and Beth Vangel for word-processing support; and Chitrita Abdullah and Robin Anne Floyd for their permissions and proofreading work.

Special thanks go to the people who reviewed sections of the book and made valuable suggestions—Astronomy: Alan Hirshfeld; Biology: Lynn Margulis, R. Dana Ono; Brain and Behavior: Donald Gash, Michael Phelps, Redford Williams; Chemistry: Robert Birge, Richard Lerner, Richard Smalley, Ioannis Yannas; Computers: George Harrar; Earth Sciences: Carl Gable, Lucia Lovison-Golob, Michael Montgomery, William Stuart; Environment: James MacKenzie, Walter Reid, Mark Trexler, Robert Winterbottom; Medicine: Louis Kunkel, Ken Mayer, James Muller, Richard Rifkind; and Physics: Edward Farhi.

Needless to say, the editors accept final responsibility for the selection of the material and the text as it stands.

Our additional thanks go to PageWorks of Cambridge, Massachusetts, for transforming our diskettes into typeset pages, and to the staff of Harcourt Brace Jovanovich for their help in making the book a reality.

Finally, we thank our friends and families for tolerating our absence during many, many evenings and weekends over the past several years as we stayed current with new science and learned more about old. And we thank Lucia Lovison-Golob, who encouraged and pushed, watched over and cared for; thank you for convincing us that we could finish the book.

Introduction

The Narrowing View

If any one word has gained importance in the late 20th century, it is "specialization"; virtually every area of knowledge has been fragmented into a number of specialties and subspecialties. Nowhere is this more evident than in the sciences, which now include fields of study that did not even exist a decade ago. While this narrowing of focus in science has been fruitful, it has also made it difficult for scientists to stay current with research outside their specialty, and generalists in the tradition of Galileo, Newton, and Franklin are becoming a rare breed. A molecular biologist, for example, may be unaware of startling new theories in cosmology. Likewise, the cosmologist may not know about significant discoveries in medical research. The danger of this "tunnel vision" is the loss of new ideas that often come about through crossfertilization.

This trend toward specialization has also affected science writing. Today's books and magazine articles for the general reader often focus on specific events or advances in the subspecialties, such as innovative cancer treatments, the ozone hole over Antarctica, or the discovery of subatomic particles. But these threads of research are often described in isolation, without considering how they fit into the rich tapestry of scientific inquiry.

The Almanac of Science and Technology addresses this problem by putting together in one volume a unique collection of articles that describe both the breakthroughs that make news and the incremental advances that fuel scientific progress. Each article describes key theories, discoveries, and events within each discipline and provides the background general readers need to gain a firm understanding of the underlying principles.

A New Breed of Science Book

How does the *Almanac* differ from the many books on science already on the bookstore shelves? The *Almanac* is the first general book to provide people with a concise, yet comprehensive, guide to the major scientific disciplines. We carefully selected each article in the *Almanac* to provide windows into key fields. Each article creates a specific knowledge base that readers can use to understand future scientific discoveries and events.

For example, while the *Almanac* was being written, a new generation of computer chips was introduced. The *Almanac's* Computers chapter also provides a framework that will enable readers to readily understand how future generations of microprocessors function.

Similarly, when we began writing the *Almanac,* high-temperature superconductivity was an impossible dream pursued by a handful of chemists and physicists. Today it is the object of a multimillion-dollar research effort involving dozens of government, industry, and university labs around the world. Our "Superconductivity" article in the Physics chapter explores the evolution of the field and gives readers a context for understanding the breakthroughs that will undoubtedly occur in the near future.

In this way, the design of *The Almanac of Science and Technology* fulfills our final goal: to promote awareness of the role played by science and technology in shaping our daily lives and the future of the entire planet. "Science literacy," like "cultural literacy," is critical in establishing a sense of who we are and where, as a world, we are headed. To this end, the *Almanac* and its subsequent editions will serve as essential guides to scientific pursuits in the 1990s and beyond.

Grand Tour of Science

While no single volume can possibly cover all new developments in science and technology, the articles in the *Almanac* represent a balanced cross-section of science today. Here's a sampling of the types of articles that you'll find:

Astronomy: A grand tour of the solar system from the Sun to Pluto... international space programs — manned flight to deep-space exploration... recent discoveries about Halley's Comet and Uranus, with its enigmatic moons and ring system... a look at the celestial bestiary, including pulsars, black holes, and supernovas... the Hubble Space Telescope and other new tools of astrophysical research... the search for extraterrestrial intelligence.

Biology: A review of human evolutionary theory, from creationism to Darwin... the latest hypotheses about why the dinosaurs disappeared 65 million years ago... new techniques in biotechnology... the growing array of genetically engineered products for medicine and agriculture... human genetics research and its potential for curing inherited illnesses.

Brain and Behavior: Brain-wave studies, positron emission tomography, and other methods to explore the workings of the brain and the mind... the mysteries of sleep research... the effects of stress on health... the promise and ethical dilemmas of brain grafting research for Parkinson's disease and other neurological conditions.

Chemistry: A primer on catalysis research and how it is changing the world of chemistry... new approaches to synthesis and their applications... innovative methods for creating artificial skin, blood, and body parts... "antirubber," radar-absorbing paint, and other exotic products from the lab bench... the brave new world of designer proteins.

Computers: A concise history of computers from room-sized behemoths to notebook-sized personal computers... the basics of computer anatomy — a guide to computer hardware and software... the evolution of computer chips and chip-manufacturing techniques... new media for storing data... recent advances in supercomputers and their applications... the fundamentals of artificial intelligence and how computers are used to mimic human thought... the expanding role of computers in the workplace.

Earth Sciences: The causes of earthquakes and volcanic eruptions and the new science of predicting these phenomena... recent geological disasters in Soviet Armenia, Colombia, Mexico, and Cameroon... exploring the rich world of the ocean floor... the effects of El Niño on global weather... the future of weather forecasting... novel techniques for probing and modeling the Earth's interior.

Environment: An in-depth look at the Exxon Valdez oil spill... the problem of hazardous waste — its dangers and potential solutions... a report on acid rain and steps to reduce its toll... the ozone hole and its frightening implications... the latest evidence of global warming... shrinking forests and growing deserts — the twin threats of deforestation and desertification... the loss of species diversity and what it means.

Medicine: The AIDS epidemic and the difficulties in developing effective treatments for this disease... heart disease causes, prevention, and therapy — from cholesterol to the artificial heart... recent strides in the diagnosis and treatment of cancer... the challenges of mapping the human genome and what genetic maps might tell us.

Physics: An introduction to theoretical physics and the world of subatomic particles... how particle accelerators work and what they are revealing about the nature of matter... the frontiers of cosmology from the Big Bang to the ultimate fate of the universe... breakthroughs in high-temperature superconductivity and progress toward practical applications... fusion — the energy technology of the future?

How to Use the Almanac

You can use the *Almanac* in a variety of ways, depending on your needs. You might, for example, scan it as a quick guide to modern science, reading an article or two from each section to sample the flavor of current research in the major disciplines. To facilitate this approach, the *Almanac* is organized as a random access guide. Although the text is cross-referenced where appropriate, each article stands essentially alone and does not presume prior knowledge of the field.

You can also use the *Almanac* to get a sense of the major advances in a discipline. The articles within each chapter contain many subsections that cover the foundations of the field as well as recent developments that may one day lead to major breakthroughs. When you finish a chapter, you'll understand the key theories and developments that drive scientific research. This basic knowledge will also make it possible for you to better understand late-breaking developments and put them into proper perspective.

However you choose to use *The Almanac of Science and Technology,* we hope that it conveys a sense of the unfolding mystery of science and gives you a taste of the excitement and wonder of scientific discovery. Above all, we hope that it broadens your understanding and appreciation of the thoughts, theories, and breakthroughs that directly affect the lives of everyone on the planet today, tomorrow, and into the 21st century.

Richard S. Golob
Eric Brus
World Information Systems
September 1989

About the Editors

World Information Systems

World Information Systems is an international publishing and environmental consulting company located in Cambridge, Massachusetts. Since its founding in 1980, WIS has undertaken numerous consulting assignments for industry and government to compile and analyze technical information, especially in the environmental field. In addition to publishing the *Hazardous Materials Intelligence Report* and *Golob's Oil Pollution Bulletin,* WIS sponsors and conducts the Hazardous Waste Business Conferences—annual summit meetings for business leaders in the environmental field. WIS has completed consulting projects for the Norwegian government, the United Nations, and several U.S. government agencies and Fortune 500 companies.

Richard Golob

Richard Golob is the founder and president of World Information Systems in Cambridge, Massachusetts. He has also served as director of the Center for Short-Lived Phenomena, formerly part of the Smithsonian Institution. He is currently the publisher of the internationally recognized *Golob's Oil Pollution Bulletin* and *Hazardous Materials Intelligence Report* and chairman of the Hazardous Waste Business Conferences. Mr. Golob's analyses of the environmental field have appeared in the *New York Times, Wall Street Journal, Washington Post, Business Week, Scientific American,* and many other leading newspapers and periodicals. He holds a degree in biochemical sciences from Harvard College and received the Newsletter Association of America's award for editorial excellence.

Eric Brus

Eric Brus has served as editor and contributing editor of the *Hazardous Materials Intelligence Report* and *Golob's Oil Pollution Bulletin* and has collaborated with Mr. Golob since 1978. He has written numerous articles on the environment, medicine, and technology and has also produced weekly audiotex features on health and fitness. Mr. Brus recently completed the Journalism Fellowship for Advanced Studies in Public Health, sponsored by Harvard University. He holds a degree in geology and English from Williams College.

Contributors

Astronomy: J. Kelly Beatty, Eric Brus, Alan Hirshfeld, David H. Smith

Biology: Richard S. Golob, Thomas H. Maugh, II

Brain and Behavior: Eric Brus, Thomas H. Maugh, II

Chemistry: Richard S. Golob, Thomas H. Maugh, II

Computers: Steven J. Bennett, Robert Hirshon, Peter G. Randall

Earth Sciences: Eric Brus, Allan Chen

Environment: Eric Brus, Richard S. Golob, Elizabeth F. Mason, Thomas H. Maugh, II, Mary Paden

Medicine: Eric Brus, Richard S. Golob, Richard Knox, Jonothan Logan, Thomas H. Maugh, II, Joe R. Neel

Physics: Richard S. Golob, Jonothan Logan, Eugene F. Mallove

The Almanac of Science and Technology

What's New and What's Known

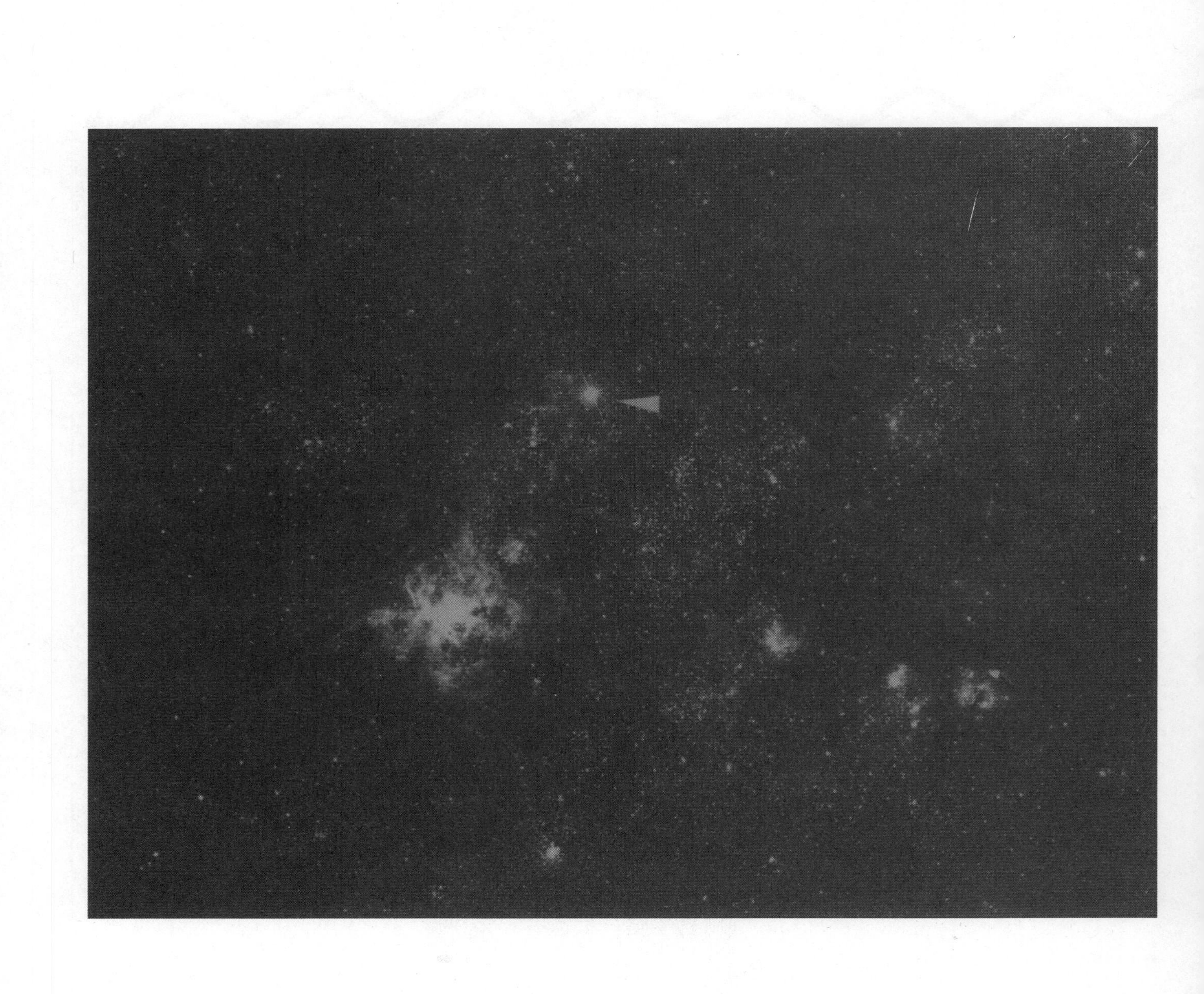

Chapter 1

Astronomy

Solar System Update

The Sun

The Sun is the nearest star to Earth and the only one on which astronomers can see surface details. Yet despite the Sun's proximity to Earth, several important scientific questions about its nature remain unanswered. Chief among these is to what extent the Sun's brightness varies. Since life as we know it is possible only within a narrow temperature range, its continuity on Earth, as evidenced by the fossil record, supports the view that the Sun's output has not varied greatly during its 4.6-billion-year history. This evidence, however, has not ruled out small, short-term variations in brightness.

For 150 years, scientists have tried to determine the solar constant, the amount of solar energy that reaches the Earth. Yet, even in the most cloud-free regions of the planet, the solar constant cannot be measured precisely. Gas molecules and dust particles in the atmosphere absorb and scatter sunlight and prevent some wavelengths of the light from ever reaching the ground.

With the advent of satellites, however, scientists have finally been able to measure the Sun's output without being impeded by the Earth's atmosphere. The National Aeronautics and Space Administration (NASA) Solar Maximum Mission satellite, or Solar Max, has been measuring the Sun's output since February 1980. Although a malfunction in the satellite's attitude-control system limited its observation for a few years, the satellite was repaired in orbit by astronauts from the U.S. space shuttle in April 1984. The results of

Solar Max's observations indicate that the solar constant is not really constant after all.

The satellite's instruments have detected frequent small variations in the Sun's energy output, generally amounting to no more than 0.05 percent of the Sun's mean energy output and lasting from a few days to a few weeks. According to solar researcher Richard Willson of the Jet Propulsion Laboratory (JPL) in Pasadena, California, these fluctuations coincide with the appearance and disappearance of large groups of sunspots on the Sun's disk. Sunspots are relatively dark regions on the Sun's surface that have strong magnetic fields and a temperature about 2,000°F cooler than the rest of the Sun's surface. Particularly large fluctuations in the solar constant have coincided with sightings of large sunspot groups. In April 1980, for example, Solar Max's instruments registered a 0.3 percent drop in the solar energy reaching the Earth. At that time a sunspot group covered about 0.6 percent of the solar disk, an area 20 times as great as the Earth's surface.

Long-term variations in the solar constant are more difficult to determine. Although Solar Max's data have indicated a slow and steady decline in the Sun's output, some scientists have thought that the satellite's aging detectors might have become less sensitive over the years, thus falsely indicating a drop in the solar constant. This possibility was ruled out, however, by comparing Solar Max observations with data from a similar instrument operating on NASA's Nimbus 7 weather satellite since 1978.

Independent checks were also obtained by solar measurements from instruments carried by balloons in the Earth's upper atmosphere and by sounding rockets, which fly briefly in space. Taken together, the satellite, balloon, and rocket observations showed a drop in the solar constant of 0.019 percent per year from 1980 to 1985. The most recent data indicate that this downward trend has reversed.

Most researchers believe that this decline in the solar constant could not have been occurring for more than a few decades; otherwise, changes in the Earth's climate would have been observed by now. Astronomer John Eddy of the National Center for Atmospheric Research in Boulder, Colorado, has calculated that a 1 percent decline in the Sun's output would cool the Earth by an average of almost 2°F. At the solar constant's former rate of decrease, the Sun's output would have declined 1 percent after only about 50 years.

Researchers have been particularly interested in learning whether the solar constant's downward trend is correlated with sunspot cycles. The number of spots on the solar disk rises and falls periodically, reaching a peak — called the sunspot maximum — on average every 11 years. The most recent sunspot maximum occurred in 1980. The 11-year sunspot cycle is one manifestation of a 22-year cycle of magnetic activity on the Sun. Approximately every 22 years, the magnetic north and south poles of newly formed sunspots are reversed relative to those found earlier.

The recent decline in the Sun's output cannot be explained by a change in the number of sunspots. If sunspots were responsible, their numbers should have increased between 1980 and 1985, but in fact, the opposite occurred. The Sun was brighter during the years when more of its disk was covered with dark sunspots. At a meeting of the American Geophysical Union in 1987, Peter Foukal of CRI, Inc., in Cambridge, Massachusetts, and Judith Lean of the Applied Research Corp. in Landover, Maryland, proposed an explanation of this behavior. They reported that data from Solar Max and Nimbus 7 showed a pattern in the variations in the solar constant. The constant fluctuated 0.04 to 0.07 percent over periods of 4 to 9 months.

Foukal and Lean discovered that these fluctuations corresponded to changes in the area of the Sun covered by features called plages, which are found near sunspots. Plages are diffuse bright areas many times larger than the sunspots nearby. Like sunspots, plages are regions of enhanced magnetic fields, and their numbers wax and wane with the 11-year sunspot cycle. The number of plages alone, however, cannot explain the decline in the solar constant. In fact, the dimming of the Sun caused by the relatively dark sunspots offsets the brightening caused by bright plages. Foukal and Lean, however, found the cause of the Sun's fading when they investigated another type of surface feature called the solar network, which is composed of regions of enhanced emission. The network covers larger areas of the Sun's disk than the plages. Like sunspots, the network is magnetic in origin, and it is most extensive at times of sunspot maximum.

According to Foukal and Lean, as magnetic activity on the Sun decreased from 1980 to 1985, the declining number of sunspots tended to brighten the Sun. The area of the solar disk covered by plages and the network also declined, however, more than compensating for the effect of the sunspots. The net effect was a decline in the solar output. If this hypothesis is correct, the constant should start to increase again as the Sun approaches the next sunspot maximum in 1990 or 1991. In fact, the most recent data from Solar Max indicate that the downward trend in the Sun's energy output has flattened out and a slight upturn is occurring.

In April 1988, Willson and Hugh S. Hudson of the

Figure 1. Solar Max. NASA's Solar Maximum Mission satellite, or Solar Max, makes detailed measurements of the Sun's energy output. These measurements have shown that this output — the so-called solar constant — actually varies over time. Solar Max became the first satellite to be repaired by astronauts in orbit, during space shuttle mission 41-C, in 1984. *Courtesy: NASA.*

University of California in San Diego, California, reported that data from Solar Max had confirmed a connection between the solar constant and Sun's 11-year sunspot cycle. As the number of sunspots and other manifestations of solar activity reached a minimum in 1986, the solar constant also attained the lowest values yet measured. But as solar activity began to rise again in late 1987, the solar constant began to rise, too. These results from Solar Max were soon confirmed using data from Nimbus 7 and the Earth Radiation Budget satellite.

Since Solar Max has provided so much valuable information about the solar constant and other facets of the Sun's workings, astronomers and space scientists were saddened to learn in December 1988 that NASA had abandoned plans to launch a second rescue mission to save Solar Max. For some time, NASA scientists have known that the satellite's orbit is decaying, with reentry into the Earth's atmosphere due in the next few years. Ironically, the satellite's slow spiral to destruction is due, in part, to an expansion of the Earth's outer atmosphere caused by the very solar activity that Solar Max was designed to measure.

Mission scientists were hoping that NASA would be able to boost Solar Max into a higher orbit and thus prevent reentry. But because of the backlog of satellite launches that occurred after the crash of the space shuttle

Challenger on 28 January 1986, NASA could not schedule a space shuttle flight to Solar Max before its reentry, which could occur by 1990. Researchers hope that Solar Max will remain aloft long enough to record the peak solar activity of the current cycle, expected in 1990 or 1991.

Solar neutrinos. Our knowledge of the Sun is limited because we can only observe its outermost layers. What little we know about the Sun's deep interior has been inferred from observations of surface conditions and theoretical models of the structure of stars. One of the most unusual "telescopes" in the world, however, has provided a window into the Sun's innermost regions. Since 1968, physicist Raymond Davis of the Brookhaven National Laboratory in Upton, New York, has been monitoring a 100,000-gallon tank containing a compound called perchloroethylene, which is commonly used as a dry-cleaning fluid. The tank is located a mile underground in the Homestake gold mine near Lead, South Dakota. This instrument is designed to detect solar neutrinos — subatomic particles emitted from the core of the Sun.

Astrophysical theories predict that neutrinos should be generated in predictable numbers by nuclear reactions within the Sun. Neutrinos interact so little with other particles that they easily escape from the Sun's interior and fly out into space at or near the speed of light. Although the neutrinos that reach the Earth pass entirely through its rocky mass, these particles can induce nuclear reactions in certain elements, including the chlorine in the cleaning fluid. The tank "telescope" is placed far underground and covered with water to shield it from cosmic-ray particles that can cause particle interactions that might be mistaken for those caused by neutrinos. A few times a week, a solar neutrino is absorbed by a chlorine atom in Davis's tank and converts the chlorine into an atom of argon gas. By repeatedly measuring the number of argon atoms in the fluid, Davis has been able to calculate the number of neutrinos entering the tank, and in turn estimate the number of neutrinos emitted by the Sun. To date, Davis has detected only one-third the number of solar neutrinos that theories predicted for a star with its observed energy output. Since the reliability of Davis's detector has been tested and confirmed, researchers have suggested that current theories of neutrino physics or solar structure must be flawed.

Over the last 15 years, astrophysicists have proposed dozens of explanations for the Sun's "missing" neutrinos, but none has been proved, much less met with widespread acceptance from the astrophysical community. One of the more popular and enduring hypotheses is that nuclear reactions in the Sun's core have temporarily slowed down. As radical as this idea seems, it is consistent with both Davis's results and some astrophysical theories. Since Davis's experiment cannot distinguish the source of the few neutrinos it actually detects, some researchers have claimed that they do not originate in the Sun at all.

Soviet physicist S. P. Mikheyev and A. Yu. Smirnov of the Soviet Academy of Sciences Institute for Nuclear Research in Moscow have proposed another idea that has attracted attention because it does not require a major change in current theories about the Sun's interior. According to the model, solar neutrinos are converted into other particles before emerging from the Sun. There are three varieties of neutrinos — electron neutrinos, muon neutrinos, and tauon neutrinos — each of which interacts in a slightly different way as it passes through dense matter like that in the Sun's interior. Mikheyev and Smirnov have proposed that some electron neutrinos may be converted into muon neutrinos on their way out of the Sun's core. Since Davis's instrument can only detect electron neutrinos, the apparently missing neutrinos may actually exist but remain undetected because they are in the muon form.

The recent detection of a pulse of neutrinos from the exploding star Supernova 1987A has added a new twist to the story of solar neutrinos. If the Soviet physicists' hypothesis is correct, then a proportion of the electron neutrinos generated in the supernova's core should also have been transformed into undetectable muon neutrinos. In other words, fewer neutrinos should have been detected from Supernova 1987A than were predicted by current astrophysical models. The number of neutrinos actually observed, however, agreed closely with theoretical predictions, a finding that casts doubt on the validity of the Mikheyev-Smirnov hypothesis.

Many of the uncertainties about Davis's solar neutrino measurements arise because his detector is sensitive only to neutrinos produced by a rare set of reactions that are secondary to the Sun's main energy-generating nuclear reactions. Ideally, researchers would like to detect the neutrinos from the primary reactions, since this would provide a direct indication of the amount of energy production in the Sun. Neutrino detectors now being built in several countries may provide new answers to the enduring solar neutrino mystery.

Solar oscillations. Solar neutrinos are not the only window to the Sun's interior. In the early 1960s solar physicists discovered that the Sun's surface oscillates and that each oscillation takes about five minutes. These pulsations, known as the five-minute oscillations, appear as rising and falling motions on small parts of the Sun's surface. Measurements of the Doppler shifts of solar spectral lines indicate that the oscillations have velocities of a few hundred meters per second (about 400 miles per hour).

At first, researchers thought that the motions on different parts of the surface were independent. Then, about 15 years ago, solar physicists at the National Solar Observatory in Sunspot, New Mexico, discovered that the random appearance of the five-minute oscillations is due to interference among a million or more different oscillating frequencies, or modes, with periods generally between 2.5 and 11 minutes. In other words, the Sun is somewhat like a bell ringing at a million different frequencies at the same time. Observing individual oscillation modes is extremely difficult. Even when one mode is filtered out from the thousands of others, it is still difficult to observe because the motions it produces are typically less than 20 centimeters per second (0.4 mile per hour).

Over the past decade, astrophysicists and solar physicists have founded a new branch of astronomy called helioseismology. Helioseismologists study solar oscillations to make inferences about the internal structure of the Sun, much as geophysicists analyze the vibrations triggered by earthquakes to understand the Earth's interior. Different solar oscillation modes penetrate the Sun's interior to varying depths. By studying the frequencies and strengths of thousands of different oscillation modes, researchers can determine how the temperature, density, and other physical properties of the Sun vary with distance from its center.

Although helioseismology is still in its infancy, astrophysicists have already made significant discoveries. Their findings have provided new support for Einstein's theory of general relativity. In the past, some researchers had suggested that the Sun's core is rotating much faster than its surface layers. If this were true, general relativity's description of gravitational fields would not correctly predict the orbit of the planet Mercury, but some rival theories would give the correct answers. Careful analyses of solar oscillations have vindicated Einstein's theory by showing that the Sun's interior actually rotates slightly more slowly than the surface.

Two major problems must be overcome if helioseismology is to continue advancing rapidly. Helioseismologists have calculated that they need to observe the Sun once every minute for two years to collect the data needed to map the Sun's internal structure in detail. Gaps in the observations due to bad weather or even night would prevent an accurate determination of the driving force causing the oscillations and the factors determining how strong each is.

Preliminary observations lasting for several days have been carried out at the South Pole to take advantage of the continuous daylight of the Antarctic summer. Some of the most recent solar observations performed at the U.S. Amundsen-Scott South Pole Station were undertaken in November 1988 by John W. Harvey and his colleagues at the National Solar Observatory. To gather 2 full years of uninterrupted data, however, the U.S. National Solar Observatory is building six solar telescopes that will be placed around the equator to observe the Sun continually.

The second problem is more formidable. To test the validity of their solar models, researchers need to observe oscillations on other stars like the Sun. These observations are much more difficult because even the brightest stars appear 10 billion times fainter than the Sun. By the late 1980s, observations by French and American teams had shown preliminary and somewhat controversial evidence of 5-minute oscillations on Alpha Centauri, Procyon, and Epsilon Eridani.

The Moon

In 1961, when President John F. Kennedy established the goal of landing American astronauts on the Moon within a decade, most planetary geologists believed that studies of material from the lunar surface would clarify such questions as how and when the Moon formed. When the final crew of astronauts returned from the Moon aboard Apollo 17 in 1972, a total of 843 pounds of lunar rocks and soil had been collected during six missions. In addition, the U.S.S.R. successfully obtained small samples of the Moon with three robot landers: Luna 16 in 1970, Luna 20 in 1972, and Luna 24 in 1976.

No spacecraft has been to the Moon's surface since Luna 24, but the precious lunar samples continue to be examined intensively by scientists around the world. Apollo samples are preserved and catalogued at NASA's Johnson Space Center, near Houston, Texas. A 120-pound "insurance set," chosen as representative of the entire collection, is stored at Brooks Air Force Base north of Houston in case a mishap damages the original collection. The Luna samples are kept in Moscow at the Vernadskiy Institute of Geochemistry and Analytical Chemistry, a

Figure 2. Sampling and Analyzing the Lunar Surface. In six manned landings on the Moon between 1969 and 1972, Apollo astronauts collected a total of over 800 pounds of lunar soil and rock. On the left, astronaut James Irwin gathers a sample during the Apollo 15 mission in 1971. Samples such as these were examined in enclosed sterile environments like the one shown on the right, at the Manned Spacecraft Center in Houston, Texas. *Courtesy: NASA.*

branch of the Soviet Academy of Sciences. The U.S. and the U.S.S.R. have exchanged small amounts of lunar material from their respective collections.

The Apollo and Luna samples have given scientists the opportunity to learn a great deal about the Moon. Sophisticated laboratory techniques allow researchers to perform complete chemical analyses of tiny rock crystals that weigh less than one-thousandth of an ounce. These analyses have shown that lunar rocks are generally similar in composition to the Earth's crust, except that they are deficient in certain metals, such as iron and nickel. In addition, none of the samples contains any water whatsoever — either as free molecules or as components of hydrated minerals. While scientific studies of lunar material have placed strict constraints on theories about the Moon, this research has not yet revealed how the Moon was formed. This and other unanswered questions must await future lunar missions.

Meteorites from the Moon and Mars. Tektites are small blobs of glass that are abundant in deposits on certain parts of Earth. Despite their abundance, tektites are so unlike most terrestrial rocks that as little as 20 years ago some astronomers believed that the glass blobs had been knocked off the Moon by a giant meteorite. In the years since, however, chemical analyses of true lunar samples have virtually eliminated the possibility that tektites originated on the Moon. Instead, most researchers now believe that tektites are bits of lava that were thrown high into the atmosphere by volcanic eruptions and solidified while still airborne.

Nonetheless, some meteorites may still have come to Earth from the Moon. In fact, this idea is now widely accepted among planetary scientists, thanks to studies of a small meteorite weighing only 1.1 ounces. The meteorite, designated ALHA 81005, was discovered in January 1982 by geologists searching for meteorites in a part of Antarctica called the Allan Hills. In certain Antarctic regions, including the Allan Hills, glaciers flow in such a way that meteorites that have fallen over a wide area are pushed to the surface and become concentrated near the bases of mountains. Once exposed to air, the meteorites are generally protected from deterioration by the dry Antarctic climate.

The ALHA 81005 meteorite is what geologists call a breccia, a mass of dust and rock fragments that have become fused together. Such assemblages are found quite often at impact craters on Earth. Chemical analyses show that some of the rock fragments in the Allan Hills meteorite consist almost entirely of a mineral called anorthosite, which is rare on Earth but quite common in many lunar rocks. Further tests of the meteorite showed that its ratio of the three naturally occurring oxygen isotopes was different from that of Earth rocks, but matched that of the lunar samples. In addition, the meteorite's ratio of manganese to iron is more like that of Moon rocks than rocks on Earth.

Since planetary geologists realized that the ALHA 81005 meteorite came to Earth from the Moon, they have become very interested in a group of eight other meteorites with unusual compositions. One of these was found in Chassigny, France, in 1815. Four others, called shergottites, are named after Shergotty, India, the site of a meteorite discovered in 1865. The other three unusual meteorites were found in Nakhl, Egypt, in 1911.

Most meteorites consist of rock condensed from hot gas and dust when the solar system formed 4.6 billion years ago. But the shergottite, nakhlite, and chassignite stones are only 1.3 billion years old. They also contain a ratio of the three naturally occurring oxygen isotopes unlike that found anywhere on the Earth, Moon, or on other meteorites. The shergottites have a crystal structure that has been deformed by mechanical shock, which geochemists believe took place 180 million years ago. The shergottites also contain trapped nitrogen gas with an isotope ratio like that found in Mars's atmosphere by the Viking landers. Many scientists have reached the conclusion that the stones probably formed during a volcanic eruption on the surface of Mars 1.3 billion years ago. The rocks were blasted intact off the surface 180 million years ago, and eventually fell to Earth. For this scenario to work, the meteorites must have been blasted off Mars without melting, because melting would erase the chemical evidence of their 1.3 billion-year age.

A number of scientists are studying how an object might strike the Moon or Mars hard enough to accelerate debris to escape velocity, yet not shock the ejected debris so much that it melts or disintegrates. In 1986, John D. O'Keefe and Thomas J. Ahrens of the California Institute of Technology (Caltech) in Pasadena reported how such impacts could occur. They calculated that if a large object struck Mars at a speed of 5 miles per second, with an oblique trajectory that made it hit the surface at an angle, a jet of vapor would spurt away from the impact at about 3 miles per second. This jet's speed is equal to the escape velocity of Mars, and any rocks caught in the vapor jet would be thrown clear of the planet. The O'Keefe-Ahrens mechanism would lift material even more effectively off the Moon, which has an escape velocity of only 1.5 miles per second.

Origin of the Moon. Although there is no historical record of when humans first gazed at the Moon and wondered where it came from, people must certainly have pondered this question for many thousands of years. After the invention of the telescope and the discovery of moons around neighboring planets, astronomers realized that our satellite is unlike others in the solar system. In particular, it is relatively large compared to the planet it circles, with a mass more than 1 percent that of Earth. With the exception of tiny Pluto and its moon, Charon, no other planet has a moon as close to its own mass. For example, Jupiter is 10,000 times more massive than all its satellites combined.

The Moon's mean density (adjusted to eliminate the compressional effects of gravity) is 3.32 grams per cubic centimeter, compared with the Earth's value of 4.45. This difference implies that the Moon does not have the same proportion of dense metals, such as iron, that Earth has. Chemical analyses of lunar material have also shown that, relative to Earth, the Moon's crust has a significantly lower proportion of volatile elements, such as hydrogen, chlorine, and mercury, which tend to melt or vaporize at low temperatures. The Moon is also depleted in what are termed siderophile elements, like nickel and cobalt, which tend to be found with metallic iron.

Before Apollo astronauts returned to Earth with samples of the lunar surface, three general hypotheses had been proposed for the Moon's origin: (1) gravitational capture, which postulates that the Moon formed elsewhere in the solar system and then passed near enough to Earth to be captured and become its satellite; (2) fission from Earth, which postulates that part of our planet was somehow separated from the main mass early in the solar system's history and later reformed in orbit; and (3) binary accretion, which postulates that the Earth and Moon formed close together simultaneously.

While each of these hypotheses explains some characteristics of the Earth-Moon system, none is consistent with all of them. Since the gravitational-capture hypothesis allows the Moon to form elsewhere in the solar system, it would explain why the Moon's composition is different from the Earth's. But astronomers consider this model to be the least likely of the three. It would be virtually impossible for the Earth's gravity to capture an object the size of the Moon. Even if it could, the resulting orbit would probably be very eccentric, not the near-circular orbit followed by the Moon.

George H. Darwin, the son of naturalist Charles Darwin, first proposed the fission hypothesis in 1879. He suggested that soon after Earth formed it began to spin so rapidly — about once every 2.6 hours — that a large chunk of matter tore free of the main mass and became Earth's satellite. Although scientists believe it is very unlikely that our planet ever spun fast enough to launch a satellite, some still find the fission model attractive because it allows the Moon to form almost exclusively out of material from Earth's mantle, which like the Moon has an average density of 3.32 grams per cubic centimeter (adjusted for compression). Moreover, the composition of the mantle is considered a good match with that of the Moon, because it contains relatively little iron and the associated siderophile elements.

In the binary accretion or "double-planet" hypothesis, the Earth and Moon formed close enough together to share some compositional characteristics, such as similar ratios of oxygen isotopes. But this model does not satisfactorily explain why the Earth has a much higher percentage of iron than the Moon. Also, it is weakened by the fact that none of the other rocky planets — Mercury, Venus, and Mars — have large satellites.

In the mid-1970s two teams of researchers independently proposed a fourth hypothesis — collisional ejection — that overcomes some of the difficulties of the earlier models. One team, William Hartmann and Donald Davis of the Planetary Science Institute in Tucson, Arizona, pointed out that collisions between planet-sized worlds probably occurred early in the solar system's history. The other team, Alastair Cameron and William Ward of Harvard University, proposed that a glancing collision between a Mars-sized object and the Earth might create a large satellite.

A more recent study by theorist George Wetherill of the Carnegie Institution in Washington, D.C., concluded that collisions between worlds should have been both violent and relatively frequent early in the solar system's history. In 1985, Wetherill noted that soon after the planets formed, the inner solar system probably contained an abundance of large objects, including some the size of Mercury and Mars. Over time, some of these passed near more massive planets and were ejected from the solar system through gravitational interactions. Wetherill calculated that others probably collided with the planets, a conclusion that helps support the collisional-ejection hypothesis.

In 1986, computer simulations performed by Willy Benz and William Slattery, both at Los Alamos National Laboratory, and Cameron confirmed that a Moon-sized satellite could have formed if the early Earth was hit obliquely by an object with the right mass and velocity. In a related computer study also completed in 1986, Marlin Kipp of Sandia National Laboratory and Jay Melosh of the University of Arizona concluded that a massive jet of vaporized rock would be ejected from the point of impact. Once in orbit around Earth, the vaporized rock would cool and coalesce into a single object.

Many planetary scientists favor the collisional-ejection hypothesis because the material thrown out from the impact would come almost entirely from Earth's iron-poor mantle, rather than from its iron-rich core. Also, a major glancing impact could provide the Earth–Moon system with the large amount of angular momentum it possesses today. Future computer simulations will focus on how sensitive the Moon's formation is to the mass, velocity, and trajectory of the impacting object. The model could be verified if future missions to the Moon show that its overall composition matches that of Earth's mantle.

Venus

Radar studies. In size, mass, density, and gravity, Venus is the planet most like our own. Yet the climate of Venus is radically different from ours. At the planet's surface, the atmospheric pressure is about 90 times that at sea level on Earth, and the temperature is 860° to 900°F — hot enough to melt lead. The reason for the extreme heat is the so-called greenhouse effect: The dense atmosphere and clouds allow sunlight to enter, while they prevent most surface heat from escaping into space.

Human eyes have never seen the surface of Venus because the entire globe is perpetually hidden under a thick, unbroken layer of clouds. Our knowledge of the planet's general topography has come from radar studies performed from Earth and from three space probes: the U.S. Pioneer Venus Orbiter (PVO) and the U.S.S.R.'s Venera 15 and 16 orbiters. Upon reaching Venus in December 1978, the PVO fired a small rocket that allowed it to be "captured" by the planet's gravity. Thereafter, as the spacecraft circled Venus, it directed a radar beam through the clouds and onto the planet's surface. The returning radar echoes allowed scientists to distinguish individual topographic features as small as 60 miles across. Over a two-year period the spacecraft mapped more than 90 percent of the planet. Although the maps revealed many previously unknown features, they were limited by the radar's relatively low resolution. Viewed at the same resolution, the Earth's continents would stand out clearly, but major mountain peaks would barely be discernible and details of even the largest river basins would be completely invisible.

On 10 October 1983, the Soviet spacecraft Venera 15 settled into orbit around Venus, followed four days later by its twin, Venera 16. Together the probes mapped all of Venus north of 30°N latitude, including some regions around the planet's north pole that could not be observed by the PVO. The Veneras carried more sophisticated radar systems than had been used on the American spacecraft. These devices allowed scientists to identify surface features as small as one mile across — comparable to the best results obtained from Earth with the 1,000-foot-diameter radar dish at Arecibo, Puerto Rico.

According to the PVO and Venera radar data, three-fifths of Venus's surface is covered with rolling terrain at an elevation near the mean radius of the planet. (The term "sea level" has no real meaning for Venus, since the planet has no oceans; therefore, geologists compare the elevations of topographic features on Venus with the value adopted for the planet's mean radius, about 3,760 miles.) About one-sixth of the terrain consists of valleys and basins well below the mean radius, and highland areas make up the remaining quarter. The International Astronomical Union, which establishes the official names for extraterrestrial objects, has decided to name most prominent features on Venus after famous women and goddesses.

Venus has two large highland regions. Ishtar Terra, located in the northern hemisphere, is about the size of the continental U.S. It contains the planet's highest mountain ranges (called Maxwell Montes, Freyja Montes, and Akna Montes), as well as a vast plateau, Lakshmi Planum, situated about 11,000 feet above the mean radius. Maxwell Montes honors the nineteenth-century British physicist James Clerk Maxwell. Its highest peak rises about 38,700 feet above the planetary mean radius and would tower more than 9,000 feet above Mount Everest.

The other major highland region, Aphrodite Terra, lies near the equator in Venus's southern hemisphere. It is somewhat larger than Ishtar Terra, covering an area comparable to South America. On the eastern end of the Aphrodite Terra, gaping rift valleys — some more than 9,000 feet deep — cut into the surface for distances of up to 3,000 miles.

Beta Regio, a mountainous region so prominent that radar stations on Earth had detected it several years before the arrival of the space probes, lies to the southwest of Ishtar Terra. Its twin peaks rise 13,000 feet above the surrounding terrain. Planetary geologists believe these peaks may be volcanoes. Together they measure 1,300 miles across at their base — about the distance from New York City to Dallas, Texas.

Venus's changing surface. Since Venus is comparable in size and bulk composition to Earth, scientists have speculated that the geologic evolution of the two planets may have been similar as well. The Earth's crust is not a single, static layer of rock, but rather a series of irregularly shaped plates. Due to heat and pressure, the layer below the plates is not rigid but behaves instead like a fluid that carries the overlying plates along as it slowly moves. In so-called tectonic activity, the plates drift slowly, colliding in some places and spreading apart in others. Mountain ranges are often created where plates collide, and both volcanic eruptions and earthquakes commonly occur in such regions. Great valleys, such as the rift valley in Ethiopia, can form where plates separate.

The PVO's low-resolution radar maps of Venus did not reveal any obvious indications of plate tectonics, leading some researchers to conclude that Venus is, and always has been, a geologically dead world. That view has recently changed as a result of the higher-resolution Venera radar studies and new Earth-based observations. According to Soviet geologists, the Venera radar maps show evidence of long, straight ridges and fissures that may have been caused by horizontal motion in large sections of the planet's crust.

The size and numbers of impact craters, created by the collision of small asteroids and comets, indicate the approximate age of planetary surfaces. Old surfaces on Venus should bear a great number of large craters, but the Venera radar mappers revealed relatively few such craters scattered across the landscape. Soviet geologists have concluded that this lack of large craters indicates that much of the surface of Venus is between 500 million and 1 billion years old. This is rather young by solar-system standards; for example, virtually all of the Moon's surface is at least 3 billion years old.

After studying all of the available radar data, geologist James Head and his colleagues at Brown University in Providence, Rhode Island, have proposed that Ishtar Terra formed during a pair of plate collisions. They suggested that the first collision created the Maxwell Montes range and a nearby region of straight, parallel ridges resembling the Appalachian Mountains in the eastern U.S. (The Appalachians were first formed millions of years ago during a collision between two plates.) The scientists speculate that the second collision created two more

Appalachian-like mountain ranges, Akna Montes and Freyja Montes.

The Brown University geologists also believe that Aphrodite Terra bears characteristics similar to the Earth's mid-ocean ridges, where new crust forms as material from the mantle wells up to the surface. In the process, old seafloor and adjacent continents are pushed aside, creating a distinctive pattern of faulted ridges and valleys. If the Brown geologists' interpretation is correct, then Aphrodite Terra is one of the youngest and most geologically active areas on Venus. Harold Masursky of the U.S. Geological Survey, however, believes that just the opposite is true. In his view, Aphrodite Terra is among the most ancient landforms on Venus, and the deep rifts observed by radar mapping are simply the scars left by ancient fault systems.

Planetary scientists have not yet reached a consensus about whether Venus experienced crustal motion in the past, largely because their knowledge of the surface is still too limited to allow a definitive interpretation of its evolution. Such analyses might become possible once the forthcoming Magellan radar-mapping spacecraft begins orbiting Venus in 1990. The craft's sensitive radar should resolve features as small as 1,000 feet across.

Volcanoes and lightning. Even before the recent Venera missions, planetologists suspected that Venus has experienced extensive volcanism in the past. The PVO's data on topography, gravity, and the atmosphere provided evidence of such activity in at least two locations — the twin peaks of Beta Regio and the eastern end of the Aphrodite Terra highland. Radar echoes from both areas resembled those from fresh lava flows on Earth, and a circular surface feature, perhaps a volcanic crater, was found on Beta Regio.

Volcanoes may still be erupting on Venus. One line of evidence comes from the study of the PVO's radio-frequency observations of Venus's atmosphere. In 1983, a research team led by Frederick Scarf of TRW in Redondo Beach, California, discovered what appeared to be "whistlers" — low-frequency radio bursts produced by lightning flashes — concentrated over the equator and especially over Beta Regio and eastern Aphrodite Terra. Scarf's team speculated that the apparent lightning may have been caused by turbulent clouds of ash and dust from a volcanic eruption.

Data from the PVO's ultraviolet spectrometer provided additional evidence of recent volcanic activity, according to atmospheric specialist Larry Esposito of the University of Colorado in Boulder. Esposito found that the amount of sulfur dioxide gas in Venus's upper atmosphere decreased steadily after the PVO began its observations in December 1978. He suggested in 1984 that a major volcanic eruption had occurred on Venus shortly before the PVO arrived. The proposed eruption would have discharged a huge quantity of sulfur dioxide gas into the planet's atmosphere, and the gas would have slowly dissipated thereafter.

Harry Taylor of NASA's Goddard Space Flight Center in Greenbelt, Maryland, and Paul Cloutier of Rice University in Houston, Texas, disagree with the active-volcano hypothesis. In a 1986 study, these researchers argued that the low-frequency radio-wave bursts are not caused by lightning accompanying a volcanic eruption but rather by the normal flow of charged particles in Venus's upper atmosphere. This alternative explanation for the radio-wave bursts is supported by the fact that instruments carried through Venus's atmosphere in 1985 did not detect any lightning in the planet's atmosphere.

Surface and balloon studies. The latest chapter in the exploration of Venus began in June 1985, when the U.S.S.R.'s Vega 1 spacecraft, which was then headed for Halley's Comet, released an 8-foot-wide capsule into Venus's atmosphere as it flew by the planet. The capsule split into a 1.5-ton lander that dropped to the surface and a small, balloon-borne atmospheric probe that remained aloft. Also in June 1985, an identical spacecraft, Vega 2, flew by Venus, releasing another lander and atmospheric probe.

The Vega 1 lander set down in a lowland region near the equator and transmitted data for 56 minutes before succumbing to the heat. The Vega 2 lander touched down about 1,000 miles farther south on the flank of Aphrodite Terra and survived for 57 minutes. Each lander was equipped with a drill designed to obtain a soil sample from just below the surface. Although Vega 1's drill malfunctioned, Vega 2's drill successfully obtained some soil for analysis. An x-ray fluorescence spectrometer determined the sample's approximate chemical composition by bombarding it with x-rays and analyzing the resulting x-ray emissions from the soil. The Vega 2 sample has an overall composition more similar to the basalt found on the Earth's seafloors than to its continental rocks. This finding supports the idea that the ground around the lander was covered with a deposit from some type of volcanic flow.

The Vega atmospheric probes, a collaboration of Soviet, French, and American scientists, consisted of a 15-pound payload suspended from a helium-filled balloon 12 feet in diameter. Each instrument package contained a battery-powered radio transmitter and an antenna, temperature and pressure gauges, a wind-speed indicator, and a light sensor to record any lightning flashes. As the atmospheric probes floated above Venus's surface, they were precisely tracked from Earth by six Soviet radio antennas, three antennas in NASA's Deep Space Network, and 11 other radio dishes worldwide.

Both probes transmitted data for about 46 hours before exhausting their battery power. By then, each probe had floated more than 7,000 miles — nearly one-third of the way around the planet — at an altitude of 33 to 34 miles. The probes encountered horizontal winds ranging from 135 to 155 miles per hour, along with unexpectedly strong vertical currents averaging about 2 miles per hour. The Vega 2 probe experienced a sudden 8-mile-per-hour downdraft as it passed over the Aphrodite Terra highland, an indication that Venus's surface topography affects wind patterns even at high altitudes. On Earth, such abrupt topography-induced currents are not encountered at altitudes greater than a few miles.

The Vega probes may have also discovered another new aspect of Venus's atmosphere. The Vega 1 probe, floating along a path about 460 miles north of the planet's equator, registered temperatures consistently 12°F warmer than did its twin, which traveled along a path about the same distance south of the equator. To explain these temperature differences, some researchers have proposed that Venus's atmosphere contains large circulating wind systems analogous to those on Earth.

Neptune and Triton

Neptune's day. Ever since Johann Galle first observed Neptune from Berlin Observatory in 1846, astronomers have attempted to measure one of its most basic characteristics — the length of its day. Normally, a planet's rotation period can be measured by timing the motion of some feature on its surface. At a distance of about 2.7 billion miles, however, Neptune had always appeared as a dim, nearly featureless disk even in the world's largest telescopes. Crude estimates of its rotation period, based on periodic light variations or theoretical calculations, have ranged from about 15 hours up to 22 hours. Recently, astronomers have finally discerned markings that have allowed them to determine the planet's rotation period more accurately.

Bradford Smith of the University of Arizona and JPL's Richard Terrile photographed Neptune in May 1983 with a sensitive electronic light detector attached to the 100-inch-diameter telescope at Las Campanas Observatory in Chile. They then used computers to bring out details that would normally be rendered indistinct by the blurring effects of the Earth's atmosphere. Images with similar cloud features were obtained by Heidi Hammel of the University of Hawaii in July 1988. These sets of photographs revealed that Neptune's upper atmosphere has bright patches that circle the planet in about 17.7 hours, with an uncertainty of only a few minutes.

This new rotation period measurement has forced astronomers to reconsider their ideas about Neptune's interior structure. According to accepted planetary theories, if Neptune's mass were highly concentrated toward its center, as in the other giant outer planets — Jupiter, Saturn, and Uranus — then the planet's rotation period should be about 15 hours. Neptune's nearly 18-hour day implies that most of the planet's mass does not lie near the center.

Radio-frequency observations, however, have led to the opposite conclusion about Neptune's mass distribution. These studies, which have measured the amount of energy that Neptune radiates into space, have indicated that the temperature in the planet's core must be at least several thousand degrees Fahrenheit. In such a high-temperature environment, lighter elements would diffuse up toward the surface and denser ones would sink toward the center. This mass distribution is much different from that expected from the 18-hour rotation period.

University of Arizona astronomer William Hubbard has suggested that the rotation rate of the bright patches observed in Neptune's upper atmosphere may be representative only of the upper atmosphere and not the planet's deeper layers. In fact, the atmosphere may even rotate at different speeds, depending on latitude. Other recent determinations of the planet's rotation period, based on high-quality observations like those of Smith, Terrile, and Hammel, have ranged from 17 hours and 42 minutes to 19 hours and 36 minutes. Planetary researchers expect to gain a definitive answer to the question of Neptune's mass distribution when the Voyager 2 spacecraft arrives at Neptune in August 1989.

Ring for Neptune? For more than three centuries after Dutch astronomer Christiaan Huygens's discovery in 1656 that Saturn was girdled by a bright ring, astronomers believed that Saturn's ring system was unique in the solar system. Then, in 1977, rings were discovered around Uranus, and two years later, Jupiter, too, was found to have rings. Some scientists now believe that Neptune also has a ring — or at least parts of one.

The first credible indication of a ring around Neptune

appeared in 1982, when astronomers from Villanova University in Pennsylvania announced that they had found evidence for a ring after reexamining data from a stellar occultation 16 years earlier. (In a stellar occultation, a planet, moon, or other object passes in front of a star.) The astronomers found that during the event the star's light was not only blocked by Neptune's disk, but may also have dimmed shortly before and after the occultation. On the basis of these observations, Villanova astronomers Edward Guinan, Frank Malone, and Craig Harris calculated that a ring or a ring system may lie between 2,200 and 4,900 miles above Neptune's cloud tops. But three other occultations observed in 1981 and 1983 failed to confirm the existence of the ring.

On 22 July 1984, Neptune appeared to "graze" a star, giving astronomers yet another opportunity to check for the presence of rings. Shortly before the event, a team of astronomers working at the European Southern Observatory on Cerro La Silla, northeast of Santiago, Chile, noted that the star dimmed by about 35 percent for less than two seconds. Simultaneously, about 60 miles to the south, a group of University of Arizona astronomers working at the Cerro Tololo Inter-American Observatory recorded a nearly identical decrease in the star's intensity. The observations indicate that the star was blocked by an object just 6 to 12 miles wide and 32,000 miles above Neptune's cloud tops. If a uniform ring does indeed exist, the star's light should have dimmed again briefly as it was occulted by another section of the ring. Neither team observed a second dimming.

Some astronomers have proposed that if Neptune has a ring, it may be broken into fragments, with most of the material concentrated in one part of the ring and very little elsewhere. During an occultation, the light of a star would dim only when it passes through the dense portion of the ring. Alternatively, the width of the proposed ring could vary significantly in different parts. If such were the case, material in relatively wide parts of the ring might be spread so diffusely that it would not significantly dim the star, while material in some other parts of the ring might be concentrated into such a narrow band that any dimming of starlight would be too brief to be noticed.

Triton. In 1846, the year of Neptune's discovery, the British astronomer William Lassell discovered the first moon orbiting the planet. Triton, as this moon became known, moves around Neptune in a direction opposite to the planet's spin and opposite to that normally found in the solar system. In fact, Triton is the only large satellite with such a retrograde orbit. Triton travels in an almost perfectly circular orbit 220,650 miles from Neptune's center. Because Neptune is massive, it exerts enough gravitational force to damp out any greater eccentricity that may have once existed in the orbit. Neptune's gravity has probably also locked Triton into synchronous rotation; that is, the satellite completes one rotation during each orbit, with the same hemisphere facing Neptune at all times. While synchronous rotation is strongly suspected, astronomers have not yet made sufficiently detailed observations of the satellite to enable them to confirm their suspicions.

With an estimated diameter of 2,160 miles, Triton is roughly the size of Earth's Moon and thus massive enough to raise a significant tide inside Neptune. This tidal bulge is not directly below Triton but lags somewhat behind the satellite because of Neptune's rotation. In turn, the bulge exerts a gravitational pull on Triton and gradually slows the satellite down in its orbit. Consequently, Triton will eventually lose so much orbital energy that it will fall into Neptune.

Before actually reaching the planet, however, strong tidal forces will tear Triton apart. By contrast, our own Moon, which revolves in the same direction as the Earth's rotation, is slowly being pushed ever farther outward. At one time, astronomers thought Triton's remaining lifetime would be no more than 100 million years, but JPL researcher Alan W. Harris calculated in 1984 that the rate of orbital decay is much slower and that Triton will remain intact for roughly 10 billion years.

Astronomers are not sure why Triton has a retrograde orbit. During the 1930s, astronomers in England and Japan proposed that both Triton and Pluto were once satellites of Neptune, traveling in normal, prograde orbits. Then a close encounter between the two giant moons ejected Pluto from the system altogether and reversed Triton's direction of motion around Neptune. In 1984, however, William McKinnon of Washington University in St. Louis showed this scenario to be dynamically impossible — Pluto would have been thrown from the solar system entirely. Rather, he believes that Triton and Pluto are independent objects that remained intact after the formation of the planets. At some point Triton passed near enough to Neptune to be pulled into an orbit by the planet's strong gravitational field.

In 1978 astronomers Dale Cruikshank at the University of Hawaii and Peter Silvaggio, then at the NASA-Ames Research Center, obtained the first crude spectrum of Triton's surface. They were surprised to find no spectroscopic evidence for water ice, but instead saw a pattern characteristic of methane. The discovery of methane strongly implies that Triton has an atmosphere, because methane is so volatile that some of the compound should exist as gas even at Triton's frigid temperature of about –330°F. In 1984, Cruikshank found another spectroscopic signature that he attributed to liquid nitrogen. Since then, theorists have concluded that the nitrogen may be either liquid or frozen, but in either case it must cover a large portion of Triton's surface. Thus, Voyager 2 may find large polar caps of nitrogen ice — or a vast liquid-nitrogen sea — when it visits Triton and Neptune in August 1989.

Pluto and Charon

Pluto's moon. During July 1978, astronomer James Christy of the U.S. Naval Observatory in Washington, D.C., photographed what appeared to be a strange bulge on Pluto's side, prompting speculation that the planet might have a moon. Astronomers soon confirmed the moon's existence and named it Charon, after the mythological boatman who ferried the souls of the dead across the river Styx to the underworld. Further observations showed that Charon orbits Pluto once every 6.39 days at a distance of about 12,200 miles. Planetary scientists believe that both Pluto and Charon are composed of ices (of water, ammonia, methane, and other compounds) and rock in roughly equal amounts. Such a compositionn is typical of the moons of Jupiter, Saturn, Uranus, and Neptune.

Pluto's atmosphere. On 9 June 1988, several teams of astronomers watched the planet Pluto occult a faint star. In effect, Pluto cast a circular "shadow" on Earth that moved quickly across the Southern Hemisphere. To anyone directly within the path, Pluto appeared to cover the star for about 90 seconds. It was the first time since Pluto's discovery that astronomers had witnessed such an occultation, and numerous observing teams recorded the event from widely scattered sites in Australia, New Zealand, and Tasmania. One group from the Massachusetts Institute of Technology (MIT), was aboard NASA's Kuiper Airborne Observatory (KAO), flying over the South Pacific. This plane, a converted C-141 cargo transport, is equipped with a 36-inch telescope that peers out the side of the fuselage.

The star's light was not cut off abruptly, as would have been the case with an airless body such as an asteroid. Instead, the light disappeared and reappeared gradually over 10 to 20 seconds (depending on the observer's location). This gradual fade-out is similar to the apparent dimming of the Sun at sunset, which occurs because the Sun's light must travel through a great deal of the Earth's atmosphere. Astronomers have therefore concluded that Pluto must have an atmosphere as well. Its composition is not known, but only a few candidate gases appear possible, given the planet's frigid temperature (about –360°F). Besides methane, other possibilities include argon, nitrogen, oxygen, carbon monoxide, and neon. Pluto's atmosphere is extremely tenuous, yielding a surface pressure about 100,000 times less than at sea level on Earth. Furthermore, the occultation record obtained aboard the KAO by James Elliot, Edward Dunham, Jr., and their MIT team indicates that the atmosphere also has a tenuous haze suspended within it. This is probably a photochemical smog of the general type that envelopes Saturn's moon Titan.

A few months before the occultation, Laurence Trafton of the University of Texas, Alan Stern of the University of Colorado, and Arthur Whipple of the U.S. Naval Observatory proposed on theoretical grounds that a single, thin atmosphere envelops both Pluto and its nearby moon, Charon. Based on the occultation data, their model now seems plausible.

Eclipses of Pluto and Charon. The great distance and extreme faintness of Pluto and Charon have made it virtually impossible for astronomers to distinguish the two bodies from one another, much less discern details on their surfaces. Fortunately, astronomers have recently had the opportunity to learn a great deal more about these distant bodies because of a rare astronomical alignment. Charon's orbit is oriented so that, for about a 6-year period occurring once every 124 years, Charon appears to pass alternately in front of and then behind Pluto as it orbits the planet. That is, once in each 6.39-day orbit, Charon partially transits Pluto and is then either partially or totally eclipsed by Pluto. Although Charon and Pluto entered their eclipse season in 1985, the starting date of that 6-year period was uncertain, largely because the orientation of Charon's orbit with respect to Pluto was not known precisely.

The task of detecting these events is complicated by the fact that Pluto's brightness varies by 30 percent as it rotates on its axis, because some parts of its surface are more reflective than others. In anticipation of the eclipse season, astronomers carefully monitored these periodic changes in brightness during the early 1980s, so that they would be able to distinguish the change in brightness caused by the overlapping disks of Pluto and Charon from that due to Pluto's uneven reflectivity.

On 16 January 1985, JPL researchers Edward Tedesco and Bonnie Buratti became the first astronomers to see Charon, as viewed from Earth, appear to graze the edge of Pluto's disk. The event caused the pair's total brightness to drop just perceptibly — by only four percent. Similar events were detected about one month later, on 17 February, by Richard Binzel, then at the University of Texas, and three days after that by University of Hawaii astronomer David Tholen. By the end of 1986, Charon's orbital plane had become sufficiently aligned with Earth that, instead of grazing Pluto's disk, it began passing completely in front of and behind the planet. According to Tholen's calculations, the eclipses and transits will end in October 1990.

The previous eclipse season took place during the American Civil War, and the next one will not occur until the twenty-second century. Thus, professional astronomers around the world have been trying to learn as much as possible before this rare opportunity ends. By comparing the observed drops in brightness during particular events to the predictions of astronomical models, astronomers have already determined many characteristics of the Pluto–Charon system.

For example, by carefully monitoring the times that the eclipses begin and end, researchers can calculate the diameters of the two bodies and trace Charon's orbit more precisely. This new data, in turn, can be used to refine calculations of the mass and density of Pluto and Charon, information that reveals much about their composition and structure. Finally, since each eclipse covers a different portion of Pluto, sensitive monitoring of variations in Pluto's brightness allows scientists to infer the distribution of light- and dark-colored features on the planet's surface.

Worlds revealed. For astronomers engaged in studies of the Pluto-Charon system, 1988 was an extremely productive year. Because both Pluto and Charon are relatively small and remote from Earth — about 2.7 billion miles away — these objects' disks could not be resolved even in the world's most powerful telescopes. Thanks to observations made during recent eclipses and transits, however, astronomers have narrowed their estimate of the diameter of Pluto to between 1,410 and 1,430 miles, and that of Charon to between 720 and 760 miles.

Since Charon's orbit has been well defined, Pluto's mass can be calculated directly. Astronomers have used this mass value, in combination with the updated estimates of Pluto's diameter, to determine that the overall density of the Pluto-Charon system is approximately 2.0 grams per cubic centimeter. This density is relatively low compared to the density of the rocky inner planets, but higher than most of the outer planets' satellites, suggesting that Pluto and Charon probably contain a great deal of ice and rock, but virtually no metal. By comparison, Earth's bulk density is 5.52 grams per cubic centimeter (or 4.45 when the compressional effects of gravity are taken into account).

Since Charon is smaller than Pluto, the satellite covers only part of the planet's disk. At various times in recent years, Charon has covered Pluto's north polar, equatorial, and south polar regions. By carefully monitoring the system's total brightness during these passages and allowing for Pluto's uneven reflectivity, many astronomers have concluded that the planet has relatively bright polar caps and a dark equatorial region. In addition, the equatorial region must have one or two large spots that may be either light- or dark-colored, depending on how the data are interpreted.

When Charon is completely hidden behind Pluto, what is seen in Earth's telescopes is no longer the combined light of two objects but only the light of Pluto alone. Spectrographic information gathered during eclipses and transits has provided information on the composition of both Pluto and Charon. According to Robert Marcialis of the University of Arizona and two colleagues, some or all of Pluto appears to be covered with methane, either frozen on its surface or as a gas in a tenuous atmosphere. The astronomers also believe that parts of Charon, in contrast, are covered with water ice.

Even more details about the Pluto-Charon system should emerge before the eclipse season ends in 1990. People alive today, however, may never know about Pluto and Charon in great detail. These objects are so distant that no spacecraft is likely to visit them for many decades to come.

Rendezvous with Uranus

On 24 January 1986, 8 years after its launch from Earth, the U.S. spacecraft Voyager 2 swept by Uranus, the seventh planet from the Sun, and gave scientists their first closeup views of that planet. The encounter marked another milestone in Voyager 2's journey through the outer solar system; along with its sister craft Voyager 1, the space probe had already visited Jupiter in 1979 and Saturn in 1981. As in these earlier flybys, the Voyager 2 data resolved longstanding puzzles while at the same time posing new ones. The spacecraft is now speeding toward remote Neptune for a rendezvous on 24 August 1989.

British astronomer William Herschel discovered Uranus by accident in 1781 while scanning the sky for double stars. Uranus is about 19 times farther from the Sun than Earth is, and is about 4 Earth diameters across. It circles the Sun every 84 years. The planet is surrounded by 5 major moons, 10 smaller moons, and at least 11 discrete rings. Uranus rotates about its own axis every 16 hours. Mysteriously, the rotation axis, which is fixed with respect to the stars, lies almost in the plane of the solar

Figure 3. Voyager 2 Spacecraft before Launch. Voyagers 1 and 2 were launched in 1977 toward the outer planets. Both spacecraft have completed their Jupiter and Saturn encounters, and Voyager 2 flew by Uranus in 1986 on its way to an encounter with Neptune in 1989. *Courtesy: NASA.*

system; that is, if the other planets are likened to tops spinning nearly upright as they orbit, Uranus is lying on its side. As a result, as Uranus orbits around the Sun, each of its poles remains in sunlight for 42 years at a time. Astronomers speculate that Uranus's peculiar orientation may have resulted from a collision with an Earth-sized object in the planet's distant past.

Voyager mission background. The combined objectives of Voyagers 1 and 2 represent a scaled-down version of the so-called "Grand Tour" of the planets. In the late 1960s, NASA had proposed the Grand Tour project to send space probes past each of the five most distant planets — Jupiter, Saturn, Uranus, Neptune, and Pluto — but budget cuts later forced NASA to de-emphasize the visits to Uranus and Neptune and abandon the planned trip to Pluto altogether. Still, the Voyagers represented a technological leap over earlier interplanetary probes.

Voyager 1 was launched from Cape Canaveral, Florida, on 5 September 1977, shortly after the departure of its twin Voyager 2 on 20 August 1977. Although Voyager 1 left the Earth after Voyager 2, it arrived first at both Jupiter and Saturn because it followed a shorter trajectory to those planets. It passed no other planets afterward and is now headed out of the solar system. Voyager 2's flight path allowed it to visit Jupiter and Saturn and then continue on to Uranus and Neptune before heading out of the solar system. Together, the two probes relayed to Earth more than 70,000 photographs of Jupiter and Saturn.

Voyagers 1 and 2 flew by Jupiter in March and July 1979, respectively. The photographs and scientific data from these flybys revealed subtle details about the chemistry, structure, and flow patterns of the planet's clouds and storms. The two spacecraft also measured the size and intensity of Jupiter's magnetic field, discovered three tiny moons, and made precise measurements of the size, mass, density, and composition of Jupiter's four largest moons — Callisto, Europa, Ganymede, and Io.

Mission scientists used Jupiter's gravity to deflect Voyager 1 toward Saturn, where the spacecraft arrived in November 1980; Voyager 2 followed in August 1981. Voyager photographs revealed that the planet's broad rings actually consist of thousands of narrow ringlets. The three rings visible from Earth, the A, B, and C rings, span over 40,000 miles from their innermost to their outermost edges, but the Voyagers found them to be less than 160 feet thick. Segments of the F ring, which was discovered by Pioneer 11 in 1979, located just outside the A ring, had an unusual braided pattern, apparently due in part to the gravitational influence of a newly discovered moon just outside the ring. In addition, the Voyagers found unexpectedly that dark "spokes" extended across the width of the B ring. Scientists concluded that the spokes were accumulations of powder-sized, electrically charged pieces of dust and ice that, for reasons unknown, had been levitated above the B ring.

During their flybys of Saturn, the Voyagers discovered six small moons, some of which are only a few dozen miles across. The opaque atmosphere of Saturn's largest moon, Titan, was found to consist primarily of nitrogen with traces of methane and other simple organic compounds. Photographs of Saturn's medium-sized moons — Mimas, Enceladus, Tethys, Dione, Rhea, Hyperion, Iapetus, and Phoebe — revealed the presence of closely packed craters, huge crevasses, and smooth plains.

After Voyager 2's rendezvous with Saturn in 1981, electronic systems aboard the spacecraft were extensively reprogrammed from Earth. Mission engineers at the JPL were able to fix a camera-pointing mechanism that had jammed as Voyager 2 flew past Saturn and also to communicate more predictably with the craft's unstable radio receiver. In addition, the engineers modified Voyager 2's computers to make them record, process, and transmit scientific data more efficiently. Because of these changes, for example, images of Uranus could be sent every 4 minutes instead of every 13 minutes, as originally planned.

The cameras on Voyager 2 required relatively long exposure times because the sunlight at Uranus is nearly 400 times weaker than at Earth and about four times weaker than at Saturn. Since the spacecraft was to pass by the planet at over 45,000 miles per hour, fine details in the images would normally have been smeared out during the lengthy exposures. To cope with this problem, mission scientists instructed Voyager 2 to rotate slightly as it recorded its pictures in order to keep the camera pointed at the target. As planned, the technique greatly increased the clarity of the images.

Uranus Observed

At 12:59 P.M. EST on 24 January 1986, Voyager 2 sped by Uranus at a distance of 50,679 miles above the planet's uppermost clouds. After a 2-billion-mile, 8.5-year trip, the spacecraft was only 10 miles away from its intended path by Uranus. (This accuracy is comparable to shooting an arrow across the United States and missing the center of a target by one inch.) To the disappointment of mission scientists, however, Uranus revealed little to the camera;

even when Voyager 2 was closest, its photographs showed what appeared to be a pale aquamarine billiard ball. An impenetrable haze of hydrocarbon "smog" particles in Uranus's upper atmosphere blanketed much of the sunlit side of the planet.

Mission scientists had hoped that the Voyager 2 images would help confirm the current estimate of Uranus's rotation period — about 16 hours. Because of the covering of haze, the period was difficult to determine from Earth. (In order to measure the rotation period of a planet, astronomers search for distinctive surface features that they can clock as the planet spins. The visible "surfaces" of the giant, outer planets, including Uranus, are the upper parts of their atmospheres where astronomers track clouds, whirls, and other gaseous features. Because winds can move clouds around, rotation rates derived from repeated cloud observations have a great deal of uncertainty.)

During Voyager 2's encounter, only four clouds could be distinguished through the haze. One was found to spin about the planet in 16.9 hours, while another — closer to the sunlit pole — completed the circuit in only 16.0 hours. Voyager's atmospheric scientists attribute the difference to zones of wind within Uranus's atmosphere that travel at different speeds, depending upon latitude.

By recording the spectrum of Uranus's light, Voyager confirmed that the planet's atmosphere was composed of hydrogen (the most abundant gas), helium, methane, and ammonia. The concentration of helium was only about 15 percent, thereby resolving a controversy that began in 1985 when JPL atmospheric specialist Glenn Orton deduced from Earth-based infrared observations that the helium concentration should be about 40 percent. This helium level would have been much higher than the concentrations observed in the Sun, Jupiter, and Saturn. Explaining why the helium level was so high at Uranus but not elsewhere would have required extreme changes in existing theories about the formation of the solar system.

In addition, Voyager 2's records of infrared emissions from Uranus have indicated that, surprisingly, the very top of the planet's atmosphere is hot: nearly 900°F on the sunlit side and, inexplicably, even hotter — 1,340°F — on the dark side. The higher temperature on the dark side is even more surprising since parts of that hemisphere have not seen the Sun for decades, due to Uranus's unusual tilt.

The infrared studies also showed that the temperature falls rapidly with increasing depth into the atmosphere. The temperature reaches a minimum of about 370°F at a depth where the pressure is about one-tenth of that at sea level on Earth. Below this depth, the temperature begins to rise again because of the pressure of the overlying layers. Scientists have suggested that a liquid water-methane-ammonia layer may exist beneath the atmosphere, and that a rocky core several times more massive than the Earth may exist at the planet's center.

Magnetic field. The existence of a magnetic field — or magnetosphere — around Uranus was suggested in 1982 by three independent research teams when the International Ultraviolet Explorer (IUE) satellite observed an ultraviolet glow around Uranus, a phenomenon then thought to be auroral activity in the planet's atmosphere. Auroras on Earth, the so-called northern lights or southern lights most commonly seen near the poles, occur when high-energy charged particles emitted by the Sun strike atoms in Earth's atmosphere, causing them to fluoresce (emit light) at ultraviolet wavelengths. Voyager 2 confirmed the existence of a magnetic field around Uranus a few days before the spacecraft's closest approach to the planet, when its instruments detected radio emissions from solar particles trapped in the magnetosphere.

Uranus's magnetic field is at least as strong as Saturn's. The axis of Uranus's magnetic field is tilted 59° from the planet's rotation axis; by comparison, Earth's magnetic axis is tipped only about 12° from its rotation axis. As Uranus rotates, its magnetic axis, and thus its magnetic field, are spun around. Simultaneously, outward-racing charged particles in the solar wind interact with this spinning magnetic field and its trapped particles, pushing them back into a long, corkscrew-like tail behind Uranus.

Both the planet's rocky core and at least some of its overlying liquid layers may be involved in the production of its magnetic field. The magnetic field of a planet is the result of the motion of material in its interior. Periodic variations in the radio emissions of particles trapped in Uranus's magnetic field have allowed scientists to fix the rotation period of the core at 17.24 hours. Thus, the interior of Uranus spins more slowly than its atmosphere, a condition that planetary scientists had not expected.

Electroglow. Since no magnetic field had yet been detected around Uranus at the time the IUE satellite first observed an ultraviolet glow around the planet, some scientists speculated that the light was instead a phenomenon known as "airglow." Unlike an aurora, which is produced by high-energy particles, airglow is produced when solar ultraviolet light strikes atmospheric atoms.

Voyager 2 proved, however, that a different process caused the ultraviolet glow, which was dubbed "electroglow" by mission scientists. First, sunlight splits hydrogen atoms in Uranus's upper atmosphere into their constituent protons and electrons; the electrons then strike nearby hydrogen molecules, which emit an ultraviolet glow characteristic of the hydrogen atom. Because this mechanism depends on sunlight, it operates only on the sunlit side of the planet and would occur even if Uranus did not have a magnetic field. The electroglow phenomenon was also observed by the Voyagers on Jupiter and Saturn.

Moons of Uranus. Even in the largest Earth-based telescopes, the five major moons of Uranus appear as only points of light. Until the Voyager flyby, astronomers could estimate only their size, mass, density, and composition — primarily water ice — from the sparse data available. From outermost to innermost, the five major moons are Oberon, Titania, Umbriel, Ariel, and Miranda. Voyager 2 also discovered 10 new moons, all closer to Uranus than Miranda. They range in diameter from about 100 miles to only 10 miles. Since Uranus is tipped on its side with its south pole currently pointing toward the Sun, a tracing of the orbits of Uranus's moons would appear from Earth to resemble the rings of a target with Uranus itself at the center. Voyager 2's trajectory carried it relatively straight through this target, and all the closeup photography and monitoring of the moons took place in a brief, 6-hour span on 24 January 1986. Voyager 2's encounters with the moons of Jupiter and Saturn took several days in each case, since the spacecraft flew parallel to the plane of the moons' orbits, rather than perpendicular to it.

Oberon. Voyager 2 passed within 300,000 miles of Uranus's outermost moon, Oberon. The spacecraft's cameras detected a mountain on Oberon over 12 miles high — more than twice the height of Mount Everest. In addition, the Voyager photographs revealed a system of long, steep-sided cliffs and fissures, suggesting that Oberon once underwent vigorous geologic activity that created the torn landscape. The moon's surface is also marked by many meteorite impact craters surrounded by bright, spoke-like rays.

According to the report of Voyager's imaging scientists, published in July 1986, most of Oberon's craters were formed early in the solar system's history. The imaging scientists suggested that the bright rays surrounding the craters are probably water ice ejected from beneath Oberon's surface by the impacts that formed the craters. The deepest parts of several of the largest craters, which are up to about 60 miles in diameter, are covered with patches of very dark material. The Voyager researchers speculated that these dark patches may have been deposited when a fluid — perhaps dirty water — erupted through cracks in the crater floors and then froze.

Titania. When Voyager 2 flew within about 227,000 miles of Titania, it found many impact craters with diameters of up to about 120 miles. Like Oberon, Titania has an intricate network of scarps and fissures, including one fissure that extends for nearly 1,000 miles. Voyager scientists have speculated that Titania's scarps and fissures were produced when water froze deep in the moon's interior, causing the layers above to expand and rupture the surface. Since these scarps and fissures have few impact craters superimposed on them, the Voyager scientists have suggested that the features have been exposed to meteorite bombardment for a relatively brief time and that they are among the youngest features on Titania's surface.

Umbriel. When Voyager 2 passed Umbriel at a distance of about 200,000 miles, the moon appeared noticeably darker than Uranus's other major moons. Spectroscopic studies by the spacecraft confirmed that Umbriel's surface is rich in water ice and that the ice is covered by a thin layer of dark material. Although the origin of the dark material remains a mystery, some mission scientists have proposed that Umbriel has gradually accumulated the material as the moon has orbited Uranus. Other researchers have suggested that a large meteor or comet struck a pocket of dark material on Umbriel; the proposed impact would have ejected a huge amount of dark debris, which would have then settled uniformly over the moon's surface. Still others have proposed that Umbriel's dark material originated in the moon's interior and was ejected onto the surface by volcanic activity. Whatever the cause, the Voyager scientists believe that the major features on Umbriel's surface have remained largely unchanged since they were formed billions of years ago and that they were covered by the dark material much later.

Ariel. At its point of closest approach, Voyager 2 passed just 79,000 miles from Ariel. The spacecraft discovered that Ariel's surface is etched with deep, and sometimes interconnecting, canyons. Most of the canyon floors are covered by an icy material that apparently flowed from the moon's interior after the canyons were formed. The composition of the material on the canyon floors has not yet been determined, although water ice is probably the most likely candidate. Mission geologists have pointed out, however, that at Ariel's extremely cold surface temperature of about –315°F, water ice would be rock hard

Figure 4. Ariel. This mosaic of the four highest-resolution Voyager 2 photographs of Uranus's moon Ariel is the most detailed view of that moon ever obtained. Voyager 2 was about 80,000 miles from Ariel on 24 January 1986 when these photographs were taken. The smallest details seen here on Ariel's surface are about 1.5 miles across. *Courtesy: NASA/Jet Propulsion Laboratory.*

and immobile. Caltech planetologist David Stevenson has proposed that ammonia, methane, or carbon monoxide might have helped lubricate the water ice, thereby allowing it to spread over the canyon floors.

Miranda. Voyager 2 flew much closer to Miranda — 17,400 miles — than any of the other moons. Although it is the smallest of Uranus's five major satellites, Miranda nevertheless proved to be by far the most unusual. In the 28 February 1986 *Science*, planetary geologist Laurence Soderblom of the U.S. Geological Survey in Flagstaff, Arizona, described Miranda's surface as "a bizarre hybrid of every kind of exotic terrain in the solar system."

One unusual region is a dark rectangular feature measuring 100 miles by 70 miles. It frames a bright, V-shaped patch, dubbed "the Chevron" by mission scientists. Two adjacent sides of the rectangle are aligned with what appear to be long, straight faults that extend at least halfway around Miranda, while a third side falls off precipitously into a deep canyon. Voyager 2 also revealed an unusual series of concentric, oval grooves on Miranda that resembles a racetrack. Mission scientists informally named this feature the "Circus Maximus," after the famous chariot race course in ancient Rome. Closer inspection of the photographs showed that concentric, oval grooves also surround the Chevron.

This jumbled landscape has led mission scientists to speculate that Miranda may have suffered a catastrophic collision with an asteroid or large meteor billions of years ago, causing the moon to shatter completely. Such collisions, once considered highly unlikely, are now believed to have occurred often early in the solar system's history. According to calculations by the Voyager scientists, the shattered pieces of Miranda would not have dispersed into interplanetary space but would have instead coalesced to re-form the satellite. Over time, denser fragments would have tended to settle toward Miranda's center, forcing the lighter fragments toward the surface. This movement might have created enough heat and stress to induce geologic activity that could have produced Miranda's peculiar surface features. Even if no settling occurred, tides caused by the gravitational pull of Uranus may have generated enough heat to trigger geologic activity.

Rings of Uranus

Uranus's system of rings was discovered accidentally in March 1977 by astronomer James Elliot (then at Cornell University in Ithaca, New York) and colleagues while over the Pacific Ocean aboard NASA's Kuiper Airborne Observatory, a C-141 jet transport equipped for astronomical use. The scientists were recording the passage of Uranus in front of the star SAO 158687. Unexpectedly, the star dimmed briefly and repeatedly about 30 minutes before the star was blocked out by Uranus itself. Based on these observations, Elliot and his colleagues concluded that Uranus has a series of nine narrow rings.

The three rings closest to Uranus are designated by the numbers 6, 5, and 4, whereas the outer six are called alpha, beta, eta, gamma, delta, and epsilon. This discrepancy in naming stems from the nearly simultaneous discovery of the rings by Elliot's group and another group of astronomers led by Robert Millis of the Lowell Observatory in Flagstaff, Arizona, who observed the occultation from Perth, Australia. Elliot's group assigned Greek letters to the six rings found initially in their data, ordering them roughly from the inside out, while Millis's group assigned numbers to the nine rings they saw. The Uranian rings are difficult to observe directly because they reflect only five percent of the sunlight that strikes them. The ring particles, which are as black as charcoal, are among the darkest materials in the solar system.

The outermost ring is no more than about 16,000 miles from the top of the planet's atmosphere — a distance equal to about one Uranian radius. Together, they are only 6,000 miles across (including the spaces between the rings), compared to a breadth of roughly 40,000 miles for Saturn's three major rings. The eta, gamma, and delta rings are nearly circular, while the other six rings — the epsilon ring, in particular — are eccentric. The epsilon ring is also the widest of the original nine, ranging from about 15 to 90 miles across, while the gamma ring is only 2,000 feet from its inner to outer edges.

Voyager 2 found that the delta ring splits into separate strands at several points. The spacecraft also discovered two new rings. The narrower of these, designated 1986U1R, is located about 15,100 miles above Uranus's clouds, between the two outermost rings, delta and epsilon. The second newly discovered ring, 1986U2R, is closer to Uranus than any of the other known rings and is much broader — more than 1,500 miles wide.

As Voyager 2 receded from Uranus, it viewed the ring system backlit by the Sun. At this angle, sunlight illuminated previously unseen dust particles between the narrow rings. Although the particles in the spaces between the rings are very small, the rings themselves consist of relatively large particles. After Voyager 2 had passed by Uranus, it transmitted signals through the planet's rings. By studying the characteristics of signals passing through the rings, researchers concluded that the rings consist of boulder-sized chunks of debris.

Mission scientists believe that the gravity of small, unseen moons keeps most of Uranus's rings narrow. Astronomers Peter Goldreich of Caltech and Scott Tremaine of MIT proved in 1977 that a ring of debris can be kept narrow by the gravitational influence of a pair of so-called shepherding moons. A moon orbiting just outside the ring and another just inside the ring confine the debris to a relatively narrow region. In the Voyager Saturn flybys, scientists observed two such shepherding moons orbiting on either side of the narrow F ring. Only two of Uranus's 15 known moons, Cordelia and Ophelia, which straddle the epsilon ring, are shepherds. More shepherding moons probably exist around Uranus, but they were apparently too small to be recorded by Voyager 2's camera.

Voyager 2: 1989 and Beyond

According to the present flight plan, Voyager 2 will fly over Neptune's north pole on 24 August 1989, passing as close as 3,000 miles above the planet's clouds at a speed of 17 miles per second. On that day, the 12-year-old spacecraft will be about 3 billion miles from Earth. A few hours later, it will pass 25,000 miles from Neptune's satellite Triton and make observations to determine whether its surface is dotted with lakes of liquid nitrogen, as Earth-based observations suggest. Then, with its four planetary missions completed, Voyager 2 will coast outward through the solar system, monitoring interplanetary space beyond Neptune and relaying the results back to Earth. Once Voyager 2 leaves the solar system, it will head in the general direction of the bright star Sirius.

Return of Halley's Comet

Every 76 years, a familiar celestial visitor emerges from the blackness of space and comes into view: Halley's Comet. Once feared as a harbinger of disaster, the comet is now awaited eagerly by astronomers because of the clues it offers about the solar system's past. In 1985 and 1986, thousands of professional astronomers and millions of others sought out its pale glow in the sky. Before Halley retreated into the outer reaches of the solar system, a series of space probes pierced its luminous shroud of dust and gas and revealed for the first time the nucleus of a comet.

Astronomers believe that comets formed with the rest of the solar system around 4.6 billion years ago. Comets are especially interesting to researchers because, unlike planets, they lack chemical, meteorological, and geologic processes that have altered the planets' primitive composition. Thus, comets hold key information about conditions in the early solar system. Variations in a comet's form also reveal much about how gas and dust in these celestial bodies interact with the Sun's light and magnetic field. This information may provide clues about the way the Sun influences Earth.

Although only about 1,000 comets have been recorded during human history, probably at least a trillion exist in our solar system. Dutch astronomer Jan Oort hypothesized in 1950 that comets exist in a vast swarm, since named the Oort cloud, about 50,000 astronomical units from the Sun. An astronomical unit, abbreviated A.U., is the average distance between Earth and the Sun — about 93 million miles. Alpha Centauri, the star system nearest to the Sun, is about 270,000 A.U. distant.

Every 100,000 years or so a star will pass near the Oort cloud and gravitationally deflect a few of the comets toward the Sun. More significantly, some researchers believe that about once every 250 million years, the gravitational field of a giant interstellar gas cloud will deflect huge numbers of comets out of the Oort cloud. Others have speculated that a dim and as-yet-undiscovered companion star to the Sun or a distant planet may periodically disturb comets' orbits and send them into the inner solar system. However they are deflected, some comets, such as Halley, may pass close to Jupiter, whose gravity alters their paths further. (The other known planets have only a minor impact on comets.) The deflected comets settle permanently into elliptical orbits around the Sun, which brings them periodically back to Earth's vicinity.

Comet Structure

In 1950, Harvard University astronomer Fred Whipple devised the "dirty snowball" model to describe the structure of comets. According to this model, a comet consists of three major components: the nucleus, coma, and tail. The nucleus, Whipple's dirty snowball, is located in the comet's head and is composed primarily of water ice and dust, with smaller amounts of other ices, such as ammonia, carbon dioxide (dry ice), and methane.

Whipple's model explains why a comet, unlike a planet, often deviates from its predicted smooth path through space. When sunlight shines on the nucleus, the ices there sublimate, passing directly from the solid to the gaseous state. The newly formed vapor surges outward in discrete jets somewhat like those produced by geysers on Earth. Each jet exerts a force on the nucleus — just as the exhaust of a jet engine exerts a force on an airplane — that over time can push the comet off course.

By studying the rate at which comets release vapor into space, astronomers have deduced that the typical nucleus must be less than a few miles across, which is too small and too obscured by gas and dust to be observed from Earth. In addition, studies of comet orbits have revealed that a typical comet nucleus has a mass probably less than one-billionth that of Earth. As a result, particles that sublimate from the nucleus can easily escape the nucleus's weak gravitational pull and be swept away to form the coma and the tail.

The coma — the visible head of the comet — attains a maximum size ranging from tens of thousands to hundreds of thousands of miles in diameter. It consists of dust particles and a wide variety of molecules and atoms expelled by the nucleus. The coma glows partly because sunlight is reflected by the dust particles and partly because the gas in it fluoresces — absorbs sunlight and reemits it at different wavelengths. Since both processes depend on the Sun's light, the coma usually does not become visible until the comet has approached to within about 3 A.U. of the Sun. Images of comets taken at ultraviolet wavelengths have revealed that a vast cloud of hydrogen gas — often a million or more miles across — surrounds each coma. The hydrogen atoms in these clouds are derived from molecules of cometary water vapor that are split apart by sunlight.

The third main component of the comet is the tail, which consists of particles originally released by the nucleus into the coma. The tail always extends away from the Sun, regardless of the comet's direction of motion, and can be tens of millions and sometimes 100 million miles in length. Astronomers recognize that comets usually have two distinct tails — a dust tail and a plasma, or ion, tail. The yellowish dust tail consists of tiny grains of dust that are swept back from the coma by the pressure of sunlight. The bluish plasma tail consists of ionized atoms and molecules, which glow by fluorescence. The plasma tail elongates as its charged particles are "blown" back by the solar wind, the torrent of charged particles expelled by the Sun. The interaction of the plasma tail with the Sun's extensive magnetic field can create rapid changes in the tail's appearance.

Halley's Comet

Earth-based observations. The first documented sighting of Halley's Comet was made by Chinese observers in 240 B.C. Since that time, the comet has returned to Earth's vicinity 29 times, most recently in 1986. The comet was named for British astronomer Edmond Halley (1656 – 1742), who recognized its 76-year periodicity in 1705 and predicted its next passage around the Sun. Halley is considered a "short-period" comet, a class of comets that orbit the Sun in less than 200 years. Its highly elongated path extends out to just beyond the orbit of Neptune and is tilted 18 degrees to the plane of Earth's orbit.

Halley's orbit is retrograde; that is, it moves around the Sun in a direction opposite to that of Earth and the other planets. Halley's reappearance in 1985 and 1986 was among the faintest in recorded history. The comet came no closer to Earth than 0.42 A.U. — three times farther than its closest approach to Earth in 1910 and ten times farther than its nearest recorded passage in the year 837. Northern Hemisphere observers had their best view of Halley in late 1985 before the comet reached perihelion, the comet's closest approach to the Sun; Southern Hemisphere observers saw the comet best after perihelion in March and April 1986, when the tail was fully developed.

While most observers did not see the comet until 1985, astronomers David C. Jewitt and G. Edward Danielson at Palomar Observatory in southern California were the first to sight the comet when they photographed it on 16 October 1982. The two astronomers recorded Halley's Comet with an electronic camera placed at the focus of Palomar's 200-inch telescope. At the time, Halley was located beyond the orbit of Saturn and was 50 million times fainter than the faintest star seen by the unaided eye.

The comet became dimly visible to the unaided eye by November 1985. By mid-December, Halley's coma had expanded to more than a million miles in diameter, far larger than during the early stages of the 1910 return. Halley also brightened much more rapidly than anticipated toward the end of 1985, fueling hopes that its coming display would not be as mediocre as predicted.

Figure 5. Halley's Comet in 1986. Halley's comet was still approaching the Sun on 9 January 1986 when this photograph was taken with a 36-inch telescope. The comet, which will not return to Earth's vicinity until 2061, reached its peak brightness during early April 1986 when it was best seen from the Southern Hemisphere. *Courtesy: Lick Observatory.*

The comet's brightening rate leveled off, however, and at its peak brilliance in March and April 1986, Halley left many observers disappointed.

To obtain the maximum scientific return from both professional and amateur observations of Halley's Comet, the worldwide astronomical community organized the International Halley Watch (IHW). IHW's purpose was to coordinate all observations made of the comet and to collect and preserve the best data obtained. A similar effort had been organized during Halley's 1910 return, but it failed completely. As a result, much of the data collected in 1910 has been difficult to find, scattered in dozens of observatory archives and hundreds of different scientific journals.

IHW was a resounding success. Over 1,000 professional astronomers from 51 countries and as many as 10,000 amateurs around the world contributed their observations to IHW clearing centers at JPL and the University of Erlangen-Nurnberg in West Germany. IHW organizers Ray Newburn and Jurgen Rahe estimate that their Halley archive now contains some 30,000 pages of data, which they intend to publish on compact disks.

Earth-based observers registered a number of firsts in their observations of Halley's Comet. In February 1985, Susan Wyckoff of Arizona State University in Tempe, Arizona, and her colleagues detected the telltale spectrum of cyanogen gas — a combination of carbon and nitrogen atoms — escaping from Halley's icy core. This was the first time that astronomers had ever witnessed the onset of sublimation from a dormant cometary nucleus. Another first came in December 1985, when a team of NASA researchers using an airborne infrared telescope obtained the first direct terrestrial observation of water vapor in a comet.

First comet rendezvous. Halley is one of the few bright comets that produces gas and dust at a high rate and has a well-determined orbit. These properties made it an attractive and relatively easy target for close-range study by spacecraft. The U.S.S.R., Japan, and the European Space Agency (ESA) took advantage of this opportunity and sent five space probes to explore the comet. These were not, however, the first spacecraft to make close-up studies of a comet. Six months before the Halley encounters, a veteran space probe successfully passed through the tail of Comet Giacobini-Zinner (G-Z).

The probe was originally the third spacecraft in a joint NASA-ESA program called the International Sun-Earth Explorers. It was launched in 1978 to monitor the solar wind far beyond the Moon, well outside Earth's magnetic field. In 1982, NASA began radioing commands that sent the spacecraft through a complex series of maneuvers to retarget it to Comet G-Z. To reflect its new objective, the probe was renamed International Cometary Explorer (ICE). On 11 September 1985, ICE passed through G-Z's tail at a distance of 4,880 miles behind the nucleus.

In an unexpected finding, ICE encountered electrically charged cometary particles over a million miles away from G-Z's nucleus, indicating that comets are much larger than photographs imply. In ICE's 20-minute passage through G-Z's tail, the spacecraft also detected water vapor, supporting Whipple's dirty-snowball model of cometary nuclei, as well as carbon monoxide gas and a small amount of dust. In addition, ICE's magnetometer confirmed the hypothesis that the plasma tail is confined by the interplanetary magnetic field draping around the comet's head as the comet moves through space.

Halley probes. In March 1986, five spacecraft passed near or through the coma of Halley's Comet. The U.S.S.R. launched two identical Vega probes, Japan sent Sakigake and Suisei, and ESA launched Giotto. Because of budgetary constraints, the U.S. did not send any probes directly to the comet but instead monitored it from the Pioneer Venus Orbiter 27 million miles away and from ICE 20 million miles distant.

The data returned by the five Halley probes are complementary, since each mission had a different focus. Giotto was sent within a few hundred miles of the nucleus to take close-up photographs and collect and analyze particles directly. The Vega probes, which entered the coma but remained several thousand miles away from the nucleus, made measurements of the comet's inner regions. The two Japanese spacecraft did not approach the nucleus and concentrated on studying the outer portions of the coma. By agreement among the research teams involved, the arrival times of the five spacecraft were staggered by several days so that variations in the comet's properties could be studied over time.

Giotto was ESA's first interplanetary space mission and also the first scientific spacecraft launched by ESA's Ariane rocket. The Giotto probe was named after Italian painter Giotto di Bondone (c. 1266 – 1337) who, in 1303, included Halley's Comet in one of his frescoes. Giotto was launched in July 1985 from ESA's Kourou Space Center in French Guiana. It was targeted to fly within 300 miles of the sunlit side of Halley's nucleus.

The main part of the spacecraft is a reinforced cylinder 6 feet in diameter, containing the rocket thrusters, fuel tanks, and a communications antenna, as well as a set of instruments and a camera with a periscope-type mirror system. The instruments were designed to analyze the composition of cometary particles, measure the coma's brightness, and study the effects of the solar wind on the comet. A dust-impact detector was mounted on the outside of the cylinder together with two magnetometers to measure the strength of the magnetic fields.

A two-layer shield protected Giotto against cometary dust impacts as the craft hurtled through the comet at 43 miles per second (about 50 times the speed of a bullet). A thin aluminum plate backed up by a thicker layer of polyurethane foam and Kevlar (a material used in bulletproof vests) could withstand impacts by dust grains as large as one gram. To keep its communications antenna pointing at Earth, the orientation of Giotto's cylinder had to be stabilized by spinning on its axis at a rate of 15 times per minute. For this reason, a special-purpose computer was needed to keep Giotto's camera fixed on its target.

The Soviet twin Vega spacecrafts are modified versions of the highly successful Venera Venus probes. The name Vega is an acronym formed from the Russian words "Venera" (Venus) and "Gallei" (Halley). Vega 1 was launched atop a Proton rocket from the Baikonur launch complex in Kazakhstan, on 15 December 1984, and its sister ship Vega 2 was launched 6 days later. On their way to Halley's Comet, Vega 1 and Vega 2 flew by Venus on 11 June and 15 June 1985, respectively. (For more information about the Vega's exploration of Venus, see the article "Solar System Update" in this chapter.)

Each 5,500-pound Vega craft consists of a central cylinder with a 30-foot-long, rectangular solar-cell array and, at each end, a conical skirt. Unlike the standard Venera probes, the Vegas were fitted with dust shields to protect them against the hazardous cometary environment. The Vegas also carried gyroscopes to stabilize them and keep their instruments fixed on target. Each Vega was equipped with a television camera to observe the nucleus directly and instruments to detect dust particle impacts, determine the gas composition within the coma, and study the interaction between the coma and the solar wind.

One of the important tasks of the Vega mission was, in fact, to provide targeting data for the Giotto probe. Dubbed the Pathfinder project, this unprecedented joint venture of the U.S.S.R., NASA, and ESA sought to pinpoint the location of Halley's nucleus in space more accurately than could be done from Earth. NASA tracking stations carefully followed the Vega spacecraft so that their positions in space were known precisely when they encountered Halley's Comet in March 1986. Using Vega observations, mission scientists were able to calculate the nucleus's location in space to within about 25 miles. Using this improved data, ESA ground controllers made final corrections to Giotto's trajectory only two days before its closest approach to Halley.

Japan sent two probes — Sakigake and Suisei — to study Halley. Sakigake (formerly designated MS-T5), an engineering prototype for Suisei, was launched from the Kagoshima Space Center in January 1985 atop the newly developed Mu-3S-II rocket. Suisei (formerly Planet A) followed in August 1985.

Sakigake and Suisei differ only in the scientific experiments they carry. Each spacecraft has a cylindrical main body 4.6 feet in diameter and 2.3 feet high with a parabolic antenna on top. During the missions, each spacecraft generally spun on its axis 6 times per minute for stability, but this spin- ning rate could be slowed to 0.2 revolutions per minute, as required for some experiments. Suisei's principal instrument, an ultraviolet camera, was used to record changes in the appearance of the comet's coma and hydrogen cloud. Sakigake's payload was devoted to plasma experiments.

Halley's nucleus revealed. On 6 March 1986, Vega 1, traveling at a velocity of 49 miles per second, approached within 5,523 miles of Halley's nucleus. Three days later, Vega 2 closed to within 4,990 miles of Halley's nucleus. Giotto, guided by the Vegas' pathfinder observations, swept to within 360 miles of Halley's nucleus on 13 March 1986. Two seconds before its closest approach, though, dust impacts caused Giotto's antenna to lose its lock on Earth, and data transmissions were lost for the next 34 minutes. Although it was not targeted at the nucleus, Japan's Suisei came within 94,000 miles of Halley's nucleus on 8 March. Its companion, Sakigake, passed 4.4 million miles from Halley three days later.

Giotto recorded over 2,100 high-quality pictures before its camera malfunctioned 12 seconds before the closest approach. The spacecraft's final image of the nucleus was sent when it was 1,020 miles away. The photographs revealed that Halley's nucleus — half in sunlight and half in shadow — is potato-shaped and roughly 9 miles long and 5 miles wide, or about twice as large as had been predicted. Its surface is very rough and covered with what appear to be craters, valleys, and hills.

Imaging-team leader Horst Keller of the Max Planck Institute for Aeronomy in Katlenburg-Lindau, West Germany, reported that Halley's nucleus reflects only 3 or 4 percent of the sunlight striking it. This means that it is one of the darkest objects in the solar system. At a press conference following Giotto's encounter, Keller said that "the true color of the nucleus is black, absolutely black, blacker than coal, almost like velvet." The low reflectivity is thought to be due to a coating of black dust.

Two bright dust jets — appearing like geysers spewing both dust and gas — were observed near the middle of the sunlit side of the nucleus. Each jet emanated from a closely packed collection of circular features that may be craters. Five other fainter jets were identified closer to the edges of the sunlit side. Mission scientists concluded that the comet gives off most of its dust and gas in jets and not uniformly over the nuclear surface, as had been believed.

During Giotto's encounter, 12,000 dust impacts were registered by the craft's impact detector. Researchers expected the dust to be similar in composition to carbonaceous chondrites, a rare class of stony meteorites containing some carbon. What they found was material rich in carbon, hydrogen, oxygen, and nitrogen — a mixture given the acronym CHON. This discovery solved one mystery. Ground-based observations had revealed far less carbon in cometary gases than expected from models of cometary composition. Giotto revealed that most cometary carbon is locked up in CHON grains.

The dust particles observed in the coma were also smaller than expected. None of the particles detected were larger than about 10^{-3} gram, and small particles weighing less than 10^{-10} gram were far more numerous than previously thought. The smallest grains detected were only 10^{-6} centimeter across and weighed 10^{-17} gram. Despite the high number of recorded impacts, the spacecraft intercepted at most one gram of cometary dust.

Giotto's instruments also found that Halley's coma was about 80 percent water vapor by volume. Carbon dioxide, carbon monoxide, and numerous other gases were also detected. In August 1987, two independent research groups announced that Giotto data revealed that a compound called polyoxymethylene was also present in Halley's coma. This long chain of interlocking formaldehyde molecules could be responsible for binding Halley's dust grains together on the nucleus and could explain why the nucleus is so dark.

Unlike Giotto, which observed Halley's nucleus for only a few hours, the Vega spacecraft observed the nucleus for a period of several days during their flybys. During this time they transmitted 1,500 photographs back to Earth. The Vega photographs also showed distinct jets on the sunlit face of the nucleus, even though the spacecrafts were much farther from their target than Giotto.

Infrared studies by the Soviet probes revealed that the temperature of the nucleus and its surrounding environment was between 80°F and 260°F — significantly hotter than would be expected if the nuclear surface were predominantly ice. Vega researchers have suggested that solar energy is absorbed by a surface dust layer and then transmitted as heat to the ice below. This proposed warming of subsurface ice and its subsequent sublimation into gas could explain what powers Halley's jets of gas and dust. Like Giotto's scientists, Vega researchers also discovered complex hydrocarbons in Halley's dust. Jochen Kissel and F. R. Krueger reported in April 1987 that when these substances react with water, they can form other complex organic compounds, such as sugars. These findings strengthened the view held by some researchers that the chemicals necessary for life may originate outside Earth.

During 5 months of observations, the Japanese spacecraft Suisei took more than 1,000 ultraviolet images of Halley. These showed that the brightness of the hydrogen cloud surrounding Halley's nucleus varies rhythmically with a period of about 52 hours. If the areas of sublimation are restricted to specific sites, the rotation of the nucleus would periodically expose these sites to the Sun's light, thus activating them and explaining the regular brightness variations.

These and other observations from Giotto and Vega, together with some ground-based data obtained in 1910, 1985, and 1986, seemed to suggest that Halley's nucleus rotates once every 52 hours. However, observations made by the Pioneer Venus Orbiter and ground-based studies by American astronomers Robert Millis and David Schleicher showed that Halley also appears to have a 7.4-day brightness cycle. These observations posed a problem, since the comet can obviously have only one rotation period.

Klaus Wilhelm of the Max Planck Institute for Aeronomy proposed a solution in May 1987. Wilhelm suggested that the 52-hour period is Halley's true rotation rate, but that the comet may also be precessing — wobbling like a top — once every 14.8 days. This precession would explain the longer period, because it would mean that rotational stresses on the nucleus would peak twice during each precessing cycle. These forces could trigger venting of material from the comet's surface and hence brightening of the hydrogen cloud every 7.4 days. Wilhelm's solution is not universally accepted. Some researchers believe that we will discover Halley's true rotational rate only by sending a space probe to rendezvous with it and make observations over a period of many months.

Stars

About 3,000 stars are visible to the naked eye under a dark moonless sky. These are, however, only a minute fraction of the several hundred billion stars that make up the Milky Way Galaxy. Stars are believed to form when clouds of gas and dust collapse under their own gravity. Just as the temperature of any gas rises when it is compressed, the contraction of a gas cloud causes the temperature in the cloud's core to rise. Once a temperature of about 10 to 20 million degrees Fahrenheit is reached, thermonuclear fusion begins, converting hydrogen into helium in a series of nuclear reactions called the "proton–proton chain." When the energy generated in the star's core is sufficient to prevent further collapse, the star is then termed a "main-sequence star."

A star's color is a consequence of its surface temperature which, in turn, depends on the star's mass. Stars more massive than the Sun generally have higher surface temperatures and are typically blue or white; those less massive than the Sun are generally cooler and are typically orange or red.

A star's main-sequence lifetime is determined by the amount of nuclear fuel (hydrogen) available. Our Sun has enough hydrogen fuel to supply its needs for another 5 billion years; it is already 5 billion years old. Blue and white stars exhaust their fuel in a few tens or hundreds of millions of years because they are hundreds to thousands of times brighter than the Sun. Orange and red main-sequence stars are often much less luminous than the Sun and burn for hundreds of billions of years.

Once the hydrogen in a star's core is exhausted and nuclear reactions cease, there is nothing to counteract the crushing force of its own gravity. The core collapses slowly until the temperature is high enough to ignite the unused hydrogen located in a shell around the star's helium-rich core. Subsequent contraction raises the core temperature even higher. When these changes take place in the star's interior, its outer layers cool and swell perhaps a hundredfold. Once this has occurred, the star is called a "red giant." The inside of the red giant may become hot enough to convert helium in the core into carbon via a series of nuclear reactions.

Stars more massive than the Sun can attain core temperatures sufficient to cause fusion reactions involving carbon, neon, and other heavier elements. What happens when all the nuclear fuel is exhausted depends on the star's mass. The cores of the most massive stars are many times denser than the center of the Sun and can collapse catastrophically when available nuclear fuels are exhausted, triggering a supernova explosion and leaving a neutron star or black hole. Stars less massive than the Sun lose their outer layers after the red giant phase, leaving a hot, dense core depleted of nuclear fuel — a white dwarf — that eventually cools into a dark stellar ember called a black dwarf.

Star Formation

Among all the stages in a star's life, the least understood is its birth. The reason is simple: astronomers cannot see the stellar nurseries in which stars form because they are embedded in opaque clouds of gas and dust. Although light cannot penetrate these clouds, astronomers can detect infrared radiation, or heat, emitted from these star-forming regions.

Many infrared sources that may be nascent stars have been located in gas and dust clouds, but there is as yet no definitive evidence that any are protostars — that is, stars still forming. Proof that an infrared emitter buried in an interstellar cloud is really a protostar would require observations that the object is actually contracting to form an object as dense as a star. Unfortunately, such events occur only in the densest and most opaque portions of interstellar clouds, from which even infrared radiation cannot escape.

Fortunately, astronomers have another way to observe star formation — radio emissions from molecules. Interstellar clouds are rich in many molecules such as carbon monoxide, water, and carbon monosulfide. These molecules emit characteristic radio signals at wavelengths ranging from a few millimeters to a few tenths of a millimeter. The signatures of these molecules appear as lines in the radio spectrum of interstellar clouds.

Many of these radio emissions show Doppler shifts caused by motions in the cloud. In other words, the emissions from the parts of the cloud moving toward Earth and away from Earth have slightly shorter and slightly longer wavelengths, respectively, than emissions from parts that are not moving along the line of sight. Astronomers can thus use these shifts in molecular radio emissions to deduce how material inside a candidate protostar is moving.

When astronomers have studied candidate protostars with their radio telescopes, they have often found that the clouds are expanding rather than contracting, the implications of which are not yet clear. The expansion could mean that the candidates are not protostars at all, but are instead very young stars whose radiation is blowing away the remnants of its protostellar cloud. It could also mean that star formation involves the simultaneous collapse of the central part of a cloud and the expansion of its outer regions. The latter possibility is supported by observations made in 1986 by astronomers from the University of Arizona and the University of Missouri. The researchers discovered evidence that the central parts of a candidate protostar called IRAS 16293 –2422 are contracting while the outer regions appear to be expanding.

Stellar Companions

Search techniques. The detection of planets around stars other than our Sun is one of the primary goals of astronomy. Current theories postulate that planets form from a disk of debris left over from the birth of a star. However, given the many uncertainties about star formation, these theories have not been proved. Even so, the majority of astronomers believe that planets exist around most if not all single stars. Planets may not exist in binary or multiple star systems because the complex gravitational fields may render their orbits unstable.

The direct detection of planets around other stars could help answer many questions about how planets form. Detecting planets is, however, a daunting task. In principle, astronomers know of four different ways in which planets could be detected. The most obvious method is to use a powerful telescope to directly observe a planet orbiting another star. Unfortunately, planets are thousands to millions of times fainter than their parent stars. While detecting the light from a planet against the background glare of its star is virtually impossible, making observations at infrared wavelengths can increase the planet's contrast against its star. In 1987, a team of astronomers from the University of Arizona announced that they had successfully detected a stellar companion by using this technique.

Another method is to observe periodic changes in the brightness of a star as its planets eclipse it. This technique is also difficult, however, because the brightness change will be very small and occur only infrequently. Eclipses will not be seen at all unless the star and planet are oriented such that the planet passes directly in front of its parent star as seen from Earth.

A more promising method is to look for slight periodic shifts in a star's position caused by its motion around the center of gravity of the star-planet system. In other words, the planet does not move in an orbit about the exact center of its star, but rather about a point in space displaced slightly from the star's center. The more massive star also follows a much smaller orbit around the center of gravity. As seen from Earth, the star's shift is very small, perhaps only one one-hundred-thousandth of a degree, but observable in principle. Several teams, most notably George Gatewood and colleagues based at Allegheny Observatory near Pittsburgh, have reportedly achieved success with this technique, but many astronomers are not convinced by their results.

The fourth technique is a variation on the third. Instead of looking for periodic shifts in a star's position, astronomers can look for periodic changes in its velocity caused by its motion around the star-planet center of gravity. The velocities involved are relatively small — only a few tens of meters per second — but detectable. In June 1987, Bruce Campbell and his colleagues from the Dominion Astrophysical Observatory in Victoria, British Columbia, reported discovering several stellar companions with this method.

Brown dwarfs. One of the most unusual incidents in the long search for stellar companions began in July 1983. A group of U.S. Naval Observatory scientists led by Robert Harrington reported measuring a slight wobble in the motion of a faint star named Van Biesbroeck 8 (VB 8), which is about 21 light-years away in the constellation Ophiuchus. The wobble could be explained only as the result of the gravitational pull of a large planet or a small star orbiting VB 8; however, this companion, which the researchers designated VB 8B, was not visible. By measuring the size of the wobble and calculating VB 8's gravitational pull, the Naval Observatory group determined that the companion is lightweight by stellar standards — perhaps only a few times the mass of Jupiter.

In 1984, University of Arizona astronomer Donald McCarthy and his colleagues confirmed and extended the Naval Observatory results. They calculated that an object of VB 8B's mass would radiate more energy at infrared wavelengths than in the visible-light portion of the spectrum. By using high-resolution infrared detection techniques, the astronomers claimed to detect VB 8B about 600 million miles from the star, a distance somewhat greater than that of Jupiter from the Sun. They estimated that VB 8B may be up to ten times as massive as Jupiter

— the largest planet in our solar system — but slightly smaller than Jupiter in diameter.

Although VB 8B is very massive compared to the planets in our solar system, the accepted theories of star formation indicate that its mass is well below that necessary to trigger the self-sustaining nuclear fusion reactions that generate energy in stars. With a surface temperature of about 2,000°F and an inherent brightness only about 0.003 percent of the Sun's, VB 8B falls between the least massive stars and the most massive known planets. Since experts disagree about whether to classify the object as a small star or large planet, some researchers have introduced the term "brown dwarf."

An interesting twist to the VB 8B story occurred in 1987. Seeking to confirm McCarthy's results, C. Perrier and J. M. Mariotti of Lyon Observatory in France, and another group of American researchers at Cornell University and the University of Rochester in New York, made their own infrared observations of VB 8, but neither group saw the brown dwarf. In an attempt to confirm his own findings, McCarthy made more observations, but he also failed to observe VB 8B again. The evidence also seemed to indicate that VB 8B was not hidden behind its companion. The final blow came when Harrington announced that additional observations at the Naval Observatory failed to find any evidence for a wobble in VB 8's orbit.

The consensus now seems to be that VB 8B does not exist. Why did two research groups find similar evidence for a nonexistent body? Most astronomers believe that the VB 8B affair was a fluke of the type that has bedeviled planet searches for decades. Nevertheless, the case of VB 8B did not put an end to searches for brown dwarfs. In June 1987, Bruce Campbell and colleagues from Canada's Dominion Astrophysical Observatory announced the results of their 7-year search for planets around other stars. They found subtle shifts in the velocities of 7 of the 16 stars that they had observed. Of these 7, two stars — Epsilon Eridani and Gamma Cephei — showed large velocity shifts that could only be explained if they were accompanied by objects perhaps eight to ten times more massive than Jupiter.

In November 1987, Ben Zuckerman of the University of California at Los Angeles (UCLA) and his colleagues reported their infrared observations of a possible brown dwarf in orbit around a white dwarf called Giclas 29-38. But additional observations of Giclas 29-38 performed by Zuckerman and others failed to confirm or disprove the presence of a brown dwarf companion. Zuckerman and Eric Becklin of the University of Hawaii in Honolulu had better luck with another white dwarf called GD165, which they observed at NASA's Infrared Telescope Facility at Mauna Kea Observatory in Hawaii. Their observations revealed a very faint companion, with a mass of 0.06 to 0.08 Suns, adjacent to the white dwarf. Unlike the case of Giclas 29-38, where the brown dwarf companion was never actually seen but only inferred from excess infrared emissions from the region around the white dwarf, Becklin and Zuckerman were able to obtain actual images of GD165's companion. From their observations of GD165, Becklin and Zuckerman concluded that very low-mass stars and even brown dwarfs could be quite common in our galaxy. If they are correct, brown dwarfs may account for a significant amount of the so-called missing mass in the universe. (For more information about missing mass see the article "Cosmology" in the "Physics" chapter.)

Disks around stars. The Infrared Astronomical Satellite (IRAS), a joint project between the U.S., the Netherlands, and the U.K., surveyed 95 percent of the sky at infrared wavelengths during its 10-month mission in 1983. One of IRAS's most surprising discoveries involved the star Vega, the fifth-brightest star in the sky. Because Vega is bright and has other well-known properties, project scientists George Aumann at JPL and Fred Gillett of Kitt Peak National Observatory in Tucson, Arizona, had been using the star to calibrate the IRAS telescope.

As they studied the star, Aumann and Gillett found that the emissions were much greater than expected at some infrared wavelengths. With further observations, they determined that the energy emanated not from Vega itself but from a disk, or perhaps a broad ring, of particles around the star. The particles, each at least about one millimeter in diameter, are located at an average distance of about 7.9 billion miles from Vega — about twice as far as Pluto's average distance from the Sun. The IRAS scientists also calculated that the amount of matter in the disk is nearly 300 times the mass of Earth, or roughly equivalent to the total mass of the planets in the solar system.

Since Vega is much younger than the Sun, some astronomers speculated that the disk may actually be a protoplanetary system — a planetary system in the early stages of formation. JPL comet expert Paul Weismann suggested in 1984 that Vega's dust cloud is in fact a ring of icy bodies that resembles the solar system's Oort cloud — a distant swarm of icy objects, some of which occasionally journey into the inner solar system as comets.

After a second bright star, called Fomalhaut and located in the constellation Piscis Austrinus, was found to be surrounded by a similar disk, IRAS scientists sifted through the satellite's observations of the 9,000 brightest stars in the sky. They found that about 90 of these stars had unexpectedly high infrared emissions. Approximately half of those had companion stars that could account for their excess infrared energy; the remainder — over 40 stars — might have protoplanetary systems.

IRAS researchers have cautioned that their observations do not prove that there are other planetary or protoplanetary systems. Excess infrared emissions might be caused, for example, by the presence of a cool, extremely dim companion star that has not yet been detected optically. Unfortunately, IRAS could not distinguish between faint companion stars and particle clouds.

In 1984, Bradford Smith of the University of Arizona and Richard Terrile of JPL reported the first Earth-based observation of a circumstellar disk. They identified a disk around the star Beta Pictoris, which was one of IRAS's protoplanetary system candidates. Smith and Terrile used sophisticated imaging techniques to block light coming from the star itself in order to record the faint glow of the surrounding disk. The photograph, taken with the 100-inch telescope at Las Campanas Observatory in Chile, clearly showed a flat particle disk extending outward some 40 billion miles from Beta Pictoris — about ten times Pluto's average distance from the Sun.

The disk is nearly edge-on as viewed from Earth; hence, Beta Pictoris is actually seen through the disk and is therefore slightly dimmer than other stars of its type and distance. Smith and Terrile said that the amount of dimming is less than what would occur if the disk extended all the way in to the star. In other words, the central part of the disk is dust-free. They calculated that the cloud of particles begins at a distance from Beta Pictoris about equal to that of Neptune from the Sun. Within the inner edge of the dust disk, the researchers suggested, the particles may have already coalesced to form planets.

ESA scientists Francesco Paresce and Christopher Burrows reported additional observations of Beta Pictoris at the January 1987 meeting of the American Astronomical Society in Pasadena, California. By observing the disk at several different wavelengths, Paresce and Burroughs were able to calculate that dust grains one micron (one-millionth of a meter) in diameter are present in the Beta Pictoris disk. Since typical interstellar dust is only one-tenth this size, they concluded that the dust around Beta Pictoris must be "snowballing." Thus, processes similar to those believed to have occurred during the formation of the solar system might now be occurring around Beta Pictoris.

UCLA's Zuckerman and his colleagues reported at the same meeting that the dust grains in the disk are similar to those produced by comets. In their view, the disk is not a nascent planetary system, but the debris from collisions between comets formed around Beta Pictoris in the distant past.

The hypothesis that Beta Pictoris is surrounded by a planetary system in the process of forming was bolstered by new evidence on the frequency of circumstellar disks presented at the January 1989 meeting of the American Astronomical Society in Boston, Massachusetts. Karen Strom and her colleagues from the Five College Radio Astronomy Observatory in Amherst, Massachusetts, reported that of a sample of 100 stars less than 3 million years old, almost 60 percent showed the characteristic infrared emissions from dust grains, indicating the presence of a circumstellar disk. Stars 10 million years old, however, showed no signs of dust emissions. According to team member Stephen Strom, these observations indicate that as a star ages, "collisions among dust grains may have led to the build-up of small planetesimals — mile-wide rocks which are the stuff out of which planets may be assembled. When that happens, there are no more small dust grains around, and the infrared signal from the young stars disappears."

Black Holes

In the last decade, astronomers have devoted much attention to objects they can never hope to see directly — black holes. These objects are regions of space where gravity is so strong that nothing, not even light, can escape. Stephen Hawking, a theoretical physicist at Cambridge University in England, overturned the established view of black holes in the mid-1970s. Until then, physicists had regarded them as being totally black and incapable of emitting radiation. Hawking, however, showed that black holes should have a characteristic temperature that is determined by their mass; low-mass black holes should have higher temperatures than high-mass black holes. If this idea is correct, then like all hot objects, black holes must emit some radiation. Since low-mass holes are hotter than their heavier counterparts, they must be the stronger emitters.

If Hawking's theories are correct, then they have profound implications for the properties of the three types of black holes thought to exist in the universe. So-called mini-black holes may have been created in the violent conditions that prevailed during the Big Bang. These objects are unstable and eventually destroy themselves in a flash of radiation. Hawking calculated that black holes less massive than an asteroid would have decayed by now, because their rate of energy expenditure is so great that their lifetimes are less than the current age of the universe.

Supermassive black holes, which may exist in the cores of quasars and galaxies, have masses ranging from millions to billions of times that of the Sun. Such black holes have low enough temperatures that they can survive for many times the current age of the universe. They are probably formed by collisions between stars in the dense innermost regions of a galaxy where stellar encounters are common. Astronomers have gathered evidence suggesting that supermassive black holes may exist in the core of our Milky Way Galaxy and several of its neighbors.

The most concrete evidence of black holes is available for those with masses similar to that of a large star. Like their supermassive counterparts, such black holes could survive for many times the current age of the universe. They would form when the core of an old star, perhaps eight or more times as massive as the Sun, collapses after depleting all available nuclear fuel because it can no longer resist the inward pull of its own gravity. These stellar black holes would be virtually impossible to detect if they existed in isolation. However, some black holes are thought to be part of binary star systems. In such cases, the black hole's intense gravity may pull matter off its companion. As this matter spirals into the black hole, it is heated by friction to a temperature of several million degrees Fahrenheit, hot enough to emit x-rays.

Astronomers have detected dozens of x-ray binary systems. In most cases, observations show that the emitting object is an ultradense neutron star. In a few instances, however, the x-ray emitter seems to have too much mass to be a neutron star. (According to current astrophysical theories, neutron stars more than about three times as massive as the Sun are unstable and collapse into black holes. Thus, any binary system whose x-ray emitting component is more than three times the Sun's mass is a prime black-hole candidate.)

The first x-ray binary system suspected of harboring a black hole was identified in 1971 and is known as Cygnus X-1. The second candidate, LMC X-3, was announced by University of Michigan astronomer Anne Cowley and her colleagues in 1983. A third example, called A 0620-00, was discovered by Jeffrey McClintock of the Harvard-Smithsonian Center for Astrophysics and Ronald Remillard of the MIT during 1986.

Conclusive proof that any of these systems contains a black hole is lacking. The evidence for Cygnus X-1 and LMC X-3 is good, but it is possible to construct theoretical models in which these binaries contain three, rather than two, stars. If these systems really are triples, then no black hole is needed to explain the observations. These models are, however, somewhat contrived, and most researchers discount them. The evidence for A 0620-00 is less certain. Its x-ray-emitting component is at least 3.1 solar masses and probably much higher. Even if the lower mass limit is correct, this star would still be massive enough to have collapsed into a black hole.

Some researchers are still not convinced that black holes exist. If they are correct, then what happens when a star too massive to become a neutron star collapses after all the nuclear fuel in its core is depleted? The answer, according to F. Curtis Michel of Rice University in Houston, Texas, lies in a family of exotic subatomic particles called hyperons. In 1988, Michel proposed that when the density inside a collapsing neutron star reaches 10 to 20 times that of an atomic nucleus, reactions begin between protons, neutrons, and hyperons (believed to be plentiful in such circumstances). The energy released by these nuclear reactions would be so great that the entire star could explode as a supernova before it could become a black hole.

Supernovae

On 24 February 1987, an event occurred which astronomers had been waiting generations to observe. A star in the Large Magellanic Cloud, a satellite galaxy of the Milky Way, died in a titanic explosion. Astronomers had observed hundreds of such supernovae in distant galaxies, but the last one close enough to be seen with the naked eye occurred in 1604, before the invention of the telescope. Observing a supernova only 160,000 light-years away would provide far more information than observing hundreds of supernova tens of millions of light-years away.

Supernovae are the most violent stellar phenomena ever observed. In a matter of hours the brightness of a single star increases until it exceeds that of a galaxy containing 100 billion Suns. Over the next few months to years, the star fades, leaving an expanding cloud of debris with perhaps a neutron star or even a black hole at its center.

Besides being cosmic beacons, supernovae play a vital role in the evolution of the universe. Without supernovae, everything we see around us, ourselves included, would not exist. All of the chemical elements except hydrogen and helium were created by nuclear fusion reactions in stars. (Astronomers believe that hydrogen and helium were created in the first few minutes after the Big Bang.)

Many of the elements — principally the lighter ones such as carbon, nitrogen, and oxygen — are created in the energy-generating reactions that power normal stars. Astrophysicists believe that most heavier elements, particularly those heavier than iron, can be formed only in the violent conditions that prevail in a supernova explosion.

Whether formed in ordinary stellar reactions or supernovae explosions, the elements are dispersed into interstellar space as stars explode. In this way, the material is carried to the clouds of gas and dust and recycled as new stars are formed. Thus, each successive generation of stars contains a higher proportion of heavy elements than its predecessor because of the supernova enrichment of interstellar matter with these elements. All of the iron in our blood, calcium in our bones, and carbon in our flesh was created by nuclear reactions in stars that died in titanic explosions in the distant past.

In addition to dispersing the chemical elements that would otherwise be trapped in stellar cores, supernovae can trigger star formation. As a star dies in a supernova explosion, it generates a blast wave that travels through space. If this wave encounters any dense interstellar clouds, they are compressed, possibly triggering the collapse of clumps of gas and dust that leads to the formation of new stars.

Astronomers believe that supernovae come in two varieties, called Type I and Type II. Type I supernovae are thought to occur in binary systems containing a white dwarf and a normal star. White dwarfs are stable only if their masses are less than 1.4 times that of the Sun. If a white dwarf pulls so much matter from its companion that its mass exceeds this limit, it collapses uncontrollably because the outward forces that normally prevent a star from contracting are overwhelmed by gravity. The heat generated by the collapse raises the star's temperature to such a level that residual nuclear fuel ignites in a cataclysmic explosion that completely destroys the star. Type II supernovae occur in stars eight or more times as massive as the Sun. When a star has depleted all the nuclear fuel in its core, there is nothing to support the star against the inward pull of its own gravity. The star's core collapses, triggering the blast that blows its outer layers into space. Astrophysicists believe that if the star's core contains less than three times the Sun's mass, it may survive as a neutron star. If it is more massive than three Suns, the core turns into a black hole.

Whether a collapsing stellar core stabilizes as a neutron star or a black hole depends on the detailed physics of stellar interiors. In a star like the Sun, pressure exerted by atomic nuclei prevents gravitational collapse. In a white dwarf, the outward force exerted by electrons is the dominant means of preventing collapse. In dense neutron stars, neutrons provide the outward force. At higher densities, even the pressure exerted by neutrons is insufficient to prevent the core from collapsing into a black hole. The ultimate fate of a star like the Sun (and other stars less massive than the Sun) is to stabilize as a white dwarf without undergoing a supernova explosion.

Discovery of 1987A. The supernova in the Large Magellanic Cloud was dubbed Supernova 1987A because it was the first supernova discovered in 1987. Like most other supernovae, it was discovered accidentally. On 24 February 1987, Ian Shelton, a young Canadian telescope operator at Las Campanas Observatory was making routine photographs of the Large Magellanic Cloud. Just before dawn he developed his last picture and discovered a bright star not visible on a photograph of the same part of the sky that he had photographed earlier. When he told the observers using another telescope at Las Campanas of his discovery, he learned that Oscar Duhalde, a Chilean telescope operator, had spotted the exploding star in the Large Magellanic Cloud with the naked eye a few hours earlier but had not realized its significance.

Word of Shelton's and Duhaldes's discovery soon reached the Central Bureau of Astronomical Telegrams, the international clearinghouse of astronomical discoveries, located at the Harvard-Smithsonian Center for Astrophysics in Cambridge, Massachusetts. From there, astronomical institutes and observatories around the world were notified of the discovery.

Because Supernova 1987A is in the far southern sky, it is always below the horizon of most of the world's major observatories in the Northern Hemisphere. Only observatories located in Australia, New Zealand, South Africa, and southern South America can see the supernova.

Six hours after Shelton's discovery, astronomer Robert McNaught of the Anglo-Australian Observatory in New South Wales, Australia observed Supernova 1987A and discovered that its position coincided with a star called Sanduleak –69° 202. (The name identifies the star as number 202 in a band located 69 degrees south of the celestial equator in a catalogue compiled by Nicholas Sanduleak of Warner and Swasey Observatory in Cleveland, Ohio.)

The fact that the supernova was a catalogued star was very significant. Among all the supernova observed before 1987A, only one — Supernova 1961V — was a previously known star. To learn more about why supernovae occur, astronomers must know something about the stars

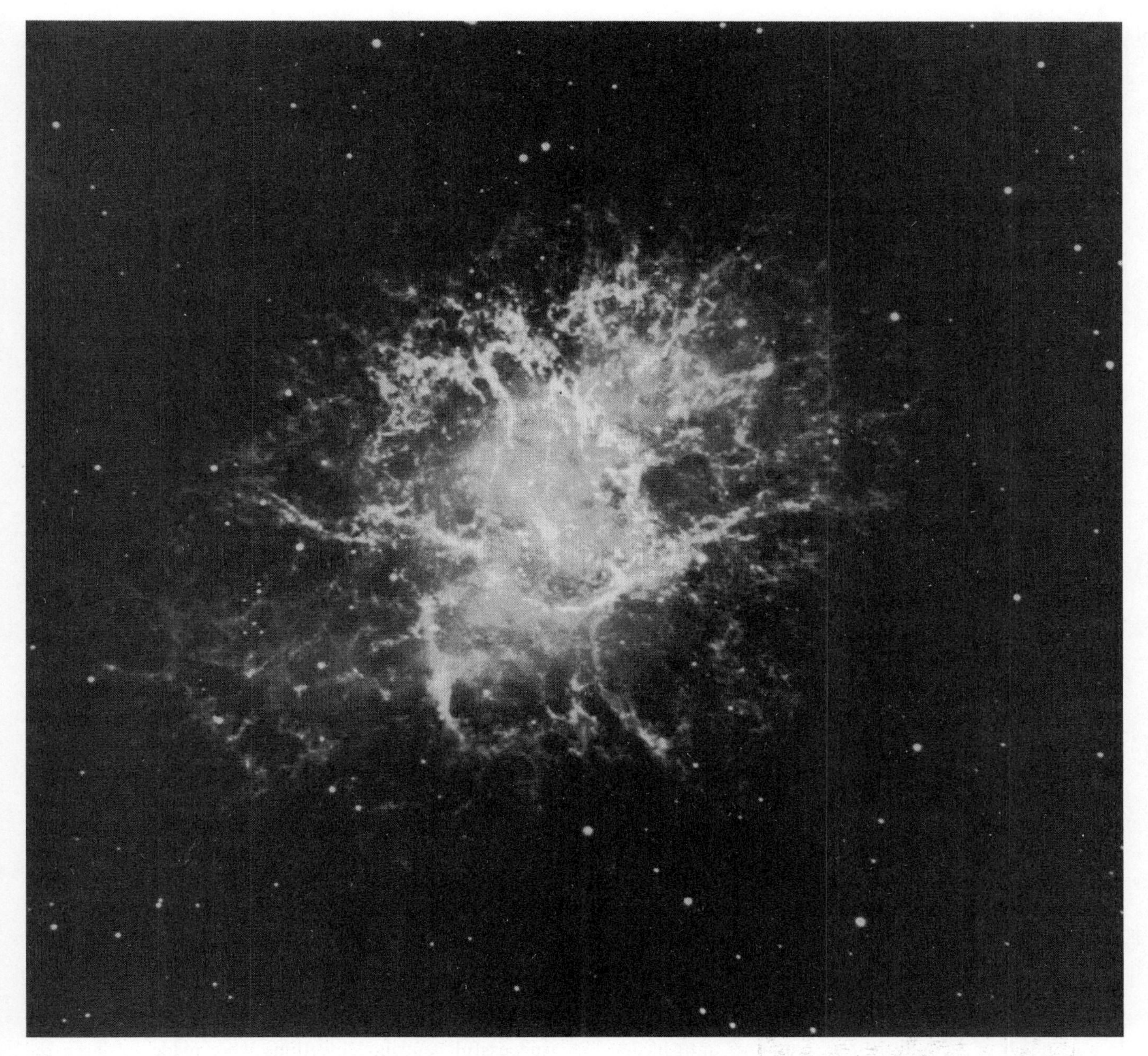

Figure 6. Crab Nebula. **The Crab nebula, in the constellation Taurus, is the gaseous remnant of an exploding star, or supernova, recorded by Chinese astronomers in the year 1054. A fast-rotating pulsar is located within the nebula.** ***Courtesy: Palomar Observatory photograph.***

before they explode. The existence of observations, and spectra in particular, for Sanduleak −69° 202 greatly helped astronomers understand how and why the star exploded.

Less than a day after Supernova 1987A was discovered, astronomers in South Africa were able to obtain the supernova's spectrum — the pattern of different component wavelengths that make up a star's energy emissions. The spectral pattern proved conclusively that this event really was a stellar explosion and suggested that it was a Type II supernova. At about the same time, the International Ultraviolet Explorer satellite obtained the first ultraviolet observations of the supernova. Japanese space scientists also directed their recently launched Ginga satellite to search for x-ray emissions from the supernova.

Stellar collapse. Supernova 1987A's brightness peaked in late May 1987, and then started to fade. Soon thereafter, astronomers began to piece together exactly what caused

Sanduleak –69° 202 to explode. Sanduleak –69° 202 was a blue supergiant star with a surface temperature 3 times greater and a mass between 10 and 20 times greater than that of the Sun. Such a star is extremely bright, perhaps 100,000 times brighter than the Sun, and burns its nuclear fuel at prodigious rates, giving it a lifetime of only 20 million years or so. By comparison, the Sun has an estimated lifetime of 10 billion years.

The key to the cause of the explosion lies in the nuclear reactions that took place in the core. All stars are in a constant battle between the inward pull of their gravity and the outward pressure caused by energy generated by nuclear reactions in their cores. At first, a star is stabilized against the inward pull of its own gravity by the outward gas pressure exerted when hydrogen is fused into helium in its core. When the central supply of hydrogen is exhausted, the core contracts and heats up sufficiently for helium to fuse into carbon. Then, when the helium is exhausted, the core contracts once again until conditions are suitable for carbon to fuse into oxygen and neon.

Each new fuel supply provides the star with less energy than its predecessor and halts the star's inexorable collapse for a shorter and shorter period. For example, the carbon fuel is sufficient to forestall the collapse for only a thousand years. The fusion of neon into silicon can delay the collapse for just a year or so. The final crisis occurs when the core has heated up sufficiently for silicon to fuse into iron. At this point, the star's structure resembles an onion: The star's central core is iron, surrounded by successive layers rich in silicon, neon, oxygen, carbon, helium, and hydrogen.

The silicon fuel is exhausted after a few days, and the iron core, weighing perhaps 1.6 solar masses, starts to contract. Since iron is the most stable element, all of its nuclear reactions are accompanied by the absorption rather than the release of energy. When the star has exhausted all of its fuel supplies, there is nothing to stabilize it. In a matter of a second, the core collapses until its density equals that of an atomic nucleus — about 50 billion tons per cubic inch. The star has finally lost its long battle with gravity.

Neutrino detection. When the star's iron core collapses, the protons and electrons in the individual iron atoms are crushed together to form neutrons, and a flood of subatomic particles called neutrinos is released from the star's core. Neutrinos are massless and electrically neutral, and they interact only rarely with other subatomic particles. What happens next is not entirely clear. Some theorists believe that the collapsing core rebounds after reaching a state of maximum density. Exactly why the rebound occurs is not fully understood, but astrophysicists believe it must occur if the collapse of the star's core in the early phase of a supernova explosion is to be converted into the rapid outward motions seen in later stages of the explosion. Some theorists think that the core "bounce" triggers a shock wave that runs out through the star ripping away its outer layers and generating the outburst that Shelton and Duhalde observed. Other theorists believe that the flood of neutrinos streaming out from the core carries energy that is deposited in the star's outer layers, generating the outward-moving shock wave that shatters the star.

In either case, most of the energy locked up in the star's gravitational field is carried away by neutrinos. In all, researchers estimate that neutrinos released some 2×10^{46} joules of energy — equivalent to the Sun's energy output for more than one trillion years — from 1987A. Only 0.3 percent of this energy is needed to blow away the star's outer layers and power the optical display seen by Shelton and Duhalde. (When the blast wave reaches the star's outer surface approximately an hour after the core collapses, the star's energy output at ultraviolet wavelengths alone is about 100 billion times the Sun's total energy output.) Spectacular as Supernova 1987A's visual pyrotechnics were, they accounted for only a tiny fraction of the total energy released in the explosion of Sanduleak –69° 202.

Some astrophysicists have estimated that a total of 10^{53} neutrinos were emitted by Supernova 1987A. Even though the Large Magellanic Cloud is approximately 160,000 light-years away, some 60 billion neutrinos from the supernova passed through every square inch of the Earth's surface. When researchers operating neutrino detectors in underground laboratories around the world heard of 1987A's discovery, they began to search their data to see whether their instruments had detected any of these neutrinos.

Within days of the discovery of 1987A, four claims of successful neutrino detections were made by large research groups in Europe, the U.S.S.R., Japan, and the U.S. The consensus among researchers is that the apparent detections by the European and Soviet groups were not valid. The claims by American and Japanese researchers, on the other hand, are generally believed to be the first detections of neutrinos from a supernova. Between 2:35:35 and 2:35:47 EST on 23 February 1987, researchers at the Kamiokande II neutrino detector in Japan's Gifu prefecture registered 11 neutrinos coming from the direction of the Large Magellanic Cloud. Between 2:35:41 and 2:35:47 EST, another eight neutrinos were observed with the Irvine-Michigan-Brookhaven (IMB) detector located near Cleveland, Ohio.

Both the Kamiokande II and IMB detectors consist of tanks containing several thousand tons of water monitored by hundreds of light-sensing photomultiplier tubes. As neutrinos pass through the water they emit characteristic flashes of light, which are registered by the photomultipliers. Researchers can reconstruct the track of a particular neutrino from the times that different photomultipliers were triggered.

The number of neutrinos detected by the IMB and Kamiokande II detectors, and the energies they carried, agreed almost exactly with predictions of the supernova neutrino emissions made by many different astrophysicists years before Supernova 1987A was observed. The neutrino observations, which provided direct evidence of events occurring at the center of Sanduleak –69° 202, confirmed that the supernova explosion was powered by the collapse of a star's core and revealed that temperatures in excess of 90 billion degrees Fahrenheit were reached at the center of the collapsing star.

Aftermath. As the cloud of debris from the explosion of Sanduleak 69° 202 expands and cools, astronomers are eagerly awaiting several events. The decline in the supernova's brightness is following almost exactly the predictions of a model postulating that the decay of the radioactive isotopes nickel-56 and cobalt-56 (radioactive elements derived from the vast amounts of iron-56 liberated by the supernova explosion) energizes the debris once the initial pyrotechnics have died down. As the cloud thinned out, astronomers hoped to detect directly the gamma rays emitted by these isotopes as they decayed. Since gamma rays cannot penetrate the Earth's atmosphere, a number of research groups are using balloon-borne instruments carried above the absorbing layers of the Earth's atmosphere to search for the radiation. Another form of high-energy radiation, x-rays, has been observed from 1987A. Instruments on the Japanese satellite Ginga and on the Kvant astrophysics module attached to the U.S.S.R.'s Mir space station detected x-rays for the first time in August and September 1987.

A major puzzle surrounding the supernova concerns the nature of a mysterious spot of light seen close to 1987A by researchers from the Harvard-Smithsonian Center for Astrophysics and Imperial College in London, England. First detected in March 1987, the spot is one-tenth the brightness of the supernova and located only 0.0000158 degrees from it. Many hypotheses have been advanced to explain the spot, but all have been rejected. Surprisingly, when the Harvard-Smithsonian group searched for the spot in August 1987, they could find no sign of it. Their failure led some astronomers to doubt if the spot ever existed at all.

Another eagerly awaited discovery as the debris thinned out was the first detection of a neutron star at the heart of the supernova. Radio and x-ray observations of this object, in particular measurement of its rotation rate and magnetic field, should shed new light on the origin and evolution of pulsars.

In February 1989, John Middleditch of the Los Alamos National Laboratory in New Mexico and colleagues reported the discovery of optical pulses from Supernova 1987A. Their observations were made with the 160-inch telescope at the Cerro Tololo Inter-American Observatory in Chile on 18 January 1989. Computer analysis of their data revealed extremely strong pulses from the expanding debris cloud coming 1,968.629 times a second. Middleditch and his colleagues interpret their data as long-expected evidence of a neutron star remaining after the supernova explosion. If the research team's observations are confirmed and the signals they detect really are from a neutron star, the star must be spinning at the unprecedented rate of once every 0.5 millisecond — three times faster than the fastest-spinning star previously known. Since Supernova 1987A continues to be active at gamma, x-ray, and infrared wavelengths and is eventually expected to "switch on" at radio wavelengths, the star will be a prime target for every major telescope in the Southern Hemisphere for many decades to come.

Supernova search programs. Most supernovae, including 1987A, are discovered by accident. Astronomers are particularly interested in observing supernovae in their very earliest stages, while they are still brightening. For the past 20 years, researchers have been trying to design automatic supernova search programs using computer-controlled telescopes. The basic strategy is for a telescope to take photographs of a preprogrammed list of galaxies every few nights. The photographs would be electronically compared with reference images stored in a computer. If any difference between the two images was found (such as a new relatively bright star), an "alarm" would go off, signaling the discovery of a possible supernova.

A pioneering version of such a system was established in the mid-1960s by Northwestern University astronomer J. Allen Hynek at Corralitos Observatory in New Mexico. Although his system used a computer-controlled telescope, it relied on observers to visually compare television images of galaxies with reference images in real time as the observations were being made. Hynek's system led to the discovery of 14 supernovae, but it exerted such a mental and physical strain on observers that it was discontinued after 2 years.

A modern version of Hynek's system was recently established at Leuschner Observatory in California by astronomer Richard Muller of the University of California at Berkeley. Operator fatigue has been reduced by storing the galaxy observations on magnetic tape so that the comparisons with reference images can be made at a more leisurely pace. By mid-1987, Muller's program had discovered three supernovae.

The goal of a fully automated search has been pursued by Stirling Colgate and his colleagues at the New Mexico Institute of Mining and Technology for 20 years. Writing computer software that can reliably distinguish between supernovae and changes in a galaxy's appearance caused by, for example, the unstable nature of the Earth's atmosphere had always been the stumbling block to this approach. Colgate's system finally went into operation in February 1987.

Muller's and Colgate's search programs have a formidable human rival, Australian amateur astronomer Robert Evans. Since November 1980, Evans has conducted a highly successful supernova search using just his memory, keen eyesight, and a homemade telescope in his garden. In his first 5 years of searching, Evans made 50,403 observations of 1,017 galaxies and discovered 11 supernovae. Evans, who works in close cooperation with professional astronomers in Australia, has become the yardstick against which all professional supernova search projects are judged.

The rivalry between Evans and the professional supernova researchers was highlighted by the discovery of Supernova 1989B. On 30 January 1989, Evans reported the discovery of a supernova (1989B) in the bright galaxy M66. His discovery was confirmed the same night by Robert NcNaught at Siding Spring Observatory in New South Wales, Australia. The following day, the Berkeley Automated Supernova Search group reported that they too had sighted Supernova 1989B on images taken by their automatic telescope on 21 and 30 January. Although the Berkeley astronomers had made the earliest observations of the exploding star, honors for its discovery were given to Evans for his prompt reporting of the event.

Pulsars

First discovered in 1967 by Cambridge University graduate student Jocelyn Bell, pulsars are rapidly spinning neutron stars that emit sharp pulses of electromagnetic energy at regular intervals ranging from a fraction of a second to several seconds. Neutron stars concentrate a mass greater than that of the Sun in a body only about 10 miles in diameter. They are so compact that they can be as dense as 50 billion tons per cubic inch.

Neutron stars are formed when an old star more massive than the Sun exhausts its nuclear fuel. With no source of energy, the star can no longer support itself against the inward pull of its gravity and its core collapses. The result is a supernova explosion in which the star's outer layers are blown off into space as the protons and neutrons in the star's core are crushed together to form neutrons.

Neutron stars spin so rapidly because angular momentum — a measure of the amount of spin — is conserved during their collapse. Since angular momentum depends on both a star's radius and its rotation rate, the rotation rate must increase as the radius shrinks if its angular momentum is to remain constant. The process is similar to the spin-up experienced as spinning ice skaters draw their outstretched arms inward. In addition, the laws of electromagnetism require that the total number of magnetic lines of force threading the star's surface also remains constant as the star collapses. Thus, a neutron star's magnetic field may be a trillion times more concentrated than it was before the star collapsed.

The combination of rapid spin and ultrastrong magnetism means that a pulsar is, in effect, a vast dynamo capable of generating immense amounts of electromagnetic energy, although the mechanism for this phenomenon is not fully understood. Energy emissions seem to be concentrated at the pulsar's north and south magnetic poles (which do not necessarily coincide with the north and south rotational poles). Astronomers detect pulses of electromagnetic energy when the star's rotation carries the magnetic poles across their line of sight. In other words, pulsars act as celestial lighthouses. All of a pulsar's energy output is ultimately derived from its rotation. Thus, as the pulsar ages, it must pay for its energy expenditure by gradually slowing down. Astronomers suspect that the intensity of the pulsar's magnetic field also falls over time. The combination of declining rotation rate and magnetic field strength may explain why many neutron stars do not pulse; if astronomers' ideas on energy generation in pulsars are correct, then the electromagnetic dynamo of old neutron stars is too weak to generate detectable amounts of energy.

Millisecond pulsars. Until 1982, the fastest pulsar known was found at the center of a glowing cloud of gas called the Crab Nebula in the constellation Taurus. Spinning 33 times every second (about once every 30 milliseconds), the Crab pulsar was born during a supernova explosion witnessed by Chinese observers in 1054. In

November 1982, a research team led by Donald Backer of the University of California at Berkeley discovered a pulsar in the constellation Vulpecula that shattered the Crab pulsar's record, spinning 642 times a second (once every 1.56 milliseconds). This object — PSR 1937 +214 — has become known as the "millisecond pulsar."

In 1983, a second ultrafast pulsar was detected by Cornell University astronomer V. Boriakoff and his colleagues. This pulsar, called PSR 1953 +29, has a period of 6.1 milliseconds and, significantly, is a member of a binary star system. In December 1985, astronomer David Segelstein and his colleagues from Princeton University discovered yet another millisecond pulsar, PSR 1855 +09. Located in Aquila, it has a period of about 5.4 milliseconds and is also a component of a binary star system. As of early 1989, eight ultrafast pulsars had been identified.

Of the more than 400 pulsars known, only 11 are found in binary systems. Yet six of the eight fastest pulsars known are binaries, a fact that many astrophysicists believe is significant. Astrophysicists began to suspect that the presence of a companion might play a crucial role in the creation and evolution of ultrafast pulsars. Another significant clue to what causes such ultrafast rotation rates is that all fast pulsars seem to have magnetic fields approximately 1,000 times weaker than those of slower pulsars.

According to the most popular model, ultrafast pulsars are very old neutron stars that have been rejuvenated by drawing matter from a companion star. As part of the evolution of a binary star, a normal pulsar forms in orbit about a less massive companion. After 50 million years or so, the pulsar "dies" because its magnetic field has become too weak, or its spin has become too slow, to generate pulses.

The neutron star remains dormant until its companion swells as it evolves into a red giant. At this point, the neutron star may be able to draw matter from its distended companion. The matter falling onto the neutron star carries angular momentum, so the neutron star starts to spin faster and faster. Eventually it may spin fast enough to start pulsing again even though it still has a relatively weak magnetic field. The result is a rapidly spinning pulsar and a very lightweight companion star. In some cases, the companion may lose so much mass that it is no longer detectable. This situation might explain the apparent lack of a companion star associated with the first millisecond pulsar. This model received a notable boost in May 1988, when Andrew Fruchter and his colleagues at Princeton University discovered a new millisecond pulsar called PSR 1957 +20, later dubbed the "black widow" pulsar.

The Princeton researchers' observations with the 1,000-foot radio telescope at Arecibo, Puerto Rico, demonstrated that this pulsar is in a binary system with an orbital period of about 9 hours. Moreover, their data indicated that the companion star has an extremely low mass, as little as one-fortieth that of the Sun. Optical images obtained by Caltech's Shrinivas Kulkarni and Jeff Hester, using the 200-inch telescope on Palomar Mountain in California, show that a plume of matter is streaming away from the binary system as the intense radiation from PSR 1957 +20 excites the surrounding interstellar gas. The intense radiation is also believed to heat the side of the companion facing the pulsar to such an extent that it evaporates into space and streams away from PSR 1957 +20 like the tail of a comet. If this interpretation is correct, then the companion star was originally much more massive and has been gradually "eaten away" by the "black widow" pulsar.

Pulsars in globular clusters. Observations of globular star clusters also support the hypothesis that ultrafast pulsars form when an old neutron star draws matter and angular momentum from a companion star in a binary system. (Globular clusters are spherical systems containing typically 100,000 to one million very old stars.) Because two of the first three ultrafast pulsars were members of binary systems, David Helfand of Columbia University and colleagues initiated a search of suitable binary systems for more ultrafast pulsars.

They confined their attention to globular star clusters where binary systems consisting of an old neutron star and a less massive companion are thought to be 100 times more common than in the rest of the galaxy. Of the 12 clusters that they searched, only M28 in Sagittarius contained a promising pulsar candidate. Unfortunately, they could not detect any pulses from this object.

Four years after Helfand's search, pulses were finally detected from his candidate object by Andrew Lyne and his colleagues at Manchester University in England. Observations with the 250-foot radio telescope at Jodrell Bank near Macclesfield, England, revealed that the object called PSR 1821 –24 pulses every 3.05 milliseconds (327 times a second). PSR 1821 –24 does not seem to be a member of a binary system. In October 1987, Lyne's group discovered another fast pulsar, which pulses every 11.08 milliseconds, in the globular cluster M4 in the constellation Scorpius. Since then, a slowly rotating pulsar has been discovered in the globular cluster M15 in Pegasus and two additional millisecond pulsars have been found in the globular cluster 47 Tucanae in the southern constellation of Tucana.

The discovery of ultrafast pulsars in globular clusters has caused much excitement because it provides new information about the origin and evolution of these rapidly rotating objects. The concentration of stars at the center of a globular cluster is so great that the average distance between stars is only about 0.1 light-year. (For comparison, the Sun's nearest neighbor, Proxima Centauri, is 4.1 light-years distant.) In these circumstances, close encounters between stars are not uncommon.

Close encounters play a pivotal role in the formation and evolution of objects such as PSR 1821 –24, according to three papers published in late 1987. In the dense central regions of a globular cluster, binary systems can form and break up relatively easily. The consensus among pulsar experts is that PSR 1821 –24's neutron star formed in isolation and was later captured by another star. After the neutron star was rejuvenated by drawing matter from its new companion, the binary system broke apart as a result of another close encounter. Roger Romani and colleagues from Caltech estimate that radio astronomers can expect to find as many as 10 ultrafast pulsars in binary systems and another three as isolated objects in each of the Milky Way's 100 dense globular clusters. As of early 1989, five pulsars were known in globular clusters, four of which seem to be members of binary systems.

Pulsars as clocks. For many centuries, the Earth's rotation was the basis of all time-keeping systems. In the last few decades, however, researchers have discovered that the Earth's rotation keeps time too irregularly for modern applications such as the navigation of interplanetary space probes. Today, all time systems are based on the atomic clocks located at timekeeping institutes around the world, including the National Institute of Standards and Technology (formerly the National Bureau of Standards) in Boulder, Colorado. Together, these atomic clocks define a time system known as UTC (the French acronym for Coordi-nated Universal Time). Several years ago, the idea of readopting an astronomically based time system was unthinkable. With the discovery of more millisecond pulsars like PSR 1937 +214, the possibility of a change is much greater.

The regular radio pulses emitted by pulsars suggests that they would be ideal clocks. The ultimate source of a pulsar's energy is its rotation. As pulsars emit energy, their spins gradually slow down. The younger the pulsar, the more energy it emits and the faster it slows down. Since most pulsars are young, this rapid slowdown, plus slight unpredictable jumps observed in their rotation rates, means that a pulsar clock would gain or lose about one second in 10,000 years. But impressive as this may appear, it is only one-tenth as precise as the atomic clocks currently used to define UTC.

The time maintained by the pulsing of the millisecond pulsar PSR 1937 +214 is another matter. Since it is such an old object and it spins so rapidly, its rotation is very stable. Its measured slowdown rate amounts to only one second in a trillion years — which means it is 10 million times as precise as current atomic clocks. Every other week since November 1982, a team led by Joseph Taylor of Princeton University has been monitoring the millisecond pulsar with the 1,000-foot-diameter radio telescope at Arecibo, Puerto Rico. Their results, published in 1987, reveal that the millisecond pulsar is a timekeeper at least as good as any existing atomic clock.

Measuring the arrival times of pulses from millisecond pulsars holds great promise as a means of accurately timing celestial phenomena. One area where the ticks of a pulsar clock would be important is in the measurement of important cosmological parameters. For example, careful timings of the pulses from millisecond pulsars may reveal evidence of gravitational waves generated during the Big Bang. Motions caused by the passage of gravitational radiation through the solar system would cause small, but detectable, changes in the times pulses from fast pulsars arrived on Earth.

Galaxies

The Milky Way

Viewed from a dark location far from city lights, our galaxy — the Milky Way — is strikingly apparent as a luminous band of seemingly countless stars across the night sky. The Milky Way Galaxy, which is about 100,000 light-years across, contains several hundred billion stars. Our solar system is about two-thirds of the way out from the center.

If our galaxy could be viewed from a great distance outside, it would appear to be disk-shaped. The central region, or "nuclear bulge," is, however, almost spherical and contains predominantly old, red stars. Surrounding the center is a disk of dim, red stars and some younger, yellower stars like the Sun. The disk, which is about 2,000 light-years thick, also contains clouds of gas and dust illuminated by the hot blue and white stars embedded in them. The hot stars, which are very young, and the clouds of dust and gas in which they formed are arranged in a series of spiral arms extending out from the nuclear bulge.

Surrounding the disk is our galaxy's halo, a roughly spherical region of space sparsely filled by several hundred globular star clusters. In orbit about the Milky Way are two satellite galaxies, the Large and Small Magellanic Clouds.

Probing the depths of space with large telescopes, astronomers have discovered that most galaxies fall into one of three basic types. Those with prominent spiral arms, such as the Milky Way, are known as spiral galaxies, while those that lack spiral arms and have an elliptical or spherical appearance are called elliptical galaxies. Finally, galaxies that have no ordered structure at all, such as the Magellanic Clouds, are known as irregular galaxies. About 77 percent of all known galaxies are spiral, 20 percent are elliptical, and 3 percent are irregular.

Heart of the Milky Way. The center of the Milky Way Galaxy is located at a distance of about 24,000 to 30,000 light-years in the direction of the constellation Sagittarius. Astronomers studying this region work under a major handicap; opaque clouds of gas and dust block visible light from the center of the Milky Way, hiding it from view. Infrared, radio, x-ray, and gamma-ray emissions can penetrate the dust, however. Recently, numerous studies carried out at these wavelengths have revealed that the Milky Way's nucleus is a dynamic, highly energetic environment, rather than the relatively inactive place once thought. Since the early 1980s, much circumstantial evidence has accumulated that a supermassive black hole surrounded by an extensive disk of gas and dust lies at the galactic center, although hard proof for this idea has been difficult to obtain.

A black hole's gravity is so strong that it prevents even light from escaping. The region surrounding a black hole can, however, emit large amounts of electromagnetic radiation as interstellar matter spiraling turbulently into the black hole heats up and radiates energy. Black holes are relatively compact celestial objects; one with a mass equal to that of the Sun would be only about four miles across.

Infrared observations show that the galaxy's center is surrounded by a very dense star cluster about 1,000 light-years across, consisting of old orange and red stars. The core of the cluster is very bright and at its center lies a bluish object called IRS 16. The total amount of infrared energy emitted by the objects in the central 10 light-years is equivalent to that of about 6 million Suns.

At radio wavelengths, two prominent sources are visible at the galactic center. One, known as Sagittarius A East (Sgr A East), is believed to be the expanding debris from a supernova that exploded a few hundred to a few thousand years ago. The other source, Sgr A West, is centered on IRS 16. A detailed map of Sgr A West, produced by Caltech radio astronomers Kwok-Yung Lo and Mark J. Claussen, revealed that gas within 5 light-years of the galactic center appears to form a four-armed spiral. At the center of the spiral is a point-source of radio emissions, called Sgr A* (Sgr A star) and coincides almost exactly with IRS 16.

A detailed study of the velocities of individual clouds in the spiral, carried out by astronomer John Lacy and colleagues at the University of California at Berkeley, suggests that one and possibly two of these arms form part of a ring of gas orbiting the galactic center. The other arms may be filaments of matter falling into the galactic center. In 1985, astronomer Michael Crawford and his colleagues, also at the University of California at Berkeley, calculated that the motions of the individual clouds in the ring and the filaments were best explained if a black hole about 4 million times as massive as the Sun were at the center of the gas ring.

The case for a supermassive black hole was strengthened when Lo and coworkers reported in 1985 that their radio observations revealed Sgr A* was to be no more than 2 billion miles across — approximately equal in diameter to the orbit of Saturn. This finding was particularly significant because the only object that could be this small and yet emit so much radiation is a supermassive black hole. Another piece of important evidence has come from several satellite and balloon observations, which revealed the presence of a variable source of gamma rays near the galactic center. These gamma ray emissions are characteristic of the annihilation of electrons and positrons and the decay of the radioactive isotope aluminum-26 to magnesium-26. According to West German astrophysicist Peter van Ballmoos of the Max Planck Institute for Extraterrestrial Physics in Garching and his colleagues, such emissions are evidence for either an exotic physical process in the region around a supermassive object at the center of the galaxy or the occurrence of some 1,000 novae or supernovae in the galactic center in the last 100,000 to one million years. If the latter explanation were correct, the West German researchers would have expected to observe gamma rays from the decay of radioactive titanium-44 in addition to those from the decay of aluminum-26. No such emission was detected, however.

The most convincing argument against the existence of a vast black hole at the galactic center comes from infrared observations first made more than a decade ago. Much of the energy emitted by matter falling into a black hole would be in the form of ultraviolet light. This energy, which would be transformed to infrared radiation by

Figure 7. Galaxy in Andromeda. This galaxy in the constellation Andromeda is over 2 million light-years from Earth and resembles the appearance our galaxy — the Milky Way — would have if it were viewed from the outside. The diffuse, oval patches to either side of the Andromeda Galaxy are smaller galaxies. *Courtesy: Palomar Observatories.*

multiple scatterings, absorptions, and re-emissions as it passed through gas and dust clouds near the galactic center, has not been detected in the amount that would be expected if a central black hole exists. Some astronomers have suggested that instead of having a black hole, the hub of the Milky Way contains nothing more exotic than a cluster of very bright stars.

Another alternative to the black hole model for activity in the Milky Way's core was proposed by MIT's Charles Engelke in 1988. Engelke believes the activity in the nucleus of our galaxy and some of its neighbors is due to normal pulsars whose emissions are smothered by surrounding high-density gas. Engelke's model has the advantage of simplicity; only one quality — the gas density — has to be varied to match the size, spectrum, and luminosity of the radio emission from Sgr A*. Nevertheless, the evidence for or against the existence of a supermassive black hole at the galactic center has remained inconclusive, and the debate may continue for years.

Quasars

In 1960, research carried out by radio astronomer Thomas Matthews of Caltech and optical astronomer Allan Sandage of Mount Wilson Observatory near Pasadena, California, led to the identification of a blue, starlike object in the constellation Triangulum with a powerful source of radio signals called 3C 48. Sandage obtained a spectrum of this "star" with the 200-inch telescope on Palomar Mountain and found that it did not resemble the spectrum of any known object. Three years later, the strong radio source 3C 273 was found to coincide with another dim, blue star. Because of their starlike appearance, both 3C 48 and 3C 273 were called quasi-stellar radio sources, or quasars.

In 1963, Caltech astronomer Maarten Schmidt obtained a spectrum of 3C 273. He found that the quasar's spectrum contained the characteristic pattern of spectral lines emitted by hydrogen atoms; however, instead of being at the usual wavelengths, the hydrogen lines were displaced toward the red end of the spectrum by about 16 percent of their usual wavelength. This shift in the positions of the spectral lines is called a red shift.

Red shifts are usually interpreted as a consequence of the Doppler effect, the apparent change in frequency of the light or sound emitted by a moving object. On Earth, the Doppler effect is most frequently encountered with sound waves. For example, the whistle of a receding train sounds lower in pitch — has a longer wavelength — than it would if the train were stationary. The faster the train is moving away, the more pronounced the lowering of the pitch.

If an astronomical body is receding from Earth, the wavelengths of the object's light as observed on Earth are lengthened or, equivalently, shifted toward the red part of the visible spectrum. The greater the extent of the red shift, the faster the object is moving away. In an expanding universe, the more a galaxy's or quasar's light is red-shifted, the farther the galaxy or quasar is from Earth. (For more information about the theory of an expanding universe, see the article "Cosmology" in the "Physics" chapter.)

American astronomer Edwin Hubble derived a relationship between red shift and distance in the 1920s. Using this relationship, Schmidt calculated that the red shift he measured for 3C 273 corresponded to a distance from Earth of at least 3 billion light-years. At about the same time, astronomer Jesse L. Greenstein of Caltech reanalyzed the spectrum of 3C 48 and discovered that all its hydrogen spectral lines showed a red shift of about 37 percent. By applying Hubble's relationship between red shift and distance, Greenstein calculated that 3C 48 is more than twice as distant as 3C 273.

If quasars are as distant as their red shifts suggest, then they must be extremely luminous. For example, if quasar 3C 273 is really 3 billion light-years away, then it emits as much energy in a second as the Sun emits in 300,000 years. Quasars have another strange property: their brightness varies erratically over periods ranging typically from several days to several months. In particular, the brightness of 3C 466 at visible-light wavelengths has been observed to double within just two days. (Faster changes have also been observed at other wavelengths — the x-ray output of the quasar H 0323 +022 has been observed to increase by an amount equal to the energy output of a trillion suns in less than 30 seconds.) Astronomers believe that these variations indicate that the energy-emitting regions of quasars are very small. The emitting region of 3C 466, for example, can be no more than 2 light-days (about 30 billion miles) across. (This limit arises because the fastest an object can vary in brightness is the time it takes light to travel from one side of the object to the other side.) The small size of quasars has been confirmed by numerous studies using the radio-astronomy technique known as "very-long-baseline interferometry." (For more information about interferometry, see the article "Astronomical Facilities" in this chapter.)

Since astronomers do not yet understand how an average quasar can emit 100 to 10,000 times as much energy as a normal galaxy from a region not much larger than the solar system,they still aren't completely sure what quasars are. Most researchers believe that quasars are the nuclei of relatively normal galaxies that contain a supermassive black hole. As described above, matter spiraling into a black hole heats up and gives off huge amounts of energy, which could account for quasars' extreme brightness.

Most distant quasar. Among the approximately 2,000 quasars catalogued by astronomers by the late 1980s, the one with the greatest red shift — 443 percent — is Q 0051 –279, which is located in the constellation Sculptor. This large red shift was discovered in a spectrum taken on 12 September 1987 by Stephen Warren, a graduate student at Cambridge University in England. Warren was using the 155-inch Anglo-Australian Telescope on Siding Spring Mountain in New South Wales, Australia, as part of a coordinated search for high-red-shift quasars.

The program, devised by astronomer Paul Hewlett of Cambridge University, made use of photographs from large areas of sky obtained with the 48-inch U.K. Schmidt Telescope, also located on Siding Spring Mountain. A single photograph taken with the Schmidt Telescope records the images of tens of thousands of stars and galaxies and some quasars. To select quasars that could have large red shifts, Hewlett scanned objects with a sophisticated computer-controlled measuring machine programmed to recognize their characteristic colors. Once a number of promising objects had been identified, Warren studied their spectra in detail.

This search technique has been highly successful. In addition to locating Q 0051 –279, Warren also discovered another distant quasar in the same observing run. Its spectral lines showed a red shift of 407 percent. The red shift of Q 0051 –279 implies that this quasar is receding from Earth at over 92 percent of the speed of light.

Converting Q 0051 –279's red shift to a distance is a difficult task. To do this, astronomers need to know the value of the so-called Hubble constant and the deceleration parameter, which tell them the rate at which the universe is expanding and the extent to which space is warped (or curved) by the gravitational field of all the matter it contains. Since both factors are known only imprecisely, assigning a particular distance to a given object is not yet possible. (For relatively nearby quasars like 3C 273, the warping of space is a minor factor, but uncertainties in the value of the Hubble constant still render distance estimates imprecise by a factor of two.) Astronomers have estimated, however, that the light they now see from this quasar started its journey when the universe was about 10 percent of its present age — about 9 billion to 18 billion years ago.

Q 0051 –279 edged out the previous red shift recordholder, Q 0000 –26, by a relatively small margin. It was discovered in 1987, only about a month before Q 0051 –279, by researchers Cyril Hazard of the University of Pittsburgh in Pennsylvania and Richard McMahon and Mike Irwin of Cambridge University in England. Like Hewlett and his colleagues, Hazard's group made use of photographs taken with the U.K. Schmidt Telescope. However, unlike Hewlett, Hazard laboriously scans the tens of thousands of images on each photograph by eye using a microscope. His systematic search techniques and keen eyesight have led to the discovery of several very distant quasars. Q 0000 –26 has a red shift of 411 percent and is moving away from Earth at a velocity of over 92 percent the speed of light.

In the last few years, red shift records have been broken almost as soon as astronomers have announced them. These fast changes are a consequence of improvements in the techniques used for searching for high-red-shift quasars. Between 1973 and 1982, the greatest red shift known was 353 percent. Since then, approximately 20 quasars with larger red shifts have been found.

Nature of quasars. When Maarten Schmidt first explained the red-shifted spectrum of quasars in 1963, many astronomers considered quasars to be an entirely new celestial phenomenon, quite unlike anything ever seen before outside our galaxy. In the 25 years since, however, attitudes have changed; optical, infrared, radio, and x-ray studies have proved that the nuclei of many apparently ordinary galaxies are far more energetic than previously thought. Even the nuclei of the Milky Way and several of its neighbors display activity that may be a scaled-down version of what occurs in quasars.

As a result of these discoveries, astronomers have come to the conclusion that the nuclei of galaxies display a wide range of activity. Some galaxies are relatively calm and peaceful, while others, such as the Seyfert galaxies and BL Lacertae (BL Lac) objects, have nuclei that are much brighter than normal but less bright than quasars. A few galaxies have luminosities rivaling those of quasars and seem to be undergoing violent bursts of star formation. The differences in the energy emissions of galaxies may be explained by the existence of black holes in their nuclei; the greater the rate at which matter falls into a black hole, the higher its energy output.

Many astronomers now believe that quasars represent a violent phase that many, and perhaps all, galaxies pass through in their youth. If quasars are the superenergetic nuclei of galaxies, astronomers should be able to see a faint glow of stars surrounding the bright region. Observations performed by astronomer Jerome Kristian at

Palomar Observatory in the early 1970s gave the first indications that quasar images are "fuzzy"; that is, point-like quasar images are surrounded by a faint glow that might be starlight from the quasars' host galaxies.

The first major study to indicate that the so-called quasar fuzz represents the light of a quasar's parent galaxy was reported by Arizona State University astronomer Susan Wyckoff and her colleagues in 1981. Since even the closest quasars are too far away for structural details, such as galactic spiral arms, to be clearly recognized in photographs, the astronomers studied the fuzzy areas spectroscopically to see whether they could find spectral lines characteristic of the stars in a galaxy. Wyckoff's group and others since have found that the fuzz surrounding 3C 48 and many other quasars bears the spectral signatures of galaxies at the same red shift as their associated quasars. In addition, the color, surface brightness, and size of the fuzzy areas are within the range of what would be expected in galaxies.

In 1985, astronomers Paul Hintzen of NASA's Goddard Space Flight Center and John Stocke of the University of Arizona reported additional evidence that quasars are galaxies. A long-exposure photograph of the region around the quasar 3C 275.1 revealed an extremely faint collection of galaxies around the quasar itself. The finding is significant because most galaxies are found in clusters; thus, if quasars are merely galaxies with high-energy nuclei, they should also be found within clusters.

Another link between quasars and galaxies was reported in 1985 by astronomer Bruce Campbell of Canada's Dominion Astrophysical Observatory and his coworkers. During a routine study of quasar fuzz in May 1983, the researchers recorded an image of the quasar QSO 1059 +730 that showed a curious bright spot within the fuzz. The spot was not present in photographs taken the year before or the year after. After ruling out various explanations for the spot, Campbell's team realized that they had captured a fleeting glimpse of a supernova — an exploding star — the first such event ever observed in a quasar. The discovery therefore confirmed that the fuzz surrounding quasars does in fact contain at least some stars. In addition, when Campbell and his colleagues compared the supernova's brightness with that of supernovae studied in nearby galaxies, they were able to calculate the distance to QSO 01059 +730. The value they obtained was consistent with the distance suggested by the quasar's red shift.

While many researchers now believe that much galactic activity is ultimately related to the presence of a supermassive black hole in a galaxy's core, they have not been able to explain why, for example, some galaxies are powerful sources of radio emissions while others show no radio emissions. Likewise, there is no explanation for why BL Lac objects have few or only weak emission lines in their spectra, while Seyfert galaxies have very strong emission lines. For many years astronomers have tried to devise a single theory to explain the many different types of activity seen in the nuclei of galaxies.

In 1989, Peter Barthel of the University of Groningen in the Netherlands outlined a model that explains how some of these apparently different manifestations of galactic activity might arise. Barthel believes that the amount of activity observed in galactic nuclei is determined by the direction from which they are observed. For example, what appears from one direction to be a quasar might look like a BL Lac object if seen from another vantage point.

Barthel came to this conclusion after carefully analyzing the hypothesis that the existence of supermassive black holes explains galaxies' activity. In the core of a galaxy, the immense gravitational field of a supermassive black hole will draw matter from surrounding stars and gas clouds into an immense disk or doughnut-shaped cloud several hundred light-years in diameter. As the matter in this accretion disk spirals into the black hole, it heats up and gives off vast amounts of energy. By some mechanism still not clearly understood, some of this energy is channeled into beams, or jets, of radiation emitted perpendicular to the disk's orbital plane; that is, along the rotation axis of the disk.

As early as the beginning of this century, the American astronomer Herbert Curtis discovered a jet emanating from the center of the galaxy M87. Since then, astronomers have found many similar structures. In particular, radio astronomers have discovered powerful radio sources which consist of two immense clouds, or lobes, of radio emission straddling a much weaker, and smaller, central object. In some cases, jets are seen linking the central object to the lobes. These observations suggest that energy generated in the central object travels outward along the jets and into the lobes where it interacts with surrounding clouds of intergalactic matter. The total energy emitted by these radio galaxies is similar to that of a typical quasar.

Barthel believes that the vantage point from which the accretion disk and jets are seen determines how it will be classified. For example, when the disk is viewed head-on, one jet appears to be pointing directly at the viewer. The observed properties of such an object closely match those of a BL Lac object. If the disk is viewed edge-on, then the region immediately surrounding the black hole is obscured by the accretion disk, but both jets are plainly visible. According to Barthel, such an object would be classified as a radio galaxy. When the disk is viewed from an intermediate angle, its observed properties match those of a quasar.

Although Barthel's model has not yet been widely accepted, it has gained the attention of many astronomers. Not only does it seem to explain many different facets of galactic activity, but it also makes predictions that can be tested observationally. Whether Barthel's model is ultimately proved or disproved, astronomers may learn a great deal about active galaxies by testing it.

Gravitational Lenses

Einstein's theory of relativity predicts that the gravitational field of any massive object, such as a galaxy, can bend and focus passing light rays. This effect was first seen in 1919 in starlight just grazing the Sun's surface. For the Sun, the deflection is small, about 0.0005 degrees. The more massive the so-called gravitational lens, however, the greater the deflection. If the distant object and a massive intervening body — such as a galaxy — are exactly oriented on the same line of sight, gravitational lensing results in a ring-shaped image of the distant object being visible to an Earth-bound observer. If the alignment is not exact, a number of individual ring segments are seen rather than a complete ring.

In 1979, a team of American and British astronomers found that two closely spaced quasars in the constellation Ursa Major have nearly identical brightnesses, spectra, and red shifts. The researchers concluded that they were actually observing two images of a single quasar, which was located behind a gravitational lens. A complete ring — the so-called Einstein ring — was not seen because the alignment between the quasar and the gravitational lens was not exact. The lens in this case, a massive galaxy, was discovered in 1980 by University of Hawaii astronomer Alan Stockton. By late 1987, the number of multiple-imaged quasars had risen to over a dozen.

The first example of a complete Einstein ring was announced in 1988 by MIT radio astronomer Jacqueline Hewitt and her colleagues. They discovered this unusual object while using the Very Large Array radio telescope in New Mexico to investigate a radio source called MG 1131 +0456 in the constellation Leo. Their subsequent radio and optical studies ruled out several alternative explanations of this structure, leaving gravitational lensing as the only plausible explanation. Since this first Einstein ring was found, two more have been discovered by researchers from MIT and elsewhere.

If Einstein rings are the most striking examples of gravitational lensing yet discovered, the so-called giant galactic arcs are close runners-up. The first of these unusual features was found in January 1987, when groups of American and French observers independently announced the discovery of a number of unusual arcs, or filaments, of glowing matter partially encircling distant clusters of galaxies. The filaments' apparent dimensions — 300,000 light-years long but only a few thousand light-years across — would make them some of the largest objects known in the universe.

Subsequent observations made by the French team in October 1987 revealed that the giant arc surrounding the cluster called Abell 370 has the spectral characteristics of a normal, but distant, galaxy. Yet its red shift is twice that of the host cluster of galaxies. In other words, the arc seems to be twice as distant as the clusters within which it appears to lie. The researchers concluded that the giant arc is a highly distorted, stretched-out image of a galaxy lensed by Abell 370. To form such well-defined arcs rather than a series of point images, the alignment between the distant galaxies, the clusters, and the Earth must be close, but not close enough to form an Einstein ring.

Since the discovery of the giant arc in Abell 370, several more have been found. Arcs are currently being studied in Abell 963, Abell 2218, CL 2244 – 02 and CL 0500 –24. Giant arcs are more than an astronomical curiosity, because they offer a promising means of calculating the masses of their parent clusters. The degree of lensing needed to form arcs is directly related to the amount of matter in the cluster of galaxies. At a meeting on gravitational lenses held at MIT in 1988, Genevieve Soucail of Toulouse Observatory in France explained how initial calculations showed that the arcs require the presence of much more matter in the clusters than can be accounted for by luminous objects such as galaxies. She added that a detailed comparison of masses determined from the arcs and those derived from traditional methods should reveal much about the nature of the so-called missing mass in the universe. (For more information about missing mass, see the article "Cosmology" in the "Physics" chapter.)

Gravitational lenses may also help scientists determine the size of the universe as well as the amount of matter it contains. The light from a distant quasar can be split into different components by the gravitational field of a massive galaxy or cluster of galaxies. Since the paths traversed by the light from the different images of a single quasar are, in general, of different lengths, changes in one will be seen in the others at different times, due to the finite speed of light. The time delay is thus a measure of the difference in path lengths. If this difference can be measured accurately, cosmologists can use it to determine the Hubble constant — a measure of the universe's expansion rate, which can be used to infer its size.

Measuring the interval between changes appearing in one quasar image and another is relatively simple because the energy output of quasars is highly variable over time. R. Florentine-Nielsen of Copenhagen University Observatory made the first measurement of this type during 1984. He noticed that a variation in the brightness of one component of the gravitational lens system 0957 +561 seen in 1982 was matched by a similar change in another component 1.55 years later. From this delay, the Danish astronomer deduced that the Hubble constant is 77 kilometers (48 miles) per second per megaparsec — almost exactly midway in the range of currently accepted values of 50 to 100 kilometers per second per megaparsec. (A megaparsec is 3.26 million light-years.)

Since Florentine-Nielsen's initial determination, other astronomers have used the same technique with better observational data. At the Fourteenth Texas Symposium on Relativistic Astrophysics held in Dallas during December 1988, Edwin Turner of Princeton University reviewed the latest measurements of gravitational-lens time delays. From a variety of observations obtained by different observers, he estimated that the Hubble constant has a value of about 86 kilometers per second per megaparsec. As more observations of the brightness variations of gravitationally lensed quasars accumulate, even more precise estimates of the Hubble constant will be possible.

Astronomical Facilities

Tools of the Astronomer

Just 40 years ago, optical telescopes were the only tools astronomers had to probe the universe. Today, scientists have a large variety of instruments that can explore the entire electromagnetic spectrum. Astronomical instruments can be divided into three groups. The first group consists of optical, radio, and certain infrared telescopes. These detect radiation that can penetrate the Earth's atmosphere. The second group consists of instruments including x-ray, gamma-ray, and ultraviolet telescopes that can detect radiation to which the Earth's atmosphere is opaque. Such telescopes must be carried above Earth's atmosphere on rockets or placed in orbit aboard satellites. The final group consists of devices designed to detect nonelectromagnetic emissions. The most developed detectors are those that register the high-energy atomic nuclei called cosmic rays and the elusive subatomic particles called neutrinos. Still in their infancy are detectors designed to register the gravitational waves predicted by Albert Einstein's general theory of relativity.

Optical Telescopes

Ever since the Italian scientist Galileo Galilei turned his telescope toward the sky in 1609, astronomers have been involved in a quest to build ever-larger optical instruments that can see ever-fainter objects. In particular, astronomers have labored to increase the aperture (diameter) of telescope optics, since the more light they gather, the more clearly they are able to observe faint objects.

Galileo's telescopes and most others in the two centuries that followed used glass lenses to gather and focus light. Large lenses are, however, very difficult to make and, more important, are prone to becoming distorted by their own weight because they can be supported only at the edge. In the middle of the nineteenth century, astronomers perfected the techniques for making concave mirrors out of glass. Not only are mirrors easier to make than lenses, but, because they can be supported from behind as well as from the side, they are less prone to distortion.

The evolutionary sequence that began with Galileo's telescope culminated in April 1936, when the 20-ton, 200-inch-diameter mirror of the Hale telescope at Palomar Observatory in Southern California was delivered to Caltech. The telescope, which was the brainchild of American astronomer George Ellery Hale, was not fully completed until 1948 because World War II caused a delay in its construction. Although many large optical telescopes have been built since then, only one is larger — the U.S.S.R.'s 236-inch reflector at the Special Astrophysical Observatory on Mount Pastukhov near Zelenchukskaya in the Caucasus Mountains. Although this giant telescope has several novel features, its image quality is generally acknowledged to be inferior to that of the Palomar instrument.

Although world-class optical telescopes have not increased in size for many years, the performance of the instruments has improved significantly. For example, the photographic emulsions in use when the 200-inch Hale telescope was commissioned in 1948 recorded at best only one or two percent of the light that fell on them. These emulsions have now been replaced by electronic detectors that register up to 80 percent of the light falling on them. Equipped with these new detectors, a relatively modest telescope with a 36-inch mirror can now make observations that would have taxed the abilities of the 200-inch Hale telescope 40 years ago.

While telescopes have remained essentially constant in size, the quality of auxiliary instruments — spectrographs and cameras — has increased so greatly that they are now approaching the limits of their capabilities. With no major advances in auxiliary equipment likely in the near future, astronomers have begun to plan new telescopes that will dwarf the 200-inch Hale.

Between 1950 and 1980, five telescopes with mirrors about 150 inches across were designed and built. Larger telescopes were not built because most researchers were convinced that they would be prohibitively expensive. Two developments, however, have challenged this assumption. First, Aden Meinel of the University of Arizona spearheaded the design and construction of a radically new telescope — the Multiple Mirror Telescope (MMT). Second, Leo Goldberg, director of Kitt Peak National Observatory in Tucson, Arizona, initiated the Next Generation Telescope project.

The MMT on Arizona's Mount Hopkins was revolutionary because it demonstrated that a large telescope need not be exceedingly expensive. The cost of a telescope depends on the cost of its three major components: the primary and secondary mirrors; the mounting that holds the mirrors in place and allows them to point anywhere in the sky; and the enclosure that protects the telescope from the weather.

Instead of using a single mirror, the MMT used 6 circular 72-inch mirrors arranged in a hexagonal pattern. The light-gathering power of these 6 mirrors is equivalent to that of a single mirror 176 inches across, but together they weigh and cost only a fraction as much as a 176-inch mirror. By reducing the weight of the mirrors, the MMT designers were also able to use a lighter (and less expensive) mounting to hold the mirrors. In addition, the smaller mounting made it possible to use a more compact and less expensive structure to house the telescope.

The Next Generation Telescope (NGT) project went a step beyond the MMT to design a telescope with a light-gathering power equivalent to a single reflector with a diameter of 1,000 inches. The NGT study in the late 1970s concluded that there are basically three ways to build telescopes larger than the Hale reflector. One method is to build a scaled-up version of the MMT. Another method is to use a single, large mosaic mirror; that is, one made up of many small segments. The third technique is to build an array of independent telescopes that bring their light to a common focus. Each of these approaches has been embodied in a project planned or under construction in 1989.

The first of these projects began in December 1984, when Caltech and the University of California received a $70-million grant from the William M. Keck Foundation of Los Angeles. The 400-inch-diameter Keck telescope, under construction in 1989 atop the 13,800-foot mountain Mauna Kea on the island of Hawaii, will begin operating in late 1991. The total cost of the project will be about $87 million. Instead of using a single mirror, the Keck telescope will use a honeycomb mosaic of 36 hexagonal mirrors, each 72 inches across, which together will form the broad reflecting surface. Each hexagonal element, weighing 880 pounds and measuring 3 inches thick, is under computer control so that it works together with its neighbors as a single mirror 400 inches across.

The second project is the European Southern Observatory's Very Large Telescope (VLT). European astronomers have begun work on an array of four separate 300-inch telescopes that can work either independently or together as a single mirror. When working together, the light of the four mirrors is combined, giving the instrument the light-gathering power of a single telescope mirror 600 inches across. Because the VLT's individual telescopes will be placed side by side in a line, they will under some circumstances be able to combine their light and simulate a single telescope 6,000 inches in diameter. To accomplish this increase, they will use the interferometry technique, which is described in the "Radio Telescopes" section below. The VLT project is expected to cost about $130 million and will begin operating at a site in Chile in 1993. Final approval for the VLT was given in December 1987.

The third project is the National New Technology Telescope (NNTT), designed at the National Optical Astronomy Observatories (NOAO) in Tucson, Arizona. The NNTT is basically a scaled-up version of the MMT. It will use four mirrors, each 300 inches across, on a common mounting to equal the light-gathering power of a single mirror 600 inches across. The individual mirrors will probably be constructed with the revolutionary spin-casting method developed by University of Arizona astronomer Roger Angel.

In traditional mirror-casting methods, molten glass is poured into a stationary, cylindrical mold to create a solid disk whose top surface is flat. The disk is allowed to cool over a period of months or years to prevent cracking or warping and then is ground and polished to a precise paraboloidal shape. The polished glass is later covered with a thin layer of reflective material, such as aluminum.

In Angel's spin-casting technique, chunks of inexpensive Pyrex glass are melted in a mold within a rotating oven. (The mold is attached to the floor of the oven, and as the oven rotates, the mold rotates within it.) The top surface of the molten material takes on a concave shape as it rises up the inner wall of the spinning container, just as coffee being stirred rises up the inside of a cup. The concave surface remains when the glass cools, and with a relatively small amount of grinding and polishing, a precise paraboloidal shape is obtained. The mold in which the glass is melted also imparts a honeycomb structure to the back of the mirror. These interconnecting ribs stiffen the mirror. In April 1988, Angel successfully cast a 140-inch mirror, proving that his spin-casting technique works.

A second 140-inch mirror was cast by Angel's team in December 1988. In early 1989, the spinning oven was being rebuilt to accommodate mirrors up to 320 inches in diameter. Once the upgrade is completed, Angel plans to cast a third and final 140-inch mirror. Sometime in 1990, the University of Arizona team will attempt to spin-cast a 260-inch mirror to replace the six 72-inch mirrors currently used in the MMT.

At a 1986 meeting of the American Astronomical Society in Houston, Texas, Jacques Beckers, director of the advanced projects division of the NOAO, said that the NNTT may not be built until after completion of the Very Long Baseline Array (VLBA), a continent-spanning array of radio telescopes (see below) that is currently scheduled to begin operating in the mid-1990s. In 1988, however, unexpected funding problems forced NOAO to suspend the NNTT project.

In early 1989, NOAO released a plan to resurrect the NNTT project in a scaled-down form. The plan calls for the construction of two 320-inch single-mirror telescopes, one each at a site in the Northern and Southern hemispheres (probably Mauna Kea on the island of Hawaii and another somewhere in Chile). To further reduce the cost of building these giant telescopes, NOAO has been searching for international partners. A number of nations, Canada and the U.K. in particular, are eager to construct large telescopes and may invest in the NNTT project.

Although the NNTT project is on hold, U.S. astronomers may still build very large telescopes using Angel's large mirrors. In April 1985, the University of Arizona joined with Ohio State University in Columbus and the Arcetri Astrophysical Observatory in Florence, Italy, in the $50 million Columbus Project. This project involves using two of Angel's 300-inch mirrors in a binocular telescope — twin telescopes on a binocular mounting — located on Mount Graham, an isolated, 10,700-foot peak about 70 miles northeast of Tucson. In October 1986, University of Arizona astronomers also agreed to join with researchers at Johns Hopkins University in Baltimore, Maryland, and the Carnegie Institution in Washington, D.C., in a project to build a single 300-inch telescope in Chile. Astronomers in a number of European nations also plan to use Angel's mirrors in large telescopes of their own.

Radio Telescopes

In 1930, Karl Jansky, an engineer at Bell Telephone, discovered that mysterious static that was interfering with transoceanic radiotelephone conversations originated in the Milky Way. Neither Jansky's discoveries nor those of his successor Grote Reber made a major impression on the astronomical community at the time. It was not until the end of World War II and the application of radar technology to scientific research that the branch of radio astronomy became established.

Radio astronomers have one major disadvantage compared with their colleagues using optical instruments: radio telescopes do not have very high resolution, which limits their ability to observe fine detail or locate celestial objects with high precision. In the 1950s and 1960s, optical astronomers could not identify optical counterparts for the radio sources discovered with radio telescopes. Without precise positions, astronomers faced the sort of needle-in-the-haystack quest of trying to locate the optical counterparts of radio sources among the millions of objects visible to their telescopes.

The larger the aperture of a telescope, the finer the detail it can discern. Engineering considerations, however, limit the size of the telescope's radio dish; giant radio dishes would collapse under their own weight. The world's largest radio telescope in Arecibo, Puerto Rico, has a diameter of 1,000 feet. It is fixed in place, however, and can study only objects within 20° of directly overhead. The largest movable dish is about 330 feet in diameter and is located at Effelsberg, West Germany.

To see the radio sky in finer detail, astronomers have to use a technique known as interferometry. By combining the signals from two or more individual telescopes some distance apart, radio astronomers can produce an image equivalent to that of a single dish with a diameter equal to the separation between the two telescopes.

At first, astronomers combined the signals from their radio telescopes by physically linking them with cables. The largest system of this kind is the Very Large Array operated by the U.S. National Radio Astronomy Observatory based in Charlottesville, Virginia. Its 27 82-foot radio telescopes are arranged in a Y-shaped pattern stretching for almost 30 miles across the Plain of San Augustin in central New Mexico. In the 1970s, researchers found that the telescopes do not need to be physically connected in interferometry. By observing the sky with a number of radio telescopes simultaneously and recording the data on magnetic tape together with timing signals from an atomic clock, researchers can later synchronize the tapes and construct their radio maps. This technique is known as very-long-baseline interferometry (VLBI). By using the VLBI method with radio telescopes on different continents, astronomers can simulate a telescope with a size equivalent to Earth's diameter.

Intercontinental VLBI link-ups between the U.S. and Europe are now common. To extend the baseline to sizes greater than Earth's diameter, astronomers have conducted experiments using NASA's Tracking and Data Relay Satellite in orbit in conjunction with earthbound radio telescopes. The ESA and the U.S.S.R. both have plans to launch dedicated VLBI satellites in the 1990s.

Even in early 1989, however, the radio telescopes employed in VLBI experiments were used for other purposes such as radio astronomy observations, and were not available for VLBI observations all the time. The VLBA is being built to overcome this problem. The array will consist of 10 custom-designed, 82-foot radio telescopes distributed across the U.S. from Hawaii to Massachusetts, and from the U.S. Virgin Islands to Washington state. When completed in the mid-1990s, the VLBA will be able to map celestial objects with a resolution 1,000 times greater than that of any existing optical or radio telescope.

Hubble Space Telescope

Even before the beginning of the Space Age, astronomers dreamed about using telescopes free from the distorting influence of the Earth's atmosphere. This dream may be realized as early as 1990 when, according to current schedules, the U.S. space shuttle carries the Hubble Space Telescope (HST) into orbit. Named after the U.S. astronomer Edwin Hubble, the 22,500-pound orbiting telescope will observe the visible-light and ultraviolet emissions of celestial objects from an orbit about 300 miles above Earth's surface.

The HST's lightweight, 94-inch mirror is much smaller than that of many ground-based telescopes; however, free from the effects of the Earth's turbulent and murky atmosphere, it will see seven times farther into space and survey a volume 350 times larger than any other telescope. HST's mirror is perhaps the most perfect optical surface ever manufactured, according to project scientists. Its shape is accurate to one-fortieth the wavelength of red light — about one millionth of an inch.

The $1.4 billion HST, which is a joint project of NASA and ESA, is about 14 feet in diameter and 43 feet long (excluding two large arrays of solar cells that supply the telescope's power). It is equipped with two cameras that will be able to see objects a billion times fainter than can be seen with the naked eye and 10 to 15 times fainter than can be seen with any existing telescopes, two spectrographs to study the chemical composition of celestial objects, and a photometer to measure the brightnesses of celestial objects.

To use all of these instruments effectively, HST must track the same object in space for many hours while astronomers make their observations. To track accurately, HST is equipped with a high-precision on-board guidance system that will enable it to remain pointed to within two-millionths of a degree of a target for up to 24 hours. The "fine-guidance sensor" used to keep HST on target will also function as a sixth scientific instrument that will be used to make precise measurements of the positions of stars.

Construction of HST was essentially completed in January 1986. Since then, it has undergone a series of tests to monitor its operation. Among the problems encountered was the discovery that the solar cells did not generate as much electricity as expected. To remedy this problem, ESA has been building a replacement solar array and NASA has been equipping HST with more powerful batteries that will be used to operate the telescope when the solar cells are not in sunlight.

The Challenger disaster severely affected the space telescope project because it was designed with the presumption that it would be launched by the space shuttle. In addition, the HST is the first satellite designed to be serviced by astronauts carried into space on the shuttle. According to current plans, HST will be visited periodically by astronauts who will conduct routine maintenance and replace any defective equipment with spares carried from Earth. NASA already plans to replace one or more of the scientific instruments with improved versions after HST has been in orbit for a few years. If the telescope suffers any major malfunctions, it can be brought back to Earth on the shuttle for repairs.

In a far-reaching report issued in 1981, the Astronomy Survey Committee of the U.S. National Academy of Sciences recommended that the U.S. build a series of major astronomical satellites to complement the capabilities of HST and study the universe across the electromagnetic spectrum from x-ray to radio wavelengths. First

Figure 8. Hubble Space Telescope. Orbiting 300 miles above the Earth's surface, the Hubble Space Telescope (HST), with its 94-inch-diameter mirror, will be able to observe celestial objects free of the distorting effects of the atmosphere. The HST had been scheduled for a 1986 launch by the space shuttle, but the launch was postponed several years because of the shuttle Challenger accident. *Courtesy: Lockheed Missiles & Space Company, Inc.*

among these proposed satellites was the Advanced X-ray Astrophysics Facility (AXAF).

AXAF will be about the same size and shape as HST and use a special system of mirrors to focus x-rays on an array of cameras and spectroscopes. X-rays cannot be focused like light or radio signals because they generally pass directly through conventional mirrors and lenses. They can, however, be reflected if they strike a mirror at a glancing angle, just as a stone can skip over water if it strikes at a shallow angle. AXAF will be equipped with a series of nested cylindrical mirrors arranged in such a way that x-rays will always hit their reflecting surfaces at very small angles.

The initial study phase of the AXAF project has already been completed, after many years of waiting for final budget approval. This was granted in 1988, when NASA decided to fund AXAF instead of another probe. U.S. astronomers, who had watched their commanding lead in x-ray astronomy erode in the wake of advances by their European, Japanese, and Soviet colleagues, will by the turn of the century have the most powerful x-ray telescope ever developed.

In addition to AXAF, the only major space astronomy mission approved as of 1989 is the Gamma Ray Observatory (GRO) satellite, which is scheduled to be launched by the space shuttle in 1990. GRO is designed to study very-high-energy (gamma) radiation emitted by neutron stars, black holes, quasars, and violent astronomical phenomena. Other missions that U.S. astronomers hope will be approved before the end of the century include the Space Infrared Telescope Facility and a large infrared-radio dish called the Large Deployable Array.

Searching for Alien Life

The development of advanced radio telescopes has allowed astronomers to attempt to answer a question that has long intrigued scientists, philosophers, and laypersons alike: Do other forms of life — and intelligent life, in particular — exist elsewhere in the universe? The legends and religions of many cultures hold that divine beings created the heavens and control such cosmic events as eclipses. But the idea that other planets might harbor life similar in development and intelligence to our own did not become popular until the nineteenth century. A few scientists of that period considered ways in which Earthlings might establish contact with other beings. One plan envisioned building huge canals in the desert in the shape of easily recognized geometrical symbols; when filled with gasoline and ignited, the canals would signal the presence of life on Earth to neighboring worlds. In 1899 the brilliant but eccentric physicist Nicola Tesla actually built a huge tower in Colorado topped by a copper ball, into which he sent powerful surges of electricity. He reasoned that the electric surges would alter the Earth's magnetic field in a way that could be used to transmit messages into space. Although he was unsuccessful, Tesla's device caused light bulbs to glow in towns up to 25 miles away.

Since then, many astronomers have become seriously interested in the search for extraterrestrial intelligence (SETI). They assume that alien beings elsewhere in the galaxy will probably try to contact us, using flashes of light or other portions of the electromagnetic spectrum to carry their message. In an article published during 1959, Cornell astronomers Giuseppe Cocconi and Philip Morrison showed that radio transmissions with wavelengths between about 1 and 10 inches were best suited for interstellar communication. This portion of the spectrum is relatively free of natural sources of noise, and transmissions at these wavelengths would travel great distances through space without distortion. Cocconi and Morrison also noted that, in space, neutral hydrogen and the hydroxyl radical — the components of water molecules — emit microwave energy within this wavelength range. Because the hydrogen and hydroxyl emissions occur naturally all across the galaxy, another civilization might preferentially select one of these so-called magic wavelengths for interstellar communication.

In 1960, astronomer Frank Drake, then at Cornell University, performed the first true search for extraterrestrial life, which he called Project Ozma. He turned a sensitive radio telescope in the direction of the nearby stars Tau Ceti and Epsilon Eridani but found no signs of transmissions that might be beacons from extraterrestrial civilizations. Since Project Ozma, about four dozen other searches have been conducted, almost all trying to detect emissions at radio wavelengths. Although no one has yet received an unambiguous signal from an extraterrestrial civilization, numerous false alarms have been caused by interference from radio-wave sources (such as military radar) here on Earth.

Harvard University physicist Paul Horowitz has undertaken the most comprehensive search to date. He developed an electronic receiver capable of monitoring a continuous band of 65,536 wavelengths simultaneously, and he attached the receiver to an 84-foot-diameter radio telescope near Harvard, Massachusetts. On 7 March 1983, Horowitz used this equipment to begin Project Sentinel, a search for alien radio beacons at the hydrogen-emission wavelength of about 8.4 inches. Every 6 months, his telescope-receiver combination scans most of the sky visible from Massachusetts, at which point a new wavelength band can be chosen and the search repeated. Funding from the Planetary Society and film producer Steven Spielberg permitted Horowitz to upgrade his electronic receiver, and Horowitz's improved detector began monitoring 8.4 million wavelengths simultaneously in September 1985.

NASA has also been involved in SETI. Between 1982 and 1987, scientists at the NASA-Ames Research Center in Mountain View, California, developed prototypes of hardware and computer programs for a major effort. NASA is currently seeking funding from the U.S. Congress to build a complete receiver and to undertake a 10-year search program. Once built, NASA's electronic receivers will be able to monitor 8.4 million wavelength channels simultaneously, as does Horowitz's receiver, but they will analyze not just individual wavelengths but whole groups of them at the same time; the idea is to listen for "chords" as well as single "notes."

One half of the NASA search plan calls for using the radio telescopes in its Deep-Space Network to scan the entire sky repeatedly. The other half would use the 1,000-foot radio telescope at Arecibo, Puerto Rico, to listen to nearby stars similar to the Sun, which may have Earthlike planets orbiting them. Roughly 800 sunlike stars exist within 80 light-years of our solar system. Although scientists involved in SETI realize that their survey will be far from complete, they believe the search must begin with small efforts. "We're new at this business," Horowitz explained in the May 1983 *Sky & Telescope*, "and when you walk into a dark, unfamiliar forest, you should probably listen before you shout."

International Space Programs

More than two dozen nations have participated in space exploration, either through their own programs or through those of the European Space Agency (ESA), the U.S.S.R., and the U.S. For example, the U.S.S.R. has cooperated with the nations of eastern Europe — Bulgaria, Czechoslovakia, East Germany, Hungary, Poland, Romania, and Yugoslavia — and a number of other countries. ESA includes 13 nations in western Europe — Austria, Belgium, Denmark, France, Ireland, Italy, the Netherlands, Norway, Spain, Sweden, Switzerland, the U.K., and West Germany. Finland is an associate member of ESA, and Canada agreed in 1981 to contribute to some of ESA's projects. Japan and the People's Republic of China have also pursued largely independent space programs of growing sophistication. The civilian space activities of ESA, the U.S.S.R., and the U.S. are described in detail below, with particular emphasis on manned and scientific missions.

European Space Agency

The European Space Agency, formerly the European Space Research Organization, was formed in 1975 to allow its member nations to pool resources for space studies and to coordinate joint research and exploration projects. ESA's total budget for 1985 was approximately $860 million. That same year, ESA administrators approved a program that would increase annual spending to about $1.35 billion by 1990.

Ariane program. On 24 December 1979, ESA's first Ariane 1 rocket successfully lifted off from the Guiana Space Center in Kourou, French Guiana. The Ariane was designed to provide launch services for ESA, non-ESA nations, and private customers, and also to lessen European dependence on U.S. boosters. Unlike the space shuttle, the Ariane rocket carries no crew and is not reusable. Ariane 1 was capable of placing payloads of up to 4,000 pounds into a geosynchronous transfer orbit. A geosynchronous transfer orbit has a low point relatively near the Earth's surface and a high point about 22,300 miles up. Satellites in a geosynchronous transfer orbit can later be boosted by small, attached rockets into a nearly circular "geosynchronous" orbit about 22,300 miles high, an altitude at which a spacecraft in an equatorial orbit remains stationary over a fixed point on the Earth below.

Arianespace, a private French-based company, markets Ariane's services and manages its launches. In the decade since the launch of the first Ariane 1, ESA has developed a series of increasingly powerful rockets. In June 1988, ESA launched the first of its third-generation boosters, Ariane 4, which can place up to 9,250 pounds into geosynchronous transfer orbit. By 1990, the Ariane 4 will be used exclusively for launches, and earlier versions of the booster will be phased out.

ESA has also approved the development of an even more powerful fourth-generation booster, the Ariane 5, by the mid-1990s. The launcher will combine an advanced liquid-fuel rocket with two solid-fuel boosters, a configuration much like that used to launch the U.S. space shuttle. Ariane 5 will be able to carry payloads weighing up 35,000 pounds to low-altitude orbits, a capability that will allow it to compete directly with the space shuttle for launching heavy satellites. The booster will also be able to launch ESA's planned Hermes vehicle, a reusable piloted spacecraft similar to the U.S. shuttle but somewhat smaller in size.

Spacelab. ESA spent about $1 billion in a joint venture with NASA to develop the reusable Spacelab, a self-contained research laboratory carried inside the payload bay of the U.S. space shuttle. The first Spacelab, which held 38 experiments developed by scientists from 13 nations, traveled into space on the shuttle's STS-9 mission on 28 November 1983. Three other Spacelab missions were flown in May, July, and October 1985. The Spacelab is actually a series of pressurized crew modules and unpressurized instrument platforms that can be assembled in various configurations, depending on the scientific requirements of particular flights. Spacelab researchers have studied the effects of weightlessness on plant growth, animal physiology, and human blood chemistry. They have also grown ultrapure crystals that will be used in scientific and industrial x-ray detectors.

In addition, Spacelab can carry telescopes above the clouds and turbulence of the Earth's atmosphere. Such telescopes have been used to observe the visible-light, infrared, ultraviolet, and x-ray emissions of the Sun, stars, and other celestial objects. Photographs of the Sun taken with one of these telescopes have shown the solar surface with unprecedented clarity. Other Spacelab detectors have monitored the flow of energy from the Sun and the influx of high-energy particles from outer space, as well as the concentration of air pollutants in the Earth's upper atmosphere.

Scientific satellite programs. NASA launched six ESA scientific satellites into Earth orbit between 1975 and 1983. Three of these — Geos 1, Geos 2, and one of three International Sun-Earth Explorers — studied the Earth's magnetic field and the field's effect on solar particles streaming past the planet. The other three satellites were devoted to astronomical studies. Cos-B, launched in 1975, monitored high-energy gamma rays from outer space for 7 years. The International Ultraviolet Explorer, which observes the ultraviolet emissions of celestial objects, reached orbit in 1978 and was still operating a decade later. Exosat, launched in 1983, studied x-ray emissions from celestial objects until it stopped operating in April 1986.

One of the most celebrated successes of ESA's scientific program has been the Giotto spacecraft, the agency's first interplanetary probe. Launched in July 1985, Giotto passed about 360 miles from the nucleus of Halley's comet on 14 March 1986. The spacecraft's cameras transmitted detailed photographs of the nucleus, and its other detectors studied the comet's composition, density, and associated magnetic field structure. (For more information about the Giotto mission, see the article "Return of Halley's Comet" in this chapter.)

Future space missions. ESA has planned an active program of scientific studies in space through the early twenty-first century. The Hipparcos astronomical satellite was scheduled to reach orbit during 1989. During its anticipated 2.5-year lifetime, Hipparcos will measure the distances, positions, and motions of approximately 100,000 stars in our galaxy with greater precision than ever before. These measurements will be the basis for the most accurate reference map of the sky to date. Scientific institutions from over 20 countries are participating in the project.

Another upcoming mission, Ulysses (formerly called the International Solar Polar Mission), is a cooperative ESA/NASA venture that originally involved sending two spacecraft toward the Sun, one provided by each agency. The pair of probes were to have flown simultaneously over the Sun's north and south poles, monitoring the solar wind — the stream of charged particles from the Sun — and solar magnetic activity in these regions. Cuts in NASA's budget forced the cancellation of the U.S. probe in 1981, but ESA decided to proceed with its part of the mission. The U.S. will fulfill its other program commitments, however, such as providing the launch vehicle, tracking, and some data analysis. Ulysses will be carried into Earth orbit by the space shuttle and then boosted deep into space by an attached rocket. After the destruction of the space shuttle Challenger, Ulysses's launch was delayed from 1986 until at least 1990.

ESA has also begun work on the Infrared Space Observatory (ISO). This Earth-orbiting spacecraft will provide observations complementary to those made by the Infrared Astronomical Satellite (IRAS) in 1983. ESA expects that ISO will be orbited by 1992. It will carry a 24-inch-diameter infrared telescope approximately 100 times more sensitive than that of IRAS. To maintain this level of sensitivity, the telescope must be cooled to a temperature of −440°F by liquid hydrogen and helium to reduce its own infrared emissions. During its anticipated 18-month operating life, ISO will study a variety of targets, including nearby stars in the process of forming and distant, faint galaxies.

Horizon 2000. In 1985, the ESA council approved substantial increases in funding for space-science programs, to about $130 million per year by 1989. These funds will enable ESA to pursue its Horizon 2000 program, which calls for the launch of numerous scientific spacecraft during the next 20 years. The Horizon 2000 plan contains four major study areas or "cornerstones" — the solar system, Earth-Sun interactions, x-ray emissions from celestial objects, and short-wavelength radio emissions.

For solar system exploration, ESA has placed a high priority on the study of asteroids and comets. These objects are believed to contain material that is largely unchanged from when the solar system formed about 4.6 billion years ago. In late 1988, the ESA science program committee officially endorsed Cassini (a joint mission with NASA) as ESA's next interplanetary venture. Cassini will leave the Earth in 1996 and take 6 years to fly to Saturn. There the main spacecraft will begin orbiting the ringed planet, and an atmospheric probe (ESA's contribution) will enter the dense, smoggy atmosphere of Saturn's large moon Titan to transmit data from its surface. A second ESA mission under study, called Rosetta, would involve sampling cometary dust and returning the dust samples to Earth. Other objectives in the solar system include observations of Mercury, Venus, and Jupiter's moon Io.

ESA has already approved two missions as part of Horizon 2000's second cornerstone. The Solar and Heliospheric Observatory (SOHO) and the Cluster mission are to be part of an international effort with the U.S. and Japan to study Earth-Sun interactions during the 1990s. Once placed in orbit between the Earth and the Sun, SOHO will monitor the Sun's outer layers, sample the solar wind, and study tiny pulsations of the solar surface. The Cluster mission calls for four spacecraft to be placed in different orbits around Earth. These spacecraft will study the plasma (ionized gas) surrounding our planet and map the structure of the magnetosphere — the roughly spherical bubble of electrically charged particles trapped by the Earth's magnetic field. SOHO and the Cluster satellites are expected to be launched in 1994 and 1995, respectively.

Horizon 2000's proposed x-ray mission consists of an orbiting, multiple-telescope observatory that will record x-rays at many wavelengths from a variety of astronomical objects. ESA's observatory will complement NASA's Advanced X-ray Astrophysics Facility, which will provide higher resolution images of celestial x-ray sources than will the ESA observatory, but will not yield as much spectral information.

The fourth cornerstone of the Horizon 2000 program is the launch of a large telescope to study the spectra of objects at radio wavelengths ranging from about 0.1 millimeter to one millimeter. Spectra at these wavelengths, which have not yet been studied in detail, would provide data on objects at temperatures ranging from a few degrees above absolute zero to several thousand degrees Fahrenheit. Likely targets for such a telescope would be the planets, evolving stars, and clouds of dust and gas.

U.S.S.R. Space Program

On 12 April 1961, Soviet cosmonaut Yuri Alekseivich Gagarin was launched into orbit aboard the Vostok 1 spacecraft and became the first human being in space. By the late 1980s, the Soviet space program had amassed an impressive series of achievements, including the longest human space flight, detailed atmospheric and surface studies of the planet Venus, and a successful mission to Halley's Comet. Considering the increasing complexity of its recent missions, the Soviet space program is now "poised for a technological quantum jump and a renaissance in the exploration and exploitation of outer space," according to U.S. space expert Nicholas L. Johnson, author of the annual report, *Soviet Year in Space*.

Manned missions. By the conclusion of the Soyuz/Apollo joint mission in July 1975, the U.S. had accumulated more than twice as many crew-hours in orbit as had the U.S.S.R. — about 22,500 hours versus about 10,700. From the end of that flight until the first launch of the U.S. space shuttle on 12 April 1981, however, the U.S.S.R. was the only country sending people into space. Between 1976 and 1986, the U.S.S.R. launched 33 Soyuz flights with crews aboard. By the end of 1986, Soviet cosmonauts had accumulated a total of 104,400 crew-hours since Gagarin's historic 108-minute voyage in 1961 — the equivalent of nearly 12 years in space by a single person. In contrast, U.S. astronauts had spent about 42,500 crew-hours in space during the same period.

Soviet missions conducted since 1978 have also included cosmonauts from Bulgaria, Cuba, Czechoslovakia, East Germany, France, Hungary, India, Mongolia, Poland, Romania, Syria, and Vietnam. Furthermore, the duration of Soviet flights has been increasing. A Soviet crew remained in space for 96 days during a mission that began in late 1977, breaking the American space duration record of 84 days set in 1974 by the Skylab 4 astronauts. Since then, Soviet cosmonauts have broken their own record several times. As of 1989, the space flight endurance record was held by Vladimir Titov and Musa Manarov, who were launched into space on 21 December 1987 and returned to Earth exactly 1 year later.

These long missions have provided Soviet scientists with extensive data on the effects of prolonged weightlessness on the human body. For example, Soviet doctors have learned that after very long periods in space, some cosmonauts experience a 15-percent decrease in the volume of their tibias, the weight-bearing bones in the lower legs. Knowledge of the health effects of space flight will be essential for preserving the health of humans on the permanent space stations and extended flights envisioned by both the U.S.S.R. and the U.S.

Soviet spaceplane and shuttle. In June 1982, the U.S.S.R. launched an unmanned subscale prototype of a reusable spaceplane on a flight that lasted 109 minutes and covered slightly more than one orbit around Earth. The winged vehicle landed in the Indian Ocean about 350 miles south of the Cocos Islands and was successfully recovered by a group of seven Soviet ships. The U.S.S.R. conducted three more single-orbit test flights of its prototype spaceplane in March and December 1983 and December 1984. The first of these missions ended in the Indian Ocean, and the others ended in the Black Sea. Australian reconnaissance photographs of recovery operations following the March 1983 flight show that the subscale spaceplane has a windshield and crew cabin, as well as thermal protection tiles to shield it from the high temperatures experienced during reentry into the Earth's atmosphere.

Some analysts believe that the full-scale, piloted version of the Soviet spaceplane will weigh only about 20 tons, much less than the 85-ton U.S. space shuttle. They expect that the spaceplane will be used primarily as a manned transport and resupply vehicle for the planned large Soviet space stations of the 1990s.

In 1986, the U.S.S.R. began landing tests of another vehicle — a space shuttle orbiter — similar in concept, appearance, and payload capacity to the U.S. shuttle. The Soviet shuttle was carried into the Earth's atmosphere atop a Bison Myasishchev Mya-4 jet bomber and then released to take a brief flight before landing on a runway at the Tyuratam launch complex. Unlike the U.S. shuttle, which glides unpowered back to Earth and must land on the first attempt, the U.S.S.R.'s test vehicle had a pair of jet engines that allowed it to maneuver during its landing attempts.

The first Soviet space shuttle, named Buran ("Snowstorm"), began its maiden flight with a dramatic night launch on 15 November 1988. Space analysts were surprised that its two-orbit shakedown mission was flown with no cosmonauts aboard — controllers managed the entire flight from the ground. Its touchdown took place smoothly on a 2.8-mile-long runway near its launch site. Buran looks remarkably like the U.S. space shuttle in size and shape. It is 118 feet long and has a wingspan of 79 feet. Approximately 38,000 ceramic tiles protect its outer skin from the heat of atmospheric reentery, using the same general scheme (white tiles on the spacecraft's sides and top, black ones on the underside) as its American counterpart. Buran's cylindrical cargo bay, 60 feet long and 15 feet across, can accommodate a payload of up to 66,000 pounds.

Space analysts expected that Buran's first mission with a crew would take place during 1989. A second orbiter is under construction, and Soviet program officials have stated that the orbiter fleet will eventually consist of three or four vehicles.

Space-science missions. While the U.S. planetary exploration program had studied six different planets (besides Earth) by early 1989, the Soviet effort has focused primarily on Venus. The U.S.S.R.'s first successful Venus probe, Venera 2, flew by Venus in 1966. In 1970, Venera 7 landed on the planet's surface and determined that its temperature is nearly 900°F. Another lander, Venera 9, relayed the first photographs of Venus's surface to Earth in 1975. Veneras 15 and 16, placed into orbit around Venus in 1983, mapped all of the planet's surface above 30°N latitude, including geologic features as small as a mile across. In December 1984, the U.S.S.R. launched the Vega 1 and Vega 2 spacecraft on dual-purpose missions. When the Vegas passed Venus in June 1985, they dropped instrumented capsules into Venus's atmosphere and then continued on for encounters with Halley's Comet in March 1986. (For more information about the Vegas, see the articles "Return of Halley's Comet" and "Solar System Update" in this chapter.)

Despite its past emphasis on Venus, the Soviet planetary program has undergone a radical change. The U.S.S.R.'s most ambitious recent planetary mission was called Project Phobos, a two-spacecraft mission to Mars and its moon Phobos. Unfortunately, errors and mechanical problems prevented the project from achieving its major objectives. The first spacecraft was launched on 7 July 1988, and the second one left the Earth 5 days later. An incorrect command radioed from Earth caused Phobos 1 to lose contact with Earth about 6 weeks after its launch, and it was not heard from thereafter. Phobos 2, however, began to orbit Mars as planned in January 1989. It gathered data for 2 months, mapping the temperature of the Martian surface, measuring the planet's magnetic field, and gathering information on the composition of the atmosphere. In late March 1989, however, Soviet space scientists lost contact with the Phobos 2 before the spacecraft could complete its main mission, which was to take photographs of Phobos from very close range and send two small landers to the moon's surface.

Other Soviet missions to Mars are in early stages of planning or design. In the Mars 1994 mission, a spacecraft will enter a low-altitude orbit around Mars and then release a balloon-borne instrument package that will float over the planet during the Martian day and settle to the surface each night. Soviet plans also call for a 1998 mission, in which robotic landers will obtain samples of the Martian surface and return them to Earth for study.

The U.S.S.R. plans to launch automated spacecraft to orbit the Moon (1993); fly past the large asteroid Vesta (1994); fly past Jupiter and then over the Sun's poles (1995); return samples from the Moon's far side (1996); orbit Jupiter and Saturn and land on Saturn's moon Titan (1999); establish a lunar base (2000); and make a long-term trek across the surface of Mars (2002). (These dates are estimates for each mission's launch date, not when the spacecraft will reach the object to be studied.)

A joint Soviet-American manned mission to Mars is now being discussed among space scientists, although it has not yet been officially endorsed by either the U.S.S.R. or the U.S. In such a mission, Soviet expertise in long-duration space flights would be combined with American experience in lunar landings. The large spacecraft necessary for sending crews to and from Mars would probably be carried into orbit in pieces and then assembled at a space station. The Planetary Society, a private organization that promotes solar system exploration, and astronomer Carl Sagan, the society's president, have been particularly active in promoting Soviet-American cooperation in space. The society has estimated that a Mars mission would cost about $17 billion, although NASA officials believe the cost would probably be much higher — $50 billion to $65 billion. With such a high price tag, most space analysts expect that the landing of humans on Mars — even in a cooperative Soviet–American venture — is still at least several decades away.

In the next few years, the U.S.S.R. plans an aggressive program to study the physics of near-Earth space and to observe the universe with astrophysical observatories in orbit. A series of spacecraft to study the Earth's magnetosphere and its interaction with the solar wind are to be launched within the next few years. These Soviet spacecraft will carry instruments provided by scientists in Eastern Europe and some Western nations. The Gamma spacecraft, scheduled for launch in 1989, represents a collaboration of the U.S.S.R., France, and Poland. It will carry a large telescope for detecting the high-energy gamma rays emitted by many types of celestial sources. Also scheduled for a 1989 launch is the Granat satellite, which will occupy a highly elliptical orbit around Earth. Granat will carry Soviet, French, and Danish instruments to study x-rays and gamma rays.

After the Gamma and Granat missions, Soviet scientists plan to launch a series of orbiting observatories with modular construction to survey the universe at ultraviolet, x-ray, and gamma-ray wavelengths. Other spacecraft missions, which would take infrared and radio-wavelength measurements of objects, are also being considered for launch in the 1990s.

U.S. Space Program

Manned space program. The space shuttle, also known as the space transportation system (STS), and the orbiting space station are the two main projects in NASA's manned space program. Flights of the space shuttle occurred regularly from the first mission on 12 April 1981 until the destruction of the shuttle Challenger on 28 January 1986. NASA resumed the shuttle flights more than two and one-half years later — on 29 September 1988 — after the successful testing of redesigned hardware for the shuttle's solid-fuel rocket boosters and other components.

The concept of a reusable STS emerged in 1970, after the first successful Apollo lunar landings. NASA asked the U.S. Congress for $10 billion to finance the space-shuttle project, but the request was rejected. After making design modifications and obtaining financial backing from the Air Force, however, NASA received a commitment from Congress in 1972 for the reduced sum of $5 billion. The target date for the first test flight was 1979; however, many problems occurred during the developmental phase, increasing the program's cost and delaying the maiden test flight for 2 years. The shuttle's actual development cost was $14 billion (in 1985 dollars).

NASA built four shuttle orbiters between 1977 and 1985 — Columbia, Challenger, Discovery, and Atlantis. A fifth orbiter, proposed in NASA's original plan, was not built due to reductions in funding. To recover most, if not all, costs associated with the shuttle program, NASA planned to fly as many as 60 shuttle missions per year and more than 570 missions in the first 12 years of the shuttle's operation. These estimates, however, soon proved to be overly optimistic, with NASA completing only nine shuttle missions in 1985, the last full year of shuttle operations before the loss of the Challenger.

Challenger accident. Mission 51-L of the shuttle Challenger was to have been its tenth flight into space and the twenty-fifth shuttle mission overall. Worldwide attention focused on the mission, not so much because of its objectives, but because one of the seven crew members was New Hampshire high-school teacher Christa McAuliffe. NASA had chosen the 37-year-old McAuliffe in July 1985 from among 11,000 applicants to be the first teacher to ride into space. She was scheduled to conduct two televised classes from orbit for schoolchildren across the U.S.

Minor mechanical problems and bad weather delayed Challenger's launch until the morning of 28 January 1986. In what appeared at the time to be a normal lift-off, the shuttle rose from its launch pad at 11:38 A.M. EST; however, it soon erupted into a massive fireball. Radio transmissions from Challenger ceased 73.6 seconds after launch. The explosion ripped the shuttle apart but left the Challenger's crew cabin intact. Debris from the explosion rained down over several thousand square miles of the Atlantic Ocean east of Cape Canaveral.

During the months following the accident, U.S. Navy ships recovered the remains of the seven astronauts as well as pieces of the shuttle. As Joseph Kerwin, director of life sciences at NASA's Johnson Space Center in Houston, Texas, stated during a NASA press conference on 28 July 1986, "The cause of death of the Challenger astronauts cannot be positively determined." Kerwin added, however, that "the forces on the orbiter at breakup were probably too low to cause death or serious injury." Therefore, the astronauts may have died either during the fire that engulfed their cabin or when the cabin slammed into the Atlantic Ocean.

Besides Christa McAuliffe, Challenger's crew consisted of the following six people: commander Francis Scobee, 46; pilot Michael Smith, 40; electrical engineer Judith Resnik, 36; physicist Ronald McNair, 36; aerospace engineer Ellison Onizuka, 39; and electrical engineer Gregory Jarvis, 41. McAuliffe, Smith, and Jarvis were all heading into space for the first time. The disaster was the worst ever in the U.S. space program and the first time in 56 missions that American lives were lost in an airborne spacecraft. (Apollo astronauts Virgil Grissom, Edward White, and Roger Chaffee were killed in a launch pad fire during January 1967.)

In addition to the loss of life, the Challenger explosion destroyed the $1.5-billion shuttle, a $117-million Tracking and Data Relay Satellite (TDRS) and the Spartan spectrometer system that was to have observed Halley's Comet.

Space shuttle program after Challenger. As a result of the Challenger disaster, the shuttle's future role in the U.S. space effort has become uncertain. The vehicle is far more costly and complex to operate than had been envisioned by its early proponents. The shuttle also exposes astronauts to the dangers of space flight in situations where unmanned boosters could serve equally well. In a far-reaching recommendation — one contrary to the Reagan administration's declared policy — a presidential commission appointed to investigate the accident urged the U.S. not to rely on the shuttle as its sole launch vehicle, but rather to develop large unmanned boosters to complement the shuttle.

By the time space shuttle flights resumed in 1988, NASA had spent more than $2 billion to correct problems with the shuttle and improve its operation. An article by William J. Broad in the 28 June 1988 *New York Times* noted that, "Although the most important changes were in the design of the flawed solid-fuel rocket, whose leak triggered the Challenger explosion, hundreds of other changes were made to the shuttle itself, its systems, and support equipment on the ground." The modifications included 8 changes to the external fuel tank, 145 to the solid-fuel rocket boosters, and 250 to the orbiter itself. Discovery, the first shuttle orbiter to undergo this exhaustive upgrading, was rolled to the launch pad in July 1988 in preparation for mission STS-26, the shuttle's first launch after more than two and one-half years of inactivity.

The 4-day flight began on 29 September 1988 and was extremely successful, releasing in orbit a much-needed TDRS communications satellite. The orbiter Atlantis made the next flight, STS-27, lifting off on 2 December 1988 with a secret military payload aboard. Although this flight was also a success, technicians discovered considerable damage to the orbiter's insulating tiles after its landing 4 days later. Problems with pumps in Discovery's main engines caused the delay of its next flight, STS-29, but that postponement was not expected to jeopardize the remaining 1989 shuttle flights.

A flight schedule released by NASA in January 1989 called for a total of seven shuttle flights during 1989, nine in 1990, and nine in 1991. Thereafter, the agency expects to launch between 12 and 14 shuttle missions annually. Meanwhile, NASA and the Department of Defense have returned to the concept of a "mixed fleet," employing expendable launch vehicles in addition to the space shuttle. A new orbiter to replace Challenger is under construction, but it will not be ready for flight until 1992.

NASA's space-science missions. During the 1970s, the U.S. launched 13 space probes toward the Sun, Mercury, Venus, Mars, Jupiter, and Saturn. The development of new interplanetary probes, however, slowed considerably during the 1980s. In fact, before 1989, the most recently launched U.S. planetary spacecraft — the Pioneer Venus orbiter, or Pioneer 12 — left the Earth on 8 August 1978.

As of early 1989, the U.S. expected to launch two interplanetary probes later in the year. The Galileo spacecraft will be carried aloft aboard the space shuttle and then boosted by a solid-fuel rocket out of the Earth's orbit. Galileo will circle the Sun three times and require almost 6 years to reach Jupiter, looping past Venus once and Earth twice en route to gain enough velocity to move into the outer solar system. The anticipated arrival date at Jupiter is now late 1995.

Between its two flybys of Earth, the spacecraft will travel deep into the asteroid belt between Mars and Jupiter, and NASA controllers may direct Galileo to pass within 600 miles of the asteroid 951 Gaspra in October 1991. This encounter would provide space scientists with their first close-up views of an asteroid's surface. Another asteroid, 243 Ida, could be visited by Galileo in August 1993 as the spacecraft heads outward to Jupiter.

The Galileo spacecraft consists of two major components: an orbiter and an atmospheric probe. About 150 days before arriving at Jupiter, the orbiter and the atmospheric probe will separate. The 260-pound probe will follow a trajectory that will take it into Jupiter's turbulent, gaseous envelope. As it descends, slowed by a parachute, the probe will relay data on Jupiter's atmosphere to the orbiter, which in turn will transmit the information back to Earth. The probe is expected to function for about an hour. Meanwhile, the Galileo orbiter will be slowed down and captured by Jupiter's gravity, becoming that planet's first artificial satellite. The orbiter will continue to circle the planet and make close passes by its four large moons — Callisto, Ganymede, Europa, and Io.

Another planetary mission — Magellan, formerly called the Venus Radar Mapper — is designed to provide a high-resolution topographic map of over 90 percent of Venus's cloud-covered surface. Magellan will travel around Venus in a 3-hour orbit at an altitude varying from about 155 miles to 5,000 miles above the surface. Magellan's radar will be used to compile images of the planet's terrain whenever the spacecraft comes closer to the surface than about 2,000 miles; the data will then be transmitted back to Earth while the spacecraft is orbiting above this altitude. Magellan radar will record surface details as small as 800 to 1,000 feet across.

One other interplanetary mission now under development is the Mars Observer, formerly called the Mars Geochemistry and Climatology Orbiter, which is to be launched in 1992 and reach the red planet about one year later. The spacecraft will perform detailed remote-sensing studies of the surface and atmosphere below, extending the knowledge already gained by Mariner and Viking spacecraft during the 1960s and 1970s.

Three instruments will provide information about the chemical and mineral composition of the Martian soil. Another device will measure the atmosphere's composition, pressure, temperature, and water content, as well as determine the characteristics of any dust suspended within it. A camera on the spacecraft will photograph strips of the surface below, revealing details as small as 5 feet across. All the information gathered by the Mars Observer will provide data to be used in selecting future landing sites for automated landers, roving vehicles, and human explorers.

Planetary exploration in the 1990s. The decision by NASA to develop the Mars Observer for the relatively low cost of about $200 million represents a policy shift away from large, expensive missions such as the $1.4 billion Galileo project and the earlier Viking Mars and Voyager multiplanetary programs. In an effort to develop an active U.S. planetary exploration program within current budgetary constraints, NASA's Solar System Exploration Committee (SSEC) conducted a 2-year study of the possible planetary missions that could be undertaken through the end of the century.

The SSEC report, released in 1983, recommended that NASA undertake a core program of low-cost and moderate-cost missions by the year 2000. The low-cost programs would fall within the range of $100 million to $150 million each, while those of moderate cost would each require about $250 million to $300 million. The total expense of the core program, including spacecraft development, mission operations, and data analysis, would be about $300 million to $350 million per year in 1984 dollars, about one-third of NASA's funding level for interplanetary spacecraft in the mid-1970s. The SSEC further suggested that, whenever possible, projects should rely on proven technologies and existing equipment to keep costs to a minimum.

The SSEC placed the highest priority in the core program on the Magellan, Mars Observer, a comet rendezvous and asteroid flyby (CRAF) spacecraft, and a Saturn/Titan scientific spacecraft. The proposed CRAF vehicle would be launched in the 1990s toward one of the comets that passes periodically near the Earth. The specific comet would depend on the launch date, which in turn would depend on the funding and development schedule. Unlike the European, Soviet, and Japanese space probes that observed Halley's Comet for a relatively short time during 1986, CRAF will follow its chosen comet for several years, orbiting within 20 miles of the nucleus. Another high-priority project in the SSEC core program — the Saturn/Titan space vehicle called Cassini — would be a joint undertaking of NASA and ESA. (For more information, see "Horizon 2000" on page 54.)

Astronomy from orbit. No astronomy research spacecraft in NASA's 30-year history has matched the complexity and the potential for fundamental discovery of the Hubble Space Telescope (HST), which is scheduled to be launched from the space shuttle in 1990. During its 15-year to 20-year lifetime in orbit, the HST's 94-inch-diameter mirror and sophisticated instruments will observe astronomical objects in much greater detail than ever before. (For further information about the HST, see the "Astronomical Facilities" article in this chapter.)

NASA has several other astronomy research payloads that are awaiting launch or are under construction. A Delta rocket was scheduled to launch the Cosmic Background Explorer (COBE) in 1989. From its 560-mile-high orbit, COBE will scan the universe to detect radiation at very long wavelengths. Its detectors will be cooled by liquid helium to within a few degrees of absolute zero, and astronomers hope that COBE will be able to observe a faint thermal emission, called the cosmic background radiation, which is in effect the afterglow of the universe's formation.

Astro-1, a quartet of telescopes sensitive to ultraviolet light, will observe the universe from the cargo bay of the space shuttle in 1990. Another spacecraft, the Gamma-Ray Observer (GRO), is also scheduled to be carried into orbit aboard the space shuttle in 1990. GRO will study cosmic sources of gamma rays, the most energetic form of radiation known.

International Space Station

During the 1980s, flights of astronauts and cosmonauts demonstrated that humans are able to work efficiently and for extended periods of time in the weightless environment of space. The success of these missions has bolstered support for permanently inhabited space stations circling the Earth. Since 1971, Soviet cosmonauts have occupied a number of small orbiting space stations with increasing frequency, and the U.S.S.R. now appears to have the capability to let cosmonauts remain aboard its newest station, called Mir, continuously. Although NASA had studied the feasibility of a space station for years, President Ronald Reagan first officially announced the goal of establishing such a facility during his State of the Union address in January 1984. After the Challenger disaster in 1986, he reaffirmed his commitment — as has President George Bush — to the establishment of a manned space station in the 1990s.

The proposed space station will be called *Freedom* and is to be built in at least two phases. According to space station plans, the Phase I portion will consist of a 508-foot-long horizontal beam, on which clusters of pressurized modules and power-generating equipment will be mounted. The space station will also have a docking area for the space shuttle and a movable remote manipulator arm for handling satellites and equipment. The 44-foot-long pressurized modules will be situated near the center of the structure and interconnected to achieve the maximum possible work area. The interior of each module can be changed to accommodate new experiments and advancing technology.

The space station's power will be supplied by eight 96-foot by 32-foot conventional silicon solar-cell panels, with four at each end of the horizontal beam. Together, they will generate at least 75 kilowatts of electricity. The station will ultimately need more electricity, but larger panels would introduce substantial drag on the station as it moves through the traces of the Earth's atmosphere that are still present in space near the planet; this drag would cause the station's orbit to decay. Solar reflectors may eventually be attached to each end of the beam to focus sunlight on a pair of heat engines that will drive electrical generators. This second system is necessary to keep the area of the solar panels to a minimum.

Ultimately, NASA hopes *Freedom* will serve as a base for studying the Earth and outer space, conducting scientific research, and developing new manufacturing technologies. It could also be used to assemble large space structures, maintain and repair satellites, and test new systems and equipment for future space operations. The shuttle will be used to carry into orbit the materials and equipment for building the space station and later to transport supplies to the facility. It will also allow for periodic crew changes once the station is functioning. The facility will be assembled in space and orbit the Earth at altitudes between 210 and 290 miles.

NASA officials initially estimated that the space station's development and construction would cost $8 billion (in 1984 dollars). After the Challenger's loss, however, concerns arose about the availability of space shuttles to bring station components into orbit and the risk to the astronauts involved in assembling the structures. By September 1986, NASA engineers had modified the station's initial "dual-keel" design (which included a huge rectangular truss measuring 310 by 150 feet) and construction plan. In the modified plan, a simpler structure will be deployed in 1995 and occupied by 1997, followed by a second phase of construction to add the dual keel's rectangular structure. Thus, the space station will be relatively small at first, but its design will allow for future expansion.

In late 1987, NASA estimated that the dual-keel station will cost $14.6 billion (in 1987 dollars), but expenditures for its long-term maintenance and the scientific equipment it will carry will raise the total investment to about $24 billion. By the end of 1987, the agency had selected four contractors to develop the hardware needed for the station's construction: Boeing Aerospace Co., McDonnell Douglas Astronautics Co., General Electric Co., and the Rocketdyne Division of Rockwell International. Each contractor will be responsible for one of the station's four work sections, which will be managed by different branches of NASA.

International participation. In his 1984 State of the Union address, President Reagan invited Canada, Japan, and the nations of western Europe to participate in the development and operation of the space station. In March 1985, the Canadian government formally accepted Reagan's invitation. Canada is conducting studies on the space station's construction and servicing facility, remote-sensing system, solar-cell power arrays, and robotic manipulator arm. (Canadian scientists developed the robot arm used on the space shuttle.) These studies will cost the Canadian government an estimated $1 billion. Japan signed an initial agreement with NASA in 1985 and will develop and pay for a pressurized laboratory module and an unpressurized work deck for the station at a cost of approximately $2 billion. (In the first phase of construction, the U.S. will build two such modules.)

In 1986, ESA formalized its agreement with NASA to build a fourth pressurized module, called Columbus, for the space station. It will be an automated experiment platform to be placed in a polar orbit, and a separate mini-station involving a pressurized laboratory module and an attached equipment module. The Columbus program will cost ESA an estimated $3.6 billion, and France, West Germany, and Italy will be its major participants. The mini-station could be carried into orbit on a single shuttle flight and assembled in space. It would contain standardized fixtures that would enable experimental payloads and propulsion and power modules to be easily plugged into and removed from the structure. The platform could be flown in the vicinity of the space station or, if necessary, diverted to a different orbit.

During September 1988, the U.S. and its international partners reached a formal agreement detailing the responsibilities of each participant. The station is to be operated by a consensus of the partners, with disagreements about using a given section of the station to be settled by the partner that produced the section.

But not all of the political implications of an international space station have been worked out. For example, will national interests preclude certain areas of cooperation? Will members of one nation have only restricted access to the modules of another nation? Whose laws will govern the inhabitants? From this starting point, future space stations may eventually evolve into truly autonomous entities, not bound by the policies of any particular nation. The space station of the 1990s may become more than just a facility devoted to scientific endeavors; it may be an experiment in international cooperation.

Chapter 2

Biology

Evolution and Creationism

Evolution

More than a century has passed since the death of Charles Darwin, the British scientist who, more than anyone else in history, increased our understanding of the mechanism responsible for the ever-changing diversity of life on Earth. Darwin developed the theory that all living things have evolved by natural selection and described that theory in detail in his famous 1859 book *On the Origin of Species by Means of Natural Selection*.

According to Darwin, chance variation in organisms, now known to be the result of genetic mutation, provides new characteristics that the organisms may pass on to their offspring. Most variations are not beneficial, and many are debilitating or deadly. A few, however, give the organism an advantage over competitors in the struggle to survive and reproduce. Darwin theorized that the organisms with advantageous mutations are likely to out-compete the original forms, gradually outnumbering and replacing them in the population. In this process of natural selection, advantageous mutations accumulate over time and ultimately produce organisms so different from their ancestors that they must be considered an entirely new species.

Despite the evidence in favor of this theory of evolution, no consensus has yet been reached among scientists concerning the precise mechanisms by which evolutionary change occurs. The recent resurgence of the creationism movement, along with the adoption of legislation in some states requiring creationism to be taught side by side with evolutionary theory, is perhaps the most striking example of the controversial nature of Darwin's theory and its theological and philosophical implications. Most scientists today, however, do accept a modern synthesis of Darwin's theory of natural selection and currently held principles of genetics and biochemistry.

Gradual or punctuated? Since Darwin's day, most scientists have believed that evolution is a continuous, gradual process, with natural selection resulting in the accumulation of minor favorable variations. The fossil record, however, is generally too fragmented to illustrate such slow, gradual change. The "gradualists" do not interpret the gaps in the fossil record between species as evidence of abrupt change. They believe that intermediate forms of species must have existed, but that the remains either were destroyed or have not yet been found.

In the early 1970s, however, a few scientists, notably paleontologists Stephen Jay Gould of Harvard University in Cambridge, Massachusetts, and Niles Eldredge of the American Museum of Natural History in New York City, presented evidence that evolution actually takes place in fits and starts. Their "punctuated equilibrium" theory postulates that evolution normally proceeds through periods of equilibrium, when existing species do not change significantly and only a few new species arise. These long periods of stability are "punctuated" by relatively brief periods when new species arise at a much faster rate. This theory implies that the key steps in the development of new species occur during brief bursts of heightened evolutionary change.

In his 1986 book *Time Frames,* Eldridge argued that small populations of organisms isolated from other members of their species by geological changes, such as the disappearance of land bridges connecting continents, may undergo rapid change in response to adverse environmental conditions, such as a rise or fall in temperature. He and Gould believe that the gaps in the fossil record support their theory because, if evolutionary change occurs relatively rapidly in isolated populations, the chances that intermediate forms will be preserved in the fossil record is relatively small. During 1987, two major studies shed new light on the question of punctuated versus gradual evolution, with one favoring each side. But the intense debate had begun several years earlier with the publication of a study involving mollusks in central Africa.

Lake Turkana fossil record. The first detailed, continuous record of one species evolving into another was reported in 1981 by Harvard University paleontologist Peter Williamson. Williamson visited a fossil-rich site east of Lake Turkana (formerly Lake Rudolf) in northern Kenya and examined a sequence of sediments about 1,300 feet thick that was deposited over a period of 3 million to 4 million years. This sediment sequence contained thousands of fossils of freshwater mollusks that lived during the Cenozoic Era 60 million to 70 million years ago.

Because the sediments were so clearly stratified and the lake basin, surrounded by mountains, was virtually a self-contained ecosystem, paleontologist Anthony Hallam of the University of Birmingham in the U.K. described the fossil record as the "nearest thing you could get to an evolutionary experiment." Williamson analyzed 3,300 mollusk fossils from 13 different species lineages and was able to identify the rare intermediate forms that linked the oldest (ancestral) species and the newer (daughter) species in these lineages. One of Williamson's central findings was that some present-day mollusks of nearby Lake Turkana still look exactly like their ancestors did millions of years ago and thus have apparently not been changed by evolution.

Within this framework of apparent stability, however, Williamson identified two relatively brief periods when many species underwent rapid change. These periods, each of which lasted between 5,000 and 50,000 years, corresponded with intervals in which the water level in Lake Turkana fell precipitously. The drop in water level caused the channels connecting Lake Turkana and nearby lakes to dry up, separating the mollusk species in Lake Turkana from those in the smaller lakes. When the water returned to its normal level, Lake Turkana was once again connected with the nearby lakes, and the mollusk species that had evolved separately intermingled. Williamson found that, in some cases, for reasons still unclear, the ancestral species became dominant and caused the daughter species to become extinct, while in other cases, the daughter species dominated.

According to Williamson's analysis, the organisms that arose and became established during periods of physical isolation and environmental stress were distinct species, not the subtly different forms that might arise through gradual change. He concluded that speciation occurs when a population of organisms is isolated under stress, which is a very different mechanism from that proposed by the gradualists.

Williamson's findings have been the subject of heated debate between "punctuationists," who believe that the study provides compelling evidence in support of punctuated equilibrium, and gradualists, who question Williamson's interpretation of the data and remain unconvinced. Evolutionary biologist Ernst Mayr of Harvard University, for example, has noted that the ancestral and daughter species that Williamson claimed to have found in the fossil record might be nothing more than different phenotypes — organisms of the same species having a different external appearance. Mayr explained that individual mollusks of a single species can have shells of different shapes, depending on local environmental conditions, such as the depth and chemical composition of the surrounding water and the amount of exposure to waves and currents. Consequently, as the conditions in Lake Turkana changed, the shell shapes could have changed in response to the new conditions even though the mollusks remained the same species.

Williamson countered, however, that his observations rule out this possibility, because some of the daughter organisms appeared to coexist with their presumed ancestral forms. This situation would not occur if the new organisms were merely different phenotypes of the same species, since only one phenotype should exist under a given set of environmental conditions. Williamson also said that the changes during periods of environmental stress seemed to follow a regular progression over 1,000 to 10,000 generations. This number of generations is much greater than that required for a phenotype change. Finally, he claimed that the extent of the changes he documented was generally much greater than the phenotypic changes that occur in even the most variable of the modern African mollusks in response to environmental stress.

Williamson has noted that the gradualists who disagree with his interpretation of the rapid change in organisms have yet to adequately explain the stability of the organisms' structures during the majority of the 3- to 4-million-year fossil record studied. Williamson claimed that the evidence of such long-term structural stability is by itself enough to call gradualism into question.

Gradualists have typically responded to such criticisms by noting that simply because an organism's bones, shells, and teeth — the parts usually preserved in fossils — might have appeared nearly constant over a long period of time, this does not necessarily mean that gradual changes were not occurring. Substantial changes in soft body parts such as the digestive system, for example, could occur without major changes in hard body parts. Consequently, the periods of "stasis" claimed by punctuationists may actually have been times of genetic change.

Gradualism and human evolution. Further support for the gradualism theory was offered in 1985 by anthropologist Milford Wolpoff of the University of Michigan at Ann Arbor. Wolpoff studied *Homo erectus*, the direct ancestor of modern humans. Gould and Eldredge have frequently cited *H. erectus* as an example of punctuated equilibrium, on the basis of a 1981 study by paleoanthropologist G. P. Rightmire of the State University of New York at Binghamton. Rightmire studied 65 skulls classified as *H. erectus* and concluded that skull size did not change during the nearly 1.5 million years before the appearance of *Homo sapiens*. He argued that the larger skull characteristic of modern man first appeared at about the same time as the new species arose.

Wolpoff traveled around the world to take a variety of skull, jaw, and dental measurements from the 92 known specimens of *H. erectus*. He divided these fossils, which range from about 1.4 million to 400,000 years old, into three groups: early, middle, and late. He found pronounced differences between the early and late groups: cranial capacity had expanded while jaw and tooth size had shrunk.

Wolpoff also studied 13 skulls classified as either late *H. erectus* or early *H. sapiens* and found a continuation of the trends observed for *H. erectus*, but no major changes. He thus concluded that *H. erectus* provides strong evidence for gradual evolution. In addition, he argued that up to 16 of the specimens Rightmire studied were not actually *H. erectus*, and that Rightmire's statistical analysis was inadequate to uncover subtle evolutionary changes.

Clams and trilobites. Paleobiologist Steven Stanley of Johns Hopkins University in Baltimore, Maryland, a proponent of punctuated equilibrium, has found support for the theory in a study of thousands of clam fossils gathered off the coast of eastern North America, Italy, and the U.K. The oldest fossils were more than 17 million years old, but most were between 1 million and 4 million years old.

In 1987, Stanley reported that he and graduate student Xiangning Yang had studied 24 different characteristics of each clam fossil and compared them to modern clams collected in the same areas. They found little change within species; differences among clams even 17 million years ago were no greater than the differences among clams today. Had the clams evolved gradually, Stanley said, more evidence of change within species should have been apparent. He concluded that the study provided strong support for punctuationism.

Geologist Peter R. Sheldon of Trinity College in Dublin, Ireland, reached the opposite conclusion after analyzing more than 15,000 trilobites, a now-extinct marine invertebrate common during the Paleozoic Era. Trilobites had a characteristic three-lobed shell and were generally 1 to 2 inches across, although some grew as large as 24 inches. Sheldon studied fossils that had been deposited over a 3-million-year period in a sedimentary formation called the Builth inlier in central Wales. He studied many characteristics, but concentrated on the number of pygidial ribs, bands across the center of the shells.

Other researchers had previously studied specimens from the beginning and the end of the 3-million-year period and concluded that new species had arisen, based primarily on the fact that the trilobites had significantly different numbers of these ribs. But when Sheldon studied the trilobites throughout the sediment, he found a steady increase in the number of ribs in each species, although periodic reversals in the number of ribs did occur. When he compared fossils from the beginning and end of the period, which had been classified as different species, with the intermediate fossils, he found no clear points at which the early species could be clearly distinguished from the later ones. He claimed that his results provide the strongest proof yet that evolution is gradual rather than punctuated.

The conflicting conclusions of these studies illustrate that evolutionary biologists have not yet reached a consensus about the manner in which evolution occurs and the way in which even a single set of fossil data should be interpreted. After more than a decade of debate, neither the gradualists nor the punctuationists appear close to convincing the other side of their theory, and the controversy appears likely to continue for years to come.

Other opinions. Some scientists think that the gradualist-punctuationist debate may be just a semantic quarrel. Geneticists, for example, would not consider Williamson's 5,000 to 50,000 years of rapid evolutionary change to be a "brief" period. About 20,000 mollusk generations would be born over the course of 50,000 years; in the laboratory, geneticists are able to produce widely varying populations of mice, corn, and fruit flies within only 50 generations.

In addition, some researchers believe that considerable evolutionary change could have taken place gradually over a relatively short period, even in hominids—modern humans and their apelike ancestors. According to biologists G. Ledyard Stebbins of the University of California at Davis and Francisco Ayala of the University of California at Irvine, the 50 percent increase in brain size between the hominids *Homo erectus* and Neanderthal man, for example, could have occurred gradually in only 540 generations, or about 13,500 years, assuming natural mutation rates comparable to those observed in studies of laboratory animals.

New Techniques for Studying Evolution

For centuries, scientists categorized life on Earth only by analyzing the anatomical structure and behavior of various organisms. While structure and behavior are useful criteria for classifying living organisms, they are less valuable for constructing evolutionary "trees," branching sequences that show the ancestors of each species and how long ago related species diverged from a common ancestor. An important limitation on the use of anatomical structure to classify organisms is that distinct species can share anatomical features without having any close evolutionary relationship. Conversely, two closely related species may be anatomically quite different. Advances in the fields of biochemistry and genetics, however, have solved this problem by enabling researchers to use biochemical techniques to establish evolutionary relationships among species. The ultimate identifying feature of an organism is the nucleotide sequence of its deoxyribonucleic acid (DNA). (Nucleotides are the building blocks of DNA. For more information on the structure and function of DNA, see the article "Biotechnology Techniques" in this chapter.) As two species diverge from a single ancestral line, subtle changes take place in the DNA of each species and in the proteins coded for by that DNA.

Biologists use several techniques to compare DNA and proteins from different organisms and derive evolutionary trees: (1) biochemical analyses of the extent to which the amino acid sequences of similar proteins vary among different species; (2) biochemical comparisons of variations in DNA sequences coding for similar proteins in different species; (3) visual analyses of the light and dark patterns on the chromosomes of different species; and (4) biochemical comparisons of variations in the entire DNA complements of different species.

In general, analyses of DNA sequences are thought to be more sensitive than analyses of amino acid sequences because there are three nucleotides for each amino acid, so that the possibility of detecting changes is greater. Also, analyses of the entire DNA complement are more sensitive than analyses of shorter DNA segments. By using all of these analytical techniques, scientists are able not only to assess the relationships among species more reliably, but also to estimate how long ago the evolutionary paths of related species diverged.

Man and apes. In 1981, geneticist Allan Wilson and his associates at the University of California at Berkeley reported on an analysis of DNA from chimpanzees, gorillas, humans, and orangutans — all of which are believed to have evolved from a common ancestor. The DNA was taken from cell structures called mitochondria, which govern a cell's energy balance. Each mitochondrion has a piece of autonomously replicating DNA, which has been found to change five to ten times faster than DNA in the cell's nucleus. The researchers found 42 sites in the mitochondrial DNA where slight differences in the sequences among the species had evolved, and used the extent of these differences to construct an evolutionary tree.

Evolutionary theories based solely on anatomical structure had previously indicated that humans had diverged from the chimpanzee, gorilla, and orangutan ancestral lines before these three types of apes themselves split into separate species. The work of Wilson and his colleagues showed, however, that the orangutan had been the first to split off. Their results also suggested the unlikely possibility that the chimpanzee, gorilla, and human species had all diverged at about the same time. Biologist Alan Templeton of Washington University in St. Louis, Missouri, later performed a statistical analysis of the Berkeley group's data and found that the human line split off first, and concluded that the chimpanzee and gorilla are more closely related to each other than either is to humans.

In 1982, pathologists Jorge Yunis and Om Prakash of the University of Minnesota Medical School in Minneapolis reported on a genetic analysis of the chromosomes of humans and the three species of great apes. The researchers matched up the pattern, thickness, and staining intensity of the dark-colored "bands" in the chromosomes of the four species and studied differences among them. Although identical banding patterns do not necessarily imply identical DNA sequences, Yunis and Prakash were surprised to find that 18 of the humans' 23 pairs of chromosomes were extremely similar to chromosomes of the three species of great apes.

They found, however, that the banding patterns of chimpanzee chromosomes were more similar to human chromosomes than to gorilla chromosomes. This observation challenged the claim by Templeton that humans diverged from an ancestral line before the gorillas and chimpanzees split. Yunis and Prakash concluded that gorillas diverged from the human-chimpanzee line before humans and chimpanzees split, making the chimpanzees a closer relative of humans than of gorillas — just the opposite of what Wilson had found.

The contradiction between the Berkeley and Minnesota findings may have been resolved by more recent work reported by zoologists Charles Sibley and Jon Ahlquist of Yale University in New Haven, Connecticut, in 1984. Using techniques they had previously developed in studies of the evolutionary relationships of birds, Sibley and Ahlquist compared the entire DNA complement of humans, gibbons, Old World monkeys, gorillas, chimpanzees, and pygmy chimpanzees.

The researchers first isolated the complete genetic material of each species, separated the double-stranded DNA into single strands, and then removed repetitive DNA sequences. For comparisons between any two species, they sheared the single-stranded material from each into 500-nucleotide-long segments, mixed them together, and allowed them to reassociate, forming what are called heteroduplexes. The strength of the bonding holding the heteroduplexes together is directly related to the number of nucleotides the two sequences have in common.

Even DNA strands that are complementary — homoduplexes — will dissociate when they are heated to a high enough temperature, usually near 212°F. Strands with a few nucleotide differences will begin to dissociate at lower temperatures, and those with more differences will dissociate at still lower temperatures. Sibley and Ahlquist observed the temperature at which half of the heteroduplexes became dissociated and subtracted that number from the temperature at which half of the homoduplexes dissociated. The result is called delta $T_{50}H$. Delta $T_{50}H$ is thought to be proportional to the time that has elapsed since the divergence of the evolutionary paths of the two species being compared. For humans and chimpanzees, for example, delta $T_{50}H$ is 1.9. This value indicates a greater genetic distance than that between the chimpanzee and the pygmy chimpanzee, whose delta $T_{50}H$ is 0.9.

In their previous work with birds, Sibley and Ahlquist had been able to associate a specific time interval with delta $T_{50}H$ by linking a particular evolutionary branching point to a specific geological event. For example, the evolutionary divergence of flightless birds, such as the ostrich and the rhea, can be linked to the separation of the African and South American continental masses some 80 million years ago. By making the reasonable assumption that nucleotide mutations in hominid DNA occur at about the same rate as mutations in birds, they were then able to estimate the following times for divergence from the chimpanzee lineage: Old World monkeys, 27 to 33 million years ago; gibbons, 18 to 22 million years ago; orangutans, 13 to 16 million years ago; gorillas, 8 to 10 million years ago; humans, 6.3 to 7.7 million years ago; and pygmy chimpanzees, 2.4 to 3 million years ago. Sibley and Ahlquist's results thus support the conclusion of Yunis and Prakash that humans are more closely related to chimpanzees than to other primates. By early 1989, Sibley and Ahlquist had accumulated more than 10 times as much data as had been available when their 1984 paper was written, and the new data reinforced their original conclusions.

The absolute dates for evolutionary divergence determined by Sibley and Ahlquist, however, are typically about 50 to 75 percent older than dates determined by other techniques. They found, for example, that the lineages of humans and gorillas split 8 million to 10 million years ago, while other techniques typically date this split at 4 million to 6 million years ago. Berkeley's Wilson speculates that such differences can be attributed to Sibley and Ahlquist's calibration points. According to Wilson, even though the African and South American continents split apart 80 million years ago, they undoubtedly remained close enough together so that even flightless birds could have passed from one to the other until much more recently, perhaps only 40 million years ago. Wilson suggested that, if Sibley and Ahlquist's calibration data were assumed to be in error by a factor of about two, all the age calculations could be reconciled.

Mother of us all. Wilson and his colleagues reported in 1986 that all people alive today may be a descendent of a single female ancestor who lived in Africa 140,000 to 280,000 years ago. Their study supports the view that human beings originated in Africa and later migrated throughout the rest of the world. Using the same techniques with which they had studied the divergence of primate species, Wilson and his colleagues examined DNA in the mitochondria of 147 placentas from women in Africa, Asia, Australia, Europe, and the U.S. Calculating from the observed similarities in DNA sequences and the known rate of mitochondrial DNA changes — about 2 to 4 percent every million years — they estimated that all of the women had descended from a single individual, or at least a small group of people, who originally lived in Africa. The results, however, do not suggest that only one couple was alive at that time — a patently ridiculous conclusion. Rather, they indicate that the descendants of all the other breeding couples eventually died out.

Later in 1986, geneticist Douglas Wallace of Emory University in Atlanta, Georgia, reported a similar conclusion about the ancestry of modern humans. Wallace studied mitochondrial DNA from 800 women around the world and concluded that all were descended from a single individual or a small group of people who lived more than 100,000 years ago. Unlike Wilson, however, Wallace concluded from the DNA data that the "mother of us all" lived in Asia rather than Africa.

Extending Human History Backwards

New evidence reported during 1988 has extended further back into prehistory the times when the first humans began using fire and tools. The evidence suggests that close relatives of early man, in lineages that later became extinct, also were able to use tools. That discovery has cast doubt on the long-held belief that it was primitive humans' ability to use tools that enabled them to outcompete other lineages.

A skilled thumb. Tools and hand bones excavated from the Swartkrans cave complex in South Africa suggest that a close relative of early man known as *Australopithecus robustus* may have made and used primitive tools long before the species became extinct 1 million years ago. They may even have made tools before humanity's direct ancestor, *Homo habilis* or "handy man," began doing so, according to anatomist Randall L. Susman of the State University of New York School of Medicine at Stony Brook. *H. habilis* and its successor, *H. erectus*, coexisted with *A. robustus* on the plains of South Africa for more than a million years.

The Swartkrans cave, one of the best sources of *A. robustus* remains in South Africa, has been under excavation since the 1940s by members of the Transvaal Museum in Pretoria, most recently by C. K. Brain and his associates. The earliest fossil-containing layers of sedimentary rock in the cave date from about 1.9 million years ago and contain extensive remains of animals, primitive tools, and two or more species of apelike hominids. The key recent discovery involved bones from the hand of *A. robustus*, the first time such bones have been found.

The most important feature of the *A. robustus* hands was the pollical distal thumb tip, the last bone in the thumb. The bone had an attachment point for a "uniquely human" muscle, the flexor pollicis longus, that had previously been found only in human ancestors. That muscle gave *A. robustus* an opposable thumb, a feature that would allow them to grip objects, including tools. The researchers also found primitive bone and stone implements, especially digging tools, in the same layers of sediments.

A. robustus was more heavily built — more "robust," in anthropological terms — than ancestors of man. *A. robustus* had large, broad faces, heavy jaws, and massive crushing and grinding teeth that were used for eating hard fruits, seeds, and fibrous underground plant parts. They walked upright, which would have allowed them to carry and use tools. Most experts had previously believed that *H. habilis* was able to displace *A. robustus* because its ability to use tools gave it an innate superiority. The discovery that *A. robustus* also used tools means that researchers will have to seek other explanations for its extinction. Perhaps their reliance on naturally occurring plants led to their downfall as the climate became drier and cooler, or perhaps *H. habilis*, with its bigger brain, was simply able to make more sophisticated tools.

Tool use in Asia. A new discovery of primitive stone tools in Pakistan suggests that tool-using hominids spread from Africa as much a million years before the date previously believed. The discovery suggests that *H. habilis* was the first tool-user to migrate away from the continent where humanity's ancestors had originated, and not its successor, *H. erectus*.

The oldest known hominid fossil in Europe is a 500,000-year-old jawbone found near Heidelberg, West Germany. The oldest undisputed evidence of hominid occupation is at Soleilhac, in the Massif Central of France, where 800,000-year-old tools and animal remains have been excavated. The recently discovered tools, excavated at Riwat, Pakistan, by archaeologist Robin W. Dennell of the University of Sheffield in the U.K., were found below a 1.6-million-year-old layer of volcanic ash, suggesting that the tools are close to 2 million years old.

Dennell reported in 1988 that he and his colleagues had found 26 potential hominid tools at one location and that six showed strong evidence of being tools. The key features of these specimens included flaked surfaces running in two or more directions and ripple marks, cracks, and protrusions clearly caused by the impact required to chip flakes from the stones. Neither the exact use of the tools nor the type of hominids that produced them has yet been determined.

Earliest use of fire. C. K. Brain of the Transvaal Museum has found the earliest evidence for the use of fire by hominids at the same Swartkrans cave where the *A. robustus* hand bones were discovered. Before the discovery, the earliest direct evidence for the use of fire was found at a Chinese site dating back only 500,000 years. In 1981, archaeologist John A. J. Gowlett of the University of Liverpool in the U.K. had reported the discovery of magnetic changes in charred clay that had been caused by a fire, but that fire could have been ignited by lightning. The new evidence, in contrast, suggests direct use of the fire by hominids.

Brain and anthropologist Andrew Sillen of the University of Cape Town in South Africa found 270 pieces of burnt wood and bone from numerous layers of sediment in the cave dating from more than 1 million years ago, but not in layers older than 1.8 million years. They concluded that the fires were made by burning stinkwood, a tree common around the Swartkrans area. The researchers are not sure how the hominids used fire, but Brain speculated that one of the first uses might have been to frighten away wild animals at night.

Earliest humans. French and Israeli researchers have found evidence that anatomically modern humans lived in an Israeli cave in lower Galilee about 92,000 years ago. The best previous evidence of anatomically modern humans had dated from only 35,000 to 40,000 years ago. The new results thus suggest that humanity in its modern form has existed much longer than researchers had previously believed. The discovery also supports the idea that anatomically modern humans evolved in Africa and spread throughout the world, rather than evolving independently at more than one location.

The researchers reported in 1988 that they had found numerous fossil fragments in the Israeli cave. About 20 burned flints found among the fossil remains were dated to about 92,000 years ago, plus or minus 5,000 years, indicating that the humans had lived in the cave at least that long ago. In a written commentary about the find, archaeologist Chris Stringer of the British Museum in London noted that the dating of the discovery raises several questions, the most important being why no early evidence of the anatomically modern humans has been found in Europe. He suggested that the climate in Europe may have been too cold for the early humans, or that perhaps the Neanderthal people, who lived in Europe from 200,000 to 35,000 years ago, were too well established for the human ancestors to move in.

Modular Genes

New biochemical evidence suggests that some types of evolution may proceed one million to 100 million times faster than scientists had previously believed possible.

Biochemists Thomas Sudhof, Joseph Goldstein, Michael Brown, and David Russell of the University of Texas Health Sciences Center in Dallas reported in 1985 that the genes coding for certain proteins theoretically could have been produced by combining segments obtained from several other genes. This process would be considerably faster than producing proteins by repetitive mutations of individual nucleotides in a given gene — the mechanism by which biologists previously thought proteins evolved. It could also be the mechanism underlying major changes in a species.

Genes in most living organisms are composed of two different types of DNA sequences, called exons and introns. Exons are DNA sequences that provide the blueprint for segments of the protein coded for by the gene. Introns are much longer "nonsense" sequences, interspersed among the exons, whose purpose is not yet known. The introns are edited out of the gene before a protein is actually produced. (For more information about gene structure and protein production, see the section "Biotechnology Techniques" in this chapter.)

The Texas group determined the DNA sequence of the gene coding for a receptor protein that binds to low-density lipoprotein, which transports cholesterol through the blood. They found that the gene is composed of 45,000 nucleotides divided by introns into 18 exon sequences. Each of the exons corresponds to a structural subunit of the protein. They then used a computer to compare the amino acid sequences coded for by the 18 exons to the sequences in other proteins whose structures had been previously determined.

The researchers found that 13 of the 18 exons were very similar to exons in other proteins. Eight of the 13 exons were closely related to exons in a precursor to a protein, epidermal growth factor, that stimulates the growth of skin cells. Three of these exons are also shared by three proteins that participate in blood clotting. The other five exons are present in an immune system protein called C9 complement. The fact that five of the exons were not found in other proteins may simply be a reflection of the limited number of proteins that have been sequenced so far.

The possibility that such genetic recombinations could occur was first suggested in 1977 by Harvard University molecular biologist Walter Gilbert to explain the presence of introns. Gilbert noted that the "nonsense-containing" introns were much longer than exons. He proposed that when genes break apart, as frequently occurs in nature, the break is more likely to occur in the introns, leaving functional exons on each side. When the fragments then recombine — which also happens naturally through the action of enzymes — the new gene produces a protein formed from segments that have already coded for other proteins. The Texas research was the first experimental confirmation of Gilbert's hypothesis.

Creationism

Ever since Charles Darwin's book *On the Origin of Species* was first published in 1859, his theory of evolution by natural selection has remained the subject of heated controversy. As soon as Darwin's book was published, biologists and religious leaders alike attacked his theory. Religious leaders, in particular, objected to the theory of natural selection on the grounds that it challenged the Biblical account of Creation described in the Book of Genesis. Darwin's theory contradicted the prevailing view of man's special relation to God; if man and the apes had a common ancestor, then man could no longer be considered to have been created in God's image.

Most scientists have since come to agree with the broad tenets of Darwin's theory of evolution through natural selection. Most mainstream religious leaders, as well as scientists and lay people who are religious, have also reconciled their belief in evolution with their faith. Nonetheless, fundamentalist religious groups have remained strongly opposed to Darwinian theory. Christian fundamentalists believe that the Bible is the word of God and is therefore literally true, making evolution incompatible with Christianity.

In the early 1980s, a resurgence of fundamentalism in the U.S. led to the passage of laws in several states requiring the concept of divine creation to be taught alongside evolution in public schools. But in June 1987, the U.S. Supreme Court rejected as unconstitutional a 1981 Louisiana law requiring that "creation-science" be taught in biology classes along with evolution. That decision, most observers agree, probably sounded the death knell for attempts to require that creationism be given equal time in U.S. classrooms.

Arkansas creationism trial. The Louisiana law that the Supreme Court struck down was modeled after an Arkansas law, Act 590, which was signed in March 1981 by Governor Frank White. Act 590 mandated that public school teachers in Arkansas teach the "creationist" theory of the origin of life — a theory that closely parallels the Biblical account of Creation — side by side with evolutionary theory.

The American Civil Liberties Union (ACLU) challenged Act 590 in the U.S. District Court in Little Rock, Arkansas. Like the famous 1925 trial in which biology teacher John Scopes was convicted of violating a state law by teaching Darwinian evolution in a classroom in Dayton, Tennessee, the trial that resulted from that challenge was the focal point of the long-standing battle between creationists and evolutionists.

Spectators at the trial heard a number of unusual statements by defense witnesses, including testimony that unidentified flying objects are evidence of a "satanic manifestation" in the world. Geologist Harold Coffin of the Geoscience Research Institute at Loma Linda University in Riverside, California, maintained that, because fossilized fish are often found with their mouths open, the fish must have died of suffocation in a worldwide flood like that described in the Book of Genesis. In addition, the state's star witness, astronomer Chandra Wickramasinghe from University College in Cardiff, Wales, suggested that life began and evolved with the help of a divine influence. To the surprise of the creationists who had flown him to the Little Rock trial, however, Wickramasinghe also testified that most of creation-science was "claptrap" and that no "rational scientist" could believe in an Earth less than 1 million years old.

Key arguments. The proponents of creation-science argued that the failure of modern science to produce life in the laboratory or to transform one species into another is evidence of a divine Creator. According to the creationists, random processes could not have produced the extraordinary complexity and variety of life today. The creationists claimed that evolution would violate the second law of thermodynamics, which prevents the accumulation of "order" through a random process. For example, a cup can accidentally fall off a shelf and break, but a pile of broken glass is not likely to reform by chance into a complete cup. Wickramasinghe testified at the trial that the chance of life evolving from nonliving substances through random processes is on the same order of magnitude as the chance that a hurricane passing through a junkyard would produce a Boeing 747.

The evolutionists countered that natural selection is precisely the process that increases the probability of the unlikely occurring and being preserved. Individual mutations that allow a new organism to compete successfully with established organisms may be extremely rare; over time, however, they do occur and then spread through a population, gradually becoming the rule rather than the exception. Through natural selection, these new organisms continue to evolve into more complex forms.

The creationists also contended that either evolutionism is true or creationism is true, and that no other alternative exists. They therefore reasoned that any evidence not supporting evolutionary theory must necessarily support creationist theory. The substantial differences of opinion among scientists regarding the actual mechanisms of evolutionary change, they said, were evidence of the failings of evolutionary theory and thus supported creationism.

Finally, the creationists insisted that creation-science is not necessarily a religion and, therefore, that teaching it in public schools does not violate the First Amendment. Creation-science, they argued, explains creation in nonreligious terms through the existence of a divine creator who may be distinct from the God that is worshipped. District Court Judge William Overton noted in his judgment, however, that theologically sophisticated Christians would be unlikely to accept the concept of a "creator God distinct from the God of love and mercy."

Verdict. In January 1982, Judge Overton ruled Act 590 invalid because it violated the separation between church and state specified in the First Amendment. Overton held that creation-science is religion, inextricably linked with the Book of Genesis and faith in God. By defining creationism as belief in both the sudden appearance of living things from nothing and the later occurrence of the Biblical Flood — a combination of events that he said was unique to Genesis — the creationists had implied the existence of a divine Creator. Since Act 590 would, in effect, force teachers to use the Book of Genesis to explain the creationist concepts, it would be forcing the public schools to proselytize a religious belief.

Overton also concluded that creationism is not a valid scientific belief because, unlike evolutionary theory, "it is not explanatory by reference to natural law, is not testable, and is not falsifiable." He said that expert witnesses for the creationists had not attempted to publish their theories in refereed scientific journals. Overton characterized Act 590's definition of evolution as "a hodgepodge of limited assertions, many of which are factually inaccurate." Finally, he ruled that the U.S. Constitution protects the religious freedom of the minority irrespective of the wishes of the majority, and teaching religion in public schools intrudes on the freedom of that minority.

Creationism in Louisiana. On the same day that Overton made his ruling, the Mississippi Senate passed a bill nearly identical to Act 590. This bill, however, did not make it out of committee in the Mississippi House of Representatives. Although similar bills have been filed in at least 24 other states, they have not passed in any state except Louisiana.

In July 1981, Louisiana Governor David Treen signed into law a bill designed to avoid the pitfalls that led to the invalidation of Act 590. The Louisiana bill mandated the "balanced treatment" of creation-science and evolution in classroom lectures, textbook materials, library materials, and other educational programs. The ACLU filed suit against the Louisiana law in the U.S. District Court in New Orleans in December 1981, again on the grounds that it violated the separation of church and state mandated by the U.S. Constitution. The suit was deferred briefly while Louisiana courts decided that the legislature did have the right to dictate the curriculum of schools within the state.

The ACLU filed for a summary judgment (a judicial decision without a trial) in September 1984 because no facts were in dispute. Judge Adrian Duplantier granted the motion in January 1985, ruling the Louisiana law unconstitutional. In his 10-page decision, Duplantier wrote: "Because it promotes the beliefs of some theistic sects to the detriment of others, the statute violates the fundamental First Amendment principle that a state must be neutral in its treatment of religions." Furthermore, he wrote that it was not necessary to repeat the lengthy trial conducted in Arkansas: "Whatever 'science' may be, 'creation,' as the term is used in the statute, involves religion, and the teaching of 'creation-science' and 'creationism,' as contemplated by the statute, entails teaching tailored to the principles of a particular religious sect or group of sects."

Duplantier specifically refused to consider the more than 1,000 pages of evidence assembled by the state and proponents of creation-science to demonstrate that the dogma is, in fact, a science: "A trial could not affect the outcome... [and] we decline to put the people of Louisiana to the very considerable needless expense of a protracted trial." Moreover, the fact that the ruling was issued as a summary judgment rather than as the result of a trial effectively thwarted attempts to pass similar laws in other states. Most legal experts agree that such laws would be struck down in a similar manner.

In June 1985, a three-member panel of the Fifth Circuit Court of Appeals in New Orleans upheld Judge Duplantier's ruling that the law was unconstitutional. The panel noted: "We seek simply to keep the government...neutral with respect to any religious controversy." In December 1985, the full court rejected by a vote of eight to seven a motion to hear an appeal of the panel's ruling. Louisiana appealed that ruling to the Supreme Court, which heard the case in December 1986.

The Supreme Court upheld the lower court ruling by a seven-to-two vote in June 1987. Writing for the majority, Justice William J. Brennan, Jr., said the law was unconstitutional because "the pre-eminent purpose of the Louisiana legislature was clearly to advance the religious viewpoint that a supernatural being created humankind." The law's primary purpose, he wrote, "was to change the science curriculum of public schools in order to provide persuasive advantage to a particular religious doctrine that rejects the factual basis of evolution in its entirety." Brennan wrote that the Louisiana legislature's claim that the law's intent was to foster academic freedom was a "sham."

Textbooks. Opponents of creationism achieved a different kind of victory in 1984 when Texas attorney general Jim Mattox ruled that long-standing school board guidelines requiring equal treatment of evolution and creationism in textbooks were unconstitutional. In an opinion solicited by State Senator Oscar Mauzy, Mattox stated that "The rules of the State Board of Education, concerning the subject of evolution, fail to demonstrate a secular purpose and are therefore in contravention of the First and Fourteenth Amendments to the United States Constitution."

The Texas rules had had a strong influence on publishers of biology textbooks. Because Texas represented nearly 10 percent of the nation's textbook market, some publishers had edited their books to conform to the Texas guidelines. Although the Texas rules had required that evolution and creationism receive equal treatment in texts, most publishers responded by deleting virtually all references to either evolution or creationism.

The creationist cause was dealt another blow in September 1985, when the California Board of Education voted to reject every science textbook submitted for use in seventh- and eighth-grade classes the following year. The board charged that the books' publishers had "watered down" sections on evolution by "systematically omitting" thorough discussions of the theory, hoping thereby to avoid controversy. Members of the board invited publishers of the seven best texts to submit revised versions that included more information about evolution. This action also had nationwide repercussions, because California purchases about 11 percent of U.S. textbooks. Revised versions of the texts were accepted at a December meeting of the board, but neither creationists nor evolutionists were pleased. The former were upset that new material about evolution had been added to the texts, while the latter argued that the hasty revisions had introduced errors into the books.

Man and dinosaur. In support of their arguments that the Earth is only 6,000 years old, creationists have long insisted that man and dinosaur coexisted during the earliest years of the planet's existence. This belief was based largely on the existence of fossilized footprints in the seasonally dry Paluxey Creek near Glen Rose, Texas. The tracks were originally exposed by a 1908 flash flood. One of two sets of footprints in the creek bed dating from the Cretaceous Period about 120 million years ago was clearly that of a dinosaur; the second set of prints was claimed to have been made by a human, even though the prints are 15 to 20 inches long. In a 1980 creationist book called *Tracking Those Incredible Dinosaurs and the People Who Knew Them*, geologist John Morris of the Institute for Creation Research in El Cajon, California, argued that the prints were made by Biblical giants.

In 1986, however, computer programmer and amateur archaeologist Glen Kuban presented convincing evidence that the second set of tracks was also made by dinosaurs. Biologists had been misled about the prints, Kuban said, because the prints did not appear to show the characteristic triple toeprint of dinosaurs. Kuban found that shallow grooves at the front of the tracks were undoubtedly the toeprints of dinosaurs and that the prints widened at the front more than human prints would. Paleontologists now believe that some material may have sifted into the toeprints and hardened, and that the tracks have been distorted by erosion. Significantly, Morris announced at a 1986 conference that he agreed with Kuban's then unpublished findings and that his book and an accompanying movie, *Footprints in Stone*, would be withdrawn.

View from the Academy. The battle between creationists and proponents of evolution is likely to continue for years. In response to the creationist movement, the National Academy of Sciences (NAS) stated during 1982 that the Academy "cannot remain silent" on the issue of teaching creation-science, which it believed endangers the "integrity and effectiveness" of education in the U.S. In accordance with this position, the NAS published in 1984 a booklet entitled "Science and Creationism," which it sent to school superintendents, high school science teachers, and religious groups throughout the U.S., explaining why the NAS believes evolution is science and creationism is not.

In the preface to the NAS booklet, Frank Press, NAS president, noted that "the theory of evolution has successfully withstood the tests of science many, many times. Thousands of geologists, paleontologists, biologists, chemists, and physicists have gathered evidence in support of evolution as a fundamental process of nature." In sharp contrast, according to the booklet, "creationism, with its accounts of the origin of life by supernatural means, is not science.... No body of beliefs that has its origin in doctrinal material rather than scientific observation should be admissible as science in any science course. Incorporating the teaching of such doctrines into a science curriculum stifles the development of critical thinking patterns in the developing mind and seriously compromises the best interests of public education."

Mass Extinctions

Scientists have long puzzled over the abrupt disappearance of dinosaurs from the face of the Earth. Until fairly recently, they assumed that dinosaurs either had somehow evolved in ways that reduced their capacity to survive or had simply been outcompeted for the available ecological niches by the mammals that appeared around the time of their extinction.

Recently, however, researchers discovered that dinosaurs are simply one group of hundreds of thousands of species that have disappeared abruptly during major extinction events in prehistoric times. In the light of these discoveries, scientists have been forced to search for reasons to explain extinction events that go beyond looking for inherent frailties in the organisms themselves. By the late 1980s, a substantial number of scientists still believed that these events were the result of changes in the Earth's climate or the global effects of sustained volcanic activity. A growing number, however, were beginning to accept an extraterrestrial explanation for the extinctions, such as collisions of comets or asteroids with the Earth.

Age of Dinosaurs

Dinosaurs dominated the world for 160 million years, from the Triassic Period, about 225 million years ago, until their extinction at or near the boundary between the Cretaceous and Tertiary periods (known as the K/T boundary, because K is the geologic symbol for Cretaceous) about 65 million years ago. The dinosaurs were an enormously diverse group of reptiles. They ranged in length from about 22 inches in the case of *Compsognathus*, a terrestrial carnivore, to about 100 feet for *Diplodocus*, an herbivore. Since the dinosaurs successfully competed with other reptile species for resources, they were able to dominate a wide range of environments.

Yet today, all that remains of the dinosaurs are fossilized bones, eggs, and skin impressions embedded in sediments. Because few dinosaur fossils have been found in sediments above the layers deposited at the end of the Cretaceous, most scientists believe that dinosaurs became extinct around the end of that period. The dinosaurs did not die alone, however. In sediments deposited before and after the K/T boundary, the diversity of organisms preserved in the fossil record changes markedly. In fact, perhaps half of the species found at certain locations did not survive past the K/T boundary.

Scientists have developed many different hypotheses to explain the K/T boundary extinctions. Some have argued that the dinosaur extinctions were the result of biological changes that weakened them or made other organisms more competitive, that they died from rickets, constipation, or other illness, or that their eggs were eaten by mammals. Others believe that environmental changes were responsible, such as continental drift, a warming or cooling of the climate, changes in the nutrient content of the oceans, or changes in the Earth's axis, orbit, or magnetic field. Finally, some scientists think that the extinctions were the result of extraterrestrial events, such as the explosion of a nearby star or the impact of a large meteorite, comet, or asteroid.

During the early 1980s, proponents of the various hypotheses could generally be divided into two major camps. The "gradualists" contended that the extinctions occurred over a period of thousands of years, while the "catastrophists" believed that a sudden event was responsible. But by late in the decade, a compromise position was rapidly evolving: In this view, climatic change and other environmental factors sapped the vitality of dinosaurs and many other species, and intermittent extraterrestrial events, such as comet showers, provided the coup de grâce.

Extraterrestrial Impact

Iridium anomalies. The extraterrestrial impact hypothesis of dinosaur extinctions arose in 1980 after a research team from the University of California at Berkeley, headed by the late physicist Luis Alvarez and his geologist son Walter, presented evidence that a comet or other body struck the Earth about 65 million years ago. The team based this conclusion on the presence of an abnormally high concentration of the element iridium in marine K/T boundary sediments in Gubbio, Italy, and in Stevns Klint, Denmark.

The researchers argued that almost all the Earth's native supply of this heavy metal sank deep into the Earth billions of years ago when the planet was still molten. Only the impact of an extraterrestrial body containing higher levels of iridium than the Earth's surface, they said, could have introduced this "iridium anomaly" into the sediments. Other researchers have since reported finding high concentrations of iridium and other heavy metals in marine K/T boundary sediments at more than 50 sites scattered around the globe.

The Alvarez team attributed the K/T extinctions to the enormous clouds of dust that would have been hurled into the atmosphere by the impact of an extraterrestrial body. As these dust clouds spread through the atmosphere and circled the Earth, they would have prevented most sunlight from reaching the surface for weeks, thereby destroying numerous plant species, all of which depended on sunlight for photosynthesis. The loss of plants would eventually have led to the extinction of the dinosaurs, which depended either on plants or plant-eating animals for food.

Physicists John D. O'Keefe and Thomas Ahrens of the California Institute of Technology in Pasadena have calculated that the impact of a large comet or asteroid could have caused severe environmental effects. They constructed a mathematical model to simulate the effects of a 6-mile-wide "bolide" (scientists' preferred term for meteors and similar objects) striking the Earth with an energy equal to 100 trillion tons of TNT. According to the model, the heat released by the impact would have caused a global temperature increase of perhaps 55°F for a period of days. This short-term increase would have been great enough to kill many organisms, including dinosaurs weighing more than about 55 pounds.

The model also indicated that dust and debris weighing 100 times as much as the bolide itself would have been injected into the atmosphere at altitudes greater than 6 miles. The dust and debris would have caused a drastic reduction in the amount of solar energy reaching the Earth's surface, and thus a decrease in photosynthetic activity and surface temperature. In addition, the high-temperature shock waves caused by passage of the bolide and its impact debris through the atmosphere would have generated nitric oxide. This compound might have reacted with atmospheric ozone, which would have led to a prolonged depletion of the ozone layer and allowed damaging ultraviolet radiation from the Sun to reach the Earth's surface.

More recently, geologists Ronald Prinn and Bruce Fegley, Jr., of the Massachusetts Institute of Technology (MIT) in Cambridge have speculated that nitrogen oxides produced by the shock waves would also have dissolved in water in the atmosphere to produce extremely acidic rain, perhaps 10,000 times more acidic than the worst acid rain now observed. This acid would have killed many plants and animals directly, destroyed eggs exposed to the atmosphere, and acidified the upper layers of the oceans, causing many organisms there to perish.

In 1985, chemists Edward Anders, Wendy Wolbach, and Roy Lewis of the University of Chicago provided further support for the extraterrestrial impact hypothesis when they discovered large quantities of soot in the same K/T layers where iridium was found. The soot was in the form of fluffy, black particles of the type produced by forest fires. The authors speculated that the impact of the bolide that released the iridium also touched off widespread forest fires that killed much of the plant life on the surface of the Earth. Perhaps more important, the fires produced large quantities of soot that, combined with dust from the impact, would block sunlight for months rather than weeks, since soot is washed out of the atmosphere more slowly than dust. The researchers concluded that a combination of events related to the impact was probably responsible for the massive extinctions of organisms at the K/T boundary.

Microtektites and shocked quartz. In 1981, geologists J. Smit and G. Klaver of the Geological Institute in Amsterdam, the Netherlands, reported that they had found another unusual feature — in addition to the iridium anomaly — in K/T boundary sediments from Caravaca, Spain. The sediments contained crystallized feldspar globules, commonly known as microtektites, in a form that only extreme heat and pressure from volcanic activity or an explosive impact could have produced. According to Smit and Klaver, the lack of volcanic rock in the Caravaca region rules out volcanic activity as the source of the feldspar globules and suggests that the globules may have been formed by the impact of an extraterrestrial body. Researchers have since discovered microtektites at dozens of K/T boundary sites.

In 1984, geologist Bruce Bohor and his associates at the U.S. Geological Survey (USGS) in Boulder, Colorado, reported additional evidence for a bolide impact in the form of "impact-shocked" quartz grains. In preparing samples obtained from the iridium-rich claystone deposited in the K/T boundary at a location near Brownie Butte in east-central Montana, the researchers used hydrofluoric acid to dissolve clay away from small quartz grains. When the USGS geologists examined individual quartz grains under a microscope, they found that the acid had etched the surfaces of some of the grains in a distinct pattern of parallel grooves unlike any pattern ever observed in volcanic rocks etched in the same manner.

The researchers later showed that such etched features are normally found in rocks that have been subjected to pressures greater than 150,000 times atmospheric pressure — much higher than the pressures generated during a volcanic eruption. The grooved pattern has also been observed in etched mineral grains taken from the sites of nuclear explosions and known meteorite impacts. In 1987, the USGS team reported that it had found similar patterns in etched quartz grains taken from K/T samples around the world. Geologists and volcanologists agree that the grains are too large to have been dispersed around the world by winds after being ejected from a volcano. Interestingly, the largest grains, about 0.2 inch in diameter, have been found in western North America, while those from the rest of the world are between 0.04 and 0.08 inch in diameter. Since the largest grains would be transported the shortest distances, this finding suggests that the meteorite struck somewhere in western North America.

Other anomalies. The search for iridium anomalies in other boundary sediments has had decidedly mixed results. In 1982, two teams of researchers reported anomalies in two sediment layers deposited only a few tens of thousands of years apart at the boundary between the Eocene and Oligocene epochs. The end of the Eocene (about 36 million years ago) has been associated with a minor mass extinction in which a large proportion of the known species of radiolarians, silica-encased microorganisms, became extinct.

The layers were described independently by chemist R. Ganapathy and his colleagues at the J. T. Baker Chemical Co. in Phillipsburg, New Jersey, and by nuclear chemist Frank Asaro of the Lawrence Berkeley Laboratory in California and other members of the Alvarez team. One layer contained microtektites, but no excess iridium. The second layer, which had first been described in 1974 by geophysicist Billy Glass of the University of Delaware in Newark, contained both microtektites and a 20-fold enrichment of iridium. Paleontologists Gerta Keller and Steve D'Hondt of Princeton University in New Jersey have also found microtektites in three distinct sediment layers deposited at the end of the Eocene. The iridium anomaly at the boundary between the Eocene and Oligocene epochs is thus the best documented anomaly other than that at the K/T boundary.

Asaro reported the discovery of a third major extinction event associated with an iridium anomaly at an October 1988 meeting, "Global Catastrophes in Earth History," held at Snowbird, Utah. Asaro reported that his group had found an iridium anomaly associated with sediments dating from 10 million to 12 million years ago during a time known as the Middle Miocene. Roughly one-quarter of the species then in existence disappeared during that period. The team found the iridium anomalies in sediments from two sites more than 6,000 miles apart, in the Weddell Sea near Antarctica and in the Tasman Sea north of New Zealand. "As the iridium was deposited at the two sites at about the same time, and they are one-quarter of the way around the world from each other, it seems likely that the deposition was worldwide," Asaro told the conference.

Chinese scientists have identified high concentrations of iridium and other trace elements in the Permian-Triassic boundary layer, formed about 248 million years ago — a time when 90 percent of species then living on Earth became extinct. Xu Dao-Yi of the State Seismological Bureau in Beijing and colleagues at other institutions found the anomalies at two separate locations in China. To the dismay of other paleontologists, however, iridium-rich layers at the Permian-Triassic boundary have not yet been found in other parts of the world.

More recently, geologists Carl Orth and Moses Artrep of the Los Alamos National Laboratory in Albuquerque, New Mexico, found two small iridium anomalies associated with Cenomanian-Turonian boundary sediments, which were deposited 91 million years ago. Both anomalies, discovered near Pueblo, Colorado, were associated with relatively small extinction events.

In 1987, a team headed by Paul E. Olsen of the Lamont-Doherty Geological Observatory in Palisades, New York, reported finding fossils in Nova Scotia that indicate a mass extinction occurred near the boundary between the Triassic and Jurassic periods about 200 million years ago. The researchers found no evidence of an iridium anomaly associated with the boundary, but they note that a large meteorite crater called the Manicouagan Structure lies about 450 miles northwest of the fossils. The 60-mile-wide crater has been dated to the same period as the extinctions. The meteorite that created the structure, they said, might have been carbon-rich rather than iridium-rich.

Gradual Extinction?

Some scientists have questioned the idea that a catastrophic event triggered the extinctions at the K/T boundary. They have cited countervailing evidence suggesting that the extinction of dinosaurs did not occur as abruptly as the catastrophists had first proposed.

Dinosaurs after the K/T boundary. A team headed by paleontologist Robert Sloan of the University of Minnesota in Minneapolis, for example, reported in 1986 that they had found bones of "the last dinosaurs known" in the Hell Creek section of eastern Montana. Their studies showed that the number of species of dinosaurs in that area had declined from 30 to 12 over a period of 8 million years preceding the K/T boundary, and that some of the species had survived for more than a million years after the boundary. Critics of Sloan's discovery have argued that the sediments in Hell Creek have been disturbed periodically by runoff from storms that jumbled the fossils and confused their interpretation. But Sloan and others have subsequently found dinosaur fossils of the same age in China, India, Peru, and elsewhere, and many scientists now agree that the dinosaurs did not completely die out at the end of the Cretaceous.

Researchers have also begun to question whether a prolonged period of dark and cold would have brought about the dinosaurs' extinction. Paleontologist William Clemens of the University of California at Berkeley reported in 1987 that he had found dinosaur fossils on the Alaskan North Slope. At the time the fossils were deposited, the area was dominated by evergreen and broad-leaf trees and had a "mild to cold-temperate" climate. Because of its high latitude, the area would have had extreme seasonal variations in sunlight, temperature, and vegetation. "This challenges the hypothesis that short-term periods of darkness and temperature decrease resulting from a bolide impact caused dinosaurian extinction," he wrote in the journal *Science*.

The gradualist view, as summarized by geologist Anthony Hallam, is that the patterns of plant species extinctions in different climatic zones favor a long, slow process of extinction. Hallam and others argue that, if a single global catastrophe was responsible for the extinctions, a large proportion of the plants on Earth would have simultaneously become extinct. The tropical plants, which are generally more vulnerable to environmental stresses than other plants, would have been most severely damaged. The fossil record shows, however, that tropical plant species were the least affected by the changes at the end of the Cretaceous Period. Gradualists interpret this finding to mean that the plant extinctions in the fossil record were due to gradual local or regional environmental changes rather than a global catastrophe.

On the basis of all this evidence, Clemens and Leo Hickey of Yale University in New Haven, Connecticut, formulated what is probably the most widely accepted gradualist hypothesis: During the late Cretaceous, the slow process of continental drift gradually changed the locations of the continents and the relative sizes and circulation patterns of the oceans. The resulting changes in global weather patterns caused a general cooling of the Earth's climate and a gradual extinction of the dinosaurs as vegetation died. Such climatic changes might have been greater in the higher latitudes than in the Tropics, leading to the pattern of extinctions observed.

Support for Clemens and Hickey's view was provided in 1988 by geologists from the University of California at Santa Barbara. Lowell Stott and James Kennett reported that studies of deep sea cores drilled in the Weddell Sea off Antarctica indicate that the Earth's climate began cooling 200,000 years before the K/T boundary. By studying the ratio of oxygen isotopes in the cores, they concluded that the average temperature of the Antarctic waters dropped as much as 4°F during that period, well before any bolide impact or volcanism occurred. They postulated that the climatic trend was driven by the gradual isolation and cooling of the Antarctic continent that had occurred as a result of continental drift. This cooling may have set the stage for a mass extinction caused by a large-scale disaster such as a bolide impact or extensive volcanism.

Volcanism. Palynologist Dewey McLean of Virginia Polytechnic Institute in Blacksburg has proposed yet another explanation for the extinctions. He believes that they are the result of a gradual greenhouse warming triggered by "Deccan volcanism," a series of massive volcanic eruptions that occurred about 65 million years ago — about the same time as the Cretaceous extinctions. These eruptions covered about 600,000 to 1 million square miles of the Earth's surface with a thick layer of lava. Even after extensive erosion, the lava layer is still about 2 miles thick near Bombay, India.

According to McLean, the world's oceans were warmer during the Cretaceous Period than they are today, averaging 57°F compared with 36°F to 38°F today. Since warm water absorbs less gas than cold water, the warm ocean waters of the Cretaceous Period would not have absorbed the vast quantities of carbon dioxide released into the atmosphere during the Deccan eruptions. Instead, the atmospheric concentration of carbon dioxide would have increased, causing a gradual global warming. Carbon dioxide allows radiation from the Sun to strike the Earth's surface, but helps prevent heat emitted from the surface from radiating into space. It thus acts much like the glass in a greenhouse.

McLean believes that plankton extinctions occurred at the K/T boundary because some of the atmospheric carbon dioxide would have dissolved into the water at the ocean surface, making it more acidic and less habitable for plankton. Without surface plankton to extract carbon dioxide from the atmosphere, the greenhouse effect would have accelerated, eventually driving the dinosaurs to extinction over a period of hundreds of thousands of years.

Geologists Charles Officer and Charles Drake of Dartmouth College in Hanover, New Hampshire, have also supported the volcanism hypothesis. Their studies of the Kilauea volcano in Hawaii have shown that high concentrations of iridium from deep within the Earth can be released into the atmosphere during an eruption. They also found that the relative concentrations of gold and iridium and of platinum and iridium in the K/T boundary sediments resemble the relative concentrations in material from the Earth's mantle. The intense emissions of dust and smoke from volcanoes, they said, could produce the same atmospheric effects as a bolide impact.

In 1986, physicist Ilhan Olmez of MIT used measurements of the iridium concentrations in emissions from Kilauea to estimate that the total amount of iridium emitted by the Deccan volcanism was almost exactly equal to the amount calculated to be present in the K/T iridium anomaly. Several groups of researchers have also found iridium anomalies of terrestrial origin at other boundaries, including the Precambrian-Cambrian (590 million years ago), Ordovician-Silurian (438 million years ago), Frasnian-Famennian (367 million years ago), and Mississippian-Pennsylvanian (320 million years ago). These findings strongly suggest that volcanism had some role in many of the extinction events.

Some recent evidence suggests that a bolide impact and Deccan volcanism may have been linked. Geologist Asisch Basu of the University of Rochester in New York reported at a 1988 meeting of the American Geophysical Union in San Francisco that he had found impact-shocked quartz grains in cores drilled in Indian rock deposits called the Deccan Trap. Significantly, the grains were found immediately below the lava that was deposited during the eruption at the K/T boundary. Basu concluded that a bolide may actually have impacted at the Deccan Trap, triggering the eruption.

That conclusion is supported by geologist David Alt and his colleagues at the University of Montana in Missoula. They reported in 1988 that their study of the Deccan Trap's geology suggests that a bolide impact opened a crater 400 miles across and triggered massive volcanism.

Lava flows then filled in the crater to produce a deep lava lake, which overflowed to spill enormous amounts of lava across much of central India. The combined effects of the impact and the volcanism, they said, would have been more than enough to account for the mass extinctions that occurred at that time.

Stepwise extinctions. Other evidence suggests that the mass extinctions that have occurred at various times in the Earth's history actually represent a series of small extinctions rather than a single large event. Marine paleontologist Peter Ward of the University of Washington at Seattle, for example, has studied an outcrop in Zumaya, Spain, that was submerged beneath the sea at the end of the Cretaceous. This rock formation, many scientists believe, is the most complete, continuous, land-based marine fossil record of the late Cretaceous and early Tertiary periods.

Ward carefully traced the history in the rock formation of ammonites, a once-populous group of marine invertebrates that bore a strong resemblance to today's sea nautilus. He found that the ammonite species had begun to undergo a serious decline more than 6 million years before the K/T boundary, and had all but disappeared 300,000 years before the boundary. Close analysis showed that the ammonite species did not die out gradually, one by one, but disappeared in three or four sharp bursts separated by hundreds of thousands or even millions of years.

In separate research, paleontologists Gerta Keller of Princeton and Erle Kauffman of the University of Colorado in Boulder have found evidence for this stepwise process in the K/T boundary at other locations. Kauffman has also discovered evidence of a similar extinction pattern at the Cenomanian-Turonian boundary, about 90 million years ago.

Multiple impacts. A first step toward reconciling the views of the catastrophists and the gradualists occurred in 1985 when chemist Gerhart Graup and his colleagues at the Max Planck Institute for Chemistry in Mainz, West Germany, reported that they had found three distinct iridium anomalies in the K/T boundary in the Bavarian Alps. The major anomaly occurred in the boundary sediments themselves, another was about 1 yard below the boundary, and a third was about 14 inches above it. Alvarez and others subsequently reported finding multiple iridium anomalies in the K/T boundary at other sites, and Keller has found them at the Eocene-Oligocene boundary. The data thus seem to suggest the possibility of multiple bolide impacts around the time of the boundary, each of which contributed its own burst of extinctions.

Volcanologists and astronomers will probably debate for years to come whether the source of the extinctions was terrestrial or extraterrestrial. "The irony," said paleontologist Donald Prothero of Occidental College in Los Angeles, "is that the predictions devolving from the volcanism model are almost indistinguishable from those arising out of the comet scenario. Both fit the spaced, stepped extinctions seen in the [geological] record."

"We don't need impact," says Keller, "but the fact is, we have impact, and the coincidence cannot be ignored." And what role did it play? "Populations already in decline and teetering on the brink of extinction because of climatic pressures may be pushed over the edge into oblivion by a minor perturbation — a collision with a comet, for example."

Cyclical Extinctions?

"Virtually all plant and animal species that have ever lived on the Earth are extinct," geophysicist David Raup of the University of Chicago has noted. "For this reason alone, extinction must play an important role in the evolution of life." Five major extinction events have occurred over the past 600 million years, with the most notable occurring during the Permian Period about 240 million years ago, when up to 96 percent of all marine species disappeared. Many smaller extinction events have also taken place. In 1984 Raup and geophysicist J. John Sepkoski, Jr., of the University of Chicago reported that these mass extinctions appear to follow a cyclical pattern.

Using data about the appearance and extinction of 3,500 families of marine animals, Raup and Sepkoski found an apparent "background" extinction rate of about three to five families per million years. They identified 12 instances, however, when the extinction rate was much higher — up to 20 families per million years — during the period from 253 million to 11.3 million years ago. These peaks of extinction appeared to occur periodically, about every 26 million years. According to Raup and Sepkoski's analysis, the Earth is currently about midway between mass extinction events, because the last mass extinction occurred about 11 million years ago. Although they originally reasoned that this cycle was too long to be caused by earthbound biological or physical processes, they have since modified their views to conclude that either terrestrial or extraterrestrial events, and perhaps a combination of both, have been responsible for the periodicity of extinction events.

In 1986, Raup and Sepkoski reported that they had used more rigorous criteria to reduce the number of reliably

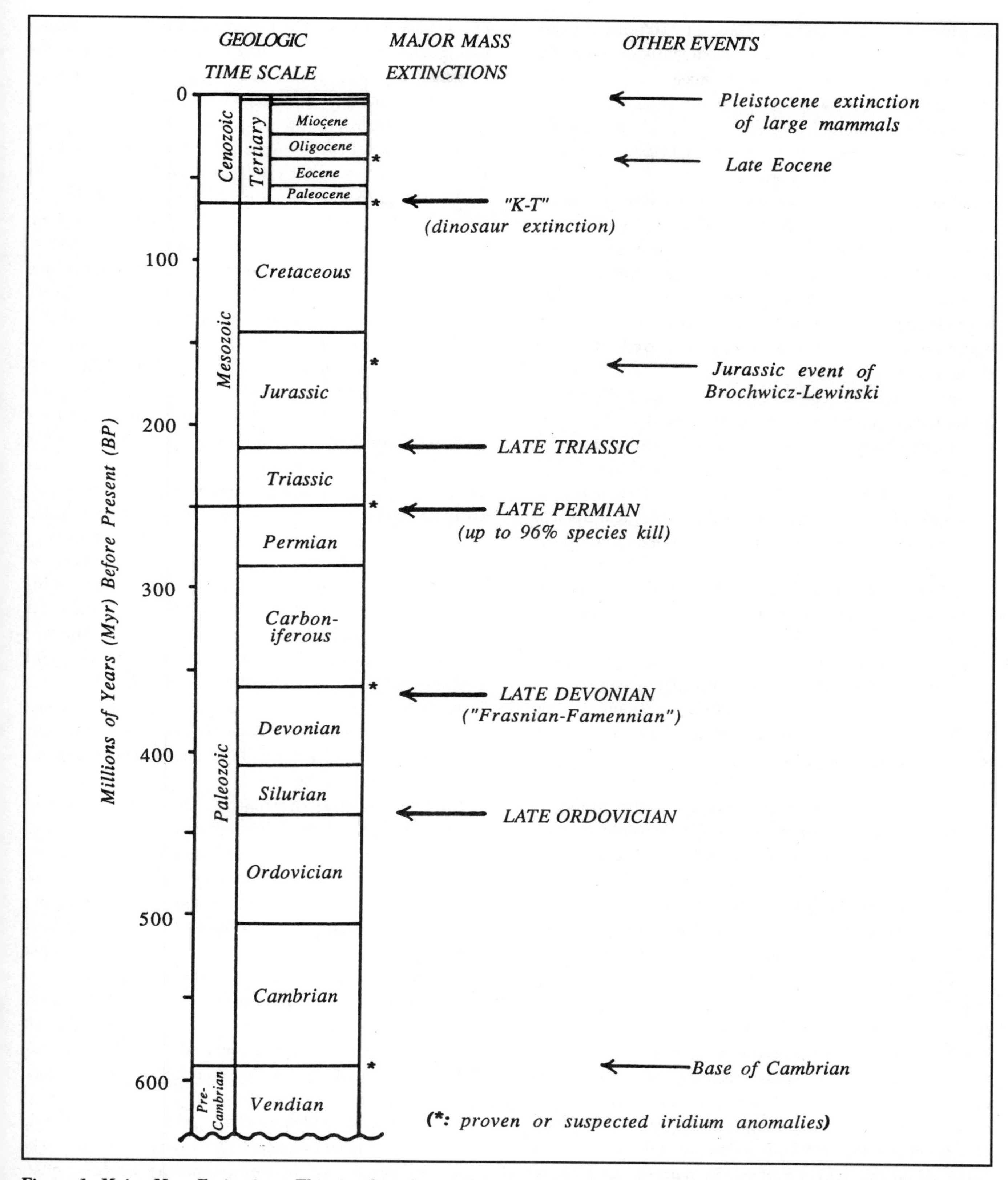

Figure 1. Major Mass Extinctions. This time line shows when the five major mass extinction events of the past 600 million years occurred. The greatest of these events took place in the late Permian Period, about 240 million years ago, when up to 96 percent of all marine species became extinct. *Courtesy: Reproduced from* The Nemesis Affair, A Story of the Death of Dinosaurs and the Ways of Science, *by David M. Raup, by permission of W.W. Norton & Company, Inc. Copyright © 1986 by David M. Raup.*

identified mass extinctions in the past 250 million years from twelve to eight. An independent statistical analysis of Raup and Sepkoski's data by physicists Scott Tremaine and Julie Heisler of MIT, however, indicated that the new data showed periodicity in extinctions at a confidence level of only 50 percent; that is, the data have only a 50 percent chance of being correct.

Other investigators have also argued that the periodicity is overstated and is, perhaps, illusory. Evolutionary biologist Leigh Van Valen of the University of Chicago has found that alternative techniques for computing extinction rates tend to obscure the apparent periodicity. Using such methods, he found that only the extinctions at the end of the Permian and Cretaceous periods stand out above background rates.

Paleontologist Steven Stanley of Johns Hopkins University in Baltimore and others have suggested that random catastrophic events could generate a periodic pattern of mass extinctions. Support for this hypothesis was offered by Michael McKinney of the University of Tennessee at Knoxville at the 1987 meeting of the Geological Society of America in Phoenix. McKinney developed a computer model that simulates the diversity of species through time, then incorporated a random series of environmental upheavals. As the catastrophists have predicted, he found that a single event severely reduces the number of living species.

But the model also showed that, after the catastrophic event, the ecosystem was relatively immune to the effects of further catastrophes. In effect, the model suggests that all the species susceptible to a disaster are killed by it, leaving behind only hardier species. Thus, after each extinction event the biological community requires a long recovery period before it is again susceptible to another catastrophe. Nonetheless, catastrophists have continued to search for a direct cause of periodic extinctions. Their search has been buoyed by the fact that the four extinction events that involve iridium anomalies occurred in a fairly regular sequence (11 million, 39 million, 65 million, and 91 million years ago), with about 26 million years between events.

Impact craters. Walter Alvarez reasoned that if an extraterrestrial source is responsible for periodic extinctions, then the bolide impacts should have left impact craters. Water and wind erosion and continental drift have erased most such craters, but about 100 of the largest craters survive in some form and many of these have been dated. Alvarez and his colleague Richard Muller selected craters between 5 million and 250 million years old, with dating uncertainties of 20 million years or less and diameters greater than 6 miles. They reported in 1984 that the impacts responsible for the 13 craters they dated occurred with a periodicity of about 28.4 million years.

Researchers Michael R. Rampino and Richard Stothers of the Goddard Institute for Space Studies in New York City used less stringent criteria to select 41 craters for evaluation, and they reported in 1984 that their data showed a 31-million-year periodicity. When Raup applied the same statistical techniques that had been used to identify the periodicity of mass extinctions, he found evidence of a periodicity of 28.5 million years for 19 selected impact craters. The USGS's Eugene Shoemaker independently created his own list of 20 craters that were accurately dated. He reported in 1985 that if the crater ages are assumed to be periodic, the most likely period is about 32 million years, and the last event would have occurred 2 million to 4 million years ago. But he added that no more than half of the craters would be included in the periodic pattern; the remaining craters were apparently produced by random impacts unrelated to a cycle.

Ideally, researchers would like to identify craters associated with specific extinction events, but that has proved to be difficult. The most work has been devoted to finding a crater left by the bolide thought to have struck the Earth at the K/T boundary. The distribution of impact-shocked quartz grains, discussed earlier in this chapter, has led the USGS's Bohor to speculate that the bolide struck near western North America. One possible candidate, Bohor says, is the 20-mile-wide Manson crater near Manson, Iowa, which was formed roughly 62 million years ago and since buried by sediments.

But work by other researchers, particularly geophysicist Frank Kyte of the University of California at Los Angeles, indicates that the sediments at the K/T boundary contain material from both continental crust and ocean crust, suggesting that the bolide may have struck just offshore in an area where sediments from land had been deposited. Some evidence suggests that a bolide may have struck off the Pacific Coast of North America. If that is true, the crater may never be found: In the past 65 million years, nearly 20 percent of the eastern floor of the Pacific Ocean has been subducted under the North American Plate by tectonic forces, so the crater may have vanished without a trace.

An alternative explanation for Kyte's findings would be multiple impacts during a short period of time. The foremost proponents of this possibility are geologists Virgil Sharpton and Kevin Burke of the Lunar and Planetary Laboratory in Houston, Texas. In 1987, they reported that they had identified five craters that all appear to be about 65 million years old: the Manson crater and two sets of twin craters, Kara and Ust-Kara in the Soviet Arctic, and Gusev and Kamensk near the northern shore of the Black Sea.

These five craters all lie within 12 degrees of a great circle extending across the Arctic, Sharpton and Burke said, suggesting that the required ocean crater may be found on the floor of the Arctic Ocean. The possibility that six bolides struck the Earth within a short period, they suggested, would be consistent with the increasingly accepted idea that the K/T event was exceptional, if not unique. Although the possibility of six bolides striking in such a regular pattern may seem small, they suggest that the individual bolides could have been formed when a large comet nucleus broke up as it rounded the Sun, a commonly observed event.

The University of Delaware's Billy Glass reported in 1987 that he had found both microtektites and tektites — larger glass globules — in association with impact-shocked quartz in sediments from the Eocene-Oligocene boundary off New Jersey. These findings suggest that the bolide responsible for the anomaly struck nearby. One candidate for the impact is a 27-mile-wide crater on the North American continental shelf about 120 miles southeast of Nova Scotia. Geologists Georgia Pe-Piper of St. Mary's University in Halifax, Nova Scotia, and Lubomir Jansa of the Bedford Institute of Oceanography in Dartmouth, Nova Scotia, have reported that the crater lies beneath 370 feet of water and is about 9,000 feet deep. The age of the crater has not yet been determined.

Astrophysical mechanisms. Even before Raup and Sepkoski's hypothesis about periodic extinctions was published, researchers who had heard of their work began proposing astrophysical mechanisms that might explain the apparent extinction cycle. Although several possible explanations have been proposed, the most favored extraterrestrial hypothesis is that periodic comet showers have been the primary cause. The evidence for such an occurrence was summarized in 1987 by a team of scientists that included Alvarez, Kauffman, Keller, Shoemaker, and Piet Hut of the Institute for Advanced Study in Princeton. The researchers believe that the most likely source of the comets is the Oort cloud, which contains billions of comets and extends far beyond the orbits of the known planets. A disturbance of the Oort cloud could send many comets traveling into the inner solar system, where one or more could strike the Earth. Hut and his colleagues have suggested that such disturbances could occur randomly as the result of gravitational attraction from stars or large clouds of interstellar dust and hydrogen passing nearby. Others have suggested, however, that the disturbances might be caused by a dwarf star circling the Sun or by a tenth planet in the solar system.

Dwarf star. In 1984, two research groups — Hut, Muller, and Marc Davis of the University of California at Berkeley, and physicist Daniel Whitmire of the University of Southwestern Louisiana in Lafayette and Albert Jackson of Computer Sciences Corp. in Houston — independently proposed the existence of a small, dark companion star traveling around the Sun in a highly eccentric orbit with a period of 26 million years. Such a star would reach a maximum distance of about 2.5 light-years from the Sun, about half the distance to the nearest star. Such an unusual orbit has never been observed for binary stars (two stars orbiting each other), even though more than half the stars in our galaxy are thought to be in binary systems.

Both groups hypothesized that, during its closest approach to the Sun, the diminutive star — which might have no more than a tenth the mass of the Sun and whose nuclear fires might never have ignited — would pass through or near the Oort cloud, sending a dense "shower" of perhaps a billion or more comets into the inner solar system. During the 100,000 to 1 million years after the star's closest approach to the Sun, some of the perturbed comets would collide with the Earth and cause mass extinctions. The number of extinctions would vary from cycle to cycle, depending upon the sizes and total number of comets that struck the Earth. Davis suggested that the proposed star should be named Nemesis, after the Greek goddess of vengeance.

The Nemesis hypothesis presents several problems, however. One of them, according to Shoemaker, is that any stable orbit that would disturb enough comets to produce intense comet showers would probably have disrupted most of the Oort cloud during the hundreds of orbits it would have completed since the formation of the solar system. Perhaps a more formidable problem is that any star in the proposed orbit would be so tenuously held by the Sun's gravity that it would have been stripped away from the Sun within a billion years. Hut suggested, however, that Nemesis originally could have been in a much closer orbit that has since grown larger.

Many astronomers are currently searching for the proposed companion star. Early in 1985, astrophysicist Armand

Delsemme of the University of Toledo in Ohio announced that he might have found the best place to look for it. After plotting the paths of 126 comets, Delsemme discovered, to his great surprise, that they journey around the Sun in oddly skewed orbits. He concluded that some large object must control the motion of the comets, and that object could be Nemesis. By the end of 1988, Muller and his colleagues at Berkeley had narrowed the search for Nemesis to about 3,000 visible stars. These stars will be observed repeatedly to determine which, if any, displays the orbital motion that would identify it as a companion to the Sun.

Planet X. Perhaps the most intriguing possibility is that an undiscovered tenth planet could somehow cause comet or asteroid showers at regular intervals. Astronomers have suspected the existence of another planet since the late nineteenth century, when they observed discrepancies in the orbits of Neptune and Uranus. The discovery of Pluto in 1930 resolved some, but not all, of those discrepancies, and the search for a tenth planet continues.

Years ago, astronomer Robert Harrington of the U. S. Naval Observatory in Washington, D.C., described what such a hypothetical tenth planet should look like. He predicted that it would be three to five times the size of the Earth, gaseous like Jupiter, with an elliptical orbit inclined to the plane of the solar system at an angle of 30° or more. Its year would be 800 to 1,000 Earth years long.

Although the University of Southwestern Louisiana's Daniel Whitmire still thinks that his proposed companion star is possible, he believes that a tenth planet, which he calls Planet X, is more likely to be responsible for the periodic comet showers. Whitmire and his colleague, astrophysicist John Matese, calculated the characteristics that such a planet should have if it is responsible for disturbing the comet cloud and triggering mass extinctions. They reported in 1985 that Planet X would have an orbital plane that slowly rotates around the Sun, completing its cycle once every 56 million years. Twice during that cycle (once every 28 million years), Planet X's orbit would carry it through a disk of comets lying just beyond Neptune, dislodging many of them. To meet all of these requirements, Planet X would need to have an elliptical orbit highly inclined to the plane of the solar system and a mass one to five times that of the Earth. In short, Planet X's characteristics are very close to those described by Harrington. The beauty of this hypothesis, according to Whitmire, is that it is based on a planet that was originally proposed for reasons having nothing to do with mass extinctions. But proof of this hypothesis, like proof of Nemesis, awaits observation of the celestial body.

Implications of the cyclical extinction hypothesis. In his 1986 book, *The Nemesis Affair*, Raup noted that if the finding of a cyclical pattern in mass extinctions is upheld, then the implications for evolutionary biology are "broad and fundamental." Although few paleontologists and evolutionary biologists are ready to entirely abandon the hypothesis that evolution occurs gradually as a result of natural selection, the growing evidence of a Cretaceous impact has convinced many scientists that catastrophes, such as impacts, have at times played an important role in the Earth's history and the development of life.

Organisms that are capable of competing and adapting under normal conditions and that would normally survive if natural selection were the only force at work in evolution might nevertheless become extinct during a catastrophe. After the effects of such a catastrophe had diminished, the surviving species would diversify, filling in the ecological niches left vacant by the extinct species and changing the direction of evolution.

Man the Killer?

One of the most puzzling mass extinction events occurred 11,000 years ago at the end of the last Ice Age. Around that time, 70 species of large mammals became extinct in North America and most vanished altogether. The list included mammoths, mastodons, saber-toothed tigers, ground sloths, native camels, armadillo-like glyptodonts, giant peccaries, giant beavers, dire wolves, and native lions. All species weighing more than 2,200 pounds vanished, as well as 75 percent of those weighing between 220 and 2,200 pounds, and 41 percent of those weighing between 12 and 220 pounds. But only about 2 percent of the species weighing less than 12 pounds disappeared.

Many anthropologists believe that the extinctions were caused by humans. Evidence from a number of archaeological sites around North America indicates that the first people arrived on the continent about 11,500 years ago. Most scientists believe that the first residents were wandering hunters who lived in the region surrounding the Bering Strait for perhaps centuries, honing their hunting and survival skills while the Arctic climate winnowed out the weakest and most disease-ridden. As the glaciers retreated at the end of the Ice Age, their movement opened a passage down the eastern flank of the Canadian Rockies.

As the ancestors of the American Indians moved down into what is now the U.S., they encountered a virtual Garden of Eden, a warm land populated by large animals

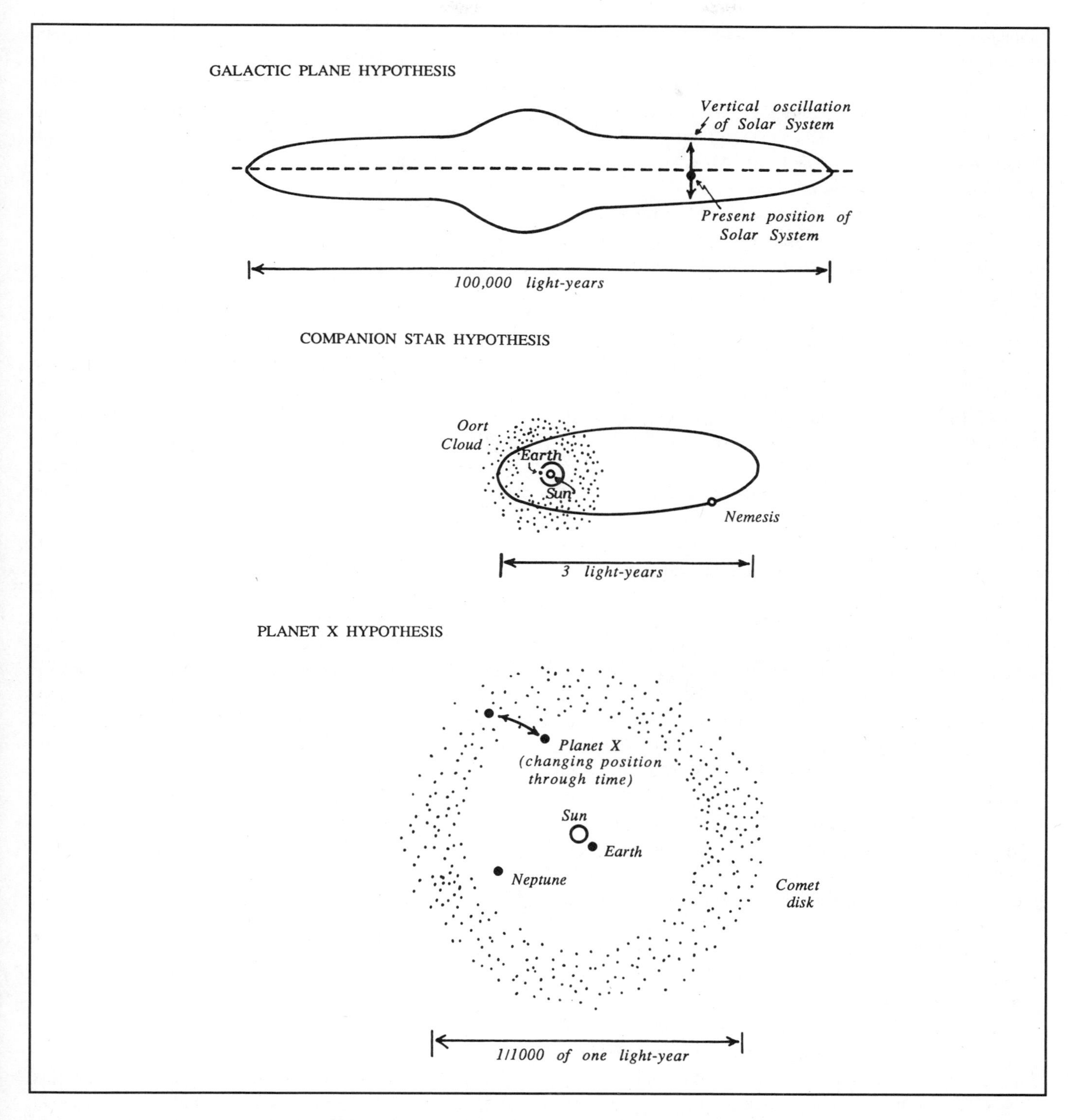

Figure 2. Three Astronomical Hypotheses of Mass Extinctions. Researchers have developed several hypotheses to explain the apparent periodicity of mass extinctions on Earth. In the galactic plan hypothesis (top), the vertical oscillation of the solar system through the galactic plane every 31 million to 33 million years would disturb the orbits of some comets and send them into the inner solar system, where one or more might strike the Earth. In the companion star hypothesis (middle), the orbits of comets in the Oort cloud would be periodically disturbed by a small, dark companion star orbiting the Sun in a highly eccentric orbit. In the Planet X hypothesis (bottom), an as-yet-undiscovered planet traveling around the Sun in an elliptical orbit would periodically disturb a disk of comets lying beyond the orbit of Neptune. *Courtesy: Reproduced from* The Nemesis Affair, A Story of the Death of Dinosaurs and the Ways of Science, *by David M. Raup, by permission of W.W. Norton & Company, Inc. Copyright © 1986 by David M. Raup.*

that had never been hunted and thus had developed no fear of man. The population of humans exploded over a period of about 500 years and the huge, lumbering beasts were all killed. The people who are thought to have carried out the hunting are called Clovis because their remains were first discovered at a site near Clovis, New Mexico. The Clovis produced an unusual, fluted spear point that has been found embedded in the remains of a variety of large animals throughout the continental U.S. Similar spear points have also been found in the area around the Bering Strait.

Some scientists question the hypothesis that humans hunted the species to extinction, arguing that many of the species that disappeared were probably never hunted, including small rodents and birds. "A fair number of animals disappeared that are not large and didn't attract much attention," paleontologist Ernest Lundelius of the University of Texas said at a 1987 symposium at the Smithsonian Institution in Washington, D.C. "We must beware of spurious correlations.... We should have some evidence beyond the coincidence in time." Lundelius noted that at least six earlier extinctions had occurred in the previous million years without human intervention, each of them at the end of Ice Ages. "I would relieve our predecessors of that responsibility," he concluded.

The disappearance of small animals not hunted by man has been the major stumbling block to the acceptance of the idea that man hunted many of the larger animals, called megaherbivores, to extinction. But in 1987, Norman Owen-Smith of the University of the Witwatersrand in Johannesburg, South Africa, offered an explanation that invoked a kind of domino effect to explain the observed extinctions. He suggested that the feeding habits of the megaherbivores had a tremendous impact on vegetation patterns, opening up habitats in which smaller herbivores could thrive. If the megaherbivores were wiped out, the vegetation would return to its earlier state and the smaller herbivores would be eliminated as well.

According to Owen-Smith, the great appetites of megaherbivores such as elephants and rhinoceroses in modern Africa have helped transform wooded savannas into open, short-grass savanna dominated by grasses and shrubs. This in turn provides more nutritious food sources for other herbivores. He believes that if the megaherbivores were eliminated, the savanna would revert to a tall grassland. More frequent and fierce lightning-ignited fires would further clear away remaining trees and shrubs, reducing food sources for the smaller herbivores. The fossil pollen record suggests that similar types of vegetation changes occurred at the end of the last Ice Age. Owen-Smith concludes that the elimination of the megaherbivores may be what separates vegetation changes at the end of the last Ice Age from those accompanying previous glacial meltdowns.

What If...?

Dale Russell, curator of fossil vertebrates at Canada's National Museum of Science in Ottawa, Ontario, has left the details of the Cretaceous extinction to other researchers and has instead focused his attention on what might have happened if the dinosaurs had not disappeared 65 million years ago. He suggested that, if the dinosaurs had continued to evolve until the present day, the descendants of at least one dinosaur species might now be a dominant form of life on Earth.

While examining fossils of a 5-foot-tall, 90-pound carnivore known as Stenonychosaurus, which lived 75 million years ago, Russell calculated that the estimated ratio of brain weight to body weight for this creature was fairly high, making the Stenonychosaurus at least as intelligent as today's opossums. The dinosaur also stood partially upright and had binocular vision and an opposable digit (a sort of thumb) on its hand — all attributes of today's more intelligent species, such as apes, chimpanzees, and humans. Russell speculated that this creature would have developed a fully upright posture if it had not become extinct.

Biotechnology Techniques

Scientists frequently find it hard to believe that less than 40 years have passed since biochemists James Watson and Francis Crick at Cambridge University in the U.K. proposed the double-helix model for the structure of DNA, the genetic blueprint of life. Their landmark paper, published in the 25 April 1953 *Nature*, concluded with the understatement: "It has not escaped our notice that the...

Figure 3. If Dinosaurs Existed Today. . . The creature in the foreground is a "dinosauroid," a product of the imagination of Dale Russell of the National Museum of Science in Ottawa, Canada. Russell extrapolated the evolutionary trends of one line of dinosaurs that included the Stenonychosaurus (shown in the background) and then built a model illustrating what he believed a modern descendant of the Stenonychosaurus might look like if dinosaurs had not become extinct. The Stenonychosaurus, with its binocular vision, opposable fingers, relatively large brain, and partially upright posture, could have evolved into the humanlike reptile pictured here. *Courtesy: National Museum of Natural Sciences, Ottawa, Canada.*

[structure]. . .we have postulated immediately suggests a possible copying mechanism for genetic material." In 1962, Watson and Crick, along with their collaborator Maurice Wilkins, were awarded the Nobel Prize for Medicine for their pioneering work in deciphering the structure of DNA.

The lines of research that emerged from that beginning have changed the face of biology forever. Perhaps most important, Watson and Crick's discovery led to the development of genetic engineering, which is the manipulation of genes themselves through cutting, splicing, and other forms of alteration. Genetic engineering is a rapidly growing field within biotechnology, a term coined to describe the manipulation of living organisms for the benefit of humans. Animal genes are now routinely inserted into microorganisms for the large-scale production of proteins, including some drugs, that were previously available only in minute amounts. Human insulin made in bacteria, for example, is now supplanting insulin derived from animal pancreases for the treatment of diabetics. Developments in biotechnology have also enabled researchers to identify gene sequences in DNA and have provided the first insights into the chemistry of gene regulation—insights that may lead, among other things, to new therapies or eventual cures for genetic diseases.

In addition, genetic engineering has given scientists the ability to modify genes in larger organisms, ranging from mice to cattle, and thereby change the characteristics of the recipients. Biologists have, for example, used genetic engineering to create mice with human diseases, thus providing new tools for the study of disease development and the testing of new drugs and therapies. Researchers are also developing livestock that use feed more efficiently and are more resistant to disease. In short, for the first time in history, humans have the potential to deliberately alter genes to produce entirely new organisms and substances.

As with any new technology, genetic engineering has prompted concern about its potential risks. In the early 1970s, critics of genetic engineering often raised the possibility that mutant bacteria might escape from a laboratory and infect people in surrounding cities or towns with hitherto unknown diseases. Proponents, however, regarded these fears as unfounded visions of monsters that would never appear, arguing that engineered microorganisms are benign and that proper precautions and laboratory techniques would safely confine them to laboratories in any case.

Although the original protests subsided after public debate, they re-emerged in the mid-1980s as investigators began to release the first genetically engineered microorganisms into the environment and to consider applying the technology to humans to cure genetic diseases. Some critics regard the manipulation of genes as the modern equivalent of opening Pandora's box which, according to Greek myth, unleashed all that is evil to mankind. Biologists argue, however, that nature has been conducting its own genetic experiments for millions of years without producing the unstoppable mutant bacteria feared by biotechnology's critics.

Genetic Engineering

Birth of molecular biology. Humans have long had an intuitive understanding of the process by which characteristics are transmitted from parent to offspring, and they have used this understanding to selectively breed plants and animals with desirable traits, even without precise knowledge of the exact mechanism by which inherited characteristics are transmitted. For example, selective breeding has produced dairy cows with higher milk yields, sheep with more and better wool, and corn with greater resistance to disease.

Gregor Mendel, an Austrian monk, laid the foundations for the modern science of genetics with his plant breeding experiments in the 1860s. By cross-breeding peas and their offspring, Mendel discovered that certain characteristics, such as flower color and seed shape, were inherited independently of one another in a nonrandom manner. For the next 80 years, however, the exact chemical nature of these units of inheritance, called genes, was unknown.

The first recognition of DNA as the primary genetic component appeared in the 1 February 1944 *Journal of Experimental Medicine*. Oswald Avery, Colin MacLeod, and Maclyn McCarty of the Rockefeller Institute in New York City reported that certain harmless bacteria known as pneumococci could be made virulent simply by mixing into their nutrient broth DNA isolated from a harmful variant. Nine years later, Watson and Crick described the molecular basis for this genetic transformation and in so doing invented the discipline of molecular biology.

DNA structure and protein synthesis. DNA consists of a linear chain of nucleotides bonded to one another. In most organisms, two strands of DNA are bonded together to form the famous double-helix structure discovered by Watson and Crick. Each nucleotide is composed of a sugar group, a phosphate group, and a nitrogen-contain-

Figure 4. Watson and Crick with the DNA Model. In 1962, scientists James Watson and Francis Crick were awarded the Nobel Prize for their discoveries about the structure of the deoxyribonucleic acid (DNA) molecule, which contains the information needed for life, growth, and reproduction in living organisms. In 1953, Watson and Crick proposed that the DNA molecule is composed of two nucleotide strands that coil around each other and bind together to form a double helix, which looks somewhat like a spiral staircase. *Courtesy: J. D. Watson,* The Double Helix, *Atheneum, New York, 1969.*

ing compound referred to as a base. In DNA, the sugar is called deoxyribose, and the bases are adenine (A), cytosine (C), guanine (G), and thymine (T). Within each nucleotide, the sugar group is bonded directly to the base and to the phosphate group. Within the DNA chain, the sugar group is also bonded to the phosphate groups of the two nucleotides immediately adjacent to it.

The double helix is maintained by the pairing of bases in the two DNA strands. This pairing is achieved through hydrogen bonding, an attraction between molecules that is weaker than the covalent bonding that holds atoms together in molecules, yet sufficiently strong to ensure specific pairing between the bases. Adenine and thymine always bind together, as do cytosine and guanine; these bases are therefore said to be complementary.

The DNA strands of complementary base sequences bind tightly together in a structure similar to a ladder, with the sugar and phosphate groups forming the vertical supports, and the hydrogen bonds between the bases forming the rungs. A DNA strand with a base sequence A-G-C-T, for example, will bind to a DNA strand with the complementary base sequence T-C-G-A. The DNA ladder is not straight; instead, it is twisted into a helical configuration.

Scientists divide living organisms into two major groups, prokaryotic and eukaryotic, on the basis of how their nuclei and DNA are structured. Prokaryotes, which include all bacteria, do not have a distinct nucleus surrounded by a membrane. In these microorganisms, virtually all the genetic information (the genome) is contained in a large circle of double-stranded DNA.

Eukaryotes, from the Greek for "true nucleus," have a well-organized nucleus and a distinct nuclear membrane; this group includes all living organisms other than prokaryotes. In eukaryotes, the DNA binds to thousands of proteins to form what is called a chromosome. Humans have 46 chromosomes; in other species, the number of chromosomes ranges from 4 in one species of flower to 82 in the turkey, 254 in the hermit crab, and thousands in some single-celled microorganisms.

The fundamental unit of genetic information is the gene, a segment of DNA that codes (serves as a blueprint) for the production of a specific protein by cellular machinery. Within the gene, information is encoded in the specific order of nucleotide bases in the DNA chain. Units

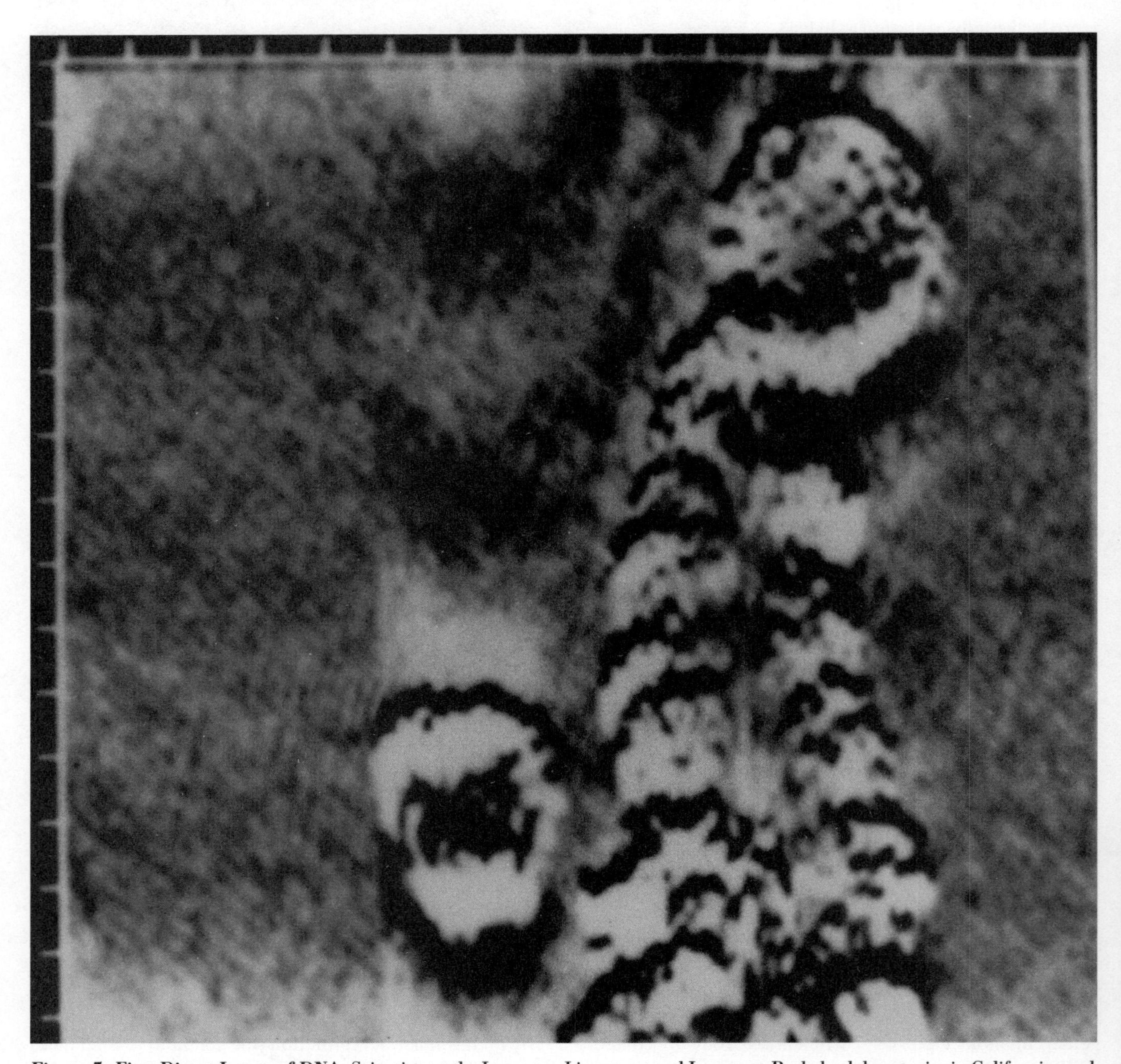

Figure 5. First Direct Image of DNA. Scientists at the Lawrence Livermore and Lawrence Berkeley laboratories in California used a scanning tunneling microscope to produce the first direct images of the DNA molecule, which were published in early 1989. The well-known double helix of DNA is composed of two intertwining strands of repeating sugar and phosphate units held together by steplike connections formed by complementary pairs of nitrogen-containing molecules known as bases. The unique specificity of DNA is based on the fact that the bases adenine and thymine pair only with each other and that the bases cytosine and guanine also pair only with each other. If the two strands are separated, this complementary pairing allows two new double helices to form, each identical to the original. *Courtesy: Rod Balhorn, Lawrence Livermore Laboratory.*

of three successive bases constitute a codon; most codons specify the order in which amino acids are combined to form proteins, but some mark the beginning and end of the gene. A typical prokaryotic gene contains about 1,000 bases, whereas a eukaryotic gene contains as many as 200,000. Human cells contain about 100,000 genes, whereas a bacterium may contain 2,000 to 3,000 genes and a virus as few as six.

The process of converting the genetic information in the DNA chain into a protein involves two steps: transcription and translation. Transcription is the synthesis of a molecule of messenger ribonucleic acid (mRNA) complementary to the DNA gene. RNA is structurally similar to DNA, but has two important chemical differences. The sugar group in RNA is ribose, and the base uracil (U) is present instead of thymine. While DNA is the master blueprint of the cell, RNA serves as a kind of working blueprint that is used by cellular components in the production of proteins, a process known as translation.

Proteins are composed of amino acids bonded together in a linear fashion. The production of a protein from a gene is called expression of the gene. Proteins are the primary building blocks of the cell. Many proteins, called enzymes, speed up chemical reactions in living creatures.

Recombinant DNA. The key to the development of genetic engineering was the discovery in the early 1970s of two families of enzymes known as restriction enzymes and ligases. Restriction enzymes cleave sugar-phosphate bonds between nucleotides in a DNA strand. The point of cleavage is specific for each type of restriction enzyme. Ligases, which are less specific than restriction enzymes, join any two DNA fragments together by forming new sugar-phosphate bonds. Restriction enzymes and ligases make it possible to take genes apart in a specific manner and then to reassemble them.

In 1973, the father of genetic engineering, biochemist Paul Berg of Stanford University in California, used restriction enzymes to remove one of the seven genes from the circular double-stranded DNA of a monkey virus known as SV40 and to cut open the circular double-stranded DNA in another virus known as lambda phage. In a laborious process, he then linked a short single strand of chemically synthesized DNA to one of the two DNA strands at one end of the opened phage DNA and an identical short strand to the second strand at the other end of the phage DNA. In addition, Berg linked to each end of the excised SV40 gene a short strand complementary to the strand he added to the phage DNA.

He might, for example, have added the base sequence G-A-A-T-T-C to the end of the phage DNA and the same sequence in reverse order, C-T-T-A-A-G, to the end of the SV40 gene. These short sequences are called "sticky ends" because they bring the phage and SV40 DNA fragments together for splicing by the ligase by the binding of complementary base pairs. Berg had succeeded in artificially combining DNA from different organisms for the first time, ushering in the era of recombinant DNA research. For this accomplishment, he shared the 1980 Nobel Prize for Chemistry.

If the process of combining genes could only have been carried out by the time-consuming chemical methods used by Berg, recombinant DNA technology would probably have advanced very slowly, if at all. Later in 1973, however, the technique was simplified dramatically by molecular biologists Stanley Cohen of Stanford University and Herbert Boyer of the University of California at San Francisco. Boyer discovered a restriction enzyme, called EcoR1, that not only cleaves DNA at precisely determined points but also produces the sticky ends needed for combining DNA fragments. Cohen found a way to use it.

When EcoR1 encounters an intact DNA double helix with one DNA strand containing the base sequence G-A-A-T-T-C, it cleaves the strand at the sugar-phosphate bond between the G and the A. The other DNA strand complementary to the G-A-A-T-T-C strand contains the same sequence in reverse order, C-T-T-A-A-G, and is cleaved by EcoR1 at the same site. Researchers can then separate the original DNA molecule into two strands, each of which has the sequence C-T-T-A-A on one end and the base G on the other end. These sequences are the "sticky ends." When EcoR1 is used on another piece of DNA with the same sequence, sticky ends are produced on it that allow the two original pieces of DNA to be joined together. More than 100 similar restriction enzymes that cleave other DNA sequences have since been isolated.

Also in 1973, Cohen found a new tool—the plasmid—for inserting DNA into a bacterium. The majority of the genetic information in a common bacterium such as *E. coli* is contained in a large, circular structure called a genophore in its nucleus. The bacteria also have small rings of double-stranded DNA, called plasmids, outside the nucleus. Plasmids may contain as few as two or as many as 100 genes and normally replicate along with the genophore when the bacteria proliferate. Plasmids are also exchanged by bacteria during sexual encounters, thereby ensuring the spread of plasmid-borne characteristics through the bacterial population. Under certain conditions, however, the plasmids can be stimulated to replicate rapidly even though the cell is not proliferating.

For example, a common gene found in plasmids of many bacteria gives them resistance to the antibiotic

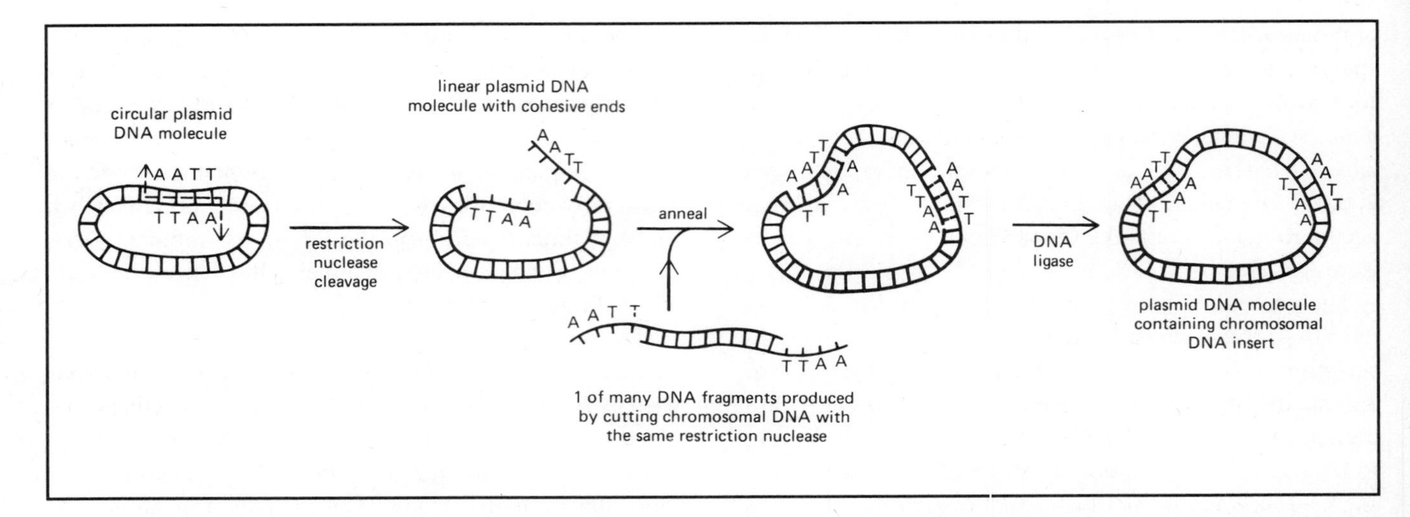

Figure 6. Gene-Splicing Technique. **This diagram illustrates the steps involved in gene splicing. The sugar-phosphate backbone of the circular DNA plasmid (left) can be cleaved with a restriction enzyme at the beginning of the A-A-T-T sequence in each strand; disruption of the complementary bonding between each A-T pair then leaves a DNA fragment with "sticky ends." A gene or gene fragment to be inserted into the plasmid is treated with the same enzyme to produce identical sticky ends. Complementary base-pairing between the new fragment and the cleaved plasmid holds the two fragments together so they can be joined by an enzyme called a DNA ligase to form a new, hybrid plasmid (right). This new plasmid can then be inserted into a bacterium, where it may replicate rapidly.** ***Courtesy: Garland Publishing, Inc.***

tetracycline. When the bacteria are exposed to the antibiotic, the plasmid increases its rate of replication, in effect producing a large number of blueprints for the synthesis of an enzyme that destroys tetracycline. The enzyme is then produced by the plasmid-bearing bacterium in large amounts, protecting it.

Cohen reasoned that the restriction enzyme EcoR1 could be used to cleave plasmids for the subsequent insertion of foreign genes, creating what he termed "chimeras," after the mythological beast that is part lion, part goat, and part serpent. In late 1973, Cohen and Boyer reported that they had done just that.

Cohen and Boyer first isolated plasmids from a strain of *E. coli* that was not resistant to tetracycline. They then used EcoR1 to cleave the plasmid DNA and open the rings, producing sticky ends. They also isolated plasmids from a second strain of *E. coli*, this one with resistance to tetracycline, and cleaved those plasmids with the same restriction enzyme. When they mixed the cleaved plasmids together, the sticky ends brought the DNA from the two plasmids together so that a ligase could be used to splice them into one large circle about twice the size of the original plasmids. The resulting chimeras were then inserted into the *E. coli* strain that lacked antibiotic resistance.

Boyer and Cohen found that the previously susceptible *E. coli* strain could no longer be killed by tetracycline, indicating that the foreign genes were "expressed"— that is, the bacteria produced the tetracycline-destroying enzyme specified by the genes. Cohen and Boyer coined the term "recombinant DNA" to describe this process of gene splicing.

Cohen and Boyer had thus created a way to excise specific DNA sequences or genes from one organism and splice them into plasmids from a second organism. Under appropriate conditions, the plasmids with the foreign DNA could be induced to replicate at high rates, thereby producing large numbers of genetically altered plasmids, giving researchers the ability to mass-produce or "clone" specific DNA sequences or genes. Alternatively, the bacteria containing the cloned plasmids could be used to produce large quantities of the protein coded for by a foreign gene. The techniques of gene isolation and cloning are so useful that they have become the foundation of the biotechnology industry. Today, these techniques have been refined to the point that they can easily be performed by a student in a high school laboratory.

Gene structure. The development of recombinant DNA techniques also provided methods for the direct study of the molecular structure of genes. The first discovery resulting from the use of these techniques was that the genes of eukaryotes are organized in a manner far different from that of bacterial genes. In 1977, biochemist Pierre Chambon and his associates at the University of Strasbourg in France began studying the gene that codes for ovalbumin, the major protein constituent of egg-whites. Chambon's group was looking in the DNA molecule for "control sequences," which determine when genes are turned on and off — that is, when gene sequences are transcribed into mRNA and when transcription is repressed.

To their surprise, Chambon's group discovered that the ovalbumin gene was not a continuous sequence of ovalbumin-coding DNA. Rather, it was a segment of DNA containing a series of eight ovalbumin-coding sequences of approximately equal size, separated from each other by sequences that did not code for ovalbumin. The total length of the intervening sequences was more than ten times greater than the total length of the ovalbumin-coding sequences. Since then, virtually all animal genes examined have been found to be split into protein-coding segments separated by much larger noncoding DNA segments.

In the late 1970s, molecular biologist Walter Gilbert of Harvard University coined the term exon, for "expressed codon," to refer to any gene segment that codes for a portion of a protein, and the term intron, for "intervening codon," to refer to any noncoding gene segment located between two exons in the same gene.

Reverse transcriptase. Because geneticists knew that mRNA is not divided into exon and intron segments, the presence of introns in eukaryotic DNA raised an important question: How do the coding segments get spliced together to make the mRNA molecule? Researchers found that genes, complete with exons and introns, are transcribed into very large mRNA molecules. The mRNA is then selectively "edited," a process in which the introns are deleted in several steps until the mRNA without introns is ready to be translated into protein.

The differences between prokaryotic and eukaryotic genes presented a significant problem in the mass production of eukaryotic proteins in bacteria, which are prokaryotes. Bacteria do not have the enzymes necessary to edit introns from the large mRNA molecules that eukaryotic genes produce. Therefore, intact eukaryotic genes cannot be used directly in bacteria.

To solve the problem, researchers made use of an enzyme called reverse transcriptase, which is found in retroviruses, viruses with genetic material composed of RNA. This enzyme can make DNA copies, called cDNA, which are complementary to its own RNA. Since protein-producing mRNA has no introns, a DNA copy of the mRNA made by reverse transcriptase also lacks those introns and therefore can be transcribed and translated by bacteria. Molecular biologists Howard Temin of the University of Wisconsin in Madison and David Baltimore of MIT discovered reverse transcriptase. In 1975, these two researchers shared one-half of the Nobel Prize for Medicine for their discovery.

DNA probes. To mass-produce specific proteins in bacteria, researchers must first isolate the appropriate mRNA from eukaryotes. The most efficient way to do this is to begin by isolating the mRNA's corresponding protein from a cell or tissue by means of chemical and physical separation techniques. Once the protein is purified, the sequence of a short segment of 20 to 30 amino acids at one end of the protein molecule can be determined. By using this amino acid sequence, along with the triplet nucleotide codes for the specific amino acids, researchers can deduce the nucleotide sequence of the mRNA that produced the protein segment, as well as the complementary nucleotide sequence of DNA. Next, an automated DNA synthesizer is used to make part or all of that DNA sequence.

The short DNA strands created by the synthesizer are used as probes to identify the intact mRNA molecules that code for the desired protein. This identification is accomplished by isolating all the mRNA molecules from cells that produce the protein and subjecting them to electrophoresis, a technique that separates molecules according to differences in their mobility in a gel matrix under the influence of an electric current.

When the gel matrix containing the separated mRNA fragments is blotted with a piece of filter paper, the fragments stick to it. The filter paper is washed with a solution containing the DNA probes that have been labeled with radioactive phosphate. The DNA probe will bind only to those mRNA molecules that have a complementary sequence of bases. In almost all cases, the mRNA with such a complementary sequence proves to be the mRNA that codes for the desired protein. The complex of the mRNA and the probe can be removed from the filter paper and separated, and the mRNA can then be reverse-transcribed into cDNA. The cDNA can be replicated to form double-stranded DNA, which is spliced into a plasmid and cloned. In this way, any protein for which a small section of its amino acid sequence has been determined can, in principle, be mass-produced by bacteria.

Monoclonal Antibodies: Biotechnology Comes of Age

A major advance in biotechnology has been the production of monoclonal antibodies by the fusion of two kinds of cells to make special cells called hybridomas. Antibodies are proteins that identify foreign molecules or microbes and bind to them, thereby marking them for destruction by the immune system. This action provides the body with protection against some infections and foreign substances. Monoclonal antibodies are highly spec- ific antibodies that are produced by a single cell and its descendants and that recognize only one type of molecule.

Before the development of monoclonal antibody technology, antibodies were made commercially by injecting chemicals, bacteria, viruses, cells, or tissues into rabbits, goats, horses, and even humans. This technique produced not a single antibody, but rather a blood sample, called an antiserum, that typically contained 1,000 to 10,000 different antibodies. The supply of an antiserum was limited to the amount produced by the animal during its lifetime. Furthermore, the composition of the antiserum could change during the animal's lifetime as a result of sickness or the aging process.

In contrast, hybridomas yield only one type of antibody and, since the artificially created cells are essentially immortal as long as their growth requirements are met, the supply of monoclonal antibodies from a specific hybridoma is virtually unlimited. All the monoclonal antibodies produced by a specific hybridoma, moreover, bind to the same foreign molecule.

By the late 1980s, biotechnology companies were using monoclonal antibodies for the purification of such biological products as the anticancer agent interferon and the anticlotting agent Factor VIII. The American Red Cross uses monoclonal antibodies routinely for typing blood, and clinical laboratories and physicians use them for identifying pathogenic bacteria and viruses in blood samples and for measuring blood levels of therapeutic drugs. Consumers also use monoclonal antibodies in over-the-counter kits for detecting ovulation, pregnancy, and hidden blood in stool. The U.S. Food and Drug Administration (FDA) has been monitoring clinical trials in humans of nearly 150 different monoclonal antibodies. In 1988, the magazine *Chemical Week* estimated that sales of monoclonal antibodies would reach $300 million in 1988 and $2 billion per year by 1990.

The first monoclonal antibody was prepared by immunologists Georges Kohler and Cesar Milstein at the Medical Research Council Laboratory in Cambridge, U.K., in 1975. This achievement earned them a share of the 1984 Nobel Prize for Medicine. Monoclonal antibodies are prepared by injecting mice with a foreign material to stimulate the production of conventional antibodies by lymphocytes, a type of white blood cell. After about a week, the animals are killed and their spleens removed and chopped into small pieces to harvest the individual lymphocytes.

Next, the lymphocytes are placed in solution and mixed with a suspension of mouse cells isolated from a bone marrow tumor called a myeloma. Before being mixed with the lymphocytes, however, the cancer cells, which are immortal as long as they receive nutrients, are chemically mutated to render them incapable of surviving if certain critical nutrients are not present in their growth medium.

Kohler and Milstein devised a technique to fuse antibody-producing lymphocytes with the mutated myeloma cells, making hybrid cells called hybridomas. Hybridomas have certain characteristics of both the lymphocytes and the myeloma cells. They produce the specific antibody associated with the lymphocyte, survive in a growth medium that lacks the critical nutrients, like the lymphocyte, and are effectively immortal, like the cancer cell. After fusion is achieved, the mixture of cells is grown in a medium lacking the critical nutrients so that the hybridomas survive and all other cells die. The surviving cells are then separated into individual wells, or depressions, in a plastic culturing tray. The colony, or clone, that grows in each well is said to be monoclonal because it is the progeny of a single cell.

The antibodies produced by the cell line in each well are then tested to determine their characteristics. Antibodies that demonstrate a high affinity (meaning that they bind very tightly to one of the molecules in the original mixture) and/or a high specificity (meaning they bind to a particular molecule in the mixture but not to other structurally similar molecules) are selected for large-scale production.

Properties of monoclonal antibodies. High affinity is an important characteristic if the antibodies are to be used in a purification procedure. For example, the protein interferon is currently isolated from cultured mammalian cells or engineered bacteria by using a monoclonal antibody that has a high affinity for interferon. A solution containing the interferon is passed through a bed of plastic beads to which the monoclonal antibody is chemically bonded. The interferon is retained in the bed when the monoclonal antibody binds to it. The beads are then washed with high concentrations of a salt solution, which breaks the binding between the interferon and the antibodies, and pure interferon is obtained.

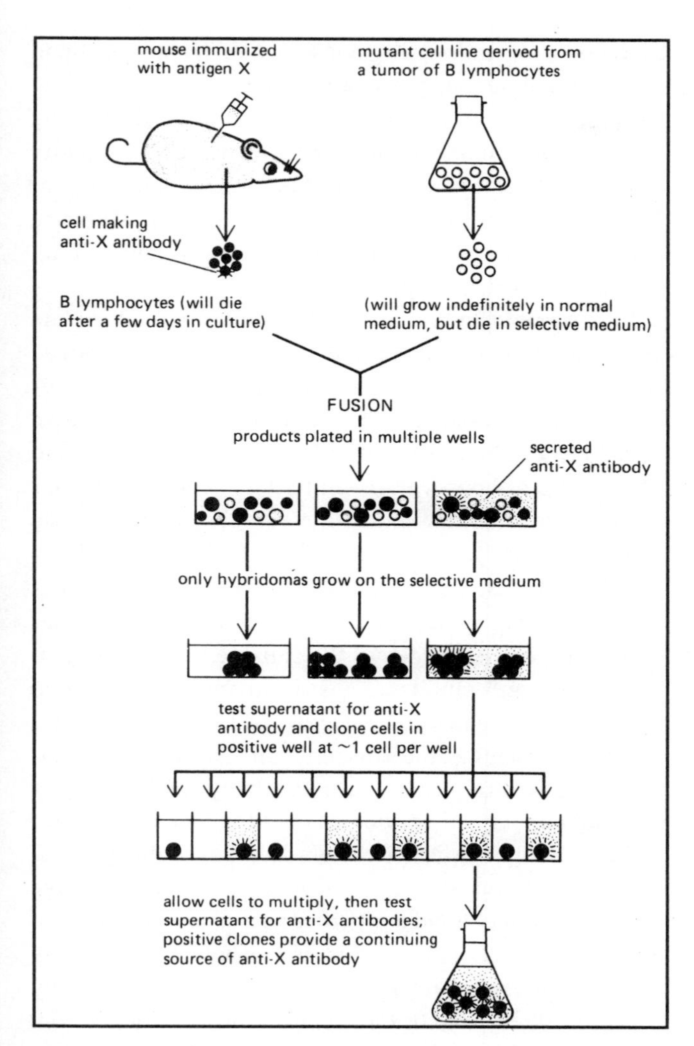

Figure 7. Hybridoma Production. When a mouse is injected with an antigen, cells called lymphocytes begin producing antibodies to that substance. These lymphocytes can be collected and fused with certain types of cancer cells which, unlike lymphocytes, can grow indefinitely in a normal growth medium. The cancer cells can be mutated, however, so that they cannot survive without certain critical nutrients normally produced by the lymphocytes themselves. When the fused cells are grown on a medium lacking these critical nutrients, cancer-cancer hybrids die because of the lack of the critical nutrients and lymphocyte-lymphocyte hybrids eventually die because they cannot grow indefinitely like the cancer cells can. The only cells that survive are the lymphocyte-cancer hybrids, called hybridomas. Individual hybridoma cells are isolated and allowed to multiply; the cells are then screened to determine which produce the desired antibodies. *Courtesy: Garland Publishing, Inc.*

Specificity is also important in any purification procedure similar to the one outlined above. If an antibody cannot distinguish between structurally similar molecules, it cannot be used to separate those molecules in a mixture. Monoclonal antibodies that have a particular specificity have also been used to detect chemicals, bacteria, viruses, and other substances in complex mixtures. The amount of antibody binding to a particular substance in a mixture can be measured, thereby indicating, for example, the presence and amount of hormones, drugs, bacteria, and viruses in blood, urine, and other body fluids.

Other species. Kohler and Milstein were either very wise or very lucky in choosing to work with mouse cells to produce hybridomas. Many other animals commonly used for biological experiments, such as cats, cattle, and horses, do not have the type of myeloma cells needed to produce hybridomas. Nevertheless, treatment of these animals typically requires monoclonal antibodies produced in cells from their own species because antibodies produced in mouse cells are frequently rejected.

A possible way to circumvent this problem was announced in October 1984 by biologists Stephen Kaatari and Barry Nevin of Oregon State University in Corvallis. Kaatari and Nevin were attempting to produce bovine monoclonal antibodies for diagnosis and treatment of infections caused by the bacterium *Streptococcus agalactiae*, which produces udder disease in cows and is a leading cause of mortality in human infants.

Monoclonal antibodies made by fusing mouse myeloma cells with bovine lymphocytes retain many of the properties of mouse cells and are thus recognized as foreign by the cow's immune system. Kaatari and Nevi found, however, that if the resulting hybridomas were themselves fused with bovine lymphocytes, the new hybridomas were more cowlike, even though they retained the immortality characteristic of mouse myeloma cells. By performing this fusion process repeatedly, the Oregon investigators found that they could make hybridomas that produced monoclonal antibodies that were not recognized as foreign by the cow's immune system. Preliminary studies suggest that this approach should be successful with other species as well.

Investigators contemplating the use of monoclonal antibodies in humans have always assumed that only hybridomas made from human cells could overcome rejection problems. Techniques for producing human-human hybridomas were developed shortly after hybridomas were first discovered, but they have never been completely satisfactory. Not only are human-human hybridomas much more difficult to make than mouse-mouse hybridomas, but for some unknown reason, they often stop producing antibodies.

The problems with human-human hybridomas may be overcome with a new technique developed by biologists Vernon Oi of the Becton Dickinson Monoclonal Center, Inc., in Mountain View, California, Sherie Morrison of Columbia University, and Leonard Herzenberg of the Stanford University Medical Center. In 1984, they produced mouse-human hybridomas that apparently are not rejected by the human immune system. They created these special hybridomas by splicing together mouse and human genes coding for different parts of a hybrid antibody molecule. The mouse portion of the hybrid molecule is the part of the antibody that targets and binds specific antigens. The human portion is the part of the antibody that tells the immune system how to destroy the undesirable cell or substance to which the antibody binds. Herzenberg believes the new antibody-producing cells have great potential for use in treating human disease because they are relatively easy to prepare.

Therapy with Monoclonal Antibodies

The natural function of antibodies is binding to molecules that the immune system recognizes as "foreign." When an antibody binds to a foreign molecule, cell, or virus, it marks the foreign body as an invader. Such invaders are eliminated in two primary ways: (1) the antibodies serve as "handles" so that macrophages, large amoebalike white blood cells, can attach themselves to the foreign materials and then ingest and destroy them; and (2) the antibodies, when bound to foreign molecules on the surfaces of invading cells or viruses, trigger a sequence of enzymatic reactions, resulting in the rupture of the surfaces to which the antibodies are attached and the destruction of the invading agent.

If a person's immune system is unable to produce a natural antibody or is overwhelmed by infection, monoclonal antibodies might be able to provide a form of artificial immunity. Preliminary evidence suggests, for example, that monoclonal antibodies can protect against bacteremia and endotoxic shock, poisoning of the bloodstream by chemicals secreted by pathogenic (disease-causing) strains of bacteria such as *E. coli* and *Klebsiella pneumoniae*. Infections by these and related bacteria strike as many as 200,000 hospitalized patients in the U.S. each year and kill 50,000 to 75,000. Such infections are normally treated with antibiotics, but killing the bacteria does not remove toxins already released into the blood, and can even lead to the release of more toxins as the dead bacteria break apart.

Many scientists believe monoclonal antibodies can be used to remove toxins from the bloodstream. In the early 1980s, physician Elizabeth Ziegler of the University of California School of Medicine at San Diego showed that it is possible to cut the mortality rate of patients with bacteremia or endotoxic shock in half by giving them conventional human antisera isolated from patients who have recovered from bacterial infections. This finding suggested that monoclonal antibodies would be equally effective in treating endotoxic shock, and Xoma Corp. of Berkeley, California, and Centocor Inc. of Malvern, Pennsylvania, accordingly began testing monoclonal antibodies for that purpose.

Xoma president Steven C. Mendell reported in late 1988 that clinical trials in more than 500 patients looked "promising," and the company planned to apply for FDA marketing approval in 1989. Centocor started its own clinical trials about 6 months after Xoma, and planned to file for FDA approval in mid-1989. Both companies estimated that the cost of treating a single patient with the monoclonal antibodies will be about $1,200 to $1,500.

In 1985, researcher Nelson Teng of Stanford University and his colleagues reported that monoclonal antibodies against a specific antigen of the pathogenic bacteria protected mice and rabbits against bacteremia. During that year, Zeigler began clinical testing of a human monoclonal antibody against the same antigen. The trial is expected to take several years, but preliminary results suggest that the monoclonal antibodies could reduce mortality.

Bacteriologist Mark Lostrom and his colleagues at Genetic Systems Corp. of Seattle, Washington, have developed monoclonal antibodies against seven strains of *Pseudomonas aeruginosa* that are responsible for many bacterial infections in hospital patients. The company hoped that the antibodies would prevent the patients from becoming infected, and it began clinical trials in late 1986. Pathologist Steven Foung of Stanford has developed similar monoclonal antibodies against varicella zoster virus, a form of herpesvirus that causes chicken pox and shingles; chicken pox is normally a relatively mild disease, but the virus can be fatal in children with weakened immune systems.

Foung has also developed monoclonal antibodies that could help prevent the blood disease caused by Rh-factor (blood type) incompatibility between a mother and her fetus. The anti-Rh antibodies destroy fetal blood cells that escape into the mother's bloodstream, thereby preventing the mother's immune system from mounting an attack against the fetus. Clinical trials of both antibodies began in 1987.

Magic bullets. Researchers have long hoped to develop antibodies that would specifically bind to and destroy the cells of a growing tumor without affecting other tissues in the body—the so-called "magic bullet" first envisioned by the German physician Paul Ehrlich nearly a century ago. Many scientists now believe that cancer cells are continuously generated in the human body, but are normally destroyed by the immune system. Occasionally, however, these cancer cells may evade the immune response, either by mimicking the appearance of healthy cells or by secreting molecules that suppress the immune system. Once the cancer cells have evaded the immune system, they divide and form a tumor.

If antibodies could be produced that would bind to and destroy only the cancerous cells, then a true "magic bullet" would be available. Monoclonal antibodies make such cancer-specific antibodies possible. In the early 1980s, immunologists Richard Miller and Ronald Levy of the Stanford University Medical Center reported that they had used monoclonal antibodies produced from mouse hybridomas to treat six humans with leukemia, a cancer of the blood-forming organs. The patients all underwent "quite remarkable changes" and one patient had "a prolonged cure."

In 1985, Levy and his colleagues used monoclonal antibodies to treat 11 patients with lymphoma, a cancer of the lymphatic system, who had not responded to conventional therapy. One of the 11 patients had a complete remission, and four others showed improvement. Five of the patients who did not respond to the monoclonal antibody treatment had developed antibodies against the mouse antibodies used in therapy.

By late 1988, clinical trials testing monoclonal antibodies against melanoma, colon cancer, lymphoma, and pancreatic cancer were under way at several institutions, including the University of Maryland, Johns Hopkins University, Stanford University, and the University of California at Los Angeles.

Scientists have even greater expectations for the future because of the development of the new mouse-human hybridomas described earlier. Levy and others were also beginning to use "cocktails" of antibodies against several different tumor antigens, which theoretically improve the therapeutic efficiency because they attack different portions of the tumor's surface. By late 1988, Levy had treated several dozen lymphoma patients with such antibody cocktails. The treatment may relieve pain, reduce the size of the tumors, and prolong life, but it rarely produces a complete remission. Progress has been steady, however, and Levy hopes for better results in the future as researchers gain a better knowledge of how the immune system functions in cancer patients.

The efficacy of monoclonal antibodies against cancer could be improved as much as 100,000-fold if they are chemically linked to toxic chemicals, such as ricin, a toxic peptide produced in seeds of the plant *Ricinus communis*, or the diphtheria toxin produced by the bacterium *Corynebacterium diphtheriae*. These toxins are extremely potent: a single molecule of ricin, for example, can kill a cell. A summary of the use of such toxins was presented in the 20 November 1987 *Science* by microbiologist Ellen Vitetta and her colleagues at the University of Texas Southwestern Medical Center in Dallas. They reported that ricin attached to a monoclonal antibody could cure a particular form of leukemia in mice; most important, the antibody-ricin combination attacked only tumor cells.

In 1986, immunologist Lynn Spitler of Xoma Corp. disclosed that the company had sponsored trials of a ricin-containing monoclonal antibody in 22 patients with malignant melanoma, an often-fatal type of skin cancer. The tumors regressed in five of the patients and stopped spreading in five others. By late 1988, several other companies, including Cetus Corp. of Emeryville, California, and Hybritech, Inc., of San Diego were also working on the development and production of such antibody-toxin combinations.

Another approach to treating cancer is to attach conventional chemotherapeutic agents to monoclonal antibodies. Most anticancer drugs harm other cells in the body as well, producing such side effects as nausea and hair loss. Many of these side effects may be avoided if the drugs are attached to antibodies that target the drugs toward cancer cells. Centocor, Inc., of Malvern, Pennsylvania, has made combinations of monoclonal antibodies with the chemotherapeutic agent doxorubicin, and Cytogen of Princeton, New Jersey, has made combinations with methotrexate.

Most combinations of monoclonal antibodies with anticancer drugs are still being studied in animals. In 1985, however, the Scripps Clinic and Research Foundation in La Jolla, California, began studying combinations of monoclonal antibodies with methotrexate in humans with lung cancer. During the following year, oncologist Robert Baldwin of the Cancer Research Campaign laboratories in Nottingham, U.K., began using the same combination to treat patients with colon cancer.

Yet another approach to treating cancer is to attach radioactive isotopes such as bismuth-212 to monoclonal antibodies. Bismuth-212 decays by alpha radiation, which involves the emission of a helium nucleus (alpha particle). The energy of the alpha particle is high enough to kill any cells it passes through. Because the particle can only penetrate a thickness of about 10 cells, however, the killing effect of bismuth-212 attached to a monoclonal antibody is restricted primarily to cells near the site where it decays.

Arnold Friedman and his colleagues at Argonne National Laboratory in Chicago, Illinois, originally

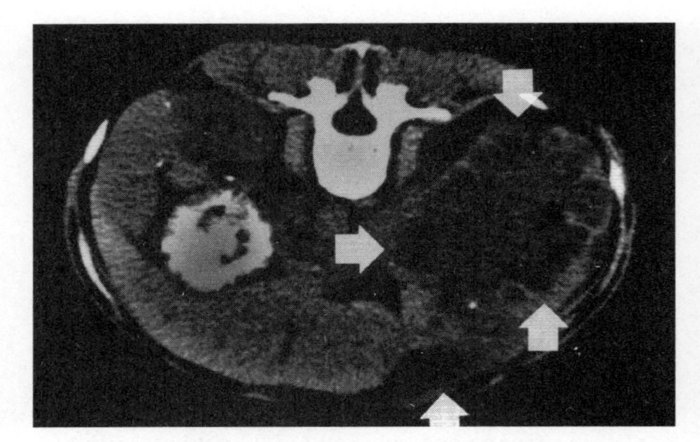

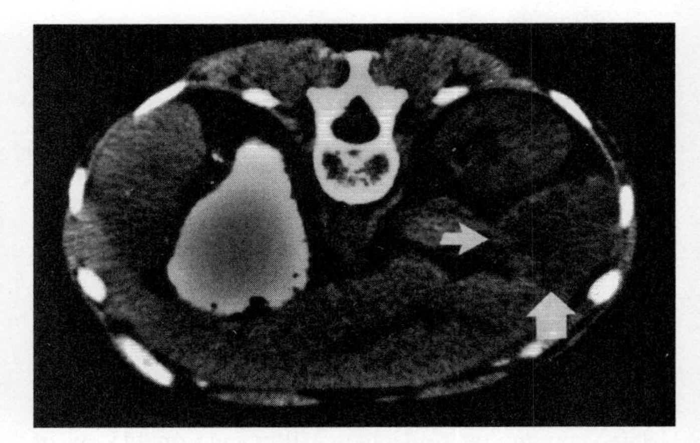

Figure 8. Liver Tumor Shrinkage After Radioactive Antibody Treatment. These two CT scans show how a massive liver tumor (indicated by the arrows) shrank after being treated with a combination of conventional radiation therapy, chemotherapy, and the use of antibodies linked to radioactive iodine-131. Before treatment (left), the tumor was too large to be removed by surgery; after about 8 months of treatment (right), however, the tumor's volume decreased nearly 70 percent to a size small enough for surgical removal. ***Courtesy: Stanley Order, The Johns Hopkins Hospital.***

developed the combinations of monoclonal antibodies and bismuth-212. They reported in 1984 that a single injection of the combinations into mice with a virus-induced leukemia increased the animals' life expectancy by 50 percent. Friedman believes that because of the limited range of the alpha emissions, the technique is probably best suited for treatment of metastases, clusters of cancer cells that have broken away from the primary tumor. In 1986, Friedman, in cooperation with oncologist James Ryan of the University of Chicago Medical Center, began clinical trials of bismuth conjugates in human patients with liver metastases.

In 1985, radiologist Stanley Order and his colleagues at the Johns Hopkins University School of Medicine in Baltimore, Maryland, revealed that they had found "the first effective treatment" for liver cancer, also known as hepatoma. Order's group used monoclonal antibodies linked to iodine-131 to treat 104 patients with inoperable hepatomas. The combination shrank the tumors by 30 percent or more in 50 of the patients and produced total remission in seven. Order found that several of the patients had also reached a "plateau" in which the tumors were neither shrinking nor growing. In contrast, with conventional therapy, a response is observed in less than 15 percent of the patients, and the disease is always fatal within 5 years. By late 1988, Order had treated more than 300 patients with monoclonal antibodies. Seven had complete remissions of their liver tumors, and nearly half had partial remissions in which the tumors shrank by at least 30 percent. In several of the latter cases, the tumors could then be removed surgically.

Imaging tumors and hearts. Radioactive isotopes attached to monoclonal antibodies are proving useful for detecting and imaging tumors. When the antibodies bind to a tumor, the emitted radiation can be detected by conventional x-ray monitors to show where the tumor is located or to identify tumors too small to be detected by other techniques. By late 1988, for example, radiologist David Goldenberg of the Center for Molecular Medicine and Immunology in Newark, New Jersey, had imaged tumors in more than 650 patients, primarily with monoclonal antibodies derived from mouse cells and labeled with iodine-131. He had successfully detected tumors of the colon, rectum, ovary, liver, and lungs, which were then treated by conventional techniques.

Similarly, Baldwin had evaluated about 500 patients with colon and rectal cancers using monoclonal antibodies labeled with iodine-131 or indium-111. In 1986, he announced that he had successfully detected about 80 percent of tumors that were too small to be imaged with other techniques. Scientists hope that such imaging will prove useful for screening patients at increased risk of developing cancer because of their genetic heritage or occupation.

Damon Biotech, Inc., of Needham Heights, Massachusetts, announced in 1988 that it had applied to FDA for permission to market a monoclonal antibody assay for melanoma. This cancer often spreads rapidly through the body, and the detection of the metastases, especially small ones, is not only difficult but currently requires expensive and time-consuming diagnostic tests.

Damon uses a melanoma-specific antibody that is linked to the radioactive element technetium-99M. When injected into the patient's bloodstream, the antibody travels to the metastases, and the attached technetium-99M emits gamma radiation that is detected by a special camera. The imaging process can be completed within a few hours. Damon reported that the assay had been successfully used on more than 200 patients in trials at medical centers around the country, including the University of Washington, the Mayo Clinic, Baylor University, and the University of Southern California.

Researchers from the National Cancer Institute (NCI) in Bethesda, Maryland, and Johns Hopkins have found that they can use monoclonal antibodies to detect lung tumors 2 years before they can be detected by conventional techniques. The team, headed by NCI oncologist James L. Mulshine, reported in 1988 that they had developed a test in which a patient's sputum is mounted on a microscope slide and exposed to a monoclonal antibody specific for lung cancer cells. If the sputum contains tumor cells, the antibodies adhere to them, allowing them to be readily seen when a special stain is applied.

The researchers tried the test on 69 sputum specimens that had been frozen at Hopkins as part of another study conducted in the 1970s; 26 of the patients who donated the samples had gone on to develop lung cancer. They found that 20 of the specimens from individuals who later developed lung tumors stained positive. Mulshine said that, on average, the sputum samples had been collected 20 months before signs of lung tumors had been detected by conventional techniques. In contrast, only five of 40 specimens from the patients who did not develop lung tumors tested positive. (The rest of the samples had deteriorated too much to yield good results.) Mulshine said the group would begin a 5-year study monitoring sputum samples from patients who have had a lung tumor removed surgically in order to detect recurrences of the tumors.

Cardiologist Edgar Haber of Massachusetts General Hospital in Boston reported in 1986 that he and colleagues at three other hospitals had been using a monoclonal antibody labeled with indium-111 to study cardiac damage resulting from heart attacks. The antibody, developed by Centocor, is specific for myosin, a structural component of the heart that comes into contact with the bloodstream only when heart tissues die. In about 150 patients, the researchers found that the labeled antibody could accurately reveal the extent of heart damage much more quickly than was possible with other imaging procedures, such as angiography. They also studied 17 patients who were suspected of having heart disease; the antibody showed that five of these patients had damage to heart tissue, which was later confirmed by other methods, while conventional imaging techniques revealed damage in only one.

Autoimmune disease. Monoclonal antibodies also have potential for treating autoimmune diseases, such as multiple sclerosis, diabetes, and rheumatoid arthritis, in which a person's immune system attacks the body's own tissues. For example, a team headed by neurologist Larry Steinman of the Stanford University School of Medicine reported in 1985 that injections of monoclonal antibodies into mice stopped the development of a disease that resembles multiple sclerosis (MS).

In people with MS, the target of the immune system's attack is the insulating sheath that surrounds nerve cells. Destruction of the sheath leads to neurological damage and the disease's characteristic symptoms, which may include weakness, paralysis, tremors, and disturbances of speech and vision. An MS-like disease called experimental allergic encephalitis (EAE) can be induced in mice by injecting the animals with nerve tissue from other mice. The mice's immune systems overreact to the foreign tissue and begin to attack their own nerve cells. Within days, the mice become weak and develop hind-limb paralysis, after which they often become completely paralyzed and die.

Steinman and his colleagues developed monoclonal antibodies that bind to and destroy a set of immune cells known as helper T-cells, which, in MS and EAE, attack the nerve sheaths. Steinman's group found that symptoms of EAE did not develop in any mice that were given the monoclonal antibody before being injected with the nerve tissue. In contrast, EAE developed in more than 90 percent of the animals that were not given the antibodies. In 1988, the group reported that administering the antibody to mice after the appearance of early signs of EAE halted the disease in more than 90 percent of the animals, even when the antibody was given to mice in the paralytic stages of the disease. The investigators began clinical trials of the monoclonal antibodies in MS patients in 1986, but abandoned them in early 1987 when the patients developed severe reactions to the mouse-based antibodies. They have since been working to develop other antibodies that will not provoke such reactions.

In contrast, neurologist Howard L. Weiner and his colleagues at the Harvard Medical School have reported that they observed no ill effects when similar antibodies were used to treat 20 MS patients. Why the Stanford antibodies provoked an immune reaction and the Harvard ones did not remains a mystery. The Harvard study was designed only to show the safety of the antibodies, and by late 1988 Weiner had begun research to determine whether they were effective at combatting MS.

Transplant rejection. The capacity of monoclonal antibodies to bind to human T-cells and target them for destruction has also proved useful in preventing transplant rejection. Researchers at Ortho Pharmaceutical Corp. of Raritan, New Jersey, have developed a monoclonal antibody called Orthoclone OKT-3 that binds to the T-cells that participate in kidney rejection. A team of transplant specialists at hospitals around the U.S. tested the antibody in 63 patients who were rejecting their transplanted kidneys, and compared the response to that obtained in 60 similar patients treated with conventional drugs. The researchers reported in 1985 that rejection was reversed in 94 percent of the patients receiving the antibody, compared to only 75 percent of those treated with conventional drugs.

By mid-1986, the antibody had been successfully used in hundreds of other patients, and at that time, FDA approved the sale of the antibody for treating transplant rejection. The antibody is expected to be widely used, because about 7,000 patients receive kidney transplants in the U.S. each year, and at least 60 percent of them suffer a rejection episode during the first year. Ortho officials have cautioned, however, that the antibody treatment can be used for only one rejection episode; since the monoclonal antibodies are foreign proteins, the body makes its own antibodies against them.

In 1987, physician C. Garrison Fathman and his colleagues at Stanford announced a breakthrough that could have major implications for diabetics and others in need of transplants. Fathman developed a new way to transport insulin-producing cells from one strain of mouse to another. Normally, such cells are rejected quickly. But Fathman injected the mice with a monoclonal antibody at the time of the transplant, triggering the destruction of 95 percent of the T-helper lymphocytes that initiate rejection.

When the mice's immune systems later replaced these T-helper lymphocytes with new ones, the new cells did not recognize the transplanted cells as foreign. The transplants survived for the rest of the animals' lives without any further treatment. Fathman predicted that the monoclonal antibody treatment could be attempted in people within 5 years.

Biotechnology Industry

The development of recombinant DNA technology during the 1970s made possible the establishment of many companies whose existence and products could not have been imagined 20 years ago. In the early 1980s, more than $1 billion per year in venture capital was raised to support the operation of 350 biotechnology companies and the commercialization of their techniques. Investment in those companies is still considered risky, however, since only a handful have yet shown a profit.

Biotechnology Products

In November 1987, FDA approved the sale of tissue plasminogen activator (tPA), a genetically engineered enzyme that dissolves heart attack-causing blood clots, by Genentech, Inc., of South San Francisco, California. The drug was the fifth genetically engineered product approved for human use, the first four being insulin, alpha-interferon, human growth hormone, and a hepatitis B vaccine. FDA has also approved the first clinical trials of a genetically engineered vaccine against acquired immunodeficiency syndrome (AIDS) produced by MicroGeneSys, Inc., of West Haven, Connecticut. The U.S. Department of Agriculture (USDA) has approved a genetically engineered viral vaccine against pig pseudorabies, as well as several other products for use in animals, including growth hormones and vaccines containing only parts of viruses.

But this small number of products is merely the first wave of what many experts predict will be a deluge. A 1987 report, *New Developments in Biotechnology*, from the U.S. Congressional Office of Technology Assessment, noted that more than 100 different potential products for human use were under development by biotechnology companies. According to Consulting Resources Corp. of Lexington, Massachusetts, sales of biotechnology-derived health care products, such as drugs and vaccines, were expected to approach $900 million in 1988.

tPA. Every year, 1.5 million Americans suffer heart attacks and more than 500,000 die from them. Most heart attacks, also known as myocardial infarctions, occur when a blood clot lodges in one of the coronary arteries and impairs blood flow to the heart muscle, killing heart muscle tissue by depriving it of oxygen. By early 1989, several studies had suggested that damage to the heart can be minimized in perhaps half of the 1.5 million victims by a process known as thrombolysis, in which enzymes are used to dissolve blood clots. The two most important clot-dissolving enzymes are streptokinase, which is produced naturally in certain bacteria, and tPA, which was originally isolated from uterine tissues but is now produced by genetically altered bacteria.

Streptokinase, which is sold in the U.S. by Hoechst-Roussel Pharmaceuticals, Inc., of Somerville, New Jersey, and SmithKline Beckman of Philadelphia, Pennsylvania, is injected into the bloodstream after a heart attack to dissolve the clot. The two companies conducted studies in the early 1980s on more than 24,000 patients and found that the drug reduces the risk of dying from a heart attack by 20 percent. Unfortunately, streptokinase dissolves clots throughout the body, which can produce internal bleeding. In contrast, tPA appears to act almost exclusively at the site of the clot in the heart, causing little bleeding elsewhere.

In 1984, the National Heart, Lung, and Blood Institute in Bethesda organized a tPA study called the Thrombolysis in Myocardial Infarction trial. The $31 million study was originally scheduled to study 4,000 patients at 25 U.S. medical centers. Preliminary results from the study, reported in 1985, showed that intravenous tPA removed the blockages in 66 percent of 118 patients, while streptokinase relieved only 36 percent of similar blockages in 122 patients. These results were so impressive that researchers terminated the study and gave all the patients tPA.

By September 1987, tPA had been tested in more than 4,000 patients around the world, and FDA approved it for general use. Sales of the drug, trade-named Activase, during 1988 were estimated to total about $180 million, the largest first-year sales for any new drug. But Genentech and many analysts had predicted first-year sales of $400 million, so that figure was viewed as a disappointment. By the end of 1988, Activase accounted for 60 percent of the market for clot-dissolvers in terms of patients treated. Streptokinase, with sales of $5 million, accounted for 35 percent.

tPA is the first large-scale product of the biotechnology industry. Genentech says that as many as 400,000 Americans could benefit from the drug every year, compared to the 120,000 that are now receiving either Activase or streptokinase. Some analysts had predicted that annual sales of the drug could reach $1 billion in the early 1990s, but that estimate is being revised downward. One problem is that the drug is expensive. In 1988, one dose cost about $2,200, compared to roughly $200 to $300 for a similar dose of streptokinase. Many physicians have also been reluctant to use either tPA or streptokinase because of the risk of internal bleeding. Other doctors argue, however, that the benefits of Activase in preventing heart damage far outweigh its increased cost and slight risk.

Genentech may not be the only company to reap the benefits of tPA. In October 1988, a British appeals court rejected Genentech's sweeping patent claims for tPA, clearing the way for another tPA patent by Wellcome PLC of London. Several other companies are also developing second-generation forms of tPA that persist in the blood longer and that have fewer side effects. Some analysts believe such products may capture a significant share of the market during the 1990s.

Insulin. The hormone insulin, which is normally produced by the pancreas, stimulates cells to absorb the sugar glucose from blood, thereby maintaining blood sugar at normal concentrations. Diabetics are unable to naturally control the level of glucose in their blood, because they cannot produce enough insulin, which enables body cells to use sugar for energy. When cells must rely heavily on fats for energy instead of sugars, they produce toxic chemicals that can cause coma and death. Poor control of blood sugar leads to long-term effects, including failure of the kidneys, loss of eyesight, and the death of nerves in the hands and feet.

Diabetics, who now number about 11 million in the U.S., have been treated with insulin from animals since the 1920s. But the worldwide demand for insulin from animals is projected to outstrip the supply by the year 2000. Furthermore, about 5 percent of the diabetics in the U.S. are allergic to animal-derived insulin or to the small amounts of other proteins that frequently contaminate the animal preparations.

In 1978, Genentech announced that its scientists had used recombinant DNA techniques to produce insulin. In October 1982, insulin became the first genetically engineered protein to be approved by FDA. Eli Lilly & Co. of Indianapolis, Indiana, markets the bacterially derived insulin, called Humulin, under a license from Genentech. At the end of 1988, the drugstore price of Humulin was

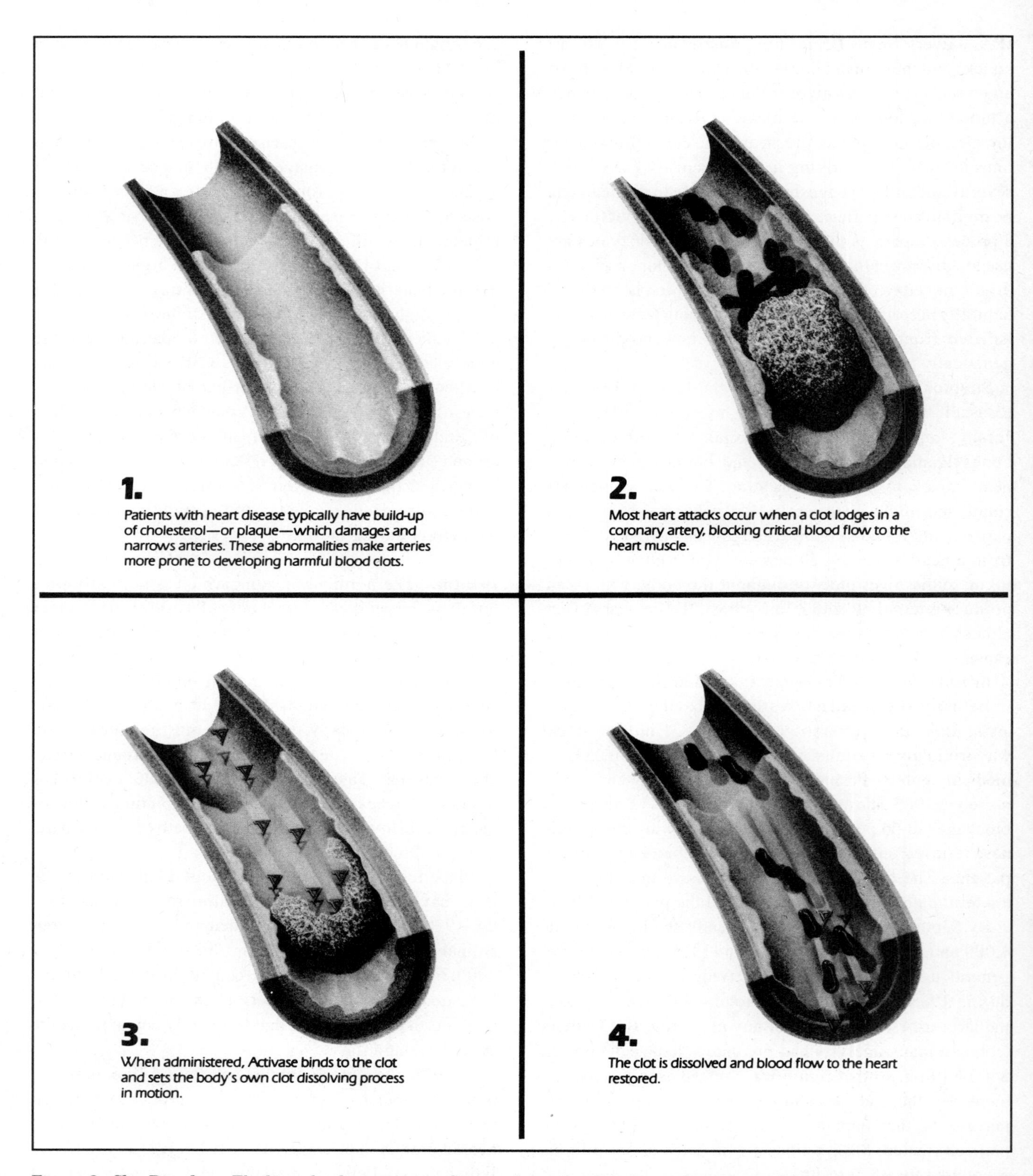

Figure 9. Clot Dissolver. The biotechnology company Genentech has developed a synthetic form of the enzyme tPA (trade-named Activase), which can dissolve the blood clots responsible for a large portion of the 1.5 million heart attacks that occur in the U.S. each year. *Courtesy: Genentech, Inc.*

about the same as the price of the highest quality insulin isolated from animals, and it accounted for about half of all insulin sold in the U.S. *Genetic Technology News* estimates the current U.S. market for human insulin to be $100 million to $125 million per year.

Human growth hormone. One out of every 30,000 children in the U.S. suffers from hypopituitary dwarfism, a genetic disease in which a defect in the gene for human growth hormone (hGH) prevents the hormone's normal synthesis. Unless these children receive regular injections of hGH, usually obtained from human cadavers, they will not reach normal adult height. hGH extracted from the pituitary glands of cadavers is, however, scarce and expensive; 50 cadavers provide enough hGH to treat only one child for one year, at an average cost in the early 1980s of about $10,000.

The hormone appears to have other therapeutic benefits as well; several experiments have suggested that hGH could speed the healing of burns, fractures, ulcers, and wounds, delay the progression of amyotrophic lateral sclerosis (Lou Gehrig's disease), assist in weight loss, and stop the bone degeneration characteristic of osteoporosis.

The overall hGH supply, sufficient to treat only about 4,000 children per year, had been too low to explore other uses for hGH. Furthermore, the distribution of pituitary-derived hGH was halted by the FDA in April 1985 because of evidence that three patients who had received the preparation later developed a rare and incurable viral infection called Creutzfeldt-Jakob disease. By the late 1980s, distribution of cadaver-derived hGH was still forbidden in the U.S., but Japan and the U.K., where it is produced by a different process, permitted its use.

The availability of hGH has since changed dramatically, however. In October 1985, the FDA approved the sale of Genentech's version of hGH, Protropin, produced in genetically engineered bacteria. This approval marked a milestone for the biotechnology industry, because it represented the first time that a biotechnology firm had developed, manufactured, and begun to market a pharmaceutical produced by genetic engineering. Before Protropin, biotechnology companies had licensed the rights to new products to major pharmaceutical companies. Genentech's decision to market Protropin itself put the company in direct competition with the larger firms. By 1988, the cost of one year's treatment with Protropin was about $15,000 to $20,000, and the cost was typically covered by health insurance. In 1987, sales of Protropin totaled $87 million.

In March 1987, Lilly received FDA approval to market its own version of hGH, called Humatrope, which is also made through genetic engineering. Humatrope is chemically identical to naturally occurring hGH, while Genentech's Protropin has an extra amino acid on one end — a characteristic of many proteins produced by genetic engineering. Lilly claims that because of this difference, fewer patients experience an immune reaction to Humatrope than to Protropin. Genentech has developed a second-generation Protropin that, like Humatrope, is identical to natural hGH. At the end of 1988, this compound had not yet been approved by FDA.

Meanwhile, Genentech has been trying to keep Lilly's product off the market. Genentech received approval for Protropin under terms of FDA's 1983 Orphan Drug Act, which gives special incentives to manufacturers of drugs designed to treat diseases with fewer than 200,000 victims. Such orphan drugs are rarely profitable because the market for them is so small. One provision of the 1983 act entitles manufacturers of orphan drugs to an exclusive 7-year marketing period before competitive pharmaceuticals can be introduced. In 1987, Genentech filed suit against FDA arguing that it had not only approved Humatrope in violation of this provision, but also had awarded Lilly orphan drug status for Humatrope, preventing Genentech from bringing out its own second-generation product. By the end of 1988, that suit had not been resolved. In the meantime, at least three other companies were developing their own versions of hGH.

The increased availability of hGH has enabled scientists to begin testing the hormone to confirm or disprove its efficacy in treating many other ailments. Some experiments have shown evidence that the hormone can be used to accelerate development in teenagers who are late to mature sexually and who reach adult height 4 or 5 years later than most of their peers. Other experiments show that the hormone can help reduce body fat in obese people.

But the wider availability of hGH has also led many researchers to fear that it may be abused. Critics charge that practitioners of "cosmetic endocrinology" will use hGH to increase athletic prowess or to increase the height of normal children in the belief that it will improve their social and economic prospects. In particular, hGH is thought to increase muscle mass like anabolic steroids, but without their detrimental side effects. There are persistent rumors of a black market in hGH in the U.S. and an even larger market in Europe, and at least one $250,000 shipment of Protropin has been stolen from a Genentech distributor. Most researchers believe, however, that the great potential of hGH far outweighs any negatives that might occur as a result of abuse.

Interferons. In response to viral infections, mammalian cells normally produce interferons, a family of at least three types of proteins. Each of these types, in turn, includes several variants. The specific mechanism by which interferons work is unknown, but they apparently stimulate infected cells to synthesize proteins that inhibit viral replication. They also seem to stimulate virally infected cells and some cancer cells to synthesize another type of protein that increases the chance these cells will be recognized and eliminated by the immune system.

The three types of interferon are designated alpha, beta, and gamma, depending on their origin. Alpha-interferon is produced by lymphocytes, the type of white blood cells responsible for initiating and carrying out immune responses. Beta-interferon is produced by fibroblasts, cells associated with the production of connective tissue. Gamma-interferon is also produced by lymphocytes. Because interferons have the potential for treating diseases as diverse as cancer and the common cold, genetic engineers have set a high priority on developing techniques to obtain the proteins in pure form.

In January 1980, Biogen became the first firm to report the use of recombinant DNA techniques to produce human alpha-interferon in *E. coli*. Then, in late 1981, a team of scientists from Genentech and the University of Washington in Seattle announced that they had produced human alpha-interferon in another type of genetically engineered organism, the yeast *Saccharomyces cerevisiae*. Yeasts have some advantages over bacteria in the production of genetically engineered proteins, primarily because their metabolism is more similar to human cells than that of bacteria. Yeasts are more suitable for carrying out the sophisticated reactions necessary for producing complex biomolecules such as interferons.

In April 1981, investigators at Cetus announced that they had successfully produced beta-interferon in *E. coli* and *Bacillus subtilis*. The company began the first phase of clinical trials with beta-interferon in humans during 1984. By late 1987, Cetus had begun the second phase of clinical trials to test the effectiveness of beta-interferon in combating hepatitis, influenza, certain other viral infections, and several types of cancer.

Researchers at Genentech and NIH reported producing the third type of interferon, gamma-interferon, in *E. coli* and yeast cells during early 1982. In 1985, Biogen announced that it was beginning clinical trials in the U.S. using gamma-interferon in the treatment of renal cell carcinoma, a type of kidney tumor. The company also confirmed rumors that its researchers in West Germany had achieved preliminary success using the drug against rheumatoid arthritis. In 1986, Biogen received approval from the West German government to begin marketing the drug for use against the disease. In 1987, Genentech began clinical trials to study gamma-interferon's effectiveness against cancer and certain viral infections.

In June 1986, the FDA gave permission to both Schering-Plough Corp. of Madison, New Jersey, and Hoffmann-La Roche, Inc., of Nutley, New Jersey, to market alpha-interferon for treatment of hairy-cell leukemia, a cancer of B-lymphocytes that affects about 10,000 people in the U.S. each year. At the time of the approvals, oncologist Howard Ozer of the University of North Carolina at Chapel Hill reported that alpha-interferon had produced significant improvement in more than 90 percent of the patients who received it. "Many patients who once required lengthy hospitalization and very frequent blood transfusions can return to work and lead normal lives after just a few months of alpha-interferon therapy," Ozer said.

In November 1988, FDA granted both companies permission to market alpha-interferon for the treatment of Kaposi's sarcoma, a disease that strikes many AIDS victims. Kaposi's sarcoma causes disfiguring reddish-purple or brown skin lesions, and can attack the lungs, brain, and gastrointestinal tract. Although the disease can cause death when it affects vital organs, most AIDS victims die of other causes. According to FDA, several human studies have shown that high doses of alpha-interferon can reduce Kaposi's tumors in 40 to 45 percent of patients in the early stages of AIDS.

Alpha-interferon has great potential for treating disease. At the time the drug was approved for hairy-cell leukemia, researchers in the U.S. were also studying its use against multiple myeloma, a cancer of white blood cells called plasma cells; malignant melanoma, a skin cancer; venereal warts, a sexually transmitted viral disease; laryngeal papillomatosis, warts of the throat; non-Hodgkin's lymphoma, a cancer of the lymph glands; and the common cold. Roferon-A and Intron A have been approved for use against many of these diseases in several countries, including Canada, the U.K., Ireland, and the Philippines.

Other Pharmaceuticals

The advent of genetic engineering has given scientists an unprecedented opportunity to develop new therapies for a wide variety of human ailments. During the 1970s and 1980s, researchers discovered a large number of naturally occurring hormones, proteins, antibodies, and other biological products that are produced by the human body in minute quantities. Many of these materials have the potential for treating disease if they could be produced in significant quantities. Genetic engineering has made such production possible, and experts predict that a large number of new biological drugs will be approved for marketing within the next 5 years.

Interleukins. Interleukins are a recently discovered family of glycoproteins (proteins linked to sugar molecules) that alter the activity of white blood cells and that may prove effective in combating cancer and a variety of other diseases. Interleukin-2 (IL-2), one of the most important members of the family, is produced by white blood cells in response to viral infections. It stimulates the growth of cells that control and regulate the immune system. As a result, many scientists have speculated that it may be useful in the treatment of certain illnesses caused by, or related to, a malfunctioning of the immune system. Such diseases include AIDS, cancer, MS, and rheumatoid arthritis.

In March 1982, Tada Taniguchi at the Cancer Institute in Tokyo, Japan, announced that, while working with Ajinomoto, Inc., also of Tokyo, he had succeeded in splicing the human gene for IL-2 into bacteria. At the same time, investigators at Hoffmann-La Roche and Immunex Corp. of Seattle, Washington, announced that they had purified IL-2 from white blood cells and determined its structure. All of these researchers then worked together to mass-produce IL-2 in bacteria by recombinant DNA techniques. At about the same time, investigators at Cetus announced that they had also been able to produce IL-2 in bacteria.

Clinical trials of IL-2 in humans began in early 1984. In December 1985, NCI oncologist Steven Rosenberg disclosed that IL-2 had reduced tumor size by more than 50 percent in 11 of 25 patients with various types of cancer. IL-2 is used in a much different way from other cancer drugs. Rosenberg extracts white blood cells from the cancer victim and incubates them with IL-2 for 3 to 4 days. By an as-yet-unknown mechanism, IL-2 converts the white cells into lymphokine-activated killer (LAK) cells that attack the tumor very aggressively when they are reinjected into the victim. Vincent DeVita, Jr., director of the institute, called the results "the most interesting and exciting biological therapy we've seen so far."

By the end of 1988, some of the initial enthusiasm about Rosenberg's approach had abated. After treating more than 300 patients over a 3-year period, Rosenberg concluded that the success rate was only about 30 percent and that the tumor shrinkage was temporary in many cases. The treatment also caused serious side effects in some patients, such as massive fluid retention and heart and lung problems. Although Rosenberg was attempting to refine the use of LAK cells, he and others had also been investigating the use of killer cells called tumor-infiltrating lymphocytes (TILs) harvested directly from a patient's tumor and activated by IL-2. In May 1988, oncologist Richard Kradin of Harvard Medical School announced that TILs had produced at least a partial remission in about half of a small number of patients with kidney tumors. TIL and LAK cells produced with IL-2 are being studied in more than 60 clinical trials around the U.S. involving more than 900 patients.

IL-2 can also be administered in the same way as a conventional drug, and Cetus Corp. has sponsored clinical trials of its form of IL-2, called Proleukin, in more than 2,000 patients suffering from more than a dozen types of cancer. The best results have been observed in metastatic renal cell carcinoma (kidney cancer) and malignant melanoma. In October 1988, FDA granted Cetus orphan drug status for Proleukin for the treatment of kidney cancer, and the agency was expected to grant marketing approval in 1989.

Preliminary studies in humans have also shown that IL-2 can restore some functioning of the immune system in people with AIDS, thereby increasing their resistance to potentially fatal infections. Researchers at Stanford University Medical Center are also testing the protein on victims of chronic hepatitis B infections to see whether it can stimulate the body's immune system to better combat the disease. Analysts predict that the market for IL-2 will total more than $100 million per year in the U.S. during the 1990s.

In 1984, three groups independently reported that they had cloned genes coding for the related protein interleukin-1. Earlier work had suggested that IL-1 is important in the activation of the immune system, that it works in the brain to induce fever, and that it may mediate the headaches and body pains that commonly accompany infections. The three groups' studies indicated that at least two forms of IL-1 exist, and perhaps many more. By late 1988, all three groups were working to produce sufficient quantities of IL-1 for further experiments.

Some evidence also suggests that the development of certain autoimmune diseases, such as rheumatoid arthritis and diabetes, may be caused at least in part by excess production of IL-1. The hormone presumably exerts its effects by binding to specific IL-1 receptors on the surface of cells. The binding initiates chemical reactions inside the cells. Researchers speculate that it might be possible to impede the development of such diseases by using chemicals, such as the receptor itself, that can bind to IL-1 so that it is removed from circulation. In 1988, researchers from Immunex Corp. of Seattle reported that they had cloned the gene for the IL-1 receptor and used it to produce small amounts of the receptor. This achievement opens the door to animal studies of the inhibition of IL-1.

In 1986, biologist Steven Clark and his colleagues at Genetics Institute, Inc., in Cambridge announced that they had discovered a third form of interleukin, IL-3. This protein stimulates the production of disease-fighting white blood cells, and Clark predicted that it would prove useful in treating bone marrow failure, in which the body's ability to produce blood cells is destroyed. Researchers at Immunex reported in 1988 that they had cloned the gene for another type of interleukin that stimulates the production of white blood cells that attack tumor cells. Investigators now believe that the interleukins may work best when two or more are used together or when they are used in conjunction with other biological or chemotherapeutic agents.

Factor VIII. When the body is injured, blood clotting occurs to limit the loss of blood. The clot formation itself is the end result of a complex series of enzymatic reactions. Hemophiliacs lack one of the enzymes involved in the clotting process, Factor VIII, and are thus unable to form clots normally. This deficiency places hemophiliacs at high risk of bleeding to death unless they receive supplemental doses of Factor VIII. Drug companies process large quantities of donated human blood to produce preparations enriched in Factor VIII.

Factor VIII processed from donated blood can be contaminated with the hepatitis and AIDS viruses. At least half of the 50,000 hemophiliacs worldwide who receive Factor VIII — 25,000 of them in the U.S. — have contracted hepatitis, and at least 80 percent of them have been exposed to the virus that causes AIDS. Processed Factor VIII is also quite expensive; the average cost of treating a hemophiliac with Factor VIII from human blood is about $10,000 per year, but emergency surgery can easily require $75,000 worth of the enzyme. Because of this high price, an estimated 450,000 individuals worldwide who could benefit from Factor VIII do not now receive it. Robert Carpenter of Genzyme in Boston has estimated that the current market for Factor VIII is about $275 million per year, but that market could increase substantially if the price could be reduced sufficiently to allow people in Third World countries to receive it.

In 1983 and 1984, Genetics Institute and Genentech announced that they had cloned the gene for Factor VIII. Several other companies are also believed to be working on Factor VIII. This enzyme is composed of more than 2,300 amino acids, making it by far the largest protein that has been produced by recombinant DNA techniques. In 1987, Genetics Institute announced that Baxter Travenol Laboratories, Inc., of Deerfield, Illinois, would begin the first clinical trials of Factor VIII. By 1989, several companies had nearly completed their clinical trials and were planning to seek FDA approval for the drug.

Tumor necrosis factor. One of the most intriguing compounds now being produced through biotechnology is a protein called tumor necrosis factor (TNF), which is made by white blood cells called macrophages, and perhaps by the adenoids, lymph glands, spleen, and tonsils. As the name suggests, TNF is effective at killing cancer cells. But unlike interferons, which stimulate the immune system to attack tumors, TNF attacks the tumor cells directly, rupturing their cell membranes so that individual cells die and the tumor shrinks.

Genentech has successfully used TNF against tumors in animals and the company began clinical trials in humans during 1985. Cetus Corp. also started clinical trials of TNF in June 1986. The companies are particularly enthusiastic about TNF because in at least some types of cancers, its effects are synergistic with those of gamma-interferon. That is, when TNF and gamma-interferon are administered simultaneously, their antitumor effect is between 10 and 100 times as great as would be expected if the effects of each were simply added together. Meanwhile, investigators at the City of Hope Research Institute in Duarte, California, and at Asahi Chemical Corp. in Tokyo, Japan, have developed a different technique for producing human TNF and have been conducting their own clinical trials at eight Japanese hospitals. Biotechnology analysts expected TNF to receive FDA approval by 1990.

TNF has a wide range of other effects in the body that may limit its usefulness as a therapeutic agent. For example, the protein has been linked to the profound weight loss (cachexia) associated with cancer and chronic parasitic or bacterial diseases. It has also been implicated in the circulatory collapse and shock associated with acute bacterial infections, the death of animals whose brains have been attacked by malaria, and the development of graft-versus-host disease, in which transplanted bone marrow cells attack the recipient's tissues.

Erythropoietin. An estimated 90,000 people in the U.S. and more than 250,000 worldwide require dialysis of their blood to remove toxic waste products because their kidneys have failed. Most of these people develop some degree of anemia — a loss of oxygen-carrying red blood cells (erythrocytes) — and at least 25 percent become sufficiently anemic to require frequent blood transfusions. They also lack the energy to lead a normal life. Part of the problem is that dialysis accelerates the normal breakdown of the red cells. Perhaps more important is that kidney failure reduces production of the hormone erythropoietin (EPO), which stimulates bone marrow cells to grow into red blood cells.

Transfusion is not an ideal therapy for dialysis-induced anemia because it carries the risk of iron overload, development of an immune reaction to blood products, and exposure to infectious agents such as the viruses responsible for hepatitis and AIDS. Now, however, physicians have a new tool to treat anemia: EPO produced by genetic engineering techniques.

EPO was first manufactured in genetically engineered, cultured mammalian cells in 1983 by molecular biologist Fu-Kuen Lin and his associates at Amgen, Inc., of Thousand Oaks, California. Clinical trials in dialysis patients began in 1985. Results from one such trial were reported in 1987 by nephrologists Joseph W. Eschbach and John W. Adamson of the University of Washington. During the 7-month study, the researchers gave EPO to 25 dialysis patients. They found that the number of red blood cells in the recipients' blood was directly related to the amount of EPO received. The well-being of all the patients improved, and 12 patients who had previously required blood transfusions no longer needed them.

In 1988, EPO was being studied in about 1,000 dialysis patients worldwide. In most cases, the patients had much more energy after receiving EPO and few required blood transfusions. Dialysis patients are not the only ones who could benefit from EPO. An estimated 2 million people in the U.S. have anemia as a result of cancer chemotherapy, surgery, or AZT treatment for AIDS, and experts believe many of these individuals could be helped by EPO. Some preliminary evidence suggests that the hormone would also be useful for treating sickle-cell disease. It might even be used to stimulate blood replacement in patients who have their own blood stored before elective surgery so that it can be used if a transfusion is needed.

During August 1988, Johnson & Johnson Co. of New Brunswick, New Jersey, received approval to market EPO produced by Amgen in Switzerland and France. Amgen received approval to market the compound in the U.S. in June 1989. Genetics Institute also produces EPO, but analysts believe that company is about 1 year behind Amgen in its efforts to market the hormone. Amgen and Genetics Institute have been locked in a patent dispute over EPO. The stakes are high. Like tPA, EPO is expensive: a 1-year course of about 150 treatments costs as much as $9,000 in Europe, but the price is expected to be about $6,000 in the U.S. Analysts predict that the annual sales of EPO could eventually reach $1 billion.

Other genetically engineered proteins on the horizon. By late 1988, more than 80 other drugs produced by genetic engineering techniques were being tested in humans. Among the most important were:

- Atrial natriuretic factor, a hormone produced by the heart. This hormone is one of the most potent drugs known for reducing high blood pressure, and researchers believe it will be extremely valuable for reducing blood pressure and water retention in people suffering from congestive heart failure and hypertension. American Home Products of New York City, California Biotechnology of Menlo Park, California, and SmithKline Beckman all began clinical trials of the drug in 1987.
- Human superoxide dismutase, produced by Bio-Technology General of New York City and Chiron Corp. of Emeryville, California, has been shown in animal experiments at Johns Hopkins Hospital to dramatically reduce damage to tissues deprived of blood flow by blockage of an artery or by interruptions of blood flow during kidney transplants. A crude form of the enzyme isolated from cattle is also used in Europe on a limited basis to treat some forms of arthritis. By 1988, clinical trials of the drug were under way in the U.S. and Europe in heart attack patients, kidney transplant patients, and arthritis victims. Two U.S. hospitals had also begun studying its use to prevent respiratory problems in some premature infants.
- Alpha-1-antitrypsin, which is manufactured by CooperBiomedical, Inc., of Durham, North Carolina, is normally produced in the liver. It protects lung tissue from an excess of elastase, an enzyme the lung secretes to destroy bacteria, smoke particulates, and other foreign matter. If insufficient alpha-l-antitrypsin is present, elastase attacks the lungs, producing emphysema. In 1985, CooperBiomedical began clinical trials of the drug in people with a genetic defect that causes emphysema.
- Protein A, produced by Repligen Corp.of Cambridge, Massachusetts, binds tightly to the proteins in antibodies; it can thus be used to purify monoclonal antibodies and immunoglobulins. Investigators have also speculated that protein A could be used to remove excess antibodies from the blood of people with autoimmune diseases.

- Human renin is produced by California Biotechnology. Renin — not to be confused with rennin, which is used in making cheese — is produced in minute amounts by the kidney and secreted into the bloodstream; it is themost potent peptide known for raising blood pressure. The availability of significant amounts of renin produced by biotechnology techniques may make it possible to develop an inhibitor that would block the substance's hypertensive effects.
- Müllerian inhibiting substance is a protein produced naturally in male embryos to cause tissue that normally grows into female reproductive organs to shrink and disappear. Preliminary results of research conducted at Biogen and Massachusetts General Hospital suggest that the protein can cause shrinkage in tumors of the female reproductive system.

Vaccines Made from Surface Proteins

Before vaccines were routinely used, immunity to a specific disease could be acquired only by exposure to the pathogens that cause the disease. This initial exposure produces what is called an immunological memory, which enables the body to mount an accelerated immune response when exposed to the disease again.

Vaccines are designed to protect the body by stimulating it to produce antibodies against particular diseases. For example, a weakened, noninfectious form of the poliomyelitis virus is used in a vaccine to cause the body to produce antibodies that provide immunity to polio. Similarly, proteins derived from the surface of the hepatitis B virus — collected from the blood of people with persistent hepatitis B infections — may be used in vaccines to protect the body from being infected by the virus itself. The proteins stimulate the production of antibodies that will bind to identical proteins on the surface of invading hepatitis viruses, thereby marking the viruses for destruction.

These vaccines, however, do entail some risk: the polio vaccine may contain the infectious form of the poliomyelitis virus, and the hepatitis B vaccine may contain some intact hepatitis B viruses. One way to get around these problems is to use genetic engineering techniques to produce only the viral proteins in bacteria, so that the risk of viral contamination is eliminated.

Hepatitis B. In August 1981, biochemist William Rutter and his associates at the University of California at San Francisco, in collaboration with geneticists Benjamin Hall and Gustave Ammerer at the University of Washington, reported that they had succeeded in producing a specific surface protein of the hepatitis B virus in yeast cells. Vaccines containing the surface proteins produced in yeast cells contain no intact viruses and thus have no risk of causing hepatitis. In July 1983, Biogen reported that it had used hepatitis B surface proteins to produce immunity against the disease in two chimpanzees.

Further trials in animals were successful, and clinical trials in humans were begun in early 1985 by Merck Sharp & Dohme using a vaccine developed by the San Francisco researchers and Chiron Corp. Those trials showed that a three-dose regimen of the vaccine stimulated the production of protective antibodies in about 90 percent of the people vaccinated — about the same efficacy as the vaccine prepared from the blood of infected individuals. The FDA approved the vaccine, called Recombivax HB, for general use in July 1986. By late 1988, at least four other companies were also testing hepatitis B vaccines produced in this manner.

A genetically engineered hepatitis B vaccine has tremendous potential because hepatitis B is endemic in many parts of the world, particularly in developing countries. The conventional vaccine costs about $100 per immunization, which is far too expensive for use on a large scale. Recombivax HB sold initially for $110 for a three-dose series, but researchers predict that it will eventually be much less expensive.

Malaria. Perhaps the best example of the advantages and the problems of using genetic engineering to produce vaccines is represented by the partially successful efforts to develop a vaccine for malaria. Malaria is one of the most widespread infectious diseases, threatening health in areas inhabited by almost half the world's population. Every year, more than 300 million people develop the disease. In Africa alone, a million children under the age of five die from its effects each year. Malaria has been very difficult to combat because of the complicated life cycle of the *Plasmodium* protozoa that cause the disease.

The *Plasmodium* life cycle has four stages, but two are most important: the sporozoite and merozoite stages. Sporozoites are the form of the protozoan that infected mosquitos inject into the bloodstream. They quickly invade the liver, where they are converted into merozoites, which invade red blood cells and proliferate, eventually rupturing the red cells and releasing more merozoites into the bloodstream. The major symptoms of malaria, which include weakness, periodic fevers, and chills, and often coma, delirium, and death, are associated with this rupturing.

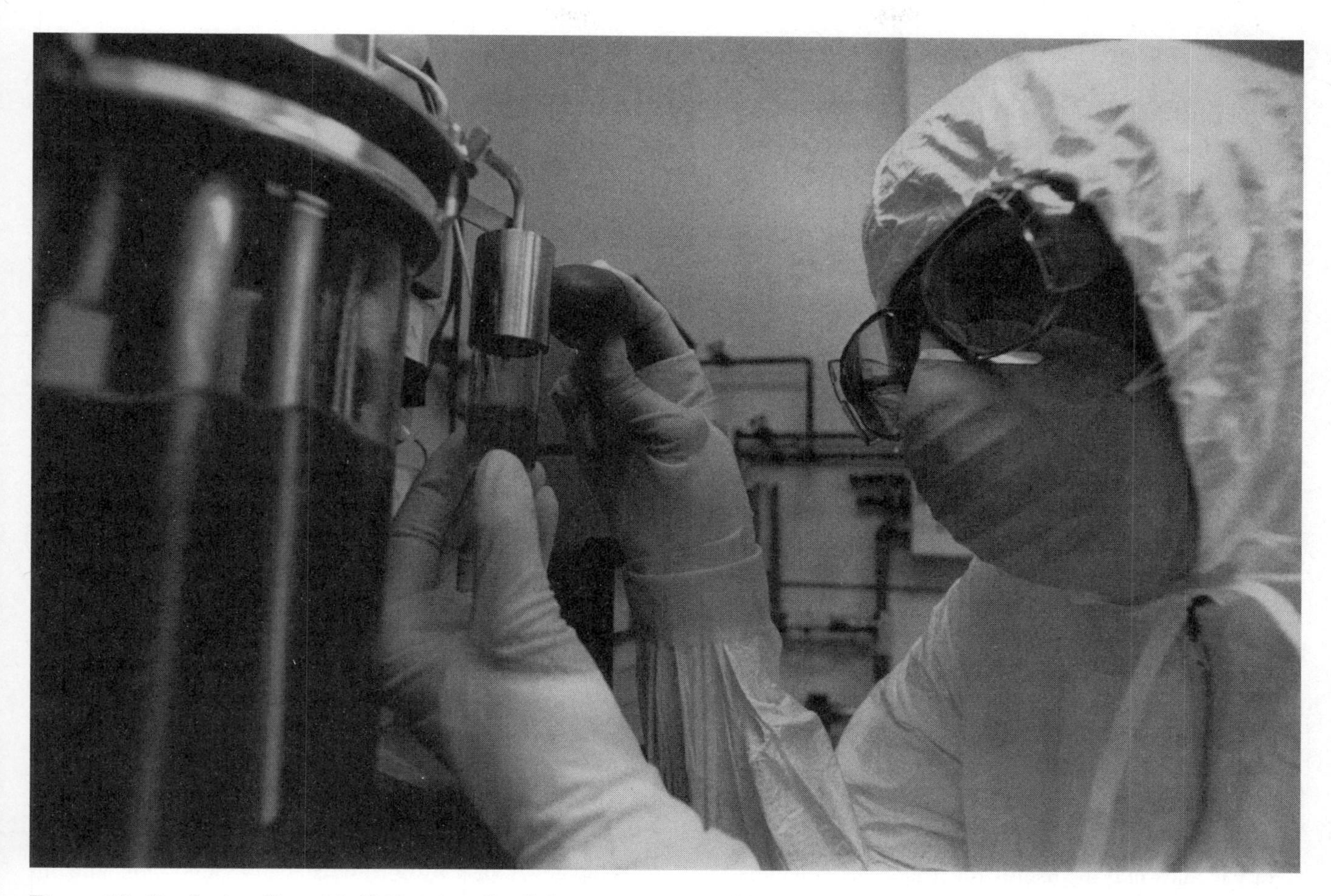

Figure 10. Producing Hepatitis B Vaccine. Small fermentation units like the one shown here were used in the development of Recombivax HB, the first genetically engineered vaccine approved in the U.S. for human use. If the cost of hepatitis B vaccines can be substantially reduced, they could have great potential, particularly in developing nations where the disease is endemic. *Courtesy: Merck Sharp & Dohme.*

Scientists have had great difficulty growing sufficient quantities of either sporozoites or merozoites in the laboratory to produce experimental vaccines. That approach to producing a vaccine may no longer be necessary, however. A team of 11 researchers from five institutions reported in 1985 that they had induced *E. coli* to produce surface proteins of sporozoites — analogous to the surface proteins used in the hepatitis vaccine — from the most important *Plasmodium* species, *P. falciparum*. Injection of these proteins into mice stimulated the production of antibodies that, when isolated, blocked the infection of cultured human liver cells by *P. falciparum* sporozoites.

To prevent malaria, the antibodies stimulated by a vaccine would have to be 100 percent effective in destroying sporozoites. If even one sporozoite were to evade the immune system and reach the liver, it would be converted to the merozoite form and would no longer be vulnerable to attack. Because no genetically engineered vaccine using surface proteins has been shown to be 100 percent effective, investigators believe that a surface-protein vaccine would initially be most valuable in protecting tourists, military personnel, and other visitors to malaria-infested parts of the world. Residents of those areas would probably be best protected by a vaccine that contained both sporozoite and merozoite proteins. The first tests to determine whether the sporozoite vaccine will stimulate production of antibodies in humans were begun in April 1986 at the Walter Reed Army Research Institute in Washington, D.C.

Unfortunately, those tests were not as successful as scientists had hoped. The researchers reported in 1987 that the vaccine protected only one of six volunteers who were immunized with it and then deliberately infected by exposure to mosquitos carrying *Plasmodia*. The Walter Reed team hopes that using different concentrations of surface proteins will increase the vaccine's effectiveness.

Researchers may also be able to synthesize surface proteins without using microorganisms. Biologists have speculated that a sequence of 10 to 20 amino acids may be sufficient to induce immunity when injected into animals. A team headed by biologists Ruth and Victor Nussenzweig of New York University reported in 1985 that they had produced antibodies against sporozoites in rabbits by injecting them with a synthetic DNA fragment (peptide) containing twelve amino acids linked to an inert protein. The peptide mimicked a portion of a sporozoite surface protein. Antibodies from the immunized rabbits prevented sporozoites from infecting cultured cells. Trials of the synthetic vaccine in humans began in the summer of 1986, but their results were also disappointing: Only one of three individuals who developed antibodies to the proteins was protected against infection.

The failure of the vaccines in the human studies may have less to do with the vaccines themselves than with the nature of malaria, according to a report published during 1987. A team headed by immunologist Stephen L. Hoffman of Walter Reed studied 83 inhabitants of western Kenya who had naturally occurring antibodies to *Plasmodium* sporozites similar to those induced by the new vaccines. The team found that despite the presence of antibodies, 60 of the 83 developed malaria during the peak exposures of the malaria season. Hoffman concluded that surface-protein vaccines may be useful for protecting travelers and residents of Central and South America and Southeast Asia, but that other approaches may be necessary to control the disease in Africa, where the degree of exposure is often much higher.

Other diseases. Many investigators fear that the problems encountered in attempting to produce a malaria vaccine may also hinder the development of a vaccine against AIDS. People with AIDS develop antibodies against the human immunodeficiency virus (HIV) that causes the disease, and the presence of these antibodies is how infection is detected. But the antibodies do not seem to provide protection against the disease. Even so, many researchers have been attempting to develop surface-protein vaccines against the disease.

MicroGeneSys produced the first such vaccine to be approved for human trials. In November 1987, the company began tests of a surface-protein vaccine on a group of male homosexuals. It may be many years before the efficacy of the vaccine can be assessed, because the researchers cannot ethically expose volunteers to the virus itself.

Another approach to combating the AIDS virus was announced in 1987 by molecular biologists Alan and Susan Kingsman and their associates at Oxford University and British Bio-Technology Ltd. in Oxford, U.K. The group used genetic engineering techniques to assemble an imitation virus, called a pseudovirus, that has all the HIV surface proteins, but none of the core proteins essential for infection and virus reproduction. The biologists believe that this pseudovirus will provoke a much stronger immune reaction than vaccines based on single proteins, and they hope to begin clinical trials by 1990.

Researchers are attacking many other infectious diseases with surface-protein vaccines. At Genentech, for example, scientists have developed a vaccine that protects against herpes infections in guinea pigs. Herpes is the most common cause of venereal disease in the U.S.; as many as 5 million infections are believed to occur each year. The vaccine consists of antigens from herpes simplex type 1 virus, which mainly causes cold sores in the mouth and on the lips; it also provided protection, however, against infection by herpes simplex type 2 virus, the type most often responsible for genital infections. None of 15 guinea pigs immunized with the vaccine developed symptoms of the disease after intravaginal exposure to herpes type 2, whereas 13 of 14 guinea pigs that received a placebo developed herpes lesions.

A similar vaccine that prevents pyelonephritis, a kidney infection caused by a virulent strain of *E. coli*, in mice has been developed by virologist Gary Schoolnik and his colleagues at the Stanford University Medical Center. Schoolnik prepared the vaccine by inserting the genes that code for proteins on pili — hairlike appendages by which the bacterium attaches itself to cells — of the pathogenic *E. coli* into harmless strains of the same bacterium. The harmless strains then produced large amounts of the proteins, which could be used as a vaccine. Schoolnik began clinical trials at Stanford in 1988. He is also developing a similar vaccine against gonorrhea.

Live Vaccines

For reasons that are not yet clear, vaccination with a live virus provokes a stronger immune response than vaccination with viral proteins or protein fragments alone. Vaccines of this type, such as the polio and influenza vaccines, normally use a weakened form of the virus that does not produce disease. Some researchers have speculated that genetic engineering techniques might be used to modify live viruses to produce more effective vaccines than are possible with surface proteins.

AIDS. In November 1987, the FDA approved the first clinical trial in humans of an AIDS vaccine using a live virus. The vaccine was developed by researchers at Oncogen, Inc., of Seattle and licensed to Bristol-Myers Co. of New York City, which is sponsoring the trials. Oncogen incorporated genes coding for HIV surface proteins into the genome of the vaccinia virus. The vaccinia virus, also known as the cowpox virus, is a safe, easily grown virus now widely used to immunize against smallpox. It does not need to be weakened, because it does not normally produce disease in humans.

When the modified vaccinia virus was injected into rabbits and chimpanzees, it stimulated the formation of significant amounts of antibodies against HIV, as well as against smallpox. The vaccine is being tested in healthy male homosexuals by virologist Lawrence Corey of the Pacific Medical Center in Seattle. In July 1988, Oncogen's president George Todaro said that the first group of 25 volunteers had suffered no ill effects from the vaccine and had developed antibodies to the AIDS virus.

Several other groups are also investigating AIDS vaccines based on the vaccinia virus. Virologist Daniel Zagury of the University of Pierre and Marie Curie in Paris, France, is thought to be taking such an approach, although he has revealed few details of his work. But he disclosed in 1987 that he had tested an experimental vaccine on himself and on several volunteers in Zaire. Zagury has not yet reported any results of the test.

Other diseases. Biologist Richard Young of MIT's Whitehead Institute for Biomedical Research and virologist Dennis Panicali of Applied bioTechnology, Inc., in Cambridge are using a similar approach to develop vaccines against leprosy and tuberculosis.

Leprosy, also known as Hansen's disease, is a stigmatizing chronic skin and nerve disease caused by *Mycobacterium leprae*. The disease afflicts as many as 15 million people, primarily in Asia, Africa, and South America. Victims lose pain sensation in their hands and feet and do not notice when they are injured, so that wounds are not treated properly and easily become infected. Through repeated injuries, the hands and feet of lepers gradually become deformed. The difficulties of producing large quantities of *M. leprae* have hampered the development of an effective conventional leprosy vaccine. Consequently, researchers have turned to genetic engineering.

Tuberculosis, caused by *Mycobacterium tuberculosis*, strikes approximately 10 million people annually, killing 3 million. It is rare in the U.S., but has become more common during the AIDS epidemic, because AIDS reduces the body's immune response. It is one of six illnesses targeted by the World Health Organization for universal childhood vaccination. The current tuberculosis vaccine is based on a harmless bacterium called bacille Calmette-Guérin. This vaccine has become ineffective in India and parts of Asia, where the tuberculosis bacterium has mutated so that it has different proteins on its surface.

Young has isolated the genes that code for surface proteins on both the leprosy and tuberculosis mycobacteria and, with Panicali — who played a key role in the development of the vaccinia technology during the early 1980s — is integrating them into the vaccinia genome. The researchers hope that when the altered vaccinia are injected into animals, and perhaps eventually into humans, the viruses will stimulate immunity to both leprosy and tuberculosis.

Panicali and others have already shown that protein genes from several microorganisms can be engineered into vaccinia viruses. This process can produce so-called multivalent vaccines, which have great potential because they could provide an easy and relatively inexpensive way to immunize large populations against several diseases at once. Vaccinia immunizations could be especially useful in developing countries, because they can be given by untrained health workers with just a scratch of the skin — no syringe is needed.

In December 1985, microbiologist Bruce Stocker of the Stanford University Medical Center disclosed the development of a genetically engineered form of the typhoid bacterium, *Salmonella typhi*, which could potentially protect people against typhoid fever. The World Health Organization estimates that typhoid fever affects as many as 12 million people worldwide each year, particularly on the Indian subcontinent.

Stocker and his colleagues deleted genes from the *S. typhi* bacterium so that it required two critical nutrients unavailable in the human body. The modified bacteria can survive in the body for several days — long enough to stimulate immunity — but without the critical nutrients, they cannot reproduce and cause disease. Initial trials of the vaccine were conducted in 1985 at the University of Maryland School of Medicine in Baltimore. Because typhoid fever is rare in the U.S., however, efficacy trials were later begun in Santiago, Chile.

Genes that code for proteins from other bacteria and viruses might be "piggybacked" on the altered *S. typhi* just as they are on the vaccinia virus. For example, virologist John Clements of the Tulane University School of Medicine in New Orleans, Louisiana, has incorporated genes from the bacteria that produce the most common form of traveler's diarrhea. Microbiologist Derrick Rowley and his colleagues at the University of Adelaide in Australia have incorporated cholera genes in *S. typhi*. Other researchers are attempting to piggyback genes that would provide protection against malaria and schistosomiasis, a parasitic disease for which there is currently no preventive agent.

DNA Probes

The growth of genetic engineering has led to the development of a new way to identify infectious microorganisms, genetic predisposition to disease, and even criminals who have left behind biological "fingerprints" in their blood or semen. The tests are based on the use of DNA probes, short pieces of DNA that have been radioactively labeled. A probe for an infectious microorganism, such as *Salmonella*, would be composed of a characteristic segment of DNA from the *Salmonella* genome. When this probe is added to DNA from a food sample, for example, it will bind only to complementary DNA from *Salmonella* in the food, forming double-stranded DNA that can be separated and identified. If no *Salmonella* is present, no double-stranded DNA will be formed, and the test will be negative.

DNA probes are already being used in screening for genetic diseases in fetuses. According to the consulting firm Robert S. First, Inc. of White Plains, New York, the market for prenatal genetic screening could total $48 million by 1990. Other analysts estimate that the total could reach $1 billion per year by the end of the century. The use of DNA probes and monoclonal antibodies in diagnostic tests may represent the largest initial market for biotechnology. Martin Nash of Molecular Biosystems, Inc. in San Diego, California, has calculated that the annual sales of DNA probes for detecting infectious microorganisms could total $500 million by 1990.

Disease diagnosis. The most important application of DNA probes could be diagnosis of infectious diseases and identification of pathogenic organisms, processes that are now most often conducted by culturing microorganisms so that they can be identified visually. This market is already estimated at $950 million per year, and biotechnology-derived tests could capture a sizable fraction because of their high specificity for particular microorganisms and the relative speed with which they yield results. One typical use is screening for *Salmonella* bacteria, which cause severe diarrhea that may last 3 to 5 days. According to Renee Fitts of Gene-Trak Systems, U.S. food companies perform at least 5 million *Salmonella* tests per year, and at least three times that many are performed worldwide.

A typical microbiological assay for *Salmonella* requires 5 to 7 days to complete. But according to Fitts, Gene-Trak Systems has developed DNA probes that can detect all 352 of the most common *Salmonella* strains in less than 48 hours. By the end of 1988, the probes were being used widely by U.S. food companies. Gene-Trak Systems is developing similar tests for *E. coli* and for *Campylobacter* pathogens. *E. coli* is not normally pathogenic itself, but its presence in food is normally taken to be a sign of contamination by animal feces. *Campylobacter* cause diarrhea and are thought to be as prevalent in some foods as *Salmonella*.

The speed with which DNA probes can identify microorganisms makes them exceptionally valuable for diagnosing diseases. A conventional assay for *Mycobacterium tuberculosis*, for example, may require as long as 8 weeks. A physician who suspects his patient has tuberculosis must thus begin administering drugs for the disease before the diagnosis is confirmed, thereby exposing some potentially uninfected individuals to the side effects of the powerful antibiotics now used for the disease. But Gen-Probe, Inc., of San Diego makes a DNA probe test kit that can confirm the diagnosis in just 2 hours, allowing the drugs to be given only to patients who have the disease. Other diseases for which probes are available include *Legionella*; several microorganisms that cause pneumonia; herpes types 1 and 2; hepatitis B; adenoviruses, which cause respiratory infections; and *Chlamydia trachomatis*, a bacterium that causes a chronic eye infection and a sexually transmitted disease that can produce sterility.

Genetic fingerprinting. When a 57-year-old Alzheimer's disease victim came home from day-care at a Tacoma, Washington hospital, her disheveled condition made her daughters suspect that she had been raped. Their fears were confirmed when a medical examination of the woman, who remembered nothing, showed semen. Law enforcement officials immediately had a suspect: Alan J. Haynes, the driver of the day-care center's van. Although Haynes was the only man who had been alone with the woman, there were no witnesses or fingerprints to connect him to the crime.

Authorities sent a semen sample taken from the woman and a blood sample from Haynes to Lifecodes, Inc., of Tarrytown, New York, for a radical new type of analysis called genetic fingerprinting. Using DNA probes that bind to specific regions of human DNA, forensic scientists at Lifecodes compared DNA in the sperm to DNA in Haynes's blood and found them to be identical. Confronted with the evidence, Haynes pleaded guilty in December 1987.

Haynes's case is one of only a handful in the U.S. in which genetic fingerprinting, developed in 1985, has been used to prosecute a suspect. But the number of cases is increasing steadily, and experts believe that the technique's unique capability to identify a perpetrator with virtual certainty will ensure its wide use in the future. The technique is a significant advance over fingerprinting or matching blood proteins. Good fingerprints are often unavailable at the scene of a crime, and blood typing can only exonerate a suspect or show that he or she may have committed a crime.

Genetic fingerprinting was developed in 1985 by geneticist Alec Jeffreys and his colleagues at the University of Leicester in the U.K. Jeffreys found that human DNA contains many short regions that are unique in composition for every individual except identical twins. He developed a set of DNA probes that could bind to these unique regions after a sample of a person's DNA has been broken into short pieces. When the combined DNA fragments and probes are separated by a technique known as electrophoresis, they form a unique pattern that serves as a genetic "fingerprint."

The electrophoretic pattern from blood or semen at the scene of a crime can then be compared to the pattern derived from a suspect's blood. The odds that two such genetic fingerprints will match range from one in 200,000 to one in 30 billion, depending on the number of probes used. In comparison, the odds that conventional fingerprints from two individuals will match are about one in 64 billion. But a suspect can always argue that fingerprints were left under noncriminal circumstances. Genetic fingerprinting can also be used to prove maternity or paternity in civil cases. Genetic fingerprinting takes about 2 weeks and costs between $110 and $300, depending on the size and nature of the sample.

The first uses of genetic fingerprinting were in immigration cases in the U.K. One such case involved a Ghanaian boy who was born in the U.K. and then immigrated to Ghana to live with his father. When the boy later attempted to return to the U.K. to rejoin his mother, British authorities denied him permission, arguing that he was either unrelated to the mother or was the son of one of her sisters. Jeffreys was able to prove by genetic fingerprinting that the boy was, in fact, the mother's son, and he was granted residence. Cellmark Diagnostics of Abingdon, near Oxford, currently performs about 250 such tests per week for criminal cases and civil suits.

Genetic fingerprinting has also been used in about 20 criminal cases in the U.K. The most famous case involved the rape-murder of two teenaged girls near the village of Enderby. Police used genetic fingerprinting to exonerate

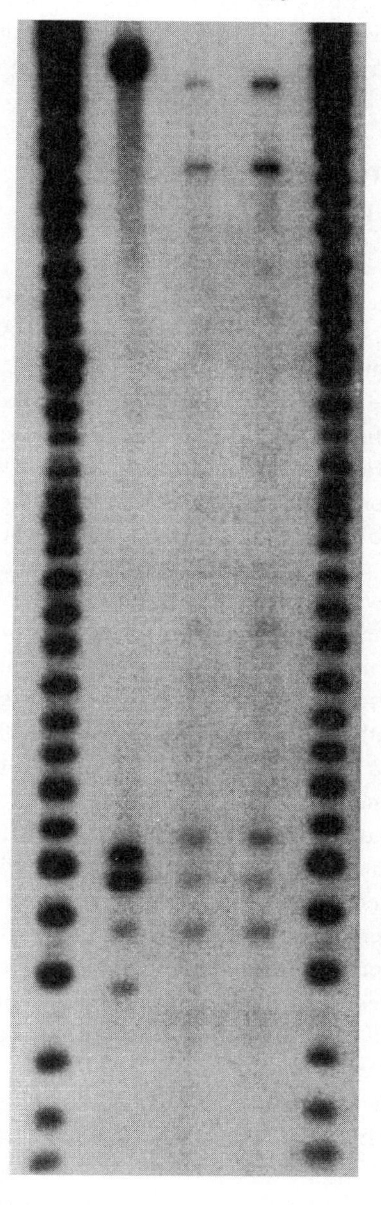

Figure 11. Genetic Fingerprinting. Certain DNA probes can be used with a sample of genetic material to produce a unique pattern — a "genetic fingerprint" — on an electrophoretic gel matrix. By comparing the electrophoretic pattern derived from a sample of the suspect's blood, it is possible to determine with near certainty whether the samples come from the same person. In this photograph, the genetic fingerprint obtained from the evidence matches that of the suspect. Note that the matched patterns are much different from the victim's pattern. *Courtesy: Lifecodes Corp.*

a youth who had been charged with one of the murders and asked 4,000 local men to voluntarily submit blood samples for testing. Police arrested a 27-year-old baker, Colin Pitchfork, when he sent a friend, Ian Roy Kelly, 23, to provide a blood sample for him. Pitchfork and Kelly were both charged with obstructing justice in the case, and in late 1988 Pitchfork was awaiting trial for the murders. His blood and semen have been tested, but British police may not legally release the results of the tests before a trial.

Genetic Engineering in Agriculture

The application of genetic engineering techniques to agricultural production is on the verge of becoming an immense business. *Genetic Technology News* has estimated, for example, that the potential annual market for microorganisms capable of converting atmospheric nitrogen into a form usable by plants is nearly $1.2 billion in the U.S. alone; if this characteristic could be engineered directly into crops, the annual U.S. market for the seeds of such plants would be $4.3 billion, while the world market would be two to three times as large.

At the same time, however, the release of genetically engineered microorganisms into the environment has been a highly controversial issue, initially focusing on research by plant pathologist Steven Lindow of the University of California at Berkeley and researchers at Advanced Genetic Sciences, Inc., of Oakland, who designed genetically engineered microorganisms to prevent frost damage to crops. The formation of frost on plants at near-freezing temperatures is often triggered by proteins secreted by bacteria, such as *Pseudomonas syringae*, that live on the surface of plants. When these bacteria are absent, plants can survive temperatures as low as 25°F for several hours. Frost is estimated to cause as much as $1.6 billion in crop damage in the U.S. each year.

Lindow and AGS used genetic engineering techniques to remove from *P. syringae* the gene that codes for the protein that triggers ice formation. They hoped that this mutant bacteria would grow on the leaves of plants and displace the naturally occurring bacteria, thereby retarding frost formation. After years of delay, Lindow was finally able to test the altered bacteria on potatoes in 1987 at the University of California's agricultural field station at Tulelake in northern California. AGS tested their bacteria at the same time on strawberries in Brentwood, California, a San Francisco suburb. Those initial tests and followup tests the next year demonstrated that the altered bacteria reduced frost damage and that they did not escape from the test plots.

Those and other successful open-air experiments with altered microorganisms eventually eased public fears about the release of genetically engineered organisms. In May 1988, a long-awaited report by the U.S. Office of Technology Assessment (OTA), *Field-Testing Engineered Organisms: Genetic and Ecological Issues*, concluded that such small-scale releases posed little threat: "With the appropriate regulatory oversight, the field tests and introductions planned or probable in the near future are not likely to result in serious ecological problems."

Caterpillar killer. Several research groups have been developing genetically engineered microbes designed to protect plants from certain types of cutworms, including corn earworm, tobacco hornworm, soybean podworm, and black cutworm. Each year, farmers use as much as $400 million worth of pesticides — some of them among the most toxic used in agriculture — to fight such pests. To minimize the need for pesticides, researchers from Crop Genetics International of Hanover, Maryland, inserted a gene from *Bacillus thuringiensis*, which has been used for more than 20 years to control pests on certain types of agricultural crops, into a bacterium, called Cxc endophyte, that normally lives inside bermudagrass. The gene codes for the production of a toxic protein called Bt toxin. Laboratory and greenhouse tests showed that if the endophyte was inoculated into corn seed, it would multiply and carry the toxin throughout the corn plant's stalks, leaves, and roots. When the toxin is in plant tissues, it can kill such pests as the European corn borer that feed on the plant. The borer infests most of the 67 million acres of corn in the U.S. and is a serious economic problem on at least 40 million of those acres. American farmers spend $50 million on chemical insecticides each year to control the borer. Crop Genetics estimates that the pest still reduces crop income by $400 million annually.

In May 1988, EPA approved field tests in Maryland of the engineered pesticide, which Crop Genetics calls InCide. The tests were successful and Crop Genetics hopes to have the new pesticide on the market by 1990. The company is developing similar pesticides for use on other crops, including wheat, rice, soybeans, and cotton.

Fertilizing alfalfa. Although molecular biologists are still far from achieving the elusive goal of giving plants the ability to fix nitrogen from the air, scientists at Biotechnica International in Cambridge, Massachusetts, have made an important step in that direction: They have engineered a naturally occurring nitrogen-fixing microbe to make it fix more nitrogen for the plants it lives on.

The common soil bacterium *Rhizobium meliloti* forms nodules on the roots of leguminous plants such as alfalfa and converts atmospheric nitrogen into nitrates that can be used directly by the plants. For many years, farmers have applied this microorganism to alfalfa seeds to increase the crop's yield. About 80 percent of the U.S. alfalfa crop is now treated in this way, according to Biotechnica vice president David Glass. The enzyme nitrogenase is essential for nitrogen fixation in the bacterium. Biotechnica researchers cloned the gene for nitrogenase and inserted extra copies of the gene into *R. meliloti*.

Tests in greenhouses demonstrated that the altered *R. meliloti* could increase the yield of alfalfa by as much as 17 percent. In 1988, the company successfully tested the engineered bacterium in Pepin County, Wisconsin. Biotechnica said the engineered microorganism could be marketed by 1991. U.S. farmers currently plant about 25 million acres of alfalfa worth an estimated $7 billion. A 15 percent increase in output would add another $1 billion to the alfalfa's value.

Genetic engineering in plants. In May 1986, researchers at Agracetus of Middleton, Wisconsin, planted 200 genetically engineered tobacco seedlings in a small plot just outside Middleton. The experiment marked the first time that a genetically modified plant had been grown outside the laboratory and only the second time that a genetically engineered organism had been released into the environment. In the summer of 1986, researchers at Calgene, Inc., of Davis, California, planted a different type of genetically engineered tobacco plant in a field near their research laboratory.

By the spring of 1988, researchers in five countries had conducted about two dozen field tests of genetically engineered plants and many others were planned, according to a May 1988 OTA report. To date, plant biotechnology research has focused largely on two main goals: giving crops the ability to survive in the presence of the herbicides used to control weeds, and endowing crops with the ability to resist insects, viruses, and fungi. *Genetic Technology News* projected that these traits could increase the U.S. yield of vegetables alone to about $6.5 billion per year from the current level of $5.5 billion.

Agracetus scientists developed tobacco plants resistant to crown gall disease by inserting a yeast gene called alcohol dehydrogenase, which disrupts the pathway by which the bacteria infect tobacco cells. The modified tobacco plants were tested in greenhouses for 2 years (six generations) before field-testing began in May 1986. The Agracetus team found that the altered plants "did not show any significant differences from unmodified plants other than an increased resistance to disease." The research did not have any commercial applications because crown gall disease is not a problem for tobacco growers. Instead, Agracetus scientists have been using resistance to the bacterium as a tool to learn more about how foreign genes are integrated into plant cells.

Other researchers have been using genetic engineering to develop plants resistant to diseases caused by viruses. Surprisingly, some experiments have shown that plants can be given a sort of artificial immunity to viral infection similar to that induced in humans, even though plants do not have an immune system. Viral diseases in plants are a major agricultural problem. According to *Genetic Engineering News*, these diseases cause between $1.5 billion and $2 billion in crop damage in the U.S. every year.

Biologist Roger Beachey and his colleagues at Washington University and Monsanto Co., both in St. Louis, Missouri, have used genetic engineering techniques that take advantage of this phenomenon. They inserted a gene that codes for a surface protein of the tobacco mosaic virus (which causes disease in many commercial crops) into cells from tobacco and tomato plants. Beachey's group found that intact plants grown from the altered cells produced the viral surface protein coded for by the transferred gene and were resistant to infection by the virus.

After completing small-scale studies in a research plot near St. Louis, the researchers conducted a much larger field study in the summer of 1987 in Jersey County, Illinois. Both the genetically engineered tomatoes and ordinary tomatoes planted at the same site were deliberately inoculated with the tobacco mosaic virus. In 1988, the team reported that the engineered plants were protected against three different strains of tomato mosaic virus, and the conventional plants produced 26 to 35 percent less fruit than those that were genetically engineered.

Perhaps the greatest amount of research on plant genetics in the mid-1980s has been devoted to developing plants resistant to herbicides. The world's major chemical companies, whose global herbicide sales total more than $4 billion per year, hope to increase, or least maintain, their shares of the herbicide market by developing seeds for plants that will tolerate their brands of weed killers.

In 1985, Luca Comai and his colleagues at Calgene reported that they had achieved the first successful expression of a commercially significant gene inserted in a major crop species. The gene provides resistance to the herbicide glyphosate, which is marketed as Roundup by Monsanto and accounts for one-quarter of the world herbicide market. Glyphosate is not toxic to mammals and degrades to harmless substances fairly quickly.

Glyphosate kills plants by inhibiting an enzyme called EPSP synthase, which plays a crucial role in plant growth. Most weeds have the enzyme and are therefore susceptible to the herbicide; unfortunately, many crop plants, and vegetables in particular, also have the enzyme. Comai and his colleagues found that the bacterium *Salmonella typhimurium* has a slightly different EPSP synthase that is inhibited to a lesser extent by glyphosate; they modified the gene for this enzyme so that the enzyme was even less sensitive to glyphosate, then inserted the gene into tobacco plants. They found that the plants displayed a significantly higher tolerance to glyphosate than ordinary tobacco. Field trials in Yolo County, California, in August 1987 and in Gila Bend, Arizona, in January 1988 confirmed this result.

Monsanto researchers have also inserted the EPSP synthase gene into canola, a nutritionally superior form of rapeseed developed by Canadian researchers. Rapeseed oil is used in margarines, shortening, and cooking and salad oils. In May 1988, researchers from Monsanto and the Alberta Wheat Pool, which represents cooperatives of wheat producers, began two separate field tests of the genetically engineered canola in west-central Saskatchewan and Wheatland County, Alberta.

Other chemical companies have been working to develop plant strains resistant to their own brands of herbicide. Ciba-Geigy Corp., for instance, is developing soybeans resistant to atrazine, the most widely used herbicide in the U.S. for corn. Although corn is naturally resistant to atrazine, soybeans — which are frequently grown in rotation with corn — are not. This can cause problems, because atrazine used on a corn crop during one season can persist in the soil and kill soybeans planted during the following season.

A major research effort has also been devoted to developing plants that are resistant to insect pests. Farmers now spend more than $3 billion per year worldwide for pesticides — with over $400 million spent in the U.S. alone for control of pests such as the cotton bollworm, the tomato hornworm, and the tobacco budworm. Plants engineered to resist insects could save farmers money and reduce the environmental damage associated with pesticide use.

In January 1985, Marc Zabeau and his colleagues at PGS in Brussels announced that they had transferred the gene for Bt toxin into tobacco plants. Rohm and Haas of Philadelphia, Pennsylvania, who sponsored PGS's research, conducted field trials of the insect-resistant tobacco in Mississippi and Florida which showed that the altered plants were not attacked by caterpillar pests. Similarly, molecular biologist David Fischoff and his colleagues at Monsanto have inserted the Bt gene into tomatoes to protect the plants from hornworm larvae. They found both the first generation of the plants and their offspring produced the Bt toxin, which provided protection from hornworm larvae.

Genetic engineering in animals. To date, the primary application of recombinant DNA techniques to animal husbandry has been the use of microorganisms as miniature factories for the production of drugs, hormones, and other chemicals that can help animals grow and protect them from disease. *Genetic Technology News* has estimated that sales of such chemicals could total $400 million per year in the U.S. by the end of the century. An even greater potential market exists for the application of genetic engineering techniques to the animals themselves.

Geneticists are particularly interested in learning how to insert growth hormones into animal genes to stimulate growth and increase milk production. During the early 1980s, for example, biologists Ralph Brinster of the University of Pennsylvania School of Medicine in Philadelphia and Richard Palmiter of the University of Washington in Seattle produced mice that were 50 percent larger than normal by giving mouse embryos the gene for a human protein called growth hormone releasing factor, which stimulates the release of growth hormone. Brinster and Palmiter found that these transgenic mice were able to reproduce naturally and pass the gene to their progeny.

Fish are good candidates for genetic engineering because they are a major food source in many nations around the world. Furthermore, researchers can work with fish eggs more easily because they are larger than those of livestock. U.S. scientists revealed in June 1988 that they had transferred a growth hormone gene from rainbow trout into common carp, obtaining fish that grow about 20 percent faster than normal. Biologist Dennis A. Powers and molecular biologist Thomas T. Chen of Johns Hopkins University in Baltimore and geneticist Rex Dunham of Auburn University in Auburn, Alabama, injected the gene into 10,000 eggs, obtaining 20 transgenic carp. The scientists ate one of the transgenic carp and found that the inserted gene did not alter the taste or quality of the flesh.

Genetically engineered veterinary products. Biotechnology companies have already begun marketing animal

Figure 12. **"Supermouse."** The so-called supermouse on the left is the product of a gene transfer from a rat to a mouse. The supermouse's cells contain a new gene that codes for a rat hormone fused to a mouse gene that accelerates the transcription of the gene adjacent to it. Although the two male mice pictured are from the same litter, the supermouse weighs about 1.5 times as much as his smaller, normal brother. According to University of Washington researchers who conducted the experiment, the altered gene is sometimes passed on to the supermice's offspring, which grow larger than normal mice. *Courtesy: R. L. Brinster, University of Pennsylvania. Reprinted by permission from* Nature, *Vol. 300. Copyright © 1982 Macmillan Magazines Ltd.*

health products because that is the easiest agricultural market to enter. Drugs and vaccines for animals typically require much less testing before approval than do genetically engineered plants and animals. Perhaps even more important, the microorganisms now used to produce such products are much easier to modify than plants and animals.

The market for animal health products is already huge. According to the Animal Health Institute in Washington, D.C., U.S. sales of animal health products reached $2.1 billion in 1986. That total included $233 million for antibiotics; $611 million for such products as hormones, antiseptics, antiparasitic drugs and vitamin and mineral supplements; and $193 million for biologicals, which include vaccines, toxoids, immune sera, and blood products. Business Trends Analysts, a consulting group in Commack, New York, estimates that the use of such products now increases livestock production by 4 to 32 percent and saves farmers $3 for every dollar spent. Business Trends projects that sales will grow at least 5 percent per year through the rest of the century.

Pets represent another large market. Although data on drug sales to pet owners are not available, the American Veterinary Medical Association in Washington estimates that Americans spent a total of nearly $5 billion in 1985 on veterinary services, including $3 billion for dogs and $1.1 billion for cats. To date, biotechnology companies have developed only a few products for pets — a vaccine against feline leukemia and several diagnostic tests. Many companies are now working, however, to develop other pet products and obtain a share of this market.

In January 1986, the USDA gave Biologics Corp. of Omaha, Nebraska, permission to market a pseudorabies vaccine based on a genetically engineered virus. Pseudorabies, occasionally called "the mad itch," is a severe herpesvirus infection characterized by intense skin irritation and sores. In genetically susceptible piglets, the disease can cause death within 48 hours. The virus can also spread to sheep, cattle, and household pets, but not to humans. Since the mid-1970s, the incidence of the disease has been increasing rapidly. USDA veterinarian Leroy Schnurrenberger has estimated that in 1974, only one pig in 200 in the U.S. showed evidence of exposure to the pseudorabies virus. By 1984, that figure had grown to one in ten. Pseudorabies virus is estimated to cost the U.S. pork industry as much as $60 million per year in lost or weakened animals.

Biologics licensed the genetically engineered pseudorabies vaccine from virologist Saul Kit of Baylor University in Waco, Texas. Kit produced the vaccine by identifying the gene that enabled the herpesvirus to spread among livestock and using genetic engineering techniques to remove that gene from the virus. When the altered virus is injected into a piglet, it invades the central nervous system like a normal herpesvirus, but it does not produce pseudorabies and it cannot escape the nervous system to infect other animals. Furthermore, once the engineered virus is ensconced in the pig's nervous system, the virulent pseudorabies virus cannot infect the animal. In January and May 1988, USDA licensed two other pseudorabies vaccines manufactured by genetic engineering. The vaccines are produced by the Upjohn Co. of Kalamazoo, Michigan, and by Syntro Corp. of San Diego, California.

Molecular biologist Hilary Koprowski of the Wistar Institute in Philadelphia, Pennsylvania, has produced a rabies vaccine by inserting the gene that codes for a large rabies virus protein into the vaccinia virus. Preliminary tests in the laboratory demonstrated that the vaccine stimulated immunity to the rabies virus without causing ill effects.

In the summer of 1988, Wistar was field-testing its rabies vaccine on wild foxes in Belgium and had applied to USDA for permission to test it on raccoons living on uninhabited islands off the Atlantic coast of the U.S. Wistar planned to soak small sponges with the vaccine, seal them with wax, and coat them with a bait attractive to wild raccoons, which are major carriers of rabies in the U.S. Drawn by the bait, raccoons would bite the sponges and absorb the vaccine. The islands being considered for the tests were Parramore Island, Virginia, and North, Cedar, and Murphy Islands off the coast of South Carolina. The researchers hope that immunization of the raccoons will help stop the natural spread of rabies, a disease that kills more than 20,000 people worldwide each year.

Several companies have developed vaccines to combat colibacillosis, more commonly known as scours, which weakens or kills millions of piglets and calves, as well as smaller numbers of other farm animals, each year. The disease is a form of diarrhea that sometimes leads to fatal dehydration. It is caused by at least three different pathogenic strains of the normally harmless bacterium *E. coli*. Cetus Corp. of Emeryville, California, has estimated that the annual U.S. market for scours vaccines is between $20 million and $40 million. In their efforts to produce a scours vaccine, researchers have focused on two specific aspects of the pathogenic *E. coli*: the hairlike pili by which the bacteria attach themselves to cells lining the intestines of animals, and a toxic bacterial protein that injures cells lining an animal's intestine, thereby producing diarrhea.

Researchers at Cetus isolated from one *E. coli* strain the gene that codes for the toxic protein and then modified the gene to make it code for a harmless protein that is structurally very similar to the toxic protein. When injected into piglets, this altered protein stimulates the animals' immune systems to produce antibodies that bind to and destroy the toxic protein. Cetus researchers also isolated the gene that codes for one pilus protein, called adhesin, that helps the bacterium bind to intestinal cells. Antibodies against this protein prevent the bacterium from binding to the intestinal cells. The researchers then inserted the genes for adhesin and the altered protein into the DNA of a harmless strain of *E. coli*, where the two proteins were produced in substantial quantities.

At the same time, researchers at Norden Laboratories — which was cooperating with Cetus — used the same technique to manufacture two other proteins found in the pili. The four proteins were then combined into a single vaccine that was injected into pregnant sows. The proteins stimulated immunity to the pathogenic *E. coli* in the sows, and the sows passed on that immunity through milk to nursing piglets. In April 1983, Norden began marketing the vaccine in the U.S. under the name LitterGuard.

Researchers at Salsbury Laboratories in Charles City, Iowa, have developed a scours vaccine by adopting a different approach. They isolated the genes that code for pilus proteins in all three scours-causing strains of *E. coli*. The researchers then inserted multiple copies of each of these genes into the DNA of one strain of pathogenic *E. coli* so that the bacteria produced unusually large amounts of the proteins. The bacteria were grown in culture and then subjected to a mechanical process that sheared off their pili. The pili could then be collected, purified, and injected into piglets to stimulate immunity against all types of scours-causing bacteria. In May 1983, Salsbury began marketing such a vaccine for pigs under the name Ecobac. By mid-1988, the company was also conducting trials of related vaccines to prevent scours in sheep and other farm animals.

Animal growth hormones represent the largest potential market for veterinary products produced by recombinant DNA techniques, and perhaps the largest market for any genetically engineered agricultural product. According to *Genetic Engineering News*, the U.S. market for such hormones could eventually total 200,000 pounds per year with a value of $900 million.

Today, the greatest amount of growth hormone research has been devoted to bovine growth hormone (bGH). At least six biotechnology companies have isolated the gene for bGH and used genetic engineering techniques to make bacteria that produce the protein. As of early 1989, these companies were testing the bacterially produced hormone in cattle. The first use of bGH has not been to increase the size of the cattle, but instead to increase cow's milk production. In one 1985 study, for example, biochemist Dale Bauman of Cornell University in Ithaca, New York, obtained a 40 percent increase in milk production during a 6 month period in which cows were injected with 40.4 milligrams of bGH per day. Studies by other researchers have shown increases of 20 to 30 percent. Several companies hope to begin selling bGH sometime in 1990.

This potential use of bGH has been harshly criticized, particularly by owners of small dairy farms, because it could produce a dramatic restructuring of the U.S. dairy industry. The farmers' chief concern is that U.S. dairies already produce about 10 percent more milk than consumers purchase. Largely as a result of falling demand for dairy products and increasing productivity per cow, the number of dairy farms dropped from 1.1 million to about 286,000 between 1965 and 1985. Critics of bGH use argue that the number of dairy farms will decrease an additional 25 percent by the year 2000 if bGH becomes widely used. They believe that most of this decrease will occur among small dairy farmers with 100 to 300 cows. Critics claim that these farmers will not be able to afford bGH, which is expected to cost between 40 and 60 cents per dose, and thus will be unable to compete with larger farms having 2,000 to 3,000 animals.

An October 1987 analysis by the USDA, however, suggested that the fears of critics may be groundless. The study found that bGH could be sold at commercially attractive prices. The report also predicted that bGH would be available to all dairy farmers and that its use would require little additional capital and few operational changes. "Thus, bGH use should reinforce — but not fundamentally change — structural trends already under way," according to the USDA report.

Pet products represent a major market that the biotechnology industry is just beginning to tap. The 52 million dogs and 56 million cats in the U.S. are linked to their owners by strong emotional ties, and those owners are much more likely than farmers to purchase expensive health-care products. The first genetically engineered product for pets is a vaccine to prevent feline leukemia and related diseases. Norden Laboratories began marketing the vaccine in 1985.

The feline leukemia virus, first discovered in 1974, is responsible for the deaths of about 1 million cats in the U.S. each year. It is transmitted among cats through saliva and possibly through urine and mother's milk. An infected cat has an unusually large number of white blood cells, and its immune system is suppressed. This immune suppression is similar to that observed in humans with acquired immunodeficiency syndrome (AIDS). In fact, this resemblance helped lead to the identification of the human T-cell virus that causes AIDS.

Several researchers first tried to make an antileukemia vaccine by using the whole feline leukemia virus, either dead or in a weakened form. In each case, however, the vaccine suppressed the animals' immune systems; although the cats did not develop leukemia, they became highly susceptible to many other diseases. In the early 1980s, virologist Richard Olsen of Ohio State University in Columbus used genetic engineering techniques to produce bacteria that make a membrane protein of the virus. This protein makes cats immune to the feline leukemia virus without suppressing the immune system. In 1985, Norden began U.S. marketing of a vaccine, called Leukocell, composed of this protein.

Gene Therapy

The practice of inserting animal genes into bacteria and then harvesting the proteins synthesized from those genes is the basis for much of the biotechnology industry. Inserting foreign genes into humans, however, is a much more difficult task and carries far-reaching medical and ethical implications. The ability to genetically engineer animals is already leading to the creation of laboratory animals with diseases that mimic human diseases, thereby providing ways to test new drugs and therapies. Researchers believe it will ultimately be possible to insert new genes into people to correct genetic defects.

Curing genetic defects in animals. By 1989, many gene transfers between species had been carried out in laboratories throughout the world. In 1985, for example, biologist David Baltimore of the Massachusetts Institute of Technology (MIT) in Cambridge altered a mouse gene that codes for antibodies so that he could readily identify the antibodies produced by the gene. He then inserted the altered gene into embryos of healthy mice. Baltimore found that the gene was present in all cells of the so-called transgenic mice he had created and that the gene was expressed only in white blood cells where antibodies are normally produced. If the antibodies had been produced in other cells, they could have caused disastrous effects. He is now attempting to determine why a gene is preferentially expressed in particular tissues, information that could reveal how cells diversify.

Biochemists Kiran Chada and Frank Constantini of Columbia University in New York City have found that they can use gene transfers to cure a genetic defect in mice that is similar to beta-thalassemia in humans. Beta-thalassemia is a form of anemia caused by a defect in the gene coding for the protein beta-globin, which is part of the oxygen-carrying hemoglobin in red blood cells. The defective gene causes defective hemoglobin to be produced, and oxygen transport through the body is impaired. About 100,000 new cases of the disease are reported worldwide each year, primarily in Mediterranean nations. In 1986, Chada and Constantini reported that when they inserted a gene coding for human beta-globin into embryos from mice with the defect, the gene produced functional beta-globin in many of the recipient mice and their descendants. The beta-globin was produced only in the appropriate location — red blood cells.

Biochemist W. French Anderson of the National Heart, Lung, and Blood Institute (NHLBI) in Bethesda, Maryland, has introduced a human gene coding for the enzyme adenosine deaminase (ADA) into seven rhesus monkeys. A genetic deficiency of ADA causes a disorder known as severe combined immune deficiency (SCID), in which a person's immune system is unable to fight infections. Anderson reported in 1987 that five of the seven monkeys produced small amounts of human ADA after the gene transfer. Unfortunately, the quantity of ADA produced was only about 10 percent of the amount that would be needed to cure SCID in humans.

A team of researchers headed by biologist Leroy Hood of the California Institute of Technology in Pasadena has used genetic engineering techniques to cure mice of an inherited neurological disease that causes them to shiver uncontrollably and die prematurely. The disease, called shiverer mutation, involves a deficiency of myelin basic protein, which accounts for about 30 percent of the insulating material that sheathes nerves in both humans and mice. The mouse disease has no exact human counterpart, but at least 40 human disorders, including multiple sclerosis, involve defects in nerve sheaths. The shiverer mice appear normal at birth, but within 2 weeks, their hindquarters begin shivering and they walk with a rolling gait. By the age of 1 month, they start having convulsions, and they generally die by the age of 3 to 4 months.

Hood and his colleagues isolated the gene for myelin basic protein, along with the genes that control its expression, and injected copies of it into 350 fertilized eggs from female shiverer mice. The eggs were then implanted in females and grown to maturity. Although the defect was corrected in just one of the offspring, two generations of mice have been produced from the cured mouse, and they are also cured, indicating that the gene has been incorporated into their DNA. "The important thing about this experiment is that the inserted gene is expressed at the right time during development and in the right location — the brain," Hood said.

In an experiment closer to the conditions that might be encountered in human gene therapy, a team headed by molecular biologist Richard Mulligan of MIT's Whitehead Institute for Biomedical Research has cured a form of beta-thalassemia in adult mice. The group was the first to demonstrate that a gene injected into animals after birth can produce significant quantities of the protein it codes for, and only in the proper body tissues.

Mulligan and his colleagues inserted the gene for human beta-globin into a specially engineered retrovirus and allowed the virus to infect bone marrow cells from the mice. The infection allowed the added genes to be incorporated in the DNA of the marrow cells. The researchers then injected the cells back into the mice's bone marrow. Mulligan and his colleagues reported that 18 mice had the human gene in their marrow. In eight mice selected for closer study, the team found that beta-globin was being produced by red blood cells, which normally make it, and not by other blood cells that would not normally make the protein. This result indicates that the added gene is under the same type of control as a native beta-globin gene, a critical requirement for the success of human gene therapy.

Human gene transfers. In the summer of 1980, Martin Cline of the University of California Medical School at Los Angeles shocked the scientific community by attempting to perform genetic therapy on two young women, one in Italy and the other in Israel. Both women suffered from beta-thalassemia and were severely ill from complications of the disease.

Cline isolated normal beta-globin genes from human cells. To insert the normal gene, he extracted bone marrow from each woman's hip and incubated the marrow cells in the presence of DNA containing the functional gene in the hope that the new gene would be incorporated into the cells. He then returned the marrow cells to the women, believing that the cells might begin producing functional hemoglobin. Although the women were neither helped nor harmed by his attempts at therapy, Cline's scientific reputation was severely damaged.

Many scientists condemned the experiment as ill-conceived, primarily because Cline had not yet been able to get the gene transplant procedure to work even in mice. Not only had he failed to demonstrate the safety and effectiveness of the technique in animals before using it in people, but he had violated National Institutes of Health (NIH) regulations by using recombinant DNA techniques in humans without permission.

For much of the 1980s, scientists had anticipated the first sanctioned attempts in human gene therapy. However, these experiments were delayed until 1989, largely because of difficulties in making transplanted genes function in animal cells. Most scientists believe some of the first applied gene therapy experiments will probably involve one of three extremely rare, untreatable genetic diseases: Lesch-Nyhan syndrome, SCID, or deficiency of the enzyme purine nucleoside phosphorylase (PNP).

Lesch-Nyhan syndrome is a devastating disease marked by progressive brain damage, compulsive self-mutilation, and death. It affects about one in every 50,000 males born in the U.S. and is caused by a defective gene that codes for the enzyme hypoxanthine guanine phosphoribosyltransferase (HPRT). Genetic therapy for Lesch-Nyhan is being studied by Theodore Friedmann and his colleagues at the University of California at San Diego, in collaboration with Inder Verma at the Salk Institute in La Jolla, California, and by C. Thomas Caskey and his colleagues at Baylor University in Waco, Texas.

SCID, which is associated with ADA deficiency, and PNP deficiency leave people without a functioning immune system, rendering them susceptible to massive infections and premature death. One well-known case of such an immune deficiency involved a Houston child who lived in a protective plastic bubble until he died in late 1984 at the age of 12. Only about 100 cases of SCID are known worldwide. Treatment for ADA deficiency is being studied by David Williams and Stuart Orkin of Harvard Medical School in collaboration with MIT's Mulligan. PNP deficiency is rarer still, with only nine known cases. David Martin and his colleagues at the University of California at San Francisco are currently studying the possibility of genetic therapy for the disorder.

The key step in curing a genetic disease is inserting a normal, functioning gene from a healthy individual into bone marrow cells from the victim. Bone marrow cells are used because they contain stem cells, which are the parent cells for blood cells. If the DNA in a stem cell can be altered, then the DNA in all the blood cells produced by that stem cell should contain the alteration. In this way, the defective blood cells would eventually be replaced by cells carrying the normal DNA. When these engineered bone marrow cells are placed back in the patient, they will presumably start manufacturing the protein coded for by the added gene and thereby alleviate the symptoms of the disease. Most scientists have assumed that the agent (called the vector) used to insert the normal gene into the bone marrow cells will be a retrovirus.

Normally, a retrovirus uses the enzyme reverse transcriptase to make a DNA copy of its own RNA, and this DNA is then integrated into the host cell's DNA. The retrovirus genes that researchers expect to use for gene therapy would normally transform the infected cells into cancerous cells. Many scientists have altered such retroviruses by removing the genes that enable the viruses to cause cancer. They have then added a healthy gene, such as that for ADA, to the retrovirus on the assumption that the virus will integrate the healthy gene into the bone marrow cells along with its own genes.

Williams, Orkin, and Mulligan reported in 1986 that they could "cure" white blood cells obtained from patients with ADA deficiency by using a retrovirus vector to insert a healthy ADA gene into the cells. Unfortunately, when they tried to implant their genetically engineered marrow cells into animals such as mice, ADA was not produced. For reasons still unclear, the parts of the retrovirus RNA that control expression of the inserted genes did not work properly.

In late 1985, biochemist Eli Gilboa of Princeton University in New Jersey devised a new retrovirus vector that was more effective at inserting genes into animals. Biochemists Randy Hock and A. Dusty Miller of the Fred Hutchison Cancer Research Center in Seattle, Washington, reported in 1986 that they had used the Gilboa vector to insert two different genes coding for antibiotic-resistance proteins into human bone marrow cells. Research teams around the world have since used the Gilboa vector or one very similar to insert an antibiotic-resistance gene into the bone marrow of mice. The transferred gene entered the bone marrow stem cells, which made significant amounts of the antibiotic-resistance proteins. Unfortunately, when NHLBI's W. French Anderson used the Gilboa vector to insert the ADA gene into marrow cells of monkeys, the cells did not produce ADA.

In May 1988, Anderson and oncologist Steven A. Rosenberg of the National Cancer Institute in Bethesda sought permission from the NIH's Recombinant DNA Advisory Committee to perform a genetic engineering experiment in which humans would be exposed to a retrovirus. Rosenberg has been working with an innovative therapy for malignant melanoma, a deadly form of skin cancer. In the treatment, white blood cells are removed from the patient, exposed to the immune-stimulating protein called interleukin-2, and then reinjected.

Only about half the patients respond to this therapy. In hopes of understanding why some patients do not respond, Anderson and Rosenberg have proposed tagging the treated cells with a nonfunctional "marker" gene that would allow the treated cells to be readily identified in blood samples removed for analysis. The tagging would allow the researchers to learn how long the cells survive and where they go in the body, clues that might be used to improve the efficacy of the treatment. The NIH advisory committee approved the proposal in October 1988, but the institute's director James B. Wyngaarden temporarily blocked the experiments until more information was provided to the advisory panel. The experiment was approved in early 1989 and began during the summer.

Control of gene therapy. More than 1,600 diseases are known to be caused by single-gene defects. These include cystic fibrosis, hemophilia, muscular dystrophy, sickle-cell disease, and thalassemia. All of these conditions are potential targets for genetic engineers, but some are more likely targets than others. A major consideration in selecting a genetic defect for treatment is how well the expression of the newly inserted gene must be controlled. In a disease such as Lesch–Nyhan syndrome, close control of the gene's expression is probably not important. Studies of naturally occurring variants of the syndrome have shown that production of even a small amount of the enzyme HPRT will help the patient and that excess amounts will not cause harm.

In contrast, in the case of beta-thalassemia, the globin genes must all work together to produce one product, hemoglobin, and control of the inserted gene's expression is therefore more important. If it should ever be possible to cure diabetes by gene therapy (probably a very difficult task since more than one gene would need to be inserted), control of the genes' expression would be crucial because insulin must be produced in precise amounts at precise times, or the patient will become seriously ill.

The prospect of genetic engineering in humans has been the subject of heated debate. Some critics have compared it to "playing God" or trying to "improve on nature." Supporters see it instead as an attempt to ease the pain and despair of people suffering from inherited diseases. In general, however, public attitudes toward genetic engineering in humans depend on the type of therapy being considered. Attempts to correct devastating genetic defects such as SCID provoke the least outcry — as long as the techniques have been proved effective in animal experiments. Such techniques are called somatic-cell therapy, from the Greek "soma," meaning body, and are distinguished from germ-line therapy, which is designed to correct genetic defects in reproductive cells.

The 1982 report, *Splicing Life*, by the President's Commission for the Study of Ethical Problems in Medicine and Biomedical and Behavioral Research, concluded that somatic-cell therapy in humans should not be prohibited and perhaps should be encouraged. This view was echoed in a 1984 report, *Human Gene Therapy*, from the U.S. Office of Technology Assessment (OTA) in Washington, D.C. The OTA report noted that "gene therapy of this type is quite similar to other kinds of medical therapy and does not pose new kinds of risks.... The factor that most distinguishes it from other medical technologies is its conspicuousness in the public eye; otherwise it can be viewed as simply another tool to help individuals overcome an illness."

Correcting genetic defects in reproductive cells is much more controversial. Many critics argue that humans should not tinker with genetic traits endowed by nature. In response, supporters argue that if a genetic defect were corrected in the reproductive cells, then a child conceived from those cells would not have the inherited defect. Furthermore, that child's descendants would also inherit the corrected gene and thus would never suffer from the disease. Over a period of time, defective genes such as that for Lesch-Nyhan syndrome might be eliminated from the population.

In 1983, leaders of virtually every major church group in the U.S. issued a resolution calling for a ban on all genetic engineering of human reproductive cells, even genetic engineering to correct inheritable diseases. If this capability were developed, the resolution warned, "the temptation to perfect humanity" might lead instead to the extinction of the species. This resolution was initiated by Jeremy Rifkin, an author and political activist who heads the Foundation on Economic Trends in Washington, D.C. Other lawsuits filed by Rifkin have blocked the release of genetically engineered organisms into the environment.

In Rifkin's 1983 book *Algeny*, he argues against most types of genetic engineering: "Once we decide to begin the process of human genetic engineering, there really is no logical place to stop. If diabetes, sickle-cell disease, and cancer are to be cured by altering the genetic makeup of an individual, why not proceed to other 'disorders': myopia, color blindness, left handedness. Indeed, what is to preclude a society from deciding that a certain skin color is a disorder?"

In a statement published in the 1984 OTA report, Rifkin argues further that "In return for securing our own physical well-being, we are forced to accept the idea of reducing the human species to a technologically designed product. Genetic engineering poses the most fundamental of questions: Is guaranteeing our health worth trading away our humanity?"

The same report also quoted Ola Mae Huntley, the mother of three children with sickle-cell disease, who holds a much different opinion, arguing that the critics are "playing God" every bit as much as the researchers. She said, "I am very angry that anyone would presume to deny my children and my family the essential genetic treatment of a genetic disease.... I see such persons as simplistic moralists who probably have seen too many mad scientist horror films."

Concerns from both sides of the ethical debate were reflected in guidelines for human gene therapy issued in September 1985 by NIH's Recombinant DNA Advisory Committee (RAC). Each case in which gene therapy is proposed would have to be approved first by review committees at the researchers' own institutions and then by RAC. Other points in the guidelines were presented largely in the form of questions: What kind of preliminary research had been or would be conducted? What would be done to safeguard the patients' interests? What obligations would the patients be expected to accept, such as long-term followup and autopsies in case the patients died? The committee noted that an accurate determination of the precise cause of a patient's death would be "of vital importance to all future gene therapy patients."

Surprisingly, the guidelines have provoked little outcry from opponents of genetic engineering. For example, none of the religious leaders who signed the 1983 petition replied to RAC's request for comments on the original draft of the guidelines, even though each was sent a copy. Even the generally outspoken Rifkin questioned only the composition of the group that developed the guidelines rather than the guidelines themselves. Perhaps the consensus was reflected best by Alexander Capron, executive director of the president's commission: "Do we want to ask these people who suffer the ill effects of the genetic lottery to bear the heavy, and sometimes lethal, effects of our unwillingness to find a finely tuned means of avoiding potential abuses of genetic alterations?"

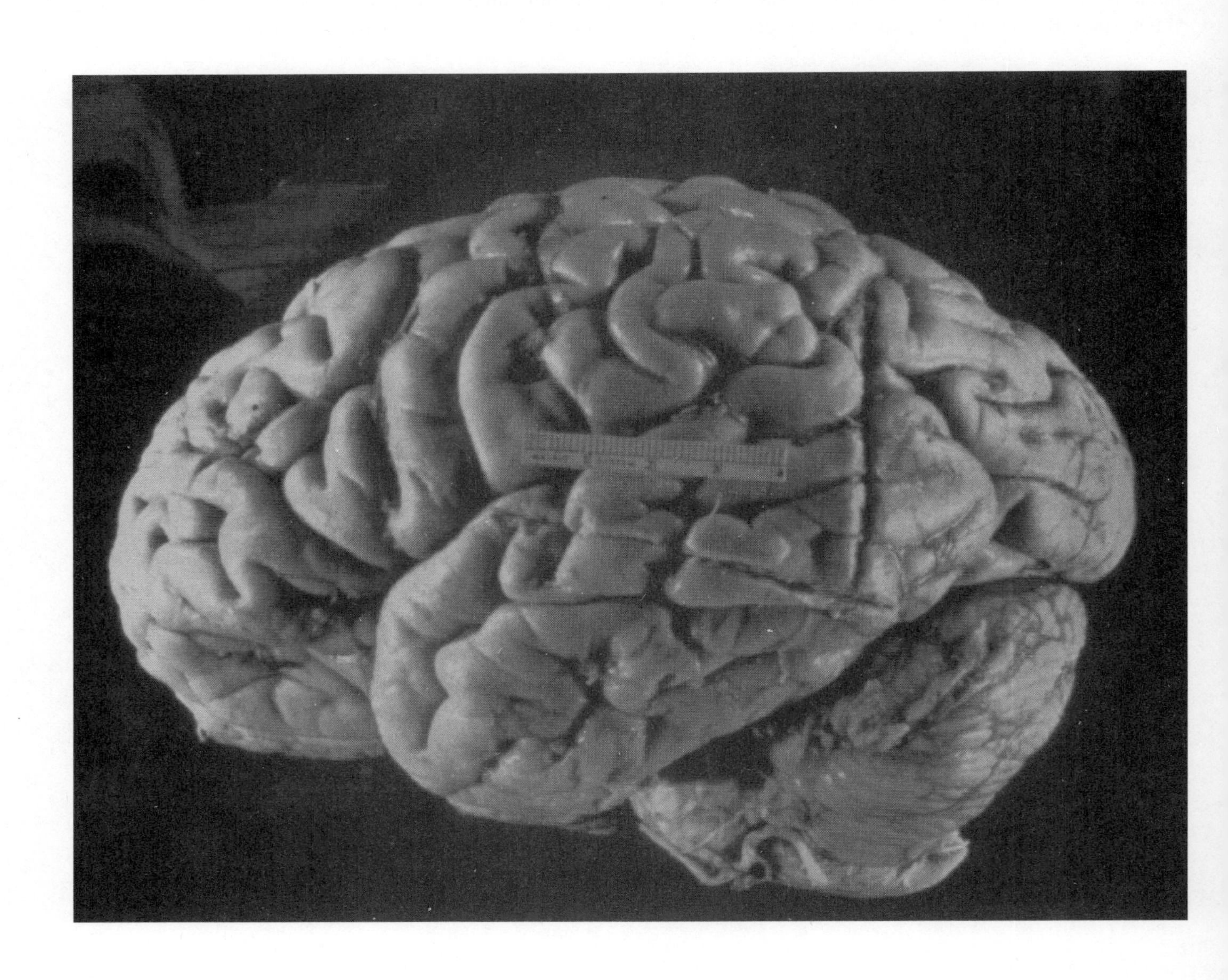

Chapter 3

Brain and Behavior

Imaging the Brain

The new techniques that scientists have developed for imaging the brain are revolutionizing the study of the brain's structure and function. Perhaps the most important of these are computed tomography (CT), magnetic resonance imaging (MRI), positron emission tomography (PET), and brain electrical activity mapping (BEAM), a variant of the old technique of electroencephalography (EEG). The first two tools are used to "see" the brain's structure, while the latter two measure the activity of cells in specific parts of the brain. CT and MRI are used routinely for detecting brain tumors and the effects of trauma, and the less widely available PET and BEAM techniques are also used to a limited extent for these purposes. For researchers studying brain function, however, all of these new techniques, both individually and together, have provided powerful new ways of understanding how the brain works.

New Techniques

PET scanning, developed in the 1970s by biophysicists Michael Phelps and Edward Hoffman of the University of California at Los Angeles (UCLA) School of Medicine, may offer the greatest promise for brain studies. The technique allows scientists literally to look inside a living brain to watch and measure its chemical activity without making a single incision. PET scanning derives from computed tomography, also known as CT scanning, a procedure that involves rotating an x-ray source around a patient's body and then using a computer to analyze the patterns of x-rays after they have passed through the body. In this way, the computer can construct cross-sectional images of internal organs, allowing physicians to detect and monitor certain diseases. Although CT scans do provide detailed information about the structure of organs, they do not provide information about metabolic activity, chemical processes, or energy use within the organs.

PET scans go further, shedding light on the chemical processes that are the basis of cell function. Positron-emitting radioactive atoms attached to chemicals similar or identical to ones normally present in the brain are first injected into the blood and then carried in the bloodstream to the brain. As the radioactive atoms decay, they emit positrons; the positrons combine with and annihilate electrons in neighboring atoms to produce gamma radiation. The PET scanner uses a ring of gamma ray detectors surrounding the patient to detect the radiation, which a computer can then integrate into a cross-sectional picture showing the activity of chemical systems throughout the brain.

In early research, PET scanning used a chemical with a structure similar to glucose (sugar), which is labeled with positron-emitting fluorine-18 atoms. Although human body cells absorb the tagged molecules just as they would ordinary glucose, they cannot metabolize the tagged molecules. The cells accumulate the radioactive molecules in proportion to the amount of glucose they use. Thus, by analyzing the emitted radiation, a researcher obtains a picture of both where the brain is using glucose and how much it is using. Because the brain uses glucose as an energy source, the amount of glucose used at any site in the brain reflects the level of the person's brain activity.

State-of-the-art PET scanning also uses many other positron-emitting atoms, such as carbon-11, nitrogen-13, and oxygen-15. For example, a researcher can inject water molecules containing oxygen-15 into a subject's bloodstream to monitor blood flow through the brain. By the late 1980s, more than 300 biological molecules and drugs labeled with positron-emitting atoms had been synthesized by researchers and used for PET scanning. Because operators deliberately select radioactive atoms with short half-lives for safety purposes, the isotopes must be prepared in a cyclotron shortly before the PET scan. As the price and complexity of PET scanning devices have declined, the number of medical centers using the technology has increased. By 1989, over 100 centers worldwide were using PET scanners.

Physicians use PET scanners to identify and assess the severity of malignant tumors and to help diagnose developmental problems and degenerative brain disorders, including Alzheimer's, Huntington's, and Parkinson's diseases. PET scanners are also used to determine the extent of brain damage after a stroke and to identify the part or parts of the brain where seizures originate in conditions such as epilepsy. Perhaps the greatest potential value of the machines, however, is in basic research, where they permit neuroscientists to study energy use, blood flow, and the distribution of neurochemicals in both normal and diseased brains without risk to the subject.

Visualizing thought. At the 1989 annual meeting of the American Association for the Advancement of Science (AAAS) in San Francisco, Phelps described how he and neurologists John Mazziotta and David Kuhl used PET scanning to study changes in energy use in the visual cortices of the brains of healthy test subjects viewing a variety of patterns and scenes. To obtain a baseline measure of the brain's energy use in a visually unstimulating environment, the researchers first obtained PET scans when the subjects had their eyes closed in a quiet laboratory with low-level background lighting.

Phelps and his colleagues then obtained PET scans as the subjects looked either at white light, at a black-and-white checkerboard pattern, or at a park outside the laboratory window. As the researchers expected, the PET scans showed that the brain's visual cortex consumed more energy as the scenes increased in complexity. In addition, the PET scans of subjects viewing complex visual scenes showed the greatest amount of activity in the associative visual cortex — the part of the cortex that interprets and forms associations with viewed objects. Compared to baseline levels, the subjects used about 6 percent more glucose in the associative visual cortex when they viewed white light and nearly 60 percent more when they viewed the visually complex park scene.

The UCLA team also used PET scans to study subjects who tried to differentiate between two sequences of audible tones. In this case, the PET scans showed that the patterns of energy use in the brain depended on the thinking strategies of the subjects. Those with little formal musical training, who simply listened and tried to remember the tones, relied more heavily on the right hemispheres of their brains, which scientists generally believe controls nonanalytical thinking in most people. Those with more musical training, who used such memory devices as picturing the notes on a musical scale, relied more on their left hemispheres, generally thought to control analytical thinking. The visual cortices of listeners who visualized the music also consumed more glucose, indicating that the centers of vision participate in the visualization process even when people are not actually "seeing" anything.

In all of their studies of the brain's response to complex visual and musical stimulation, the UCLA researchers noted high glucose consumption in the frontal cortex, the area of the brain generally believed to be the center of higher thought processes, such as logic, language, and planning. These studies not only provided insights into the ways in which brains process information, but they also demonstrated that patterns of a brain's energy use reflect individual differences in thinking strategies and training.

Phelps and pediatric neurologist Harry Chugani have also turned to PET scanning in their effort to understand brain development in infants and children. In 1986 they reported that the primary brain metabolic activity of infants 5 weeks of age and younger occurs in those areas that control primitive sensory and motor activities. Although metabolic activity in other areas of the brain is low at 5 weeks of age, it increases slowly until the infants reach an age of about 7.5 months, by which time the patterns in their PET scans are virtually identical to those of adults. According to Phelps, these results suggest that newborn infants begin life with a limited capacity for thought and other high-order functions, and that this capacity increases slowly as they grow.

On another front, the work of psychiatrist Richard Haier of the University of California at Irvine has suggested that high intelligence may be the result of an efficiently organized brain. Using PET scanners, Haier found that the brains of people who perform well on intelligence tests use less energy than the brains of poor performers. He monitored 3 groups: 8 people taking a test called Raven's Advanced Progressive Matrices, 13 people taking a visual vigilance test, and 9 others simply watching flashing stimuli. The difficult, nonverbal matrices test measures a person's abstract reasoning ability, and the results have correlated well with other tests of general intelligence. The test requires that the subject recognize a pattern within a matrix of abstract designs and select a design that completes the pattern. The visual vigilance test requires that the subject monitor a TV screen and push a button when a particular signal appears on the TV.

At the 1988 AAAS annual meeting in Boston, Massachusetts, Haier reported that the brains of subjects who performed best on the Raven's matrices test used the least energy in the cortical areas, where abstract reasoning takes place, while those who did poorly used much more energy. His results suggest that the high scorers did not work harder than the other subjects, but simply worked more efficiently. In the other two tests, which only required the subjects to pay attention to flashing lights, Haier observed no differences in the patterns of energy use in the brains of the good and poor performers.

Scanning for brain disorders. Researchers have also used PET scanning to identify abnormal brain metabolism in people suffering from such disorders as dementia, epilepsy, schizophrenia, stroke, and Parkinson's and Huntington's diseases. CT scans of these patients' brains may have little value, since the scans often appear normal. PET scans of people who suffer from certain types of seizures that originate in the visual cortices reveal abnormally low use of glucose in the visual cortices during seizure-free periods when the patients usually experience restricted vision. But during seizures, when some patients have vivid hallucinations, the PET scans show dramatically high levels of glucose metabolism in these parts of the brain. By pinpointing the areas of abnormally low energy use with PET scans, Mazziotta and his associates have identified the sites in the brain where seizures originated in more than 500 patients. Subsequent surgery confirmed that the areas were, in fact, abnormal.

Psychiatrist Monte Buchsbaum and his associates at the University of California at Irvine have used PET scans to compare energy use in the brains of schizophrenic patients to that of normal subjects. At an October 1987 symposium at Irvine, Buchsbaum reported that the schizophrenics use an abnormally low amount of glucose in their brains' frontal cortices. As the brain's "chief executive," the frontal cortex plays an important role in planning and carrying out goal-directed behavior. Schizophrenics, who number 2 million to 3 million in the U.S., often cannot distinguish fantasy from reality, an indication, in Buchsbaum's view, that the "executive" is not working properly.

Buchsbaum used PET scans to monitor the brain activity of schizophrenics performing tasks that they found difficult — in particular, tasks that required them to pay close attention. He concluded that his results support the idea that inability to pay attention is a fundamental symptom in schizophrenia.

In related studies, radiologist Henry N. Wagner, Jr., and his colleagues at the Johns Hopkins School of Medicine in Baltimore, Maryland, detected a neurochemical abnormality in the brains of schizophrenics. Researchers had long speculated that the schizophrenics' brains might not use the neurotransmitter dopamine in a normal way. At an October 1986 seminar at Johns

Hopkins, Wagner reported that studies of 25 schizophrenics showed that they had an average of twice as many dopamine receptors in their brains as were found in normal people.

Some researchers working with animals had thought that neuroleptic drugs (which block dopamine receptors) used to treat schizophrenics might be responsible for the higher number of receptors. Wagner found, however, no significant difference in the number of dopamine receptors between the 11 of his subjects who had never received neuroleptic drugs and the 14 who had. This finding suggested that the increased number of receptors was a result of the disorder rather than the therapy.

Wagner, radiologist Dean F. Wong, and their colleagues have also found abnormal numbers of dopamine receptors in patients with affective disorders and psychoses. Affective disorders are mental illnesses characterized by extremes of mood, such as depression or elation. Psychoses are characterized by hallucinations, delusions, disordered thoughts, and disorganized behavior. At the July 1987 meeting of the Society of Nuclear Medicine in Toronto, Wong reported that five patients who suffered from both manic-depressive disorder and psychoses had more dopamine receptors in the caudate nuclei of their brains than did either nonpsychotic patients or normal individuals. Because the researchers found the abnormality only in patients with one subtype of manic-depressive illness — that is, those with psychoses — Wong speculated that PET scans will help physicians differentiate among the various types of the disorder.

In PET-scan studies of autistic adults (people who are typically withdrawn from the world, communicating and empathizing with other people only to a limited extent), Buchsbaum and psychiatrists Peter Tanguay and Robert Asarnow of the University of California at Irvine found significantly less activity in the right halves of the brains of autistic people than in those of people without the illness. The right hemisphere is generally thought to control many aspects of personality, including social interaction, emotion, and the emotional inflection of speech.

The three researchers studied 15 autistic adults who had partially recovered from the disorder and found less activity in the right sides of all their brains. In a report scheduled for publication in 1989, they noted that seven had almost completely recovered and scored relatively high on IQ tests. "They may represent a more pure group of autistics," Buchsbaum said. "My suspicion is that [the remaining patients and] a number of other people are brain damaged during birth, and the irregular nature of the symptoms makes them seem like autistics ... The adults in our study were better able to communicate than they had been as children, but they still have residual problems, and none were really fully employed. Our support and care for them will be better when we really understand what is happening in their brains."

PET scans of patients with Alzheimer's disease, an affliction of the elderly characterized by progressive loss of memory and other mental functions, have shown significantly less energy use in their brains than in those of normal subjects. Psychiatrist Thomas Chase of the National Institute of Neurological and Communicative Disorders and Stroke in Bethesda, Maryland, reported in 1983 that total glucose use in the brains of 20 patients with Alzheimer's disease averaged 28 percent below normal. While glucose use appeared nearly normal in many areas of the patients' brains, it was as much as 40 percent below normal in their posterior parietal lobes, the parts of the brain generally thought responsible for integrating information gathered by the senses and for performing complex thinking and reasoning. Decreased glucose use in the parietal lobes is seen even in Alzheimer's disease patients with very mild symptoms. This finding indicates that PET scans may be useful in the early diagnosis of this condition.

Down's syndrome, a congenital form of retardation, has also been studied with PET scans. Psychiatrist Neil Cutler and his colleagues at the National Institutes of Health (NIH) in Bethesda have monitored glucose use in one middle-aged and four young patients with Down's syndrome, comparing their findings to glucose use in healthy volunteers of the same ages. The NIH group reported in 1983 that glucose use in the young Down's syndrome patients was 25 to 41 percent above normal, while that of the middle-aged patient was only slightly above normal. Other investigators have reported abnormally high energy use in the brains of autistic adults. The NIH team suggested that Down's syndrome and autism may belong to a class of disorders in which the brain must use excessive amounts of energy just to maintain minimal thought processes, perhaps because of neurochemical imbalances or incomplete metabolism in the brain.

Researchers are now turning their attention to other mental disorders and the effects that commonly abused drugs have on the brain in the hope of obtaining new insights into the biochemical abnormalities that may cause or contribute to mental illness and drug abuse. The use of PET and other techniques may eventually make possible a much more precise diagnosis of complex mental disorders, and help neurologists and psychiatrists decide which therapies are most appropriate.

Electroencephalograph Studies

Research into the brain's electrical activity has gradually helped scientists identify the parts of the brain that are essential for such complex mental processes as reasoning, feeling, and remembering. To measure the brain's electrical activity, scientists place electrodes on the subject's scalp, then connect the electrode wires to a machine called an electroencephalograph, which records the brain's activity, or brain waves, and produces a tracing called an electroencephalogram (EEG). When brain waves occur in response to a specific stimulus, such as a flash of light or a sudden noise, scientists call the resultant EEGs evoked potentials (EPs). Changes or abnormalities in EPs can be signs of brain disease.

A major problem with conventional EEGs is that each electrode attached to the skull produces its own trace on graph paper, making EEG analysis complicated and time-consuming. An important breakthrough in the use of EEGs was achieved in the early 1980s when Frank Duffy of Harvard Medical School in Boston, Massachusetts, developed a technique called brain electrical activity mapping, or BEAM. Using sophisticated computer technology, BEAM combines the signals from each electrode to produce a continuous, color-coded map of electrical activity in the brain. In December 1987, Nicolet Instrument Corp. of Madison, Wisconsin, began selling BEAM instruments commercially.

In the BEAM system, a standardized test protocol is used to acquire data. This standardized system greatly simplifies use of the equipment. In the protocol, both regular EEG data — for measurements of resting-brain activity — and EP data are collected. Because researchers have measured and cataloged BEAM activity from a large number of healthy people in different age groups, they can instantly compare this data, stored in the computer's memory, to a specific patient's BEAM patterns. In just one example of the use of BEAM, Duffy and his colleagues have achieved 80 percent accuracy in identifying patients with dyslexia (a learning disorder) and schizophrenia, both of which can sometimes be difficult to diagnose with other techniques.

Treating epilepsy. By 1988, physicians were beginning to use EEG and related techniques to identify epileptics who could benefit from surgery and to pinpoint the parts of their brains where surgery would be effective. By some estimates, about 2 million Americans suffer from epilepsy, a condition that may cause unpredictable seizures and convulsions, sometimes leading to unconsciousness. These seizures and convulsions are symptoms of underlying problems such as tumors, head injuries, or any one of a variety of other potential causes. A person may develop epilepsy months or even years after suffering a brain injury. As many as 400,000 epileptics suffer from seizures that cannot readily be controlled with conventional epilepsy drugs, which sedate the brain. In up to one-quarter of those people, the seizures start in a small region of the brain that could be reached and treated with surgery.

A small number of treatment centers in the U.S. are now providing comprehensive examinations and surgery for epileptics. The process typically begins with MRI or CT imaging to identify any obvious lesions or abnormalities in the brain, although in most cases, none is found. The patient is hospitalized and then undergoes simultaneous BEAM studies and videotaping for periods as long as 2 weeks to correlate seizures and other symptoms with patterns of brain activity. At some centers, such as the Minnesota Comprehensive Epilepsy Program in Minneapolis, the electrodes are glued directly onto the head. At others, such as the University of Florida's Epi-Life Center in Gainesville, they are implanted on the skull beneath the skin for greater sensitivity.

About one-third of the patients monitored in this way are found to have a lesion in the left or right temporal lobe, the area of the brain where, among other things, memories are processed. If the BEAM patterns indicate that the seizures originate in a single lobe, neurosurgeons may consider removing the entire lobe. According to neurosurgeon Robert J. Gumnit of the Minnesota Center, at least 75 percent of patients who have a temporal lobe removed suffer no more seizures and have little loss of function.

If the neurosurgeons conclude that the seizures originate in both lobes, they may elect to implant long, thin needles directly into the lobes through holes drilled in the skull over the ears. These needles each contain several electrodes that can help identify the site of the most important source of seizures, which can then be removed with confidence.

In many cases, however, the source of the seizure lies deeper in the brain or higher up in the frontal lobe, where both speech and movement are controlled. Because large-scale removal of tissue would be destructive in such cases, physicians must pinpoint the source of the seizures more precisely. To do so, neurosurgeons remove a large portion of the top of the skull and implant a grid containing between 64 and 100 small electrodes, which remain in place for up to a week. Because these electrodes receive electrical signals directly from the brain rather than after the signals have been diffused by the skull, they provide a much clearer picture of the exact location of the lesion, thus permitting neurosurgeons to remove only the diseased tissue.

Of patients diagnosed and treated this way in Minnesota, 63 percent were free of seizures and 22 percent had at least a 90 percent reduction. Similar results have been obtained at Florida's EpiLife Center, according to neurosurgeon Steven Reid. "Most medical centers using long-term, brain-wave monitoring and surgery of this type report a success rate of 80 to 90 percent with either complete cure or bringing the . . . seizures under control medically," said Reid. "That's really an accomplishment because these are people who have had incapacitating seizures for years that have been completely uncontrolled with medicine, so surgery is literally their last hope."

Harry Chugani and Michael Phelps of UCLA have used PET scanning to identify epileptic lesions in very young children, who have an extremely poor prospect for recovery if they are not treated surgically. Their seizures begin shortly after birth, and the combination of seizures, medication, and the underlying disease slows the children's development. These children are often institutionalized during the second decade of life and have a short life expectancy. Phelps reported at the 1989 annual AAAS meeting, however, that PET imaging can accurately identify the part of the temporal lobe that is defective in these children. According to Phelps, with surgery, the children can become seizure-free, require no medication, and develop normally. Even if a whole hemisphere is removed, the amazing plasticity of the child's brain allows the remaining lobe to progressively take over the functions that would have been controlled by the lobe that has been removed.

Interpreting brain waves. Researchers are slowly beginning to match specific EEG and BEAM patterns to particular cognitive or functional processes in the brain. This represents, in a crude way, the first step toward the development of techniques for using computers to interpret the meaning of brain waves. If such interpretation becomes possible, revolutionary computer applications might become practical. Quadriplegics, for example, might be able to control computers with their minds, and perhaps even use the computers to control mechanical prostheses.

Physiologist Walter Freeman of the University of California at Berkeley has found that the olfactory bulbs, which are involved in interpreting smells, produce specific EEG patterns when animals encounter known smells. Freeman implanted arrays of 64 electrodes (similar to those described above in epilepsy research) in the olfactory lobes of rabbits. The electrode arrays produced computer tracings of regions of high and low amplitude that looked like contour maps.

When Freeman exposed the rabbits to odors, such as banana and clove oils, their olfactory bulbs initially exhibited chaotic patterns. But after a conditioning session of 50 to 100 sniffs, a repetitive pattern emerged. That pattern reappeared when the animals were later exposed to the same odors. Because each animal exhibited a different pattern, Freeman concluded that the brain stores information about an odor by making physical connections between neurons to produce the specific EEG patterns. He predicted that this new understanding may lead to the development of computers that will store information in a similar way. Freeman has already written equations which create and process chaotic patterns that eventually merge into regular patterns, thereby mimicking the activity of the olfactory bulbs.

In prosthetic research, neuroscientist Apostolos P. Georgopoulos and his colleagues at Johns Hopkins have taken the first step toward neuronal control of artificial limbs by mapping the electrical activity involved in the movement of a monkey's arms. They found that when an animal anticipated moving an arm, that movement was foreshadowed by changes in the pattern of electrical activity across the animal's brain.

The researchers trained a monkey named Lambda to respond to a series of lights arranged in a circle like the hours on a clock face while they monitored the activity of the individual cortical cells involved in arm movement. At the beginning of each experiment, Lambda was sitting in front of the circle and a light came on in its center. After a variable period of time ranging from 0.75 to 2.25 seconds, the center light went off and one of the lights around the edge flashed on in either a dim or bright mode. Lambda was trained with a banana juice reward to turn a handle in the same direction as the light when it came on dim, or perpendicular to and counterclockwise from the direction of the light when it came on bright. Lambda could eventually perform the correct motions 80 percent of the time.

As Lambda considered her movements, electrodes that had been placed on her head picked up the electrical activity of her cortical cells and relayed them to the computer, which interpreted them in terms of vectors (directional forces). In 1989, the researchers reported that when the light was dimly lit, the vectors aligned in the same direction as the light within milliseconds after it came on. When the light flashed brightly, the vectors again aligned in the direction of the light, but then they rotated until they pointed perpendicular and counterclockwise to the direction of the light. In each case, the vectors formed before the arm moved. Georgopoulos noted, "We've used a technique that lets a mental process be seen as a physical one." He speculated that a similar combination of electrodes and computers could help amputees control artificial limbs.

Physicist Erich Sutter of the Smith-Kettlewell Eye Institute in San Francisco has developed a prototype device that uses electrodes implanted under the skull's surface to help severely disabled people communicate by using a computer. The electrodes pick up the person's brain waves, which are analyzed to determine what part of a video screen is being viewed. When the person focuses on squares containing letters, numbers, or words, a computer attached to the electrodes generates words and sentences that appear on a video screen while being spoken by a synthesizer.

The first patient to have the electrodes implanted was Dr. Lance Meagher, a 42-year-old internist who had been afflicted with amyotrophic lateral sclerosis (Lou Gehrig's disease) for 12 years and who had been severely disabled for 6 years. Despite his illness, Dr. Meagher treats other victims of the disease with the assistance of computer technology. When Meagher and Sutter demonstrated the new device at a November 1988 meeting at California State University at Northridge, Meagher said that he preferred it to the computer device he had previously used.

State-of-the-art devices designed to help the handicapped use computers often employ video cameras that monitor eye movements as the person looks at a video screen. These devices have limitations, Sutter said, including the fact that users must hold their heads relatively still in the range of the camera — a difficult feat for many victims of muscular disorders. The devices must also be recalibrated frequently, and reflections from eyeglasses can interfere with the camera's tracking. "The ideal would be to interface with the brain directly, but that hasn't been possible so far," Sutter said. "But it is possible to use brain waves on a much more modest scale."

Sutter is using brain-wave monitoring to overcome many of the limitations of current devices. His system includes a video screen divided into 64 checkerboard-like squares called keys, and a letter or a commonly used word is assigned to each key. In order for the system to determine which key the user is viewing, each key must generate a specific brain signal. Sutter achieves this by having each key change color, flicker, or change patterns at its own characteristic frequency. When the user focuses on a particular key, his visual cortex emits brain waves that vary at the same frequency. Although conventional electrodes attached to the scalp could detect these brain waves, Sutter has obtained much greater sensitivity and convenience by implanting the electrodes just beneath the skull, outside the membrane that encloses the brain.

The electrodes are connected to a thin cable that passes under the user's scalp, exits the skin on the neck below the hairline, and plugs into a miniature radio transmitter that sends the signal to the computer. When the video screen is operating, some keys display commonly used words, while others display letters. When the user focuses on a letter, the screen then displays words beginning with that letter. This system allows users to quickly pick any of 600 to 800 words with just one or two "keystrokes." If the desired word does not appear on the list, the user can spell it out by picking further letters. By the end of 1988, Sutter was planning similar implants in other patients to refine the system.

Combinations of Techniques

University of Texas researchers have found the first concrete evidence that stuttering and another speech disorder called spasmodic dysphonia are caused by biochemical abnormalities rather than emotional disturbances. The discoveries could remove much of the stigma associated with the disorders based on the belief that they have a psychological origin. The findings also provided the basis for new types of therapy that are already being used experimentally.

Stuttering affects about one in every 100 people in the U.S. Researchers have long wondered why stutterers do not stutter all the time. Country singer Mel Tillis, for example, stutters when he speaks, but not when he sings. Other people do not stutter under specific conditions, such as when they are speaking to children or involved in certain tasks. Because of this variability, many psychiatrists have believed that the condition is caused by a psychological trauma that occurred during childhood. This assumption had made many parents feel unnecessary guilt, according to speech scientist Frances J. Freeman of the University of Texas Center for Vocal Motor Control in Dallas.

Spasmodic dysphonia affects perhaps one-tenth as many people as stuttering. It typically strikes people in their thirties and forties, but can begin as early as the teens or as late as the sixties. Typically, spasms close the larynx prematurely, chopping words off in the middle; in some cases, however, the larynx muscles relax so completely that words cannot be formed at all. Because the disorder is progressive, most people with this condition eventually completely lose the ability to speak. Some psychologists have assumed that the victims of spasmodic dysphonia unconsciously wish to punish someone in their lives and do so by not speaking. The conventional therapy for spasmodic dysphonia has thus been psychotherapy, but it has not been effective.

The University of Texas researchers made their new discoveries through the combined use of MRI, BEAM, and a variant of PET called single photon emission computed tomography, or SPECT. None of the techniques used individually could demonstrate that biochemical abnormalities were the cause of the disorder, but the combined evidence was persuasive, neurologist Kenneth Pool of the Dallas center reported at the 1989 AAAS meeting. MRI, for example, showed that about one-quarter of 51 patients with stuttering or spasmodic dysphonia had abnormalities in the structure of three specific regions of the brain, two of which are "classically associated with speech function," according to Pool. BEAM showed that up to about two-thirds of 43 patients studied had abnormal electrical activity in the same areas. And SPECT studies of 49 patients revealed that about three-quarters had abnormally low blood flow in those regions.

"We can say with great confidence that 84 percent of the people we have studied with these two disorders have a neurological defect of the central nervous system that causes the disorder," said radiologist Terese Finitzo, the project director. Although the affected areas are not precisely the same in the two disorders, they are close enough together to suggest that both impairments might be subtly different manifestations of similar underlying damage, according to Finitzo. Concerning the other 16 percent, Pool said, "Failure to find neurological dysfunction does not preclude its existence. It just means that our techniques aren't sophisticated enough yet to see it."

Once the University of Texas group determined which parts of the brain were involved, they began looking at other bodily functions controlled by the same brain regions to see whether those functions were also affected. One region of the brain involved in spasmodic dysphonia, for example, also helps control the digestive system. The researchers found that victims of the speech disorder often produce insufficient stomach acid and frequently suffer from gastrointestinal upsets. Other areas of the brain implicated in spasmodic dysphonia are also involved in motor control of the body. Here, researchers found evidence of reduced bodily function; people with spasmodic dysphonia and many stutterers have more trembling in their hands than normal, and they often have diminished coordination and longer reaction times.

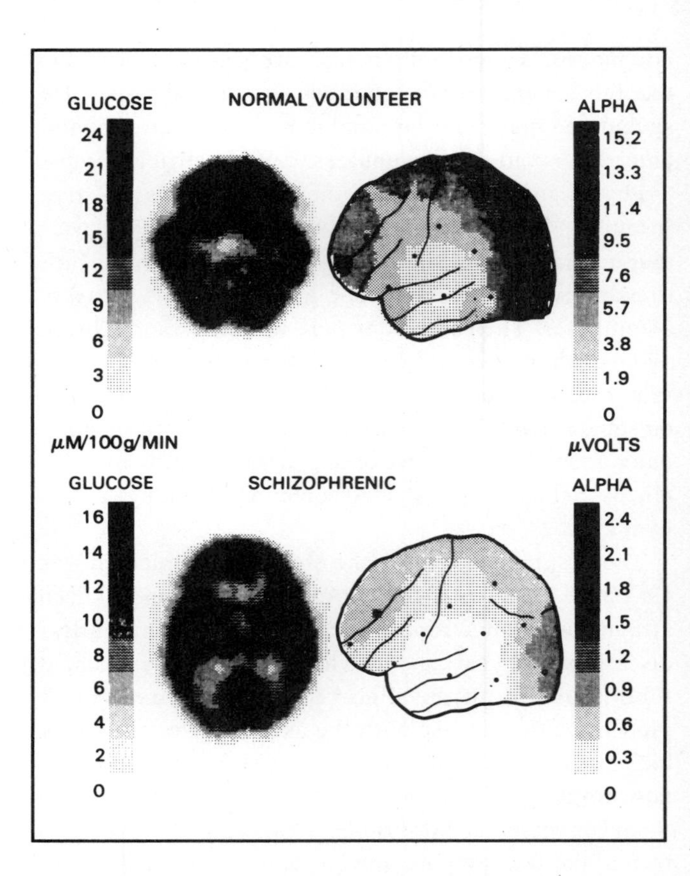

Figure 1. Simultaneous PET Scans and EEG Maps of Brain Activity. These PET scans (left) and EEG maps (right) show a consistent pattern of reduced energy use and electrical activity in a schizophrenic patient compared to a normal person. The lightly shaded areas on the PET scans represent regions of low glucose use, while the lightly shaded areas on the EEG maps represent regions of low electrical activity. The darker areas represent regions of high glucose use on the PET scans and high electrical activity on the EEG maps. These patterns reflect the schizophrenic's characteristically low level of mental activity and lack of response to the environment. *Courtesy: Monte Buchsbaum, NIMH.*

Such impairments were often overlooked by other physicians and the patients themselves because they were most concerned about the vocal problems. The presence of these other impairments strongly confirms the idea that the disorders spring from a biochemical abnormality. The research, however, still does not answer the question of why some stutterers can sometimes speak clearly.

These speech disorder findings have suggested new approaches for treatment. For spasmodic dysphonia, the Dallas center now extensively uses conventional speech therapy, a treatment option that was previously ignored when the cause of the disease was thought to be psychological. Since the imaging techniques showed that people with these speech disorders sometimes had impaired blood flow to some regions of the brain, conventional drugs have been used to dilate blood vessels and increase blood flow. Therapists are also using drugs normally used to treat movement disorders caused by impairments in the same parts of the brain as those involved in the speech disorders. Although it is too early to tell if any of these approaches will produce long-term benefits, the University of Texas team is confident that one or more of the techniques will prove successful.

Sleep

The decision to launch the ill-fated Challenger space shuttle in January 1986 despite abnormally cold weather conditions was made in the early hours of the morning when most of the decision-makers would normally have been asleep. The disastrous accidents at the Chernobyl nuclear power plant in the U.S.S.R. and the Three Mile Island nuclear facility in Pennsylvania, as well as many less serious incidents at other nuclear power plants, occurred between 1 A.M. and 4 A.M. Less dramatic, but more significant in terms of lives lost, is the fact that the largest number of single-vehicle accidents involving trucks occurs between 3 A.M. and 6 A.M.

In the case of airplane accidents, the National Transportation Safety Board has cited pilot fatigue as either a cause or a contributing factor in 69 crashes between 1983 and 1986 in which 67 people died. Finally, a 1988 study by the Institute for Circadian Physiology in Boston, Massachusetts, found that sleeping problems in the workplace cost American companies an estimated $70 billion annually in lost productivity, increased medical bills, and industrial accidents.

Sleep — or, more precisely, the lack of it — has become a major problem in the U.S. and other industrialized countries. As technological developments and increasing round-the-clock operation of industrial facilities such as power plants have required a greater degree of alertness on the part of human operators, scientific knowledge about sleep cycles and the metabolic changes during sleep has not kept pace. Researchers still have not found a definitive answer to the most basic questions about why people need sleep and how much sleep they actually need.

Conventional wisdom has held that adults need 7 to 8 hours of sleep each day to maintain their physical and mental health. Many sleep researchers now dispute this idea, claiming that even though most adults sleep an average of 7 or 8 hours a day, many people regularly get less sleep with no apparent ill effects.

Whether sleep is necessary or not, researchers have clearly proved that the average amount of sleep an individual gets generally declines with age. Newborns may sleep as few as 11 or as many as 22 hours a day and often average 12 to 16 hours a day in the first half-year of life. For the average person, daily sleep time declines to about 12 hours by age 2, 11 hours by age 5, and then gradually drops to 7 or 8 hours by late adolescence. After the age of 60, the average drops further, to about 6.5 hours a day.

But the amount of sleep needed can vary widely from person to person. "The basic function of sleep is to help you stay awake," said psychiatrist Wallace B. Mendelson of the State University of New York at Stony Brook. "To a great extent, both the amount and quality of your sleep are normal when you believe them to be so. When you believe that lack of sleep is hampering your daytime efficiency, it is, no matter how 'normal' that sleep might appear on a tape recorder."

Sleep Cycles

Scientists have learned most of what they now know about brain activity during sleep since the 1930s, when sleep researchers began to systematically use EEGs to record the electrical activity of the brain and the electro-oculograph to monitor eye movement. In the 1950s, researchers discovered that two distinct sleep states repeat in cycles roughly 90 minutes long. The two states are called REM (rapid eye movement) sleep and NREM (non-rapid eye movement) sleep. During REM sleep, the sleeper's eyes move rapidly under his eyelids, while during NREM sleep, little eye movement occurs. An individual generally goes through four or five 90-minute sleep cycles during the night, progressing from NREM sleep into REM sleep and back again to NREM sleep to begin the next cycle.

Scientists further divide NREM sleep into four separate stages. As people fall asleep, their muscles relax and their heart and breathing rates gradually decrease until they drift out of conscious awareness of the surrounding world

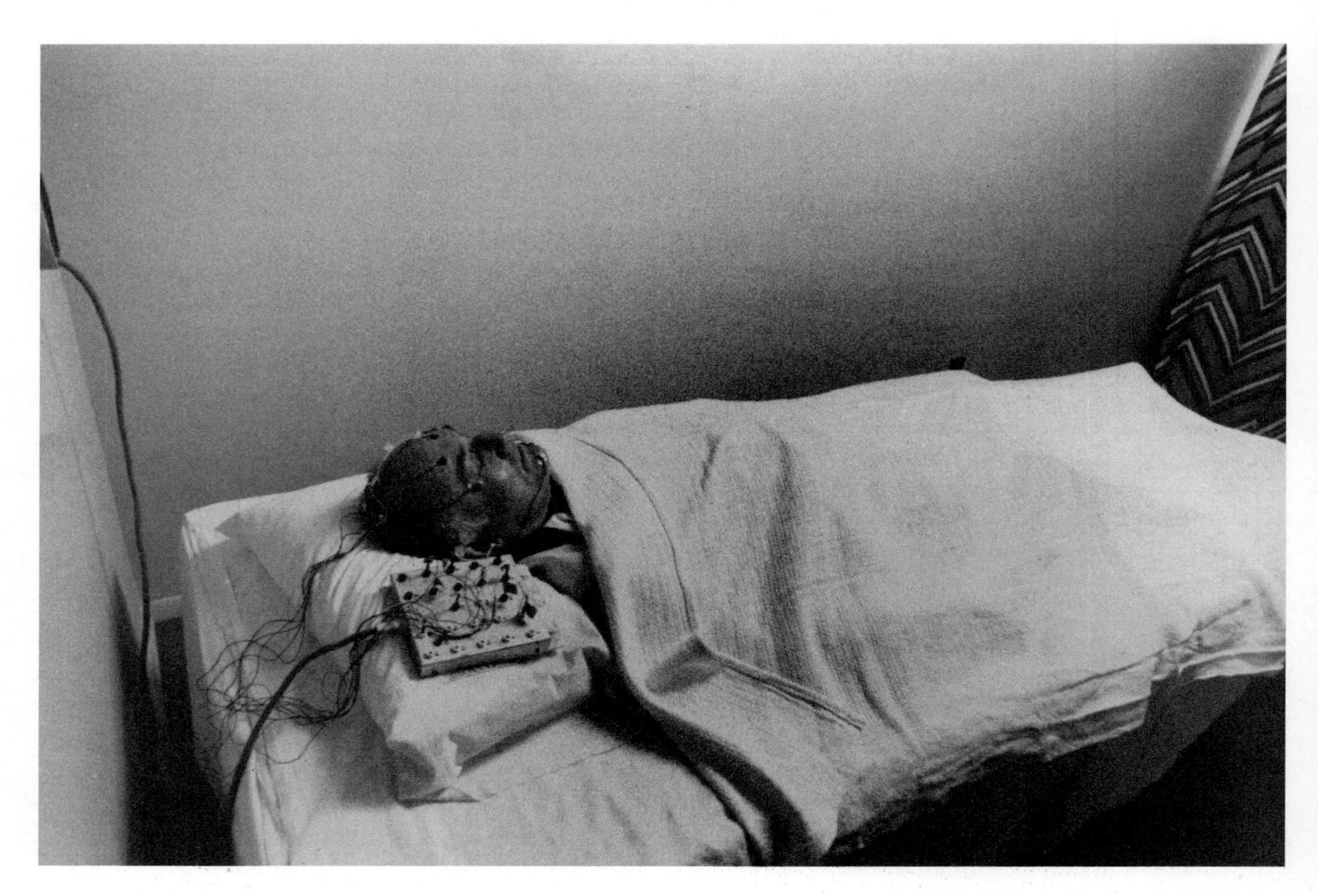

Figure 2. Sleep Disorders. The most commonly reported problem at sleep clinics in the U.S. is not insomnia — trouble beginning or maintaining sleep — but rather excessive sleepiness during the daytime. In a recent study of 3,900 patients with sleep problems, 51 percent were diagnosed as suffering from excessive sleepiness, while 31 percent had insomnia, 15 percent had parasomnia — conditions such as nightmares and sleepwalking that disturb sleeping — and the remaining 3 percent had disorders of the sleep-wake cycle, in which a person's cycles of sleeping and waking are out of phase with the usual hours for those activities. *Courtesy: Sleep-Wake Disorders Center, Montefiore Medical Center.*

into stage one sleep. Typically, about 15 minutes after falling asleep, people enter stage two, which can be distinguished from stage one by differences in the sleepers' EEG patterns. While people awakened from stage one will become alert almost immediately and scarcely realize they have fallen asleep, a person disturbed during stage two will usually take several seconds to become fully awake.

Stage three — which, together with stage four, is known as deep sleep — begins about half an hour after people fall asleep. As sleepers pass through stage three to stage four, their EEG patterns change again as their brains begin to produce slow, rhythmic patterns known as delta waves. People disturbed during deep sleep have drifted so far from waking consciousness that they may take several minutes to awaken completely.

In studies of EEG maps of brain electrical activity, Mendelson and his colleagues discovered that, as sleep deepens, delta waves not only become stronger, but also are produced by more parts of the brain. During stage two sleep, delta-wave activity spreads down from the vertex or top of the brain and, by stage four, covers much of the parietal and frontal regions. Because the map of delta-wave activity during deep sleep resembles a nightcap covering much of the upper portion of the brain, the researchers refer to this effect as a "delta-wave nightcap."

After spending perhaps 40 minutes in stages three and four, sleepers gradually drift back through stages two and one and into REM sleep. REM sleep has also been called paradoxical sleep because it is such an active physiological state that it almost appears as if the brain is awake. During

REM sleep, sleepers' eyes move rapidly under the eyelids, their pulse and respiration rates speed up, and their blood pressure rises. Muscles twitch briefly, and sleepers may become sexually aroused. Most sleep researchers believe that nearly all dreaming occurs during REM sleep. After completing the first REM period, sleepers descend through NREM stages to deep sleep and then ascend back up to REM sleep once again.

The sleep cycles that occur early in a night's sleep are usually dominated by stages three and four, with a total of about 80 to 90 minutes of deep sleep in the first 3 hours of the sleep period. As the night progresses, sleepers spend less time in deep sleep and more in REM sleep until, just before waking, REM sleep generally dominates. According to researcher Laverne Johnson of the San Diego Naval Health Center in California, young adults spend on average about 6 percent of their sleep time in stage one, 50 percent in stage two, 7 percent in stage three, 16 percent in stage four, and 20 percent in REM sleep.

Circadian rhythms. Most organisms follow circadian rhythms — cycles of about 24 hours in length — that govern both metabolic processes and behavior. In mammals, as many as 50 different rhythms exist, including those for sleeping and waking, body temperature, liver function, cell division, and resistance to drugs. The circadian rhythms of humans or other animals placed in an environment without any external clues to day and night will become what psychologists call "free-running," dissociated from normal periods of light and darkness. Humans' circadian rhythms will expand to about 24 ¾ hours. In other words, their "day" will be 24 ¾ hours long and they will go to sleep about 45 minutes later each day.

Because light synchronizes circadian rhythms with a 24-hour day, individuals who travel from one time zone to another require 2 or 3 days to resynchronize themselves with the light-dark cycle at the new location. Different circadian cycles become resynchronized at different rates, producing the disorientation known as "jet lag." Researchers once thought that the light-dark cycles themselves set the rhythms, but new research has shown that the rhythms are established internally.

In animals, two organs produce the circadian rhythms. Research has shown that the source of circadian rhythms in most birds is the pineal gland, a light-sensitive organ in the brain. In mammals, however, the situation has not been as clear. Researchers long believed that a section of the brain called the suprachiasmatic nucleus was the source of circadian rhythms, since surgical removal of the nucleus abolished circadian rhythms and transplants restored them. But that evidence did not necessarily mean that the nucleus created the rhythms; it may, in fact, have simply allowed them to be expressed.

In 1988, however, biologists Michael Menaker and Martin Ralph of the University of Virginia in Charlottesville showed conclusively that the nucleus does, indeed, generate the rhythms. Menaker and Ralph identified a strain of hamsters with abnormally short circadian rhythms. While hamsters usually have a free-running period of 24.2 hours, the new strain's cycle was only 20 hours long. Since the tendency to have a short circadian cycle could be passed on to the hamster's progeny, the researchers concluded that the trait was tied to a specific gene. Cross-breeding experiments showed that if a hamster inherited two copies of the abnormal gene (one from each parent), its free-running period was 20 hours; with only one copy (from one parent), its period was 22 hours.

This finding set the stage for proof that the suprachiasmatic nucleus actually generates circadian rhythms. When Menaker and Ralph transplanted suprachiasmatic nuclei among hamsters, they found that the recipient hamster always developed the free-running period of the donor, whether the donor's period was 20, 22, or 24 hours. Unfortunately, biologists still do not know how either the pineal gland or the suprachiasmatic nucleus generates a circadian rhythm. But the discovery will make it possible for researchers to compare the biochemistry of suprachiasmatic nuclei from hamsters with different circadian cycles to obtain clues about how the organ generates the rhythms. Isolation of the gene itself may also provide new insights.

Sleep Deprivation

Regardless of how much an individual may need, sleep is crucial to both humans and animals. When people lose only 1 or 2 nights of sleep, their capacity to think and behave normally is seriously impaired, and if animals are kept awake for extended periods, such as a month or more, they can actually die. Although researchers do not yet know exactly what causes these problems, new studies, particularly with animals, are beginning to reveal some of the metabolic changes that occur during sleep deprivation.

Animal studies. Experiments conducted with animals over the last three decades have revealed that prolonged sleep deprivation may cause illness or death. Some scientists have criticized these experiments, however, because the researchers failed to consider the possible effects of the stimuli used to prevent sleep. For example, when researchers force an animal to remain constantly active or administer electric shocks to prevent sleep, the stress from the stimuli themselves rather than the sleep deprivation may cause the observed ill effects.

To overcome such criticism, psychologist Allan Rechtschaffen and his associates at the University of Chicago Sleep Research Laboratory in Illinois devised a way to separate the effects of stress from those of sleep deprivation. They designed an apparatus that delivered a relatively mild stimulus to two rats in such a way that one rat was prevented from sleeping while the other was not. In each experiment, two rats were kept in adjacent cages. The floors of the cages were flooded with water, and the rats lived on a rotatable platform that extended into each cage. Whenever the researchers rotated the disk, both animals were forced to walk in the opposite direction to avoid being plunged into the water.

The researchers placed electrodes on both animals to record their brain-wave and muscular activity. A computer could then analyze the impulses from the electrodes and determine whether the animals were awake or sleeping. Whenever the rat chosen for sleep deprivation fell asleep, the computer immediately caused the disk to rotate, rudely awakening the rat when it hit the water. The rotation of the floor was then stopped until the rat, which immediately climbed out of the water, fell asleep again. Both rats thus shared the same environment and were subjected to the same floor-rotation stimulus, but one was not allowed much sleep, while the other could sleep whenever the sleep-deprived rat was awake.

In their first studies, which Rechtschaffen summarized at the 1989 annual AAAS meeting in San Francisco, California, he and his colleagues subjected eight pairs of rats to the disk-rotation experiment for periods ranging from 6 to 33 days. The group found that the sleep-deprived rats got only about one-eighth their normal amount of sleep, while the control rats still got over two-thirds their usual amount. On average, the animals were forced to walk about 0.9 mile per day — not an extreme stress, considering that rats are known to run up to 30 miles a day.

Within 33 days after the experiment began, three sleep-deprived rats died and four others were killed by the researchers because death seemed imminent. These seven rats had grown very weak, had lost muscular coordination, and showed a marked drop in brain-wave amplitude. The eighth sleep-deprived rat was killed after the electrode separated from its skull; it had also shown signs of weakness and lack of coordination, but no decline in brain-wave amplitude. Examination of the dead rats revealed that they had suffered from various ailments, including lesions on their paws and tails, stomach ulcers, scruffy, discolored fur, fluid in the lungs and trachea, and enlarged bladders. Although measurements showed that the rats' metabolic rates also increased so that they ate more food, they still lost weight.

In contrast, aside from weight loss and minor skin irritations, all of the control rats appeared healthy, active, and responsive throughout the experiments. Rechtschaffen and his associates noted that, because the only difference in the treatment of the two groups of rats was sleep deprivation, the ailments and deaths of the deprived rats could only be attributed to sleep loss or a related problem. Although they concluded that their experiments support the idea that sleep serves an essential physiological function, they still do not know why sleep deprivation is so harmful. "We've ruled out a lot of obvious things," said Rechtschaffen, "but the rats still die and we don't know why."

Human sleep-deprivation studies. Lack of sleep can have a strong effect on people's creativity and ability to cope with unfamiliar situations, according to psychophysiologist James A. Horne of Loughborough University in Leicestershire, U.K. Horne subjected 24 healthy college students to an array of tests designed to measure thinking time, fluency, flexibility, and originality of thought, and the ability to elaborate on thoughts. The tests had no "right" answers. In one test, for example, the subjects saw a picture of a person performing an apparently meaningless act. They were then asked to write down all the questions that came to mind that might help explain the situation and to list as many possible causes and consequences of the act as they could. In another test, the students were given pages of small circles, and they were asked to incorporate the circles into as many different pictures as possible.

Scorers unacquainted with the students rated the tests in four categories: flexibility, which involved the abilities to think up a variety of ideas, to shift from one approach to another, or to use a variety of strategies; originality, which involved the potential to produce ideas that are not obvious or commonplace; elaboration, which involved the ability to develop and embellish ideas; and fluency, which involved producing a large number of ideas.

When the students were initially divided into two groups and tested, their results were roughly the same. Then one group was kept awake all night while the second group slept as much as they normally would. In the December

1988 *Sleep*, Horne reported that when both groups were tested again the next day, the sleep-deprived group scored one-third to two-thirds lower than the well-rested group. To determine whether lack of sleep might be impairing the students' motivation, Horne offered them money for high scores on one of the tests, but the results remained the same. He hypothesized that one primary function of sleep is "to repair the cerebral cortex from the wear and tear of consciousness" and that lack of sleep causes something to "go wrong with cerebral function," disturbing some fundamental part of the decision-making process.

Sleep deprivation and depression. Psychologist David Sack and his colleagues at the National Institute of Mental Health (NIMH) have found that sleep-deprivation can reverse some of the effects of severe depression. Although some previous studies had obtained similar results, the effect disappeared after the patients finally slept. Sack has found, however, that mood-enhancement persists if the patients are deprived of sleep for several nights in a row.

As an example of the effect of sustained sleep deprivation, Sack cited the case of a 43-year-old teacher in Maryland who "was so disabled by her depression she couldn't enjoy anything. She couldn't go out. She was hopeless, had suicidal thoughts, no energy. She felt worse, much worse, in the morning. She had a long history of depressions and she'd had just about every treatment known to man.... She had an excellent therapist and had been in psychotherapy for many years, but she just didn't respond."

When the woman eventually went to NIMH for treatment, Sack deprived her of a few hours of sleep each night for 3 days — a technique called partial sleep deprivation. "After 3 days," he said, "her mood completely improved and stayed improved. We followed her for the next 6 months, and although she did relapse after that time, she improved again in response to the partial sleep deprivation." Given these and similar results with a few other patients, Sack, now at Los Altos Hospital in Los Angeles, California, and other researchers have begun a larger study of partial sleep deprivation to determine whether the therapy can be more widely applied in people suffering from depression.

Sleep Disorders

At one time or another, almost everyone has trouble sleeping, and large numbers of people regularly have trouble with their sleep. In a 1973 Gallup poll, 15 percent of adult Americans complained that they suffered from chronic insomnia, which sleep researchers define as difficulty beginning or maintaining sleep. Even so, according to a 1979 National Academy of Sciences report, only 6 percent of people with sleep problems seek advice or treatment from a physician. When people complain of sleeping problems, such as difficulty falling asleep, awakening in the early morning, or repeated interruptions during sleep, they usually lump them together under one catchall term, insomnia.

Surprisingly, the most prevalent sleep problem encountered in clinics is not insomnia. In 1982, a team of researchers headed by Richard Coleman at the Stanford University Medical Center reported that the most common difficulty among patients visiting sleep clinics was excessive sleepiness during the daytime. In the Coleman study, organized by Project Sleep and the Association of Sleep Disorders Centers in Washington, D.C., the researchers reviewed the medical records and polysomnograms — tracings of brain and other physiological activity during sleep — of 3,900 patients at U.S. sleep disorder centers. About 51 percent of the patients were diagnosed as suffering from hypersomnia, which is defined as "inappropriate or uncontrollable daytime sleepiness," while 31 percent were found to have insomnia.

Of the nearly 2,000 clinic patients who experienced excessive daytime sleepiness, 43 percent suffered from sleep apnea, periodic pauses in breathing that disrupt sleep at night and leave victims sleepy during the daytime. Twenty-five percent of the patients suffered from narcolepsy, a disorder causing an uncontrollable tendency to fall asleep in the daytime, regardless of the quality and length of nighttime sleep. Among the more than 1,200 patients categorized as insomniacs, the most common cause was psychiatric disorders, which accounted for about 35 percent of the cases. About 15 percent of the insomniacs had trouble falling asleep because of anxiety not associated with a psychiatric disorder, and another 12 percent suffered from insomnia as a result of alcohol or drug dependency.

Parasomnia, the clinical term for sleeplessness caused by such problems as epileptic seizures, head-banging during sleep, heartburn, nightmares, and sleepwalking, accounted for about 15 percent of all sleep complaints. About 3 percent suffered from a fourth major category of sleep disturbance: disorders of the sleep-wake schedule. Although patients with this recently recognized condition often complained of insomnia or excessive sleepiness, research has shown that the problem is actually that their sleep-wake cycles are out of phase with conventional hours of waking and sleeping. Some of the patients exhibited a disorganized sleep pattern that followed no regular 24-hour pattern, but many experienced delayed sleep phase syndrome (DSPS), which is described below.

Sleep apnea. Sleep apnea is caused by repeated constriction of tissues in the upper throat, blocking the passage of air, or by a malfunction in the brain's respiratory center. People with sleep apnea can stop breathing literally hundreds of times each night, severely reducing the quality of their sleep. Although the disorder was once thought to be rare, studies in the 1980s have shown that it affects about 1 percent of the entire population, and perhaps up to 3 percent of adult males. This condition is most common among obese middle-aged and elderly men, but it can affect women and children as well. Physical abnormalities such as oversized tonsils can predispose a person to sleep apnea, and it can be exacerbated by alcohol, sleeping pills, and tranquilizers, all of which increase the throat's resistance to air flow as one breathes during sleep.

Sleep apnea is closely linked to heavy snoring, which has been associated with an increased incidence of hypertension, strokes, and heart attacks. While some researchers believe that sleep apnea is a contributing factor in many deaths, no one has proved that hypothesis. In 1986, psychologist Donald Bliwise and his colleagues at Stanford's Sleep Disorders Center began studying apnea and mortality in 200 elderly volunteers who lived in the Palo Alto, California area. By the end of 1988, 20 of the volunteers had died from various causes. According to Bliwise, sleep apnea was more prevalent among those who died of cardiovascular disease than among those who died of cancer or other causes. He also found that more sleep apnea sufferers died between midnight and 6 A.M. than at any other time.

In another ongoing study, psychiatrist Susan Ancoli-Israel and her colleagues at the University of California at San Diego are studying 235 nursing-home patients in southern California, monitoring their sleep for 1 or 2 nights and then following the patients' health over a period of years. By early 1989, Ancoli-Israel had found that women patients with frequent episodes of sleep apnea ran a significantly higher risk of dying from all causes than did patients without the disorder. She also found that the sleep of all the nursing home patients was "enormously disturbed."

By the late 1980s, physicians were exploring a variety of possible treatments for sleep apnea. The most successful is tracheostomy, in which a surgeon creates a permanent opening in the throat to permit the free flow of air. Nearly all patients respond well to tracheostomy and no longer have symptoms of apnea, according to psychologist Frank Zorick of the Henry Ford Hospital Sleep Disorders and Research Center. But since most physicians would prefer to use less drastic measures, Zorick and other researchers have been evaluating two other therapies.

The simplest of these is nasal continuous positive airway pressure (CPAP), which opens the upper airway by delivering air under pressure through a nasal mask that patients wear during sleep. But 30 percent of sleep apnea patients cannot tolerate the mask or will not use it for other reasons. Those patients can be treated with uvulopalatopharyngoplasty (UPPP), in which surgeons widen the airway by removing the uvula and other unneeded tissues. In 1988, Zorick reported that the two procedures appeared to be about equally effective: about 80 percent of the patients receiving either therapy reported less daytime fatigue.

Behavioral therapies can also help some patients. Because most snoring and apnea episodes occur when patients sleep on their backs, some people improve when they are induced to sleep on their stomachs or sides. Psychologist Rosalind Cartwright of Rush-Presbyterian-St. Luke's Medical Center in Chicago has been able to obtain dramatic results by using an electronic device that beeps 15 seconds after patients begin snoring on their backs, awakening them so that they turn over. She found that when patients grew accustomed to sleeping on their sides, episodes of sleep apnea decreased from 50 to 80 per hour to 5 or less per hour.

Narcolepsy. Although most people think of a narcoleptic as a person who suddenly falls asleep at inappropriate times, victims of narcolepsy also share a number of other symptoms, including excessive daytime sleepiness, cataplexy (a sudden loss of muscle strength following an emotional event), hallucinations, and the sudden onset of REM sleep. Such symptoms often make it very difficult for the estimated 250,000 narcoleptics in the U.S. to keep a job.

Some scientists may have found a genetic basis for narcolepsy. During 1985, a team of researchers from Canada, France, Japan, the U.K., and the U.S. reported that 150 of 152 narcolepsy patients studied had a specific antigen — a protein on the surface of cells that plays a role in the immune system's recognition of the cells — called HLA-DR2 on their white blood cells. This antigen normally occurs in no more than one-third of the general population.

Medical microbiologist Hugh McDevitt of Stanford has suggested that the association of narcolepsy with an antigen linked to the immune system may mean that narcolepsy is an autoimmune disorder in which the patient's immune system attacks cells in the brain or that the disorder is caused by a gene that lies very close to the HLA-DR2 gene. One possibility is that the gene codes for a receptor on the surface of cells that binds to Factor S, a recently discovered chemical that induces sleep. In 1988, biochemist Manfred J. Karnovsky and his colleagues at Harvard Medical School in Boston found that white blood cells taken from 12 narcoleptics did not have receptors for Factor S, while cells from healthy patients did. Their results suggest that narcoleptics may not respond properly to naturally occurring chemicals that induce sleep.

Resetting the biological clock. When working and sleeping according to a conventional schedule, DSPS sufferers typically complain that they cannot fall asleep until much later than they would like. But when people with DSPS are allowed to follow their natural sleep-wake cycles, they sleep well and awaken refreshed. While the body temperatures of DSPS patients fall with sleeping and rise with waking in the same pattern as those of normal sleepers, their sleep and temperature cycles typically run 3 or more hours out of phase with conventional bedtimes and waking times. Before DSPS was recognized as a physical disorder, many physicians advised patients suffering from this problem to drink alcohol or take drugs before going to bed. Unfortunately, these approaches seldom worked for long and sometimes created dependency.

A team of scientists from the Albert Einstein College of Medicine in New York City and Harvard Medical School has successfully tested a new drug-free treatment for DSPS called chronotherapy, which literally resets the biological clocks of DSPS sufferers to conform to conventional waking and sleeping times. In chronotherapy, the patient simply goes to bed progressively later each day until he or she reaches the bedtime of choice.

For example, a DSPS patient who would usually fall asleep at 6 A.M. but who preferred to fall asleep at midnight might delay his bedtime to 9 A.M. on the first day of therapy, noon on the second day, and so forth, until by the sixth day his bedtime had progressed to midnight. Remarkably, this simple technique has worked even for one woman who had suffered from DSPS for more than 50 years.

Insomnia and memory loss. Many physicians prescribe sleeping pills for patients with short-term insomnia. Although these medicines help the patients get to sleep, the sleep may not be completely normal. Moreover, psychologist Cheryl Spinweber of the Naval Health Research Center in San Diego has found that the drugs can interfere with both memory and thought processes. Spinweber studied the effects of one such drug, triazolam, on 56 sailors, all in their early twenties. She reported in 1986 that triazolam users awakened during the night to memorize word lists did not remember the words the next morning as well as did volunteers who had taken a placebo. The sailors on triazolam could not easily recall, identify, and match nouns with adjectives if they had learned the word pairs after having the drug. The drugged sailors also took longer to perform a task that involved sorting playing cards by suit.

Modifying Work Schedules to Enhance Productivity

"In the past 15 years, the growth of research on sleep and the biological clocks that control it has led to significant discoveries about the many ways in which these processes influence human health and functioning," stated a 1988 report by the Association of Professional Sleep Societies' Committee on Catastrophes, Sleep, and Public Policy. "These influences can occur without our awareness of them and affect us more profoundly than we realize," concluded the report, which was published in the journal *Sleep*. Research during that 15-year period has led to a better understanding of the consequences of sleep disorders, improper sleep schedules, shift work, and daytime sleepiness.

One major discovery has been that the neural processes controlling alertness and sleep produce an increased tendency to sleep and a diminished capacity to function during the early morning hours from 2 A.M. to 7 A.M. and, surprisingly, during the afternoon hours from 2 P.M. to 5 P.M. These patterns can greatly influence a person's susceptibility to heart attacks and to vehicular and industrial accidents.

Obstetricians have long known, for example, that spontaneous labor and birth are most likely to occur in the middle of the night. New evidence obtained during the mid-1980s shows that death also occurs most often in these "sleepiness zones." More than 50 studies involving 437,000 deaths show that the largest number of deaths from all causes occurs between 4 A.M. and 6 A.M., with a smaller peak between 2 P.M. and 4 P.M. Similarly, peaks in the incidence of heart attacks are "related to the rhythmic processes underlying sleep and its timing," according to the 1988 report.

While most traffic accidents happen during daytime hours when traffic is heaviest, single-vehicle accidents such as driving off the road (accidents more likely to be related to sleepiness) follow a pattern similar to that of birth and death. Studies of 6,000 single-vehicle accidents attributed to "falling asleep at the wheel" reveal a major peak in incidence between midnight and 7 A.M., with most accidents occurring between 1 A.M. and 4 A.M. A secondary peak occurs between 1 P.M. and 4 P.M.

One study of about 500 single-vehicle truck accidents showed a peak rate between 1 A.M. and 7 A.M. Another study of 13,700 truck accidents in the state of Washington found, as expected, that the number of accidents peaked during midday, when traffic was highest. But the study also showed that serious accidents involving hazardous materials were most common between 6 A.M. and 9 A.M, usually under good driving conditions on a straight road. Finally, a study of emergency braking incidents on the German Federal Railways found that problems caused by operator errors peaked between 3 A.M. and 6 A.M. and again between 1 P.M. and 3 P.M.

The most serious accident at a commercial U.S. nuclear power plant occurred at the Three Mile Island reactor at 4 A.M. on 28 March 1979. Between 4 A.M. and 6 A.M., shift workers at the plant failed to recognize a loss of core coolant water caused by a stuck valve. Although a mechanical problem precipitated the incident, this human error of omission and the subsequent failure to take the proper corrective action caused the problem to escalate to the near meltdown of the reactor later that morning.

More recently, three other serious incidents at U.S. nuclear power plants and one abroad involved human errors committed during the early morning hours. On 9 June 1985, the Davis-Besse reactor at Oak Harbor, Ohio, went into automatic shutdown after the reactor suffered a total loss of cooling water at 1:35 A.M. The problem worsened when an operator pushed two wrong buttons in the control room, thereby inactivating an auxiliary cooling system. Although a combination of equipment malfunctions and human errors made the situation dangerous, corrective action did eventually stabilize the reactor.

On 26 December 1985, the Rancho Seco nuclear reactor near Sacramento, California, lost power to the integrated control system at 4:14 A.M. For a variety of reasons, including equipment design flaws, inadequate training, and human errors, operators were slow to regain control of the plant. In a third U.S. incident, which occurred in 1987, the Philadelphia Electric Co. was forced to shut down its Peach Bottom Power Station after investigators found control-room operators sleeping on night and weekend shifts. It should come as little surprise, then, that Soviet officials finally acknowledged that the nuclear power plant catastrophe at Chernobyl began at 1:23 A.M. as the result of human error. The limited information available, however, makes it difficult for scientists outside the U.S.S.R. to assess the role that sleep- or fatigue-related errors may have played in the disaster.

Committee's recommendations. "Sleep and sleep-related factors appear to be involved in widely disparate types of disasters," wrote the Committee on Catastrophes, Sleep, and Public Policy, noting that these factors have been given little consideration in follow-up studies of disasters such as the Challenger and Chernobyl incidents. The group suggested that both government and industry leaders pay more attention to the dangers associated with working and traveling during the hours of increased sleepiness. It also recommended that programs be developed to identify and reduce sleep-related human errors.

Managers should also limit active-duty hours for all personnel to ensure adequate time for sleep between shifts, and attention should be given to identifying poor schedules for shift work and substituting schedules that promote health and safety. Finally, laboratory and field research is needed to pinpoint the underlying biochemical processes that make people sleepy at different times. By identifying these mechanisms, researchers will be better able to develop effective countermeasures to minimize the consequences of sleep-related problems. "Such countermeasures will be essential," the committee concluded, "if society is to continue its relentless push to around-the-clock operations in all aspects of life both on the planet and as we broach the frontiers of space."

Adjusting work schedules. Physiologists and sleep researchers have already begun applying their knowledge of the human sleep-wake cycle to work schedules, designing shifts to increase employee satisfaction and productivity at companies operating 24 hours a day. Harvard Medical School sleep specialist Charles Czeisler summarized these efforts at the 1989 AAAS meeting. In the early 1980s, Czeisler had undertaken one of the first worksite studies in cooperation with psychologist Martin Moore-Ede, now at the Institute for Circadian Physiology in Boston, and Stanford's Richard Coleman. They compared the sleep and productivity of a group of men working on rotating shifts at the Great Salt Lake Minerals and Chemicals Corp. in Ogden, Utah, with those of a control group of men working on nonrotating shifts at the same company.

Figure 3. Fatigue-Related Accidents and Performance Errors. Recent sleep research indicates that a wide range of accidents and performance errors tend to occur at times of the day and night when humans are most susceptible to sleepiness — between the hours of 1:00 A.M. and 7:00 A.M., and to a lesser extent, between 2:00 P.M. and 5:00 P.M. A 20-year study of gas meter reading errors in Sweden, as well as studies in the U.S. of single-vehicle automobile accidents attributed to falling asleep at the wheel, revealed time distribution patterns shown in the top and bottom graphs, respectively. In the graphs, "N" refers to the total number of errors or accidents studied. *Courtesy: Raven Press, Ltd. Reproduced from Mitler et al., "Catastrophes, Sleep, and Public Policy: Consensus Report,"* Sleep, *Vol. 11, No.1 (1988), pp. 100–109.*

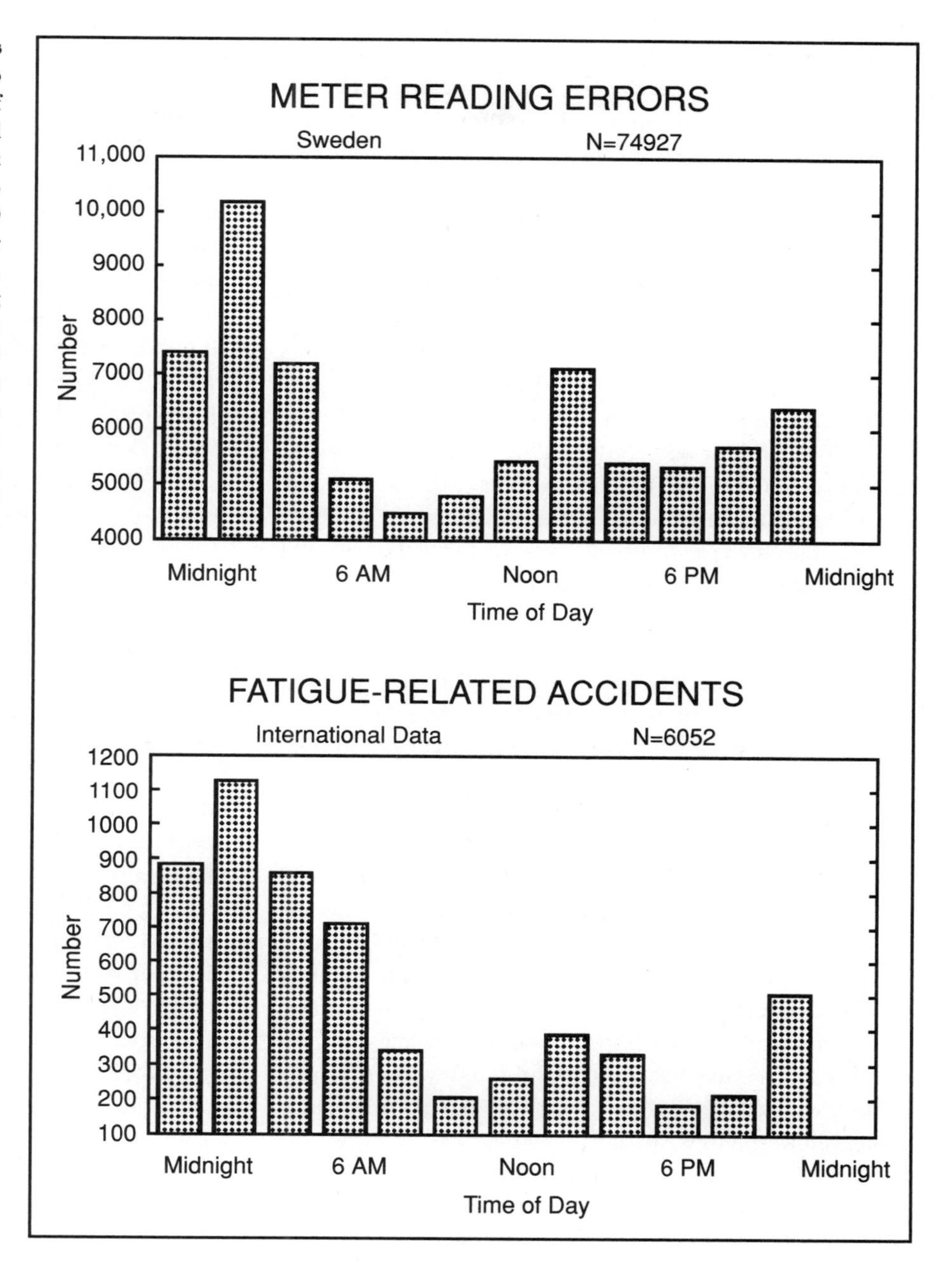

The men on rotating shifts worked on a given shift for 7 days before rotating to the earlier 8-hour shift. A worker on a rotating shift who began working on the night shift (midnight to 8 A.M.), for example, would work on the evening shift (4 P.M. to midnight) the next week and then the day shift (8 A.M. to 4 P.M.) the following week, repeating the full cycle every 3 weeks.

Czeisler and his associates found that workers on rotating shifts had significantly more problems with insomnia than those on nonrotating schedules. About two-thirds of the workers on rotating shifts complained of sleeping poorly, compared to about one-quarter of the workers on nonrotating shifts. Twenty-nine percent of the rotating-shift workers said that they sometimes fell asleep at work, compared to only 6 percent of the nonrotating day workers. Ninety percent of the rotating-shift workers complained that the shifts changed too often.

After evaluating these findings, the Harvard and Stanford team obtained permission to test another schedule at the Ogden plant, one they believed would be better adapted to the natural sleep-wake cycle. Their new schedule was based on research showing that people generally find it easier to overcome jet lag induced by westward travel, in which clocks are shifted back, than that induced by eastward travel, in which clocks are shifted ahead.

The new schedule rotated shifts to later hours, from night to day to evening (a shift in the delay direction), instead of rotating to earlier hours in the advance direction, from night to evening to day. Czeisler's group also suggested that the shifts be changed just once every 3 weeks rather than weekly; this would allow workers sufficient time to adapt fully to each shift before changing to the next. To test their new shift schedules, the researchers then divided the rotating-shift workers into two groups. Some workers were put on a delay-shifting schedule that changed each week, while others were on a delay-shifting schedule that changed every 3 weeks.

After 3 months, the workers preferred the new delay-shifting schedule by a three-to-one margin. Although most workers on the weekly delay rotation still felt that their shifts changed too often, they were much less dissatisfied than they had been with their previous advance rotation. Only about 20 percent of the workers on the 3-week schedule felt that the schedule changed too quickly. After 9 months, worker productivity, measured by the rate of mining and processing potash, was about 25 percent higher than it had been a year earlier on the old shift-rotation scheme.

Czeisler has since worked with several other companies to initiate similar changes in work schedules. Productivity per person-hour at Amax Coal Co. in Indianapolis, Indiana, for example, increased 10 percent in 1987 when the company switched to a schedule that rotated workers' shifts once every 28 days, with a 7-day break before workers changed to the new shift.

At the beginning of 1986, the Philadelphia police department began an 11-month study in which the work schedules of 300 detectives were rotated in the delay direction rather than the advance direction, as had been the previous policy. The detectives' time on each shift was also increased from 8 days to 18. During the experiment, the department's rate of on-the-job automobile accidents dropped 40 percent. Also, compared with the old schedules, half as many officers reported problems with daytime fatigue, and 25 percent fewer said they slept poorly. The city's police commissioner Willie Williams said that he favored the new schedules, but that implementing the new program citywide would be too costly. By the end of 1988, the police union, which argued that scheduling should be part of contract negotiations, had taken the matter to arbitration.

Factor S — A Sleep Vitamin?

In the early 1980s, physiologists John Pappenheimer and Manfred Karnovsky of Harvard Medical School and James Krueger of the University of Tennessee at Nashville isolated a sleep-promoting chemical called Factor S. When administered in even minute amounts, this chemical lengthens the time animals sleep. The researchers believe it will have the same effect in humans, but by the end of 1988, no experiments had yet been tried on humans. Karnovsky, Krueger, and other researchers have since found that many other similar chemicals can also induce sleep. Surprisingly, Factor S and these related chemicals can also induce fever and stimulate the immune system to increase its ability to fight infectious agents. No one knows why these things occur and what relationship, if any, these functions have to sleep.

Karnovsky and Pappenheimer first observed in the 1970s that mammals' cerebrospinal fluid (CSF) contained a sleep-promoting chemical; this chemical was presumably Factor S, but the quantities present in the fluid were so small that the researchers were not able to collect enough to determine its structure. They later found that a sleep-promoting substance is also present in human urine. Pappenheimer, Karnovsky, and Krueger were eventually able to extract 30 millionths of a gram of the substance from 4.5 tons of human urine, enough to enable them to identify its structure and to study its activity. When a dose of just a few billionths of a gram of the purified substance was injected into the brains of rats, rabbits, and cats, it produced a 50 percent increase in the length of the animals' sleep periods.

In the mid-1980s, the three researchers, along with Stephen Martin and Klaus Biemann of MIT, reported that Factor S has an unusual structure. It is composed of two sugars, *N*-acetylglucosamine and *N*-acetylmuramic acid, and four amino acids, all linked together. This structure is surprising because *N*-acetylmuramic acid and one of the amino acids, diaminopimelic acid, are normally produced by bacteria but not by humans or other mammals. In fact, Factor S bears a very strong structural resemblance to the building blocks present in the cell walls of certain bacteria, including the *Escherichia coli* normally found in the human intestine.

Some critics have argued that the evidence of a bacterial origin for Factor S indicates that the substance may actually be a urine contaminant with no normal biological function in humans. To refute such criticism, Karnovsky and Zhai Sen of Harvard looked for molecules containing muramic acid — which is not produced in mammalian cells — in rat tissues that had been sterilized to prevent bacterial contamination. They found muramic acid-containing molecules in rat brain, kidney, and liver tissues, suggesting that such chemicals have a physiological function in animals.

Krueger and Karnovsky have studied a number of synthetic muramic acid-containing peptides prepared by Edgar Lederer and his colleagues at the Université de Paris-Sud in Orsay, France, and by 1988 had found that several of these also induce sleep in animals. The sleep-inducing substances share two characteristics: each contains muramic acid, and each contains the unusual D-form of the amino acid isoglutamine rather than the naturally occurring L-form. All of the peptides that promoted sleep also produced a fever and stimulated the recipient's immune system.

Perhaps the most surprising finding is the discovery by Krueger and L. Chedid and his colleagues at the Pasteur Institute in Paris that Factor S is closely related to a family of chemicals known collectively as interleukin-1. Mammalian cells produce interleukin-1 in response to viral and bacterial infections; it stimulates the immune system to fight off the infection, causes fever in the brain, and is thought to be responsible for the muscular aches that often accompany infections. Krueger, Chedid, and their colleagues have found that interleukin-1 promotes sleep.

All of these findings seem to indicate that at least some cells of the body can use chemicals, such as muramic acid, derived from bacteria to produce chemical agents that influence sleep. If bacteria rather than human cells are its source, then Factor S can be considered a sleep "vitamin."

Stress, Temperament, and Health

Psychologists have long believed that the inability to cope with stressful events causes or contributes to illness, but until recently they could marshal little hard evidence linking stress or psychological traits to bodily changes that make a person vulnerable to disease. In 1919, Japanese researcher Tohru Ishigama undertook one of the earliest attempts to find such links, measuring the activity of phagocytes — white blood cells capable of engulfing and devouring bacteria, protozoa, and other organisms and materials harmful to the body — in tuberculosis patients. Ishigama found that phagocyte activity decreased during times of emotional distress over business, family, or personal problems, leading him to conclude that the stresses of contemporary life decrease the body's immune response and make it more susceptible to disease. While additional evidence for the link between personality traits, stress, and disease risk has been collected by other investigators over the past 70 years, most of this research has taken place since 1970.

In 1987, the Institute for the Advancement of Health, headquartered in New York City, published three large volumes edited by psychiatrist Steven E. Locke of the Harvard Medical School and summarizing all of the published studies on the relationship between the brain and disease. In an interview published in the *New York Times*, Locke noted that the quantity of the studies was impressive. He cautioned, however, that many of them suffered from severe limitations. "With the exception of a few areas, like hypertension and headaches, there are very few studies in behavioral medicine that have true randomized, well-controlled clinical trials of the sort that are standard in medical research. There are many studies that show improvement after treatment, but the problem is that some patients may come [to the physicians] when their condition is at its worst. Some of the improvement you see may be due to the normal oscillation in the condition, not the treatment itself."

Books on this subject have also proliferated. *Love, Medicine & Miracles*, by surgeon Bernie S. Siegel of Yale University in New Haven, Connecticut, *Minding the Body, Mending the Mind*, by psychologist Joan Borysenko of Harvard Medical School, and *Hypnosis and Behavioral Modification*, by psychologists Daniel Brown of Cambridge Hospital in Massachusetts and Erika Fromm of the University of Chicago in Illinois have all tried to present recent research in a clear and understandable fashion.

Taken together, such books and the massive summary compiled by Locke provide a fascinating overview of one of the most intriguing questions in medicine: What influence does our state of mind have on our physical well-being? Like so many sagas of medical discovery, the story begins with animal experiments.

Stress in Laboratory Animals

Many studies have shown that stressed animals fight disease much less effectively than unstressed animals. Psychologists Vernon Riley and Darrel Spackman at the Pacific Northwest Research Foundation in Seattle,

Washington, have presented some of the most convincing evidence of this phenomenon. After designing a special cage for mice that would provide them with an undisturbed environment and minimize the stress of confinement, Riley and Spackman mounted the cage on a modified phonograph turntable that could rotate it at rates of 16 to 78 revolutions per minute (rpm). This device allowed the researchers to vary the amount of stress inflicted on the animals in the cage.

In most of their experiments, Riley and Spackman rotated the cages at 45 rpm for 10 minutes each hour. In one test, they determined that the stressed mice's white blood cell counts — an indicator of immune function — dropped to half the normal level after just two 10-minute periods. A full day of such stress caused the mice's thymus glands to begin to shrink and continue shrinking to as little as 40 percent of their normal size. Because the thymus gland, located at the base of the neck, helps the body fight infectious diseases, its reduction in size was an indication that the animals' immune systems were impaired by the stress.

Riley and Spackman hypothesized that the combination of a low white blood cell count and thymus gland shrinkage would seriously impair the mice's ability to fight disease. To test this idea directly, they implanted lymphosarcoma cells, a malignant tumor of the lymphatic tissue, into two groups of mice. A healthy animal's body would normally reject the tumors within a few days. One test group received 10 minutes of stress every hour for 3 days, beginning on the fourth day after the tumors were implanted, while the second group was not stressed. In the unstressed mice, the tumors followed the expected pattern of restrained growth and eventual rejection. In the stressed group, however, the tumors grew rapidly and the animals died. In fact, the faster the scientists spun the turntable, the faster the tumors grew.

Another key study, designed by NIH psychologist Novera Herbert Spector, showed that the immune system can be trained to respond to external signals in much the same way that Russian physiologist Ivan Pavlov conditioned dogs to salivate in response to a bell. Immunologists Brent Solvason and colleagues at the University of Alabama School of Medicine in Birmingham, who carried out the experiment, exposed two groups of mice to the distinctive odor of camphor for nine separate 3-hour periods. The odor itself had no effect on the immune systems of the mice, but each time the researchers exposed the animals to the smell, they simultaneously injected one of the groups of mice with a chemical that stimulated the activity of natural killer cells (white blood cells that engulf and destroy foreign cells, bacteria, and viruses) and the other group with a placebo (a harmless chemical that had no effect on the immune system).

During the tenth exposure to the odor, the mice received no injections. Nevertheless, the mice that had received the chemical injections all showed a large increase in natural killer cell activity. According to Spector, such results indicate that the brain can unconsciously activate the immune system even in the absence of chemical stimuli or infectious agents that typically trigger an immune response. Immunologist John Bienenstock and his colleagues at McMaster University Medical Center in Hamilton, Ontario, obtained similar results when they exposed rats to both lights and sounds while injecting them with egg albumen. Such injections provoke an immune response similar to that which occurs in allergies. In 1989, Bienenstock's group reported that, after conditioning, the animals produced the immune response when exposed to the audiovisual cue alone.

Conditioned fear or anxiety can also induce a dramatic change in immune function, according to psychologist Donald T. Lysle of the University of Pittsburgh School of Medicine in Pennsylvania. Lysle and his colleagues trained rats to associate a clicking sound with a mild electric shock the rats received shortly after hearing the sound. The click itself soon produced fear. At a 1988 meeting of the Federation of American Societies for Experimental Biology in Las Vegas, Nevada, Lysle reported that lymphocytes isolated from the rats after exposure to the clicking alone divided less than one-third as vigorously as cells from nonexposed rats. Such reduced lymphocyte vigor would impair an animal's response to infection.

Examining a different type of stress, psychologist Christopher Coe of the University of Wisconsin at Madison found that infant monkeys separated from their mothers for 24 hours also suffered from impaired immune functions. In a 1987 meeting of the American Psychological Association, Coe reported that the impairment persisted for as long as 2 months after the separation. He cautioned that his study does not imply that children dropped off for day care will suffer a similar immune impairment. The monkey's separation, he said, was more analogous to that of children removed from a parent under highly stressful conditions, such as by a sudden hospitalization.

The effects of stress can be reduced if the animals are allowed to take steps to stop the stress, according to psychologists Steven Maier of the University of Colorado at Boulder and Mark Laudenslager of the University of

Colorado Health Sciences Center at Denver. The researchers subjected rats in a box to periodic electric shocks administered through an electrode attached to their tails. However, the rats could stop the shocks simply by turning a wheel within the box, and they quickly learned to do this. A second group of rats in an adjacent cage received electric shocks whenever the first group did, but they could not stop the shocks.

Maier and Laudenslager then took samples of white blood cells called T-cells, and measured those cells' capacity to multiply in response to a stimulus, an indicator of immune functioning. They found that T-cells from the rats who could control their exposure to stress multiplied as rapidly as T-cells from unstressed control animals, while T-cells from the rats that could not control the stress multiplied slowly. The researchers concluded that the stress of the shocks interfered with immune system response only in animals that could not control their exposure to stress.

Stress in Humans

A growing body of evidence in the emerging field of psychoneuroimmunology has established specific links between people's emotions, the body's immune defenses, and the progression of serious diseases. Stress, shock, lack of sleep, exam anxiety, depression, unemployment, marital breakup, and grief all can depress the body's immune system. Conversely, some studies have shown that a positive mental attitude can help a patient combat a serious illness.

Separation or divorce. Marital disruption, either through divorce or death, can be one of life's most stressful events. In fact, psychologist Janice K. Kiecolt-Glaser and immunologist Ronald Glaser of the Ohio State University College of Medicine in Columbus believe that divorce and death of a spouse are the most powerful predictors of future physical and emotional illness.

In 1987, for example, they studied a group of Columbus women who had been separated from their husbands for 1 year or less and found that these women had poorer immune function than did carefully matched women who lived with their husbands. Moreover, among the divorced and separated women, the researchers discovered a correlation between both shorter separation periods and greater attachment to or preoccupation with the ex-husband and greater depression, loneliness, and a poorer immune function. A 1988 study showed a similar effect in divorced and separated men.

The death of a spouse creates an even stronger effect. Since the mid-1970s, psychiatrist R. W. Batrop and his colleagues in New South Wales, Australia, have studied the emotional and physical effects of bereavement. Batrop's group was among the first to report a measurable weakening of the body's immune system following severe psychological stress. They found, for example, that surviving spouses had lower T-cell activity for as long as a year after their loss.

More recently, psychiatrist Michael Irwin of the University of California School of Medicine in San Diego studied 37 women whose husbands were dying of lung cancer. After evaluating them for symptoms of depression, including insomnia, weight loss, mood swings, and contemplation of suicide, he drew blood samples both before and after the death of their husbands. In 1987, Irwin and his colleagues reported that the women anticipating the death of their husbands and those who had already lost them had lower-than-normal blood concentrations of T-cells and natural killer cells. Furthermore, among the women whose husbands had died, Irwin found the lowest levels of T-cell and natural killer-cell activity in the women who were most depressed.

Although the disruption of a relationship can stress an individual, the mere existence of a marital partner is not necessarily beneficial to health. Data accumulated since the early 1970s suggest a relationship between the quality of a marriage and the health of the partners: unhappily married people consistently report poorer health than either divorced or happily married individuals of the same age, sex, and race. Kiecolt-Glaser and Glaser have also reported that both men and women stuck in unhappy marriages displayed poorer immune response on several different measures than did people in happier marriages.

Stress and herpes. A variety of evidence has accumulated linking stress to outbreaks of herpes among infected people. Herpesviruses — which cause cold sores on the mouth, venereal disease, and other infections — are particularly suitable for psychoneuroimmunological studies. Unlike other common viruses such as rubella (measles) or poliovirus that disappear from the body after the infection has run its course, herpesviruses remain hidden in the bodies of infected individuals for the rest of their lives, erupting periodically to produce lesions. Researchers can thus assess the activity of herpesviruses by measuring the concentration (titer) of antibodies in the blood. The higher the antibody titer, the more active the virus has recently been.

Psychologist Stanislav Kasl of Yale University in New Haven, Connecticut, has studied antibodies to the Epstein-Barr virus (EBV), a herpesvirus that causes infectious mononucleosis, among cadets at the U.S. Military Academy at West Point, New York. Over a 4-year period, Kasl and his colleagues gave all the entering cadets blood

Figure 4. Stress and Space Flight. In studies conducted during the early 1970s, NASA physicians found that following splashdown, a very stressful day for astronauts, the Skylab and Apollo crews had significantly higher numbers of white blood cells, which combat infection, than they had before their flights began — evidence that the body's immune system responds strongly to stress. *Courtesy: NASA.*

tests to determine whether they had been exposed to EBV. They also interviewed the cadets about their family histories and their expectations for future accomplishments.

Each year, the team found that about one-fifth of the entering cadets were infected with the virus, although only one-quarter of those actually had developed the symptoms of mononucleosis. Furthermore, infections were most likely to occur among cadets with high levels of motivation and whose fathers were "overachievers" but who were themselves poorer academic performers. These students spent the longest time recovering in the infirmary.

Kiecolt-Glaser and Glaser have also discovered abnormally high EBV titers among recently separated women and separated and divorced men. In addition, the men had higher titers for herpes simplex virus type 1 (HSV-1), which causes genital herpes — suggesting that they were less able then other men to fight such infections. Kiecolt-Glaser and Glaser also found that family members who care for victims of Alzheimer's disease have higher titers for EBV than do other individuals of the same age and background in the same community, again an indication that their immune systems may be weakened as a result of stress.

The Ohio State researchers' evaluation of the immune function of medical students at their university also detected high titers against EBV, HSV-1, and cytomegalovirus (which produces a mononucleosis-like disease) during final exams and low titers immediately after summer vacations. When the researchers tracked some of the medical students over an entire year, they found a cyclic pattern of titer changes, with the highest levels occurring during exam periods. They also found that T-cells and B-cells showed reduced activity during exams and that the concentration in the blood of interferons — peptides that help fight invading viruses — decreased sharply at the same time, indicating that the students were most susceptible to disease then. But for reasons still unknown, according to Kiecolt-Glaser, "These students don't get sick."

Depression. Like stress, depression can inhibit the immune system. In 1986, immunologist Steven Schleifer and his colleagues at the Mt. Sinai School of Medicine reported on their study of immune function in 16 patients hospitalized for severe depression, 10 outpatients with mild depression, 10 nondepressed men hospitalized for surgical repair of hernias, and an equal number of healthy, nondepressed individuals. Among these diverse groups, the researchers found that T-cells and B-cells from the patients hospitalized for depression multiplied much less rapidly than similar cells from the healthy individuals, while T-cells and B-cells from the mildly depressed outpatients multiplied normally. Cells from the men hospitalized for hernia operations also multiplied normally, suggesting that the suppression of the immune system resulted from the severe depression rather than from hospitalization.

An overview. Perhaps the most convincing evidence of the link between emotions and illness appeared in the June 1987 *American Psychologist*, where psychologists Howard S. Friedman and Stephanie Booth-Kewley of the University of California at Riverside described their use of a recently developed analytical technique called meta-analysis to combine the results of 101 different psychoneuroimmunological studies. By combining smaller studies into what amounts to a large research project, they were able to detect subtle effects that would not necessarily be apparent in research involving fewer subjects.

The University of California psychologists found that neurotic people are twice as likely as nonneurotics to complain of physical symptoms that suggest a wide variety of illnesses. According to Friedman and Booth-Kewley, "treatment of medical patients by health psychologists and clinical psychologists seems prudent and worthwhile. At the very least, such interventions may improve the psychological adjustment among a high-risk population. Such interventions may also have a beneficial effect on the progression or recurrence of serious chronic disease."

In addition to stress and anxiety, an individual's overall outlook on life can affect vulnerability to disease, according to psychologist Christopher Peterson of the University of Michigan in Ann Arbor. Peterson conducted intensive interviews with members of the Harvard University classes of 1939 through 1944 who were considered to be "fittest" by their classmates. He reported in the July 1988 *Journal of Personality and Social Psychology* that the students who felt the most fatalistic or pessimistic while in college were more susceptible to diseases of all kinds later in life. Although he saw little difference in all the subjects' health for more than two decades after they left college, "the turning point came during their early forties, when the more pessimistic ones began to fall ill."

Fighting Disease with Emotions

If deep-seated hostility and a pessimistic outlook on life can make a person more likely to get sick, can a good attitude help one recover more rapidly? A growing body of evidence suggests that the answer to this question is yes. In 1976, Norman Cousins, then the editor of *Saturday Review*, wrote in the *New England Journal of Medicine* (*NEJM*) that he had cured himself of spinal arthritis by adopting a healthy mental attitude, laughing a lot, and taking vitamin C. Eventually, Cousins expanded the article into a book, *Anatomy of an Illness*, in which he promoted the benefits of positive thinking in curing disease. Although many other self-help books have since proclaimed that individuals can help overcome their illnesses by adopting healthy mental attitudes, until recently, physicians have found little firm evidence to support these claims.

Emotions and cancer. By 1988, several researchers had found evidence that a patient's positive emotional outlook does indeed appear to promote recovery from cancer. For example, oncologist S. Greer and his colleagues at King's College Hospital in London, U.K., have found a strong correlation between a woman's emotional response to breast cancer 3 months after breast removal and her long-term survival. Of the 57 women the physicians monitored for their study, 10 mastectomy patients denied their cancer. Most of these women rationalized that they had agreed to the operation as a preventive measure, and only half of them were still alive after 10 years.

Another 10 of the 57 women demonstrated what Greer and his colleagues called a "fighting spirit," reacting to their breast cancer with the attitude that they would take every step possible to battle the disease. Seven of these 10 women were still alive after 10 years. The remaining 37 women either reacted with a stoic acceptance of their disease, showing no signs of distress, or felt completely hopeless about their prospects. Only 10 of the 32 stoical women were alive after 10 years, and only one of the five women who felt hopeless survived. Given the small number of women studied, the researchers noted that their results should be interpreted cautiously; yet they concluded that emotional therapy might improve the survival rates of women with cancer.

In a similar study, psychologist Lydia Temoshok and her colleagues at the University of California at San Francisco have found that among patients with melanoma — an often fatal form of skin cancer — those who freely expressed feelings of anger and distress displayed a stronger immune response to their cancers than did those who suppressed their emotions. Similarly, oncologist Ronald Grossart-Meticek and his colleagues at the Interdisziplinares Forschungsprogramme Fozialwissenschaftliche Onkologie in Heidelberg, West Germany, reported that in a study of 100 women with metastatic breast cancer (disease that had begun spreading to other parts of their bodies), the women who received a combination of chemotherapy and psychotherapy survived longer than those who received chemotherapy alone.

Psychologist Sandra M. Levy of the University of Pittsburgh School of Medicine studied 36 women with advanced breast cancer to test the hypothesis that anger and a desire to fight the disease would prolong survival. After 7 years, 24 of the 36 women had died from their disease and Levy found, to her surprise, that after the first year, anger made no difference at all. The only psychological factor that correlated with increased survival was a sense of joy about living. "Perhaps a sense of joy is a reflection of the body's resilience and stamina," she said.

Emotions and AIDS. Acquired immunodeficiency syndrome (AIDS) appears to progress more slowly in patients who handle stress well, have positive attitudes, receive strong support from friends and relatives, and otherwise cope well with stress. In a study of 18 men with AIDS, Temoshok and UCLA psychiatrist George Solomon found that men who were assertive, who exercised, and who displayed lower levels of anger, depression, fatigue, stress, and tension had higher numbers of healthy white blood cells than did men who lacked these characteristics.

Psychiatrist Karl Goodkind of the University of Texas Southwestern Medical Center in Dallas reported similar results at the 1989 AAAS annual meeting. Goodkind studied 40 male homosexuals: 13 were healthy, 10 had been infected with the AIDS virus but showed no symptoms, 4 had AIDS-related complex (ARC), and 13 had fully developed AIDS. The results showed that the men who were infected with the virus but who had not yet developed ARC or AIDS felt less stress in their lives, received more social support, and had a more confident coping style than either the noninfected group or the groups that had developed ARC or AIDS.

At the same meeting, psychologist Michael Antoni of the University of Miami reported on his research, which involved measuring various signs of psychological and immune-system functioning among 39 symptom-free gay men at various times before and after tests for the AIDS virus. Some of the men had been participating in a 10-week aerobics exercise program, while the others had not. Among those who tested positive for the AIDS virus, Antoni found, the exercisers had higher levels of T-cells and B-cells. They also suffered less anxiety, depression, and confusion or bewilderment after the positive diagnosis.

Goodkind suggested that when all of the results are considered together, they suggest that "we should encourage people [with the AIDS virus] to do things that may be helpful and that certainly can't be harmful." He cautioned, however, that failure of optimism to slow progression of the disease might wrongly lead infected individuals to blame themselves.

Opposing views. Not all physicians agree that a patient's mental attitude can control the outcome of disease. In the 14 June 1985 *NEJM*, medical sociologist Barrie Cassileth and her colleagues at the University of Pennsylvania Cancer Center in Philadelphia reported finding no relationship between the attitudes of 204 cancer patients toward life or their diseases and the patients' survival rates. The study, according to the researchers, "suggests that the inherent biology of the disease alone determines the prognosis, overriding the potentially mitigating influence of psychosocial factors."

An accompanying editorial in the same issue by *NEJM* assistant editor Marcia Angell — herself a physician — dismissed most of the studies that had previously linked emotional states and disease, concluding that "our belief in disease as a direct reflection of mental state is largely folklore." The Pennsylvania study and the editorial provoked a strong reaction, especially among psychologists, who argued that Cassileth had made a fundamental mistake by focusing on terminally ill patients.

The American Psychological Association responded with a resolution denouncing the editorial as "inaccurate and unfortunate," asserting that it ignored a substantial body of research findings linking psychological factors and health. *NEJM* later reported that it had received a greater volume of mail about the editorial than about any previous article or editorial in its history.

Two other researchers, Paul T. Cost and Robert R. McCrae of the National Institute on Aging in Bethesda, Maryland, studied 347 men who were given medical and psychological tests several times over two decades. They reported in 1988 that men who scored highest on tests of neuroticism consistently reported more medical complaints of all kinds, from respiratory problems to skin disease, but they also found that the neurotic men lived as long as men who were not neurotic.

Stress and Heart Disease

With the near-eradication of many major infectious diseases in most industrial nations, heart disease has become the leading cause of death. Epidemiologists have identified various medical risk factors for heart disease, such as high blood pressure, high blood cholesterol levels, and cigarette smoking. Over the past 20 years, researchers have built a strong case that psychological factors — personality traits and stress — may also influence the incidence of heart disease.

In a study of 7,000 adult residents of Alameda County, California, Lisa Berkman of Yale University and S. Leonard Syme of the University of California at Berkeley found that men and women with few social ties were much more likely to die of cardiovascular disease and related conditions, such as stroke, than were people with greater access to family and friends. During the 9-year study period, the death rate for women with few social ties was 2.8 times greater than normal, while for men it was 2.3 times greater than normal.

According to James Herd of the Baylor College of Medicine in Houston, Texas, several studies of Japanese emigrants who settled in California have shown that reduced support from families and a lack of social stability increased the risk of heart disease. These studies found that Japanese emigrants who maintained a traditional family-oriented lifestyle and forged relationships with the Japanese community had a much lower incidence of heart disease than individuals who abandoned their traditional way of life. Most of the studies, however, have not taken into account the differences between the traditional Japanese diet and the American diet, differences that scientists think account for the lower rate of heart disease in Japan itself.

Risks of Type A behavior. Claims of a connection between heart disease and personality factors that increase stress have been particularly controversial. In 1959, cardiologists Meyer Friedman and Ray Rosenman of the Harold Brunn Institute of Mount Zion Hospital in San Francisco, California, proposed that people with "Type A" personalities — typically achievement-oriented and often hurried, impatient, aggressive, and hostile — appear to run a higher risk of developing heart disease than "Type B" people, who are more relaxed, easygoing, and cooperative.

Friedman and Rosenman originally developed their hypothesis that Type A behavior increases the risk of heart disease after observing that the seats of upholstered chairs in their waiting room were worn most at the front; their heart patients were typically in such a hurry to conclude their visits and get back to work that they never sat back in their chairs. The idea that heart patients behaved differently from other people was eventually supported by several studies, including Friedman and Rosenman's own study of 3,000 healthy, middle-aged men over an 8.5-year period. In that group, Friedman and Rosenman found that Type A men had about twice as many cases of coronary heart disease as did Type B men.

Friedman and Rosenman's hypothesis has remained controversial for three decades, and by the late 1980s, researchers were still not in complete agreement about the risks involved. In 1981, however, the American Heart Association (AHA) reviewed all the available studies of Type A behavior and concluded that such behavior should, in fact, be classified as a risk factor for heart disease. But in the years since, some new studies have questioned some of Friedman and Rosenman's basic tenets.

In one such study led by Robert Case of St. Luke's-Roosevelt Hospital Center in New York City, a team of cardiologists analyzed the personalities of over 500 heart-attack patients and then monitored their health for up to 3 years. In a 1985 report on the study the group concluded: "We fully expected that the people who scored highest in Type A traits would have more heart attacks and die sooner. Then we analyzed the data and found no relationship at all."

Although many cardiologists have been impressed by these findings, Friedman and Rosenman have criticized the study's experimental technique. Case and his colleagues asked their subjects to fill out a questionnaire about their habits and traits; the questions asked, for example, whether they found it difficult to make time for a haircut and whether they were more hard-driving than their coworkers. Although the AHA has endorsed this type of questionnaire for diagnosing Type A personalities, Friedman and Rosenman argue that Type A traits can be measured accurately only in face-to-face interviews, because they believe that the way an answer is delivered is often as important as the content of the answer. They claim that it is crucial for an interviewer to observe telltale signs of Type A behavior, including rapid speech, frequent interruptions, fist-clenching, and grimacing.

A second 1985 study did involve interviews, but it also found no link between personality and heart disease. Epidemiologist Richard Shekelle and his colleagues at the University of Texas Medical Center in Houston studied over 3,100 middle-aged men who had not yet suffered heart attacks, but who were considered prime candidates because of their blood pressure, cholesterol levels, or cigarette-smoking habits. During the 8-year study, 129 of the men suffered heart attacks, and 62 of those died. The cardiologists observed no difference in the incidence of heart attacks or deaths between Type A and Type B men.

Some cardiologists have argued that the apparent contradiction between these findings and earlier studies may arise because Case's and Shekelle's groups were not sufficiently skilled at identifying Type A behavior and did not classify their subjects properly. Case has suggested instead that perhaps only certain components of Type A behavior increase the risk of heart disease.

Hostility. At a 1989 AHA symposium in Monterey, California, internist Redford B. Williams, Jr., of the Duke University Medical Center in Durham, North Carolina, and author of *The Trusting Heart,* said that he had conducted several studies indicating that cynicism, mistrust, and anger toward others are the "toxic core" of Type A behavior, and that these forms of hostility place workaholics at increased risk for heart disease.

In one of his studies, for example, Williams examined the case histories of 118 lawyers who had undergone psychological testing during law school. After 25 years, the death rate among lawyers with high hostility scores was 4.2 times higher than that of their cohorts. At especially high risk were those who were inclined to harbor a cynical mistrust of the motives of others and who openly and frequently expressed their anger — people, he said, who were inclined to explode over slow elevators and long lines at banks and supermarkets. Their death rate over the 25-year period was 5.5 times higher than that of the more trusting attorneys.

Williams suggested that Type A people have a weak parasympathetic branch of the autonomic nervous system — the branch that exerts a calming effect after an emergency has triggered the so-called fight-or-flight reflex. Citing a recent study involving isoproterenol, an epinephrine-like substance that stimulates the fight-or-flight reflex, Williams pointed out that the calming parasympathetic response seemed to be triggered sooner in Type B men than in Type A men. Epinephrine and other hormones stimulated by stress may, for example, trigger changes in the lining of blood vessels and thus increase the risk that arteries will become clogged. If this is so, then long periods of high epinephrine levels would be detrimental to the heart.

Meta-analysis findings. Psychologists Stephanie Booth-Kewley and Howard S. Friedman of the University of California at Riverside used meta-analysis to combine results from 150 separate studies of the link between Type A behavior and heart disease. They reported in the March 1987 *Psychological Bulletin* that meta-analysis revealed a modest but genuine link between some of the personality variables they studied and coronary heart disease. The strongest associations arose for Type A behavior and depression, but firm associations also were found for anger, hostility, aggression, and anxiety.

They also found that structured interviews of the type used by Friedman and Rosenman more effectively diagnosed Type A behavior than did pencil-and-paper tests. Furthermore, prospective studies — in which researchers identify individuals at the beginning of a study and follow them for long periods of time — showed weaker links between personality and heart disease than did cross-sectional studies, in which researchers study a selected population at a particular point in time.

"This review also revealed that information about the interrelations of personality predictors of coronary heart disease is sorely needed," the report concluded. "The picture of coronary-proneness revealed by this review is not one of a hurried, impatient workaholic but instead is one of a person with one or more negative emotions. We suggest that the concept of the coronary-prone personality be broadened to encompass psychological attributes in addition to those associated with Type A behavior and to eliminate those components that the accumulated evidence shows to be unimportant."

Survival after a heart attack. While people with Type A personalities may be more prone to heart attacks than those with Type B personalities, they may also be more likely to survive heart attacks if they live through the first 24 hours. Epidemiologist David R. Ragland of the University of California School of Public Health at Berkeley reported in 1988 that he had studied 231 men who had survived a heart attack for more than one day. The men were part of the original group of over 3,000 studied by Friedman and Rosenman and classified by them as Type A or B. Ragland found that the death rate among the Type A patients in the group was only 58 percent of the rate among the Type B patients. Ragland speculated that Type A patients may respond differently to heart attacks, perhaps complying more strictly with medical treatment or changing their lifestyles more aggressively. Alternatively, they may pay more attention to symptoms, such as chest pains, that Type B patients might ignore.

Behavioral change. Whether the risk of heart disease is linked to Type A behavior in general or to specific aspects of this behavior, however, there is evidence that Type A individuals can reduce the risk of heart attacks by modifying their personalities, according to Meyer Friedman and Carl Thoreson of Stanford University. Working at the Harold Brunn Institute, the two investigators and their colleagues recruited more than 1,000 heart-attack survivors, people statistically at high risk for future heart attacks. The scientists divided the survivors into three groups: a behavior-change group, a cardiology information group, and a control group. Members of the first group attended an exercise program and were counseled by psychologists about Type A behavior and its relationship to cardiovascular disease.

The members of the second group received advice and information on heart disease treatment from cardiologists, but they were not specifically encouraged to change their Type A behavior. The control group received no counseling of any kind. The cardiologists found that after 3 years, the group counseled about Type A behavior had only half the heart-attack recurrence rate of the group receiving only cardiology information, which in turn had a slightly lower recurrence rate than the control group. The behavior-change group achieved a recurrence rate of less than 3 percent per year, compared to a rate of about 6 percent for both the information-only and control groups.

Because most people associate Type A behavior with people who are successful in their careers, some scientists have wondered whether a change from Type A to Type B behavior might lead to reduced success. Scientists have also questioned whether Type A people who have not yet suffered a heart attack can be persuaded to modify their behavior to lower their risk. When James Gill and his colleagues at the Harold Brunn Institute studied officers who were students at the U.S. Army War College in Carlisle Barracks, Pennsylvania, they observed rampant Type A behavior among men and women training to become commanders and top staff officers. From a group of middle-aged volunteers, Gill's group selected 116 men and two women who scored exceptionally high on questionnaires and interviews designed to identify Type A behavior.

The volunteers were randomly assigned either to a test group that underwent an 8-month, 21-session counseling program to reduce their Type A behavior or to a control group that received no behavioral counseling. The researchers also matched each of the counseled participants with a classmate charged with alerting the cardiologists to changes in behavior that might undermine the effectiveness of an army officer. At the end of the counseling program, the cardiologists re-evaluated the personalities of all the subjects and even asked their spouses to assess changes in the officers' behavior and attitudes.

The study revealed a marked reduction in Type A behavior in 42 percent of the counseled group, and a significant reduction in an additional 26 percent. In contrast, only

9 percent of the officers who were not counseled (but who knew the purpose of the study) showed a marked reduction in Type A behavior and another 19 percent showed a significant reduction. The most dramatic changes in the counseled group occurred in their hostility and time urgency scores. In the September 1985 *American Heart Journal*, Gill and his colleagues wrote, "No adverse effects on the military leadership qualities of Type A counseled participants were observed by their classmates." The cardiologists thus concluded that effective counseling could reduce the Type A behavior of healthy, middle-aged individuals without reducing their leadership skills. The Army was so impressed by the changes in behavior that it incorporated similar counseling as a regular part of officer training at the War College; the program may be extended to other Army groups as well.

Other risk factors. Boston University Medical Center epidemiologist C. David Jenkins and his colleagues have associated three other behavioral factors with a high incidence of heart disease: low socioeconomic status, sustained disturbing emotion, and work overload. According to Jenkins, epidemiological studies have indicated that, in prosperous countries, heart disease is more common in people with lower social status and jobs requiring limited skills than it is in people with higher social status and skilled jobs. Although the reasons for this correlation are not yet known, Jenkins speculated that high-risk individuals may be less educated about the health dangers of cigarette smoking and poor diet and may feel more frustrated by their jobs and living situations; the frustration and the accompanying biological changes may increase the risk of heart disease.

Sustained disturbing emotion — strong negative emotions, such as anxiety or depression, experienced repeatedly or over an extended time — and work overload both place excessive psychological demands on people. Naturally competitive Type A individuals, for example, will feel great pressure to achieve in a work-overload situation and may place impossible demands on themselves. Failures at work or emotional difficulties at home could further increase the risk of heart disease.

Emotions and sudden cardiac death. Short-term stress, such as that caused by the death of a spouse, appears to contribute greatly to sudden cardiac death. This condition is twice as common in men as in women, and most of the victims are middle-aged. In addition, about 25 percent of the victims have shown no earlier symptoms of heart trouble. Death can generally be attributed to ventricular fibrillation, a condition in which the heart's muscles lapse into uncoordinated, irregular twitching. When the heart stops beating in a coordinated way, a person may quickly die.

To explore this phenomenon further, psychiatrist Peter Reich of Brigham and Women's Hospital in Boston, Massachusetts, studied people who had been resuscitated after suffering from ventricular arrhythmias. Of the 117 patients interviewed, 25 reported intense emotional experiences earlier on the same day as the arrhythmia. Although anger was the most common emotion cited, one man suffered an arrhythmia in his jubilation over a victory. As a group, the 25 patients who had experienced intense emotional states generally had fewer prior signs of heart disease than the other 92, an indication that emotions alone may have been enough to trigger some of the arrhythmias.

Parkinson's Disease and Brain Grafting

On 7 January 1988, neurosurgeon Ignacio Navarro Madrazo of the La Raza Medical Center in Mexico City reported in *NEJM* that he had transplanted brain tissue taken from a spontaneously aborted fetus into the brain of a woman with Parkinson's disease — the first reported human brain-to-brain transplant. Madrazo reported that before the September 1987 operation, the recipient of the graft, 35-year-old Leonor Cruz Bello, had stiffness in her right side, walked with difficulty, and could not prepare her own food. Within weeks after the procedure, Bello could cook and take care of her two children, although she still had a slight amount of tremor and tired easily climbing stairs.

Madrazo's apparent success in the controversial new field of brain grafting may pave the way for new treatments of Parkinson's disease. One year earlier, he had stimulated a tremendous surge of interest in the subject when he reported that the health of two Parkinson's disease victims improved significantly after tissues from their own adrenal glands were implanted into their brains to provide an alternative source of a neurotransmitter that is deficient in victims of the disease. By the end of 1988, about 200 patients around the world had received brain grafts of adrenal tissue and another 35 had received grafts of fetal brain tissue.

Despite their growing popularity, the procedures have generated heated controversy. Critics have charged that researchers should have conducted much more extensive experimental studies on primates before attempting transplants in humans, while supporters have argued that they had adequate experimental data to justify human studies. The efficacy of the operation has also been debated, with supporters claiming that up to two-thirds of the patients benefit, and critics charging that few, if any, patients have shown any lasting improvement. Finally, the growing use of fetal tissues has opened up a new controversy about whether this practice will encourage abortions and spur unethical attempts to secure fetal tissue.

In spite of these controversies, supporters of brain grafting insist that the technique holds great promise as a treatment not only for Parkinson's disease but also for Huntington's disease, Alzheimer's disease, and a variety of other conditions thought to involve a deficiency of neurotransmitters. The technique might also eventually be used to restore certain types of memory and thought processes.

Parkinson's Disease

Parkinson's disease, named after nineteenth century English physician James Parkinson, who was the first person to describe it completely, is a chronic, progressive disorder of the central nervous system affecting more than 1.5 million people in the U.S., most of them over the age of 55. Its symptoms include tremors and rigidity of the limbs, with the symptoms eventually becoming so severe that patients cannot carry out such ordinary activities as feeding and dressing themselves. About 30 percent of people with Parkinson's disease also suffer dementia, a loss of mental capacity.

Although scientists do not know the cause of Parkinson's disease, they do know that the symptoms are caused by the death of brain cells that secrete dopamine, a neurotransmitter that plays a key role in controlling body movements. The disorder can be controlled, at least in the early stages, with the drug L-dopa, which increases the brain's dopamine supply. As the disease becomes more severe, however, people gradually benefit less from the drug.

A bizarre series of incidents in Maryland and California has given researchers several clues to the cause of Parkinson's disease. The incidents began in the late 1970s when a young chemistry graduate student in Bethesda, Maryland, suddenly developed severe Parkinson's disease symptoms. The case baffled the medical community because Parkinson's disease normally strikes the elderly and usually progresses very slowly. Eventually, however, NIMH researchers found that the graduate student had been synthesizing for illegal sale a compound closely related to the narcotic meperidine, better known by the trade name Demerol. A chemical analysis of drug samples from the student's basement laboratory showed that the samples contained several contaminants, among them a compound called MPTP.

Subsequently, neuroscientists William Langston, now at the Santa Clara Valley Medical Center in San Jose, California, and Ian Irwin of the Stanford University Medical Center identified several other drug abusers with symptoms identical to those of the graduate student. In 1983, the Stanford researchers reported that, like the young student in Maryland, each of the victims had attempted to synthesize meperidine-related drugs or had illegally purchased and used drugs contaminated with MPTP. By the end of 1988, investigators had identified at least 400 victims of MPTP exposure with symptoms of Parkinson's disease. Unfortunately, the symptoms are irreversible.

Neurologist R. Stanley Burns, now at the Vanderbilt University School of Medicine in Nashville, Tennessee, discovered that injecting MPTP into rhesus and squirrel monkeys produces a condition almost identical to Parkinson's disease. This discovery gave researchers the first animal model for the disease, thereby allowing them to test new ideas about its causes and treatment, particularly with the new brain grafting techniques. As a result of such studies, researchers have learned that MPTP itself does not cause the disease. In tests with monkeys, neuroscientist Sanford Markey and his colleagues at NIMH have observed that MPTP is converted in the animals' brains into one or more toxic derivatives by a compound called monoamine oxidase. Langston has found that the effects of MPTP on the brains of squirrel monkeys could be prevented by administering a chemical called pargyline that blocks the activity of monoamine oxidase.

Several other drugs also block the activity of monoamine oxidase and could potentially be used for treatment of Parkinson's disease. In 1985, neuroscientist Joussa Youdim of the Technion-Israel Institute of Technology in Haifa reported that treatment of Parkinson's disease patients with one such inhibitor, deprenyl, extended their lives by an average of 15 months and slowed the progression of their disease. In June 1987, the National Institute of Neurological and Communicative Diseases and Stroke began a 5-year, $10-million study of the use of deprenyl and another drug, tocopherol, on Parkinson's disease victims at 28 U.S. medical centers. Preliminary results suggest that these drugs work best in the early stages of the disease. Unfortunately, however, by the time neurologists diagnose Parkinson's disease in most people, the condition has progressed to the point where the drugs may do little good.

No genetic link. Unlike many other brain disorders, Parkinson's disease does not appear to be linked to a genetic defect. In one recent study, neurologist Urpo Rinne of the University of Turku and colleagues from the University of Helsinki studied the patterns of Parkinson's disease in 16,000 pairs of identical and fraternal twins born in Finland before 1958.

If Parkinson's disease did have a genetic component, the twin siblings of people with the disease should have had a much-greater-than-normal chance of developing it. But among the 16,000 pairs of twins, the Finnish researchers found only 42 cases of Parkinson's disease, and in only one of those cases had both twins developed the disorder. The researchers speculated that this single occurrence was the result of chance, and concluded that no hereditary factors existed. Similar results have been obtained in other twin studies.

Evidence for an environmental link. The apparent absence of genetic factors in Parkinson's disease and the striking similarity between the MPTP-induced and naturally occurring disorders has led many researchers to speculate that environmental toxins similar to MPTP might contribute to the onset of Parkinson's disease. A major study supporting this hypothesis was presented at the Eighth International Symposium on Parkinson's Disease held in 1985 in New York City.

The late Andre Barbeau and his colleagues at the Clinical Research Institute of Montreal in Quebec collected data on the prevalence of Parkinson's disease in nine regions of the province of Quebec and compared the prevalence to patterns of pesticide use in those regions. The chemical structures of many pesticides are similar to that of MPTP. The researchers found a striking correlation: A major agricultural region southwest of Montreal where pesticide use was very high, for example, also had the highest prevalence of Parkinson's disease (0.89 cases per 1,000 population). Conversely, in a region where pesticide use was low, the incidence of Parkinson's disease was only 0.13 cases per 1,000. Barbeau cautioned, however, that other toxins besides pesticides might play a role in the disease, and that many industrial chemicals — particularly dyes and chemicals used in making paper — have structures similar to that of MPTP. Nevertheless, Barbeau's findings strongly implicated environmental factors in the disease.

Further support for an environmental link comes from a multinational team of researchers headed by neurotoxicologist Peter S. Spencer of the Albert Einstein College of Medicine. The team had analyzed the unusually high prevalence of Parkinson's disease and other brain disorders among the Chamorro population of the Pacific islands of Guam and Rota. In the 1950s, the prevalence of some neurological disorders was 50 to 100 times higher than in the U.S. and other developed countries, but the prevalence has since declined. The researchers linked the disorders to a neurotoxin in a plant known as the false sago palm or cycad.

The Chamorro traditionally ate large quantities of rice, but during the Japanese occupation of the islands during World War II, they relied more heavily on cycad, soaking plant tissues and seeds in water for days or weeks to remove known poisons, then drying the material into a flour for use in tortillas, soap, and beverages. After the war, the Chamorro became more westernized and their use of cycad declined — a decline that has paralleled the decline in the incidence of the neurological diseases.

To prove their speculation that the disease had been caused by toxins in the cycad seeds, the researchers isolated a neurotoxin from the seeds called beta-*N*-methylamino-L-alanine or BMAA. When they fed this toxin to macaque monkeys, all of the monkeys developed progressive neurological disorders within 2 to 12 weeks. That result convinced the researchers that BMAA in food and folk medicines was a major cause of neurological disease in Guam and Rota, and that similar materials might also cause these conditions elsewhere.

Grafting for Parkinson's Disease

During the late 1970s and early 1980s, Swedish and American researchers successfully transplanted dopamine-secreting brain cells, called dopamine neurons, from the brains of fetal rats into the brains of adult rats. By doing so, they hoped to reverse Parkinson's-like symptoms they had produced in the adult rats by injecting a toxic chemical into their brains to damage or kill dopamine neurons. In one typical class of experiments, the researchers damaged dopamine neurons in only one brain hemisphere; when these cells stopped secreting dopamine and the rats were exposed to a particular drug, they began walking in circles. When the researchers then grafted healthy dopamine neurons from rat fetuses into their brains, the animals were able to again walk in straighter lines.

In 1982, neurobiologist Anders Bjorklund of the University of Lund in Sweden and Lars Olson of the Karolinska Institutet in Stockholm reported that they had successfully grafted fetal brain tissue from one species to another, in this case, dopamine neurons from fetal mice into adult rats whose dopamine neurons had been chemically damaged to induce circling. After receiving the graft, these rats could also walk straighter.

In July 1987, Bjorklund and neurologist Patrick Brundin of the University of Lund reported that they had reversed most of the chemically induced Parkinson's disease symptoms in rats by implanting brain cells obtained from aborted human fetuses. Those grafted cells were as effective as rodent cells in restoring function to the disabled animals.

Other researchers took advantage of the fact that some cells outside the brain produce some of the same hormones as the brain. For example, the adrenal glands, a pair of walnut-sized glands that sit atop the kidneys, also produce dopamine. The dopamine produced by the adrenal glands is normally prevented from reaching the brain, however, by the blood-brain barrier, which protects delicate brain tissues from foreign chemicals and microorganisms in the blood. Neurobiologists William Freed and Richard Jed Wyatt, and their colleagues at NIMH reasoned that, if adrenal cells were grafted into the brain, they might produce enough dopamine to alleviate the symptoms of Parkinson's disease. When they damaged dopamine cells in one hemisphere of the brains of rats and then grafted adrenal cells from healthy animals into the damaged brains, they found that half the animals could walk in straighter lines.

Experiments in primates. Neuroscientist Roy A. E. Bakay of the Emory University School of Medicine in Atlanta, Georgia, reported one of the first successes with transplanted cells in primates at the 1985 World Congress of Neurosurgery in Toronto. Bakay produced a Parkinson's-like condition in two monkeys with MPTP, then transplanted dopamine-producing cells from the brainstems of 35- and 37-day-old rhesus monkey fetuses into the adults' brains. Although the animals did not fully resume their earlier level of activity, they did become more mobile and developed normal dopamine levels within 2 months after the operation.

In 1986, neurobiologists John R. Sladek, Jr., of the University of Rochester School of Medicine and Dentistry in New York and D. Eugene Redmond of the Yale University School of Medicine successfully reduced the symptoms of MPTP-induced disease in African green monkeys with grafted fetal brain cells. Two years later, they reported similar results after implanting human fetal brain cells into the monkeys, a feat subsequently duplicated by several other researchers.

Brain grafts in humans. Encouraged by the success of adrenal tissue grafts in alleviating the symptoms of Parkinson's-like conditions in animals, Lars Olson and Ake Seiger of the Karolinska Institutet and neurosurgeon Erik-Olaf Backlund of the University of Bergen in Norway reasoned that transplants of adrenal tissue into the brains of humans with Parkinson's disease might produce similar results. For their first patient, they selected a 62-year-old retired clergyman who was almost completely crippled by the disease.

In 1982, a team of surgeons at the Karolinska Institutet removed one of the patient's adrenal glands through an incision in his back. Olson and Seiger then minced the organ into small clumps of cells that could pass through the needle of a hypodermic-like device. Finally, Backlund placed the patient's head in a metal frame that secured it in place, drilled a small hole in the skull, positioned the needle point in the brain, and injected the cells.

Before the operation, the man moved with difficulty, and his condition was rapidly deteriorating, even though he was receiving large doses of L-dopa. Within days after his surgery, the man regained some mobility in his arms and fingers, and his body became less rigid. He also required less L-dopa to control his symptoms. To the researchers' disappointment, however, these gains disappeared within 2 months of the operation. Improvements were also transitory in three other patients who underwent the same procedure, and researchers became pessimistic about the potential of the operation.

Ignacio Madrazo's 1987 report that two patients had significantly improved after having adrenal tissues implanted into their brains thus both shocked and galvanized neurologists. The most dramatic recovery was made by 34-year-old Joseluis Meza. Before the surgery, Meza could neither speak clearly nor perform such tasks as walking, dressing, bathing, or feeding himself without help. Ten months after his surgery, he could do all of these tasks and more — even kick a soccer ball with his 5-year-old son. A second patient, a 39-year-old man, had also improved significantly.

Madrazo attributed his success to the use of a different procedure than that of the Swedish neurosurgeons, one that more closely mimicked the approaches taken with animals. Animal researchers such as Bakay had reversed Parkinson's disease symptoms by implanting adrenal tissue into the lateral ventricles — fluid-filled cavities in the brain. Hoping to amplify the effects of the grafted cells, the Swedish surgeons had injected the cells deep into a part of the brain called the caudate nucleus so that the secreted dopamine could more easily reach the cells that required it.

Madrazo, in contrast, implanted the adrenal cells into the ventricles, as had been done in the animal experiments. Madrazo believed that this procedure enabled the cells to survive much longer than they did in the Swedish transplants, because the cells were bathed in nutrients. But the questions of whether the implanted cells did indeed survive longer and how the recovery occurred have been the subject of much dispute. Autopsies on two of Madrazo's patients who eventually died yielded inconclusive results, with no strong evidence that the implanted cells had survived. Many investigators believe that the implanted cells may have secreted other hormones, called trophic factors, that stimulated some regrowth of the brain's own cells. Others believe that the operation itself may have stimulated some brain cells to release trophic factors. A few researchers have even argued that the apparent recoveries were illusory because Madrazo and other neurosurgeons simply did not analyze their patients carefully enough before the operations.

Nonetheless, interest in the operation spread rapidly. Within 2 weeks after Madrazo's report was published, neurosurgeon George S. Allen and his colleagues at Vanderbilt performed the first U.S. operation on 42-year-old Dickye Baggett. Baggett was only moderately impaired by the disease when Allen operated on her. She had tremors on her right side, her right foot dragged when she walked, and she had difficulty writing, although she was able to perform her job as an office clerk. Shortly after the operation she said, "I take less medicine, my foot doesn't drag, and my tremors are less severe.... I'm quite happy with the results and I would do it again in a minute."

By the end of 1988, more than 200 such operations had been performed worldwide, 50 of them by Madrazo, who had traveled extensively to demonstrate his procedure to other physicians. Because some neurosurgeons have reported good results from brain grafting while others have perceived little or no benefit, the technique remains controversial. Clearly, however, some results have been spectacular. One staunch supporter has been neurosurgeon Kemp Clark of the University of Texas Health Sciences Center, who has performed the surgery eight times. "We're not talking about amelioration of a few symptoms or a little bit of the problem," he said. "When it works, it is absolutely phenomenal."

Clark cited the case of 38-year-old Dan Covey, now a supervisor at a water company in the Dallas suburb of Plano. Before Clark operated on Covey in July 1987, Covey had been forced to quit his job because his stiffness made walking very difficult. He could no longer drive or feed himself because of the severity of his tremors, and his memory had also faded to the point where he needed a tape recorder to remind himself of what he had done during the day. But within 3 months after the operation, Covey was back at work and driving his car again. In the year after surgery, he was running 3 miles a day.

A greater number of graft recipients have shown a more modest improvement. As neurosurgeon C. Warren Olenow of the University of Tampa in Florida, who has operated on six patients, observed, "Parkinson's is often a cyclic disease in which patients are 'off' [severely impaired] or 'on' [less impaired]." Olenow found that his patients had longer "on" periods and less severe symptoms during the "off" periods. "The patients were less bad, less often," he said. "That is not a trivial improvement, especially if it persists."

But some patients, perhaps one-third, have not benefited from the surgery at all, and that finding is the focus of the controversy surrounding Madrazo and the operation itself. Many physicians argue that Madrazo claimed all of his patients had improved. According to Madrazo, however, the dispute was merely semantic. "I have never claimed that all my patients have improved. What I did say was that, in all my patients, the evolution of their disease had changed." In other words, some had not improved, but at least they had stopped getting worse. Nonetheless, the American Academy of Neurology, reflecting the disappointment of some researchers, issued a position paper in December 1988 calling for "great caution" in expanding the use of the procedure except as a research tool at highly specialized centers. The statement concluded: "It would be wise to encourage further transplantation studies in animal models before large-scale [human] trials are undertaken."

Fetal Tissues

The pace of human research, however, was already outdistancing the concerns of the American Academy of Neurology, for Madrazo's 1988 *NEJM* report had raised the stakes even higher. Madrazo reported that he had transplanted brain tissue taken from a spontaneously aborted fetus into the brain of a woman with Parkinson's disease. In another operation on the same day, he had also transplanted adrenal tissue from the same fetus into the brain of a woman with Parkinson's disease, the first time fetal adrenal tissue had been used for that purpose. The apparent success of those operations in reducing symptoms raised researchers' hopes about the potential of fetal cell transplants. Not only have animal studies shown that fetal cells reduce the symptoms of Parkinson's disease more effectively than adrenal cells taken from adult animals, but many neuroscientists believe that fetal tissues may also be used to treat other degenerative brain disorders, including Alzheimer's and Huntington's diseases. Opponents of the procedure, among them antiabortionists and some ethicists, have argued that the use of fetal cells should be forbidden because it will encourage abortions and the exploitation of the unborn.

Despite this growing controversy, researchers are continuing to study fetal tissues to learn more about human development and gain new insights into the causes of diseases. However, such studies still represent only a small fraction of government-funded biomedical research in the U.S. In 1987, the most recent year for which figures are available, about 118 U.S. research groups received $11.8 million — about 2 percent of NIH's budget — for research involving fetal cells, according to Charles R. McCarthy, director of the NIH Office of Protection from Research Risks. "Such studies are a critical factor in research in virtually every major disease category," McCarthy said. One particularly intriguing use of fetal tissues is their implantation into immune-deficient mice to produce a model of a human immune system that can be used for studying AIDS.

Only a handful of researchers are studying the potential use of fetal cells in humans. About 200 patients around the world have received fetal liver tissue — which produces bone marrow cells — primarily to restore bone-marrow loss resulting from cancer therapy, according to hematologist Robert P. Gale of the University of California at Los Angeles. Gale unsuccessfully implanted fetal liver cells in six victims of radiation poisoning from the 1986 Chernobyl nuclear accident in the U.S.S.R. The procedure has largely been abandoned, Gale said, because it has not been successful.

Fetal pancreas cells. By the end of 1988, nearly 170 U.S. and 100 Chinese diabetics had received implants of insulin-producing fetal pancreatic cells in an attempt to reverse their condition. Diabetes mellitus occurs when cells in the pancreas secrete insufficient amounts of the hormone insulin. The disease leaves other cells in the body unable to metabolize sugars properly. Insulin injections, however, allow the body to use sugars once again. Nonetheless, wide swings in blood-sugar concentrations may occur between insulin injections, and many physicians believe that these swings are responsible for the long-term complications of diabetes, including blindness, kidney failure, and nerve damage in the arms, legs, and other parts of the body.

Such complications might be avoided if insulin were released into the bloodstream only when needed. A pancreas transplant might achieve this goal, but such transplants normally provoke a strong immune response in the recipient's body, making that organ more likely to be rejected than a heart or a kidney. Moreover, only small clusters of pancreatic cells, called islets, actually secrete insulin. The rest of the pancreas, which is about the size and shape of a fillet of sole, produces digestive enzymes and plays no role in diabetes. It is very difficult to separate islets from an adult pancreas for transplantation; however, islets develop earlier in fetuses than the enzyme-producing cells and, at about the fourteenth week of gestation, they can be readily separated from the aspirin-sized pancreas itself.

Immunologist Kevin Lafferty of the University of Colorado Health Sciences Center has transplanted islet cells from electively aborted fetuses into 24 insulin-dependent diabetics who had already received kidney transplants because high blood-sugar levels had damaged small blood vessels in their own kidneys. Since these patients were already receiving drugs to suppress kidney rejection, they were less likely to reject transplanted pancreas cells. At a 1988 press conference, Lafferty said that five of the seven patients given "large amounts" of fetal cells required 30 to 50 percent less insulin daily after the surgery. The treatment provided no benefit for patients who received fewer cells. Lafferty implanted fetal cells in 10 more patients during 1988 and was planning to monitor their progress for 2 years before performing any further operations.

In similar efforts to combat diabetes, immunologist Bent Formby and his colleagues at Sansum Medical Research Foundation in Santa Barbara have implanted fetal pancreatic cells in seven insulin-dependent diabetics. In the most successful cases, in which cells from 12 fetuses were implanted, the patients' daily insulin requirement dropped by about half within days after the operation, then rose again as some of the transplanted cells died. Formby believes the operation would have been more successful if his team had implanted twice as many fetal cells, but the researchers had difficulty obtaining enough fetal tissue for their studies.

To circumvent the difficulties of acquiring fetuses, Hana Biologics, Inc., of Alameda, California, has spent 6 years refining techniques to grow fetal pancreas cells in the laboratory. By using specially developed nutrients and growth enhancers, the company can grow enough cells from one fetal pancreas to treat 20 adult diabetics. Researchers at four medical centers began implanting the cultured cells in 24 insulin-dependent diabetics in August 1987. Those tests proved successful, and in the fall of 1988, researchers at 12 medical centers around the U.S. began implanting the cells in 120 diabetics. Hana president Craig McMullen would not reveal the names of the centers because "we are afraid the centers would be swamped by diabetics demanding treatment." Hana expects to be able to provide enough cells for 15,000 transplants per year by 1991. The company has also developed similar techniques to grow fetal brain cells for use in treating Parkinson's disease and, by early 1989, was testing the cells in animals. McMullen expected to begin testing the cells in humans in 1990.

Fetal cells for Parkinson's disease. As had been the case with his use of adrenal cell transplants, Madrazo's use of fetal tissues sparked a surge of similar operations. Madrazo himself was unable to perform more of these procedures because Mexico, a predominantly Catholic country, prohibits abortions. His operations were possible only because the donor mother, with a history of spontaneous abortions, miscarried in the hospital, where physicians could collect the fetal tissue.

By the end of 1988, approximately 30 implants of fetal brain tissue had been performed, including 2 in Sweden, 10 in Cuba, 8 in the U.K., 6 in Spain, and 2 in the U.S. The results, as with adrenal tissues, have been mixed. The Swedish patients have not shown significant improvement, according to the University of Lund's Anders Bjorklund. In contrast, Bjorklund reported that the Cuban patients, who were operated on by neurosurgeon Hilda Molina of Hospital Hermanos Ameijeiros in Havana, have shown "clear improvement."

In late 1987, neurobiologist Irwin J. Kopin of the National Institute of Neurological and Communicative Diseases and Stroke applied to NIH's institutional review committee (which must approve experiments involving human subjects) for permission to implant fetal cells into one of his patients. The application eventually reached Robert E. Windom, assistant secretary of health at the U.S. Department of Health and Human Services. In March 1988, Windom announced a temporary ban on all human experiments at NIH involving tissues from aborted fetuses, a ban that was extended in May to all research involving NIH funds. Windom also directed NIH officials to form an independent advisory committee to examine the ethical implications of using tissues from aborted fetuses. The committee was chaired by retired federal appeals court judge Arlin J. Adams.

Despite the NIH ban, neurobiologist Curt Freed of the University of Colorado Health Sciences Center implanted fetal tissues into the brain of a 52-year-old man with Parkinson's disease in November 1988, becoming the first person to perform the procedure in the U.S. Although Freed's research has been supported by NIH, he used privately donated funds to cover the cost of this particular operation.

Freed was severely criticized by many researchers throughout the U.S. who feared that the timing of the operation would interfere with the deliberations of the NIH panel. Freed defended himself by arguing that "the time had simply come for human experiments" and noting that his team faced a 1-month "window" when team members, the necessary operating rooms, and fetal tissue would all be available at the same time. He said that if he had not operated then, he would have had to wait until at least February 1989. Despite the furor, he said, "the minute we finished the operation, I knew we had done the right thing .... I'd do it all over again in a minute." In December, Yale's Eugene Redmond also performed a fetal-tissue graft but, in sharp contrast to Freed's case, the operation received little attention.

Ethical issues. The debate over the use of aborted fetal tissues for human therapy encompasses a wide range of opinions. Perhaps the most extreme negative position is held by the National Right to Life Committee in Washington, D.C., a militant antiabortion group. Kay James of that group stated before a meeting of the NIH panel that 1.5 million fetuses are "deliberately killed" each year in the U.S. "We must not compound this gigantic moral and ethical lapse by weaving this slaughter into the warp and woof of modern medical management," she argued.

Less extreme critics fear that the use of fetal tissues will provide an incentive, either emotional or financial, for women to have abortions. Attorney James Bopp, Jr., of Terre Haute, Indiana, who served on the NIH panel, argued that the immense number of people who could potentially benefit from fetal therapy would create a $6 billion market for fetal tissues, a market whose gross revenues would be 30 times that now generated by abortion clinics. "The likely result is increased numbers of abortions, changes in abortion procedures, and delayed abortions to facilitate acquisition of more useful fetal tissue," he testified.

Many ethicists also worry that women could become pregnant and have an abortion either to provide fetal tissue for a close relative who is ill or to sell the tissue. Such critics fear that women in Third World countries could be exploited for the benefit of people in wealthier countries. Many researchers and physicians dismiss such fears out of hand, however, arguing that no reputable physician would tolerate either practice and noting that such problems have not developed for organ transplants. U.S. federal law, furthermore, already prohibits the sale of fetal tissues.

Most researchers and ethicists urge that the abortion and fetal-cell transplant issues be considered separately. They believe that as long as abortions are legal, the tissues should be available for medical use. "I do believe we'd be better off as a moral community without abortion," testified Thomas H. Murray, director of the Center for Biomedical Ethics at Case Western Reserve University School of Medicine in Cleveland, Ohio. "But the ethics of research on human fetal-tissue transplantation is not inextricably tied to approval of abortion any more than recovering organs from accident victims is equivalent to approving of accidents." Lee Ducat, president of the National Disease Research Interchange, which supplies human tissues to many researchers, said that if the medical community did not use aborted fetuses for research or therapy, the fetal tissues would simply be "wasted, trashed, or incinerated."

A preliminary draft of the NIH panel report, released in December 1988, concluded that the issues of abortion and use of fetal tissues for research or therapy should be kept completely separate. The panel did not address the issue of abortion, but concluded that if abortions are legal, then the use of the tissues resulting from the abortions is morally acceptable. Nonetheless, many researchers are concerned about the future of fetal-tissue research. The 1988 election of President George Bush, who has openly stated his opposition to abortion, and the impending retirements of the three most liberal U.S. Supreme Court justices suggest that the court will soon be more conservative. Many antiabortion groups hope that the court will overturn the controversial *Roe v. Wade* decision that legalized abortion in the U.S. If that happens, then the supply of fetal tissue could nearly disappear, and promising therapies for Parkinson's disease, diabetes, and a variety of other diseases might no longer be possible.

Alternative sources of cells. Researchers hope that they will eventually be able to avoid the ethical issues associated with use of fetal cells by developing alternative sources of cells. The work at Hana represents one potential solution to the problem. By growing fetal cells in culture, Hana's biologists have greatly increased the number of patients who can be treated with cells from one aborted fetus. Such an approach could greatly reduce the long-term demand for tissues obtained from abortions. Other researchers are studying approaches that would completely eliminate the use of fetal cells.

Neurobiologist Don Marshall Gash of the University of Rochester, for example, has been studying cells from a tumor of the human nervous system, called a neuroblastoma. The tumor cells can be grown in the laboratory to produce as many cells as needed and then treated with chemicals to stop their proliferation. After their growth is halted, the cells began producing acetylcholine, a neurotransmitter produced in insufficient amounts in the brains of people with Alzheimer's disease.

Gash reported in 1986 that he had grafted 320,000 chemically treated neuroblastoma cells into the brains of five African green monkeys. A significant percentage of the grafted cells survived for 270 days — the longest period during which Gash monitored the cells — and did, in fact, produce acetylcholine. Gash believes these results suggest that the cultured tumor cells "may serve as a practical source of donor tissue for implants."

Another possibility, according to Anders Bjorklund, would be to derive the graft tissue from species closely related to humans, such as monkeys. Animal experiments have indicated, however, that the graft success rate plummets when tissue is transplanted from one species to another. Nevertheless, Bjorklund and other researchers have been working to improve the success rate of interspecies transplants by using drugs that impede rejection, such as cyclosporine.

In another important development, neuroscientist Fred H. Gage of the University of California at San Diego has genetically engineered rat skin cells called fibroblasts to secrete a hormone called nerve growth factor, which helps sustain the health of certain types of brain cells. Gage and his colleagues reported in late 1988 that the engineered cells survive and secrete the hormone after being implanted in the brains of other rats. If this approach can be further refined, fibroblasts taken from a person with a brain disorder could be collected, engineered, and then implanted in the person, eliminating the need for fetal tissues. Gage said it could be 10 years before the engineered cells are tested in humans, but others think such a test might be attempted sooner because of the great potential of the approach.

Fetal-Cell Therapy for Other Brain Disorders

Brain-tissue grafting may eventually be used for a broad variety of diseases. In most cases, the grafted cells would simply act as miniature pumps that provide missing hormones or neurotransmitters. The evidence suggests, however, that the grafted cells could also become integrated into the brain, where they could become completely functional.

Biological pumps. Rochester's Don Gash and his colleagues have successfully transplanted a particular type of nerve cell, the vasopressin neuron, from the brains of normal fetal rats into the brains of Brattleboro rats, which are born without these neurons. Because the hormone vasopressin helps regulate water consumption and retention, Brattleboro rats typically consume more than their body weight in water every day and thus produce large amounts of urine.

At a 1986 symposium, Gash reported that after vasopressin neuron grafting, water consumption and urination dropped dramatically, although not to normal levels, in about 20 percent of the rats treated. When Gash dissected the rats, he found that the positive effects had occurred only in animals who had received the fetal cells at the precise sites where vasopressin-producing neurons are found in healthy rats. Precise placement of the graft within the recipient's brain thus seemed crucial.

Neurobiologists Earl Zimmerman of the Oregon Health Sciences University in Portland and Ann-Judith Silverman of Columbia University in New York City performed similar research with rats whose brains do not produce gonadotropin-releasing hormone (GnRH), a hormone that plays a role in the development of the reproductive organs. At the symposium, Zimmerman reported that grafts of GnRH-producing fetal brain cells caused the testicles of adult GnRH-deficient males to grow and begin producing sperm. Ten adult GnRH-deficient female rats who received similar treatment developed normal ovaries and uteruses. "All ten mated, and six gave birth to normal litters," Silverman said.

In other experiments, Bjorklund and Lars Olson found that transplanted brain cells can also improve coordination in old rats. The researchers monitored the agility of old rats by requiring them to walk along narrow rods. The old rats either clung tightly to the rods and remained in one place or often fell off when they did try to walk. At a 1986 meeting of the New York Academy of Sciences, Bjorklund reported that grafts of fetal cells producing dopamine and acetylcholine restored the old rats' ability to walk along the rods. According to Bjorklund, the old rats "manifested gait and posture similar to young rats and fell off less frequently."

Alzheimer's disease. Some evidence indicates that transplants of hormone-secreting cells can restore mental as well as physical agility. If, as many investigators believe, at least part of the memory loss incurred by patients with Alzheimer's disease stems from the degeneration of acetylcholine-secreting neurons in their brains, then transplants of healthy cells might be beneficial.

To test this notion, neurobiologist Gary Arendash and his colleagues from the University of South Florida in Tampa and the University of British Columbia School of Medicine in Vancouver chemically damaged acetylcholine-secreting neurons in 10 rats. They then implanted healthy fetal cells in the brains of five. After the animals recovered from surgery, they were subjected to a test requiring them to learn and remember how to avoid an electrical shock. At a 1984 conference on Alzheimer's disease, Arendash reported that untreated brain-damaged rats were slow to learn how to avoid the shocks and forgot the lesson easily. In contrast, the transplant recipients learned and retained the lessons almost as well as normal animals.

In related research, John Sladek and neurobiologist Helen Barold of the University of Rochester studied a strain of rats whose ability to learn new tasks naturally declines with age. They found that the implantation of fetal brain cells that secrete the neurotransmitter norepinephrine, which is also believed to be deficient in people with Alzheimer's disease, improved the rats' ability to remember the location of a hidden object.

Huntington's disease. At the same symposium, several researchers also presented evidence that brain grafts can alleviate the symptoms of Huntington's disease. Researchers can induce some of these symptoms in laboratory animals by injecting chemicals that kill specific brain cells, after which the animals lose some control of their limb movements.

Neurobiologist Stephen B. Dunnett of Cambridge University in the U.K. found that rats with this chemically induced brain damage had greater difficulty reaching out

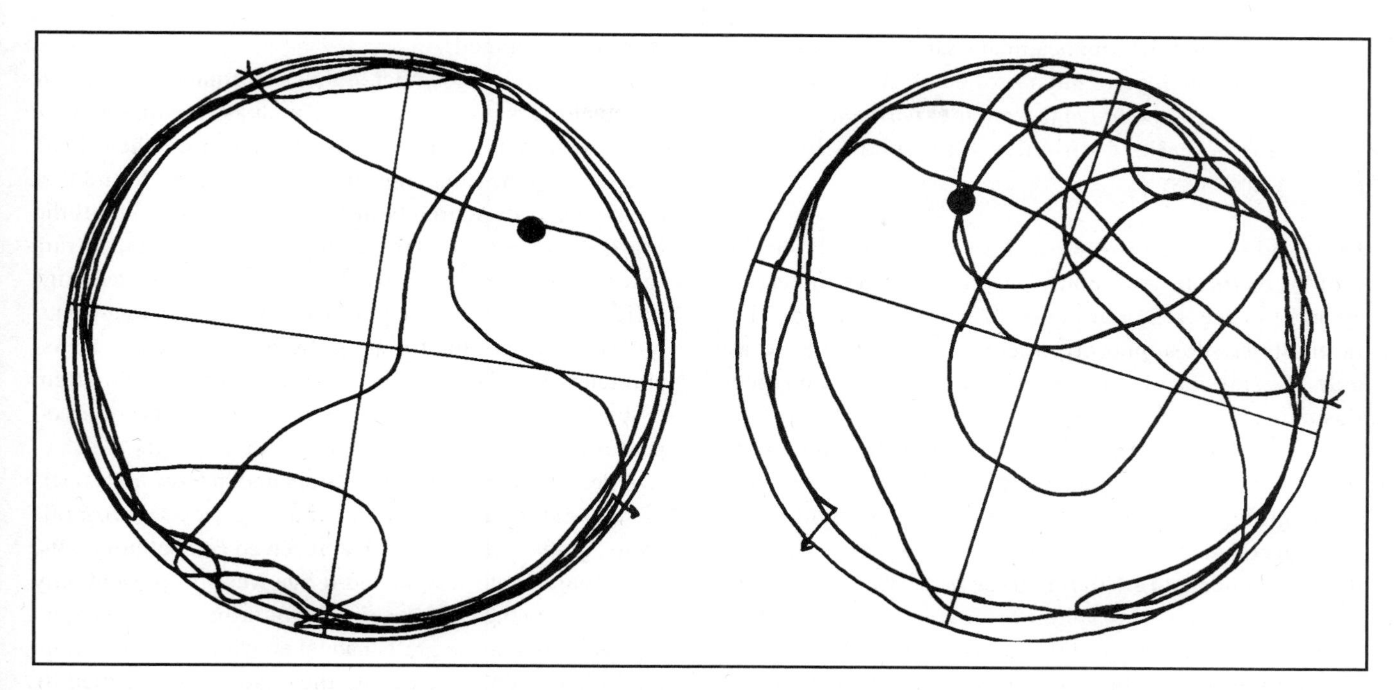

Figure 5. Alzheimer's Disease Model in Rats? Experiments conducted at the University of Rochester suggest that the memory deficits common in certain strains of older rats may be related to diminished production of norepinephrine — a neurotransmitter secreted by specialized neurons in both rat and human brains. In a series of trials, old rats swimming in a small circular pool had difficulty remembering the location of a slightly submerged clear glass platform upon which they had been able to rest during earlier trials, but which later had been removed. As shown by the paths traced in the diagram on the left, one typical older rat swam repeatedly around the wall of the pool, making only occasional excursions to the vicinity of the platform (indicated by the black dot). However, 5 weeks after fetal rat norepinephrine neurons were implanted into the older animal's cerebral cortex, its search activity was concentrated in the area near the platform location (see diagram on the right). Some researchers believe that norepinephrine deficiency plays a role in Alzheimer's disease. *Courtesy: Timothy Collier, University of Rochester School of Medicine.*

of their cages to obtain food from a nearby bowl. Before the injections, they retrieved food on 60 to 70 percent of such attempts; afterwards, they were successful only about 20 percent of the time. When Dunnett grafted fetal tissue into the animals' brains, however, the rats could eventually retrieve food nearly as successfully as healthy rats. Other researchers have reported similar results, including psychiatrist A. Wallace Deckel of the University of Medicine and Dentistry of New Jersey in Newark.

One particularly intriguing report was presented by neurosurgeon Noel B. Tulipan of Vanderbilt University, who grafted fetal tissue into rat brains before administering the chemical that induced Huntington's disease symptoms. Surprisingly, Tulipan found that the graft somehow protected the rats' brains from developing the symptoms. This finding could be significant, because many scientists believe that Huntington's disease is caused by the action of similar chemicals on the brains of humans who are genetically susceptible to the disease. Tulipan believes his results suggest that fetal-cell grafting might halt the progression of Huntington's disease. In April 1988, Tulipan and George Allen implanted adrenal tissues into the brain of a 41-year-old man with Huntington's disease, but by the end of the year, they had not yet announced whether the procedure had produced any significant effects.

Epilepsy. Gage and neurologist Gyorgy Buzsaki of the University of California at San Diego believe that fetal-cell implants may also be useful in reducing the effects of epilepsy. They simulated temporal-lobe epilepsy in 14 rats by surgically severing specific nerve pathways leading to the hippocampus, an area of the brain particularly vulnerable to epileptic activity. With the pathways severed, neurons in the hippocampus emit random electrical signals, triggering seizures that can either occur spontaneously or be triggered by stress. The two researchers implanted fetal hippocampal cells in half of the rats and fetal cells from the locus coeruleus, an area of the brain stem, in the other half. After several months, the rats that received locus coeruleus tissue had no spontaneous seizures, fewer induced seizures, and much less uncontrolled electrical activity in the brain. In contrast, the rats receiving the hippocampal tissue had many more seizures and much more uncontrolled electrical activity. The researchers concluded that their experiment demonstrated the potential of brain-cell transplants for controlling epilepsy.

Impaired thought processes. Recent studies have shown that brain-tissue grafts may partially restore impaired thought processes in animals with brain damage. The implanted cells apparently secrete trophic factors that help re-establish electrical connections in the damaged tissue.

Donald Stein of Rutgers University in Newark, New Jersey, and his colleagues have found that fetal brain-tissue grafts can improve the performance of brain-damaged rats on tests of spatial learning. In their experiments, the researchers surgically damaged a part of the frontal cortices — which play a major role in awareness, intelligence, and memory — of 21 adult rats. Seven days later, they implanted frontal cortex tissue from healthy fetal rats into eight of the brain-damaged rats and healthy tissue from fetal cerebellums into six others.

The remaining seven brain-damaged rats received no transplanted tissue, although the researchers reopened the wounds from their original operations so that all rats underwent two operations. In addition, another eight adult rats served as a control group. Incisions were made in the skulls of these animals on the same days that operations were performed on the brain-damaged rats, but their brains were not damaged.

The researchers compared the performance of the four groups of rats on spatial tasks, such as learning mazes, that researchers commonly use to determine the effects of frontal-cortex damage. As expected, they found that the control rats scored better in the tests than did all the brain-damaged rats. However, the brain-damaged rats that received frontal-cortex transplants performed significantly better than the brain-damaged rats that either received cerebellum tissue or did not receive any transplanted tissue. Overall, the rats that received cerebellum tissue (which does not normally participate in spatial tasks) performed no better than those with no grafts.

After completing the tests, Stein and his associates performed autopsies on five rats that had received frontal-cortex grafts and three that had received cerebellum grafts. They found that transplanted tissue was still present only in those rats that had received frontal-cortex tissue, an indication that those grafts had taken while the cerebellum grafts had been rejected and the dead cells absorbed by the rats' bodies. This finding was consistent with the results of the spatial tests.

No one expects fetal cells to be grafted into human brains at any time in the near future to correct memory dysfunction. But most researchers do believe that brain-cell grafting will continue to be studied intensively in both animal and human experiments. If this research is successful, then grafting could someday become an important treatment for a variety of brain disorders.

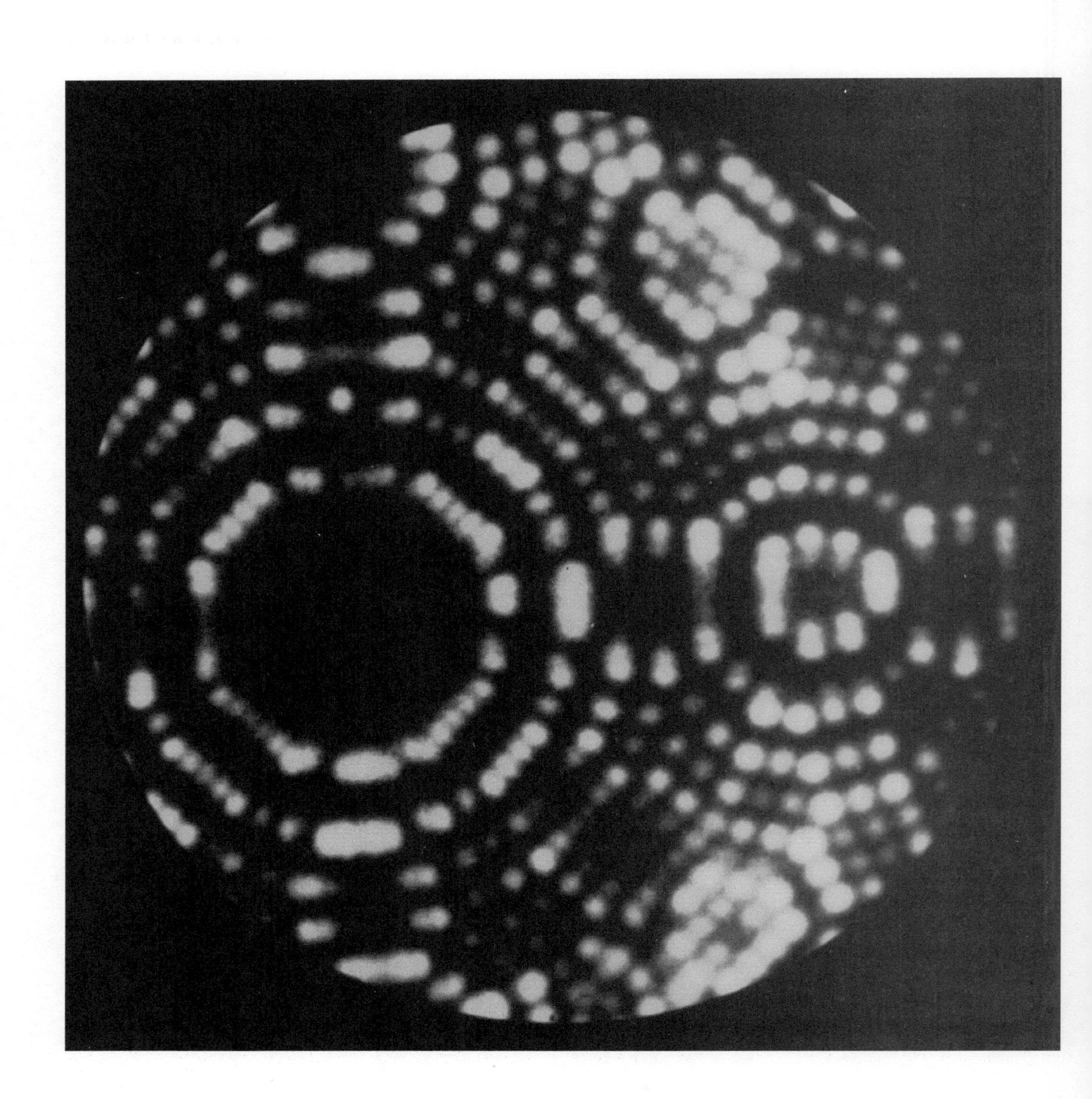

Chapter 4

Chemistry

Catalysis

The history of chemistry reveals a relentless pursuit of selectivity — the ability to make a particular chemical compound to the exclusion of all others. Whereas a chemist in the 1890s might have wanted to synthesize a ring structure containing six carbon atoms, the chemist of the 1990s is more likely to want to make a plastic containing literally tens of thousands of carbon atoms linked together in a precise manner. Perhaps even more difficult, a biochemist might want to create molecules containing only 20 to 30 carbon atoms, controlling not only which atoms are linked to which other atoms but also the three-dimensional configuration of those atoms.

The key to achieving selectivity is a technique called catalysis. Catalysts have been well-known and widely used tools since 1835, when the Swedish chemist Jons Jakob Berzelius first coined the term to describe materials that speed up chemical reactions without being consumed in the reactions. Since then, catalysis has become an essential part of industry; it now contributes directly or indirectly to products accounting for one-sixth of the value of all goods manufactured in the U.S.

Despite the importance of catalysis to industry, remarkably little has been known about how catalysts work. One of the first and, until recently, one of the few insights into the nature of catalysis came at the turn of the century, when the German chemist Friedrich Wilhelm Ostwald recognized that catalysts worked simply by changing the rate of a reaction. He also recognized that catalysts cannot force a reaction to occur if it is not chemically possible. Since then, chemists have found that a potentially important trait of catalysts is their ability to alter reaction rates selectively so that a desired reaction is enhanced while competing reactions are suppressed. Beyond these facts, so little has been known about catalysts that the development of new ones has been a trial-and-error process, part science and part art.

Renewed interest. During the 1980s, research on catalysts has undergone a renaissance. Both academic and industrial chemists and physicists are bringing a remarkable array of new instruments and techniques to bear on the study of how existing catalysts work and how new catalysts might be designed to improve selectivity and increase reaction yields. These efforts have met with great success. Although it is still an art to develop and refine new catalysts and improve on old ones, science now plays a more significant role.

Several converging trends have stimulated this growing interest in catalysis. One of the most important is a change in the types of petroleum used for fuels and chemicals. The amount of thin, low-sulfur crude oil remaining in oil fields worldwide is declining, as it is pumped from the ground. This depletion is forcing refineries to use thicker, more sulfurous crude. According to the National Petroleum Council, the percentage of low-sulfur oil in U.S. refinery feedstocks will have declined from 65 percent in 1969 to about 41 percent by 1990. Refining the new feedstocks will require entirely different catalyst systems, since the more sulfurous crudes contain larger amounts of metals that gradually block the activity of conventional catalysts and shorten their useful life.

The energy crises of the 1970s also stimulated a new interest in alternative sources of fuels and chemicals. The most important alternative to petroleum is coal, which can be burned directly for its energy or gasified to produce a mixture of carbon monoxide and hydrogen known as synthesis gas, or syngas. This gas can then be converted into various chemical feedstocks, such as acetaldehyde, ethanol, and methane, or into gasoline by a process known as Fischer-Tropsch. Sasol Technology, Ltd., has three plants in Sasolburg, South Africa, that use coal to produce chemical feedstocks, diesel fuel, and gasoline.

Other alternative fuels include the thick oils obtained from tar sands, kerogen from oil shale, methanol from biomass fermentation, and hydrogen obtained by breaking apart water molecules. Some of these fuels are already being used: Syncrude, Ltd., and Great Canadian Oil Sands, both of Edmonton, Alberta, are producing low-sulfur crude from tar sands near Fort McMurray, Alberta. In mid-1985, the New Zealand government opened a plant that converts methanol into gasoline using a new catalytic process developed by Mobil Research and Development in Princeton, New Jersey. New catalysts are required to convert each of these alternative materials into a useful product.

The chemical industry is also changing its sources of raw materials. The traditional feedstocks for chemical production have been natural gas and the light petroleum fractions, such as naphtha. Because of the potential for future shortages and price increases, the industry has been looking to other sources, particularly synthesis gas and methanol. A harbinger of this change is a Tennessee Eastman Co. plant opened in April 1984 in Kingsport, Tennessee, to produce acetic anhydride and other chemicals from synthesis gas.

Another incentive for finding new types of catalysts is the current heavy reliance on noble metals as catalysts. In 1987, according to the U.S. Bureau of Mines, the petroleum, chemical, and automobile industries in the U.S. used 690,492 troy ounces of platinum at an average price of $553 per ounce, 66,446 troy ounces of rhodium at $1,222 per ounce (nearly double its price in 1984), and 236,032 troy ounces of palladium at $130 per ounce (12 troy ounces equal one pound). Virtually all of these metals were mined in South Africa and the U.S.S.R., a situation that puts the U.S. — indeed, the whole Western world — at a potential strategic disadvantage because of the possibility of boycotts or political extortion.

Many scientists, as well as government officials, hope that the development of new catalysts based on more common metals will reduce this dependence on potentially unstable suppliers and lower the cost of manufacturing processes. An April 1987 study conducted by Battelle Pacific Northwest Laboratories for the U.S. Department of Energy estimated that the development of innovative catalysts could save $100 billion over a 15-year period. The study estimated that U.S. chemical and refining plants could save $31 billion in capital costs and $69 billion in operating costs during that period.

Heterogeneous Catalysis

Heterogeneous catalysis — catalytic reactions at the surface of a solid, typically a metal particle — is the foundation of most chemical and energy production technologies. Nearly all petroleum products and 80 to 85 percent of chemical products are created by such reactions. During 1987, sales of catalysts totaled $1.65 billion in the U.S. alone and more than $850 million in Japan and Europe. Yet until the late 1970s, remarkably little was known about the nature of catalyst surfaces, how gases and liquids interact with the surfaces, or how the reactions themselves occur.

The extensive use of analytical techniques for studying catalyst surfaces has permitted the scrutiny of catalysts on an atomic scale. For the first time, the structure and chemical composition of molecules on the catalyst surface can be determined by spectroscopic techniques and correlated with the rate of the catalytic reaction, the energy required to begin the reaction, and the chemical composition of the products. This information should soon make it possible to predict what will happen at the surface of new catalysts; in the longer run, it may also allow scientists to design catalysts to carry out specific chemical reactions.

Modeling commercial catalysts. A major problem in the past has been that heterogeneous catalysts are such complex mixtures of metal particles and "supports" (usually inert materials that hold the particles in place and keep them separated) that it is almost impossible to identify any particular intermediate or final product on the metal surface. To overcome this problem, a number of chemists, including Gabor Somorjai of the Lawrence Berkeley Laboratory in Berkeley, California, M. Albert Vannice of Pennsylvania State University in University Park, and D. Wayne Goodman of Sandia National Laboratories in Albuquerque, New Mexico, have begun using specially designed instruments to study the properties of catalytic metals. In these instruments, developed during the early 1980s, a single metal crystal — the catalyst — can be enclosed in a small pressure chamber located inside the vacuum chamber of a spectrometer, a device for measuring physical properties of the sample, such as emitted or absorbed light.

A catalytic reaction on the surface of the crystal can be run in the small chamber at the high pressures characteristic of a heterogeneous reaction. The progress of the reaction can be studied with spectrometric techniques that are not affected by the presence of reactant and product gases. The reaction can also be stopped at any point and the gases removed from the chamber to study molecules on the catalyst surface with spectrometric techniques that require a vacuum.

Many industrial scientists have criticized this approach, however, arguing that what happens on the surface of a single crystal may have no relationship to what happens on the surface of industrial catalysts, which are composed of polycrystalline metal particles. Thus, one of the most important recent findings in research on heterogeneous catalysis was the demonstration by Goodman and others during the mid-1980s that the characteristics of a catalytic reaction observed on the surface of a single crystal are identical to those of catalytic reactions carried out by industrial catalysts.

Catalytic reactions at crystal surfaces. Studies of the interaction of crystal surfaces with hydrocarbons — chemicals containing only carbon and hydrogen — have led to a growing understanding of the mechanisms involved in heterogeneous catalysis. Somorjai of the Lawrence Berkeley Laboratory has, perhaps, provided the most detailed picture of the surface of a catalyst. He has used single crystals of platinum to catalyze the hydrogenolysis, or breaking apart, of alkenes to produce methane. Alkenes are hydrocarbons with at least one double bond, in which two carbon atoms are held together by four electrons rather than the two electrons in a single bond.

Below about 212°F, alkenes undergoing hydrogenolysis are adsorbed, or bound, on the platinum surface with the double bond perpendicular to the metal surface. Somorjai observed that the interaction of the alkenes with the metal atoms stretches the double bond, indicating that dissociation has already begun during the binding of the alkenes to the surface; in effect, the formation of carbon-metal bonds weakens the carbon-carbon double bonds.

As the temperature increases, some of the carbon-carbon double bonds break apart completely, and many of the resulting hydrocarbon fragments lose hydrogen atoms. A layer containing hydrocarbon fragments and hydrogen then begins to accumulate on the metal surface. Between 212°F and 238°F, this layer has an average of about three hydrogen atoms for every two carbon atoms. Because the layer still contains a relatively high proportion of hydrogen atoms, the hydrocarbon fragments can recombine with hydrogen and break free of the surface as intact alkenes.

At higher temperatures, more hydrogen is released from the bound hydrocarbons, and two distinct hydrocarbon regions begin to form. One still contains significant quantities of hydrogen and is thus similar to hydrocarbon; the other contains very little hydrogen and is thus more like graphite, which is pure carbon. Although little hydrogen is chemically bonded to the carbon in the graphite-like layer, some hydrogen is adsorbed onto its surface.

The carbon atoms in the graphite-like layer are more tightly bound to each other and to the metal surface than those in the hydrocarbon-like layer, so they no longer recombine with hydrogen. These nonreacting carbon atoms block the approach of other alkenes to the metal surface and thus block the activity of the catalyst. But since catalytic activity is still observed, the surface must have "islands" of clean or unblocked platinum. Somorjai thus concluded that the key step of the hydrogenolysis reaction — the cleavage of the carbon-carbon double bond — must occur on these clean islands.

Somorjai also observed that, in the hydrocarbon-like layer, the hydrocarbon fragments resulting from the cleavage of the carbon-carbon double bonds are too strongly bonded to the metal surface to escape. Nevertheless, the fragments are constantly in motion on the surface and can move to the nearby graphite-like layers. The hydrocarbon fragments bind more weakly to the graphite-like layers than they do to the metal surface. As a result, they react with the adsorbed hydrogen on the graphite-like layer, forming molecules of methane (one carbon atom and four hydrogens) and ethane (two carbons and six hydrogens), which escape from the metal surface.

Robert Gomer of the University of Chicago in Illinois and others have observed that this type of movement from layer to layer occurs readily. Gomer also found that about 10 times as much hydrogen can be adsorbed onto the graphite-like layer as can be adsorbed onto the metal surface; thus more hydrogen is available on that layer to combine with the fragments than is available on the clean metal surface. This makes it easier to form methane on the graphite-like layer.

These results suggest several ways in which the activity of the platinum catalyst can be changed. An increase in hydrogen pressure in the reaction vessel, for example, will remove more of the hydrocarbon-like layer because the extra hydrogen will react with the hydrocarbon fragments in that layer. As a result, a greater proportion of the metal surface will be bare, and a larger number of sites will be available to catalyze the reaction.

When tiny amounts of certain metals, such as gold and tin, are added to the platinum crystals as they are forming, the foreign atoms alter the spacing between the platinum atoms and the number and concentration of active sites on the platinum surface, thereby changing the reaction rate. The foreign metal atoms can also catalyze the reaction between the hydrocarbon fragments and the hydrogen on the graphite-like layer, thereby increasing the rate at which methane is formed. Researchers hope that detailed studies of other metal surfaces will make it possible to modify catalysts deliberately.

Other studies with single crystals have shown that the nature of the products from a catalytic reaction is highly dependent on the structure of the crystal surface. For example, Goodman has found that the type of products made during the hydrogenolysis of butane depends on which crystal surface of iridium catalyzes the reaction. Butane is a hydrocarbon composed of four carbon atoms, connected in a straight line, and ten hydrogen atoms; three hydrogens are attached to each of the end carbons and two to each of the center carbons. When hydrogen is added to butane at high temperatures in the presence of a catalyst, such as iridium, the butane is broken down into methane and ethane.

Goodman studied two of the many possible iridium surfaces. On one face, called (111) in the complex nomenclature used by crystallographers, each iridium atom is surrounded closely by nine other atoms. On the second face, known as (110)-(1x2), each iridium atom is surrounded by only seven other atoms. Goodman found that, on the (111) surface, each butane molecule is broken down into one ethane molecule and two methane molecules; however, on the other surface, each butane is broken down into two ethane molecules, and no methane is formed at all.

Goodman also observed that, on the (110)-(1x2) face, both end carbons of a single butane molecule could bind to one iridium atom to form a five-membered ring. This ring then split apart to yield two ethane molecules. On the (111) surface, however, the iridium atoms are so close together that the two end atoms of butane cannot bind to one iridium atom. Hence, the butane molecule binds at only one end, and this configuration favors the splitting off of two carbon atoms, one at a time, to form both methane and ethane. Goodman concluded that the products made in industrial reactions can sometimes be changed by controlling which crystal faces of a catalyst are exposed to the reactants.

How catalytic converters become "poisoned." Physicist Gary L. Kellogg of Sandia has studied processes on the surface of catalysts by using techniques that allow individual atoms to be observed and identified. One of Kellogg's research goals is to determine how the rhodium catalyst in an automobile's catalytic converter is poisoned — contaminated so that it no longer functions — when too much oxygen is present. The rhodium catalyst joins toxic carbon monoxide in the exhaust with oxygen to form carbon dioxide.

For his studies, Kellogg used a field ion microscope, which consists of a vacuum chamber to hold the sample, a refrigeration unit for cooling the sample to a very low temperature, and a fluorescent screen like that on a television. The sample itself is a sharply pointed needle or whisker, called the tip, and the apex of this tip is only a few tens of atoms wide.

An image of the atoms at the apex of the tip is obtained by introducing a small amount of a gas, such as helium or neon, into the chamber and applying a high voltage to the tip. The voltage produces an electrical field at the surface of the tip, which continuously ionizes the helium or neon gas atoms directly above each surface atom by capturing electrons from them. A magnetic field accelerates the ions to the fluorescent screen, where they produce spots. The image on the screen, therefore, is a direct image of the protruding atoms on the tip's surface. When a catalytic reaction is being studied, the field ion microscope can be combined with a mass spectrometer to identify atoms or molecules that bind to the surface of the tip.

Kellogg studied a rhodium whisker in the field ion microscope under conditions that simulate the reaction in a catalytic converter. The whisker was about the same size as the rhodium particles present in an automobile catalytic converter. Before the reaction began, Kellogg had determined the whisker's precise surface structure. He found that the reaction between carbon monoxide and oxygen proceeded normally as long as the ratio of oxygen to carbon monoxide was 30 to 1 or less. As the ratio increased beyond 30 to 1, however, a stable oxide of rhodium began to form on the surface and the rate of the oxidation reaction decreased.

The formation of rhodium oxide, according to Kellogg, occurs at about the same oxygen-to-carbon monoxide ratio at which poisoning of the catalyst occurs in an automobile, so it is clear that surface oxidation of the catalyst is responsible for the damage. He speculated that the presence of the rhodium oxides on the surface keeps the carbon monoxide from binding to the rhodium long enough to react with oxygen. For an automobile catalytic converter to operate efficiently, he concluded, it must be designed so that the amount of oxygen reaching it is less than 30 times the amount of carbon monoxide.

Exploring bimetallic catalysts. Catalysts composed of two metals often function better than either metal alone. When the surface of a ruthenium catalyst is partially covered with copper, for example, the resulting bimetallic catalyst can carry out a particular reaction — the removal of hydrogen from cyclohexane to produce benzene — 40 times faster than ruthenium can alone. Copper by itself does not catalyze the reaction at a significant rate. Goodman and his colleagues at Sandia found that the copper/ruthenium catalyst acts almost as if it were an entirely new metal.

Although the copper/ruthenium catalyst is not commercially important, a number of researchers have studied it thoroughly. The reactions that it catalyzes are well known, even though the underlying mechanisms are not. Goodman's group studied the bimetallic catalyst in a small pressure chamber placed inside a larger vacuum chamber. (For a description of this apparatus, see the section "Modeling Commercial Catalysts" earlier in this article.) The researchers began with a single crystal of ruthenium and coated part of its surface with a layer of copper only a single atom thick. Then they used sophisticated spectroscopic techniques to study the material's surface.

The researchers found that the copper atoms "sit in registry" with the ruthenium atoms. That is, each copper atom occupies the same position on the crystal's surface that an added ruthenium atom would have occupied. Because the atoms of ruthenium (atomic number 44) are slightly larger than those of copper (atomic number 29), however, the bonds between the copper atoms must stretch by about 5 percent for the atoms to maintain their fit with the ruthenium crystal structure. The layer of copper atoms thus becomes what is known as a strained lattice.

The new bimetallic catalyst does not resemble either copper or ruthenium, and its other chemical and physical properties are different as well. "It's something that doesn't look like any known material," Goodman said. The change in the copper's chemical properties can be attributed, at least in part, to a change in its electronic properties. The Sandia group's studies have shown that the underlying ruthenium atoms donate electrons to the copper atoms, and the resultant increase in the electron density at the copper surface helps change the properties of that surface.

Similar effects occur with other bimetallic catalysts. Goodman found that individual atoms in a one-atom-thick layer of nickel applied to the surface of tungsten occupy the same positions that would be filled if additional tungsten atoms were added to the crystal surface. Because nickel atoms are much smaller than tungsten atoms, the density of the nickel atoms in the monolayer is thus only 55 percent or 79 percent that of nickel alone, depending on which crystal face of tungsten is used. Because of this change in density and the electronic interaction between nickel and tungsten atoms, the monolayer of nickel atoms binds either hydrogen or carbon monoxide much more weakly than either nickel or tungsten alone, so that the bonded molecules can more readily take part in hydrogenation reactions, such as the conversion of carbon monoxide into formaldehyde.

The most important aspect of this research, however, is the demonstration that such detailed experiments with bimetallic catalysts can be conducted. "We have shown that the problem of studying bimetallic catalysts is actually manageable," Goodman said. Because of this demonstration, according to Goodman, "we can now start talking about some very specific experiments to learn the precise answers [about catalytic mechanisms]. The potential is enormous.... A general understanding of these interactive effects [between metals] should result in highly specialized [catalysts]" with surface properties superior to those of existing catalysts.

Watching bimetallic catalysts grow. Sandia's Kellogg, working independently of Goodman, has used his atom-imaging technique to study nickel atoms deposited onto different surfaces of a tungsten crystal. Such a nickel/tungsten catalyst can be used for the hydrogenolysis of ethane, in which hydrogen combines with ethane, breaking the carbon-carbon bond and forming two methane molecules. Studies have shown that the rate of this reaction increases as the amount of nickel on the tungsten surface decreases. Scientists believe that an ensemble of nickel atoms containing some as-yet-unidentified critical number of nickel atoms is necessary for the reaction to occur.

Kellogg wanted to determine how many nickel atoms were required for the reaction and to see whether individual nickel atoms are mobile or stationary on tungsten surfaces. He found that the answer to the latter question depends on the surface itself. On a relatively smooth tungsten surface, the individual nickel atoms moved freely at temperatures well below room temperature; however, on a rougher surface, the atoms moved around only when the temperature was increased above room temperature. Pairs of nickel atoms behaved in the same manner.

When larger numbers of nickel atoms were applied to the surface, the shape of the nickel clusters formed was dependent on the temperature at which the atoms were applied. Atoms applied at –99°F, for example, formed linear, one-dimensional chains aligned parallel to a particular crystal plane of the tungsten. Atoms applied at –45°F or higher, however, formed two-dimensional clusters. When either type of cluster was heated to temperatures of 80°F, they remained immobile on the surface. As the temperature was increased above 80°F, the clusters began moving across the surface as units. And finally, at temperatures above 216°F, the clusters fell apart. Three-dimensional clusters were never formed, according to Kellogg, and attempts to deposit nickel atoms on top of the nickel clusters failed.

When Kellogg studied the ethane hydrogenolysis reaction under each of these conditions, he found, contrary to the commonly held belief of researchers, that the reaction requires only single nickel atoms to proceed, and that the rate of the reaction was highest when the atoms were dispersed most widely across the tungsten surface. He thus concluded that commercial hydrogenolysis catalysts should be prepared under conditions that favor the formation of single nickel atoms.

When Is a Metal Not a Metal?

The metal particles in a commercial catalyst typically have a number of different crystal planes exposed. They also have edges, terraces (places where one layer of atoms on the surface stops, so that a lower layer is exposed), defects (irregularities in the crystal structure), and adatoms (individual atoms projecting from the surface). For some types of reactions that use catalysts, the location and number of metal atoms surrounding the reaction site are not important because only one or two atoms participate in the reaction. As a result, such reactions can be catalyzed by any crystal surface.

For most reactions, however, that environment is crucial. For example, in the Fischer-Tropsch conversion of synthesis gas (carbon monoxide and hydrogen) to various hydrocarbons on one catalyst particle, the precise reaction sites are not known. One might imagine that methane would be formed only on single metal atoms at the edge of a crystal, cyclohexane (which has six carbon atoms connected in a ring) at a site near a terrace, butane (which has four carbon atoms connected in a row) at a site where four metal atoms are located in a row, and so forth. Hence, many different products might be formed on a single catalyst particle.

If the size of the metal particle could be reduced, the number of different types of reactions taking place on its surfaces might also be reduced. In principle, if the metal particles were reduced to clusters consisting of only a few metal atoms arranged in a specific shape, only one reaction might occur, and one product might be produced. Reducing the size of the particle would also reduce the amount of precious metal required as a catalyst.

Reducing the size of the particle, however, introduces a new problem. An isolated atom of a metallic element does not have the same properties as a metal atom surrounded by a large number of similar metal atoms — that is, an atom in bulk metal. As surprising as it may seem to a casual observer, a single copper atom does not constitute a metal. Its electronic properties — the characteristics of the electrons orbiting around the nucleus — are different from those of an atom in bulk metal and, more importantly, from those that characterize a metal, such as the ability to conduct electricity.

Two copper atoms together also do not constitute a metal, but 200 copper atoms together do. The question, then, is how many atoms of an element such as copper must a cluster of atoms contain to acquire the characteristics of bulk metal? The answer has become particularly important as researchers have begun to focus on the use of small metal clusters as catalysts with increased selectivity. The electrons of the metal atoms control the catalyst's adsorption of organic molecules, which in turn controls the breaking of carbon-carbon bonds in a catalytic reaction. The fact that small metal clusters have different electronic properties than bulk metals may mean that they can catalyze reactions that do not occur on bulk metals.

The point at which the electronic properties of a small metal cluster change from those characteristic of isolated atoms to those of a bulk metal is still not clear. The properties of clusters containing one to 100 metal atoms, according to Richard Smalley of Rice University in Houston, Texas, remain "an almost totally uncharted wilderness." For example, as recently as 1985, physical chemistry texts did not even list the length of the metal-metal bond in a compound containing only two metal atoms because that information was not yet known.

Chemists have had great difficulty making small metal clusters and determining their properties. The manufacture of conventional industrial catalysts generally involves heating the metal, which causes small metal clusters to aggregate into larger particles. Furthermore, determining the properties of the clusters is difficult because analytical instruments tend to measure the properties of the material that supports the clusters, which is present in much higher quantities than the clusters themselves. In the early 1980s, however, two different methods for making and studying small clusters were developed in U.S. laboratories.

Studying small metal clusters with lasers. Smalley has developed a technique in which a laser is used to vaporize a metal sample. The laser produces a localized temperature of about 10,000°F, hot enough to vaporize any metal that might be used as a catalyst. As the metal vaporizes, it forms clusters ranging in size from one to 200 atoms.

The clusters are then passed through a device that cools them to a temperature of about –457°F. This decrease in temperature sharply reduces the molecular and atomic vibrations in the clusters, simplifying the optical spectra caused by those vibrations and making it easier to determine a cluster's properties. Various types of spectrometry are carried out during the few hundred microseconds in which a cluster can be studied before its momentum carries it from the instrument.

Rice University researchers have reported that the electronic properties of clusters of copper or silver atoms depend on the number of atoms making up the cluster. They measured the ionization potentials of the clusters — the energy required to remove an electron from an atom — and found that the potential decreases sharply as the number of atoms in a cluster increases from 2 to about 10. The potential then falls more slowly until the characteristic ionization potential of an atom in a bulk metal is reached at about 80 atoms. In general, however, the researchers found that clusters with odd numbers of atoms have lower ionization potentials than the even-numbered clusters on either side of them.

These findings tend to agree fairly well with theoretical calculations performed by Roger Baetzold of Eastman Kodak Co. in Rochester, New York, and Roald Hoffman of Cornell University in Ithaca, New York. Other findings, however, are in conflict with high-level calculations, such as those conducted by William Goddard of Caltech. One of Goddard's calculations had predicted, for example, that the atoms in a cluster containing two chromium atoms would be only weakly bound, with a relatively long bond length of about 3 angstroms. Smalley found that the bond length is actually 1.68 angstroms, making it, in fact, one of the shortest and strongest bonds known. "A lot of theoretical calculations are going to have to be redone," according to Smalley.

Chemists Andrew Kaldor, Donald Cox, and Eric Rohlfing of Exxon Research and Engineering Co. in Clinton, New Jersey, also observed that the ionization potentials of certain clusters of either iron or nickel atoms deviate rather sharply from the pattern observed with clusters of copper or silver atoms, indicating that not all metals behave in the same way. Other properties of small clusters do not vary in such unusual ways, however. David Johnson of Cornell University and a team headed by Peter Edwards of Cambridge University in the U.K. found that the magnetic properties of osmium clusters vary in a more straightforward manner.

Edwards and his colleagues also observed that clusters containing two or three osmium atoms have very little paramagnetism — the ability to become magnetic in the presence of a magnetic field — in contrast to bulk osmium, which displays strong paramagnetism. The paramagnetism of the osmium clusters increases as the number of atoms in the cluster grows, until a paramagnetism very close to that of bulk metal is obtained when the cluster reaches 10 atoms. Johnson is now studying the variation in paramagnetic properties in platinum clusters.

Chemical reactivity of clusters. Other results indicate that chemical reactivity is also dependent on the number of metal atoms present in a cluster. Smalley and his colleagues at Rice University have examined the dissociative binding of molecular hydrogen and deuterium to metal clusters generated in their laser device. Such binding, in which the hydrogen-hydrogen or deuterium-deuterium bond is cleaved as the molecule attaches to a metal atom in the cluster, is an important step in many catalytic reactions. The researchers found that dissociative binding displayed "the most dramatic cluster-size dependencies ever measured for any property of metal clusters."

While bulk cobalt, for example, binds hydrogen and deuterium very strongly, Smalley and his team observed that single cobalt atoms and clusters containing 2, 6, 7, 8, and 9 cobalt atoms did not bind hydrogen or deuterium at all. Clusters containing 3, 4, or 5 cobalt atoms and those containing 10 or more atoms, in contrast, did bind both molecules. Niobium clusters showed similar variations. Individual niobium atoms and clusters of 2 atoms did not react with the hydrogen, while clusters containing 3 to 7 niobium atoms were moderately reactive. Clusters containing 8 to 10 niobium atoms were inert, while those containing more than 10 atoms — with the exception of the cluster containing 16 atoms — were extremely reactive. In copper clusters, no binding with hydrogen was observed until the clusters contained at least 20 atoms.

Chemist Stephen J. Riley of the Argonne National Laboratory near Chicago used an apparatus similar to Smalley's to study the reactivity of iron clusters with hydrogen. Riley found that an iron cluster containing 23 atoms is at least 1,000 times as reactive with hydrogen as one containing 8 atoms. Perhaps even more impressive was the 50-fold increase in reactivity when one iron atom was added to an 18-atom iron cluster. Increases in the number of iron atoms beyond 23 did not result in any increases in reactivity, suggesting that a 23-atom iron cluster behaves chemically like bulk iron.

Riley and his colleagues also studied how many hydrogen atoms can bind to the surface of iron and nickel clusters. They found that, in clusters up to 80 atoms, the surfaces of both types of clusters are fully covered; that is, one hydrogen atom is attached to each surface metal atom. For larger nickel clusters, the surfaces remain fully covered, but for iron clusters, this is not the case. For iron clusters containing 260 atoms, less than 80 percent of the surface is covered with hydrogen.

This difference in hydrogen coverage may explain differences in the catalytic activities of iron and nickel, Riley said. In the reaction of carbon monoxide with hydrogen, for example, nickel catalysts yield only methane while iron catalysts also produce larger hydrocarbons. "Perhaps a nickel surface, fully covered with hydrogen, can adsorb only single carbon monoxide molecules and convert them to methane," according to Riley. "On a less-covered iron surface, however, two or more carbon monoxide molecules may be adjacent, allowing them to form carbon-carbon bonds and thus produce larger hydrocarbons."

Scientists do not yet know why the clusters behave in this way. Smalley has speculated that the electronic structure of the cluster may change according to the number of atoms in it. Alternatively, the geometry of clusters containing certain numbers of atoms may affect the interaction of the metal atoms with hydrogen. In either case, the observations provide fertile ground for theorists attempting to predict the chemical reactivity of such clusters. "If a theory can predict these striking reactivity patterns," said Smalley, "it is likely to be a good one."

Trapping small metal clusters by freezing. Chemists Geoffrey Ozin and Martin Moscovits of the University of Toronto in Ontario have independently used a technique different from that of Smalley and Kaldor to produce small clusters. The University of Toronto researchers mix gaseous metal atoms with relatively large amounts of inert gases, such as argon and krypton, and cool the mixture to about −445°F. The inert gas freezes in a film and traps the metal atoms singly or in small clusters. The chief advantage of this approach is that the film holds the clusters in place so that they can be studied for as long as the researchers want. The inert gas film does not interfere significantly with spectroscopic studies of the structure of the clusters.

Using the inert gas technique, Ozin has found that small metal clusters can have unusual shapes and properties. Clusters containing three silver atoms, for example, can form triangles — a shape that is not seen in bulk metal. Silver-silver bond lengths are also not the same as in the bulk metal; in general, the smaller the cluster, the longer the bond. (The compounds of two chromium atoms observed by Smalley appear to be an exception to this rule.) According to Ozin, when reactivity with hydrocarbons, rather than ionization potential or paramagnetism, is measured, the transition from clusters with properties of isolated atoms to clusters with properties of bulk metals seems to occur at about 13 atoms. Baetzold and other theorists have been working to explain why changes in the different properties occur at different cluster sizes.

Meanwhile, Ozin and others have conducted several useful new catalytic reactions with small clusters. To deposit small clusters of iron or cobalt on supports, Ozin has used techniques similar to those used to trap clusters. He has found that, when the Fischer-Tropsch process (conversion of synthesis gas into products) is conducted with these cluster catalysts, it can yield a product that contains 85 percent butenes — alkenes containing four carbon atoms. Ozin said that the 85 percent selectivity for butenes is an "extraordinary" degree of selectivity, which had not seemed possible in the past. Ozin's catalysts have industrial applications, since butenes can be used to make synthetic rubbers.

Moscovits has used a somewhat different technique to produce small clusters of iron atoms. He has achieved a somewhat smaller selectivity for butenes than Ozin, but has noted that more of the synthesis gas is converted into butene in his system. He has also reported the development of nickel cluster catalysts that convert nitrogen oxide pollutants to harmless nitrogen in the presence of either hydrogen or methane. These nickel cluster catalysts work as well as expensive platinum catalysts and perform significantly better than commercial nickel catalysts. Consequently, they have the potential to replace precious metals in the catalytic converters that remove pollutants from automobile exhaust. By early 1989, a Canadian company had licensed the technology and was developing several commercial applications for it.

Moscovits has also developed a different approach for making small clusters of uniform size. Over a period of many years, he had been developing catalyst-supporting aluminum oxide substrates that have large numbers of small surface pores oriented perpendicular to the surface of the substrate. These pores are open at the surface of the aluminum oxide and closed at their other end. Moscovits uses a laser to generate small clusters of metal atoms and an electromagnetic device called a Wien filter to pick out clusters of a given size; the selected clusters are then accelerated by an electrical field onto the aluminum oxide substrate, where each cluster fills a pore.

When the metal atoms are nickel, cobalt, iron, or a mixture of these elements, the substrate/cluster combination has unusual magnetic characteristics. Each cluster can be readily magnetized by an external magnetic field, but the clusters' own magnetic fields do not interfere with one another. This property makes the substrate/cluster combinations ideal for use as magnetic memories in computers because bits of data could be stored closer together than in existing magnetic media without the magnetic field at one data storage site impinging on the field at adjacent sites. Moscovits speculated that the materials could be used to produce floppy diskettes for personal computers that would hold 20 times as much information as diskettes now available, and he is negotiating with several companies for a license on the materials. He is also developing a technique for making magnetic recording tape by applying a thin aluminum oxide film with pores to a thin plastic material.

Making thin metal films. Chemist Kenneth J. Klabunde of Kansas State University in Manhattan has developed an alternative technique for making clusters or thin films of metal catalysts. In his technique, called solvated metal atom dispersion, single metal atoms are suspended in an organic solvent, such as toluene, at very low concentrations. The solvent is then placed on a support surface, usually silica or alumina, and evaporated. This process causes the metal atoms to be deposited in clusters or a thin film. Solvated metal atom dispersion is an improvement over more conventional techniques for making such metal catalysts because it prevents the nonuniform build-up of deposited atoms, which can degrade a catalyst's selectivity.

Klabunde has developed a manganese/cobalt catalyst that is more than 100 times as effective as cobalt alone for catalyzing the hydrogenation of alkenes. Since manganese does not itself act as a catalyst for the reaction, the catalytic reactivity must arise from some unusual interaction between the two metals. Klabunde has also developed a platinum/tin catalyst that converts crude oil into high-octane liquid fuels. In early 1989, Amoco Chemicals, Inc., of Naperville, Illinois, was studying that catalyst intensively.

Trapping Atoms

Because single metal atoms may have different catalytic effects than bulk metal catalysts, scientists are interested in the atoms' chemical and physical properties. If single atoms can be cooled to temperatures very close to absolute zero (−459.67°F, the temperature at which all motion ceases), precise spectra can be obtained because slow-moving atoms produce sharper spectrographic lines than do atoms moving at high speeds. Such studies should yield a great deal of information about the atoms' electronic structure. In addition, single atoms at such temperatures scatter large amounts of light, making spectroscopy easy. According to chemist Richard Thompson of Imperial

College in the U.K., a single atom at low temperatures scatters 100 million photons per second, enough to make it visible to the naked eye.

Isolating single atoms is difficult, however. Whereas ions (atoms or molecules with an electrical charge) or elementary particles can be trapped relatively easily in electric and magnetic fields, neutral atoms (those with neither a positive nor a negative charge) are much more difficult to trap. Most neutral atoms, however, have a magnetic dipole moment; that is, they act as if they are tiny bar magnets. This slight magnetism allows them to be trapped and retained in a magnetic field, but only if they are moving very slowly — less than about 9 feet per second.

Because temperature is a measure of the average random motion of atoms, movement at less than 9 feet per second is equivalent to a temperature of a few degrees above absolute zero. At room temperature, gaseous atoms typically have a random motion averaging at least 1,500 feet per second.

Trapping neutral atoms. Two independent teams from the National Institute of Standards and Technology, one in Gaithersburg, Maryland, headed by physicist William Phillips and the other in Boulder, Colorado, headed by physicist John Hall, reported in early 1985 that they had found a way to reduce the speed of neutral atoms to such an extent that the atoms could be trapped in a magnetic field. They slowed the atoms by focusing an intense beam of laser light on the atoms. Normally, when laser light is directed into a chamber full of gas, photons (the packets of energy that make up light) collide with gas molecules and transfer energy to them. On average, this energy transfer increases the speed of the gas molecules, thereby raising their temperature.

The two research teams were, however, able to slow the gas molecules by arranging their experimental apparatus so that a beam of sodium atoms collided head-on with a beam of photons from the laser. With this technique, energy is still transferred from photons to atoms, but because the atoms and photons are traveling in opposite directions, the energy transfer slows the atoms. Each head-on collision between an atom and a photon reduces the speed of the atom by slightly more than one inch per second. An atom starting out with an initial speed of 3,000 feet per second would thus have to collide with more than 30,000 photons to come to a complete stop. The process, Phillips said, is analogous to trying to stop a rolling bowling ball by throwing large numbers of Ping-Pong balls at it.

Moreover, each atom interacts only with laser light of a specific frequency, so that the light is in resonance with the atom. But the frequency of the laser light "perceived" by the atoms depends on their speed. Those atoms traveling toward the laser at high speed perceive the frequency of the light as being significantly higher than the actual frequency. Those traveling more slowly perceive it as being only slightly higher than the actual frequency. This apparent change in frequency is known as the Doppler effect.

The Doppler effect is illustrated by the horn of a passing train. As the train approaches an observer, the pitch of its horn seems to be higher than normal; the faster the train is approaching, the higher the perceived pitch. After the train has passed, the pitch appears to be lower than normal. Similarly, as the atoms in the sodium beam slow down, the frequency of light they perceive shifts away from the resonant frequency. As a result, they can no longer interact with the light, and they are no longer slowed.

Phillips and Hall developed different ways to compensate for the Doppler effect. Hall increased the actual frequency of the laser light as the atoms slowed down, thereby ensuring that the frequency of the light perceived by the atoms remained constant. Phillips's team took advantage of the fact that placing atoms inside a magnetic field reduces the frequency at which they will interact with light. His team compensated for the Doppler effect by placing the atom beam inside a graduated magnetic field that changed the resonant frequency of the moving atoms so that it always matched the perceived frequency of the light. The key difference between these approaches is that varying the magnetic field in space makes it possible to slow atoms continuously, whereas changing the laser frequency is a process that can slow only one batch of atoms at a time and needs to be repeated for each new batch.

Both groups were able to trap neutral sodium atoms successfully. Phillips reported that atoms trapped by his technique had a speed equivalent to a temperature of 0.017°F above absolute zero. Most of the neutral atoms escaped from the magnetic trap in less than a second, however; one second is long enough for many experiments, but it is far shorter than the trapping times for charged particles. The major limit on the time that the sodium atoms remained in the magnetic trap was the number of collisions that they had with random molecules from outside the trap.

By early 1989, several other research groups had built similar magnetic traps that held larger numbers of atoms for periods of 2 minutes or more. The technique had also been applied to hydrogen and cesium atoms. The longest period of confinement of hydrogen atoms was recorded by physicists Thomas Greytak and Daniel Kleppner of the Massachusetts Institute of Technology (MIT) in Cambridge, who held 500 billion hydrogen atoms in a trap for 20 minutes at an estimated temperature of 0.04°F above absolute zero.

Optical molasses. Physicist Steven Chu and his colleagues at the AT&T Bell Laboratories in Holmdel, New Jersey, developed an alternative technique for trapping neutral atoms. They used what they call an "optical molasses" to confine as many as 500 sodium atoms in a small space, with a volume of about 1 cubic centimeter, for as long as 4 seconds.

The AT&T researchers used a laser in the same way as earlier researchers to slow down a beam of sodium atoms. Once the speed of the atoms was reduced, however, they were directed into a region where six additional laser beams intersected: One pointed north, one south, one east, one west, one up, and one down. These beams, Chu said, created a "thick soup" of photons that opposed the atoms' motions in all directions and cooled the atoms to a temperature of 240 millionths of a degree above absolute zero.

Finally, the researchers shot a powerful "trap" beam into the region where the other lasers intersected. This laser trap beam was not uniform like most laser beams; instead, it was very intense in the middle and weakened toward the edges. The electric field at the center of this beam interacted with the minute magnetic fields of the sodium atoms and, like a set of "optical tweezers," pulled the atoms into the center of the beam, thereby trapping them in a small space.

Homogeneous Catalysis

Ultimate selectivity. If decreasing the size of a metal cluster increases the selectivity of a heterogeneous catalyst, then a single metal atom should provide the ultimate selectivity. Since it is difficult to place single metal atoms on a solid support, an alternative approach is to dissolve single metal atoms or clusters of two to six atoms in a liquid with the aid of organic molecules known as ligands. Such dissolved complexes of metal atoms and ligands are known as homogeneous catalysts because all components of the reaction are in solution. With heterogeneous catalysts, the metals are not in solution.

Before 1970, the term homogeneous catalysis referred strictly to catalysis by simple acids and bases, such as hydrochloric acid or sodium hydroxide. Since then, the term has generally been used to describe catalysis by complexes of metal atoms and ligands. The use of homogeneous catalysts expanded rapidly around 1970, when six major new industrial processes based on such catalysts were implemented in chemical plants. These new processes included the use of Wilkinson's catalyst, a rhodium-containing complex, to add hydrogen to both sides of a double bond; the Shell Chemical Co. "higher olefin process," in which a nickel complex is used to convert ethylene to olefins containing 10 to 14 carbon atoms; and the Monsanto Co. process that uses rhodium and iodide ions to produce acetic acid from methanol and carbon monoxide.

The development of new homogeneous catalysts offered significant cost savings because reactions carried out with them can usually proceed under less extreme conditions — at lower temperatures and pressures — than reactions carried out with heterogeneous catalysts. Lower temperatures usually lead to increased selectivity as well as reduced fuel costs, and lower pressures obviate the construction of expensive high-pressure containment vessels, which are necessary for heterogeneous reactions.

Homogeneous processes are now used to produce chemicals that account for 15 to 20 percent of the total annual sales of the chemical industry. Since the early 1970s, however, few new homogeneous processes have been introduced because the savings were not enough to offset the costs associated with installing equipment for the new processes. One notable exception is the Tennessee Eastman plant in Kingsport, Tennessee, which uses a proprietary homogeneous catalyst system to produce acetic anhydride and other chemicals from synthesis gas.

Improving industrial processes. Albert Chan of the Monsanto Co. in St. Louis, Missouri, has developed a new homogeneous catalyst for producing ethylene glycol from synthesis gas. Ethylene glycol is the primary component of antifreeze and is a feedstock for the production of many other chemicals. According to Chan, existing catalytic reactions can produce formaldehyde from synthesis gas; his new catalyst can be used to convert the formaldehyde to ethylene glycol.

In the Monsanto process, rhodium complexes are used as catalysts to efficiently hydroformylate formaldehyde; that is, the rhodium catalyst adds an aldehyde group (which has a carbon bonded to a hydrogen and doubly bonded to an oxygen) to the formaldehyde to produce a compound that can be easily converted into ethylene glycol. In the late 1980s, the cost of petroleum — the raw material from which ethylene glycol is currently made — was not high enough to justify a switch to the Monsanto process; nevertheless, this process holds potential for the future when petroleum prices may rise again.

Monsanto has also developed a new process for converting alkenes into branched alkenes containing an additional carbon atom. The process uses a different rhodium catalyst to hydroformylate alkenes selectively to produce branched molecules, which can then be converted to hydrocarbons containing two double bonds. Such hydrocarbons can be readily linked together to form rubber molecules.

Another new homogeneous catalysis process, discovered at Argonne, could replace existing techniques for converting methanol to ethanol. The best current methanol-conversion technology uses a cobalt complex to combine two molecules of hydrogen and one molecule of carbon monoxide with one molecule of methanol to produce one molecule of ethanol and one molecule of water. This process is inefficient in its use of hydrogen because part of the hydrogen is wasted in the formation of water; more importantly, the resulting water must be separated from the ethanol by distillation, an expensive process that consumes a significant amount of energy. The new process, which was developed by Argonne chemists Jerome Rathke and Michael Chen, uses an iron complex as the catalyst to promote the reaction of one molecule of methanol with one molecule of hydrogen and two molecules of carbon monoxide. This reaction produces one molecule each of ethanol and carbon dioxide, but no water, eliminating the need for distillation.

Despite the potential of such processes, the field of homogeneous catalysis is at a very early stage of development. According to chemist George Whitesides of Harvard University in Cambridge, Massachusetts, "One of the hopes is that we will understand homogeneous catalysis well enough to design catalysts rationally. I anticipate another 20 years of trial and error before we reach that stage." (The use of enzymes to synthesize organic compounds is also a form of homogeneous catalysis and holds great promise as a synthetic technique. For more information about enzymes, see the articles "Designer Proteins" and "New Approaches to Synthesis" in this chapter.)

Imaging catalyst surfaces. Physicist David J. Smith at Arizona State University in Tempe and chemists J.O. Bovin and R. Wallenberg at the University of Lund in Sweden have developed an unusual new way to view the surfaces of crystals that might serve as catalysts. They first reported using the technique on gold surfaces in late 1985, but by early 1989 they had also studied surfaces of several metals widely used as catalysts, including platinum, rhenium, ruthenium, cadmium, and zinc.

Smith, Bovin, and Wallenburg connected an exceptionally sensitive television camera and videotape recorder to a high-resolution electron microscope and used the system to watch the movement of atoms on the surface of metal crystals. In their original studies, they used easily prepared gold crystals, each of which contained 55 atoms of gold, that had been dispersed on a carbon film and placed in the electron microscope's vacuum chamber.

When the researchers viewed the crystals under the microscope, they found them to be very unstable. The crystal structures changed almost constantly, and some of the crystals grew by capturing atoms from other crystals. "It's like watching living atoms," Smith said in *Science News*. "You can sit and look at one small particle for 10 minutes. You may get 29 different shapes in 30 seconds, and then it will sit still for a while."

The researchers also saw clouds containing several atoms above some crystal surfaces and observed the coordinated movement of several atoms from one row of atoms in a crystal to another. "It may well be," according to the researchers, "that column hopping and changes of cloud shape are an indication of how atoms locate the most favorable lattice position during crystal growth. The phenomena described here also suggest that it should be possible to monitor the way in which other atoms interact with a metal surface, and this should be of fundamental importance for studying heterogeneous catalysis."

Smith and his colleagues have extended their observations to a variety of other metals used in catalysis, and several other laboratories have adopted their imaging technique. Smith's group found that most of the metals behaved in the same way as gold and that the amount of rearrangement of the atoms was directly related to the intensity of the microscope's electron beam.

How atoms are held in place. In mid-1986, physicists Robert J. Hamers, Rudolph M. Tromp, and Joseph E. Demuth of the IBM Thomas J. Watson Research Center in Yorktown Heights, New York, announced that they had made the first pictures of both the surface atoms on crystals and the bonds that hold them in place. The IBM researchers used silicon crystals, but they believe the same principles can be applied to imaging the surface of metal catalysts. The images were made with a scanning tunneling microscope developed in 1981 by physicists

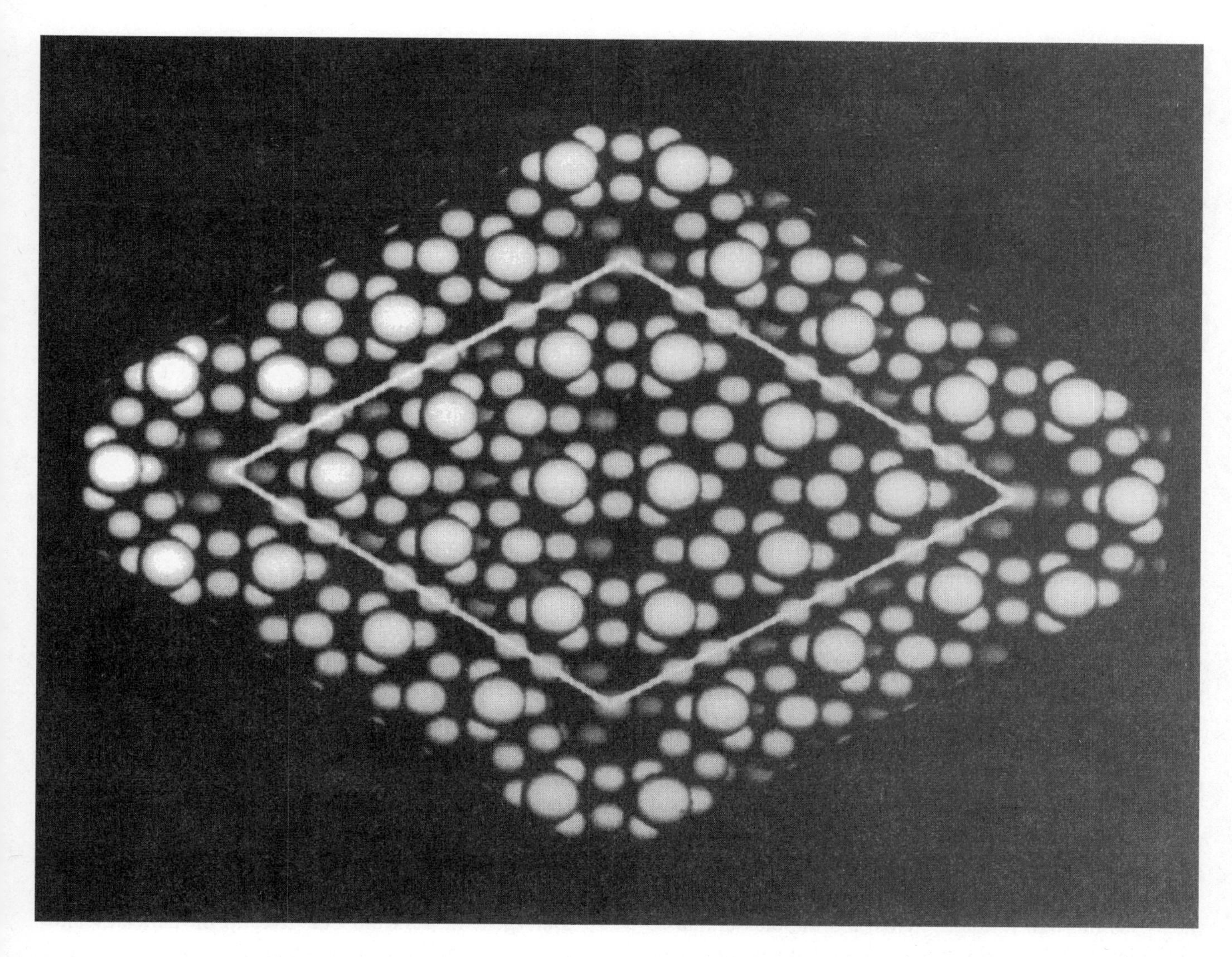

Figure 1. Atomic Surface. IBM physicists have used a scanning tunneling microscope to obtain the first images of both the atoms on crystal surfaces and the bonds that hold them in place. This image shows the surface of a silicon crystal magnified about 30 million times. The top atoms in the crystal appear as large balls; the smaller balls are atoms deeper within the crystal. *Courtesy: IBM, Thomas J. Watson Research Center.*

Gerd Binnig and Heinrich Rohrer of the IBM Research Laboratory in Zurich, Switzerland. Their work with the device won them the 1986 Nobel Prize for Physics.

In the scanning tunneling microscope, a voltage is applied across the gap between the sample and a probe tip that ends in a single atom. As the probe is moved to within a few atomic diameters of the sample, a "tunneling" current dependent on the tip-to-surface distance flows between the sample and the probe. The tip is then moved back and forth over the surface at a constant distance from the surface; this constant distance is maintained by an electronic feedback loop that senses small changes in the current and adjusts the tip's position, thereby bringing the current back to the desired value. A computer records the changes in the tip's position and produces a contour map of the surface.

Hamers and his colleagues modified their scanning tunneling microscope to include an extra step in the imaging process. As the tip scanned the surface of the sample, they periodically interrupted the feedback loop and recorded the electric current resulting from the motion of electrons around every atom. The combination of the contour and current maps reveals the relationship between the surface atoms and the electronic bonds between them.

Previously, it had been possible only to study electron clouds around atoms and infer the approximate location of the electronic bonds. This process shows them directly.

The characteristics of the electrons on the surface of the silicon determine the electronic properties of a semiconductor. If the bond characteristics can be correlated with the material's electronic behavior, Hamers said, "it should enable us to build better electronic devices." Furthermore, the technique can be used to locate small defects that produce problems in the fabrication and operation of semiconductor devices, and may make it possible to correct the problems.

Reactions Catalyzed by Sound

Although catalysis is normally associated with metal surfaces, metal atoms, and related organometallic materials, new techniques, such as the use of ultrasonic waves — sound waves that are beyond the range of normal hearing — can be used to accelerate a reaction. Chemist Philip Boudjouk of North Dakota State University in Fargo has found that ultrasonic waves can be used to improve the yields of commercially valuable chemicals. Boudjouk has studied many industrial reactions that are normally carried out at high temperatures and pressures and has found that the use of ultrasound allows them to be conducted near room temperature and at atmospheric pressure.

The ultrasonic waves apparently create tiny bubbles as they pass through a liquid. When the bubbles collapse, they release high-energy shock waves that generate temperatures of several thousand degrees Fahrenheit and pressures of 5 to 10 atmospheres for fractions of a second. By early 1989, several U.S. pharmaceutical companies were using ultrasound to catalyze a reaction called the Reformatsky reaction, which is used in the synthesis of some antibiotics.

Boudjouk also discovered that ultrasound can speed up the nickel-catalyzed synthesis of silicones — polymers containing alternating oxygen and silicon atoms — so that manufacturers can replace expensive platinum catalysts with the much cheaper nickel. In 1988, the Dow-Corning Corp. of Midland, Michigan, was studying the use of this technique for its silicone production operations. Boudjouk has found that the use of ultrasound speeds up about 25 other reactions, but none of the accelerations are as dramatic as that associated with silicone production.

Chemist James Rusling of the University of Connecticut in Storrs has found that ultrasound increases the rate of the electrical destruction of polychlorinated biphenyls (PCBs), which had commonly been used in transformers and other electrical devices before their serious health effects became known. Rusling found that PCB destruction, catalyzed by electricity and an organic complex called bipyridyl cobalt, proceeds three times faster and destroys more of the PCB molecules when ultrasound is used. In 1988, Rusling was seeking ways to make the reaction proceed even faster so that it would be useful for large-scale cleanup operations at hazardous waste sites.

Designer Proteins

Enzymes were the darling of the multibillion-dollar detergent industry in the 1960s, when it became possible for the first time to produce them cheaply enough to allow their use in commercial products. They represented a seemingly ideal way to remove food, blood, and grass stains without damaging even the most delicate fabrics. By the end of the 1960s, enzymes were included in as many as 60 percent of all detergents. As the decade progressed, however, many consumers complained of skin irritation and rashes from handling enzyme-enhanced detergents. Consumers were also frightened by reports of allergic reactions among workers in the European factories where the enzymes were prepared. As a result, by the early 1970s, enzymes had fallen out of favor and were no longer included in consumer detergents.

Detergent makers did not give up on enzymes altogether, however. When researchers developed encapsulation techniques for enzymes that eliminated skin contact, the proteins once again began to appear in detergent products. By mid-1988, encapsulated enzymes were used in 40 percent of laundry detergents in the U.S., according to Robert Petrowski, a marketing manager at Novo Laboratories, Inc., of Wilton, Connecticut, which dominates the $80-million-per-year U.S. detergent-enzyme market. Enzymes are also used in 80 percent of the detergents in Western Europe and 60 percent in Japan.

Enzyme producers still face many obstacles in the detergent market. The enzymes are degraded quickly in hot water, for example, and they lose more than 90 percent of their activity in the presence of bleach. Industrial detergents are also highly alkaline, and enzymes do not function in alkaline environments. In addition, a growing percentage of detergents are sold in liquid form. Since dissolved enzymes degrade more rapidly than those in powdered form, manufacturers must add extra enzymes to make up for the loss, at an additional manufacturing cost.

To overcome these problems, enzyme producers are turning to the new science of protein engineering to create more desirable molecules.

Protein engineering. Protein engineering involves altering the structure of enzymes or other proteins to change their properties. It has long been considered something of a stepchild of genetic engineering, in which researchers manipulate deoxyribonucleic acid (DNA) to alter the characteristics of organisms or give them new capabilities. Within the last few years, however, protein engineering has become an important field in its own right. Scientists have begun to speak of custom-made designer proteins in the same wide-eyed manner that they previously talked about designer genes.

Designer enzymes "promise to be of great use in producing chemicals, foods, drugs, and fuels," according to biochemist Alexander Klibanov of MIT. "They can be used to break down industrial waste and analyze chemicals. Indeed, virtually any imaginable chemical process can be catalyzed by some enzyme," he said. Other designer proteins may have a significant impact in medicine. Already, custom-designed forms of interferon and interleukin — two drugs used to fight cancer and other conditions — have been found to have greater effectiveness in fighting diseases than their natural counterparts. Researchers have also been studying artificially produced antibodies that may be less expensive to produce and more effective for detecting and fighting disease.

Many biotechnology companies that once devoted themselves exclusively to genetic engineering, such as Genentech Corp. of South San Francisco, California, Genex Corp. of Gaithersburg, Maryland, and Amgen, Inc., of Thousand Oaks, California, have expanded their protein engineering departments or created new subsidiaries to study designer protein technology. Industrial giants such as E. I. du Pont de Nemours & Co., Inc., of Wilmington, Delaware, have also created new research groups to study protein structure and modification. "Every major drug or chemical company involved in biotechnology is now at least putting a big toe in the water," said biochemist Kevin Ulmer, director of the new Center for Advanced Research in Biotechnology in Shady Grove, Maryland, which is itself devoted primarily to protein engineering.

Interest in protein engineering is not restricted to the U.S. In 1986, Japan's Ministry of International Trade and Industry initiated an 8-year, 30-billion-yen (over $225 million) protein engineering research program. In the U.K., several firms, including RTZ Chemicals, J and E Sturge, Celltech, Glaxo, Imperial Chemical Industries, and British Petroleum, have formed a Protein Engineering Club to finance research at selected British universities. In continental Europe, Plant Genetic Systems of Ghent, Belgium, has organized a protein engineering consortium with members from several European countries.

"My impression is that the field as a whole is where genetic engineering was in 1976," according to Ulmer. "We have numerous proofs that the principle is sound and that the technology works, but we haven't had the first big commercial breakthrough to really get things moving." When that breakthrough occurs, Ulmer added, "I think protein engineering will move forward much faster than genetic engineering did."

DNA is often called the blueprint of life. By that analogy, proteins are the bricks and mortar from which organisms are constructed — as well as the masons and carpenters who put them together and the maintenance people who keep them running. Proteins are composed of tens to many hundreds of individual amino acids strung together like beads on a necklace and then folded into intricate balls. They are the major structural components of tissues throughout the body. They are also essential for converting food into energy, and for fighting off infections and disease. About 1 million proteins are thought to occur in nature.

Major targets. Each type of protein represents an opportunity for protein engineers, but the greatest interest at the moment is focused on enzymes — catalysts that carry out specific chemical reactions without themselves being consumed in the process. A very small amount of an enzyme in yeast, for example, converts the sugar in dough into the large volume of carbon dioxide that gives bread and rolls their characteristic textures.

Industrial chemists are interested in enzymes for two reasons: enzymes typically produce only one product in any given reaction, and they operate under very mild conditions of temperature and acidity that are essentially the same as those in the human body. Virtually all conventional chemical reactions, in contrast, produce many unwanted by-products in addition to the desired product, so that valuable starting materials are wasted and expensive separation techniques must be employed. Many industrial chemical reactions also must be carried out at high temperatures and pressures, necessitating the use of expensive containment vessels and significant amounts of energy. These problems could all be reduced or avoided by using enzymes.

Chemists face a dilemma, however. While literally thousands of enzymes have been identified in nature, only 216 are produced commercially in quantities of more than

Figure 2. Semisynthetic beta-Endorphin. The late Emil Thomas Kaiser and his associates at Rockefeller University in New York City created a semisynthetic variant of beta-endorphin, a chemical that acts as a natural painkiller. The molecule, pictured in this three-dimensional model, has been designed to resist degradation by enzymes and thereby have longer-lasting painkilling effects than naturally occurring beta-endorphin. *Courtesy: Emil Thomas Kaiser.*

an ounce or so. If the desired reaction is not carried out by one of those enzymes, the synthesis costs more because the other enzymes are much more expensive. Or, as is the case with detergent enzymes, they may not function properly when used under adverse conditions.

In the short term, protein engineers are attempting to change the structure of enzymes slightly so that they can better withstand adverse conditions. The researchers would also like to take one of the readily available, inexpensive enzymes or proteins and change its structure so that it carries out a completely different reaction than its natural one. In the longer run, protein engineers would like to design enzymes from scratch, using either amino acids or completely synthetic chemicals.

Semisynthetic Enzymes

If an existing enzyme is not an appropriate catalyst for a desired reaction, chemists may be able to produce such a catalyst by synthesizing a compound that combines the catalytic activity of metal atoms with the ability of an enzyme to bind to specific molecules. Chemists Harry Gray and Ruth Margalit of the California Institute of Technology (Caltech) in Pasadena and Israel Pecht of the Weizmann Institute of Science in Rehovot, Israel, for example, have chemically linked the protein myoglobin to a synthetic organic molecule containing a ruthenium ion that catalyzes the transfer of electrons from one molecule to another; they call the new catalyst a "semisynthetic bioinorganic enzyme."

Myoglobin binds oxygen in muscles and resembles blood hemoglobin in function, but it has no catalytic activity. When three ruthenium complexes are chemically linked to its surface, however, the new material can reduce (add electrons to) oxygen dissolved in the enzyme solution, while oxidizing (removing electrons from) various organic molecules, such as ascorbic acid (vitamin C). Because it binds to ascorbic acid, the semisynthetic enzyme is about 200 times as effective at oxidizing it as the ruthenium complex alone and is nearly as effective as the natural enzyme ascorbate oxidase.

Flavopapains. The late Emil Thomas Kaiser and his colleagues at Rockefeller University in New York City successfully created several semisynthetic enzymes and proteins. Kaiser called one group flavopapains. Papain is an enzyme that breaks apart the ester bonds that link amino acids in proteins, and flavins are nitrogen-containing compounds that catalyze many oxidation reactions.

The Rockefeller University group created the flavopapains by chemically linking a flavin at a specific site on the side of the papain molecule where the protein normally binds. The group then studied the ability of the semisynthetic enzyme to oxidize dihydronicotinamides, which are closely related to chemicals that have a role in oxidation/reduction reactions in cells. They found that flavopapain oxidizes the dihydronicotinamides at a rate 1,000 times faster than the flavin alone and that the level of catalytic activity approaches that of many naturally occurring enzymes.

The researchers created two other semisynthetic enzymes that also carry out oxidations. In early 1988, Kaiser's group reported that they had coupled a flavin to the enzyme glyceraldehyde-3-phosphate dehydrogenase. This new flavoprotein was even more effective than the original flavopapain at oxidizing biological materials.

In a similar way, they linked flavin to hemoglobin — the oxygen-binding protein from red cells — to mimic an enzyme called cytochrome P-450. Cytochrome P-450 can reduce many biological compounds, but the process normally requires the participation of a second enzyme called a reductase. Kaiser and his colleagues found that flavohemoglobins can carry out the same reactions without the participation of the reductase, so that one semisynthetic enzyme can take the place of two natural enzymes. Kaiser said that the flavohemoglobin can also remove methyl groups from many molecules and thus may find widespread application in synthesizing new molecules.

Using the same approach, Kaiser's group successfully produced semisynthetic variants of two other proteins. In 1984, Kaiser reported that the group had produced new forms of calcitonin, the hormone that controls the calcium concentration in the bloodstream, and beta-endorphin, a chemical that acts as a natural painkiller. Neither protein is an enzyme, but both have binding sites that are very similar to those of enzymes.

In the case of calcitonin, Kaiser and his colleagues undertook a complete analysis of the protein's structure and shape. As they learned the function of different components of the calcitonin molecule, they synthesized new structures that could replace each component while, at the same time, retaining the component's original function. For example, they found that one segment of the amino acid folds naturally to form a structure called an amphipathic helix, in which charged amino acids line up on one side of the molecule and uncharged ones line up on the other side. Kaiser found that amino acids could be substituted in this portion of the calcitonin molecule without changing its biological activity, as long as the amphipathic helix was maintained.

After studying each portion of the molecule in this way, Kaiser's group eventually developed a semisynthetic calcitonin that differs from natural calcitonin in 60 percent of its segments but, when administered to rats, still retains about one-tenth of the activity of salmon calcitonin — the most active naturally occurring form of calcitonin. Kaiser and his colleagues reported in late 1988 that they had changed several other amino acids in the semisynthetic calcitonin to produce a molecule that is three times more active than salmon calcitonin in rats.

Kaiser's research on semisynthetic calcitonin could have medical applications. Natural calcitonin is used to treat Paget's disease, which is characterized by bone deformation, and other calcium-related illnesses; however, it must be given intravenously rather than orally because stomach enzymes destroy it before it can be absorbed into the circulatory system. Natural calcitonin can also be attacked by the immune system and degraded. The semisynthetic form was designed so that it would resist degradation by enzymes and the immune system. Similarly, semisynthetic beta-endorphins are expected to have longer-lasting painkilling effects than naturally occurring beta-endorphin because they are designed to resist enzyme degradation.

Cleaving DNA. Enzymes that recognize specific sequences of DNA and cleave the DNA at specific sites are some of the most important tools in genetic engineering. Such enzymes, called nucleases or restriction enzymes, allow genetic engineers to break DNA into segments of specific lengths that are useful in searching for defective genes. They also enable researchers to excise specific genes from a chromosome so that they can be inserted into other organisms.

Only a limited number of naturally occurring nucleases are available, and most recognize only very short sequences of the bases that compose DNA. In 1987, however, two research groups made new semisynthetic nucleases that bind to different DNA sequences, thereby expanding the number of nucleases available to biotechnologists. Their results suggest that chemists will be able to synthesize semisynthetic enzymes that will cleave DNA at any point.

Biochemists Chi-Hong Chen and David Sigman of the University of California at Los Angeles made a new nuclease from a DNA-binding protein called the tryptophan gene (*trp*) repressor, which was isolated from the bacterium *Escherichia coli*. The *trp* repressor binds to a DNA segment that contains 20 deoxyribonucleic acids in a specific sequence. They reported in late 1987 that they had chemically linked the *trp* repressor to a synthetic molecule, called phenanthroline, that binds a copper ion. The phenanthroline/copper complex, in the presence of hydrogen peroxide, cleaves DNA by oxidizing and breaking apart a base.

The new semisynthetic enzyme — a combination of the *trp* repressor and the phenanthroline/copper complex — therefore has the ability to both bind to a specific DNA segment and cleave it at a desired position, where, for example, researchers might want to insert a new gene. The new enzyme also has a different specificity than existing nucleases.

Caltech chemist Peter Dervan has made a family of semisynthetic restriction enzymes. Dervan and chemist Heinz Moser created some by using a chemically synthesized segment of DNA to bind to the DNA to be cleaved, and a complex of the organic molecule ethylenediaminetetra-acetic acid (EDTA) and iron as a catalyst. Moser and Dervan found that the semisynthetic enzyme produced from the two components cleaves DNA only at the desired location.

To create a second family of enzymes, Dervan and his associates started with a 52-amino-acid fragment of a restriction enzyme called Hin recombinase and joined it to EDTA. This new semisynthetic enzyme cleaves DNA at a different location than the original Hin recombinase. Dervan predicted that these new semisynthetic enzymes will be the first of a whole new set of restriction enzymes that will ease the way for further manipulating DNA and custom-designing proteins.

Designer Enzymes

The most common way to engineer proteins is a process known as site-directed mutagenesis, in which researchers selectively change the identity of one or more amino acids in the molecule. This can be done readily by making small changes in the DNA that codes for the protein. Biochemist David Estell and his colleagues at Genencor, a South San Francisco-based joint venture of Genentech and Corning Glass Works, have developed a variation of this technique that enables them to replace any amino acid in an enzyme with any of the other 19 common amino acids, thereby producing 19 new proteins. The process of making the new proteins takes about 3 weeks, but determining their characteristics can take several months.

Estell and his Genencor group have used the amino acid replacement technique to create nearly 300 variants of the bacterial enzyme subtilisin, which breaks down proteins and is very similar to the enzyme used in detergents. In the course of their experiments, the researchers have found several variants that are more resistant to heat and bleach than natural subtilisin and that can be stored longer and function under more alkaline conditions. In early 1989, Genencor and other companies that have made similar changes in other protein-degrading enzymes were negotiating to sell those designer enzymes to detergent manufacturers.

Intuitive process. The process of deciding which amino acids in a protein to alter and which amino acids will replace them is still very much an intuitive one, and success can be assessed only by measuring the properties of the end product. "Sometimes the experiments go almost exactly as predicted," according to Estell. "Some of them go in the opposite direction." One good example of this unpredictability is provided by the work of biochemists Charles Craik, William Rutter, and their colleagues at the University of California at San Francisco. The researchers have been using site-directed mutagenesis to produce variants of trypsin in order to learn how this digestive enzyme recognizes its substrate (the chemical that it modifies).

Trypsin preferentially binds positively charged substrates, and the San Francisco research group reasoned that this preference arises from the fact that the enzyme contains a negatively charged amino acid (aspartic acid) at the active site, where the substrate binds. To test their hypothesis, the researchers replaced the negatively charged aspartic acid with a positively charged amino acid (lysine). This modification did change the enzyme's specificity, but in an unexpected way; the new variant preferred to bind amino acids with no charge, rather than those with a negative charge, as had been expected. By late 1988, the researchers were replacing other amino acids around trypsin's active site to get a better understanding of how the enzyme binds its substrates.

Rutter and his colleagues have also been studying carboxypeptidase, another enzyme that breaks down proteins. Earlier studies of the three-dimensional structure of carboxypeptidase suggested that one amino acid — a tyrosine molecule in the enzyme's active site — plays a key part in the enzyme's catalytic activity by acting as an acid; that is, by donating a proton during one stage of the catalytic reaction.

The researchers replaced the tyrosine with an amino acid that could not donate a proton and found, to their surprise, that the enzyme's activity was unchanged. Substitution of other amino acids at the same site also had no effect. Clearly, researchers still have much to learn about the role of amino acids in proteins. Protein engineering "is an imperfect science," said biochemist Phil Whitcome of Amgen. "At best, it is like looking through a fog bank and trying to see what is inside."

To eliminate some of the uncertainties in protein engineering, chemist Arieh Warshel of the University of Southern California in Los Angeles has been developing a computer program to predict the effects of amino acid substitutions more precisely. The program calculates all the normal repulsive and attractive forces between amino acids within a protein to determine the most stable configuration, which should be the most catalytically active. Warshel has then compared the results of these calculations with results from actual experiments conducted by Estell, Rutter, and other researchers. "We are able to predict the effect of mutations on the speed with which enzymes work," Warshel said. In the future, researchers might be able to use Warshel's technique to predict the most effective amino acid changes before actually making them, thereby reducing the amount of work needed to make new enzymes.

The drawback to Warshel's approach is that the calculations for each amino acid change in an enzyme require about 10 hours on a Cray supercomputer, making the analysis a very expensive proposition. According to Warshel, however, once the usefulness of the calculations has been proved, "people will build dedicated computers only for this type of calculation, and they will be much cheaper than the Cray." Despite the high cost of the calculations, Warshel has been working with laboratory scientists to try to predict the outcomes of experiments before they are performed.

Using altered enzymes. Aside from laundry detergents, one of the first commercial applications of designer enzymes may involve the use of an altered form of subtilisin to link individual amino acids together in the synthesis of peptides (short chains of amino acids). The altered enzyme, called thiol-subtilisin, was first created during the 1960s by chemist Daniel E. Koshland, Jr., and his colleagues at the University of California at Berkeley. They used chemical methods to change a hydroxyl group in the active site of the enzyme subtilisin into a sulfhydryl group; that is, they replaced an oxygen atom with a sulfur atom. The group had hoped that the resultant thiol-subtilisin might have unusual chemical reactivity, but it did not. Thiol-subtilisin would not break down proteins very well, nor did it seem to do much of anything else.

In 1987, interest in thiol-subtilisin was revived when Rockefeller University's Kaiser began studying the enzyme's "reversibility." If an enzyme such as subtilisin breaks a peptide bond to destroy a protein, it can usually carry out the reverse reaction. In the case of subtilisin, the reverse reaction involves forming a peptide bond. Under normal reaction conditions — such as those found in the body — this reversal is not favored and occurs to only a limited extent. But if the reaction conditions are changed, the reverse reaction may sometimes be favored and dominate the forward reaction.

Kaiser and his colleagues found that changing an oxygen to a sulfur in subtilisin to form thiol-subtilisin changed the reaction conditions sufficiently to favor peptide formation. According to Kaiser's group, thiol-subtilisin is an ideal reagent for combining synthetically produced short peptides to make larger peptides. The designer enzyme carries out this reaction very specifically and without making any chemical changes in the peptides, whereas other chemicals that carry out the reaction degrade part of the peptides and reduce the yield of the reaction. By late 1988, Kaiser and Rockefeller University had patented the technique, and Rockefeller was negotiating with several companies for production of the altered enzyme. The new technique should reduce the cost of making synthetic peptides and proteins.

Cassette mutagenesis. Although site-directed mutagenesis is leading to new knowledge about how proteins work, many investigators believe that the creation of useful proteins will require much greater changes in a protein's structure than those produced by site-directed techniques. Such changes can be effected by cassette mutagenesis, in which sections of different genes are grafted together to produce molecules that are only distantly related to their progenitors.

Amgen, for example, has used cassette mutagenesis to custom-make a totally synthetic form of the antiviral agent interferon, which is normally obtained from mammalian cells. According to Amgen's Whitcome, the company's researchers first studied all of the naturally occurring forms of interferon and determined which sections of each molecule were most important for its activity. "We then combined these sections to produce a theoretical structure with the same number of amino acids [as natural interferons], but whose gene did not exist in nature," Whitcome said. "We synthesized the gene and placed it in a [bacterium], which produced the protein," called consensus interferon.

Amgen later found that consensus interferon is more effective than natural interferon in fighting some types of cancer. In addition, the company has modified the anticancer agent interleukin-2 to produce a protein that, in animals, is more effective than the natural form.

Amgen has been sponsoring clinical trials of both consensus interferon and modified interleukin-2 for cancer treatment.

Biochemist Robert Ladner and his colleagues at Genex have constructed new antibodies in a fashion similar to that used to create consensus interferon. Antibodies are composed of four separate protein chains linked together to form one large molecule. Only a small part of each chain is actually important for the antibody's function, however. Ladner's group has taken these individual sections from each chain and linked them together to form a single-chain antibody that is much smaller than the original and thus easier to produce.

The engineered antibody binds to foreign objects as well as the natural antibody does, according to Ladner, but it is much more stable and can survive for longer periods in hostile environments. This increased stability, he said, should make it very useful in biosensors, which are used for detecting small amounts of specific chemicals in blood or industrial effluent. Single-chain antibodies could also be used in protein purification systems to remove desired proteins from complex mixtures.

Chemists Greg Winter and Herman Waldmann of the Medical Research Council Laboratory of Molecular Biology in Cambridge, U.K., have also been using cassette mutagenesis to engineer antibodies. They have focused on monoclonal antibodies. (For more information about monoclonal antibodies, see the article "Biotechnology Techniques" in the "Biology" chapter.) Most monoclonal antibodies are made from mouse cells, and humans develop antibodies against them, which limits their effectiveness. Winter and Waldmann have been attempting to "humanize" mouse antibodies by using human DNA to replace the portion of the mouse antibody gene that codes for the amino acids that appear on the surface of the antibodies. They hope that these antibodies may overcome the immunologic reactions that hinder the therapeutic use of the mouse antibodies. The first application of the technique, according to Winter, will probably involve treatment of graft-versus-host disease, in which transplanted immune cells attack the recipient's body.

Abzymes

A radically different approach to the construction of designer enzymes was announced in late 1986. It involves converting antibodies into artificial enzymes, which have been dubbed "abzymes." Many researchers believe that this approach has the greatest promise of all the techniques that have been developed for producing new and different enzymes. Furthermore, the approach might yield new proteins for treating diseases. Conventional antibodies do not normally destroy target molecules themselves; instead, they trigger the destructive activities of other immune system proteins and cells. An abzyme that could both bind to a specific protein and degrade it might prove very effective for dissolving blood clots or for finding and destroying tumor cells.

The most important characteristic of an enzyme is its ability to bind to a specific substrate. The substrate fits into the enzyme just as a key fits into a lock. Antibodies possess this same ability to bind in a highly specific fashion, and biologists have speculated that it might be possible to use this specific binding ability to produce a synthetic enzyme. But the problem in producing such an abzyme was finding a way to make the antibody catalyze a chemical reaction.

The key to solving this problem was found in the way in which enzymes work. Most researchers believe that all chemical reactions proceed through an intermediate structure, called a transition state, that represents the state of highest energy through which reactants must pass before they are converted to products. For example, in the common reaction in which an iodide ion interacts with the carbon atom of a substrate to displace a bromide atom attached to the carbon, the transition state is the intermediate in which both the iodine and the bromine are simultaneously bonded to the carbon.

Scientists have not yet been able to detect transition states directly because molecules pass through them too quickly for observation. Nonetheless, researchers are confident that transition states exist and that anything that makes transition states more stable will increase the speed of a reaction. In many enzyme reactions, for example, atoms on the side chains of amino acids in the enzyme's active site bind to the transition state and stabilize it, thereby speeding up the reaction. Biochemist Richard Lerner of the Scripps Clinic and Research Foundation in La Jolla, California, reasoned that if an antibody could be prepared that would bind specifically to the transition state of a reaction, it would then catalyze the reaction itself.

Because transition states are so short-lived, however, it is impossible to produce antibodies against them directly. An alternative is to use a molecule, called a template, that has the same size and shape as the transition state. Lerner and his colleagues Alfonso Tramontano and Kim Janda at Scripps wanted to make an abzyme that would catalyze the hydrolysis (breaking apart) of common chemicals called esters. In an ester, the key carbon atom is double-bonded to one oxygen and single-bonded to another oxygen. In a typical hydrolysis reaction, the key carbon atom interacts with an oxygen atom from the enzyme or from a water molecule to form a transition state in which three oxygens are attached to it.

As their template, Lerner and his colleagues chose a molecule known as a phosphonamide, whose central atom is phosphorus rather than carbon. The phosphonamide has the same shape and approximately the same size as the transition state for ester hydrolysis. Lerner and his colleagues injected the phosphonate ester into mice, isolated antibody-producing spleen cells, and prepared monoclonal antibodies against the phosphonamide by conventional techniques. In December 1986, the Scripps researchers reported that, when they tested the antibody, it catalyzed ester hydrolysis, speeding the reaction by a factor of about 1,000. They also found that the antibody had some enzyme-like specificity, catalyzing the hydrolysis of esters with the same general shape as their template, but not others.

Also in December 1986, chemists Scott Pollack, Jeffrey Jacobs, and Peter Schultz of the University of California at Berkeley reported that they had produced catalytic abzymes in the same manner. The group was trying to design an enzyme to hydrolyze a carbonate ester (in which the central carbon atom is attached to three oxygen atoms, rather than to two oxygens and another carbon as in a conventional ester), but for their template they used a phosphorus compound similar to that used by the Scripps workers. They found that their abzyme increased the hydrolysis rate by a factor of 15,000. Nonetheless, they noted, that increase is still many powers of 10 less than the increase produced by naturally occurring enzymes.

By early 1989, both research groups were making other abzymes that would carry out important biological reactions. Among their research goals were abzymes that will form new carbon-oxygen, carbon-phosphorus, and carbon-carbon bonds. They were also trying to produce abzymes that will create optically active compounds. Tramontano and Lerner, along with chemists Andrew Napper and Stephen J. Benkovic of Pennsylvania State University in University Park, succeeded in making an abzyme that catalyzed the formation of an optically active lactone — a cyclic ester — of high purity. This technique may thus offer a major new way to make optically active materials.

The new technology for making abzymes has already spawned at least one company, IGEN, Inc., of Rockville, Maryland. The company is developing abzymes that could be used for a variety of purposes, such as the production of pharmaceuticals. The company is also attempting to produce abzymes that would disable viruses or tumor cells by binding to them and cleaving specific proteins. In addition, it is working on an abzyme that would break down the primary component of blood clots that cause heart damage. According to Jim Massey, IGEN's president, the company hopes to begin selling its first products to pharmaceutical manufacturers by 1990.

Wholly Synthetic Enzymes

A wholly synthetic enzyme is an artificially synthesized molecule with a cavity for binding a specific chemical and with one or more catalytic groups attached. "The aim is to produce new catalytic systems that might have some of the same high selectivities and high [reaction] rates characteristic of enzyme processes, but that might have special advantages over natural enzymes," said chemist Ronald Breslow of Columbia University in New York City. One such ability might be the capacity to carry out reactions that are not catalyzed by naturally occurring enzymes.

Only a few researchers are currently investigating wholly synthetic enzymes, but interest in the field is growing as the potential becomes evident. One sign of that growing recognition is the awarding of the 1987 Nobel Prize for chemistry to three scientists working with wholly synthetic enzymes: Donald Cram of the University of California at Los Angeles, Jean-Marie Lehn of Louis Pasteur University in Paris, France, and Charles Pedersen, who has retired from E. I. du Pont de Nemours & Co. of Wilmington, Delaware.

Cyclodextrins. One of the most thoroughly studied approaches to producing synthetic enzymes involves a family of doughnut-shaped molecules called cyclodextrins. If the molecules contain six glucose molecules linked in a ring, they are called alpha-cyclodextrins; if they contain seven, they are known as beta-cyclodextrins; and if they contain eight, they are called gamma-cyclodextrins. The cyclodextrins have a central cavity 7 angstroms deep and 5 to 9 angstroms in diameter; the top of the cavity is somewhat larger than the bottom, so that it is shaped like a cone rather than a cylinder.

Breslow and chemist Myron Bender of Northwestern University in Evanston, Illinois, have independently performed much of the work using cyclodextrins as synthetic enzymes. Their work has been directed toward accelerating the rate of simple catalytic reactions, with the glucose

Figure 3. Synthetic Enzymes. Cyclodextrins — a family of doughnut-shaped carbohydrates — have been used to make synthetic enzymes that accelerate some simple chemical reactions by a factor of about 6 million. The cyclodextrins, modeled here, are composed of 6 (left), 7 (center), or 8 (right) glucose molecules linked together in a ring. *Courtesy: M. L. Bender.*

hydroxyl groups on the rim of the cone serving as catalysts. One of their findings is that benzene rings bind in the cavity, as does a molecule called ferrocene. If an ester is attached to the benzene or ferrocene by a flexible hydrocarbon chain, the binding of the benzene or ferrocene to the cyclodextrin brings the ester close to the hydroxyl groups on the rim of the cyclodextrin, and the hydroxyl groups act as catalysts to break apart the ester.

In the presence of the cyclodextrin, the ester is broken apart about 1,000 times faster than in water without cyclodextrin. If the flexible hydrocarbon chain is replaced with a more rigid linkage that forces the ester closer to the hydroxyl groups, the rate of reaction is further increased. Breslow has found that, when an ester is attached to a ferrocene molecule by a highly rigid linkage, its reaction rate is accelerated by a factor of 5.9×10^6 in the presence of beta-cyclodextrin. This acceleration of the reaction rate is comparable to the rate increases observed with natural enzymes.

Another reaction catalyzed by the cyclodextrins is the chlorination of the compound anisole. Anisole is composed of a hexagonal ring of carbon atoms with a small tail of carbon atoms attached to one of the ring carbons. Normally, when chlorine reacts with anisole, chlorine atoms become attached either to a carbon atom immediately adjacent to the tail or to one across the ring from it, resulting in a mixture of products. In the presence of alpha-cyclodextrin, however, the anisole settles tail-first into the molecule's cavity, thereby shielding the carbons adjacent to the tail and preventing them from reacting with chlorine. All the chlorine atoms then add to the carbon atom opposite the tail. It is especially interesting to note that an enzyme called chlorinase is also able to chlorinate anisole, but it results in a mixture of products. The wholly synthetic enzyme thus has greater selectivity than the natural enzyme.

Cavitands. Cram and other scientists have been studying a different kind of synthetic cavity-containing molecule. Cram, for example, has synthesized many different ring compounds, most of them bowl-shaped, that have a cavity lined with oxygen atoms. He calls these molecules "cavitands." Many different small molecules can slide into the cavity of these molecules and bind to the oxygen atoms to form stable complexes.

One of the cavitands developed in Cram's laboratory has a relatively long, narrow cavity that can form complexes with such molecules as carbon disulfide, sulfur dioxide, and oxygen. Each of these molecules is composed of atoms aligned in a straight row so that the molecules can easily fit into the cavity. According to Kent Stewart in Cram's laboratory, this cavitand is the first known molecule that can bind oxygen without the participation of a metal ion. Cram believes that a similar cavitand could be used to carry oxygen in a synthetic blood substitute.

Cram has also combined pairs of the cavitands to form hollow, closed shells that he calls "carcerands," from the Latin word for prison. When the carcerands are brought together in a solution, they trap a small amount of the solvent. Carcerands allow chemists to study the differences between a small molecule in solution and the same molecule trapped within another molecule, where its movements are severely restricted.

Carcerands themselves are very insoluble in either water or organic solvents. Cram has overcome this problem by attaching as many as eight long chains of carbon atoms to each half of the shell to make them soluble in organic solvents; the chains give the molecules the appearance of an octopus. Chemical alterations in the chains, such as the addition of hydroxyl groups, can make the carcerands soluble in water. Cram predicts that the carcerands might prove useful as delivery systems for drugs or pesticides. The degradation of the carcerands by acids within the human body, for example, would slowly release their contents, providing a sustained dose of a drug. They might be used, for example, to release insulin over a long period of time in diabetics, thereby eliminating the potential for a toxic reaction when a large dose is injected. Carcerands might also be used to release anticancer drugs continuously in the immediate vicinity of a tumor.

New Approaches to Synthesis

Only about 90 chemical elements exist in nature, and all material things are built from combinations of them. Atoms of these elements are linked together in unique ways to form molecules. The job of chemists is to take these molecules apart and understand them better, to reproduce them in larger quantities, and to make new molecules that have unusual properties. The new materials synthesized range from foods and drugs to synthetic metals and building materials. Given the important role that such materials play in our world today, one of the most important challenges facing chemists is to develop improved techniques for synthesizing new molecules, whatever their ultimate use.

Mimicking Mother Nature

Physicians and biologists have become increasingly interested in using natural products (chemicals normally found in the body) and closely related compounds, because such materials are less likely than artificial chemicals to produce undesirable side effects in people. The function of such molecules is closely tied to a specific three-dimensional structure, and as a result, there has been a growing need to carry out synthetic reactions in which the spatial relationships of individual atoms are controlled. The importance of three-dimensional structure arises from the fact that a carbon atom can bind to four separate atoms or groups of atoms. If each of the four groups attached to the carbon atom is different, the molecule can exist in two different forms that are called stereoisomers; the three-dimensional structures of stereoisomers are different even though their chemical composition is the same.

Two stereoisomers are related to each other in the same way that a right hand is related to a left hand: they are mirror images. Scientists say that one stereoisomer cannot be superimposed on the other, just as a glove for the right hand cannot be worn on the left hand. Each of the two stereoisomers is asymmetric — no imaginary plane can be passed through the molecule in such a way that the portion of the molecule on one side of the plane is the mirror image of the portion on the other side, as would be the case for a symmetric molecule.

Scientists describe an asymmetric molecule as being chiral, from the Greek word "chiros" or hand; in the case of carbon, chirality results when four different atoms or groups of atoms are bonded to a specific carbon atom. That carbon atom is called a chiral center. If a molecule contains one chiral center, it has two stereoisomers, and if it contains more than one chiral center, it has several stereoisomers. A symmetric molecule exists in only one form and is termed achiral; its mirror image can be superimposed upon itself. By definition, symmetric molecules have no stereoisomers.

Stereoisomers can be differentiated from one another by one of three different systems of nomenclature, depending on the physical properties of the molecules. A pair of stereoisomers can be called (+) and (–), (*R*) and (*S*), or D and L. A mixture containing equal quantities of both members of a pair of stereoisomers is said to be racemic.

Most compounds in nature could potentially exist as two or more stereoisomers, but biological systems usually produce only one. The configuration of that stereoisomer is extremely important for its biological activity. For example, amino acids, the building blocks of proteins, occur naturally mainly in the L-form; D-amino acids are not used to build proteins, and they cannot be digested by most organisms. Similarly, most sugars found in nature occur only in theD-form, and L-sugars cannot be digested by most organisms.

Specific three-dimensional structures can be extremely important in the functioning of drugs, hormones, and many other compounds. The natural stereoisomer of the gypsy moth sex attractant (+)-Disparlure, for example, agitates males and disrupts mating when it is present in high concentrations, while the (–)-stereoisomer not only has no effect on mating, but even interferes with the activity of (+)-Disparlure. The (*R*)-stereoisomer of the drug Inderal, which is used to treat hypertension and certain heart disorders, is 100 times as effective as the (*S*)-form. The (*R*)-(+)-stereoisomer of the drug thalidomide is a safe and effective sedative, while the (*S*)-(–)-stereoisomer causes birth defects when ingested by pregnant women. Thalidomide, sold as a sedative in Europe during the late 1950s, contained both stereoisomers and caused birth defects in thousands of European children.

Conventional synthesis of chiral compounds. In the past, chemists who wanted to synthesize a specific chiral drug, hormone, or other product had only two methods for carrying out the process. The oldest and most common method is to synthesize a racemic mixture containing both stereoisomers and then to separate the desired stereoisomer. This separation is generally achieved by crystallizing the product in the presence of a different chiral compound. The added chiral compound forms a complex with the desired stereoisomer in the racemic mixture. The complex will crystallize out of solution, leaving the undesired stereoisomer still dissolved. Repeated crystallizations generally yield the complex in high purity, and the added chiral compound is then separated from the desired stereoisomer.

This process of separating stereoisomers, called resolution, is expensive because of the time required for carrying out the crystallization process. It is also inefficient because half of the starting material is converted to the undesired stereoisomer, which cannot be used. It is, however, the only way to obtain a particular chiral product when the desired stereoisomer cannot be synthesized directly.

Another method of obtaining a specific chiral product is to use a starting material that already has a chiral center identical to that of the desired stereoisomer. With this method, the chemist has to modify only those portions of the molecule that are connected to the chiral center to obtain the desired material. D-glucose, which has four chiral centers, can be used as a starting material for the synthesis of vitamin C, which has two chiral centers; in this case, two of the chiral centers in the glucose retain the same configuration in the vitamin C, and the other two are converted to achiral centers during the synthesis.

A better approach is to create a chiral center during the synthesis of the desired compound. Ideally, creation of this chiral center should be done with a selectivity of greater than 95 percent; that is, the undesired stereoisomer should account for less than 5 percent of the product. Chemists have attempted many different approaches to the problem of chiral synthesis, but only three chemical methods have produced reliable results: chiral hydrogenation, chiral epoxidation, and chiral boration.

Hydrogenating olefins. Chiral catalytic hydrogenation of olefins (certain hydrocarbons containing one carbon-carbon double bond) involves the addition of a hydrogen atom to each of the two carbons joined by a double bond. This catalytic technique was developed independently in 1968 by chemists William Knowles and his colleagues at Monsanto Co. in St. Louis, Missouri, and L. Horner and his associates at the University of Mainz in West Germany.

The catalyst for the hydrogenation reaction is a rhodium atom bound to one or more large chiral organic molecules called ligands, which allow the olefin to bind to the metal in only one configuration. Once the olefin is bound to the metal, the ligands "shield" it so that a hydrogen molecule can approach it from only one direction and thus can form only one stereoisomer. This process produces the desired stereoisomer with a selectivity of greater than 98 percent.

Chiral hydrogenation is most commonly used for the production of amino acids. Monsanto, for example, uses the process for the large-scale production of L-dopa, an amino acid derivative that reduces the symptoms of Parkinson's disease, a disorder of the nervous system. The technique is also used on a smaller scale to produce radioactively labeled L-amino acids and rare D-amino acids for use in biological research.

Making chiral epoxides. One of the most important new methods for chiral synthesis, called chiral epoxidation, was developed in the early 1980s by chemists K. Barry Sharpless and Tsutomu Katsuki, who were then at Stanford University in California. The method involves adding an oxygen atom to a specific side of the double bond of an olefin, resulting in the formation of a chiral three-membered ring (composed of two carbon atoms and one oxygen atom) called an epoxide.

The researchers used a titanium complex as a catalyst, *tert*-butyl hydroperoxide to donate the oxygen to the ring, and a naturally occurring, inexpensive chemical (either (+)- or (–)-diethyl tartrate) as a chiral ligand. As is the case in chiral hydrogenation, binding the ligand to the metal atom governs the direction from which the oxygen atom is attached to the double bond, and the three-dimensional structure of the ligand determines which stereoisomer is produced. In Sharpless and Katsuki's system, the desired stereoisomer makes up 95 percent or more of the reaction products.

Chiral epoxidation has been used for the cost-effective production of a variety of compounds. In 1980, for example, it cost about $56,000 per ounce to prepare (+)-Disparlure, the gypsy moth sex attractant, by resolution of a racemic mixture. In 1988, (+)-Disparlure produced by chiral epoxidation was available for about $7,000 per

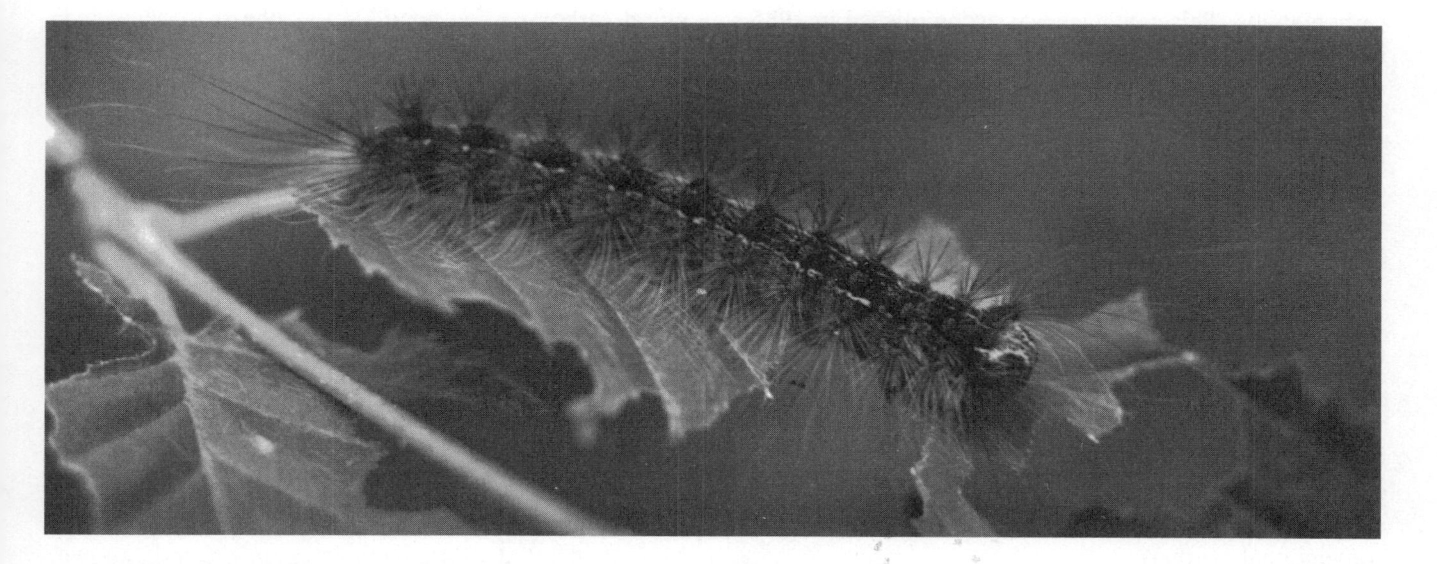

Figure 4. Gypsy Moth Control. Gypsy moth caterpillars (top), which defoliated tens of millions of acres in the Northeast and Mid-Atlantic states during the early 1980s, may be brought under control by (+)-Disparlure, the naturally occurring stereoisomer of the gypsy moth sex attractant. As chemists have developed more efficient techniques for making synthetic (+)-Disparlure, the compound's price has fallen to the point that it can now be economically used to draw adult gypsy moths (bottom) into traps. *Courtesy: U.S. Department of Agriculture.*

ounce, which made it possible for manufacturers to include the substance in gypsy moth traps for home use.

One of the most dramatic applications of chiral epoxidation is the synthesis, from two-carbon molecules, of all eight possible L-sugar molecules that consist of six carbon atoms. This synthesis was accomplished in 1985 by Sharpless, Satoru Masamune, and their colleagues at the Massachusetts Institute of Technology (MIT) in Cambridge. The method should be very useful for producing radioactively labeled sugars for biological studies and for synthesizing certain sugar-containing antibiotics.

In 1987, Takasago Perfumery Co., Ltd., of Tokyo, Japan, began using chiral epoxidation to produce about 1,000 tons of menthol per year. Menthol is used in many consumer products, such as cigarettes, chewing gum, toothpaste, pharmaceuticals, and personal-care products. Much of the 3,500 tons of menthol consumed worldwide each year is extracted from *Mentha arvensis*, grown in China, but chiral epoxidation allows the use of inexpensive starting materials and eliminates the expensive optical resolution techniques necessary for conventional synthesis.

ARCO Chemical Co. of Chicago has licensed the chiral epoxidation technique from MIT and begun developing a process for producing large quantities of glycidol, a three-carbon alcohol containing an epoxide group. Glycidol is a very versatile starting material for many chemical syntheses, and at least six pharmaceutical companies have each purchased 50 to 100 pounds of it for use in the synthesis of new drugs. At least four other pharmaceutical companies have used epoxidation for the production of potential new drugs. If one or more of these drugs is marketable, chiral epoxidation could become much more widely used in the drug and chemical industries.

Forming carbon-carbon bonds. The newest and potentially most versatile method for chiral synthesis was developed by Nobel laureate Herbert Brown and his colleagues at Purdue University in West Lafayette, Indiana, in 1984. Brown used inexpensive and readily available chemicals to convert achiral olefins into chiral alkyl boranes (compounds in which a chiral carbon center is attached to a boron atom) of almost 100 percent purity.

Since techniques are already available to transfer chiral carbon centers from the borane molecules to other molecules without making a racemic mixture, the new method should make it possible to synthesize many types of chiral compounds. The chiral carbon center can, for example, be transferred to a hydroxyl group (an oxygen atom bonded to one hydrogen atom) to produce a chiral alcohol (a compound containing a hydroxyl group). It can also be transferred to a nitrogen compound to form a chiral amine (a compound containing at least one nitrogen atom bound to two hydrogen atoms). Most importantly, the chiral carbon center can be transferred to a carbon atom to form a chiral carbon-carbon bond — a synthesis that has been particularly difficult in the past.

By late 1988, the chiral boration method was being used commercially to produce small amounts of chemicals destined for research use, but Brown predicts that the technique will find widespread use in industry for producing drugs and hormones.

Enzymes in Syntheses

Nature has approached the problem of chiral synthesis by using enzymes, which catalyze a chemical reaction with a remarkably high selectivity. Many chemists have become interested in using enzymes isolated from bacteria or other sources as catalysts to carry out specific reactions. More than 2,000 different enzymes are currently known, but because of the high costs of isolating and purifying them, only 16 are available commercially in large quantities, and about 200 are available in limited amounts. The enzymes are used in laundry detergents, cheese production, food preparation, and pharmaceutical synthesis. In 1985, U.S. sales of such enzymes were about $185 million, and the total was expected to top $260 million by 1990. Worldwide enzyme sales were about $400 million in 1985.

The small number of enzymes available commercially may be discouraging to the chemist who wants to carry out specific reactions. According to chemist Alexander Klibanov of MIT, however, most enzymes can carry out several different reactions in addition to those that they catalyze in nature. Klibanov has reported, for example, that the enzyme glucose oxidase can catalyze a variety of reactions. In cells, glucose oxidase converts the sugar D-glucose into another product while simultaneously reducing molecular oxygen — adding two hydrogen atoms to the oxygen to produce hydrogen peroxide. In the laboratory, the isolated enzyme can be used with D-glucose to reduce a number of organic materials to produce useful products.

For example, glucose oxidase can reduce benzoquinone to hydroquinone, which is used as a photographic developer. It can also convert certain other chemicals into vitamin K derivatives, which are used to treat various illnesses. Similarly, the enzyme galactose oxidase, which normally oxidizes the sugar galactose, can also oxidize many alcohols with great selectivity.

Immobilized Proteins

Enzymes are expensive catalysts and ideally should be reused many times. The best way to reuse an enzyme is to immobilize it by chemically linking it to a solid support, entrapping it in a gel, or encapsulating it in a membrane that allows reactants and products to pass through. The same methods can be used to immobilize whole bacterial cells, dead or alive, that contain enzymes.

Existing uses. Immobilized enzymes have had limited use in industry because they are more expensive than heterogeneous or homogeneous catalysts for most processes. Until recently, the most important commercial applications of immobilized enzymes were: the use of glucose isomerase to convert glucose from corn into high-fructose corn syrup, which is used in soft drinks and many packaged products such as baking mixes; the use of the enzyme aminoacylase to destroy one of the stereoisomers in a racemic mixture of amino acids, leaving a pure stereoisomer; the use of enzymes in immobilized bacterial cells to produce L-amino acids; and the use of penicillin acylase to convert penicillin molecules into 6-aminopenicillanic acid, a precursor of many antibiotics.

Those applications were introduced in the early 1970s, after which no important new processes were developed for several years. Interest in immobilized enzymes was revived in the 1980s, however, because biotechnology made it possible for large quantities of many enzymes to be produced at reasonable prices. The first new immobilized enzyme process since the early 1970s was incorporated into a yeast-manufacturing plant opened in October 1983 in Winchester, Kentucky, by Nutrisearch Co. — a joint venture of Kroger Co. of Chicago and Corning Glass Works of Corning, New York. The plant uses an immobilized enzyme called lactase to convert the sugar lactose from a cheese by-product called whey into two other sugars, glucose and galactose. These sugars are then used to grow bakers' yeast, which is later sold at Kroger grocery stores. The process also helps to solve a pollution problem, since half the whey produced in the U.S. is now washed down the drain.

Other applications of immobilized enzymes are being considered for specialized markets. Researchers at the Midwest Research Institute in Kansas City, Missouri, for example, have used an immobilized cholinesterase as a means of detecting pesticides in air, water, and soil; the pesticides inhibit the enzyme, and its reduced level of activity can be readily monitored. The Midwest Research Institute is now working under a contract from the U.S. Environmental Protection Agency to determine whether other commercially available enzymes are inhibited by chemicals on the agency's toxic chemicals list.

New applications. Bioengineer Robert Langer of MIT has used an immobilized enzyme called heparinase to destroy heparin in blood. Heparin is an anticoagulant that is used to keep blood from clotting when it is circulated outside the body in devices such as a heart-lung machine. When the heparin-laden blood returns to the body, however, it can cause internal bleeding. Langer has found that immobilized heparinase can destroy the heparin before the blood is returned to the body, thereby diminishing unwanted anticoagulant effects. Tests with dogs showed that the immobilized enzyme can destroy 99 percent of the heparin in a dog's circulatory system within 2 minutes.

In mid-1988, Langer and his colleagues reported that the immobilized heparinase has also been successfully tested in 70 sheep and that the treatment did not have any adverse effects on the animals' blood. Langer has licensed the immobilized heparinase technology to Continental Pharma Cryosan of Montreal, Quebec, and in late 1988 that company was preparing further studies in the hope of receiving approval from the Food and Drug Administration (FDA) to use the heparin-destroying process in humans.

Langer, Klibanov, chemical engineer Cynthia Sung of MIT, and pediatrician Arthur Lavin of Boston Children's Hospital have also developed a method that may prove useful for treating neonatal jaundice. Jaundice is caused by the abnormal accumulation of bilirubin as a result of kidney malfunction. Bilirubin, which is a reddish-yellow pigment formed from hemoglobin, binds to cellular and mitochondrial membranes, causing cell death in a variety of tissues. Bilirubin toxicity in an infant can lead to mental retardation, cerebral palsy, deafness, seizures, or death.

The two main treatments now used to treat jaundice are phototherapy and exchange transfusion. In phototherapy, the infant is exposed to blue light that converts bilirubin in blood vessels directly under the skin to a less toxic chemical. Since only 15 percent of the body's bilirubin can be converted by this process, it is effective only for mild cases. More severe cases require exchange transfusion, in which the infant's blood is removed and replaced with bilirubin-free blood from an adult. This process also has many undesirable side effects, such as transmission of infectious diseases.

The MIT group's method involves trapping the enzyme bilirubin oxidase in a small blood filter. The enzyme converts bilirubin to biliverdin, which is much less toxic than bilirubin and can be removed by the body. In tests with human blood, the MIT team has shown that the system removes 90 percent of the bilirubin in human or rat blood in one pass through the filter. Clinical trials of the bilirubin system were scheduled to begin in 1989.

Several other uses of immobilized enzymes are also being studied. Chemist Thomas Chang of McGill University in Toronto, Ontario, has developed a system in which three different enzymes enclosed in a microcapsule can be used to break down urea from the blood of kidney-failure patients or ammonia from the blood of liver-failure victims. Researchers at Kraft are working on processes that would use the immobilized enzymes rennin and pepsin to coagulate milk for cheese production, and immobilized lipase and esterase to break down butterfat for use in cheeses.

Chemist George M. Whitesides and his colleagues at Harvard University have developed a technique to trap enzymes inside commercially available tubing used for kidney dialysis. With Whitesides's method, organic chemicals in solution outside the tubing diffuse inside, where they are converted into the desired product by the trapped enzyme; the product then diffuses back into the solution.

Perhaps the most intriguing use of immobilization involves a protein that is not an enzyme. Joseph and Celia Bonaventura of Duke University in Durham, North Carolina, have developed a process for trapping hemoglobin, the blood component that carries oxygen, in a polymer. They call this hemoglobin-containing polymer a "hemosponge." In a patent application filed in 1983, the Bonaventuras reported that their hemosponge can extract oxygen directly from seawater. The oxygen can then be released from the hemosponge by exposing the polymer to a vacuum or by passing a weak electrical charge through it.

According to Joseph Bonaventura, a small hemosponge unit carried on a diver's back would allow the diver to work underwater indefinitely by extracting oxygen from the seawater around him; a larger unit 3 feet long and 10 feet in diameter could provide oxygen indefinitely to 150 workers on the ocean floor. Bonaventura said that an even more important application is the possible use of the hemosponge to supply oxygen for underwater combustion to power a submarine. If oxygen were available to burn gasoline or kerosene in a submarine, those fuels would provide 300 times as much energy per weight as the batteries currently used.

In late 1983, the Bonaventuras and Duke University licensed the technique to Aquanautics Corp. of San Francisco, California, and since then that company has been developing a variety of applications for the hemosponge, including submersible vehicles that extract oxygen from water, respirator units for the bedridden that extract oxygen from the air (thereby eliminating the need for oxygen tanks), and small devices that would remove oxygen from packaged food and other products, thereby minimizing spoilage.

Artificial Body Parts

Accidents and diseases have always taken a heavy toll on the human body, causing the loss or destruction of tissues, organs, limbs, and other body parts. "For as long as the history of medicine can record, there have been attempts to restore shape or function to affected parts by some kind of replacement," according to materials scientist David Williams of the University of Liverpool in the U.K. "In early days, replacements were confined to relatively crude external prostheses, notably the glass eye, the wooden leg, and gold teeth.... Now it has become possible to carry out functional and structural replacements that seemed, only a short time ago, to be a millennium away."

Surgeons now routinely implant stainless steel hip and knee joints to replace joints destroyed by arthritis, Dacron polyester blood vessels to replace vessels blocked by atherosclerosis, plastic eye lenses to replace lenses damaged by cataracts, and artificial heart valves to replace valves damaged during heart attacks. Less commonly, they implant artificial hearts, synthetic bone materials, artificial ligaments, and artificial skin.

While the body may not seem to be a hostile or aggressive environment, Williams said, metals in contact with blood plasma can be degraded 10 times as fast as metals in contact with sea water, which is normally viewed as a hostile environment. In addition, enzymes in tissues can break down plastics and metal implants, causing the release of metal ions or organic chemicals that are irritating or toxic to the body's tissues. Often, the degradation of the implants forces the removal of prostheses to maintain the patient's health.

The problem of biocompatibility is particularly acute with materials that come into contact with blood. Unfortunately, nearly all artificial materials seem to be very good at activating the clotting process. As a result, many laboratories throughout the world have been trying to develop materials that will prevent clotting on foreign surfaces. This research is a major component of the effort to develop an effective artificial heart.

One of the important steps in the blood-clotting process is the aggregation of platelets (round or oval disks found in the circulating blood) at the site of a wound or on the surface of an artificial material. The platelets become enmeshed in a framework of a filamentous, insoluble protein called fibrin to form the clot. Many researchers have therefore approached the problem of producing biocompatible materials by attempting to block the adhesion of platelets.

Anticoagulants such as heparin or prostaglandins can stop the abnormal clotting process, but it is dangerous to have them circulating freely in the blood for long periods because they also block normal clotting during an injury. One research approach has focused on binding the anticoagulants chemically to the surface of biomaterials to prevent platelet adhesion. Materials scientist Clement Bamford of the University of Liverpool, for example, successfully attached a synthetic form of prostaglandin to the surface of various plastics and found that it retained the ability to prevent clotting. Bamford is currently implanting the coated materials in animals to determine how long they retain their protective coating.

Artificial Blood Vessels

Small blood vessels with a diameter of less than 0.25 inch frequently need to be replaced after they are damaged by physical injury or diseases such as diabetes or atherosclerosis. Such small vessels are most commonly replaced with the saphenous vein from the leg — one of the veins visible through the skin on the front of the leg. This procedure, however, requires the patient to undergo two operations: one to remove the vein from the leg, and the other to implant it elsewhere. Moreover, as many as 25 percent of the 300,000 prospective patients who require a blood vessel replacement each year either do not have a usable saphenous vein or have already used their saphenous veins for previous bypass surgery.

Researchers have been working to produce artificial blood vessels that could be used instead of the saphenous veins. Tubes made of Dacron polyester are used routinely to replace large blood vessels, but are not effective replacements for small ones. Small Dacron blood vessels become blocked by blood clots much more readily than large ones, often within an hour after implantation. To avoid this problem, researchers have been developing ways to modify the interior surface of small replacement vessels to minimize clot formation.

Since the 1950s, scientists have known that coating the inside surface of a synthetic blood vessel with albumin (a protein that is found in especially high concentrations in egg whites) reduces clot formation on prostheses. The reason for this reduction remains a mystery. Two techniques for applying albumin to the inside of artificial vessels are under development. One technique, which has not yet proved successful, involves modifying the surface of existing polymers to increase their affinity for albumin. The other technique, pioneered by a research team headed by chemist Donald Lyman of the University of Utah in Salt Lake City, uses completely new polymers that have an inherent affinity for albumin. Some of Lyman's results suggest why other researchers have been unsuccessful.

Lyman's polymers have a high oxygen content, which greatly increases their affinity for albumin. For the last decade, Lyman and his colleagues have been implanting small synthetic vessels made from the polymers into dogs. The vessels were successful in that clots did not form on their interior walls, but a second problem developed. Within 200 hours after implantation, the vessels became blocked at the juncture, or anastomosis, between the natural and the synthetic vessels. This blockage phenomenon, known as the anastomotic effect, is characterized by excessive growth of smooth muscle cells from the natural blood vessels into the synthetic vessels. This growth causes impaired blood flow and, eventually, vessel blockage.

Lyman concluded that the anastomotic effect was caused by the difference in flexibility between the natural and the synthetic vessels. The diameter of natural vessels changes as much as 30 percent during the pulsation of blood, whereas the diameter of the synthetic vessels remains constant. This difference in flexibility places great strain on the natural blood vessel at the anastomosis and stimulates the abnormal growth of smooth muscle cells there. Lyman concluded that it was the anastomotic effect, rather than the chemical nature of the artificial vessels, that was responsible for the failure of modified polymers used by other researchers. This effect is not as important with larger synthetic blood vessels because the natural blood vessels do not flex as much and because the larger vessels can accommodate some ingrowth of smooth muscle cells without becoming blocked.

In the early 1980s, Lyman's group began synthesizing oxygen-containing polymers that had been altered to give them greater elasticity. More than 90 percent of the new elastic vessels that Lyman implanted in dogs remained unobstructed after 3 months, and some were still clear when the animals were sacrificed after 29 months. Lyman has successfully implanted vessels as small as 0.03 inch in diameter.

Lyman began testing the synthetic small blood vessels in humans in early 1983. Most of these tests have been successful, he said, despite the fact that the implants were performed as a last resort in patients who had already been subjected to other types of blood vessel replacement, such as a saphenous vein implant, and who were in imminent danger of losing a limb because of poor circulation. In 1985, Lyman formed a company to market the artificial blood vessels and, in September 1987, he sold it to Research Industries, Inc., of Salt Lake City. In April 1988, Research Laboratories received permission from the FDA to begin testing the synthetic vessels in the lower legs of humans.

Surgeon David Annis of the University of Liverpool uses the same approach as Lyman, but with a different type of material. He and his colleagues have been making blood vessels from plastics by knitting or weaving plastic threads together so that the vessel walls are filled with pores. As soon as blood flows through a stretch of the tubular fabric, it oozes through the pores and rapidly clots. With larger vessels, this process is self-limiting and generates a layer of tissue similar to that on the interior of a normal vessel. Such grafts are very durable.

Annis has used an oxygen-containing plastic called poly(etherurethane) and a process called electrostatic spinning to produce small blood vessels with very small pores. The polymers produced by the spinning process, much like those made by Lyman, have a flexibility that is nearly identical to that of natural small blood vessels. Annis has reported that preliminary studies in animals have been successful and that he has begun limited trials of the vessels in humans.

Synthetic Skin

On 22 October 1985, 25-year-old Mark Walsh was working at an aerosol can factory in Holbrook, Massachusetts, a suburb of Boston, when an explosion and fire burned nearly 80 percent of the skin off his body. Four others died in the fire, but Walsh was lucky. He was taken to Massachusetts General Hospital in Boston, where surgeon John F. Burke closed his wounds and, no doubt, saved his life using a new artificial skin developed by materials scientist Ioannis V. Yannas and his colleagues at MIT. The artificial skin covered Walsh's burns to protect them from infection and fluid loss and provided a scaffolding on which new cells could grow to replace the lost skin. In the absence of adequate protective coverings, at least 10,000 of the 130,000 Americans hospitalized for severe burns each year die from their injuries.

Artificial skin grafts. A graft of the patient's own skin is the ideal covering for burns, but severely injured burn victims generally do not have enough unburned skin to meet this need. Skin from cadavers or animals, typically from pigs, can be used as a covering, but it generally must be removed between 3 and 9 days after application, before it is rejected by the body. Yannas's synthetic skin provides an excellent alternative to these types of grafts because it can be produced in large quantities and is not rejected by the recipient's body.

The principal material of Yannas's artificial skin is a highly porous polymer composed of collagen fibers that are bonded to a sugar polymer, called a glycosaminoglycan, obtained from shark cartilage. Both components are found normally in skin, and research has shown that the combination in the artificial skin stimulates the growth of healthy skin cells around it. This material is covered with a sheet of silicone rubber that serves as a barrier to infection and fluid loss and provides mechanical strength so that the graft can be sewn into place. The artificial skin resembles strips of moist, smooth, elastic paper toweling.

Burn victim Mark Walsh was fortunate that some of Yannas's artificial skin was available at the time of his accident; even in 1989, it was still in relatively short supply. Four days after the fire, Walsh underwent the first of six operations that lasted a total of 24 hours over 4 days. During the operations, Burke cleaned away all burned flesh and fat and covered Walsh's neck, chest, abdomen, and arms with approximately 9 square feet of the synthetic material. The artificial skin was draped over the burned areas, sutured in place, and covered with bandages.

Once the procedure was complete, skin cells began to grow into the synthetic skin from the bottom and sides, and the polymer was slowly degraded by the body. After about 20 days, the silicone protective layer was removed, and small patches of epidermis — the outer layer of the skin — were transplanted from elsewhere on Walsh's body. This procedure is much less traumatic than a conventional graft of the patient's own skin because the two lower layers of skin are not removed. In 7 to 10 days, new epidermis grows back over the area from which the

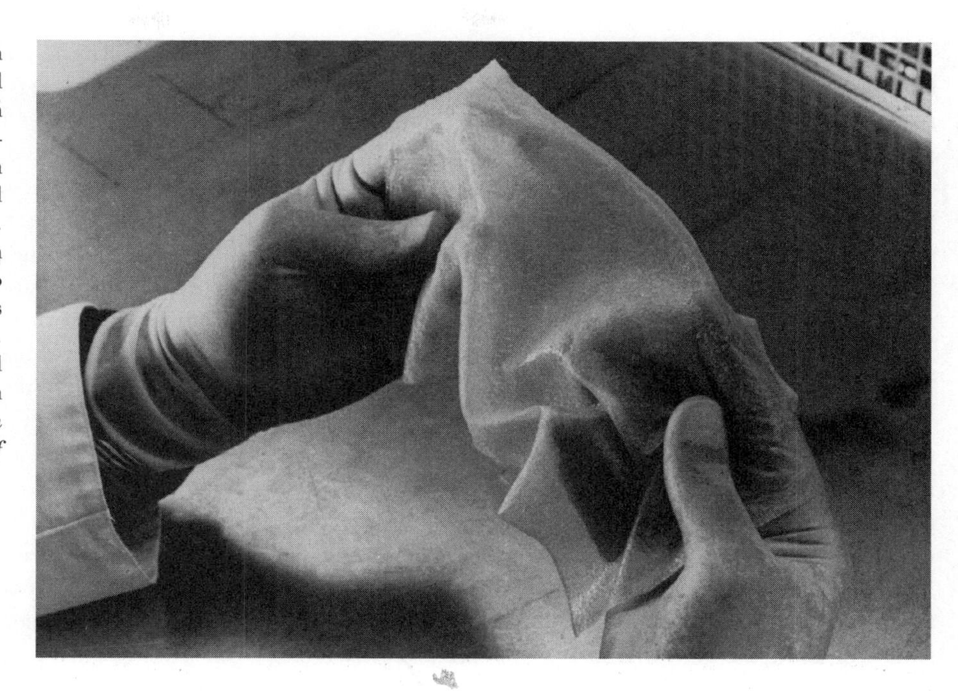

Figure 5. Synthetic Skin. Synthetic skin can help save the lives of severely burned patients and prevent disfigurement in people with less severe burns. The material consists of a very porous collagen polymer bonded to a sugar polymer and covered by a sheet of silicone rubber. After the graft of synthetic skin is sewn over the burned area, skin cells grow into it; the collagen polymer gradually breaks down as it is replaced by healthy tissue. The protective silicone is later removed and replaced with epidermis taken from other parts of the body. *Courtesy: Calvin Campbell, Massachusetts Institute of Technology.*

skin was removed — a process similar to the healing of a sunburn.

When Walsh's bandages were removed in January 1986, he saw healing areas of ruddy pink skin. He was able to distinguish hot and cold and to experience pain in the regenerated areas. The new skin could be distinguished from his own only by the fact that hair did not grow on it and it had no sweat glands. Walsh required several months of therapy to regain use of his stiffened joints, but his appearance was much as it had been before the accident. The skin had fulfilled its two primary purposes: it kept Walsh from developing a life-threatening infection in the short term and it gave him a normal appearance.

In 1980, MIT licensed the patent on the skin to Marion Laboratories, Inc., of Kansas City, Missouri, and the company has been developing techniques for producing the large amounts of material necessary for clinical trials. In late 1988, surgeon David Heimbach of Harborview Clinical Center in Seattle, Washington, reported that the artificial skin had been successfully used in 106 patients at 11 medical centers around the country. Marion expected FDA approval for wider use of the skin within 12 to 24 months.

Combining artificial and natural skin. Since 1984, Yannas and Burke have been working on a new type of skin called stage II, which will not require a graft of the patient's epidermis. The basic structure of this material is the same as that of the original material, which Yannas calls stage I, but actual skin cells are incorporated into its structure. The scientists take a small biopsy sample of skin from the potential recipient and isolate the basal cells — those cells that eventually mature into the dermal and epidermal cells of the skin. This collection and isolation process takes about 80 minutes.

In tests with animals, these basal cells are seeded into the stage II skin along with a small quantity of fetal calf serum, which stimulates the growth of the cells. The artificial skin is then sewn over the wound. This material works much like the stage I skin, but the basal cells multiply and produce epidermis as well as the underlying tissue. Near-normal skin grows over the burned area within 30 days after the synthetic skin is implanted.

Yannas said that, as of September 1988, the stage II skin had been used in six patients, but that he and Burke had encountered unexpected difficulties. They decided not to use the fetal calf serum in the skin for humans because they feared that the serum would sensitize the patients to dairy products; however, in the absence of the serum, the implanted epidermal cells proliferated very slowly. One of the six patients recovered successfully, Yannas said, but he and Burke were forced to remove the artificial skin from the remaining five patients 3 weeks after it was implanted because the wounds were not healing fast enough. The researchers are now searching for human growth factors to use instead of the calf serum.

Yannas's technique of building a polymer matrix that stimulates the growth of tissues may be applicable to other tissues. He and his colleagues have used it, for example, to restore small sections of sciatic nerves that have been excised from rat legs. Cutting the nerve leaves the rats unable to spread their toes, maneuver their feet, or walk properly. If peripheral nerves such as the sciatic nerve are simply cut and sutured back together, they will normally reconnect, restoring about 80 to 90 percent of the normal nerve function. But if a segment of the nerve is removed, leaving a large gap, repair is virtually impossible.

Yannas and his colleagues removed about 0.5 inch of the rat sciatic nerve, and then inserted the two cut ends into the ends of a small silicone rubber tube. If the tube is empty, the nerves do not grow back together; however, if it is filled with the polymer matrix, there is nerve growth. Yannas's group found that, to ensure optimum conditions for nerve regrowth, the pores of the polymer matrix should be about the same diameter as that of the regrowing nerve cells, about 1 micron, and that the pores should be oriented in the direction of nerve growth. Under these conditions, he said, about 50 percent of the nerve function can be restored within a few weeks.

Artificial Bone

Scientists at the University of Texas at Austin have developed a new type of synthetic bone that is virtually identical to natural bone in structure and composition. The new material is much stronger than previous bone substitutes and better able to withstand stress, according to University of Texas chemist Richard J. Lagow. Tests in animals have shown that once the new material is implanted in animals, it is slowly broken down by bodily processes and replaced by living bone. According to Lagow, this stimulation of new bone growth suggests that the grafts should last indefinitely. He predicted that the new material may prove useful in orthodontics, reconstruction of shattered bones, and replacement of bones or bone fragments removed during cancer surgery.

About 65 percent of living bone is composed of a mineral known as hydroxyapatite. This mineral provides strength and rigidity and acts as a porous matrix that supports bone marrow, bone-synthesizing cells, and blood vessels that sustain the cells. Tooth enamel is a pure, nonporous form of hydroxyapatite. Researchers have synthesized hydroxyapatite in the past as a replacement for bone, but the synthesized form was always brittle and broke easily. They tried to strengthen the synthetic material by reinforcing it with silica, alumina, or organic polymers, but these materials increased the risk of rejection. Surgeons have also used bones from cadavers for repairing or replacing bones, but these bones are even more likely to be rejected and they are not generally available. "No bone graft material around now is ideal," said plastic surgeon Marc Gottlieb at City of Hope Medical Center in Duarte, California.

Lagow and graduate student Paul J. Capano devised a way to synthesize pure hydroxyapatite in a strong, porous form that is "virtually identical" to natural bone, Lagow said, as well as in a denser form similar to that found in tooth enamel. When the hydroxyapatite bone is implanted in animals, natural bone covers it and grows into the pores during a 2- to 4-week period. Blood vessels also grow into the pores. Over a period of months, the synthetic bone is broken down by specialized cells and replaced with natural bone — just as the body's own bone is continually replaced. "This is the only material I know of that, if you put it in the body, capillaries vasculate it, cells grow... and real bone replaces it," says biomedical chemist Alan Davison of MIT.

The synthetic bone has so far had limited testing in humans. Oral surgeon Edward T. Farris of the Baylor Medical Research Center in Dallas has used it in several patients for jaw reconstruction and found that it worked very well. J. Lester Matthews, executive director of the Baylor Medical Research Foundation and a member of Lagow's team, is also working with several orthopedic surgeons in the Dallas area and hopes that they will obtain FDA permission to implant the synthetic bone material in humans. The first orthopedic uses, Lagow said, will probably be for replacing small bone fragments or segments.

Artificial Ligaments

The knee is one of the weakest joints in the body and highly susceptible to injury caused by sports and other strenuous activities. When the knee is traumatically bent or overstretched, the most common result is a torn anterior cruciate ligament, one of the four ligaments that give the knee joint strength and prevent it from wobbling. Ligaments are a fibrous form of cartilage whose chief characteristics are that they are strong and do not normally stretch. When torn, they typically do not heal readily and are usually replaced.

The replacement most often used is a piece of healthy

tendon taken from the side of the knee. The replacement procedure is successful most of the time, but it can fail if other areas of the knee, particularly the muscles or bones, are weak. The procedure can require as long as a year for recovery and rehabilitation, during which time the tendons may stretch, leaving the joints susceptible to weakening later in life. Researchers at W. L. Gore and Associates in Flagstaff, Arizona, have devised an artificial ligament that both reduces rehabilitation time and strengthens the knee.

The artificial ligament is made from a form of Teflon, a fluorinated polymer that is widely used as a nonstick finish for cooking utensils. Gore was founded in 1969 when Robert Gore devised a way to inject air bubbles into Teflon to make it bulkier without losing its key properties, such as the ability to repel water. The material, known as GORE-TEX, is widely used in waterproof raincoats, ski parkas, and camping gear. It is also used for making synthetic blood vessels. The key to its success, says Gore spokesman James Dykes, is that "GORE-TEX is totally inert. It doesn't react with anything.... Human tissue will grow into and around the GORE-TEX. A year later, you'll have trouble finding it because it is incorporated in your tissue."

The 6-inch-long, artificial anterior cruciate ligament is composed of more than 1,000 GORE-TEX fibers braided together like a rope. It is four times as strong as the natural ligament. To implant the artificial ligament, the surgeon makes a small incision above the kneecap and another across and below the knee. Using an arthroscope (an examining and surgical instrument), the surgeon threads a strip of flexible tape in the top incision and out through the bottom. The ligament is attached to one end of the tape and pulled through until its looped ends protrude through both incisions.

The surgeon then uses stainless steel screws to anchor the top end of the ligament to the femur (the thigh bone) and the bottom end to the tibia (the shin bone). Natural bone eventually grows into the device, providing a natural, permanent anchor. The surgery can be done on an outpatient basis, and rehabilitation takes only about 6 weeks.

During clinical testing in the early 1980s, ligaments were implanted in about 1,000 humans at 27 medical centers around the country. Based on those studies, the FDA approved the sale of the artificial ligament in October 1986, but only for use as a last resort for patients in whom a tendon implant had failed. In 1988, Gore was continuing human trials in hopes of obtaining FDA approval for use in all patients suffering from torn anterior cruciate ligaments.

Blood Substitutes

Donated blood is often needed to save the lives of accident victims or patients undergoing extensive surgery. Unfortunately, because some people contracted AIDS (acquired immunodeficiency syndrome) as a result of blood transfusions — a problem that has now been minimized by extensive screening of blood before use — many people are fearful of receiving blood, and sizable numbers are afraid even to donate it. Always in short supply in the past, blood has become even scarcer in the age of AIDS.

While AIDS has been a major incentive in the development of blood substitutes, hepatitis is also regarded as a major problem for blood banks. Of the 4 million Americans who receive transfusions each year, 250,000 contract hepatitis as a result. Whole blood, furthermore, is very expensive to collect, store, and administer. It is highly perishable and cannot be stored, even in refrigerators, for more than 3 weeks. Whole blood therefore cannot be used for routine treatment of accident victims or battlefield casualties. The military especially would like to have a blood substitute for use in combat that would remain stable for months or years at room temperature. Finally, blood exists in four major types — A, B, AB, and O — and can be either Rh-positive or Rh-negative. If blood is not matched carefully before a transfusion, the recipient can suffer a fatal reaction to it. Researchers hope that blood substitutes could be used universally, eliminating the need for such matching.

Properties of blood. The average human has between 5 and 6 quarts of blood, totaling about one-eleventh of the body weight. Blood binds to oxygen in the lungs, turning bright scarlet in the process. It carries the oxygen throughout the body, releasing it to tissues. After releasing the oxygen and becoming a bluish-red color, it picks up the carbon dioxide produced by body tissues and returns it to the lungs for expulsion. About 45 percent of blood by volume is plasma, a yellowish fluid that contains salts, sugar, hormones, fats, amino acids, proteins, and such waste products as urea and creatinine dissolved in water.

Blood also contains three main types of cells and cell fragments. Erythrocytes, or red blood cells, carry oxygen. Leukocytes, or white blood cells, are the primary components of the body's immune system. Platelets are cell fragments that bind to the wall of a blood vessel at the site of an injury and initiate the complex process of coagulation, or clotting. A cubic millimeter of blood — little more than a drop — contains about 5 million erythrocytes, 5,000 to 10,000 leukocytes, and 200,000 to 300,000 platelets.

In transfusions, the key component of the blood is the red cell. Other components of the blood can be replaced relatively quickly by the body, but without sufficient numbers of red cells to transport oxygen, tissues throughout the body will die. Erythrocytes are formed in the bone marrow. After about 120 days in circulation, during which time they typically incur substantial damage, they are broken down and removed by the spleen. The primary component of the erythrocyte is hemoglobin, an iron-containing protein that binds to one molecule of oxygen. Each erythrocyte contains about 5 billion hemoglobin molecules. Because the principal goal of a transfusion is to improve the oxygen-carrying capacity of blood, most blood substitutes are actually some type of artificial hemoglobin.

Fluorocarbons. Fluorocarbons may at first appear to be unlikely replacements for hemoglobins because they are so different structurally. Fluorocarbons are a family of organic molecules in which the hydrogen atoms have been replaced with fluorine atoms. Nonetheless, fluorocarbon-based blood substitutes are being marketed in Japan and are being studied in the U.S. for a variety of uses.

Interest in fluorocarbons was sparked in 1966 by a dramatic demonstration by pediatrician Leland C. Clark, Jr., of the University of Cincinnati College of Medicine in Ohio. Clark attached a laboratory weight to the tail of a mouse and dropped the mouse into a beaker full of a fluorocarbon called perfluorobutyltetrahydrofuran. The mouse's lungs quickly filled with the fluorocarbon as it floated just under the surface, just as a drowning man's lungs would fill with water. But surprisingly, the mouse continued to breathe normally. When it was removed from the beaker an hour later, the mouse had not been harmed.

This astonishing feat was possible because liquid fluorocarbons dissolve large amounts of oxygen. Whereas salt water or blood plasma dissolve about 3 percent oxygen by volume, and whole blood about 20 percent, fluorocarbons dissolve 40 percent or more. Moreover, carbon dioxide, which must be removed from the body through the circulatory system, is at least twice as soluble in fluorocarbons as in blood.

Fluorocarbons cannot be simply injected into the bloodstream, however, because they are immiscible with blood, just as gasoline and oil are immiscible with water. The contact with blood would produce vessel obstructions called embolisms, which are often fatal. In the late 1960s, surgeon Henry A. Sloviter of the University of Pennsylvania and nutritionist Robert P. Geyer of the Harvard School of Public Health showed that this problem could be overcome by emulsifying the fluorocarbons with detergents and albumin, thereby dispersing them into microdroplets that remain suspended in solution.

Clark and others demonstrated that mice and dogs showed no ill effects when as much as 90 percent of their blood was temporarily replaced with a fluorocarbon emulsion. The only visible effect is that the skin and eyes of the mice were no longer tinted pink by the blood underneath them; instead, they developed an eerie white luminescence. Despite their success in sustaining the lives of animals, these early fluorocarbon emulsions were still not suitable for use in humans; the fluorocarbons tended to concentrate in the liver and spleen, and researchers feared long-term harmful effects.

Clark appeared to solve this problem in 1973 when he discovered that perfluorodecalin is completely eliminated from the body through exhalation or perspiration without being metabolized. He later found, however, that the microdroplets in a perfluorodecalin emulsion tend to agglomerate into larger droplets with time. If the droplets became large enough, they could block capillaries and cause the death of tissues in the limbs and elsewhere.

Chemist Ryoichi Naito of the Green Cross Co., a pharmaceutical firm in Osaka, Japan, solved this problem by adding about 25 percent of a second fluorocarbon, perfluoropropylamine, to the perfluorodecalin emulsion. Green Cross named this translucent, milk-like emulsion Fluosol-DA. The perfluoropropylamine stabilized the Fluosol-DA emulsion, but increased its half-life (the time required for half to disappear from the body) to 65 days, which the company's researchers found acceptable.

By 1988, physicians in Japan were using Fluosol-DA during surgery, and the Japanese defense department was freezing quantities of it for potential military use. Clinical trials sponsored in the U.S. by Green Cross's subsidiary Alpha Therapeutics, Inc., of Pasadena, California, have been less than successful, however. In one key trial at the University of Chicago Medical School, for example, Fluosol-DA was transfused into 23 surgical patients who had refused blood transfusion on religious grounds. Some patients complained of dizziness and chest pains and suffered a rapid drop in blood pressure. The researchers concluded that the blood substitute was unnecessary in mild anemia and ineffective in severe anemia, and the FDA refused to approve its use.

Green Cross researchers are working on a second-generation fluorocarbon called perfluoro-4-methyloctahydroquinalidine. This compound is stable in emulsions by itself and is excreted from the body within a few weeks. Meanwhile, Alpha Therapeutics is studying other

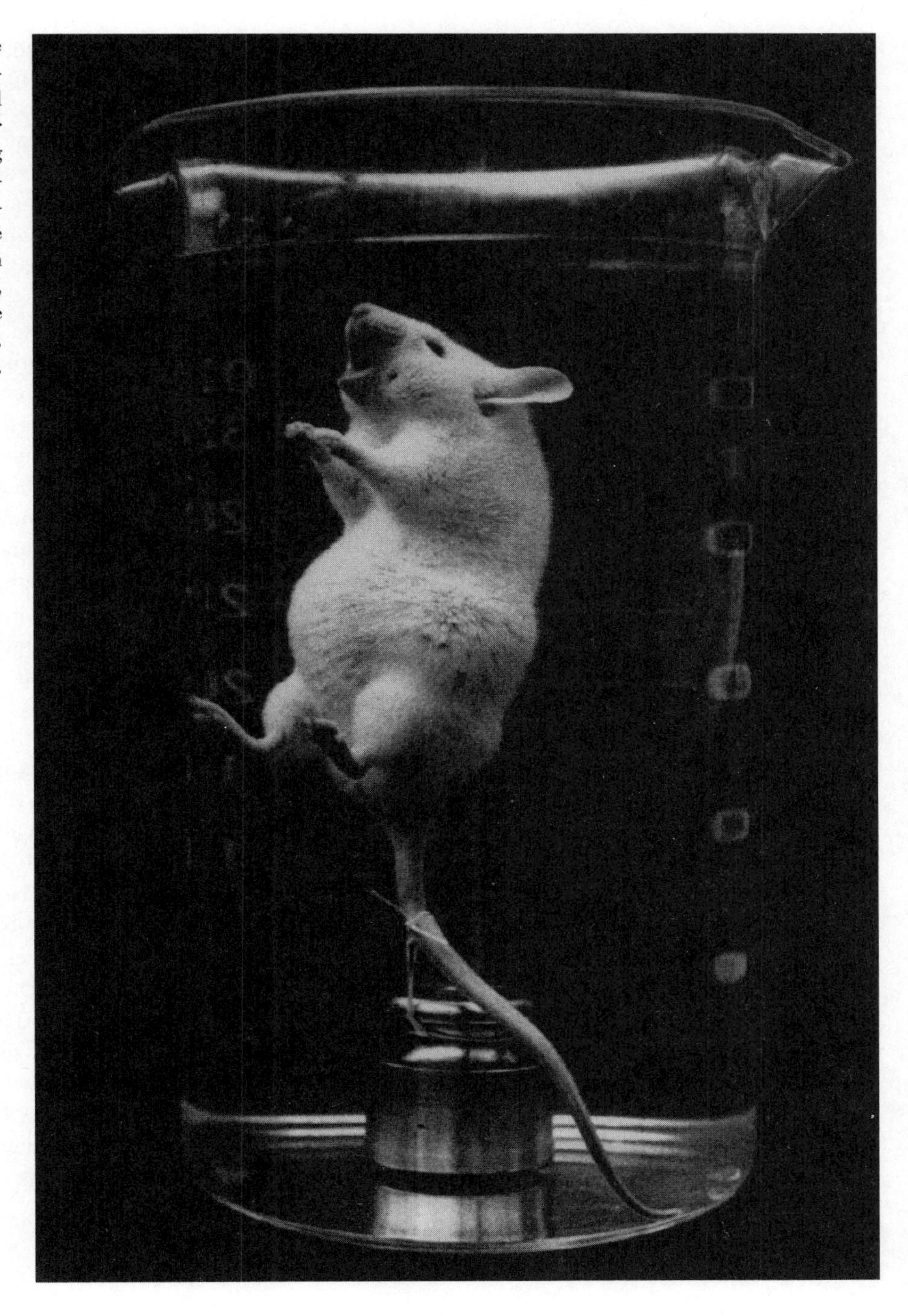

Figure 6. Liquid-Breathing Mouse. The liquid fluorocarbon in this beaker contains such a high proportion of dissolved oxygen that a mouse can breathe it for hours without drowning or suffering other ill effects. The high oxygen-carrying capacity and long shelf life of fluorocarbons may allow these compounds to be used in place of blood under certain conditions, such as on the battlefield, where the supply of whole blood may be very limited. *Courtesy: Leland Clark, Children's Hospital Medical Center, Cincinnati, Ohio.*

potential applications of Fluosol-DA. The company is particularly interested in the fact that microdroplets of the emulsion are only about one-seventieth the size of an erythrocyte and may thus be able to pass through partially blocked arteries and reach areas that the blood cells cannot.

Animal tests, for example, have suggested that Fluosol-DA could be useful during balloon angioplasty (a surgical procedure in which a tiny balloon is inserted and temporarily inflated inside a coronary artery to compress deposits that are blocking the vessel). Because the balloon itself blocks the flow of blood, it can be inflated for only a short period or the tissues beyond it will die for lack of oxygen. The animal studies have suggested that the Fluosol-DA can flow around the balloon and oxygenate the tissues so that the balloon can be kept inflated for longer periods, perhaps increasing the likelihood of the operation's success. Similarly, animal studies have suggested that the blood substitute might minimize damage after a heart attack by allowing oxygen to reach areas of the heart where the flow of blood has been blocked. Damage to the brain from a stroke might also be prevented in the same way.

In addition, Fluosol-DA may be useful for enhancing cancer therapy. Neither radiation nor chemotherapy are very effective killers of cancer cells unless those cells have a high oxygen content. "So by oxygenating a tumor with Fluosol, we are able to sensitize it to radiation or chemotherapy," said George Groveman of Alpha Therapeutics. Clinical trials of this technique, he said, have proved "very encouraging." Finally, Fluosol-DA may be useful for oxygenating donor organs while they are being prepared for transplants.

Hemoglobin. Although chemical changes in the structure of erythrocytes during storage render the blood unsuitable for use within 3 weeks, they do not affect the hemoglobin in the erythrocytes. Many researchers have thus attempted to find ways to use this hemoglobin, which can be easily obtained by rupturing the blood cells with a dilute salt solution, to produce a blood substitute. Initial studies, however, showed that this hemoglobin caused severe damage to the kidneys of laboratory animals. Researchers soon found that the toxic substance was the outer membrane of the erythrocytes, called the stroma, and devised ways to remove it.

Nevertheless, isolated hemoglobin is not a good oxygen carrier. Hemoglobin is a protein that is composed of four subunits that bind together to form the oxygen-carrying molecule. In the bloodstream, free hemoglobin breaks down into the individual subunits, which are small enough that they are removed by the kidneys and lost in urine. In addition, white blood cells called phagocytes ingest and remove small foreign particles, such as the hemoglobin subunits, from the blood. Hence, the amount of free hemoglobin present in the blood is halved every 4 hours. A good blood substitute, in contrast, should have a half-life of 24 to 36 hours in the blood.

Researchers have taken two different approaches in trying to circumvent the half-life problem. In one, pharmaceutical chemist C. Anthony Hunt of the University of California in San Francisco has attempted to make an artificial erythrocyte by encapsulating hemoglobin in a lipid membrane. Hunt emulsifies an aqueous solution of hemoglobin with a nonaqueous solution of cholesterol and phospholipids (fats that are made partially water soluble by the presence of a phosphate ion at one end of the polymer). During emulsification, the fats surround the hemoglobin, forming a water-soluble particle.

Hunt has successfully replaced as much as 95 percent of the blood of laboratory animals with the emulsified particles, which he calls neohemocytes. Unfortunately, however, the neohemocytes are recognized as foreign by the body and broken down within a few hours. Quest Blood Substitute, Inc., of Detroit, Michigan, has purchased Hunt's technology, and researchers at that company are working to extend the life of the neohemocytes.

The second approach has been to link the hemoglobin subunits together chemically so that they cannot come apart in the bloodstream; some researchers have even linked several hemoglobin molecules together to form a polyhemoglobin that they believe will be even more stable in blood. One leader in this field is surgeon Gerald Moss of the University of Chicago. Moss has used glutaraldehyde, a simple chemical that occurs naturally in the body, not only to link the subunits together, but also to link two or three intact hemoglobin molecules together. This approach has the additional advantage that the glutaraldehyde kills any viruses that might be present in the hemoglobin. Moss has licensed the technology to Northfield Laboratories, Inc., a small private research firm in Northfield, Illinois, that has successfully tested the polyhemoglobin in laboratory animals, including primates.

In September 1987, Northfield received permission from the FDA to begin clinical trials of the polyhemoglobin in humans. Northfield chairman Richard DeWoskin said that the polyhemoglobin should cost about $225 per unit when it reaches the market; regular blood costs patients about $150 in processing costs. He has estimated that the market for the product could be $200 million per year in the U.S. within 5 years after it is introduced. Researchers at the Letterman Army Research Institute in San Francisco have developed a similar form of polyhemoglobin, and they began tests on animals in the fall of 1987.

Both the Army researchers and Northfield have been relying on discarded blood as a source for their hemoglobin, but that may not be sufficient. "A very small percentage, 1 to 3 percent, of all 14 million units collected yearly are discarded," says Dale Smith, group vice-president in charge of blood products for Baxter-Travenol. "It's not insignificant, but not nearly enough to make a major impact."

To make up for the short supply, some researchers have been using animals as a source of hemoglobin. Biopure, Inc., of Boston, for example, has been working with highly purified hemoglobin from cows since 1984 and has shown that it is completely nontoxic in a variety of animal species. In fact, when converted into polyhemoglobin, it has proved to be an even better carrier of oxygen than the human protein. Quest Blood Substitute is also working with bovine hemoglobin, but it is not yet clear if the FDA will approve testing in humans.

Plastic Lenses

New types of soft plastics may soon replace those currently being used in artificial lenses implanted in the eyes of cataract patients, reducing the risk of long-term damage to the eye from such implants. Similar lenses can also be used as artificial corneas and perhaps reduce reliance on corneal transplants, which are often rejected.

The new materials, which are just beginning to be tested in humans, could have widespread use. In 1987, more than 1.2 million Americans — 90 percent of all cataract victims — received intraocular lens implants during cataract operations, and more than 2 million people received the lenses worldwide, according to materials scientist Eugene P. Goldberg of the University of Florida in Gainesville. In addition, about 30,000 Americans received corneal transplants in the same year following damage to their eyes due to chemical burns or immunological disorders, according to chemist Jean T. Jacob-LaBarre of the Tulane University School of Medicine in New Orleans. About 6,000 of those transplant recipients have already rejected or will reject the new corneas, she said. Because donor corneas are in short supply, about 5,000 people were on waiting lists for cornea transplants.

Polymethylmethacrylate is now used for intraocular lenses and, to a limited extent, as an artificial cornea, but it produces problems in both applications. The hard plastic intraocular lens is inserted in the eye just behind the pupil. Abrasive interactions between the lens and the fragile eye tissues can damage the endothelium (the cells at the back of the cornea). In some cases, the iris of the pupil can stick to the hard, dry surface of the plastic, becoming damaged in the process.

Goldberg devised a technique for chemically bonding a second plastic, similar to that now widely used in soft contact lenses, to the polymethylmethacrylate surface. This hydrophilic (water-loving) surface coating absorbs water and becomes soft and slippery. The iris thus does not stick to the lens, and other tissues slide smoothly over it. Tests in rabbits showed that the surface-modified material caused less than one-third of the damage caused by unmodified polymethylmethacrylate. Goldberg has licensed the technique for coating the lenses to Pharmacia Ophthalmics, Inc., of Pasadena, California, which is sponsoring clinical trials of the lenses at six U.S. medical centers. By late 1988, nearly 200 lenses had been successfully implanted, and the company expected to receive FDA approval for marketing the lenses by 1990.

Artificial corneas can produce some of the same problems as hard plastic intraocular lenses. Even more importantly, however, their rigidity frequently causes them to be sheared free from the more flexible surface of the eyeball. Jacob-LaBarre and Tulane ophthalmologist Delmar R. Caldwell have developed an artificial cornea that has apparently solved the shear problem. The center of their artificial cornea is a clear plastic lens similar to those now in use. They have devised a chemical technique for attaching to the rim of the lens a ring of porous, hydrophilic plastic with six "spokes" extending radially outward.

The artificial cornea is placed on the surface of the eye and each of the six spokes is sutured to the side of the eyeball. Tissue then grows around them and into the porous plastic to anchor the cornea. Caldwell successfully implanted the first artificial cornea in a human patient in December 1987 and had implanted two more by late 1988. He has received permission from the FDA to implant the lenses in a total of 12 patients.

Beyond Silicon

Organic Electronics

Chemical research may provide a radical new way to construct electronic components, such as the chips that serve as computer memories and the transistors that power electronic devices. Because it may eventually not be possible to continue reducing the size of electronic components produced from silicon or other semiconductors, many researchers have concluded that one way to make further reductions in component size would be to use organic chemicals. "What we would like to do is reduce the size of switching and memory components down to the size of one molecule," said chemist Forrest Carter of the Naval Research Laboratory in Washington, D.C. Such an achievement would produce memory cells a fraction of the size of the smallest now available.

In 1985, according to Carter, researchers at the IBM Thomas J. Watson Research Laboratory in Yorktown Heights, New York, constructed silicon memory chips that had a capacity of 1 megabyte — 1 million bytes — and were as small as a fingernail. Carter predicted that, within the next decade, conventional technology could be used to make a 16-megabyte chip, but that much higher densities may be difficult or impossible to achieve.

If conventional silicon devices are placed closer together on a circuit board to achieve increased memory capacity, according to physicist Phillip Seiden of the IBM Watson Research Laboratory, the components may short-circuit. As the components are moved closer together, minute electromagnetic fields created by each component impinge on nearby components, interfering with their operation or altering the stored data. Organic molecules are less susceptible to such electromagnetic interference, Seiden said, and can be stacked one atop another, allowing many more electronic components to be packed into a smaller space.

In the fastest supercomputers, the speed of operation is limited primarily by the amount of time required for an electronic signal to travel from one component to the next. Even though electronic signals travel at very high speeds, a finite amount of time is required for them to pass through the complex pathways of the computer. Organic electronic components, called molecular electronic devices (MEDs), would reduce that time by allowing components to be packed closer together. If 10 electronic components, such as transistors, could be compressed into the space formerly occupied by one, then an electronic signal would have to travel only one-tenth as far to carry out a given operation, and the new circuit would be 10 times faster. "Small means speed," according to Seiden.

A small, MED-based computer using sophisticated artificial intelligence programming might be used, for example, as the "brain" for an unmanned spacecraft. The artificial intelligence could make most, if not all, routine decisions for the craft, making it self-guiding on a long mission. Existing computers that could handle such programming are simply too large to fit into spacecraft, and consequently, spacecraft must now be controlled by humans on Earth. Such ground-based control introduces time lags that decrease the ability of a distant spacecraft to respond to unforeseen situations.

A new generation of MED-based supercomputers for weather forecasting would be able to acquire more meteorological observations than existing computers and process them more rapidly, thereby producing more reliable predictions. Many scientists have also noted that the Strategic Defense Initiative might require such enhanced supercomputers to track all the missiles and decoys that would be launched during an attack on the U.S. and to coordinate their destruction in the short time that would be available before they reached their targets.

The great potential of such computers, as well as of other electronic devices that could be made smaller and more efficient with MEDs, has prompted the government and major corporations, such as IBM, Westinghouse, and Hughes Aircraft, to invest more than $100 million per year on MED research. That figure is expected to grow to $1 billion annually by the end of the century. The governments of the U.K., West Germany, Italy, Japan, and the U.S.S.R. have also invested heavily in developing MED technology. Some experts have predicted that MED memory devices could be ready for commercial use within 2 to 10 years. Electronic apparatus incorporating MEDs, furthermore, could account for 10 percent of the total computer market by the year 2020, according to a 1986 forecast by Technical Insights, Inc., of Fort Lee, New Jersey. A 10 percent share of the market would amount to $30 billion in sales.

Chemicals similar to those used in MEDs may find use in other electronics-related applications. Researchers have, for example, recently discovered a variety of chemicals that change colors when an electric current is applied to them. Such chemicals could be used in display devices, such as indicators on electronic equipment, and could even find application in color television screens. Related chemicals could also be used to make so-called smart windows that, at the touch of a switch, will become opaque to let less sunlight shine through.

Memory devices. The principal characteristic that an organic molecule must display for use in a memory device is the ability to exist in two characteristic states, such as two different colors or two different electronic states, which can represent the "1" and the "0" of the binary code commonly used for manipulating letters and numbers in computers. Many researchers have been working with unusual molecules that can change color when exposed to light.

For example, chemist Robert Birge of Syracuse University in New York, has been studying an organic pigment called bacteriorhodopsin, which is part of the photosynthetic system of the purplish bacterium *Halobacterium halobium*. Perhaps fittingly, *H. halobium* was first isolated in the salt marshes of San Francisco Bay, close to

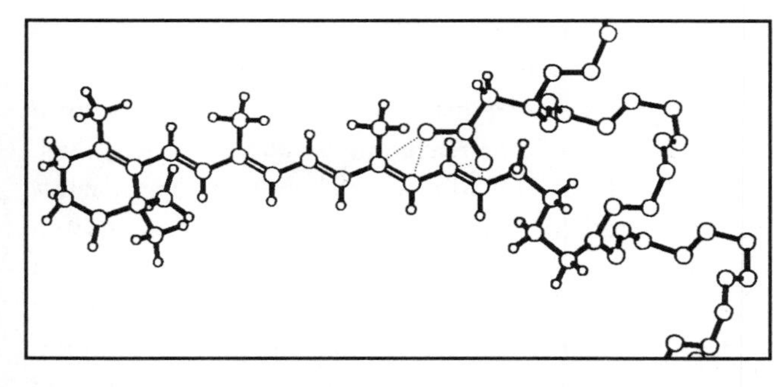

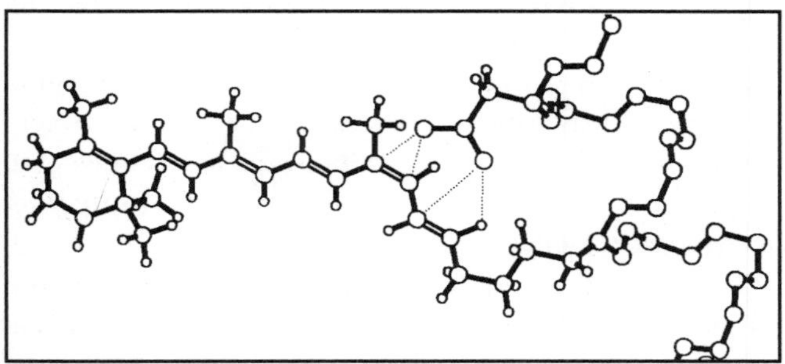

Figure 7. Promising Molecule for Optical Memory. The organic pigment bacteriorhodopsin can exist in two stable states, one that absorbs green light (top) and one that absorbs red light (bottom). Shining green light on the bacteriorhodopsin molecules in the green-absorbing state converts them to the red-absorbing state, and vice versa. Since bacteriorhodopsin can be readily converted back and forth between its two states, the molecules can be used to store data in the binary code commonly used by computers. Bacteriorhodopsin-based optical memory devices would be much faster than conventional magnetic memory devices and could store far more data in a given amount of space. *Courtesy: Robert Birge, Syracuse University.*

Silicon Valley, the home of the electronics industry. Isolated from the bacterium, bacteriorhodopsin can exist in two states: one that absorbs green light, and the other that absorbs red. When bacteriorhodopsin in the green-absorbing state is exposed to green light, the molecule's electronic configuration changes, so that it becomes red-absorbing, and vice versa.

Information can be stored in the bacteriorhodopsin molecules with two laser beams: one green and the other red. In order to store a piece of information, clusters of molecules at precise sites are irradiated with a green laser, changing the normally red molecules to green. This process is analogous to changing the magnetization of a specific site in a magnetic memory, but the cluster of bacteriorhodopsin molecules can be much smaller than a magnetic site.

The red laser can then scan the clusters to determine the color of each cluster and thereby retrieve the stored information. If the cluster is red, the red light will not be absorbed, but if the cluster is green, the light will be absorbed. A detector is used to determine each cluster's color by indicating whether or not the light has been absorbed. (Each time the red laser reads information at a site and converts the green cluster to red, the green laser is used to restore the cluster's original color so that no information is lost.)

According to Birge, the bacteriorhodopsin memory is much faster than a conventional magnetic memory. Only five trillionths of a second are required for the molecules to change color, which is about 1,000 times faster than information can be stored in or retrieved from a magnetic device.

Using this approach, Birge and physicist Rick Lawrence of Hughes Aircraft Corp. in Los Angeles, California, have succeeded in constructing a thumbnail-sized, high-density memory device; they applied 1,000 layers of bacteriorhodopsin, each layer one molecule thick, on the surface of a quartz plate. "It looks like a piece of glass with a clear, deep, rich red coating," Birge said. About 10,000 molecules are used for storing each bit of information. According to Birge, the device has a potential storage density of nearly 10 megabytes — 10 million bytes — per square centimeter, comparable to the storage density of magnetic devices used in large computers.

Birge and Lawrence said that they hope to increase the storage capacity of the device by another factor of 10, but they are faced with two primary problems: applying the molecules to the quartz plate uniformly, and controlling the position of the read and write lasers within 8 millionths of an inch — the accuracy needed to use the high storage density. The bacteriorhodopsin's two states, furthermore, are stable only at temperatures below –313°F. According to Birge, this problem can be overcome by using genetic engineering to change the molecule's structure slightly so that the red-absorbing state is more stable. Birge has predicted that the bacteriorhodopsin memories will be ready for commercial use within 2 years.

Physicist Theodore Poehler of the Johns Hopkins Applied Physics Laboratory in Laurel, Maryland, has been using a family of completely synthetic organic molecules deposited on a metal surface to construct the same type of memory device. He and chemist Richard Potember deposit a thin layer of copper or silver on a sheet of plastic or silicon, and then cover it with a layer of tetracyanoquinodimethane (TCNQ) or a TCNQ derivative and warm it gently. The warming causes a charge transfer salt to form: an atom of copper, for example, donates an electron to

TCNQ so that the copper becomes positively charged and the TCNQ becomes negatively charged. Irradiating a cluster of salt molecules with an argon laser reverses the process, forcing the electron back to the copper atom and making the molecules electrically neutral again.

If the metal is copper, irradiation causes the color of TCNQ to change from blue to pale yellow; if the metal is silver, the color changes from violet to silver. In either case, the change can be detected by determining whether or not light is absorbed by the TCNQ. The result is a memory device similar to Birge's. A memory bit can be erased by using a different laser that heats the cluster slightly, thereby causing the charge transfer salt to reform.

Seiden and IBM chemist Ari Aviram have constructed MED memories from which information is recovered electrically rather than optically. They have synthesized elongated molecules, called quinones, that undergo a structural rearrangement when a small electric field is applied; in the rearrangement, a proton moves from one end of the molecule to the other. This shift can be detected with an electrical field.

To make the memory device, the two IBM researchers deposited the molecules in a thin layer on a metal plate so that they are standing on end side by side, and attached microscopic electrical leads at the top and bottom of clusters of a few quinone molecules. When the protons are in their normal position at the bottom of the clusters, each cluster is read as a "0" in the binary code. Application of a small electric field shifts the protons to the top of the molecules, which is read as a binary "1." The presence or absence of the charge at the top of the clusters can be detected electromagnetically in much the same way that current magnetic media are read.

Both Seiden and chemist Robert Metzger of the University of Alabama have built small computer memory devices based on quinones, but they have had difficulty making them work. The primary problem that they have encountered is the difficulty of purifying the complex quinones. Unfortunately, according to Metzger, "the most exciting compounds are the hardest to purify."

Organic transistors. Chemist Mark Wrighton and his colleagues at MIT were the first to produce transistors and diodes from organic chemicals. A transistor is an electrical device that allows a current to pass from one metal electrode (the source) to a second electrode (the drain). The current flow is regulated by a third electrode (the gate), which is typically made of a semiconducting material. A semiconductor is a material whose electrical conductivity is intermediate between that of a conductor, such as copper, and an insulator, such as glass, but whose conductivity can be increased under specific conditions, such as the application of an electric field. In a transistor, the semiconductor gate normally prevents current from passing from the source to the drain; however, when a small electrical field is applied to the semiconductor, it allows a large current to pass from the source to the drain. This amplification can be used, among other things, to boost a weak radio signal so that it becomes audible.

In Wrighton's transistors, semiconducting organic polymers such as polyaniline were used as a gate between two miniature gold electrodes that serve as a source and drain. The polymers are normally insulators, so that no current passes from source to drain; they can be converted into conductors in at least two different ways. Polyanilines can be converted into a conducting state by the application of an electric field; the field initiates a chemical reaction that makes the polymer conducting, and the effect can be reversed by turning off the field. The first MED transistor that Wrighton made with this material, reported in 1984, amplified a current about 100-fold, typical of many transistors; because a chemical reaction was involved, however, the device required 10 seconds to cycle from off to on, much slower than conventional transistors.

In late 1987, Wrighton reported that he had produced similar MED transistors that cycle in about 1 millisecond (one-thousandth of a second) — a distinct improvement over his first MED transistor but still only about a millionth as fast as a silicon-based transistor. According to Wrighton, much faster MED transistors are possible, "but great speed may not be necessary for many applications. One potential use of such devices would be to interface between biological and electronic systems, and for that you don't need a lot of speed."

Many biological systems, for example, involve chemical reactions that produce a pulse of electrical current. The transistor, Wrighton said, would be ideal to amplify such currents so that they could be more readily detected and used. Such a device might be attached to nerves leading to an artificial arm so that electrical impulses from the brain could be used to control movements of the device.

The MIT researcher has developed a second type of MED transistor in which speed is less important, and which may find practical application much sooner. In this device, the gate is opened by a chemical signal rather than an electrical signal. Poly(3-methylthiophene) is normally insulating, for example, but combines with oxygen to become a conductor. When the oxygen is removed, the polymer again becomes an insulator. The conductivity of the polymer, and thus the amount of electricity that it allows to flow through the gate, is a function of the concentration of oxygen.

This device also has an on/off cycle of a few milliseconds — too slow to make a useful transistor, but sufficiently fast to make a commercially useful oxygen sensor. Wrighton and his colleagues passed an inert liquid over the transistor for 30 hours, injecting uniform pulses of oxygen into the flow at 10-minute intervals. The output from the transistor showed distinct spikes of consistent amplitude every time the oxygen was injected, and no degradation of performance was observed. The sensor produced a distinct signal even when the amount of oxygen present was too small to be detected by conventional sensors. Wrighton has developed other prototype sensors that detect carbon monoxide, acidity, and certain enzymes, and several companies have contacted him about the possibility of making commercial versions of these sensors.

Chemist Scott Rickert of Case Western Reserve University in Cleveland, Ohio, has used a slightly different technique to make sensors similar to Wrighton's. He makes thin polymer films, much like Seiden and Aviram's, that look like metal foils. "The electrical [conductivity] of these foils," Rickert said, "is very sensitive to environmental changes such as temperature, pressure, humidity, and chemical composition. As a result, they make very good chemical sensors, better than any sensor yet developed." These sensors respond to changes in the concentration of chemicals in their environment much faster than do conventional sensors, and they recover much faster than conventional sensors when the chemical being monitored is removed.

A major improvement in MED transistors was reported in late 1988 by chemist Richard Friend and his colleagues at the University of Cambridge in the U.K. Using polymer processing techniques developed at the University of Durham in the U.K., Friend made transistor gates from polyacetylene. The chief advantage of polyacetylene is that, unlike the polymers used by Wrighton, it does not have to undergo a chemical reaction to become suitable for computing. Instead, applying an electric field to the polyacetylene gate causes a rapid rearrangement of bonds in the polymer. The transistor can thus be turned on and off in a millionth of a second, which is about 1,000 times faster than Wrighton's transistor. Friend predicted that refinements could further increase the switching speed so that the devices could become competitive with silicon-based transistors.

The processed polyacetylene has an additional property that could make it useful even sooner, according to Friend. When the electric field is applied to the polymer, it becomes opaque to light. Transistorlike devices made from the polymer could thus be used to turn beams of light on and off in optical computers, which use light waves to control logic circuits.

In the most ambitious MED project, Birge has been trying to make a NAND gate from organic molecules; a NAND gate allows a computer to carry out logical functions, such as "if A and B, then C." Computers are constructed by connecting such gates together in the right sequence. Birge and his coworkers have made several such gate molecules, he said, "but they don't work reliably yet." When the devices do work, however, their switching time will be on the order of a trillionth of a second, about 1,000 times faster than comparable silicon-based gates. "A single molecular gate could be synthesized on the order of 5 nanometers (200 billionths of an inch) square," Birge said.

A NAND gate can be envisioned as a "Y." If a signal is fed into either arm of the Y, nothing happens; however, if signals are fed into both arms simultaneously, another signal is emitted from the stem. Birge's NAND gate is constructed from one complex molecule. The arms of the Y are two light-sensitive quinone derivatives, each of which absorbs light of a different color and releases an electron; one of the quinone derivatives absorbs blue light, and the other absorbs green light. The arms are chemically bonded to an electron acceptor called a cyanine dye, which is itself bonded to another dye called a protonated Schiff base. The dyes form the stem of the Y. These complex, Y-shaped molecules are very difficult to synthesize, but Carnegie Mellon chemist Jonathon Lindsey has built a machine that produces them automatically.

If Birge's NAND gate is illuminated with either blue or green light, one quinone gives up an electron to the cyanine, but nothing happens; if the NAND gate is illuminated with light of both colors, however, the cyanine receives two electrons — one from each quinone — and gives up one electron to the Schiff base, causing that dye's color to change. This color change can be monitored with a laser by shining light on the NAND gate and determining whether the light is absorbed. By early 1989, this device was still only in the experimental stage. The major difficulty, according to Birge, has been to find a way to link the individual gates together to form a complete circuit, which would be needed for use in a computer. "A single gate demonstrates the feasibility of the approach," he said, "but it doesn't do anything useful [in applications]."

Optical disks. In September 1987, Sumitomo Chemical Corp. of Osaka, Japan, and Hoechst Celanese Corp. of New York City announced a program to develop a new type of "write-once, read-many" (WORM) data disks based on dyes that vaporize when exposed to light from a laser. (For more information about WORM disks, see the article "Mass Storage Systems" in the "Computers" chapter.) As part of the program, the companies plan to build a new disk manufacturing plant in Japan by 1990.

WORM disks are used for the high-density storage of data that will be read frequently but not changed — for example, electronically recorded encyclopedias or prerecorded movies. The disks now in use are typically made from a plastic, such as polycarbonate or polyacrylate, on which a thin, inorganic coating has been deposited. Sumitomo uses primarily tellurium compounds as the inorganic coating, while other manufacturers use a combination of cobalt, dysprosium, iron, and neodymium. To write on the disks, a laser is used to etch a pit in the coating. At any given point on the disk, the presence of a pit might represent the "1" of the binary code, and the absence of a pit the "0" of the binary code. The pattern of the pits in the coating can be read with a low-power laser.

Sumitomo and Hoechst Celanese plan to use organic dyes rather than inorganic materials as a coating on the disks because they offer several potential advantages. Lasers etch a much smoother pit in the dyes than in the inorganic films, resulting in a better signal quality when the disk is read. The organic dyes can also be applied to the disks at a much lower cost and with less difficulty. Finally, the dyes permit data to be stored at a much higher density, because smaller pits can be made in the dyes. Several other companies are also developing dye-based optical disks. According to a forecast by Nomura Research Institute in Tokyo, sales of optical disks could total nearly $500 million per year in the U.S. by 1991 and $357 million in Japan, and dye-based disks could account for more than half of the total.

Erasable disks. A new type of compact disk that can be repeatedly erased and written on has been developed by researchers at Optical Data, Inc., of Beaverton, Oregon. Compact disks now in use for audio and video recordings are WORM disks because of their high storage density and signal quality. If the disks could be erased and written on, according to chemist Nancy Iwamoto of Optical Data, their use for home recordings and in personal computers could be greatly increased.

The Optical Data disks are composed of two polymer layers. The top layer is a thermoplastic — a relatively stiff material that becomes pliable when it is heated. The bottom layer is an elastomer — a rubberlike material. Both layers are dyed so that they absorb light at different wavelengths. Data are recorded on the bilayered plastic as a series of bumps that are formed by the interaction of the two layers when they are struck by pulses of heat-generating laser light. The maximum length of a bump is 1 micrometer (about 40 millionths of an inch); this small size permits very high-density storage of data.

The data are read by scanning the disk with the same laser at a lower power; the bumps scatter the light in a characteristic manner, and that scattering is detected by sensors and translated into binary code. The bumps are erased by heating them with light of a different wavelength from a second laser, which causes both layers of plastic to return to their original shapes.

The useful life of the medium is limited by a slight swelling that develops after several thousand erasures, Iwamoto said. The first applications of the new bilayered plastic should thus be in situations that require only about 1,000 read/write cycles, such as data storage in personal computers. According to Iwamoto, however, anticipated progress with the plastic should yield lifetimes far beyond the current level. (The lifetimes of magnetic media are in the millions of cycles.) Optical Data has successfully tested data-storage devices using the medium, Iwamoto said, and hopes to have commercial products available within 2 to 3 years. (For more information about optical disks, see the article "Mass Storage Systems" in the "Computers" chapter.)

Chemicals That Change Color Electronically

Imagine automobiles whose colors can be changed at the flick of a switch on the dashboard, windows that can be darkened electrically to filter out varying amounts of sunlight, or a color television screen thin enough to hang on a wall. These are some of the many potential applications of new chemicals that change color when an electric voltage is applied to them. The commercial potential of such chemicals is huge. According to the trade journal *Digital Design*, the market for flat-panel display devices alone is expected to reach $450 million in the U.S. and $2 billion worldwide by the 1990s.

Chemicals that change color under various conditions have, of course, been known for a long time. They have been used since the turn of the century for monitoring changes in the acidity of a solution, for example, or for detecting the presence of specific chemicals in analytical applications. Such chemicals typically have only two or three distinct colors. The distinguishing characteristic of the new materials is that they can pass through as many as seven different colors.

Chemist C. Michael Elliott of Colorado State University in Fort Collins has developed a color-changing polymeric compound containing the metal ruthenium. The binding of the organic polymer to each ruthenium atom stabilizes the metal in a large number of oxidation states — both those in which extra electrons have been added to each metal atom and those in which some of the electrons have been removed from each metal atom.

When the ruthenium atoms are in their normal nonionized state, the polymer is blue. When one electron is removed from each ruthenium by applying an electrical current, the polymer is purple; when two are removed, it is orange. When one electron is added to each ruthenium atom, the polymer is green-blue; when two are added, it is brown; when three are added, it is rust-colored; and when four are added, it is cherry red. By varying the chemical composition of the polymer, the color spectrum can be changed.

The color-changing polymers are very durable, Elliott said. He has deposited them on a tin oxide electrode in a layer less than one-millionth of an inch thick and cycled them through all seven oxidation states more than one million times. After one million cycles, according to Elliott, nearly 90 percent of the material remained intact, and its appearance was unchanged. The polymers also respond very fast — in only a few hundredths of a second — to changes in voltage. The principal problem with the polymers is that they are very sensitive to oxygen and water vapor, but Elliott said that this sensitivity can be eliminated.

One of the first applications of the new materials might be to replace liquid crystal displays in digital watches and similar devices, according to Elliott. Liquid crystals now have several disadvantages. For example, they can be very difficult to see if viewed from an angle, and they respond slowly to changes in voltage, thereby limiting their use in high-speed display devices. The new ruthenium polymers have none of these problems, he said. In 1988, one Japanese company was evaluating the materials for use in optical displays, and several others had expressed an interest. Many Japanese companies have been developing similar color-changing materials, according to Elliott, but have not discussed their results publicly.

Chemist Sze Cheng Yang of the University of Rhode Island in Kingston has synthesized a completely organic color-changing polymer, although the range of colors is not as broad as Elliott's polymer. Yang announced in mid-1987 that the color of a thin film of polyaniline can be switched from purple to blue to green to transparent by applying varying voltages. As with the ruthenium compounds, the sequence and identity of the colors can be altered by using different aniline derivatives to form the polymer. Yang has predicted that the polyaniline derivatives will have many of the same applications foreseen by Elliott, but he is also investigating the possibility of developing variable-colored sunglasses and ski goggles.

Industrial Materials from the Laboratory

Radar-Absorbing Chemical

The study of rhodopsin and other chemicals that enable humans to perceive light has led to the discovery of retinyl Schiff base salts that absorb radar waves. These salts may prove valuable in the production of so-called stealth aircraft, such as the F-19 Stealth fighter or the Stealth Advanced Technology Bomber now under development, which are not readily detected by radar. The discovery may be particularly valuable because it comes at a time when Soviet scientists have apparently developed techniques to defeat the radar-absorbing coatings currently used on stealth aircraft. Experts believe that, within 3 years, a family of paints using the retinyl Schiff base salts could be developed to render aircraft, missiles, ships, and tanks invisible to radar.

Chemist Robert R. Birge and his colleagues at Carnegie Mellon University discovered the retinyl Schiff base salts while trying to find relatively simple chemicals that mimic the action of pigments in the eye, such as rhodopsin. Rhodopsin is present in the rod-shaped cells in the retina and undergoes a structural rearrangement when struck by a photon. The rearrangement initiates a cascade of reactions that ultimately result in a signal reaching the brain. The retinyl Schiff base salts, which are black and physically resemble graphite, undergo a similar structural rearrangement in response to electromagnetic waves at radar wavelengths rather than at light wavelengths. In a fraction of a second, the chemicals revert to their original structure, releasing the absorbed energy in the form of heat.

A much different technology — or rather, series of technologies — is used in current stealth aircraft. First, such aircraft are designed so that they have no sharp corners or angles, which reflect radar waves intensely; all their contours are designed as smooth, continuous curves. The F-19, for example, is a manta ray-shaped aircraft in which the wings, fuselage, engine inlets, and control surfaces are smoothly merged. Second, as much of the plane as possible is constructed of plastics and other materials that do not reflect radar waves as strongly as metal. Finally, the airplane is coated with iron oxide ceramics, called ferrites, that further reduce radar reflection.

A typical antiradar coating consists of three layers: two radar-reflecting ferrite layers sandwiching an inner layer called a dielectric. Radar waves pass through the outer layer and are reflected between the innermost and outermost layers. This allows the incoming waves and the reflected outgoing waves to cancel each other, much as the ripples from two stones thrown into a pond may cancel each other when they meet. The dielectric between the two layers helps trap the radar waves long enough for cancellation to occur.

Antiradar coatings can cancel only radar of a specific wavelength, which is determined by the distance between the two ferrite layers. Multiple coatings or a combination of ferrite coatings and structures are used to defeat a range of radar transmission wavelengths, but the weight of the coatings makes it impractical to build an airplane that cancels all possible wavelengths. American experts now believe that Soviet scientists have developed variable wavelength radar units that scan through all wavelengths until they find one at which the plane can be detected. Such radar devices would render ferrite-based coatings ineffective.

A mixture of different Schiff base salts could be used to absorb radar of all wavelengths without an appreciable weight penalty, according to Birge. The principal problem with the materials is that they do not dissolve well in the solvents normally used to prepare paints. Nonetheless, Birge said that the production of a commercial paint that would absorb at least 80 percent of incoming radar waves could be accomplished at a cost of about $3 million. The cost of coating a fighter plane with the paint would be about $30,000.

Birge initially reported his discovery in a little-noticed paper in the 1 April 1987 *Journal of the American Chemistry Society*. On 18 May, *Aviation Week and Space Technology*, a highly respected weekly magazine that covers the aerospace industry, ran a story discussing the potential application of Birge's findings to the stealth aircraft program. After the story appeared, Birge told the *New York Times* that he was asked by "half a dozen government agencies" to say nothing more about his discovery. Two government agencies, which he declined to name, have decided to finance research on the radar-absorbing salts, but Birge said that he will not participate in it because he is already involved in research on the biochemistry of vision and has adequate funds to continue his work.

Stronger Ceramics

Metallurgist Mark Newkirk of the Lanxide Corp. in Newark, Delaware, has developed a new technique for manufacturing ceramic materials composed of a mixture of a metal and its oxide. The term "ceramics" originally referred to articles prepared from pliable, earthy materials that were made rigid by high-temperature treatment. In the aerospace industry, the term now refers to strong, highly heat-resistant alloys made by mixing, pressing, and then baking a combination of a metal and its oxide.

Metal oxides are commonly encountered as the iron oxide (rust) on steel surfaces or the aluminum oxide film on aluminum sheets. Such oxides, like conventional ceramics, are normally brittle and tend to fracture or shatter under stress. Newkirk's alloys, in contrast, are much tougher and lighter, as well as much less expensive to produce than the ceramics now used in the aerospace industry. The new ceramics thus appear to have potential for use in jet engines and armor plating for aircraft and military vehicles.

Newkirk said that he has produced hundreds of the new materials, which he calls lanxides, by reacting a molten metal with a gaseous oxidizing agent, such as the oxygen in air. Typically, the molten metal must be "doped" (impregnated) with trace quantities of at least two materials, and the temperature of the metal must be kept within strict limits.

For the production of aluminum lanxides — a composite of aluminum and aluminum oxide — magnesium and silicon are effective dopants, and the temperature must be about 2,300°F. Aluminum lanxide is a honeycomb structure of aluminum oxide in which the pores are filled with aluminum. The honeycomb structure forms on the metal surface and can reach a thickness of one inch. The physical properties of the material can be altered by varying the temperature or quantity of the molten metal.

Materials scientist Rustum Roy at Pennsylvania State University has evaluated some of the new materials at the request of the U.S. Department of Defense, which has funded Newkirk's research for several years. According to Roy, Newkirk's method of producing lanxides may rank higher in importance than the discovery of the transistor. Further details about the synthesis and about the properties of other lanxides were scheduled to be published during 1986, but the report never appeared and Newkirk has refused to talk about the materials in public.

In late 1987, however, Lanxide Corp. and Du Pont formed a joint venture to develop the new ceramics. Du Pont will work on applications in aerospace equipment, turbine engines, and heat exchangers, while Lanxide will work on ballistic applications such as armor. Du Pont foresees a multibillion-dollar market for the new technology.

Diamonds

Humans have had an eternal fascination with diamonds — the clear crystalline form of carbon. Few natural substances can match their beauty, and even fewer can match their durability. The diamond is one of the hardest, most chemically resistant substances known to man, and those properties make it ideal for many industrial applications. But because diamonds are rare in nature, they are too expensive for most industrial uses.

Naturally occurring diamonds are thought to have been produced by the compression of organic materials at very high temperatures and pressures far below the Earth's surface. When General Electric Co. scientists first produced synthetic diamonds in the laboratory in 1954, they achieved the feat by mimicking nature. The scientists placed carbon in 10-story-tall presses to crush it under a pressure of 1 million pounds per square inch and simultaneously heated it to temperatures as high as 2,500°F. After a few days of pressure and heat, small diamonds were formed. This process is still used in several countries to produce industrial diamonds, but the resulting diamonds are extremely expensive, and scientists continue to look for alternative ways to manufacture them.

Diamond films. A diamond film would be an ideal coating for many industrial materials. It has unmatched hardness and Teflon's slipperiness, carries heat better than gold or copper, and can be made to conduct electricity. Ball bearings coated with a thin layer of diamond, for example, could outlast conventional bearings and would work well in high-temperature environments where conventional lubricants break down. Diamond-coated cutting instruments, such as scalpels and scissors, could be brought to a finer edge and stay sharp for a lifetime. And industrial machine tools coated with a diamond film would last longer and perform better than those coated with flecks of industrial diamond.

In 1977, B. V. Derjaguin and his colleagues at the Institute for Physical Chemistry in Moscow, U.S.S.R., first described a technique for making diamond films by heating methane and atomic hydrogen in the absence of oxygen. The process produced an extremely thin film of diamond on the surface of objects in the chamber where the heating took place. Most Western scientists did not believe that the Soviets were forming true diamond, even when Japanese scientists reportedly replicated the process 5 years later.

In 1986, however, a team of researchers at Pennsylvania State University headed by chemists Russell Messier and Karl Spear reported that they too had been able to make diamond films. The team mixed a small volume of methane gas with a large volume of hydrogen and passed it through a powerful beam of microwaves to create a plasma, a hot gas in which all the atoms are ionized. When the plasma was passed over a cooler solid, the gases condensed to form a diamond film. According to Messier, with this technique, "we should be able to put a protective diamond film on virtually any solid object." A thin diamond film could be applied to the lens of a camera at only a slight increase in price, for instance, to make it scratchproof. It could even be applied to the surface of inexpensive gemstones to make them more durable and attractive.

Perhaps the greatest potential for diamond films lies in electronics. Conventional silicon chips generate large amounts of heat while they are operating, so individual components must be separated by relatively large spaces to prevent overheating and melting. Because diamond sheds heat much more quickly than silicon, electronic components layered onto a diamond film could be packed more tightly together. They could also be used in extremely hot environments, such as jet engines or nuclear reactors, where silicon does not work.

Diamond can also withstand higher voltages than silicon, so chips manufactured on a diamond film could replace conventional components in microwave communication devices, where the voltage is too high for silicon chips. Finally, unlike silicon chips, diamond semiconductors can continue operating after they have been exposed to radiation, such as that produced by atom bombs. Most communications devices that use silicon chips would not operate for days or weeks after exposure to radiation from exploding warheads. By early 1989, over 20 U.S. companies were working to develop diamond-coated products.

Rediscovering a lost process. Using a process described in a letter in the 24 August 1905 *Nature*, chemists at Virginia Polytechnic Institute and State University in Blacksburg have created microscopic particles that they believe are diamonds. The process is unusual in that it does not require the high pressures normally associated with diamond production. Chemist Felix Sebba of Virginia Polytechnic stumbled across the process in a book on gemstones published in 1908. The book cites the *Nature* letter, in which British astrophysicist Charles Burton claimed to have synthesized diamonds by an unusual method. Further checking by Sebba showed that the letter had not been abstracted in any of the existing abstract journals, and that it had not been cited in any subsequent book on carbon chemistry. In short, the paper seemed to have been both ignored and forgotten.

Burton's diamond-producing method involved dissolving carbon in a lead-calcium alloy heated to about 990°F under normal atmospheric pressure. At that temperature, the carbon is more soluble in the alloy than in lead alone. Burton then reacted the calcium with steam to form a volatile product that escaped from the alloy, leaving a supersaturated solution of carbon in lead. The carbon then precipitated from the lead in the form of diamonds. In a sense, the process is similar to dissolving large amounts of sugar in hot water to form a supersaturated solution, from which sugar crystals precipitate when the solution is cooled. Burton's method differed, however, in that it did not require cooling the solution of carbon in lead.

Sebba and his research associate, N. Sugarman, attempted to reproduce Burton's synthesis, even though the experimental details in the original paper were sketchy. They succeeded in obtaining a black powder — presumably graphite, the most stable crystalline form of carbon — in which were embedded "many transparent crystals which scintillated with considerable fire in reflected light." Preliminary evidence indicated that these crystals were diamonds. According to Sebba, "there is a strong presumption that...Burton had synthesized diamonds in 1905, half a century before General Electric." In 1989, Sebba was trying to refine the process to produce larger diamond fragments.

Diamonds from explosions. Scientists from the Center for Explosives Technology at the New Mexico Institute of Mining and Technology in Socorro announced in early 1987 that they had used explosives to create diamonds that are 85 percent as hard as natural diamonds — a new record for synthetic diamonds. Visiting professor Akira Sawaoka from the Tokyo Institute of Technology and his coworkers placed diamond powder in a stainless steel capsule and then used explosives to drive an iron plate against the capsule. The explosion produced a shock wave with a pressure of nearly 1 million times atmospheric pressure; this shock wave compacted and heated the powder, forming larger diamonds. The diamonds are of industrial quality, according to Sawaoka, and could be used for numerous applications, such as coating machine parts for increased durability.

The production of diamonds is one of many projects at the 4-year-old Center for Explosives Technology, where researchers use powerful explosives to produce unusual materials. The key to their success is fashioning explosives into a special shape that will achieve the desired results. When a conventionally shaped explosive detonates, it generates a roughly spherical shock wave that expands more or less uniformly in all directions. Such a shock wave shatters, scatters, and distorts objects in its path. To be useful for producing materials, a shock wave must be planar, much like the flat face of a swiftly moving hammer head.

Researchers at the center create planar shock waves by molding a thin cone of a fast-detonating explosive around a cone of a slower-detonating explosive. When this double cone is detonated at its tip, the shock wave passing around the outer cone must travel farther to reach the base than the wave moving straight through the inner cone. But since the explosive in the outer cone detonates faster, both waves actually reach the base at the same time. Moreover, the shock wave from the outer cone confines the shock wave from the inner cone so that all the explosive force is directed against the base of the cone, thereby generating a planar wave.

Antirubber

One of nature's most familiar and seemingly immutable laws — that materials get thinner when they are stretched or bulge out when they are squeezed — has been overturned by a biomedical engineer at the University of Iowa in Iowa City. Roderic Lakes has created what might be termed "antirubber," a perverse material that behaves in a manner contrary to intuition. When stretched in one direction, the material expands in every other direction. When

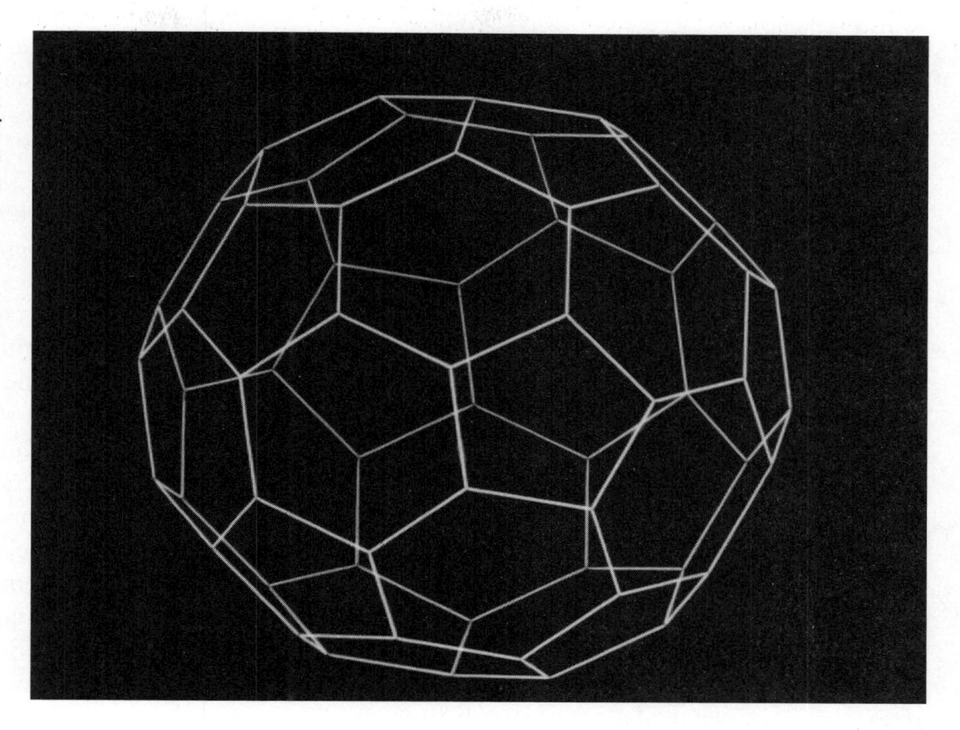

Figure 8. Buckminsterfullerene. During 1985, Rice University chemist Richard Smalley and his colleagues reported the discovery of a molecule composed of 60 carbon atoms linked together in a shape identical to that of a soccerball. The molecule, pictured here in a computer graphics model, has been named buckminsterfullerene, because the molecule resembles the geodesic domes designed by the American architect R. Buckminster Fuller. *Courtesy: Florante A. Quiocho, Rice University.*

squeezed in one direction, it contracts in every other direction. Lakes thinks that antirubber might find use in such applications as sponges, shock absorbers, air filters, and Velcro-like fasteners.

The tendency of a material to shrink in one direction when it is stretched in another is quantified by a number known as Poisson's ratio. Rubber and biological tissues have a high Poisson's ratio, meaning that they become much thinner when stretched. Metals have a somewhat lower, but nonetheless appreciable, Poisson's ratio. In contrast, the ratio for cork is nearly zero, making it an ideal material for sealing wine and other bottles — it does not bulge out while being removed with a corkscrew. If cork had a higher Poisson's ratio, pressure from the interior of the bottle would cause it to expand and jam in the bottle's neck.

Materials with a negative Poisson's ratio are theoretically possible, but were unknown before Lakes's experiments. In thinking about what would be required for a negative Poisson's ratio, Lakes envisioned a cube in which each face was puckered inward. Mechanical models demonstrated that pulling on any two faces would cause the other faces to stretch outward as well, and Lakes set out to make a real material with such a structure.

Lakes took a conventional polymer foam such as that used in mattresses, compressed it in all three directions, and then heated it to a temperature slightly above its softening point. The mold was then cooled to room temperature, and the foam was released. He found that the individual air cells in the foam each had a shape similar to his model cube, and that the foam had a large negative Poisson's ratio. Lakes has been attempting to make metal structures with similar properties.

Buckminsterfullerene

Large molecules containing only carbon atoms are known to exist around certain types of carbon-rich stars and in interstellar dust, but the structures of these molecules have not been determined. Similar molecules are also thought to be present in soot formed during the incomplete combustion of organic materials. In experiments designed to mimic the conditions under which such molecules could be formed, chemist Richard Smalley and his colleagues at Rice University have identified an unusual new molecule containing only carbon atoms; the researchers speculated that it may be fairly common throughout the universe. Other molecules identified in the experiments could also form the basis of novel materials that would have the strength of graphite but not its brittleness.

In 1985, Smalley and his colleagues reported that they had produced the large molecules by vaporizing material from a graphite surface with a laser. The vaporized material was then cooled by expansion and, while still in gaseous form, passed into a mass spectrometer for analysis. The mass spectrometric data showed two different types of molecules. The most common group had a "magic number" of carbon atoms (11, 15, 19, or 23). The second group had between 40 and 70 carbon atoms per molecule. Researchers have been especially fascinated by the molecule containing 60 carbon atoms because of its unusual structure.

According to Smalley's group, the most probable structure of the 60-atom molecule was a truncated icosahedron — a hollow, spherical object with 32 faces, 12 of them pentagons and the rest hexagons. Each pentagon is surrounded by five hexagons, and each hexagon is bordered by three pentagons and three hexagons. This structure is identical to that of a soccer ball, and Smalley's group originally considered such names as ballene, spherene, soccerene, and carbosoccer for the molecule. They eventually settled on the name "buckminsterfullerene" because of the structure's resemblance to the geodesic domes designed by American architect R. Buckminster Fuller.

How was such an apparently complicated structure formed? Smalley's group noted that graphite normally exists as two-dimensional sheets of interlinked carbon hexagons — a structure that resembles chicken wire. The sheets form stacks, but the individual layers slide across each other easily, giving graphite its characteristic slipperiness. When heated with a laser, the layers separate, and many carbon-carbon bonds within each layer are broken, leaving the atoms at the edges of the fragments highly reactive. These edges ultimately wrap around and link up with other edges on the same fragment, forming icosahedrons.

Caged atoms. Smalley and his group learned that, when they vaporized graphite impregnated with ions of elements such as lanthanum or calcium, they obtained molecules in which the carbon icosahedron enclosed a single metal ion. For reasons not yet known, icosahedrons containing iron could not be formed.

Not everyone agrees that the materials identified by Smalley are composed of a metal ion inside buckminsterfullerene, however. Chemist Andrew Kaldor and his colleagues at Exxon Research and Engineering Co. reported in mid-1987 that they had observed similar clusters containing one, two, or three metal atoms. They argued that the evidence suggests that the metal atoms are attached to the edge or face of graphite flakes, and are not enclosed inside buckminsterfullerene. By early 1989, both groups were trying to obtain further structural evidence to support their theories.

If the 60-carbon molecules can be synthesized on a large scale, according to Smalley, "the chemical and practical value of the substances may prove extremely high." If the metal ion in the center were uranium, for example, the substance would be as easy to cut as butter, he said. If each carbon atom were linked to a fluorine atom — as is the case in Teflon — the material might be a superlubricant.

Many other carbon clusters containing an even number of carbon atoms between 40 and 80 may also be stable and roughly spherical in shape, according to Smalley. He speculated, for example, that a 70-carbon molecule his group analyzed with the mass spectrometer probably has an extra band of hexagons around the middle to give the molecule an egg-shaped structure. Theoretical studies conducted by chemist H. W. Kroto of the University of Sussex in the U.K. support that possibility. Kroto has said that a 50-carbon molecule is probably also spherical. In addition, theoretical chemist A. D. J. Haymet of the University of California at Berkeley has predicted that a molecule containing 120 carbon atoms would be stable and roughly spherical; such a molecule has not yet been observed, however.

Magic numbers. The structure of the clusters containing the four "magic numbers" of carbon atoms has been more difficult to determine. But in late 1987, materials scientist James Van Vechten and chemist Douglas Kesler of Oregon State University in Corvallis proposed a structure that would explain all four clusters. They discovered the structures inadvertently when they were attempting to determine the structure of the fine "whiskers" of carbon that they had made by sputtering — bombarding a graphite target with ions. Electron microscopy had shown that the whiskers were crystalline but that they had the structure of neither diamond nor graphite.

Van Vechten and Kesler concluded that the only feasible structure for the molecules in the whiskers was an 11-atom "paddlewheel" structure consisting of two "hub" atoms along an axle surrounded by three "paddles," each containing an additional three carbon atoms. They also concluded that this structure could provide an explanation of the very prominent 11-atom clusters formed during laser vaporization. In addition, the paddlewheel structure would account for the other "magic numbers" because each addition of four carbon atoms would permit a stable, graphite-like six-membered ring to form along the side of one paddle.

Van Vechten is attempting to grow carbon crystals in which the 11-atom structure is stacked into a honeycomb pattern, interlocked by carbon chains. The interlocking structure would prevent the honeycomb planes from slipping past each other, which is the cause of graphite's brittleness as well as its slipperiness. Such a material, according to Van Vechten, could have 10 times the tensile strength of titanium with only a third of its density. The electronic properties of the molecule, he added, suggest that it might even be superconducting under certain conditions, which could make it useful for electronics applications.

New State of Matter

A new state of matter that may have unusual electronic and other properties valuable to industry has been discovered by electron microscopist Dan Schechtman of the Israel Institute of Technology-Technion in Haifa, Israel, and physicists Ilan Blech, Denis Gratias, and John Cahn of the National Institute of Standards and Technology in Gaithersburg, Maryland. The theoretical underpinning for their discovery was developed independently by theoretical physicists Paul Steinhardt and Dov Levine of the University of Pennsylvania. Both the discovery and the theoretical explanation were discussed for the first time at the March 1985 meeting of the American Physical Society in Baltimore, Maryland. By early 1989, researchers at a large number of laboratories around the world were studying the new state of matter but still had only a limited understanding of its structural properties.

Solid matter normally exists in one of two forms: amorphous or crystalline. In an amorphous solid like flour, the atoms (or molecules) are arranged in a completely random fashion, while in a crystalline solid like salt or sugar, they are arranged in a repetitive array. All crystals are composed of a basic unit, called the unit cell. In many metals, for example, the unit cell is a body-centered cubic cell, in which eight atoms are arranged at the corners of a cube and a ninth atom is located at the center. This unit cell is repeated in a regular manner in all three directions throughout the crystal.

The most important feature of the new state of matter, which Steinhardt calls "quasicrystalline," is that two different unit cells are combined in a repeating, nonperiodic way to produce a structure that is neither crystalline nor amorphous. The distinguishing characteristic of quasicrystals is the unusual symmetry that they display during electron diffraction analysis.

Unusual symmetry. While Schechtman was on sabbatical in Gaithersburg, he and his colleagues inadvertently discovered the new state of matter while trying to produce an alloy of aluminum and manganese that they hoped would be unusually strong and light. Normally, aluminum and manganese cannot be combined into an alloy, but the researchers were experimenting with a new approach that entailed melting a mixture of the two metals and then cooling them extremely rapidly to form a metallic glass.

When Schechtman studied the new metallic glass by examining its electron diffraction pattern — a pattern of dots produced on photographic film by a beam of electrons passed through the material — he found an unusual fivefold symmetry that does not occur in an ordered crystalline structure. In fivefold symmetry, the dots are arrayed as though they are at the apexes of concentric pentagons. The fact that the diffraction pattern consisted of discrete dots rather than the rings or other shapes characteristic of amorphous solids indicated that the atomic arrangement did have some order. The researchers were unable to explain their results until the diffraction pattern was observed by Steinhardt, who recognized that it matched the one that he and Levine had previously predicted for a particular theoretical quasicrystal.

The problem that confronted the scientists is easier to understand when visualized in two dimensions. Consider the task of trying to cover a floor with tiles of equal size and shape. If the tiles also have to be the same length on all sides, then only three tile shapes can be used: triangles, squares, and hexagons. No other shape will work. Pentagons, for example, cannot be fitted together without leaving spaces between the adjacent tiles.

The problem is similar when it is extended into three dimensions. Tetrahedrons, which have four triangular faces, and cubes can be used to fill a volume, but no three-dimensional shape with a fivefold symmetry can be used. The simplest such shape is an icosahedron, a highly symmetrical solid with 20 triangular faces, 12 vertices, and 30 edges. Icosahedra cannot be packed together in a repetitive manner to fill a volume because, just like the pentagon in two dimensions, spaces are left between adjoining units. In their attempts to explain the unusual three-dimensional symmetry, Steinhardt and Levine extrapolated from the solution to the two-dimensional problem.

Penrose tiling. Steinhardt and Levine's theoretical work was inspired by the work of mathematician Roger Penrose of Oxford University in Cambridge, U.K. In 1974, Penrose had shown that a two-dimensional plane could be covered completely in a nonperiodic way by using two types of polygonal structures. The result looked much like the spiraling, dovetailed patterns made famous by the painter M. C. Escher. Steinhardt and Levine demonstrated in 1984 that a three-dimensional volume could also be filled in a similar, nonperiodic way.

The Pennsylvania researchers had shown that quasicrystals can be based on many different unit cells made of two components. One such theoretical quasicrystal was constructed of icosohedra, and its theoretical electron diffraction pattern was identical to that observed by Schechtman and his colleagues. Steinhardt and Levine reported that the icosahedron cell they observed could be broken down further into building blocks of two types: one was a kind of "squashed" cube and the other was a "stretched" cube. Because of their unusual shapes, neither of these "cubes" alone could be assembled into any periodic structure; yet they could be fitted together to form an icosahedron cell.

By early 1989, as many as 30 other quasicrystals had been discovered, according to Steinhardt, and researchers at several institutions had devised at least five new methods for producing them. Some of the quasicrystals contained three or four elements, compared to the two elements — aluminum and manganese — in the original quasicrystals. Most of the newer quasicrystals had the same icosahedral structure originally observed by Schechtman, but several had previously unobserved structures. The scientists are working to improve their understanding of the properties and potential uses of these materials.

Hard Disk Drive:
Media Type: Fixed
Cylinders: 977
Heads: 5
Sectors: 17
Extended
Disable
Backlight: 0 Minutes

Chapter 5

Computers

Anatomy of a Computer

Computing Yesterday and Today

The fields of computer science and technology have grown at a remarkable pace; in just 40 years, the computer has evolved from the ENIAC, a 30-ton, 2-story-high marvel covering 15,000 square feet, to today's lightweight, portable machines that can fit inside a briefcase and yet compute many times faster than their ancestor. The cost of that computing power has also dropped dramatically as new technologies have made it possible to pack thousands of circuits and components onto chips smaller than a fingernail. Several years ago, one computer analyst calculated that if automobiles' prices and efficiencies had improved at the same rate as those of computers, a Rolls-Royce would cost $2.70 and would get 2 million miles to the gallon.

Computer historians refer to four generations of computing, a classification scheme derived from the types of electronic devices used to process data. Each generation represents an increase in computing power and a decrease in the size of the equipment.

First generation. Many history books cite engineer J. Presper Eckert and physicist John Mauchly of the University of Pennsylvania in Philadelphia as the creators of the first computer. During World War II, the two scientists approached the U.S. Army with an idea for a device that would automate the computation of gunnery coordinates, a task that otherwise required laborious hand calculations. The government quickly supported the project, and in 1946 Eckert and Mauchly demonstrated the first large-scale, general-purpose computer, ENIAC, which stands for Electronic Numerical Integrator And Calculator.

The story, however, is not so simple. In Germany during the mid-1930s, Konrad Zuse built a general-purpose computer based on electromechanical relays. He filed for a U.S. patent in 1938, but was denied one because his description of the device was not specific enough. In the U.S., physicist John V. Atanasoff claims his ABC device, built in 1939, entitled him to be credited as the father of the computer age. (ABC stands for the Atanasoff-Berry Computer, named for Atanasoff and Clifford Berry, who also contributed to the development of the device.) The courts agreed, and Atanasoff gained the legal right to be called the inventor of the digital computer. The court case has failed to resolve the matter, however, as supporters of both sides still vigorously debate the issue. But in point of fact, it was ENIAC that became the basis for the field of modern computer science.

While ENIAC was designed for military use, it was hardly a portable device that could be used on the battlefield or on ships and planes. In fact, ENIAC could have held a boxcar with room to spare, and it required 150,000 watts to power its 18,000 vacuum tubes and tens of thousands of components. Capable of performing 5,000 calculations per second, it could do in several minutes tasks that had required many hours of human effort. Still, it differed markedly from modern computers because operators had to reset switches and actually alter the wiring whenever they wanted the machine to perform a new type of calculation. This severely limited ENIAC's usefulness as a general computing device.

A solution to the problem came from mathematician John Von Neumann, who suggested that both the data and the program instructions for manipulating data could reside in a computer at the same time, thus eliminating the need to rewire the machine for each new application. Von Neumann's ideas, which presaged the concept of modern software, paved the way for more versatile and useful computers.

The first commercial computer to make use of stored programs was UNIVAC, also the brainchild of Eckert and Mauchly. Realizing the business potential of a machine that could store payroll and other data, Eckert and Mauchly formed their own company, the Electronic Control Corp., which Remington Rand purchased in 1950.

Eckert and Mauchly envisioned that corporations would use UNIVAC to automate boring and time-consuming computational tasks. Considering the computer's popularity today, the business world was surprisingly unreceptive to UNIVAC, although the U.S. Census Bureau did begin using it in 1951. The machine was at first considered something of a novelty. In 1952, for example, UNIVAC gained national attention when it predicted (less than an hour after the polls had closed) that Dwight D. Eisenhower would be the winner of the presidential elections with 438 electoral votes. Eisenhower actually defeated his opponent, Adlai Stevenson, by winning 432 electoral votes. In 1954, the General Electric Co. finally became the first major company to purchase a UNIVAC to handle its accounting functions. Later in the 1950s, other companies followed suit, using UNIVAC and other computers to automate their bookkeeping and manage their inventory.

Second generation. The first generation of computers suffered from several major drawbacks: they were enormous, used cumbersome vacuum tubes, consumed tremendous amounts of electricity, generated a great deal of waste heat (which made them require expensive air conditioning systems), and were generally unreliable. Constant breakdowns ran up costly repair bills.

The second generation of computers, which used transistors, eliminated many of these problems. Transistors, invented in 1947 by William Shockley and colleagues at Bell Laboratories, performed the same function as vacuum tubes, but took up much less space and cost much less. They consumed considerably less energy, produced far less heat, and malfunctioned less often.

Second-generation computers also introduced magnetic cores (doughnut-shaped magnets) for internal memory, replacing the magnetic drums used by first-generation computers. The magnetic cores outperformed the earlier drums, which had to be mechanically turned to access data. Second-generation computers were the first to use tape and disk media for secondary storage, a major improvement over the punched cards used to enter and store data in the first-generation machines. Other improvements included the grouping of like components, a modular design that made for easier repairs, and the use of powerful programming languages.

Throughout the 1960s, a half dozen companies — Sperry, International Business Machines Corp. (IBM), Honeywell, Inc., Control Data Corp., and others — began manufacturing all-transistor computers that purchasers could use to automate a variety of tasks, from accounting and inventory management to airline reservations.

Third generation. While the transistors of the second generation represented a vast improvement over the vacuum tubes of the first, the integrated circuits (ICs) of the third generation produced an even greater advancement. Integrated circuits, invented in the late 1950s by Jack Kilby of Texas Instruments, Inc., and Robert Noyce of Fairchild Semiconductor, combined whole clusters of interconnected transistors and other components onto silicon chips. This eliminated one of the main problems of the second-generation computers—extensive wiring between components. It also meant that a computer manufacturer could pack more computing power into much less space.

The epitome of the third-generation machines was the IBM System 360, which was introduced in 1964 as the first family of computers designed to allow users to upgrade their hardware as their data-processing needs increased. IBM's six System 360 computers were upwardly compatible, which meant that programs written for the low-end model could also run on the high-end model. Before the System 360, whenever users purchased additional computing power they had to write new programs or convert old ones, both of which are costly, time-consuming tasks. The System 360 catapulted IBM to its position as the world leader in computers.

Another significant third-generation development came from Digital Equipment Corp. (DEC), which introduced the minicomputers, also called "minis," in 1965. The computer carried an attractive price tag of only $18,000 — far less than mainframe computers. Before the advent of the minicomputer, many computer users participated in time-sharing services that permitted them to draw on the power of a central mainframe computer, thus avoiding the high cost of purchasing and maintaining their own systems. When minis became available, more users could afford their own machines. As minicomputers caught on, other companies followed DEC's lead and began making them.

Fourth generation. Whereas each of the first three generations of computers involved striking advances in engineering and capabilities, the fourth generation represents more of a refinement of the IC and a dramatic drop in price. The original ICs, classified as small-scale integration (SSI) chips, contained about 100 components. Improvements led to medium-scale integration (MSI) chips containing 100 to 1,000 components, and large-scale integration (LSI) chips containing 1,000 to 10,000 components. Very large-scale integration chips with more than 10,000 components have now become available, as well as ultra-large-scale integration (ULSI) chips that have more than 100,000 components.

The ability to pack more components onto a chip led to the development of the microprocessor, which allows a manufacturer to fit the entire circuitry of a computer on a single chip at low cost. Intel Corp. manufactured the first microprocessor chip in 1971, selling it for $200, less than half the cost of the first Texas Instruments IC, even though the new product contained more than 20 times the number of components.

The microprocessor made possible the microcomputer, an inexpensive machine that made computing affordable to individuals. In the mid-1970s, Altair, Radio Shack, and Apple Computer, Inc., began offering microcomputers, creating what evolved into the multibillion-dollar personal computer market of the 1980s. At the same time that the LSI circuits made personal computers a reality, they also enhanced the power of the mainframes, paving the way for supercomputers, such as the Cray-1, which can process 80 million instructions per second (MIPS).

Toward the fifth generation. In 1981, the government of Japan sponsored a conference on what it called fifth-generation computing. Designers of fifth-generation computers will base their products on artificial intelligence. This generation will respond to natural language and make decisions that will be useful in manufacturing and other fields. The fifth-generation effort, largely funded by the Japanese government, was undertaken to make Japan a leading developer of computer technology by the early 1990s. By 1989, however, the project had produced few concrete results.

Central Processing Unit

Today, the word computer connotes a wide variety of devices, ranging from inexpensive personal computers to supercomputers that cost millions of dollars. Regardless of their differences in size and power, however, all computers share a number of basic features. A microcomputer costs much less, occupies less space, and bears little physical resemblance to its larger cousins, but even so, like minis and mainframes, microcomputers have a central processing unit that manipulates data and devices for getting data into and out of the machine.

The Central Processing Unit (CPU) performs operations on data. In a sense, it serves as the brain of the computer, the only part that can actually alter data. At its heart, the CPU consists of an arithmetic and logic unit (ALU) and an operational control unit (OCU). The ALU contains circuits that perform arithmetic computations, such as adding and multiplying, and logical operations, such as comparing two numbers or strings of characters.

An ALU carries out instructions that define an operation performed on an operand (the data being processed), which resides at a particular address in the computer's memory. The OCU gets instructions from the memory, breaks them down into simple steps for the ALU to execute, and thereby controls the overall flow of processing.

In addition to the ALU and OCU, the CPU also contains a small amount of high-speed memory, called registers, that holds the data and instructions with which the CPU is working, as well as intermediate results of the CPU's computations.

Other components sometimes associated with the CPU include additional microprocessors programmed for specific purposes, such as supervising the operation of a printer, or for solving a particular type of math problem. Additional processing and control circuits may relieve the main processor of some of the repetitive housekeeping chores of computer operation, so that it can concentrate on the more important computations.

Several specifications describe the performance of a CPU. Processing speed is given in instructions per second. This number, however, provides only a rough estimate of computer performance, because different manufacturers use different testing procedures, different computers are often not directly comparable, and a given computer's processing speed may be higher or lower during particular applications.

The amount of memory a computer can interact with — address — also determines its power. The CPUs of most personal computers built before 1985 could address about 1 megabyte. (A byte is the equivalent of 8 bits, the minimum number of bits generally required to describe a single alphanumeric character. A megabyte is 1 million bytes.) Some new personal computers that use 32-bit microprocessors can address many times that amount. The Motorola 68000 microprocessor, which is the CPU of the Apple Macintosh personal computer, can address 16 megabytes. Some 32-bit microprocessors introduced in the late 1980s address up to 4 gigabytes (4,000 megabytes).

Evolution of microcomputers. In the world of microprocessors, success is measured by the competition between the two giants: Motorola, Inc., and its family of microprocessors based on the 68000 chip, and Intel Corp. and the succession of chips that have evolved from the 8080 chip. The key to the evolution of a family of chips is the upward compatibility of software applications. Each new chip must be able to outperform its predecessor while still running the same programs. While this constraint has, at times, slowed the advance of chip design, it is an economic necessity. A chip that has a pre-existing base of available software that it can run faster than its predecessor has far more market appeal than a higher-performance chip that cannot run available software. Furthermore, the evolutionary design allows programmers familiar with the chip family to build on their previous experience, rather than having to start from scratch every time a new chip is introduced.

The development of the Intel family of processors that has become the basis for the highly successful IBM line of personal computers (PCs) began with the introduction of the Intel 8080 microprocessor in 1974. The 8080 was not originally built for general computing applications. Intel's intention was instead to develop an intelligent device controller that could be easily programmed to run consumer appliances such as dishwashers. When early PC hobbyists started using the microprocessor and attaching it to more memory and input/output devices, such as monitors and cassette recorders, they were able to produce the early forerunners of today's personal computers.

The Intel 8080 is an 8-bit processor with an 8-bit data path. These two features, together with the clock speed at which devices are run, largely determine the operating performance of a computer built with a given microprocessor. The term 8-bit processor describes how large a piece of data the processor is able to manipulate through such operations as adding, subtracting, multiplying, or simply comparing. An 8-bit processor is able to process data 8 bits, or 1 byte, at a time. In contrast, a 16-bit processor can manipulate 16 bits (2 bytes), while a 32-bit processor can handle 32 bits (4 bytes).

A second important feature is the width of the data path into and out of the processor. While manufacturers usually set the data path at the same number of bits as the internal processor, this is not always the case. For example, in Intel's 8088 and 80386 SX processors, economic considerations led the company to retain a narrower data path than the processors' internal data capabilities. When the 8088 was introduced in 1979 as a 16-bit processor, there were relatively few 16-bit peripheral devices, such as disk controllers, memory systems, or monitor controllers. The requirement for 16-bit peripherals made true 16-bit computers more expensive than their 8-bit counterparts. By keeping the data path at 8 bits while increasing the internal processing to 16 bits, however, Intel was able to increase performance without greatly increasing the overall cost of the resulting computer, because it could still use relatively inexpensive 8-bit peripheral equipment. For this reason, the 8088 was much more widely used than its more powerful predecessor, the 8086, which was a true 16-bit chip.

The other important measure of a microprocessor's performance is its clock speed. The microprocessor chip performs one activity, such as fetching data or performing one of the steps of a mathematical computation, during each click of an external clock chip called a crystal. The clock chip contains a precisely machined quartz chip that vibrates at a constant frequency. The clock chip is used to control the step-by-step actions of the microprocessor and all the other devices in the microcomputer. In a sense, it is the beating heart of the computer. The original IBM Personal Computer used the 8088 microprocessor that ran at a speed of 4.77 megahertz (MHz). This meant that the crystal vibrated 4.77 million times per second and the processor stepped through its processing in 4.77 million discrete steps each second. In many cases, the crystal actually vibrates at a speed that is an even multiple of a particular processing speed. For example, the microprocessor might count two or more clock clicks for each of its steps.

When a manufacturer produces a chip, it assigns the chip a rated speed, which is the maximum reliable speed that the chip maker supports for its product. This rating is generally conservative, and many system manufacturers have found that they can produce faster machines just by increasing the clock speed — up to a point.

When selecting an operating speed, the manufacturer of the PC must consider the speed not only of the microprocessor itself, but also of all the peripheral chips, such as memory or video controllers. Again, higher-performance peripheral chips cost more than their slower cousins. In order to allow higher-performance microprocessors to work with slower, lower-cost peripheral chips, the manufacturers introduced the concept of wait states. The chip would perform internally at its fully rated speed, but would wait for one or more clock cycles whenever it dealt with one of the slower peripheral chips.

Returning to the history of the Intel 8080 family of microprocessors, the original chip was the 8080 with an 8-bit internal processor and an 8-bit data path. This was followed in 1977 by the 8086 microprocessor, a full 16-bit processor with both a 16-bit internal capability and a 16-bit data path. Although these were used in several of the first personal computers, the PC industry did not become well-established until IBM chose the 8088 for its first personal computers.

The 80286 introduced two new features to the world of microcomputing: virtual memory and protected-mode memory. To understand these concepts, one first needs to understand that a computer's physical memory typically consists of random access memory (RAM) chips, each of which has its own directly accessible address. (For a detailed description of RAM, see the article "Computer Chip Technology" in this chapter.) Virtual memory involves the swapping of large blocks of data back and forth between the computer's main memory and a secondary storage device, such as a hard disk or additional RAM chips. Through this rapid swapping of data, virtual memory systems make a computer appear to have a much larger main memory than actually exists. In protected-mode memory, memory allocated to one program is protected from accidental alteration by another program operating concurrently.

Theoretically, the 80286 has the capacity to address not only the 16 megabytes of physical memory in protected mode, but also to address up to 1 gigabyte (1 billion bytes) of virtual memory. When IBM first introduced its highly successful PC-AT line of microcomputers using the Intel 80286 processor, the company used a 6-MHz chip. Many buyers discovered that they could easily replace the clock chip with one that provided 8-, 9-, or even 12-MHz performance. In this way they were able to increase their performance 33 to 100 percent for only a few dollars. By the late 1980s, 80286-based computers with clock speeds of 12 to 16 MHz were fairly common, and a few machines with speeds of 20 MHz were being sold.

The Intel 80386 brought true 32-bit microprocessing to microcomputers, with both a 32-bit internal register and 32-bit data paths. In addition, the 80386 has the capacity to directly address 4 gigabytes (4 billion bytes) of physical memory and an incredible 64 terabytes (64 trillion bytes) of virtual memory, or the equivalent of over 30 billion typed pages of text. First introduced at 16 MHz, 80386 machines were available with clock speeds of 20 MHz and 25 MHz by the late 1980s, with the possibility of speeds as high as 50 MHz in the future. In April 1989, Intel introduced its 80486 chip, which was expected to run software two to four times faster than the fastest 80386 chips.

While the raw processing power of microprocessors has leapt several orders of magnitude in the first 15 years of the 8080 family, the software that runs actual computer systems has lagged behind. Almost no software was written to make use of the 80286 chip's protected-mode memory and virtual-memory system. By the time developers were ready to begin developing the next generation of software, they were already anticipating the availability of the 80386, which had far greater capabilities than those of the 80286 and other features that were easier to use. For example, the 80386 offered a virtual 8086 mode that allowed it to emulate many individual 8086 machines simultaneously, thereby permitting the simple development of multiprocessing applications while still supporting the protected mode.

Remembering the greater success of the 8088 as compared to 8086 at the beginning of the decade, Intel introduced a trimmed-down version of the 80386 in 1988. Instead of a full 32-bit data path, the 80386 SX features a 16-bit data path while retaining the 32-bit internal architecture. This allows the chip to process a full 32 bits internally, while using less expensive 16-bit peripheral components. However, as the price of 32-bit peripherals declines, the price difference between 80386 and 80386 SX machines will probably decrease, and the 80386 will probably become the pre-eminent chip in the Intel family until 80486 chips become widely available.

Is there a limit to the power single users will require? Certainly, machines operating as file servers or shared processing servers on a network will continue to require more power as more users are added to the network. But what about the true personal computer serving a single user? Could such a limit be reached or even defined? Perhaps the best way to look forward is to review the past. When the 8086/8088 chips were first introduced in PCs, the computer had only 64 kilobytes (K) of RAM for the main memory and the ability to increase that to 640K. By the time the 80286 was introduced, the minimum configuration was 512K with the ability to increase that to 16 megabytes. The 80386 and the advanced operating systems that take full advantage of its power need at least 1 to 2 megabytes of RAM, with many systems in the range of 4 to 6 megabytes. The current software is so much more sophisticated than the first software for the 8086/8088 machines that it demands much more memory and processing power to run efficiently, even for a single user. The memory requirements have grown from 256 kilobytes to 2 megabytes — an eightfold increase — in only 8 years. This trend toward greater power appears likely to continue into the future.

CPUs with multiple processors. In fault-tolerant computers, the CPU contains redundant processors to ensure that if one fails, another will take over immediately. In such machines, multiple processors do not add to the computer's power, but they do enhance its reliability.

In another type of computer design, multiple processors may work simultaneously on a single program in what is called parallel processing. In most computers, one processor fetches data and instructions from memory, performs a desired operation, then repeats the process with the next instruction. Speeding up computers with this architecture thus depends on increasing the speed at which the CPU performs each operation, but even very high-speed ICs are limited by the speed at which electronics can function. In contrast, multiple processors are capable of performing more complex operations on larger pieces of data.

Engineers have developed different types of parallel processing, including Single Instruction, Multiple Data (SIMD), and Multiple Instructions, Multiple Data (MIMD). In SIMD computers, the multiple processors all receive the same instructions, but they perform them on separate pieces of data. This approach, the closest to classic computer architecture, is used in the Cray-1 supercomputer. In MIMD computers, the multiple processors receive different instructions, which they carry out on different pieces of data. Keeping the processors working in harmony is more difficult than in the SIMD computers, but MIMD computers are better-suited for highly branched programs like those used in artificial intelligence.

Computer engineers further categorize parallel-processing computers as either coarse-grained or fine-grained. Coarse-grained machines use a small number of powerful processors, while their fine-grained counterparts use very many less powerful processors. Coarse-grained computers are generally easier to program because they more closely resemble classic computer designs. Fine-grained computers, on the other hand, incorporate a greater degree of parallelism — and thus make fuller use of the advantages of having multiple processors — but they differ so radically from conventional computers that users must learn how to custom-program them.

Parallel-processing computers also vary in terms of whether or not the individual processors share a central memory or each have separate memories, and in terms of how the program is divided among the processors. The way in which the program is divided may be specified by either the programmer or the computer.

Most research in parallel computing took place in the early 1980s, and by 1986 dozens of applications benefited from parallel architectures. One of the first parallel-processing computers, the Massively Parallel Processor (MPP), was made by Goodyear Aerospace for the National Aeronautics and Space Administration (NASA) in 1983. Still one of the most powerful parallel-processing computers in the world, the MPP uses 16,384 processors, arranged in a 128 × 128 array. NASA uses this fine-grained SIMD primarily for image processing of satellite photographs.

Reduced instruction set computing. Less impressive than parallel processing but perhaps more immediately significant was the introduction of reduced instruction set computing (RISC) in the early 1980s. Developed by IBM and now offered by two dozen different manufacturers worldwide, RISC derives from a new philosophy of computer design that aims to speed up computers by simplifying them. Every CPU accepts a certain set of instructions and processes data according to instructions. Over time, CPUs have become continually more complex and versatile, with more and more instructions involved as they operate. Proponents of RISC architectures are uneasy with this trend, because they claim that the extra instructions — even those designed specifically to speed up the execution time — ultimately slow down the CPU. They argue that the CPU spends far too much time locating the proper instruction from the many that govern how it operates. In contrast, RISC computers sacrifice some of the features of standard computers for greater speed and programming simplicity.

In addition to using fewer instructions, RISC computers are also "hard-wired," which means that instructions are built into the chip's circuitry rather than programmed into its memory. Such hard-wired instructions run much faster than programmed instructions. RISC also standardizes certain characteristics of CPU operation such as the ways in which data may be accessed and the format in which instructions may be written.

Two manufacturers, Pyramid and Ridge, produce RISC minicomputers that compete with similar-sized non-RISC machines offered by DEC and Prime Computer. IBM and Hewlett-Packard Co. have also introduced entire families of RISC computers. IBM calls its family of engineering workstations the PC/RTs, while Hewlett-Packard refers to its RISC family as Spectrum.

Inputting Data

All computers provide one or more ways for users to feed information into them. To enter the data and instructions for a single calculation, the first computer, the ENIAC, required its programmers to rewire it by hand and manually set dozens of switches. Later first-generation computers used punch cards. Since then, the keyboard has become the standard means for data entry, although various pointing devices, such as mice and digitizers, have become popular in recent years. On the cutting edge of data-entry technology are advanced optical character scanners capable of reading handwritten characters and voice recognition devices designed to further reduce the barriers between people and their machines.

Keypunch cards. A punched card, called a Hollerith or IBM card, is a piece of thin cardboard that accommodates 80 columns and 12 rows. A keypunch machine makes rectangular holes in the card, with the positions of the holes designating specific numbers, alphabetic characters, or symbols. After being punched, a card can pass through a card reader equipped with photoelectric or mechanical wire brush sensors. When a beam of light or a fine wire brush passes through a hole in the card, a circuit is completed, thus allowing the device to determine the correct number, character, or symbol. The reader then transmits the information to the computer for processing.

Modern card readers can read 600 or more cards per minute, and modern keypunch machines have various buffering mechanisms that store all the keyed information, so the operator can verify correctness before physically punching the cards. Although punched cards were the primary means of entering data into first-generation computers, they have virtually disappeared now that more convenient forms of data entry are widely available.

Keyboards. Keyboards are by far the most common device for entering data into a computer. Each key works as a switch closing a circuit and thereby informs the computer that various characters or codes are being entered. In addition to the familiar alphanumeric (letters and numbers) and symbol characters on a typewriter keyboard, a computer keyboard often includes nontext keys that perform special functions.

There are two basic styles of keyboards: "QWERTY" and "Dvorak." The QWERTY keyboard follows the traditional typewriter layout, the first six characters of the top row of letters being q, w, e, r, t, and y, respectively. Traditional wisdom held that the QWERTY arrangement of letters did not facilitate fast typing, but was instead originally designed to prevent people from typing too fast and jamming a typewriter's keys. However, in 1988, motor skills researcher Donald Genter, who studied the correspondence of Christopher Sholes, the inventor of the typewriter, found no evidence that Sholes had intentionally designed the keyboard to slow typists down. In 1932, August Dvorak developed an alternative layout by grouping the most commonly used keys together at the center of the keyboard to increase typing efficiency and speed. Despite these advantages, Dvorak's arrangement has not been widely adopted because most people learn to type with the QWERTY layout.

Mice and digitizers. For some applications, users find mice and digitizers more useful for entering data than keyboards. A mouse is a hand-held device connected to the computer by a cable or linked through an infrared beam. When the user moves the mouse, the cursor on the computer screen moves a proportional distance in the same direction. For design and desktop publishing applications, in which a user wants to move the cursor frequently to different locations on the screen, the mouse can be significantly faster and easier to use than the cursor keys on a keyboard.

Three types of mice are currently available: mechanical, optical, and optomechanical. Mechanical mice contain a metal ball that passes an electrical current to nearby sensors. As the ball turns, it rotates a pair of continuously variable resistors, one of which reads horizontal movement while the other reads vertical movement. The change in voltage passing through the resistors as the mouse

Figure 1. Programming an Early Computer. To perform a computation, the operators of ENIAC had to rewire the machine manually. In this photograph, ENIAC coinventor John Mauchly carries out the laborious process of changing switch settings. *Courtesy: The Bettmann Archive.*

moves is interpreted so that the cursor on the screen moves the appropriate direction and distance.

Optical mice use an infrared light-emitting diode, a reflective tablet with an etched or painted grid pattern, and a sensor. When the infrared beam hits the tablet it is reflected back to the sensor. As the beam moves across one of the grid lines on the tablet, the reflected beam changes, allowing the sensor to detect movement. By counting the number of lines crossed, the mouse can move the cursor a proportional amount on the screen. The high precision of an optical mouse and its lack of moving parts subject to wear, deterioration, and failure, make it a handy device. However, a person can use it only on the grid tablet.

Finally, an optomechanical mouse combines features of both optical and mechanical versions. It uses a ball to capture the movement, but instead of rotating mechanical resistors, it rotates a pair of patterned disks. Light reflected off the rotating disks is picked up by a sensor, which translates the shifting reflections into horizontal and vertical cursor movement.

Like mice, digitizing pads enable a user to move the cursor without cursor keys. A digitizing pad or tablet usually consists of a drawing board and a stylus (pen). As the user presses and moves the pen on the drawing board, corresponding lines or dots appear on the screen. Digitizing pads work especially well for creating free-form shapes or tracing existing drawings.

Optical readers. Optical readers have been used to read pencil marks on standardized tests and questionnaires since the 1960s. The technology involves passing a beam of light over a surface on which dark marks have been made, then recording areas too dense for the light to pass through or too dark to reflect the light. The data can be transmitted to a computer which then evaluates them.

Optical Character Recognition (OCR) devices operate on the same principle, scanning a page and creating a digital signal based on dark and light areas. By comparing the patterns of the scanned images against a library of known characters, the OCR can identify specific characters. Since OCRs cannot easily distinguish similar characters, they often use a special font, with distinct shapes and forms designed to minimize confusion. In fact, the original OCR readers could only read text printed in this typeface. Some of the devices now available, however, can read a variety of fonts with a high degree of accuracy.

Many retail stores use optical readers to interpret bar codes made up of vertical lines, varying in thickness, that represent specific characters and numbers. When the optical reader passes a beam of light over the merchandise label, it creates a digital signal based on the lines and the white space between them. This information can then be transmitted to a cash register, eliminating the need for a clerk to enter price data on a keypad. The same technique can also help stores maintain accurate inventory records.

Magnetic ink character recognition. Another type of character recognition technology, used primarily by banks, employs magnetic ink. When a check is processed by the bank, the dollar amount is printed with magnetic ink onto the lower corner of the check. The banks preprint their own codes and data about individual accounts in magnetic ink on clients' checks, and machines can scan dollar amounts when the check is processed. A high-speed check processor reads all the data in magnetic ink and then transmits it to the computer for posting.

Voice recognition. After World War II, some students of technology suggested that typewriters capable of understanding human speech would soon be developed while other advances, such as putting a man on the Moon, would require many years. As it turned out, the opposite occurred: In 1969 the first human set foot on the Moon, but typewriters that recognize and accurately translate ordinary human speech are still not available.

Part of the difficulty surrounding voice recognition involves dialects and accents. Not only must a device be able to distinguish different words, but it must also recognize the same word when spoken by a native of Brooklyn, New York, Tuscaloosa, Alabama, or London, England. To make matters worse, human languages abound with colloquialisms, idiomatic expressions, and syntactical oddities that make programming them extremely difficult.

During the past 5 years, developers of voice-recognition systems have made strides toward overcoming such vexing problems. One type of voice-recognition system breaks speech into units of sound, compares these units with a library of known sounds, selects the best match, and then transmits the results to the computer. Another system involves teaching the computer what to do after certain words are spoken. The user first speaks the words and then uses the keyboard to tell the computer to execute certain tasks. This teaches the computer to perform specific tasks when particular words are spoken. While this type of system generally achieves good word recognition, it only recognizes the words as spoken by a given operator, and it must be reprogrammed to decode the voice of anyone else using the system.

The first commercial voice-recognition unit was introduced in 1971 by the Threshold Co. The device, which sold for approximately $15,000, could recognize only several dozen words. In the years since, the technology has advanced significantly. During 1987, Kurzweil Computer Products, Inc. (now owned by Xerox Corp.) announced a system with a vocabulary of more than 5,000 words and the ability to adapt to a user's voice patterns over time, thereby increasing accuracy.

Modems. Modems enable computers to transmit data to each other over telephone lines. Since computers use only digital signals and common telephone lines require analog signals, the modem from the transmitting computer must first modulate the signal into an analog form. The receiving computer's modem then demodulates the signal, translating it back into digital form, hence the name "modem," a contraction of modulator/demodulator.

The speed at which modems transmit data between computers is measured in baud, an old telegraph term roughly equivalent to one bit per second (bps). In the mid-1970s, the most common transmission speed was 300 baud. While faster speeds were possible, transmission errors caused by random noise on the telephone lines often negated the savings in time. By the late 1980s, better error-checking software and transmission procedures had made transmission speeds of 1,200 and 2,400 baud routine, and 9,600 and 19,800 baud modems had already begun entering the marketplace.

Other input modes. In addition to the traditional means of inputting data into a computer, users can employ sensors, transducers, photocells, and other devices. Once data from these devices are converted into digital signals, they are transmitted to a computer for analysis.

Output Devices

Computer output falls into two general categories: soft copy and hard copy. Soft copy output devices, such as visual display terminals, produce temporary output. Hard copy devices, such as printers and plotters, produce more permanent records.

CRTs. The most common type of monitor is the cathode ray tube or CRT. Like a television set, a CRT contains an electron gun that projects an electron beam onto a phosphor-coated screen. Wherever an electron hits the screen, the phosphor coating glows, creating a dot of light. To produce complete images, the electron gun rapidly scans across the screen, pulsing the beam of electrons on and off as directed by the software. Since this technique uses patterns of dots to create images, it is referred to as the "dot-matrix" technique.

Another technique uses a process called vector scanning, in which the device generates images as continuous lines (rather than as configurations of dots), by projecting a constant rather than a pulsing beam of electrons across the phosphor screen. Because such monitors create smooth circles and curves, they are well-suited to graphic and computer-aided design applications.

Modern computers use one of three types of CRT displays: storage tubes, stroke/refresh displays, and raster images. Storage tubes maintain an image indefinitely on the screen, without having to be "refreshed" (redrawn) constantly by the beam of electrons. They also provide very high image resolution. However, users cannot partially change storage tube images; even a small alteration of the picture requires that the entire display be erased and redrawn. Storage tubes can also handle only a limited range of colors.

With stroke/refresh displays, as with storage tubes, a beam of electrons "writes" on the screen, but stroke/refresh images fade away and therefore must be redrawn many times per second (typically 60). Although this requires more electronic circuitry and limits the complexity of images that can appear on the screen compared with those of storage tubes, it does allow a user to alter an image quickly. Stroke/refresh images can be easily animated and moved around the screen, and they can be in a greater range of colors than images on storage tubes.

Most television sets and computer displays now rely on raster technology. With raster displays, the electron beam traces a fixed number of parallel lines across the screen many times each second, modulating in intensity as it moves. The changes in beam intensity result in variations of the brightness of the phosphor dots on the display screen. Full-color CRTs group red, green, and blue phosphor dots into so-called picture elements, or pixels.

As with stroke/refresh displays, raster display images fade away and must be refreshed many times per second. Unlike stroke/refresh displays, however, raster displays redraw the entire screen — every pixel — with each refresh cycle. Consequently, raster displays do not slow down with increasing image complexity. This characteristic allows users to draw complex images and whole fields of color easily. However, raster displays generally have lower resolution than either stroke/refresh or storage tube displays, and they require more computer processing power and memory to operate.

While most parts of computers have become lighter and more compact, computer displays have, for the most part, remained fairly constant in size and weight. Viewing comfort dictates their horizontal and vertical dimensions, and a CRT's depth must be a specific multiple of its screen size to give the electrons adequate room to accelerate and find their target phosphor dots. In addition, CRTs must be constructed of heavy glass, because they must maintain an internal vacuum.

A standard CRT is suitable for most applications, but some types of equipment, such as portable computers and military instruments, demand flat, lightweight displays. Flat-screen displays can also reduce the size and weight of televisions. As a result, dozens of companies worldwide have worked on developing flat-screen displays for over 30 years.

Some companies have tried placing the electron gun off to the side of the screen, using lenses to redirect the beam at the face plate. Because such systems are expensive and difficult to keep aligned, however, no such screens are yet being sold commercially. A new design developed by RCA Corp. and Matsushita does away with the electron gun entirely, and instead uses long strips of material to emit electrons. Grids placed in front of the strips break up the lines of emitted electrons into single streams, one for each pixel. Although RCA has not announced any commercial plans for the technology, Matsushita has demonstrated at several trade shows prototype televisions with 10-inch pictures and tubes less than 4 inches deep.

Non-CRT displays. While CRTs provide good resolution, their weight and size make them poorly suited for portable computers. As a result, manufacturers have developed a number of alternatives to CRTs, including liquid crystal displays (LCDs), electroluminescent displays (ELDs), and gas plasma displays. In LCDs, a liquid crystal is suspended between two pieces of glass that contain a grid of electrodes. When current passes through the grid, the molecules of the liquid crystal align themselves, which in turn causes them to reflect light differently from nonaligned crystals. In this way, patterns of dark and light form the shapes of characters and graphic elements. LCDs require relatively little power and can typically run for 4 or more hours on batteries, depending on the model.

Since LCDs emit no light and provide poor contrast, users need sufficient light and must view the screen from the proper angle to see the image clearly. This readability problem has been largely solved by the use of backlighting and so-called supertwist technology, which improves the display by adding a lighting source, a set of front and rear polarizers, an antiglare coating, and a high-density electrode grid embedded in the front and rear glass. The polarizers rotate the liquid molecules, creating light and dark areas on the screen. The backlight source produces a considerably higher contrast than that found in conventional LCDs. Most manufacturers of portable computers use backlit screens for better readability, although that feature reduces battery life because of the additional power drain.

In late 1988, Zenith Data Systems Corp. further refined the backlighting concept by using a fluorescent lighting source and a compensating layer that eliminates the bluish light characteristic of supertwist screens. This so-called page white display has a contrast close to that of CRT displays. In 1988, Mitsubishi Electronics America, Ltd., also introduced a portable computer with a page white display.

Another type of flat screen, the ELD, sandwiches a phosphor film between the rear and front plates of glass, which are impregnated with an electrode grid. When a current reaches an intersection on the grid, the phosphors emit a burst of light. Many portable computer manufacturers prefer ELDs because they produce better contrast than LCDs without the weight and size of a CRT. One important drawback is that ELDs require too much power to be run on batteries. A number of industry experts predict, however, that battery-powered ELDs will become feasible in the near future. Another drawback is the relatively high cost of ELDs compared to LCDs. Nevertheless, as cost and power consumption drop, ELDs could become the most commonly used flat displays.

Gas plasma displays are another flat-screen alternative to CRTs and LCDs. A gas plasma screen consists of a layer of neon gas sandwiched between two pieces of glass. A set of transparent electrodes is embedded in the front piece of glass, while the rear piece contains reflective electrodes. When current passes through the electrodes at selected sites, the neon gas ionizes, creating visible dots of light that can be used to create characters and images. Like ELDs, gas plasma displays provide high resolution, but since they consume a great deal of power, they require external power sources. Gas plasma displays also cost much more than CRTs and LCDs.

Impact printers. Computers can be connected to a number of different types of printers. One type, the impact printer, works like a typewriter; a fully formed raised character presses an inked ribbon or film against paper, leaving a printed character. Several types of printers are available. With impact line printers, the raised characters sit along a continuous chain or metal band rotated swiftly by two gears. As the chain or band moves, hammers located at each character position across the page press the appropriate character when it passes in front of the proper position on the page. The character in turn presses against an inked ribbon, leaving an imprint on the paper. Line printers can print up to 1,500 lines per minute (a typical single-spaced 8-1/2 by 11 inch page contains about 55 lines), and are generally used with mainframes and minicomputers.

Daisy-wheel printers, another popular type of impact printer commonly used with microcomputers, operate on a similar principle. The daisy wheel itself consists of a plastic hub with spokes radiating from its center. The end of each spoke contains an alphanumeric character or symbol. As directed by the software, the daisy wheel rotates so that a hammer can strike the appropriate character against a ribbon, which in turn makes an imprint on the paper.

A variation of the daisy wheel is the print thimble, which consists of a plastic cup made of individual prongs, the tops of which contain alphanumeric characters or symbols. To leave an imprint, a hammer on the inside of the cup presses the characters against the ribbon and paper.

Daisy-wheel printers produce excellent print quality and are fairly inexpensive. However, they are relatively slow (10 to 80 characters per second) and noisy, and they cannot produce graphics. Users must also suffer the inconvenience of stopping the printer and changing the daisy wheel whenever they want to change the typeface on a document.

Dot-matrix printers. While daisy-wheel and line printers use sets of fully formed characters to create output, dot-matrix printers create characters from a cluster of wires, or pins, that are pressed against a ribbon. Dot-matrix printers are inexpensive and quite fast, with some models attaining speeds of more than 500 characters per second. However, in the early dot-matrix printers, which contained only seven or nine printing pins, print quality was relatively poor compared to that of impact printers. Seven or nine pins can do little more than generate crude outlines of characters. More detailed characters require smaller and more closely spaced dots.

To improve the quality of dot-matrix output, some printer manufacturers have designed machines that operate in so-called correspondence mode, which entails laying down a line of characters, slightly shifting the printhead, then repeating the line of characters. This rounds out characters and fills in spaces, so that the characters look more like those produced with an impact printer or typewriter. However, since the printer must print each line twice, the correspondence mode effectively cuts the printer's speed in half. Another solution to the quality problem involves increasing the number of pins in the printhead. In correspondence mode, the 18- and 24-pin printers now sold by a number of manufacturers approach the quality of daisy-wheel printers.

Dot-matrix printers can print coarse graphics and simulate large and fancy fonts by printing a series of dots to draw the image. While this process is time-consuming, it can produce reasonably high-quality graphics and elaborate printing at a low cost.

Laser printers. The laser printer is the most significant development in output technology of the past 20 years. Unlike line, daisy-wheel, and dot-matrix printers, which print a line at a time, laser printers store a full page of characters (or dots in an image) before the printing begins. To create a page, the computer sends signals to the printer, which shines a laser at a mirror system that scans across a charged drum. Wherever the beam strikes the drum, it removes the charge. The drum then rotates through a toner chamber filled with thermoplastic particles. The toner particles stick to the negatively charged areas of the drum in the pattern of characters, lines, or other elements the computer transmitted and the laser beam mapped.

Once the drum is coated with toner in the appropriate locations, a piece of paper is pulled across a so-called transfer corona wire, which imparts a positive electrical charge. The paper then passes across the toner-coated drum. The positive charge on the paper attracts the toner in the same position it occupied on the drum. The final phase of the process involves fusing the toner to the paper with a set of high-temperature rollers.

Laser printers offer the advantages of high speed, excellent print quality, and quiet operation. Users can also combine a variety of typefaces on a single page and include graphic elements as well. Moreover, while impact printers are generally most efficient when used with continuously fed paper, laser printers can print on loose sheets of paper stacked in one or more bins. This eliminates the need to separate sheets at the end of a long print job.

High-volume laser printers designed for use with mainframes and minicomputers can generate 120 or more pages per minute. Their size and cost ($50,000 or more), however, make them inappropriate for personal or ordinary office use. Laser technology did filter down to a microcomputer level in 1984, however, when Hewlett-Packard Co. introduced its revolutionary LaserJet printer that cost about $4,000 and could print eight pages per minute. Hewlett-Packard also offered a selection of plug-in font cartridges, which added a wide variety of typefaces and sizes, enabling users to create documents with print quality approaching that of professional typesetting machines. Within 2 years after the introduction of the LaserJet, more than 50 other manufacturers offered similar printers.

While recent models of the LaserJet and similar laser printers typically contain six built-in or "resident" fonts, some laser printers from other manufacturers, such as Apple Computer, contain 30 or more. Many of these printers use the powerful PostScript page description language developed by Adobe Systems, Inc., in Palo Alto, California, and they generally cost between $4,000 and $6,000. They work especially well in desktop publishing and other applications that require a variety of typefaces and sizes.

According to two 1988 reports by the market research firm Dataquest and the industry research firm CAP International, Inc., laser printers were the fastest growing segment of the personal computer printer market. Laser printer sales jumped from 49,000 in 1984 to 600,000 in 1987. Nevertheless, dot-matrix printers still lead in total sales (an estimated 5.5 million in 1987), in part because dot-matrix printers cost only a fraction as much as laser printers and can perform one task that no laser printer can ever perform — printing on multiple-copy forms. Since laser printers make no physical impression on the paper, they can only imprint the top sheet of a multiple-copy form.

Crystal printers. In a variation on the technology used in laser printers, LCD (liquid crystal display) printers substitute a crystal display for the laser mechanism. Instead of using a laser beam to discharge a positively charged drum, crystal printers create lines of dots on an LCD. Light is transmitted through or reflected from the display (depending on whether a particular dot is off or on) to remove charge from parts of the drum. Toner particles are then stuck to the discharged areas and transferred to the paper in the same way as in laser printers. Crystal printers cost less than laser printers and eliminate a delicate laser-aiming mechanism that requires precise alignment.

Other printers. Thermal printers and electrostatic printers use a conventional dot-matrix printhead but no ribbon. Thermal printers require a special heat-sensitive paper that heats up and darkens when struck by the printing pins. The patterns of darkened areas form characters. Electrostatic printers conduct an electrical charge from printing pins to special aluminum-coated paper. The charged pins vaporize the coating, leaving a darkened area beneath each pin. The various patterns of darkened areas form the numbers, alphanumeric characters, and symbols in the same way as dot-matrix printers do. Thermal and electrostatic printers are very inexpensive and are often used in hand-held calculators and portable computer terminals.

Ink jet printers spray minute drops of ink against the paper. One type uses an ink chamber with a piezoelectric wall that flexes when charged with electricity and flings a drop of ink onto the paper to form a dot. The individual dots in a pattern form characters in the same way as the dots from a dot-matrix printer.

An alternative approach, bubblejet technology, replaces the piezoelectric wall of the ink chamber with a thermal resister that boils the ink. When a bubble of ink bursts, it forces an ink drop onto the paper. Both Canon, Inc., and Hewlett-Packard use the bubblejet approach in several of their models. The chief benefit of ink jet printers is their silent operation. They can produce extremely high quality output, with some models reaching the resolution of laser printers at much lower cost.

Plotters. Plotters move a pen or series of pens along horizontal and vertical axes in increments as small as one-thousandth of an inch. They allow users to create precise graphs, maps, architectural plans, and other types of highly detailed hard copy. There are two main types of plotters — flat-bed and moving-paper.

In flat-bed plotters, the paper remains stationary and the pens travel across its surface. Most models can accommodate large sheets of paper and are quite expensive, generally costing $4,000 or more. In moving-paper plotters, paper moves through the plotting mechanism as the printing takes place. Since the paper travels along the horizontal axis, the pens need only move up and down along the vertical axis. Although this arrangement is not as precise as when the pens move in both dimensions, it allows moving-paper plotters to be smaller and less costly than flat-bed plotters.

Microfilm and microfiche. When computer users want output mainly for long-term storage or record keeping, they may not want to incur the expense of printing and storing paper copies. As an alternative, they can produce output in the form of Computer Output Microfilm or Microfiche (COM). Each frame of microfilm or microfiche card may contain hundreds of pages of data. The data are transferred to a tape, which is run on a COM device, which in turn displays each frame or card on a CRT and then converts that image to film. Once the data reside on the film, a user can read them either manually, with a microfilm or microfiche reader, or on a computer system equipped with software to retrieve and display the images.

Programming Languages

Hardware would be useless without some type of operating instructions to tell the computer what to do and how to do it. These instructions come in several different forms and operate at different levels. At the lowest level, the so-called bootstrap loader is a sequence of programming code that literally instructs the computer about its own identity, telling it what parts it has and where to find them. This code is generally contained in the computer's read-only memory (ROM), which is included with the computer when it is manufactured. The bootstrap loader gets its name from the phrase "picking oneself up by one's bootstraps" and is the source of the verb "boot," which means to start a computer or program. The bootstrap loader contains enough information to activate the computer to read a predetermined program called the operating system into memory.

The operating system contains instructions about how the various parts of the computer and supporting equipment, such as printers, can work together. Since the operating system deals with the hardware of the computer, it is specific to the machine and often has to be modified to accommodate different hardware configurations. For example, the IBM PC was designed to operate with the DOS (Disk Operating System) operating system developed by Microsoft Corp. of Redmond, Washington.

Since the operating system and bootstrap loader work directly with the hardware of the computer, they are often written in what is called machine language, which as the name implies, deals directly with the hardware. Machine language is arcane and generally difficult to use and understand. Another programming language, called assembly language, has been designed to facilitate the development of operating systems that are more human-oriented ("user-friendly"). Although assembly language is machine-specific, it is easier for programmers to understand. Developers use assembly language when they want to make a particular piece of programming code as efficient as possible. Many programs are written in a mixture of a higher-level languages and assembly language.

The next level of software is generally called application software. This can take one of many forms. Language interpreters and compilers allow users to write their own programs, while many specific applications programs, such as electronic spreadsheets, databases, and word processing programs, are sold ready to run. Many programming languages have been developed for particular purposes. One of the earliest was COBOL, which was developed by the U.S. Navy in 1959 to handle such standard business applications as inventory control, payroll, and billing. Another early programming language still in use is FORTRAN, which was developed in the mid-1950s at IBM for scientific and engineering applications. One of the most popular programming languages is BASIC, developed in the early 1960s by T. E. Kurtz and J. G. Kemeny at Dartmouth College in Hanover, New Hampshire, specifically to teach programming to students.

The English-like vocabulary of the higher-level languages must be converted into machine instructions before it can run on a given computer. Programs that convert higher-level languages to machine code are called compilers or interpreters. A compiler converts the entire text of a program into machine instructions at one time, and then permits the compiled code to run. In contrast, an interpreter converts a single line of code, runs it, and then converts the next line sequentially as it executes the program. The advantage of an interpreter is that it helps programmers develop software more quickly and easily, because they can pinpoint any errors as they progress, rather than completely compiling the program each time they test it. The advantage of a compiler is that it can produce code that runs much faster once the program is fully debugged (made error-free).

A multitude of both general-purpose and specialized programming languages have been developed. Some popular general-purpose programming languages include C, Forth, and Pascal; some special-purpose languages include LISP and Prolog in the field of artificial intelligence.

Computer Chip Technology

The rapid progress in computer science, particularly the development of integrated circuits, has been accompanied by advances in the electronics industry. Many thousands of circuits can now be etched into a single chip smaller than a fingernail, a breakthrough that allows engineers to design powerful microprocessors a fraction of the size of their ancestors of just a few years ago. Regardless of the type of electronic device used to process data, the underlying concept of digital computing derives from Boolean logic and binary mathematics, which uses combinations of ones and zeros to represent data.

Boolean Logic, Binary Numbers, and Digital Electronics

All computers, from the smallest to the largest, rely on a special class of mathematical logic developed in the 1850s by British mathematician George Boole. Named for its inventor, Boolean logic is a system for reducing a problem to a series of true or false propositions, which can be represented by zeros and ones. A problem-solver can then work out a solution with what is called binary mathematics.

The binary number system differs from the decimal system people commonly use in that it represents all numbers with only two symbols — "0" and "1" — in contrast to the decimal system, which uses ten symbols — 0, 1, 2, 3, 4, 5, 6, 7, 8, 9. In the decimal system, counting above single digits, or "ones," requires adding another column representing "tens" and starting over with a one and a zero that represent one "ten" and no "ones." Similarly, when counting beyond 99, a third column of "hundreds" is added. Thus, each added column represents a successive power of ten. For example, the decimal number 437 can be broken down to "4 × 100, plus 3 × 10, plus 7 × 1."

The binary system, in contrast, uses only two symbols and adds new columns for every power of two. Counting above one thus means adding another column for "twos," so that the number two is expressed as "10." This could be broken down into "1 × 2, plus 0 × 1." The decimal number "3" is expressed as "11" in binary (1 × 2, plus 1 × 1), and decimal "4" becomes "100" in binary (1 × 4, plus 0 × 2, plus 0 × 1). Instead of columns representing ones, tens, hundreds, and so forth in the decimal system, the binary system uses columns for ones, twos, fours, and eights, with each successive column representing the next higher power of two.

Any decimal number may be represented as a binary number, although this often takes far more columns than in the decimal system. Forty-seven written in decimal requires only two columns: "47." In binary, it requires six columns: "101111" (1 × 32, plus 0 × 16, plus 1 × 8, plus 1 × 4, plus 1 × 2, plus 1 × 1).

The rules for solving logic problems with binary math were set forth in George Boole's 1854 paper, "The Laws of Thought." Over 80 years later, in 1937, Claude Shannon wrote his master's degree thesis at the Massachusetts Institute of Technology (MIT) in Cambridge, describing how electrical switches, which exist in either an on or off state, could represent binary numbers and, following Boole's laws, could conceivably be used to solve logic problems. He further described how such switches could be arranged to form numerical or symbolic calculating machines. An "on" switch might represent a "1" or a Boolean "true" or "yes," while an "off" switch would represent a "0" or a Boolean "false" or "no." Banks of electrical switches could then represent Boolean logic operations.

If designers of such banks of switches made the electrical output of a given switch dependent on the input it received from another switch or set of switches, they could construct *logic gates*, each of which could represent a single step in Boolean logic. As it turns out, no problem in Boolean logic requires more than three such gates: AND, OR, and NOT. An AND gate outputs "1" (true) only when both of its inputs are "1" (true); otherwise, it outputs "0" (false). An OR gate outputs "1" (true) when either of its inputs are "1" (true). A NOT gate merely reverses whatever it receives as input: a "1" input results in a "0" output, and a "0" input gives a "1" output. As Shannon pointed out, each of these three logic gates could be made from two or three discrete electronic components, and Boolean logic problem solving could then be accomplished by stringing together AND, OR, and NOT gates.

The use of electrical switches to solve Boolean logic problems represented in binary math is called digital electronics. With it, a machine using three basic circuits, the inputs and outputs of which are always either "1" or "0," can solve any logic problem, no matter how complex.

However, even simple problems require a large number of logic gates. Performing the operation "1 + 1," for example, requires 11 AND gates, 5 OR gates, and 3 NOT gates, all 19 of which consist of 2 or 3 electronic components. Although designers can minimize the number of components by letting a single component perform its unique task more than once at different stages of the circuit's operation, the circuit still requires about two dozen components, more than the number in a typical radio or phonograph. To perform meaningful work, a simple computer would need hundreds of such logic circuits.

Using bulky vacuum tube technology, a useful machine would become huge and unwieldy. In fact, without the advent of solid-state electronics based on a group of materials called semiconductors, computers would have reached their maximum practical size during the mid-1950s.

Semiconductors

Semiconductors are chemical elements or compounds that conduct electricity much more efficiently than insulators, such as rubber or glass, but not as efficiently as conductors, such as copper or aluminum. As early as the 1930s, scientists thought they might someday use semiconductors to replace vacuum tubes, then widely used in radios, phonographs, and telephone systems to amplify electrical signals and to switch signals on and off. When used as amplifiers, vacuum tubes accept a weak signal, such as one picked up by a radio antenna, then pass it between a negative and a positive electrode, thus amplifying the original signal. Radios and telephones both need such amplification.

Vacuum tubes contain a positive electrode, called the anode or collector (because it collects electrons), and a negative electrode, called the cathode or emitter (because it emits electrons). A control grid between the emitter and the collector regulates the amount of current passing between them. The anode and cathode are placed in a vacuum or oxygen-free glass chamber to prevent deterioration, hence the name "vacuum tube." Although some scientists believed that they could produce the same effects in a solid block of semiconductor material, their early efforts proved unsuccessful. Then in 1945, Bell Laboratories in New Jersey assembled a large team of scientists and engineers, led by physicists Walter Brattain, John Bardeen, and William Shockley, to develop new components useful in communications systems, specifically a solid-state amplifier. The team studied semiconductor materials in hopes of learning enough about the properties of the materials to use them to build experimental models.

To understand the characteristics of semiconductors, the scientists had to consider their atomic structure. The most stable chemical compounds have a specific number of electrons in their outer electron shells. To attain a stable state, they may share electrons with other compounds or elements by forming covalent bonds with them. Silicon provides a good example. Silicon atoms combine to form a structure known as a crystal lattice by sharing each of their four outer electrons with other silicon atoms. Since a material's conductivity depends on its number of free (unbonded) electrons, silicon in this fully bonded pure crystalline state makes a poor conductor. The only free electrons available are those released from their bonds when the silicon is heated or knocked loose by photons when exposed to light. This latter phenomenon accounts for one of the unusual properties of semiconductors — they are photoelectric, which means they produce an electric current when exposed to light. Semiconductors also become better conductors when they were heated. The photoelectric properties of semiconductors make them ideal for use in solar cells, which generate electricity from light, and for light-emitting diodes (LEDs), which generate light from electricity.

Engineers can precisely alter the ability of semiconductors to conduct electricity through a process called doping, the introduction of impurities into the pure crystal lattice. These impurities disrupt the lattice structure in predictable ways, causing either an excess of free electrons (making the material negatively conductive) or a lack of electrons (making it positively conductive).

Phosphorous, an element with five valence electrons (electrons in the outer shell or shells of an atom that participate in the formation of chemical bonds), is a good doping agent. If an atom of phosphorous is added to a silicon crystal lattice, four of the five phosphorous electrons bond with silicon atoms, with the fifth electron left over. The addition of more phosphorous atoms creates more free electrons in the crystal, making the compound more negatively conductive. Scientists call such negatively doped semiconductors n-type.

The opposite effect occurs when the semiconductor is doped with boron, which possesses only three electrons in its valence shell. If an atom of boron is added to a silicon crystal lattice, the three electrons bond with silicon atoms, resulting in a so-called hole. Even if a bonded electron near the hole breaks its bond and fills the hole, it then leaves another hole in its place. When a positive current runs through boron-doped silicon, electrons move toward the positive electrode, leaving behind a receding stream of holes, which flow toward the opposite electrode. The more boron atoms are added, the more holes open up, and the more positively conductive the material becomes. Scientists call such positively doped semiconductors p-type.

Of all the semiconductor materials, silicon works especially well in microelectronics applications because, when combined with oxygen, it forms silicon dioxide, one of the best insulators known. Therefore, depending on the additive, silicon can become positively conductive, negatively conductive, or insulative. In addition, silicon is the second most abundant element in Earth's crust, after oxygen.

However, the earliest experiments at Bell Laboratories and elsewhere focused on devices made of the element germanium, because scientists had already used it extensively in electronics and because its relatively low melting point made it easier to purify and handle. The research and development effort at Bell Laboratories proved successful within two and a half years, and in December 1947, the team demonstrated the first solid-state amplifier, later dubbed the transistor. In 1956, Brattain, Bardeen, and Shockley received the Nobel Prize in physics for their work.

Figure 2. Early Point-Contact Transistor. Although primitive compared with today's junction transistors, early point-contact transistors worked in essentially the same way as their modern counterparts, amplifying electrical signals by passing them through a solid semiconductor material. *Courtesy: AT&T Archives.*

Transistors

The first transistor was the so-called point contact transistor, which consisted of a block of germanium soldered to a metal base with the negative terminal of a battery connected to the germanium, and the positive terminal connected to the metal. Two electrodes were attached 0.002 inch apart on the surface of the germanium. One electrode carried the weak input signal to be amplified, and the other carried away the amplified signal, boosted by the high-voltage battery running between the germanium and the metal.

Because of the difficulties in manufacturing point-contact transistors, chiefly related to positioning the two surface electrodes, researchers developed two other types of devices still commonly used today: bipolar transistors (also known as junction transistors) and field-effect transistors.

William Shockley developed the bipolar transistor in 1950. It consisted of a sandwich of doped semiconductor material, either some n-type semiconductor material between two layers of p-type material (called a pnp transistor) or some p-type material between two layers of n-type material (called an npn transistor). The center region, or base, is analogous to the control grid in a vacuum tube. The two outer layers serve as the emitter and the collector, analogous to the electrodes in a vacuum tube. Also as in a vacuum tube, a small input signal passed through the base controls a large signal passing between the emitter and the collector. Bipolar transistors are one of the most popular types of transistors for applications demanding speed rather than low power or small size.

Although Shockley and several other researchers first described the principles of the field-effect transistor in the early 1950s, no one actually built one until 1962, when RCA introduced the metal-oxide-semiconductor field-effect transistor (MOSFET). Field-effect transistors (FETs) have a source, a gate, and a drain, instead of an emitter, a base, and a collector. Like the junction transistor, these form either an npn or a pnp sandwich. A plate located above the middle layer (gate) acts as the control. A small current applied to the plate induces a so-called field effect. In the case of an npn transistor, a positive charge will repel positive ions in the p-type silicon beneath it. This repellent positive force prevents the ions from crossing the channel from source to drain and, consequently, a small amount of current applied to the plate controls the large current applied between source and drain.

Early FETs were inefficient because the plate actually touched the semiconductive material below, draining current from the control circuit and thus requiring more current to operate. Researchers reasoned, however, that if a layer of silicon dioxide were used to insulate the gate from the semiconductive material, the transistor would require far less power to operate. RCA's MOSFET applied this insulated gate concept, which was used in the first commercial FETs. Though slower than bipolar transistors, MOSFETs can be found in the most complex modern microelectronics components, which demand small size and low power.

These developments in solid-state electronics made possible such battery-operated electronic devices as transistor radios, portable televisions, and transistorized hearing aids. Because transistors have many potential military uses, the U.S. government awarded millions of dollars to both vacuum-tube makers developing transistors and new companies specializing in transistors.

Integrated Circuit: Birth of Microelectronics

The term microelectronics generally applies to the integrated circuit (IC), a device that may contain many transistors and other electronic components linked together. Transistors like those developed at Bell Laboratories and widely used in radios and hearing aids in the 1950s replaced vacuum tubes on a one-to-one basis. Like individual vacuum tubes, resistors (which impede the flow of electricity) and capacitors (which store electricity), transistors are discrete components. Discrete components wired together, usually on plastic or masonite circuit boards, form complete circuits.

ICs consist of transistors, resistors, and capacitors interconnected on a single chip of semiconductor material, which itself forms a complete electronic circuit. Many ICs may, in turn, be wired together on circuit boards and combined with discrete components to form larger, more complex circuits.

Several developments led to the modern IC. The first two occurred at Bell Laboratories in 1955 as a result of efforts to improve transistor fabrication. One improved technique was diffusion, a way of doping semiconductor material by exposing it to vapors of the dopant in a furnace at extremely high temperatures. By varying the temperature and the length of time the semiconductor spends in the furnace, engineers can precisely regulate how much dopant diffuses into the semiconductor.

However, engineers still had to be concerned about where the semiconductor absorbed the dopant. The more accurately they could position doped areas of n-type and p-type silicon, the smaller they could make the transistors. This precise control was accomplished with a contact printing process called photolithography, which means literally, "writing on stone with light."

Jean Hoerni, a Swiss-born physicist working at Fairchild Semiconductor, used diffusion and photolithography techniques to make the first flat, or planar, transistors. To fabricate all the elements of a transistor on a flat surface, Hoerni used photolithography techniques and placed an insulating layer of silicon dioxide over the top of the transistor, effectively isolating different regions with interconnecting paths. Fairchild engineers found a method for vaporizing aluminum onto the top of the transistor, so that it filled in the interconnected paths with thin aluminum wires. Next, they covered the entire flat transistor with an insulating layer of silicon dioxide. The flatness of planar transistors made them more rugged than other transistors. More importantly, transistors could now be completely machine-made, substantially reducing the time and cost of manufacturing. Soon transistors in the form of silicon wafers were produced on assembly lines in large batches.

While engineers at Fairchild worked on the planar process, Jack Kilby, an engineer at Texas Instruments, developed a method for putting several components on a single piece of semiconductor, thereby creating the first IC. In 1958, Kilby demonstrated an entire circuit called a phase shift oscillator — consisting of a transistor, a capacitor, and three resistors — that he had built on a base of germanium. The components in Kilby's design, altogether smaller than a match head and more reliable than a conventional circuit, had been wired by hand. In February 1959, Texas Instruments patented Kilby's circuit on a chip and in March of that year offered it as a $450 product.

Unaware of Kilby's efforts, Fairchild had also been working on building a circuit on a chip. Though Hoerni's work on the planar transistor had absorbed the efforts of the small company's engineering staff, news of Kilby's success quickly refocused their attention. Fairchild researchers Robert Noyce and Gordon Moore applied Hoerni's planar process to the problem of developing a circuit on a chip. Using planar technology, they could fabricate different circuit elements on a single piece of silicon, isolate them from one another with silicon dioxide, and interconnect them with aluminum. While the components on Kilby's version had to be physically isolated from one another and then hand-wired together, the version for which Noyce filed a patent in July 1959 could be mass-produced. As planar transistors became the technology of choice for transistors, planar ICs became the only practical ICs. Even Texas Instruments eventually licensed the technology for the manufacture of its own ICs. The planar fabrication process was such an important step in the development of the IC that Noyce is often credited as the co-inventor, along with Kilby, of the IC, even though his patent came over 6 months later than Kilby's.

As with transistors, the U.S. military subsidized much of the early development of ICs by purchasing the devices at premium prices. In the 1960s, the Minuteman Missile project, together with the NASA manned space missions, accounted for a large percentage of the purchases of complex ICs. ICs became increasingly intricate as manufacturers developed purer semiconductor materials and refined equipment for designing and fabricating chips. Engineers refer to the period between 1960 and 1966 as the era of small-scale integration (SSI), when manufacturers put up to 100 transistors on a single chip. SSI gave way to medium-scale integration (MSI), which prevailed between 1966 and 1969, with between 100 and about 1,000 transistors on a chip.

In 1969, the era of large-scale integration (LSI) began, mainly because scientists had developed new techniques for manufacturing MOSFETs, which are considerably smaller than bipolar transistors. LSI chips contain between 1,000 and 10,000 transistors each. Despite great strides in manufacturing such dense, intricate chips, designing them still posed many problems. Not only were few engineers trained in IC design, but the time consumed by the design process threatened to make each new chip obsolete before it went into production. A further complication was that customers were putting the more complex chips to work in increasingly specialized applications, thus reducing the overall volume of chips they purchased. Nevertheless, once an IC maker developed the necessary factory equipment, it could mass-produce chips in vast quantities for only pennies apiece.

User-Configurable Logic

One answer to the problem of low-volume complex chips entailed making chips in high volumes, then letting customers configure them to fit their particular applications. The first of these was the gate-array chip, or master slice chip, called micromosaic, which Fairchild introduced in 1967. With the gate-array approach, the chip manufacturer makes a large quantity of identical chips, consisting of a pattern of AND, OR, and NOT gates, none of which has yet been interconnected. Customers then determine how they want their chip connected, so the chip maker can customize it based on those instructions, connecting some logic gates and bypassing others. Depending on how efficiently the customer makes use of the available logic, this may or may not result in an effective use of chip space. However, the approach does provide a quick turnaround time, since the chip buyer need only wait for the manufacturer to perform the logic interconnection step on an otherwise completed chip. This approach allows the chip maker to produce large volumes of chips that can be kept in stock and then quickly modified for customers.

In 1977, Monolithic Memories, Inc., developed a new kind of user-configurable logic chip, called the programmable logic array (PLA). While the PLA chip resembled the gate-array chip in that it consisted of many unconnected logic gates, its gates were accompanied by tiny fuses that, when blown, established the required interconnection pattern. By using a special computer system, chip buyers can customize their own PLAs, without having to depend on the chip maker. State-of-the-art PLAs hold up to 5,000 logic gates.

Once the gate-array or PLA chips have been completed, they cannot be changed. To overcome that limitation, erasable programmable logic devices (EPLDs) were introduced in 1983 to allow chip buyers to customize chips and then change them as many times as they wish. EPLDs couple capacitors, rather than fuses, with the logic gates. Since capacitors can store electric charge, each logic gate can be connected or bypassed at will by modifying the charge in the associated capacitor. Moreover, each capacitor is insulated with silicon dioxide, which becomes slightly conductive when exposed to ultraviolet light, thus permitting the capacitor's charge to leak away. Once customers program their EPLDs, they can erase them by exposing them to ultraviolet light and then reprogram them. This lets chip buyers try out various program combinations on a single chip and then correct any errors or implement new designs on existing chips.

Regardless of the method used to interconnect an IC's logic gates, each chip is designed to perform a particular function. For example, the gates on a particular IC might accept a numerical input and run it through a series of steps, corresponding to the steps taken to solve an equation or series of equations, and output the result. Once that chip is used in a product, such as a calculator, its function remains fixed. Chips manufactured for popular functions and sold complete to customers are called standard, or off-the-shelf, ICs. Chips such as gate arrays, PLAs, and EPLDs, along with custom chips tailored to a customer's particular specifications, are all termed application-specific ICs (ASICs).

Microprocessors and Intelligence

Microprocessors are ICs containing general-purpose groups of logic gates that users may access with stored instructions. In other words, they work like the central processing units (CPUs) of general-purpose digital computers, although they are squeezed onto a single IC. Microprocessors differ from ASICs in the same way that computers with stored programs differ from calculators. In ASICs and calculators, a fixed number of procedures may be performed on an input, determined by the way the logic gates are interconnected. With a microprocessor or computer, an input may run through one series of logic gates, and then be stored temporarily while an instruction is fetched from memory. This instruction may call for the input to go through the same logic gates again or through a different set of gates. A series of such microinstructions enables a relatively simple chip to perform extremely complex tasks. Microprocessors function somewhat more slowly than ASICs, since they must store intermediate results and wait for the next instruction. In many applications, however, the user does not even notice the delay.

Most significantly, however, the microprocessor allows almost infinite versatility. For example, the same microprocessor can be programmed to perform the work of a calculator, regulate the operation of a car engine, or act as a timer on a microwave oven. In those applications, the manufacturer programs the microprocessor. A microprocessor can also be used as the CPU of a computer, in which case the computer maker programs some instructions and the user programs others.

Since microprocessors can be programmed for so many different applications, manufacturers can produce them in huge volumes, which leads to very low costs per unit. If the price is low enough, the microprocessors can be used in low-cost consumer items — everything from music boxes to lawn sprinklers.

The microprocessor's ability to perform different operations, based on either the output of the last operation or the input from an external source, makes them intelligent. This means that they can make decisions. A robot arm with an IC chip preset to perform a certain function repeatedly is termed dumb, no matter how complex its task. In contrast, an arm with a microprocessor that allows it to alter its operation under specific conditions is called intelligent.

The first commercially available microprocessor was the Intel 4004, officially announced in 1971. Intel developed the microprocessor concept while designing a family of calculators for a Japanese company named Busicom. Rather than designing a whole set of ICs, Intel created a single chip that contained a CPU. The proper instructions, stored on another chip, could turn the microprocessor into a calculator.

Although the microprocessor could process data with all the power of a computer, it was not, strictly speaking, a computer on a chip. By definition, a true computer must contain, in addition to a CPU, some memory for storing intermediate computation results and instructions, as well as the electronics needed to convert input signals into digital data for processing and then to convert the results back into some form of output or display.

Intel offered its 4004 chip in a four chip set: one that could store instructions (the read-only memory chip); another that could temporarily store inputs and intermediate solutions to problems (the random access memory chip); one that could convert signals from the keypad into the proper form for processing; and one that could be used as an application-specific controller. The advantage of this approach was that it enabled designers to select the best combination of components for each application without having to redesign the entire processor.

Bits and Bytes

Instead of processing information 1 bit (a single one or zero) at a time, computers accept a number of bits simultaneously, then process them all as a group. Computer scientists refer to the number of bits a computer can accept at one time as the machine's word length. In general, the larger the word length, the more powerful the computer's processing capability. Today's computers most commonly use 8-, 16-, and 32-bit word lengths, though some larger machines can handle 48- and 64-bit lengths. Eight-bit processing was once so widely used that an 8-bit word became known as a byte. Although word length may vary from one computer to another, a byte always equals 8 bits.

Microprocessors, like computers, also accept specific word lengths. The 4004, for example, could handle only 4 bits. In 1972, Intel designed the 8-bit 8008 for the computer terminal manufacturer Display Terminal Corp. (now called Datapoint Corp.). As with the 4004, Intel offered the 8008 with the other chips needed to make a complete microcomputer system.

Intel had designed the blocks of logic on both the 4004 and the 8008 for customers with specific applications in mind, but it then offered the same products to other customers. Although the 4004 and 8008 were general-purpose chips in the sense that customers could program them for many different tasks, they were not designed for versatility. In 1974, Intel introduced an 8-bit, nonspecific microprocessor, the 8080. The 8080 was so versatile, so powerful, and so easy to program that Intel sold millions of the ICs and licensed their manufacturing rights to numerous other companies, which in turn sold millions more. The chip still remains popular today. Capable of performing 200,000 operations per second, the 8080 computes about 40 times faster than ENIAC and costs about $2.

Although the Intel chips and chip sets were technically computers, years elapsed before they were actually used as computers. From 1971 to 1975, electronic companies bought microprocessors and microcomputer chips to program for specific applications in their products. As a result, the users of the products did not think they were buying computers; instead, they were simply buying a cash register, appliance, or burglar alarm that contained a preprogrammed computer as a component. Today, these types of applications remain the largest market for microprocessors.

In 1975, MITS in Albuquerque, New Mexico, introduced the first computer based on a microprocessor chip set, the Altair 8800, which MITS sold to hobbyists in kit form. Driven by the Intel 8080 chip, the computer cost several hundred dollars and sold by the thousands. Confusingly, the Altair 8800 and other small computers based on microprocessors, microcomputer chips, and chip sets, soon became known as microcomputers. Therefore, when discussing microcomputers, it became necessary to specify whether the term referred to a chip, a set of chips, or a fully operating computer on a desktop.

Through the late 1970s and 1980s, microprocessors increased in power and speed as their word lengths increased from 8 bits to 32 bits and their complexity rose from 2,250 transistors on the Intel 4004, to over 400,000 transistors on the 32-bit Hewlett-Packard HP9000, introduced in 1981, and on the National Semiconductor 32131 chip, introduced in 1986. Intel's 80486 and Motorola's 68040 chips, which were introduced in 1989, each contain about 1.2 million transistors.

Even with advanced 32-bit microprocessors, the most powerful computers still use discrete logic chips on computer boards rather than microprocessors as their CPUs, because some types of IC logic chips can incorporate faster, but larger and more power-hungry, transistors that cannot be easily fabricated on a single chip. Also, the CPUs of some of the most powerful computers are too complex for even the most advanced microprocessor. Consequently, boards and sets of boards composed of high-speed logic chips still dominate the world of high-end CPUs.

Nonetheless, computers based on the latest microprocessors challenge the power of the mainframe computers of only a few years ago, setting up classification problems, since the usual term "microcomputer" seems inappropriate for a powerful minicomputer or mainframe computer. As a result, the computer world often refers to such powerful microcomputers as supermicrocomputers.

Semiconductor Memory

Another IC of immense importance to computers, the semiconductor memory chip, closely paralleled the rise of the microprocessor. To do their work, computers need two types of data: the actual data a user wants to process and the instructions that tell the computer exactly what to do. Once encoded into digital ones and zeros, both types of data can be stored in either a computer's secondary memory (also known as bulk storage or mass storage) or in its primary memory (also called main memory). A computer's secondary memory can hold immense quantities of information but requires a relatively long time to access the data. Secondary storage devices include floppy disk drives, rigid disk drives, tape drives, and optical disks.

Main memory typically stores less data than does mass storage, but it can be accessed more quickly. Computers must store the instructions that they are currently following, the data they are processing, and the intermediate results of complex calculations in their main memory, so that the CPU can perform its functions without waiting.

Computer scientists use several terms to describe computer memory. Volatile memory is erased when the user shuts off the power; nonvolatile memory is preserved even without power. Memory may also be destructive, meaning that it is erased each time it is read, or nondestructive,

Figure 3. Integrated Circuits. The first integrated circuit (top), invented by Jack Kilby of Texas Instruments in 1958, consisted of several hand-wired components and was slightly smaller than a match head. Intel's first microprocessor (bottom), the 4004, which was introduced in 1971, contained 2,250 transistors. *Courtesy: (top) Texas Instruments, Inc., and (bottom) Intel Corporation.*

meaning that it remains after being read. If the memory is dynamic, it tends to leak away and must be periodically refreshed with current; if it is static, it retains data as long as it receives constant power. Read-only memory (ROM) is nonvolatile and never added to; write-once memory allows data to be stored once, but then never changed; and read/write memory can be stored and read repeatedly.

Finally, memory may be random access, which means a user can call up any bit of information stored in the computer, or serial access, which means the data are arranged sequentially so that some bits are more quickly accessed than others. A tape record, a computer memory device similar to video tape, is a type of serial access memory, because users cannot read data at the beginning of a tape unless they rewind the tape to that point.

A computer's main memory, which must be very fast, is always random access. Until 1969, most computers used magnetic core memory, which consisted of a grid of wires with donut-shaped magnetic rings approximately one-sixteenth of an inch in diameter at each intersection. Two currents, one sent along a horizontal wire and another sent along a vertical wire, would combine at their intersections to change the polarity of the magnetic ring. While rings magnetized in one direction denoted a "1," rings magnetized in the opposite direction indicated a "0." Sensing wires passing diagonally through the cores detected the polarity of the core, and thus the resident ones or zeros.

Developed by MIT researcher Jay Forrester in 1951 and used in MIT's Whirlwind computer, magnetic core memory resembled panels of window screening arranged in walls. Such core memory was nonvolatile and random access, and it could read a bit of data in one-millionth of a second. But it was also destructive, losing its data as soon as it was read.

During the late 1960s, many companies worked on applying the IC to the memory problem. Researchers tried both bipolar and field-effect transistors, and although they developed some devices that were faster than core memories, they could not at first get them to store enough data to replace magnetic cores. Instead, engineers used them together with other chips to create a small amount of fast memory.

IBM shipped the first IC memory chips in 1968. Based on bipolar technology, each chip stored 64 bits of data in memory cells composed of transistors arranged in a configuration called a flip-flop, a circuit long used for data storage. In fact, the old ENIAC stored its data in vacuum tubes arranged in flip-flop circuits. Flip-flop circuits switch from one voltage to another, maintaining a given state as long as they receive a steady power supply. Since a flip-flop consists of two transistors and two resistors, a chip storing 64 bits needs 256 components just for storage and many others to route electrical signals coming into the chip.

IBM's memory chips were one part of a main memory system for the company's mainframe computers. Because the memory chips had a low storage capacity and were relatively expensive, IBM used them with other, more established memory technologies, such as magnetic cores. IBM restricted their use to its own products. However, in 1970, Fairchild introduced a chip called the 4100, the first commercial IC memory product. The 4100 could hold 256 bits. Although it cost more than a comparable amount of core memory, it ran faster, reading one bit in 70 billionths of a second, and was nondestructive. Unlike core memory, however, it was volatile, losing its data when power went off. The 4100, like the IBM chip, was a static random access memory (RAM) chip. The flip-flop memory cells in static RAM chips maintain their state, and thus their data, as long as they keep getting power.

Manufacturers could make memory chips using flip-flop memory cells with either fast, high-power bipolar junction transistors, or slower, lower-power FETs. In either case, they could put only a limited number of cells on each chip because each flip-flop required several components. A simpler system, called a dynamic RAM memory, uses MOSFETs coupled with capacitors, which store electrical current. The presence or absence of charge indicates a one or a zero. Dynamic RAM can store more information on a single chip, but since the charge on the capacitors tends to leak away within a millisecond unless they are refreshed with current, dynamic RAM chips must be accompanied by a secondary refresher circuit. This adds to the cost and complexity of the memory system. Nonetheless, the dynamic RAM chip does take up less space and costs less. Intel introduced the first dynamic RAM chip in 1970 and called it the 1103. The chip stored 1,024 bits of data, four times the amount stored by Fairchild's static RAM chip introduced the same year.

Memory chips are the simplest type of IC, since they perform only the single, simple function of storing ones or zeros. They are also the most dense ICs, and their capacity increased extremely quickly. As memory capacity grew, the microelectronics industry began using prefixes commonly used in the decimal system as a kind of shorthand. These included kilo-, or K, for 1,000, and mega-, or M, for 1 million. Since digital electronics works in base two, however, the industry altered the meaning of the prefixes to stand for numbers more convenient in base two. Thus, kilo- in microelectronics stands for 1,024 (2^{10}), and mega-stands for 1,048,576 (2^{20}). Today, giga-, which stands for

1 billion in the decimal system but for 1,073,741,824 (2^{30}) in microelectronics, is also used.

When referring to memory, the prefixes kilo-, mega-, or giga- precede either bit or byte (8 bits). Thus, a 1-kilobit memory chip stores 1,024 bits of information, and a 1-megabyte memory board can store 1,048,576 bytes of information, which are equal to 8,388,608 bits. Today, a reference to an IC eliminates the suffix "bit." Therefore, someone referring to a 64K memory chip means a 64-kilobit, rather than a 64-kilobyte, device. Conversely, someone referring to memory boards and secondary storage devices will drop the suffix "byte." Thus, a reference to a 256K main memory assumes 256 kilobytes.

To put this into perspective, a computer needs 1 Kbit, or 1,024 bits, to store about 20 average-length words in English. ENIAC, with its 18,000 vacuum tubes and other components, could store about 5 Kbits or 100 English language words. In 1972, Intel introduced a 4K dynamic RAM chip, which could store about 80 words. In 1975, Intel introduced the first 16K dynamic RAM chip, which could store about one and a quarter double-spaced, typed pages of text, or about 310 words. The same year, IBM announced an experimental 64K dynamic RAM, which it began producing in 1977. A 64K chip can store about 5 pages of text, or about 1,250 words.

In 1983, Hitachi described the first 1-Mbit dynamic RAM chip (DRAM), a device that could store the equivalent of about 80 typed pages. Several other companies soon followed, with the first actual shipments of 1-Mbit memory chips occurring in late 1985. At that time, several companies announced chips that could hold 4 Mbits, or about 320 typed pages. In 1987, scientists at Nippon Telegraph and Telephone's Atsugi Electrical Communications Laboratory announced a 16-Mbit DRAM. And in September 1988, IBM announced that it will be using advanced x-ray lithography techniques to create 64-Mbit DRAMs by the mid- to late 1990s. Some analysts have predicted that chips with capacities of 1 billion bits will be available by the year 2000.

ROMs, PROMs, EPROMs, EEPROMs, and EAROMs

Nonvolatile, random-access IC memory is called read-only memory (ROM). Users cannot alter this memory, which holds the fixed and permanent data a computer requires for operation. For example, the basic instructions of a microprocessor stay the same, even though they may be performed in any order. Manufacturers permanently install instructions on standard ROM chips during the fabrication process, wiring the transistors to represent the pattern of ones and zeros necessary to define those instructions.

With some microprocessor-based computer systems, the user can change the function of the device by replacing one ROM chip with another, thereby altering the instructions to the microprocessor. In fact, one company demonstrated the versatility of the first microprocessor, the Intel 4004, merely by swapping ROM chips in a product that could perform a variety of functions, from telling the time to playing music.

Programmable read-only memory (PROM) chips work like standard ROM chips, with the exception that fuses reside at each transistor. PROM-programming systems allow customers to blow fuses selectively, thus putting their own instructions on the chip. In this way, customers can buy a large number of standard memory chips and customize the chips themselves without having to rely on the IC manufacturer.

In 1971, Intel introduced the 2K-bit model 1702 erasable programmable read-only memory (EPROM). EPROMs consist of FETs that store a one or a zero, depending on the absence or presence of a charge on a capacitor that accompanies each transistor on the chip. A layer of pure silicon dioxide insulation allows the capacitors on EPROMs to retain their charges indefinitely, unlike conventional MOSFET RAM chips. As with a PROM, customers can use a special computer system to program their own memory chip.

EPROMs take advantage of silicon dioxide's tendency to become slightly conductive when exposed to ultraviolet (UV) light. As long as the chip remains unexposed to UV light, it retains the data programmed into it. When UV light strikes the chip, however, the silicon dioxide becomes conductive, and the charges on the capacitors leak away, effectively erasing programs on the chip. The resulting blank chip can then be reprogrammed.

EPROMs have become extremely popular, especially for companies developing operating software (also called microinstructions or microcode) for microprocessors, since the memory chips allow them to erase the chips when they make a mistake or a change and then start all over again. Such companies often perfect a program on an EPROM, and then mass-produce it on ROM or RAM chips, which cost considerably less than EPROMs.

The EPROM played a crucial role in the success of Intel's first microprocessor. Early potential customers worried about making a heavy investment in writing the microinstructions the ROM chip needed to control the functions of the microprocessor. Using ROM and PROM chips to do so would increase their costs, since mistakes meant throwing out the chip and starting over. However, the availability of EPROMs allowed customers to buy microprocessors without having to invest heavily in ROM memory experiments. Using a single EPROM chip, customers could develop their own microprocessor instructions, making mistakes and changes along the way and starting over, until they got the program exactly the way they wanted it. Without this freedom, far fewer companies would have purchased microprocessors as soon as they did.

Erasure of the first EPROMs required about 30 minutes of exposure to UV light of the proper wavelength. The drawbacks of these UVEPROMs include the fact that users must protect them from light and physically remove them from the system for erasure. To overcome these difficulties, people can now use another type of EPROM, called electrically erasable PROMs, or EEPROMs, which they can erase by exposing the chips to a particular electrical current while still in the circuit.

EPROMs cannot be selectively erased. Therefore, researchers have developed what they call electrically alterable read-only memory (EAROM) chips, which allow customers to change parts of the memory while leaving the rest of the chip intact. Unlike a RAM chip, the EAROM is nonvolatile. However, the EAROM's relatively high cost and its lower data densities have restricted its use to specialized applications.

Charge Coupled Devices

Scientists developed two other electronic memory technologies at nearly the same time as RAM chips: charge coupled devices (CCDs) and magnetic bubble memories (MBMs). Although both provide serial access rather than random access, these devices run much faster than flexible disk drives and tape drives. In addition, while they are not as fast as RAM chips, they may be suitable for some main memory applications.

Developed at Bell Laboratories in 1969, CCDs consist of an n-type silicon base, a layer of silicon dioxide insulator, and a metallic layer of electrodes. The n-type silicon under the insulator, which sits at the points immediately beneath the electrodes, stores a small number of positive ions in response to a voltage applied to the electrodes. The absence or presence of ions defines ones and zeros.

A steady, low voltage keeps the ions under a particular electrode. A slightly higher voltage applied to the adjacent electrode attracts the ions. When the ions receive pulsing current from the grid of electrodes, they travel along from electrode to electrode, eventually reaching the bottom of the CCD and traveling out as a sequence of ones and zeros. To prevent data loss, the ions then cycle back to the top of the array and start over. Changing a particular bit involves waiting for it to reach a particular point on the chip, then applying the proper voltage to congregate a group of ions at that point.

The amount of time required to access a particular bit depends on the size of the grid and the speed at which the bits travel. It takes only about a microsecond (one-millionth of a second) or less to shift the bit over one step, about as fast as accessing a bit on a RAM chip. If the bit lies 50 positions away, however, that time increases fiftyfold. On average, it takes about 500 microseconds (0.5 milliseconds) to access any given bit in a CCD. Although this is too slow for most computer processing tasks, it is fast enough for intermediate data storage and many other peripheral memory applications, such as video display memory. Manufacturers use CCDs extensively in video cameras, where the chips are used to store signals that represent light intensities.

Magnetic Bubble Memories

MBMs were also invented at Bell Laboratories in 1969. As in CCDs, stored data move sequentially through the storage bubbles until reaching the read/write mechanism. However, MBMs store data in magnetic domains in a thin film of a substance such as garnet rather than ions in a layer of silicon. In the presence of a strong magnetic field perpendicular to its plane, the film maintains magnetic areas (domains) polarized either up or down. By using a permanent magnet rather than an electromagnet for this field, MBM builders can make the devices nonvolatile, so that they retain their memories even when the power is off. Reading and writing to an MBM does require electricity, however. Weak electromagnets along the MBM's horizontal and vertical axes move the magnetic areas along a grid of electrodes. A write head, consisting of a loop-shaped electromagnet, alters the polarity of a domain from up to down, or vice versa, to record a one or zero. The domains, visible under a microscope, look like cylindrical bubbles in the garnet.

A user accesses the magnetic areas in much the same way as reading a CCD, and with the same problem: serial access means relatively long access times on the order of 1 to 4 milliseconds. However, unlike CCDs, MBMs do not need to be refreshed, and they have tremendous storage capacity. In 1984, Intel introduced a magnetic bubble memory chip storing 4 Mbits of information, 16 times more storage than any RAM chip sold at that time.

In the 1970s, computer experts predicted that magnetic bubble memories would take over much of the market for computer memory. At that time, RAM chips stored relatively few Kbits and had the added disadvantage of being volatile. Bubble memories seemed to promise far greater nonvolatile capacities. Their lack of speed seemed a surmountable problem, especially at a time when computer processing itself was comparatively slow.

By the early 1980s, rapid development of RAM memories combined with far greater computer processing speeds made bubble memories unsuitable for nearly any computer-related applications. One of the few instances in which they did perform well during the early 1980s was in data storage for computers operating in hostile environments. Since MBMs are rugged and immune to power interruptions, they made great sense in factory automation equipment, laboratory instruments, and field testing and measurement devices.

Developments in the technology may significantly improve the performance of MBMs. In 1985, Floyd Humphrey of Carnegie-Mellon University in Pittsburgh, Pennsylvania, discussed how structures called Bloch walls between magnetic bubbles could be used as media for storing data. Vertical lines of magnetism in these walls occur in stable pairs that researchers have used to represent data in laboratory experiments. Far smaller than the bubbles themselves, these vertical Bloch walls may allow data densities several times greater than even the highest capacity MBMs made today.

Design of Integrated Circuits

Before the advent of ICs, engineers designed circuits at their drafting tables with pencil and paper, using standard symbols to represent logic gates. Once they completed the design of the logic, they would determine how best to build gates from actual electronic components (transistors, capacitors, and so forth) by drawing them on paper as well. It was then necessary to design the physical layout of the circuit board and to determine what kind of components could best do the job. Designers also had to arrange components in a way that minimized wiring without overcrowding and overheating the circuit. Finally, the designers would build and test a prototype, or breadboard, of the circuit.

The complexity of today's advanced ICs has made designing them on paper virtually impossible. A microprocessor containing a half-million circuit elements would require a sheet of paper the size of a tennis court. These difficulties will only increase in the years ahead as the maximum number of transistors on a chip rises from about 10 million in 1988 to an estimated 1 billion by the turn of the century. In addition, testing a prototype of the circuit has become impractical, since fabricating an IC requires a huge initial investment in equipment that becomes economical only when large quantities of chips are produced. For a simple IC, an engineer might simulate the circuit with discrete components on a circuit board, but that would only permit tests of the circuit's logic design. Other factors, such as how well the IC will dissipate heat generated during operation, or how fast the chip will actually operate, cannot be tested in that way. Consequently, engineers began using computers to design and test ICs.

Fabrication of Integrated Circuits

Factories that make ICs must be cleaner than surgical suites, because the smallest particle of dust can ruin their components. Similarly, the raw materials must be very pure, because even a few atoms of contaminant will render them unusable. Given such problems, IC fabrication facilities take great pains to maintain scrupulously clean rooms. For example, the air is filtered to reduce the number of tiny particles (no more than 0.5 microns in size) to less than 100 per cubic foot. People working in these facilities must wear sterile gowns, slippers, masks, and hats. To keep the number of people in the clean areas to a minimum, modern IC factories automate much of the fabrication process, using conveyor belts to move the circuits from station to station and robot arms to manipulate the circuits at each station.

Most ICs today are made of silicon. Chip makers create single, pure crystals of silicon up to 4 feet long and 6 or more inches in diameter by slowly pulling them from vats of molten silicon. They then slice the silicon cylinder into thousands of wafers, each about 0.02 inch thick. Each of these wafers may eventually be fabricated into hundreds of rectangular ICs, each composed, in turn, of many components.

In most instances, manufacturers use photolithography to define the areas on the wafers that will receive the deposits of n-type silicon, p-type silicon, silicon dioxide, and metal conductors that make up the transistors, resistors, and other electronic components of the ICs. Other fabrication techniques include electron-beam lithography and x-ray lithography.

The photolithographic process begins with a polished wafer of pure silicon, which is coated with an insulating layer of silicon dioxide by exposure to water vapor in a 2,000°F furnace. The wafer is then covered with a photoresist, a material that reacts when exposed to UV light. Negative photoresists harden when exposed to UV light, while positive photoresists are hard when left unexposed. In either case, a manufacturer can form a pattern on the wafer by covering it with a photomask (similar to a negative used in photography), exposing it to UV light, and then etching away the soft photoresist with acid (called wet etching) or ions (called dry etching or plasma etching). The etching process washes away not only the soft photoresist, but also the silicon dioxide underneath, right down to the original silicon wafer.

Once the photoresist has defined a pattern on the wafer, it is removed completely, leaving a wafer with a pattern etched through the layer of silicon dioxide. Next, the first stage of doping occurs, generally through a process called diffusion. The wafer is placed in a furnace with the dopant gas, which diffuses into the surface of the silicon. Ion implantation offers even greater control in the doping of silicon wafers. With this method, an ion gun ionizes the dopant material, accelerates it electrically through a tube, then fires it onto the silicon wafer in precisely controlled quantities.

The entire process is repeated for another layer of the chip by using a second photomask, followed by a third, and so on, until the entire circuit has been defined. Some chips require as many as 15 separate photomasks, each delineating a layer of n-type or p-type semiconductor, insulator, or conductor. The final masks define the interconnective wiring of the IC.

Shrinking the circuit design produced by the engineer down to the size of a chip one-quarter-inch square is a major challenge. At the end of the design process, an engineer has created a diagram consisting of a pattern of symbols representing an electrical circuit. Given the symbolic nature of this diagram, it must first be converted into a pictorial representation of the actual physical pattern that will go onto the silicon and create the circuit. Since IC chips consist of multiple layers, each photoengraved one at a time, the final diagram must include multiple photomasks, each representing one layer.

Experts expected photolithography to produce ICs with features as fine as 2 microns (0.00008 inch), but they assumed they could never focus UV light precisely enough to delineate narrower lines. State-of-the-art photolithographic techniques, however, can produce line widths of less than 1 micron, thanks to improved lens and focusing mechanisms.

In light of the success of photolithography, electron-beam and x-ray lithography have been slow to enter mainstream IC manufacturing facilities. Nevertheless, some manufacturers now use electron-beam lithography to produce very detailed masks, which can then be used in conventional photolithography, or to write directly onto the wafer. Although direct-write electron-beam lithography machines produce finer line widths than do photolithography systems, they take far longer. For this reason, electron-beam lithography is almost always used in special-purpose, low-quantity applications. More often, manufacturers use electron beams to make photomasks. Such photomasks outperform optically produced masks, but the resultant ICs do not have as fine details as ICs made with direct-write electron-beam systems.

X-ray lithography can produce even finer line widths than electron-beam lithography, but the equipment is costly and the process is relatively slow compared to other fabrication techniques. It is difficult to generate x-rays of the proper wavelength and intensity, except with synchrotron radiation generated from electrons in a storage ring. The Fraunhofer Institute for Microstructure Technology in West Berlin, West Germany, in cooperation with the West German government and several electronics companies, built an IC production facility that included the first electron storage ring devoted solely to chip fabrication. With it the group hopes to develop a pilot line for a 4-Mbit RAM IC.

Heat-Resistant Chips

Conventional silicon chips can operate in environments up to about 650°F; beyond that temperature, silicon becomes fully conductive, losing the properties that make it useful in digital electronics. In late 1987, a team including Robert F. Davis and John Palmour at North Carolina State

University in Raleigh announced that they had produced silicon carbide transistors capable of functioning at 1,200°F with the same performance levels as silicon transistors operating at lower temperatures. Such chips would be useful in many high-temperature applications, such as monitoring the status of internal combustion engines.

Another approach to making heat-resistant semiconductors has come from a group of scientists at Japan's National Institute for Research in Inorganic Materials. The team, headed by Osamu Mishima, developed a diode produced from cubic boron nitride that can withstand temperatures as high as about 1,000°F. They doped the boron nitride with beryllium and fabricated it at a temperature of about 3,100°F and very high pressure. The research team believes that boron-nitride-based diodes may be able to function at temperatures over 2,300°F.

Gallium Arsenide Chips

The next generation of computer chips may be based on a compound called gallium arsenide. Chips made of this material are significantly faster than silicon-based chips, because electrons pass through gallium arsenide faster than they travel through silicon. Since gallium arsenide chips also generate less heat, they can be packed closer together. The resultant shorter distances between the components on the chips reduce the travel time of electrons.

Despite these advantages, silicon chips will probably remain popular for some time. Silicon is readily available, inexpensive, and easy to fabricate into chips. In addition, an insulating layer of silicon dioxide forms naturally during the fabrication process, protecting the chip against imperfections. In contrast, gallium arsenide is difficult and expensive to extract and purify. The compound is also brittle, which makes it difficult to fabricate. In addition, gallium arsenide does not develop an oxide layer during fabrication, making it difficult to coat gallium arsenide chips uniformly with a protective insulating layer.

To reduce the chips' cost, manufacturers apply a layer of gallium arsenide to a silicon wafer, a compromise that improves the speed of silicon chips, while avoiding some of the inherent problems of gallium arsenide. To protect the chip, chip makers can also modify the transistor structure so that an energy barrier forms between the gallium arsenide and the metal, providing a protective coating. This type of device is known as a metal-semiconductor field-effect transistor (MSFET). An alternative way to provide protection is being explored by scientists at IBM's Yorktown Heights Laboratory and Bell Communication Research, who are developing means for chemically or photochemically protecting gallium arsenide surfaces. The process is still experimental, but the results may lead to innovations in gallium arsenide chip production.

In the late 1980s, a number of chip and computer makers were pursuing gallium arsenide chip technology. Cray Research, Inc., announced plans to use gallium arsenide in its Cray-4 supercomputer, and IBM and DEC also considered developing computers based on gallium arsenide chip technology. Part of the reason for the pursuit of gallium arsenide chips stems from the U.S. Department of Defense's interest in monolithic microwave integrated circuits (MMICs), which will enable radar designers to develop compact, yet powerful units.

Mass Storage Systems

The rapid development of high-power computer systems is due in large part to advances in mass storage, which save data files and programs for long periods of time. Unlike the computer's main, or local, memory, which stores only the information being used by the CPU at a given time, mass storage can save large amounts of data essentially permanently. In general, a computer moves only the data and software with which it is currently working from mass storage to main memory, and then it sends the results to mass storage when it is finished. As computers have increased in power, so has the need to find techniques for storing and retrieving information more quickly and to develop media that can hold more data in less space.

300 Years of Information Storage

Punched cards and paper tape. The oldest means of storing data involved punching holes in cards. Punched cards were first used in the early eighteenth century to automate looms used in the weaving industry. In automatic looms, holes in the cards were "read" by a series of needles. When a needle detected a hole, it permitted the shuttle carrying the thread to pass through it. When a needle touched a part of the card without a hole, the shuttle stopped. In a similar way, cards were used to shift the pattern of stationary thread through which the shuttle had to pass. By using cards both to alter the pattern of the stationary threads and to change the movements of the shuttle, the weaver could control the pattern on the material the loom produced.

Punched cards also became popular in the late nineteenth century, when they were used in calculating machines. In early calculating machines, the holes in the cards were read by metal needles similar to those used in weaving looms. In these machines, however, the needle passed through the holes and made contact with a pool of mercury lying below the card. When the needle touched the mercury, it completed an electric circuit. The electricity passing through the circuit, in turn, rotated gears or moved levers in the calculator.

When the concept of stored computer programs became popular in the 1950s, punched cards became a convenient way to save both programs and data, with a hole or a lack of a hole representing the ones and zeros of the binary system. To write data or program code on a punched card, the operator simply used a machine to punch a series of holes through the card in the particular patterns assigned to specific alphanumeric characters or commands. The coded patterns and their physical locations needed to conform to an agreed-upon standard between the card-punching device and the card-reading device. The most common standard was devised at the end of the nineteenth century by Herman Hollerith. It is still used by some systems today, primarily as a return card containing customer information for billing systems.

Punched tape was also once popular for storing programs and data on the same medium. The tape operated in the same way as punched cards, in that the presence or absence of holes coded for ones and zeros. The main difference was that punched tapes were continuous, which made reading data simpler than in punched cards and eliminated one of the greatest fears of early programmers — dropping their deck of cards and mixing up the order.

Magnetic memory. Magnetic storage was first used in the 1950s, with the introduction of the magnetic drum. The magnetic-media-based memory systems all operate on the same principles, regardless of their size and form. Magnetic drums, fixed disks, floppy diskettes, and tapes all use a base material (a metal or plastic disk or plastic tape) coated with iron oxide or another material capable of being magnetized. The material is moved continuously past a read/write head.

When writing to the disk or tape, the read/write head produces an electromagnetic field whose polarity changes to code the data. The magnetic particles on the surface of the media are reoriented in a direction corresponding to the polarity of the electromagnetic field. The orientation is either positive, corresponding to one in the binary language of computers, or negative, corresponding to zero. As the magnetic medium is moved past the stationary read/write head, the head is able to code the series of ones and zeros that make up the data files or programs for the computer. When reading magnetic media, the same read/write head is able to detect the polarity of the magnetized surface as it passes the head. The read/write head translates this polarity into an electric current that is passed on to the computer as data.

The two specifications that define a mass storage system's usefulness are capacity (the amount of information that can be stored) and access time (the speed with which data can be obtained). Reels or cassettes of magnetic tape can store huge amounts of information, but they have relatively slow access times because they must be rewound or advanced to the location of any particular piece of data before it can be used. Reels of tape are particularly useful for sequential processing applications such as billing systems, where many data records are processed one after another. For such uses, the relatively slow access time to the first record is less important than the fact that the system knows that it can find the next record in sequence on the tape.

IBM introduced flexible, or floppy, diskettes in 1972. Floppy disks spin continuously and are accessed by read/write heads that travel radially across their surface, recording or sensing magnetic impulses as directed by a computer's software. Since the heads can move quickly to any position on the diskette, floppy disk drives have relatively fast access times; but since they have much less surface area than a reel of magnetic tape, they also have lower capacity. Nevertheless, the advent of floppy drives opened important new avenues for storing and transporting data, and was an important step toward the development of microcomputers.

Hard disk drives, also known as rigid disks or Winchester disks, were introduced by IBM in 1973. They have the same basic design as floppy disks, but the magnetic medium is coated onto a metal plate, which is sealed in a dust-free chamber. Dust-free chambers are necessary because even a single particle of dust can cause a read or write error. As a disk's storage density (the amount of data stored in a given unit of area) increases, a dust-free environment becomes even more critical, because more data are at risk. Consequently, high-performance disk drives are generally sealed. Hard disks can store 20 or more times the data per recording surface as floppy disks. They can also access data considerably faster, since they typically spin at 3,600 revolutions per minute (rpm) versus 360 rpm for floppy disks, and they have multiple read/write heads, reducing the number of times any one head must travel across the medium to access data. In addition, hard disks' higher data-transfer rates to the CPU further enhance their speed advantages over floppy disks.

State-of-the-Art Magnetic Memory

Better tape drives. Although the basic principles of magnetic data recording are the same as they were 30 years ago, each of the three forms of media — tape, floppy disks, and hard disks — have advanced considerably, especially during the past few years. In 1985, IBM shipped a magnetic tape drive called the 3480, which was the result of a 10-year development effort called Project Saguaro. Using plastic tape cartridges measuring 4 inches by 5 inches by 1 inch, the device replaced tape drives using the familiar 10.5-inch open reels. The 3480 drives' improved magnetic tape coatings and read/write heads allow them to record information at higher densities.

Since the cartridges are sealed and do not need to be threaded, they can be loaded automatically from a rack of tapes. This feature can give users ready access to hundreds of tapes containing billions of bytes of data. The cartridges also require only one-quarter the shelf space of open reels. Despite these advantages, it will probably be many years before the 3480 drives replace open-reel drives because so much data already exist in the older tape format.

Denser floppy disks. Floppy disk drives have become significantly smaller during the past 15 years. The 8-inch diskettes of the 1970s gave way to 5.25-inch minifloppy diskettes in the 1980s. In 1984, several companies, including IBM, introduced microfloppy disk drives, using diskettes measuring between 3 and 4 inches in diameter. By 1985, a 3.5-inch diskette designed by Sony Corp. of Tokyo, Japan, had become the industry standard for microfloppies, forcing the discontinuation of the other designs. Microfloppy diskettes currently have a storage capacity of more than 1 megabyte — equivalent to over 600 double-spaced typed pages. Using special diskette coatings and modified read/write heads, capacities of 5 megabytes have been achieved, but the cost of producing the high-capacity microfloppy drives, and especially the cost of the diskettes themselves, have kept these products from being adopted by computer manufacturers.

3M Corp. of Minneapolis, Minnesota, has been experimenting with the production of sheets of a flexible diskette medium stretched taut like the surface of a drum, instead of the usual floppy sheet of plastic used in ordinary diskettes. Their technique, which is called stretched surface technology (SST), allows over 20 megabytes to be stored on a surface the same size as a 5.25-inch minifloppy diskette. Several disk-drive manufacturers are also testing the medium in high-capacity disk drives, and 3M claims that removable SST cartridges are under development. The cartridges would offer both the storage capacity of hard disks and the convenience of floppy disks.

Improved hard disks. The performance of hard disks has increased more dramatically than any of the other magnetic storage technologies. Improvements in the magnetic material coatings, together with more sensitive read/write heads, greatly increased hard disk performance and capacity during the early 1980s. The introduction of a low-cost hard disk drive by Seagate Technology in 1980 brought the price of fixed disk drives within the reach of owners of personal computers. The wholesale cost of 10 megabytes of hard-disk memory fell from several thousand dollars in 1975 to $200 or less in 1988. The maximum storage capacity of fixed disks for personal computers exceeded 300 megabytes by 1987, and by the end of 1988, several companies began offering hard disks with over 500 megabytes of storage.

The ability to squeeze more data onto a given amount of disk surface has led to the design of smaller fixed disks. Micro-Winchesters, with the same surface area as 3.5-inch microfloppy drives but with 40 megabytes of storage, were first announced in 1985. By 1988, some 3.5-inch fixed drives for personal computers had capacities over 100 megabytes.

Special applications for 3.5-inch hard disks. A spin-off development of 3.5-inch hard disks was the plug-in drive card, the first of which was offered by Plus Development Corp. of Milpitas, California. The Plus HardCard contained a 10-megabyte fixed disk and all the electronics necessary to operate it, and it fit directly into the expansion slots of IBM's line of personal computers and compatibles. Before the HardCard was available, a user adding a hard disk had to disassemble the computer, install the disk unit in a drive bay, and connect it to an additional electronics board, the controller, which linked the drive to the CPU. In contrast, the HardCard could be plugged into the PC within minutes and did not require sacrificing a drive bay. By 1988, a number of hard disk cards were available, with storage as high as 60 megabytes.

The 3.5-inch hard disks also began appearing in laptop computers during the mid-1980s, enabling users to easily carry the large amounts of data previously restricted to desktop computers. Toshiba America, Inc., set a new data storage standard for laptop computers in 1986 when it introduced a 14-pound computer equipped with a 10-megabyte hard disk. By 1987, more than a dozen computer makers were offering hard disk laptop computers with storage capacities as high as 40 megabytes.

In December 1987, engineers at IBM's Almaden Research Center in San Jose, California, announced the development of a new 3.5-inch hard disk that can hold 10 billion bits of information, the equivalent of over 600,000 typed pages. The IBM engineers etched a dense pattern of magnetic cells on a film of cobalt alloy, then removed all extraneous magnetic material. This created the smallest particles (0.5 micron wide) ever used in the coating of a magnetic storage device, resulting in an exceptionally high storage density.

Optical Disk Drives

In the late 1970s, a number of companies announced their intention to develop and market optical disk drives, which would use lasers to write large amounts of data onto flexible media. Several of the companies, including Storage Technology Corp. and Burroughs Corp. (which recently merged with Sperry to form Unisys), eventually abandoned their optical disk projects after investing hundreds of millions of dollars with no return. One French company, Alcatel Thomson Gigadisc, however, did succeed in developing 12-inch optical disk drives, but could not produce enough disks to make the product economically. After nearly going bankrupt, Gigadisc turned to the 3M Co. for a supply of the materials used to make the disks.

Even when optical disks became available in the early 1980s, computer makers doubted whether many buyers would want a device that allowed them to read but not write data. In 1983, Matsushita Electrical Industrial Co. announced that it was developing an erasable optical storage system, but technical difficulties kept the system from ever being produced.

In addition to technical problems, potential makers of optical disks did not want to commit research and development funds to optical disk drive development when disk size standards had not yet been established; some disk suppliers were promoting 12-inch disks, while others were promoting 5.25-inch disks. Finally, although optical disk drives could hold a tremendous amount of data — the equivalent of hundreds of thousands of pages — their access times were little better than those of floppy disks, rendering them impractical for most applications.

By the mid-1980s, however, optical disks finally became a practical means of data storage, largely because of the success of consumer products, such as stereo compact disks (CDs), on which music is digitally encoded and read with a laser. CDs have created a greater demand for many of the components, such as semiconductor lasers, that originally made optical memory devices so costly. Mass-production of these parts for consumer products has led to dramatic reductions in the costs of the parts, which in turn has reduced the cost of optical memory devices.

CD-ROM devices. The most direct use of consumer technology in an optical memory computer product is the compact disk read-only memory, or CD-ROM, codeveloped by Sony Corp. and Philips in 1984. CD-ROM devices typically read up to 540 megabytes of data (equivalent to about 300,000 pages) from a single optical disk. Large data files, instructional manuals, or complex computer programs can be recorded onto optical disks at a factory.

CD-ROM disks are coated with a special material that reflects a laser beam back to a detector circuit. To store data on the disk, the reflective surface is broken so that the beam is not reflected. The data are coded as a pattern of reflective and nonreflective points on the disk. CD-ROM disks are all manufactured from a single master disk, much like phonograph records are produced. The master disk is made by using a high-power laser to burn small holes at the data points corresponding to ones, while leaving the areas designated as zeros untouched, or vice versa. The spots recorded onto optical disks may be extremely small, allowing very high data densities. And unlike the magnetic charges recorded onto disks and magnetic tape, the spots burned into an optical disk cannot be accidentally erased.

The master disk is used to make a production negative disk, which is used in turn to press the final distribution copies of the disk. These distribution copies are identical

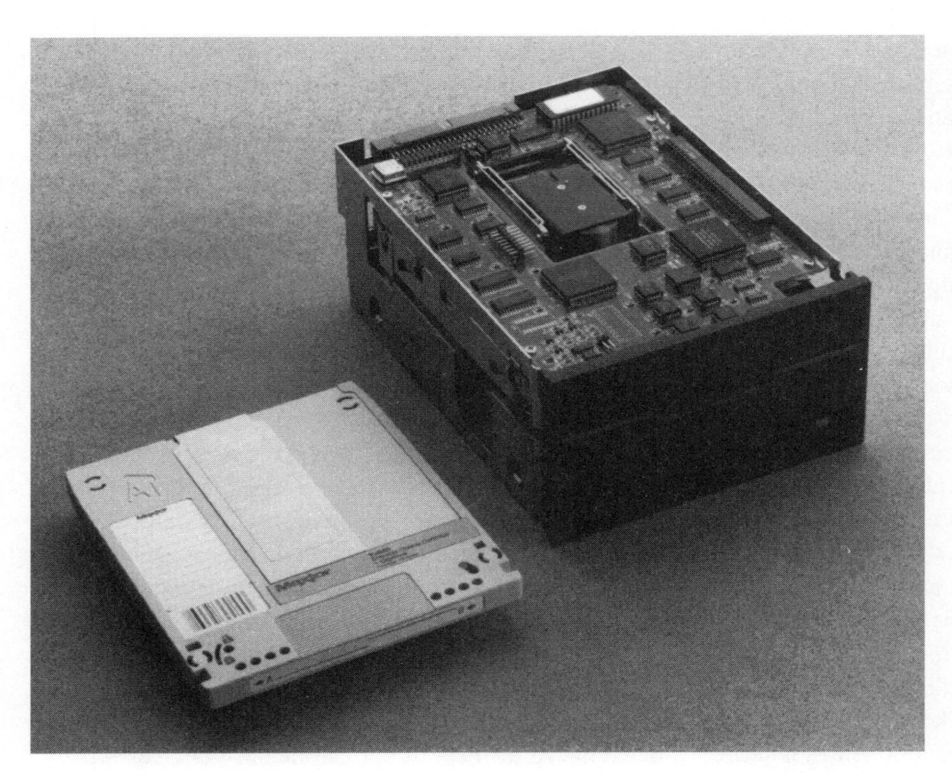

Figure 4. Erasable Optical Memory. Each removable cartridge used in Maxtor Corp.'s Tahiti I, a 5.25-inch erasable optical disk drive, holds 600 megabytes of memory, the equivalent of about 1,700 typical 5.25-inch floppy diskettes. The Tahiti I design has overcome what had been one of the major limitations of erasable optical disks — slow access to data. *Courtesy: Maxtor Corporation.*

to the master disk, containing the same pattern of holes and reflective surfaces.

CD-ROM publishing. By 1988, a number of publishers began to take advantage of the enormous storage capacity of CD-ROMs. For example, Grolier Electronic Publishing began offering an electronic version of its standard encyclopedia in 1987, and in the same year, Microsoft Corp. of Redmond, Washington, introduced a CD containing *The Chicago Manual of Style*, *The American Heritage Dictionary*, *The 1987 World Almanac and Book of Facts*, *Bartlett's Familiar Quotations*, and six other major reference works. Also in 1987, Standard & Poor began selling Compustat PC Plus, a CD-ROM database containing statistics on 7,000 companies. In 1988, Lotus Development Corp. of Cambridge, Massachusetts, released Datatext, a four-disk compendium of business information, listing financial information on more than 10,000 companies.

One of the advantages of searching for information on a CD-ROM is that the user can specify criteria (such as annual revenues, number of personnel, and so forth), and information on all the companies matching those criteria will be retrieved. This makes it possible to compile unique clusters of information that would be extremely time-consuming, and in some cases impossible, to assemble by hand. For example, with Standard & Poor's Compustat, a user could request data on all companies with annual sales between $100 million and $150 million and then refine the search with additional criteria.

CD-ROMs are already having a major impact on the sciences, where thousands of new articles are continually being published. In medicine, for example, an estimated 6,000 new articles appear in journals each day. A number of CD-ROM disks containing the abstracts or complete text of articles are available to medical libraries and physicians. For example, Elsevier's Excerpta Medica (EMBASE) abstracts a broad range of articles in the major medical disciplines. Some scholars of the English language now use the CD-ROM version of the Oxford English Dictionary, which defines about 300,000 words and contains over 40 million words.

Write-once, read-many drives. In the mid-1980s, the goal of developing optical storage devices that can both write and read data was achieved. Several manufacturers, including IBM, began offering 5.25-inch write-once, read-many (WORM) drives. WORM optical disks start with a blank disk that is completely reflective. Data are permanently written to the disk by a high-power laser that burns holes into the reflective surface of the disk in the same way that a master disk is made for CD-ROM. Once data are written to the WORM disk, it cannot be erased. This makes WORM disk drives well-suited for archival backup where permanence is important and the inability to erase and rewrite data is not a disadvantage. Each time a data file is updated, a new series of holes must be burned into the medium. Since the previous version of the file cannot be erased, it remains as a backup. In this way, the user continually backs up data without losing earlier versions of a file. Companies can also use WORM drives to write large databases to optical disks and then distribute the disks to their branch offices, which are also equipped with WORM or CD-ROM drives.

Erasable optical memory. The ultimate goal of many developers of optical storage devices is an optical disk that can be erased and reused like a conventional floppy or hard disk. Two techniques for creating erasable CDs are currently under development, those based on magneto-optic properties, and those based on crystal phase change.

Magneto-optic recording — also called optically assisted magnetic recording — uses a medium that can shift between two states of reflectivity under the combined influence of a laser and a magnetic field. This allows data to be written to the disk by altering the disk's reflectivity, and it also allows the data to be erased by returning the reflectivity to its starting state. Since the writing method does not permanently alter the surface of the disk, as is the case in CD-ROM and WORM disks, erasable optical disks can be erased and rerecorded.

In 1988, Maxtor Corp. of San Jose, California, introduced an erasable magneto-optic disk drive called the Tahiti I. The drive has an access time of 43 milliseconds, which is more than twice as fast as most optical disks and comparable to that of Winchester hard disk drives. Maxtor achieved this high speed by using a combined read/write head assembly that weighs half as much as conventional read/write heads.

The second erasable optical memory technique uses materials that change from a crystal-lattice structure to an amorphous state when exposed to specific wavelengths of light, and then back again when exposed to other wavelengths. Although phase-change crystals have been used in nonerasable disks, the introduction of erasable versions has been hampered by instability of the crystals and the material's tendency to lose its capacity to change phase after a number of reversals. Matsushita planned to introduce a product in 1989 based on crystal phase change.

In 1988, Tandy Corp. stunned the computer industry when it announced the introduction of THOR-CD, an erasable CD that can record, play back, and erase digital signals from an audio or computer system. The company first plans to apply its technology to music systems expected to reach the market by late 1989 or 1990, and then adapt it to data storage. While precise details of Tandy's system have remained proprietary, some industry experts believe that the system will use two lasers, one of which will raise small areas on optical disks coated with a dye polymer. Each raised area marks a bit of data, and a photodetector would read the data areas. The second laser would be used to flatten the raised areas, thereby erasing the disk.

Digital Paper

In early 1988, a new high-density mass storage medium called digital paper became available. Developed by the British company ICI Electronics and marketed by Creo Products, Inc., digital paper is actually made from high-density recording film and is also referred to as flexible optical tape. In the late 1980s, digital paper was the lowest-cost storage medium available, with a cost of just $0.05 per megabyte of data. In contrast, conventional magnetic tape costs about $0.20 per megabyte, and an optical disk costs about $0.25 per megabyte. Moreover, a 12-inch reel of optical tape can store 1 terabyte (1 trillion bytes) of data. Since digital paper can be cut into a round shape, it could eventually be used as a medium for flexible 5.25-inch or 3.5-inch diskettes.

In addition to its relatively low cost per megabyte of storage, digital tape offers high-speed data access. For example, the access time for a 5.25-inch floppy disk with digital tape media is 40 milliseconds, compared to 100 to 200 milliseconds for a fixed optical disk, such as a CD-ROM. Finally, digital tape has the added advantage of a long shelf life. While magnetic tape has a maximum shelf life of 10 years under ideal temperature and humidity conditions, digital paper can last up to 15 years under ideal conditions, and ICI is currently working to extend the shelf life to 20 years.

Digital Audio Tape

Digital audio tape (DAT) is a new concept in magnetic tape recording. DAT recorders produce a mathematical value for an incoming signal, based on the binary code, hence the use of the word digital in digital audio tape. When the values are reconstructed during playback, the reconstructed sound is so much like the original that the human ear cannot distinguish the difference. The reproduction is so good, in fact, that American record companies have aggressively lobbied lawmakers to prevent DAT from being sold in the U.S., claiming that its availability could encourage people to copy CDs and violate copyright laws.

While the recording industry has fought what some analysts see as the inevitable introduction of DAT stereo equipment into the U.S., some computer companies consider DAT to be the medium of the future for storing massive amounts of data. A single DAT cartridge, which would be about the width and length of a credit card, could hold 1.3 gigabytes (billion bytes) of data, equivalent to about 700,000 pages of double-spaced text.

Several companies, including Gigabyte, Hewlett-Packard Co., Sony Corp., and Archive Corp., had either introduced or planned to introduce DAT storage systems by 1989. Hewlett-Packard and Sony have joined forces to propose an industrywide standard for DAT machines, and Archive has entered into a joint partnership with the Japanese electronics manufacturer Matsushita Electric Industrial Co. to develop commercial DAT storage machines. Some analysts believe that the ultimate success of DAT storage devices will depend on whether DAT consumer products succeed, much as the development of CD-ROM systems received a much-needed boost from the acceptance of stereo CDs in the consumer market.

Supercomputing Power

Computer scientists and engineers generally work with three main types of computers: microcomputers, minicomputers, and mainframes. Microcomputers refer to machines that are designed for individual users and that may cost from as little as a few hundred dollars to many thousands of dollars. Minicomputers, which can process and store considerably more data than microcomputers, can cost as much as $200,000 or more. Mainframes, which can cost millions of dollars, are used to process massive amounts of data for applications such as scientific studies, complex statistical analysis, and large-scale business operations and planning.

A fourth type of computer, the supercomputer, can process data hundreds to thousands of times faster than the fastest computers of the 1970s. By some accounts, Cray Research, Inc., ushered in the era of supercomputers when it introduced the Cray-1 in 1976. In terms of pure speed, the Cray-1 was certainly in a class by itself at the time. But the architecture now commonly used in the most powerful supercomputers today — parallel processing, a design whose goal is to make maximum use of different parts of the machines' processing components — was pioneered by IBM in the late 1950s and used in its STRETCH computer. Parallel processing was also incorporated into the design of Control Data Corp.'s 6600 and 7600 models, which were introduced in the 1960s and 1970s. In a sense, supercomputers have been around for more than 20 years in the form of specialized mainframes. Nevertheless, the Cray-1 marked a radical departure from IBM's STRETCH and Control Data's 6600 and 7600 models, being the first computer with "massively parallel" architecture. The Cray-1 consisted of two separate processors linked to a common memory. This arrangement enabled the Cray to process 400 million floating-point operations per second (flops) — a speed nearly 10 times that of its predecessors.

Supercomputer development was proceeding rapidly by the mid-1980s, when the U.S Department of Defense began its Strategic Computing Program, which had the goal of developing extremely powerful parallel computers that could drive sophisticated artificial intelligence applications. During this period, the Japanese government also launched the National Superspeed Computer Project and the Fifth Generation Computer System Project in an attempt to make Japan a world leader in both supercomputer technology and artificial intelligence systems. By 1987, several computer manufacturers were making computers capable of processing more than 1 billion flops.

In addition to their artificial intelligence applications, supercomputers are now being used in a variety of academic and research fields for modeling and design, which require enormous processing power. For example, supercomputers are used extensively in structural and fluid dynamic analyses, as well as the simulation of electronic circuits for chip and system design. Supercomputers have even spawned an entirely new discipline called computational physics, in which computer programs are used to prove hypotheses that cannot be confirmed through actual experiments. Supercomputers can also produce highly detailed graphic representations of data, which are particularly valuable for three-dimensional modeling.

Super Speed in a Small Package

The "super" aspect of supercomputers has nothing to do with their size. In fact, the size of the main processing units of the fastest supercomputers — in the range of about 4 to 5 feet high and of similar width and depth — is often smaller than that of mainframes and minicomputers. Within that compact structure, however, manufacturers pack tens of thousands of components. For example, the

Figure 5. Cray Supercomputer. The Cray Y-MP/832 computer system was the top-of-the-line supercomputer system offered by Cray Research, Inc., in the late 1980s. It is a parallel-processing computer with eight central processors. *Courtesy: Cray Research, Inc. Photo: Paul Shambroom.*

Cray-3 supercomputer, scheduled for delivery in 1990, will incorporate 16 processors containing 88,000 ultradense, ultrahigh-speed gallium arsenide chips. The space between components has been kept to a minimum to eliminate as much wiring as possible; the shorter the distance between components, the faster the machine can operate. The Cray-3 is expected to reach peak speeds of 16 billion flops.

Although supercomputer processors are relatively small, they consume a tremendous amount of electricity and give off large amounts of heat. A high-end supercomputer with multiple processors may consume a million dollars or more worth of electricity each year and generate as much heat as a furnace. The University of Illinois Supercomputer Center actually uses the waste heat from its computers to warm a nearby underground garage. Manufacturers use various techniques to cool their supercomputer processors. For example, IBM uses a water-cooled system, while Fujitsu relies on forced air. Cray Research used a Freon cooling system with its X-MP model, but switched to another fluorocarbon system with the Cray-2. The entire processing unit of the Cray-2 is immersed in the coolant.

Parallel Processing

In a general sense, a supercomputer is just like any other computer, in that it consists of a central processing unit (CPU), a means for entering data, and a means for outputting data. The structure of a supercomputer's CPU, however, is radically different from that of its less powerful relatives. Non-supercomputers typically use a single processor that performs operations one after another in a sequence. The processor fetches data and instructions from memory, performs an operation, and then repeats the process with the next instruction. This sequential operation limits the speed at which computations can be performed. In contrast, supercomputers use multiple processors linked together so that operations can be performed in parallel, rather than sequentially. The result is an enormous boost in speed.

There are several different types of parallel processing. The two broadest categories are Single Instruction, Multiple Data (SIMD) and Multiple Instructions, Multiple Data (MIMD). In SIMD computers, the multiple processors all get the same stream of instructions, but they perform them on separate streams of data. Each processor has its own local memory unit, but is also linked to a common memory unit. Of the new designs, SIMD is the closest to conventional computer architecture.

While SIMD computers use multiple processors to handle one set of instructions, MIMD computers use independent processors to handle more than one instruction set. The processors are linked together and share a central memory source. One MIMD supercomputer, the Cray X-MP, uses two linked processors, while the Cray X-MP/48 uses four processors. Keeping the processors working in harmony is more difficult than in the SIMD computers, but MIMD computers are better suited for highly branched programs of the type used in artificial intelligence applications.

Parallel-processing computers may also be classified as either coarse-grained or fine-grained. Coarse-grained designs use a small number of powerful processors, while fine-grained computers use many smaller processors. In general, coarse-grained computers are more easily programmed. Fine-grained computers have a greater degree of parallelism, but they are so different from conventional computers that they must be custom-programmed.

Other Supercomputer Design Techniques

Pipelining. All complex arithmetic tasks can be broken into a series of steps. One way to tackle them is to complete each step, one at a time, until the task is complete. When a computer processor handles arithmetic tasks in this way, some of its logic circuits are idle after they have performed their functions. This would be analogous to a manufacturing operation in which a worker completes a task and then waits for the next item to be manufactured. If some tasks take longer than others, workers with shorter tasks are idle until they have a new unit to assemble. A more efficient alternative is an assembly-line approach, in which a worker is constantly active, completing a task and then repeating it on the next unit on the line.

A similar approach can be used to pipeline data in a computer. Rather than allowing certain logic circuits of the processor to be idle after they have executed their instructions, pipelined architectures direct more work to the circuits so that they can continually manipulate data as if they were on an assembly line. Pipelining can significantly reduce processing time, especially when the computer is performing repetitive arithmetic tasks. However, the nature of the task and the software design have a major impact on the pipeline throughput. IBM pioneered the pipeline architecture in the 1950s when it introduced its STRETCH model computer, and the technique is still being used today.

Interleaved memory. In many operations, memory access time places a major constraint on processing speed; no processing can occur until the data are retrieved from memory and, in some cases, the manipulated data have to be placed back in memory. Moreover, data passing from the processor to the memory may have to wait until newly fetched data clear the channel. In either case, the processor's functional units will be inactive while the transfer takes place.

One solution to this problem is called interleaving, which entails using separate memory units that can be addressed by the logic circuits. With separate circuits, data "traffic jams" are eliminated, and more than one parcel of data can be fetched and acted upon simultaneously. The IBM STRETCH used memory interleaving techniques to boost processing speed, and these are still used today to reduce memory access times.

Supercomputer Applications

Supercomputers are now regularly used at institutions throughout the world. By 1990, as many as 100,000 people are expected to use supercomputers in their work. Government-sponsored programs have given some researchers greater access to supercomputers, and as the price of computer power continues to decline, more institutions will be able to afford their own supercomputing facilities.

National supercomputer centers. In 1986, the U.S. National Science Foundation (NSF) launched a $143 million program to support five supercomputer centers that would give researchers in many disciplines access to high-power data-processing systems. The centers are located at Princeton University, Cornell University, Carnegie-Mellon University, the University of Illinois, and the University of California at San Diego. The work of chemist George McRae is in some ways typical of the research for which supercomputers are now being used. McRae has used the powerful computer at the Pittsburgh Supercomputer Center to study the interactions of chemicals found in automobile engine exhaust. McRae's computer analysis suggests that, contrary to common sense, decreasing certain types of emissions can lead to greater pollution problems, because the chemical reactions that take place in exhaust depend on the relative levels of the different chemicals in the exhaust. This discovery could help the U.S. Environmental Protection Agency refine its automobile emission standards.

At Princeton's John Von Neumann National Supercomputing Center, another of the five government-funded operations, Arthur Winfree is studying how life-threatening fibrillations (uncontrolled contractions) of the heart muscles occur; this research may lead to the design of more effective defibrillation devices. Other researchers at the Princeton facility are using the supercomputer for projects ranging from the operation of internal combustion engineers to the structure of galaxies.

Linking supercomputers. More than 100 information networks now link various universities, government laboratories, and private institutions in the U.S. As computer networks proliferate, some experts fear that it will become increasingly difficult to maintain a coherent national system that will allow computers in one part of the country to instantly share information with colleagues in another location. Some computer scientists and communication specialists believe that the NSF's network, called NSFNET, which links the nation's supercomputer centers, could serve as the heart of a nationwide computer network system. Others are skeptical, however, noting that the various data communication networks in the U.S. have evolved randomly, creating a tangled web of pathways with different standards.

Regardless of whether NSFNET ultimately becomes the core of a national system, the NSF has extensive plans to develop the system further. About 200 universities were expected to have access to NSFNET by the end of 1989, and links have already been built between NSFNET and the Department of Defense network ARPANET, which connects research facilities working on military contracts. In 1988, the NSF awarded the contract for upgrading and maintaining NSFNET to Merit, Inc., an Ann Arbor-based consortium of eight Michigan universities. Merit, working with IBM and the telecommunications carrier MCI, hopes to increase the data transmission speed on NSFNET. In early 1989, the network was capable of transmitting 1.5 million bits per second; however, by improving the system's hardware and software, that transmission rate could be increased to 45 million bits per second. Such improvements will be necessary to maintain NSFNET's usefulness. If demand continues to rise at the rate of the late 1980s, the capacity of the network will be exhausted by 1990. NSF has budgeted $14 million over the next 5 years to ensure that the system expands to meet user needs.

Supercomputers at NASA. The National Aeronautics and Space Administration's Numerical Aerodynamic Simulation Facility (NASF), began operating in 1987. NASF is a supercomputer capable of performing up to 172 million flops. The computer is based on a Cray-2 supercomputer that has a sustained output of 250 million flops and, with the anticipated addition of a second processor, should be able to achieve a sustained speed of 1 billion flops. NASF has been used to simulate flight, process weather information, study atmospheric conditions on Mars, and analyze the evolution of our galaxy. In addition, engineers plan to use the supercomputer to design an aerospace plane capable of flying both near the ground and in orbit.

Using computers to analyze sedimentary basins. In 1988, a group of researchers from the University of Illinois at Urbana, the Exxon Production Research Co. in Houston, Texas, and the National Center for Supercomputing Applications in Urbana reported that they had used a supercomputer to model the geological processes involved in the formation of sedimentary basins. The computer analyzed chemical and physical processes as they might have occurred across geological time and then simulated different scenarios that could account for the origin of ores and the distribution of hydrocarbons, such as oil and gas. The researchers concluded that the numerical analysis performed by the computer, combined with information on regions' stratigraphic, sea-level, and plate tectonic histories, is a powerful tool for studying how sedimentary basins evolve and change.

Other research uses for supercomputers. Oceanographers have found that supercomputers can be used to model currents and eddies in various ocean regions. These models may yield insights into how energy is transferred both horizontally across the oceans and vertically through the water column. Meteorologists are benefiting from supercomputers, which can process the large quantities of data required for accurate forecasts of global weather patterns. Astrophysicists have also put supercomputers to use in analyzing the dynamics of planets' orbits and simulating a variety of stellar and galactic phenomena. For example, in 1988, Cornell University researchers Stuart L. Shapiro and Saul Teukolsky reported that they had used Cornell's National Supercomputer Facility to solve Einstein's equations of general relativity as they apply to the dynamic evolution of a relativistic star cluster. Such clusters contain many extremely dense neutron stars moving at close to the speed of light. If the forces keeping the stars apart are disturbed, the entire cluster may collapse into a black hole.

Since it is not possible to observe the formation of a black hole directly, the phenomenon must be modeled mathematically through computer simulation. In addition, tracking the positions and velocities of thousands of stars would provide a staggering mass of data that would be difficult to interpret. The Cornell researchers solved the problem by converting the numerical output from their computations into an animated color graphic image that showed movements of stars in the cluster on a computer screen. Shapiro and Teukolsky believe their technique could be applied to simulating other stellar phenomena.

Commercial uses for supercomputers. Supercomputers are now used for design applications in many industries. Before supercomputers became available, the design of aircraft, ships, and automobiles required laborious and time-consuming calculations to determine their optimal shapes. Supercomputers can drastically reduce design time by using the principles of fluid dynamics to calculate the effect of air and water on vehicle surfaces.

For example, when supercomputers are used in conjunction with wind tunnels and other testing equipment, they can provide detailed simulations of the performance characteristics of aircraft in a fraction of the time required by earlier generations of computers. Such simulations allow engineers to explore different design options at a far lower cost than conventional design techniques. For example, the Airbus A-310 is significantly more fuel-efficient than the earlier A-300, largely due to improved wing designs developed with the aid of supercomputers. Other aircraft, such as Boeing's 757 and 767 have also benefited greatly from the use of supercomputers during their design.

Just as aircraft manufacturers have used supercomputers to speed up the design process and identify potentially costly or dangerous flaws that would be difficult to repair later during testing, boat makers have used supercomputers to help them design high-performance vessels, including the *Stars and Stripes II*, which recaptured the America's Cup in 1986 after it was lost to an Australian team in 1983. Cray Research provided the developers of *Stars and Stripes II* with access to a Cray X-MP, which was used to test various keel and hull configurations.

On land, automakers have used supercomputers to model expensive crash tests that evaluate the safety of design prototypes. The automakers are now experimenting with the use of supercomputers to simulate accidents and test various designs for their crashworthiness. While computer simulations cannot fully replace crash tests performed with actual vehicles and dummy drivers and passengers, they can reduce the number of tests needed and help engineers pinpoint design flaws that would otherwise be difficult to identify.

Supercomputer Marketplace

In the late 1980s, the total market for supercomputers was estimated at $1 billion, and many analysts believe that the market will grow by 40 percent a year. To meet the anticipated demand for supercomputers, more companies are devoting their efforts to develop a variety of machines, including new highly parallel models and single-user machines. By the end of 1987, more than 60 supercomputer models were available from 40 manufacturers. These ranged from high-end machines costing as much as $20 million for a fully configured Cray-2, to about $75,000 for an Apollo Series 10000 single-user machine (see below).

High-end machines. The top tier of the supercomputer market has been dominated by Cray Research, Inc., and Control Data/ETA Systems in the U.S., and by Fujitsu, NEC, and Hitachi in Japan. At the end of 1988, Cray had installed 220 supercomputer systems worldwide, followed by Fujitsu with 57, Control Data/ETA Systems with 43, NEC with 14, and Hitachi with 12, according to figures from Cray Research.

NEC hopes to gain a larger share of the market with its new SX-3 supercomputer series, which is expected to perform up to 22 billion flops when it becomes available in 1990. IBM has attempted to enter the lucrative high-end supercomputer market by teaming up with Supercomputer Systems, Inc., of Eau Claire, Wisconsin, a company founded by former Cray chief engineer Steve S. Chen. To date, IBM's only recent entry in the supercomputer market has been an auxiliary vector processor designed for use with its 3090 mainframe computer. In addition to funding the development of a new supercomputer, IBM will supply Chen's company with access to its disk-drive and communications technology.

The goal of the IBM/Supercomputer Systems partnership is develop a computer with 100 times the processing power of supercomputers now on the market. The computer is expected to use 64 linked processors to achieve processing speeds of 128 billion flops, the same as the Cray-4, which is currently under development at Cray Research. Industry analysts believe that both the IBM/Supercomputer Systems and Cray machines will be introduced in the early to mid-1990s.

Mid-range supercomputers. While the top tier of the supercomputer market is dominated by five companies, the market for mid-range supercomputers (computers with processing speeds of 20 million to 100 million flops and costing from $200,000 to $1.3 million) is hotly contested. Some analysts have noted that a shakeout in the industry appears to be occurring. In 1987, some 16 companies had announced that they would produce new supercomputers, but by early 1988, six of those companies failed to produce machines. In April 1989, Control Data Corp. announced that it would close its 6-year-old ETA Systems subsidiary and stop producing supercomputers.

Single-user supercomputers. During the past few years, interest in single-user supercomputers (also known as personal supercomputers, superworkstations, or graphics supercomputers) has grown dramatically. In 1988, Apollo Computer, Inc., introduced a ground-breaking supercomputer made for individual use, the Series 10000, with a peak processing speed of 140 million flops. Ardent Computer also introduced a single-user supercomputer in 1988, the Titan, with a processing speed of 64 million flops. The price range of the Apollo Series 10000 is $70,000 to $80,000, while the Ardent Titan sells for about $80,000 to $150,000, depending on the configuration. These and other single-user supercomputers are expected to be especially useful for computer-aided design (CAD) and other tasks that involve manipulating three-dimensional images or processing large quantities of data. An advantage of such graphics and processing power is that engineers can experiment with various design alternatives and see their results immediately. As prices fall in the years ahead, a growing number of companies and individuals will soon have the power of supercomputers at their fingertips.

Artificial Intelligence

Nature of Intelligence

"Artificial intelligence" (AI) has been defined in a variety of ways, a fact that has led to a great deal of confusion about the term's meaning. This is largely a result of the rapid development of the field within the past few years, with many different researchers concentrating on different areas of research. For example, Marvin Minsky, one of the first AI researchers, defined artificial intelligence as "teaching computers to do tasks that, if done by humans, would require intelligence."

Another leader in the field, Avron Barr described AI as "the part of computer science concerned with designing intelligent computer systems, that is, systems that exhibit characteristics that we associate with intelligence in human behavior." Researcher Elaine Rich defines AI as "the study of how to make computers do things which, at the moment, people are better."

The common thread in each of these definitions is that they do not so much specify an individual discipline of computer science as they define a performance expectation. That is, AI research has the goal of making machines perform more like humans. This objective can be pursued by making computers think, see, talk, understand speech, move, or do physical tasks like people.

The major difference between AI and other areas of computer science is the intention to model what we humans think of as intelligence. But what is intelligence? For the purposes of AI research and its applications, it is often easier to define intelligence by describing what it enables people to do. For example, intelligence allows people to understand ambiguous or contradictory information, by applying outside knowledge to the problem to resolve the ambiguity or contradiction. For example, imagine that you have just arrived at the airport of a city you are visiting for the first time. You have received two different sets of directions to a particular place in the city from two different people. One of these people drives a taxi in the city while the other is just visiting the city for the first time himself. You would probably be able to resolve the question of whose directions to follow by applying the knowledge that taxi drivers make their living by knowing how to get from one place to another.

In the same way, intelligence allows people to evaluate and rank various pieces of information on the basis of their relative importance. Continuing with the same example, you could rank the directions to the location you are seeking by first ensuring that you are in the right country and city, and then seeking the particular street and finally the specific address on that street. You would not consider wandering randomly from city to city and street to street, looking at each building with a particular street number until you found one that matched both the city and street you wanted. Your intelligence allows you to rank the various elements of the address — in this case, city, street, and number — by their relative importance.

One of the key elements of intelligence is that it enables people to see differences between situations despite apparent similarities and, conversely, to notice similarities despite differences. Imagine viewing the facades of apartment buildings in various cities. You might find many that are very similar. For example, many are a few stories high, built of brick, and have dozens of windows, a front entrance, and so on. Yet you would have no trouble identifying your own building on the basis of a few relatively subtle differences, such as the color of the trim around the door. Conversely, you can easily classify two buildings as apartment buildings even though one has green door trim and the other has red trim.

The final element of human intelligence required for AI is flexibility of solution. When traveling from the airport to a particular apartment building, you could follow one of a relatively few logical routes or any of many illogical routes. Not only can you usually select one of the better routes, but if you should run into an unexpected roadblock along your chosen route, you can generally manage to change the route slightly and continue on your way. In essence, you are able to develop a flexible solution to the travel problem and modify it as you gain additional information.

Artificial intelligence in all of its implementations seeks to model one or more aspects of human intelligence to solve specific problems. René Descartes summed up the essence of being human with the phrase "Cogito, ergo sum" — I think, therefore I am. With the implementation of AI, can we really expect to give computers the human capacity to think? Perhaps a better question is, how can we tell if a computer is really thinking or merely demonstrating clever programming?

Turing Test

In 1950, Alan Turing proposed that if you could engage a computer in a blind dialogue with a person (through a terminal) and fool the human into thinking he was communicating with another person, then the computer would be thinking. Specifically, the Turing test runs in the following way: A person called the interrogator is placed in an empty room except for a computer terminal, which is linked to a terminal in another room where a man (player A) and a woman (player B) are staying. The interrogator can use the terminal to ask questions of player A or player B, with the goal of determining which player is the man and which is the woman. While this game may at first seem simple, Turing added a twist: Only one of the players is obligated to tell the truth, while the other is free to lie in order to deceive the interrogator. The game proceeds until the interrogator is ready to guess which player is which.

As a variation on this scenario, you can substitute a computer for one of the people and change the objective to determining which player is human and which is a computer, with the computer's responses designed to deceive the interrogator into thinking it is a human. If the interrogator is no more successful in the computer versus human game than in the male versus female game, then the computer is thinking according to Turing's definition.

ELIZA. One of the early attempts to simulate the interaction of the Turing test was a program called ELIZA, written by Joseph Weizenbaum of MIT. ELIZA impersonates a psychotherapist who engages the patient in a dialogue. The program responds to each statement by the patient with an apparently thoughtful additional question. For example, ELIZA might respond to the statement "I am sad" with the question "Why are you sad?" After a few rounds of interaction, however, it becomes apparent that ELIZA does not really understand the statements, but is just using a series of rules to break the patient's statements into parts and then use some of the parts in new questions. While short exchanges with ELIZA may resemble some human conversations, ELIZA is not generally considered to have passed the Turing test.

Birth of Artificial Intelligence

The beginnings of AI as a discipline can be traced to a conference held in Dartmouth, New Hampshire, in 1956. This conference involved representatives of industry and

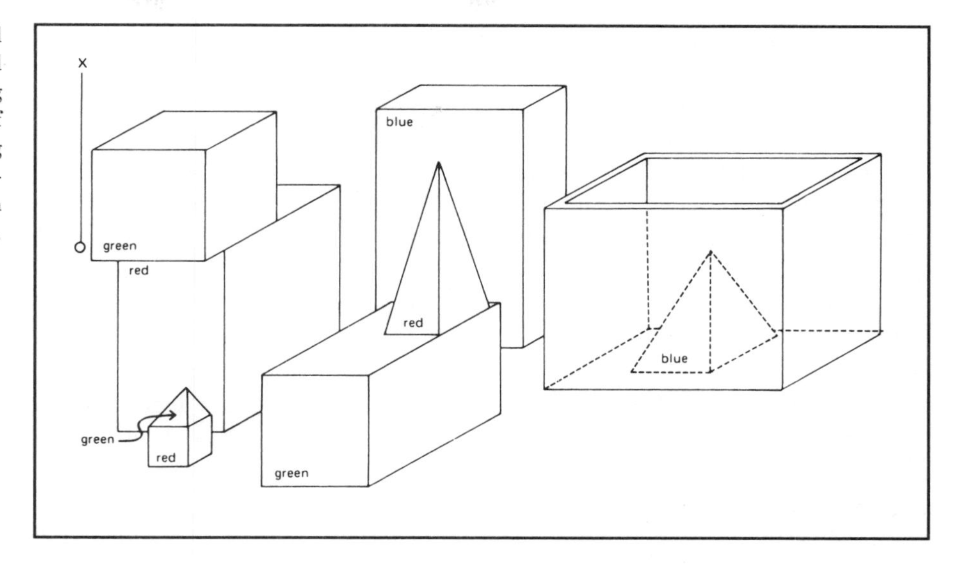

Figure 6. SHRDLU. Terry Winograd developed an innovative program called SHRDLU, which is capable of describing in English an environment consisting of various geometric shapes. People using SHRDLU can also manipulate the elements in the environment with English instructions. *Courtesy: Academic Press, Inc.*

academia and included four men now recognized as the founding fathers of AI: John McCarthy, Marvin Minsky, Nathaniel Rochester, and Claude Shannon. McCarthy was an assistant professor of mathematics at Dartmouth College, while Minsky was a junior fellow in mathematics and neurology at Harvard University in Cambridge, Massachusetts. They had been graduate students together at Princeton University in New Jersey and had worked with Shannon at Bell Laboratories. Rochester was manager of information research with IBM at the time of the conference. While the conference produced few concrete results other than the coining of the term "artificial intelligence," it is noteworthy for the people it brought together, and it served as a focal point for the varied fields of study — including mathematics, neurology, and computer science — that fall under the umbrella of AI.

Two other people at the conference, mathematician Allen Newell and economist Herbert Simon, later developed a computer program that they called the Logical Theorist designed to prove each of the theorems contained in Bertrand Russell and Alfred North Whitehead's *Principia Mathematica*. This was perhaps the first program to make use of heuristic processing rather than a fixed series of program steps; that is, the program used rules of thumb that lead toward an answer but that do not guarantee reaching it.

IBM computer scientist Arthur Samuel, who also attended the Dartmouth conference, developed a program that played checkers. An important aspect of Samuel's checker program was its ability to learn from its mistakes. This feature enabled the program to become a better checker player than its programmer, whom it soon beat easily. Another early AI application developed by a person who attended the Dartmouth conference was a chess program written by IBM scientist Alex Berstein. This program used heuristic programming techniques to search for the best moves among the multitude of possible moves in chess.

Basic Principles of Artificial Intelligence

Symbolic languages. In the years since the Dartmouth conference, several important subfields in AI have developed. These include expert systems, natural language interfaces, speech recognition and synthesis, machine vision, and robotics. While each of these specialties uses distinct programming techniques, they all share some features. All start by creating a model of the problem or the reasoning process used to solve it. The model is then translated into a computer program that uses one of the specialized programming languages developed for AI, such as LISP or Prolog. These are called symbolic processing languages, in contrast to the commonly used numerical processing languages such as Fortran, COBOL, C, or Basic. Symbolic processing languages are designed to deal with facts and their relationships in an AI application as compared to an emphasis on mathematics in numerical processing.

Symbolic processing languages such as LISP and Prolog operate by codifying knowledge into facts and rules or procedures for dealing with them. These elements are generally maintained as a list. This characteristic is reflected in the name of the program LISP, which is a contraction of the words "list processor." In LISP, facts are statements of the form BIRD(ROBINS EAGLES), which means that both robins and eagles are birds. Procedures use defined functions called primitives to operate on the facts. For example, the primitive CAR is defined to mean "take the first value of the object list." Thus, the procedure CAR(ROBINS EAGLES) would yield the list ROBINS. Programs written in LISP and other so-called procedural languages are built of such statements of facts and the procedures for dealing with them.

In contrast, Prolog is more descriptive. A Prolog program typically has many more statements of fact about the environment and relatively few rules for declaring how to process them than a LISP program. If you have adequately described the knowledge domain, Prolog will be able to find the solution without being told explicitly how to perform the processing. For example, suppose you wanted to describe the hierarchy of the animal kingdom with a series of facts based on a relationship called "member":

member(lion, mammal)
member(robin, bird)
member(mouse, rodent)
member(rodent, mammal)

The first line simply states that lion is a member of the group called mammal, the next line states that robin is a member of the group called bird, and so on. You could then ask Prolog several questions and it would search its list of facts for the result by matching the pattern of information. For example, you could use the following statement to ask what group lion belongs to:

member(lion, group)?

In this case, the variable "group" would be assigned the value "mammal."

More significantly, you could ask Prolog if a mouse is a mammal in the following way:

member(mouse, mammal).

Prolog would infer that this statement was correct and respond affirmatively despite the fact that the relationship was not explicitly stated nor was the logical hierarchy that a mouse is a rodent and a rodent is a mammal specifically explained to Prolog.

Semantic networks. The nature of the relationships in a knowledge base can usually be represented by a diagram called a semantic network. In this design, each of the facts is represented by a node and each of the relationships is represented by a connecting link. The process of searching through the semantic network can also be represented graphically as a tree of branching options. The tree is often called a game tree because it resembles the process of evaluating various options for the next move in many games such as checker or chess.

The process of searching the game tree of the knowledge base can be very time-consuming and has led to several different strategies for improving performance. These include depth first, breadth first, best first, difference reduction, hill climbing, and heuristic searches. In the depth first strategy, you (or the computer) select one branch of logic and proceed down it until you reach a conclusion. If the result is unsatisfactory, you return to the top of the game tree and proceed down a different path. In the breadth first strategy, you analyze all possible branches from the initial state down one level. If none of these is satisfactory, you move down to the next lower level and again analyze all possible branches at that level. If no solution is found, you proceed to the next level, and so on.

The best first strategy can be used if the problem allows you to evaluate the relative performance of the nodes on a given level. In this case, you seek the best option at each level, proceed down its branch, seek the best option at the next level, and so on. Difference reduction is similar to the best first strategy. In this technique, you evaluate the difference between each node and a stated goal and then select the branch that reduces the difference most. The hill climbing strategy is similar to difference reduction. In hill climbing, each branch is evaluated to determine if it moves you toward the goal or away from it. If a branch leads you toward the goal, you continue down it. If it moves you away, you move back up one level and try another branch. Heuristic searches use rules of thumb that may greatly reduce the number of options you need to evaluate or that help you concentrate on a particular group of branches first.

In addition to using different search techniques, it is sometimes worthwhile to approach the problem from a different direction. When you search from the initial state toward a desired goal, the process is called forward chaining. The process of searching from a desired goal back to the initial state is called reverse or backward chaining, and the process of searching from both directions toward the middle is called bidirectional chaining. Forward chaining is usually best when there are few initial states and many satisfactory goals, while backward chaining is best for problems where there are few satisfactory goals and many potential initial states.

AI Applications

Expert systems. Expert systems are computer programs that assemble the collective knowledge of one or more experts in a particular field (the knowledge domain) together with the procedures for applying that expertise. Once programmed, these systems can apply the collected knowledge to problems of types foreseen by the developers of the system. Expert systems are also inherently flexible enough to deal with some entirely new situations unforeseen by the developers, as long as the new situations are within the knowledge domain of the system.

One of the earliest expert systems, called MYCIN, was designed by Bruce Buchanan and Edward Shortliffe to diagnose bacterial blood infections. The expert knowledge used in MYCIN's knowledge base was reduced to a series of rules. For example, MYCIN "knows" that if the patient exhibits symptom X, the probable causes are Y and Z. MYCIN asks for information about the patient, applies rules to that information, asks additional questions if necessary, and then draws conclusions about the probable diagnoses and causes, further testing that may be needed, and proposed treatment. MYCIN also includes one additional feature now considered critical to the success of any expert system — the capacity to explain how or why the program reached its conclusions.

One of the first commercial applications of an expert system in industry was developed by DEC to help systems engineers configure its customer's VAX computer systems. This system, called XCON, analyzes a customer's system specifications and looks for incompatible equipment and other possible configuration problems. For example, if a company orders a certain number of terminals, XCON checks to see whether the company's system has enough communication ports to handle them all. If not, XCON makes recommendations for correcting the order. This expert system is able to review system configurations in a fraction of the time required by a systems engineer. By the late 1980s, DEC was reportedly saving about $70 million to $100 million a year by using XCON and other expert systems.

Another commercial application was developed by the American Express Co. to handle complex questions of credit approval for its credit card customers. Unlike most other credit-card companies, American Express does not use a fixed dollar limit for its credit policy. Instead it looks at a person's spending history and makes each credit approval based on the person's record. Implementing this credit policy had been costly, because it required human credit analysts to make decisions about each account. Now, most of American Express's credit inquiries can be answered automatically by using an expert system.

In the time since the specialized systems described above became available, several general-purpose expert systems shells have been developed by commercial software companies. These systems are called shells because they provide a general structure for information together with programming called an inference engine to analyze data. Users simply provide the information for the knowledge base, adapting the expert system to their needs.

By 1988, about 2,000 different expert systems were being used worldwide, according to Stanford University computer scientist Edward Feigenbaum. Feigenbaum has found that these systems can reduce the amount of time required to complete a task between 10- and 300-fold. A 300-fold increase in speed is comparable to the difference between strolling leisurely and flying in a jet.

Natural language interfaces. Natural language interfaces make it easier to communicate with computers by allowing the operator to use commands more like ordinary spoken or written language than formal computer languages. Ideally, users would be able to phrase questions in their own words, and the computer would then interpret the questions, perform the tasks needed to answer the questions, and then respond in the users' native language. In AI, natural language interfaces are treated separately from speech recognition and speech synthesis, which are described below.

A natural language interface requires that the computer not only be able to recognize the text of the command but also to understand it properly so that the computer can act upon it. The first step involves parsing the command to determine which words have been used for which parts of speech. The next step is to interpret the meaning of the request and translate it into a command that the computer can act upon. For example, a person who would like to search a database of clients might make the following request: "Give me a list of good clients in New Hampshire."

While the definition of "good" may be subject to further clarification, the system might interpret it as meaning better than average sales and develop a command such as the following:

```
LIST ALL clients FOR
(location="New Hampshire") AND
(sales >= AVG(sales)) TO PRINT
```

Natural language interfaces are usually associated with a particular software product and are most common in database applications. While no natural language interface can understand every possible request, several commercial products can handle a sufficient range of vocabulary and sentence structure to perform many tasks for inexperienced users.

Speech recognition and speech synthesis. An ideal user interface would also enable the computer to understand verbal commands and respond verbally. Speech recognition requires that the computer analyze the electronic signal produced when a person speaks into a microphone and then recognize the spoken words by matching them to a library of known words. This problem is extremely complex, because no two people pronounce a word in exactly the same way. In fact, even the same person may pronounce a given word quite differently in different situations.

To help simplify these problems, speech recognition systems are generally trained to understand the speech patterns of a particular user. Systems that can recognize the speech of many different users have been difficult to create and usually recognize only a small number of words. Further complicating the process of speech recognition is the human tendency to speak in continuous sentences, sometimes without detectable pauses separating each word. A computer can generally recognize isolated words much more easily than the connected words of ordinary speech.

Simply recognizing a particular word is often insufficient to allow a computer to act upon it. For example, the words two, to, and too all sound alike but have much different meanings. A sophisticated speech recognition system must be able to make a selection among the three choices based on the context in which the words are used. After recognizing all the words in the spoken phrase, the system must then be able to understand them in order to act on them. This step ties directly into the natural language interfaces discussed above.

In many ways, speech synthesis presents the opposite problem from speech recognition. Once the computer has performed the requested action, it must determine how to present the answer to the user. In many cases, determining what to say is easier than determining how to say it. Simply uttering words is not sufficient. To be like true speech, the phrase must be presented with the proper pronunciation, pacing, and intonation. In some cases, determining precisely what words to say or when to speak is the more difficult task. Each of these problems must be addressed by the speech synthesis system.

Machine vision. Computers that can "see" the world around them and act upon it have long been the subject of both science fiction and computer research. The problem of endowing a machine with the ability to correctly interpret the images it sees is in some ways similar to speech recognition and perhaps even more complex. For example, a newspaper photograph of a building is actually a collection of thousands of tiny dots, which is what the computer sees. Since the human brain is particularly adept at pattern recognition, you can immediately recognize the pattern of dots as a picture of a building and might even be able to identify what specific building it is, such as your home or the White House. However, teaching a computer to recognize even relatively simple objects is a complex problem.

Two basic techniques used for pattern recognition are template matching and edge detection. In template matching, the computer compares the image it sees to an inventory of known shapes and looks for matches. This can work well when the system is looking at objects with a limited number of known shapes. Edge detection is a more general technique, which involves analyzing the pattern of dots for edges. The computer interprets abrupt changes in color, intensity, or other features as edges. Once the edges have been determined, the relative shape of the edges is used to identify the object.

Robotics. Perhaps the ultimate products of AI would be fully mobile, self-sufficient robots capable of performing work in a home or factory and interacting with the people and things in a changing environment. Such robots are still years away, but more limited applications have been around since the beginning of the Industrial Revolution. By the loosest definition, a robot is simply any machine that performs a physical task previously done by humans. In that sense, Eli Whitney's cotton gin was a sort of robot because it emulated the way people processed cotton by pulling the fibers from the husks. Modern-day "intelligent" robots use sensors and servo-controlled actuators to detect, grasp, and orient objects correctly and then move them to particular locations, such as storage bins. Robots can also be programmed to carry out repetitive assembly-line tasks and perform some activities too dangerous for humans. In fact, as robots become increasingly sophisticated as a result of AI and improved engineering, they may well become key "employees" in some industries.

Japan has been the world's leading user of industrial robots, with a total of 141,000 robots in 1987, according to statistics compiled by the International Federation of Robotics in Stockholm, Sweden. This total is nearly five times the number of industrial robots in the U.S., which was second in robot use, with 29,000 robots in 1987. The total number of industrial robots installed worldwide increased 10-fold between 1980 and 1987, rising from about 22,000 to 230,000.

Computers in Manufacturing and Business

Second Industrial Revolution

During the past 25 years, computers have been responsible for major changes in the way goods are designed and produced. Just as the Industrial Revolution of the nineteenth century paved the way for production on an unheard-of scale, the computer revolution is reshaping manufacturing operations in industrial nations around the world. Modern manufacturing companies not only operate more efficiently than their precomputerized counterparts, but they provide a wider range of goods that meet ever-changing customer demands.

Three areas in particular stand out as examples of advances made by computers in the manufacturing arena: manufacturing resource planning (MRP), computer-aided design (CAD), and computer-aided manufacturing (CAM). Manufacturing resource planning allows companies to determine what parts should be ordered and to schedule the shipments of thousands of items, tasks that might otherwise require a large staff of full-time employees. Computer-aided design systems enable engineers to create three-dimensional models of objects and manipulate them in ways that would be impossible with traditional drawings, while computer-aided manufacturing involves the use of robots and other programmable devices to increase manufacturing efficiency.

Manufacturing Resource Planning

Before the dawn of the computer age, manufacturing companies had to devote considerable time and expense to ordering and scheduling the shipments of new parts. For example, a large company might inventory 10,000 or more individual parts. Manufacturers typically used what is called an order point system. In this system, an order would be placed when the number of each part fell to a designated level in an effort to keep adequate stocks on hand. Although this method may seem logical, it has two important drawbacks. First, it does not take into consideration when a part is actually going to be used. Consequently, a particular part may actually remain in a warehouse bin unused for many months, occupying space and tying up a company's money. A second and related problem is that the system does not indicate whether a particular item has to be matched with another item to make a finished product. Even if the manufacturing plant has 99 percent of the components for a given product on hand, if it lacks one component, the product cannot be completed and sold.

The job of coordinating ordering schedules for 10,000 or more parts is almost humanly impossible. However, with the advent of the computer, programmers have now developed software packages that help simplify this difficult task. In the 1960s, software programmers at the companies J. I. Case and Twin Disk began writing the first MRP software. Because this software was hampered by the slow access speed of the mass storage devices of the time, its scheduling functions were relatively limited. As more advanced data storage devices became available, however, true scheduling software became a reality, and the computer gained an important place on the floors of major manufacturing companies throughout the U.S.

By the 1970s, a number of commercial software companies began writing MRP programs capable of running on the most commonly used mainframe computers. These packages allowed firms to schedule orders for a vast number of parts well in advance, so that unnecessary and costly parts inventories could be kept to a minimum and production could proceed without interruption.

In the late 1970s, software called MRP II gave computers a new role in production scheduling. In this method, data from all departments within a company are fed back to the planners and schedulers, in what manufacturing expert Oliver Wight called "the closed loop." The closed loop involves feedback of information from all departments. The best schedule will be of little use if the company does not have both the capacity (people, equipment, and so forth) and the materials (components and raw materials) needed for the job. As capacity and materials plans change, this information is evaluated by the MRP II software, which makes appropriate adjustments in the ordering of parts.

Likewise, if the marketplace has suddenly changed and more or fewer products are needed, then the schedule must be altered. With the closed-loop system, information from the sales and marketing departments is fed back to the planners, so that production schedules can be adjusted accordingly. In this way, the company can theoretically tailor its output to the changing demands of the marketplace.

In the late 1980s, MRP II software cost anywhere from about $75,000 to $500,000. Although this initial investment is substantial, the pay-back can be enormous. A recent study by the Oliver Wight Companies, Inc., of Essex Junction, Vermont, determined that some companies may save as much as $250,000 per month by using MRP II software, when the costs of excess inventory and lost business opportunities are considered. Even so, only a small fraction of manufacturing companies worldwide now use MRP II, and few of the companies that have it are using the software to its full potential.

Computer-Aided Design

Computer-aided design (CAD) systems allow engineers to automate many of the drafting and computational tasks that until recently were done manually. With CAD, three-dimensional models can be produced on a video display in a fraction of the time required with pencil, paper, and drafting tools. Once a model is created, a number of analytical techniques can be applied to study its properties. In finite element analysis, the model is divided into simpler components so that its stress distribution can be more easily determined. Once the computer has determined the stress pattern of the individual components, it can determine the engineering characteristics of the entire object.

The early CAD systems available before the mid-1970s could create so-called wire frame models of the external surfaces of an object, but they could not analyze or preview the interior of a structure. Consequently, the systems were used mainly as drafting aids. Current CAD systems allow engineers not only to analyze the design and three-dimensional structure of solid objects but to easily compare different design options, a normally laborious process when manual drafting and modeling techniques are used.

CAD software is available for a range of machines from microcomputers to mainframes. Microcomputer CAD software is generally used for small-scale engineering and architectural tasks. As microcomputers become increasingly powerful and the resolution of video displays improves, however, CAD drawing and engineering applications will increase. Major manufacturers, such as automobile and aircraft makers, now typically use the powerful CAD software programs designed for minicomputers and mainframes, although this may change as the capabilities of microcomputers continue to grow.

The $4 billion CAD market is expected to expand significantly during the early 1990s as more companies and government agencies take advantage of the increasingly sophisticated software becoming available for all sizes of computers. For example, during the next 3 years, General Motors Corp. plans to spend more than $750 million on CAD equipment, the U.S. Army Corps of Engineers plans to spend $100 million, and the U.S. Navy plans to spend $1 billion to $2 billion.

Animation techniques now used by the film industry to produce photographlike images and special effects are expected to become major CAD tools in the 1990s. The computer graphics company Pixar of San Francisco, California, won an Academy Award in 1989 for a short cartoon called "The Tin Toy," which was made entirely with computer graphics. The realism possible with such sophisticated programs could have a variety of business applications. For example, architects could take their clients on animated video tours of buildings not yet constructed and ship designers could see how their new vessels would look in the water.

At the frontier of computer science is a developing technology called virtual reality, in which persons wearing special helmets and gloves can see three-dimensional artificial environments and manipulate the objects in such computer-generated worlds with their hands. Although practical and affordable virtual reality systems are probably still many years away, these systems could have an immense number of applications in business, education, and entertainment. For example, an interior designer might experiment with different furniture arrangements in a computer-generated room, a physics student might experience the world from the perspective of a subatomic particle, and persons seeking adventure might explore exotic fantasy worlds without leaving their own homes.

Computer-Aided Manufacturing

Computer-aided manufacturing systems develop approaches that programmable machines can use to carry out tasks

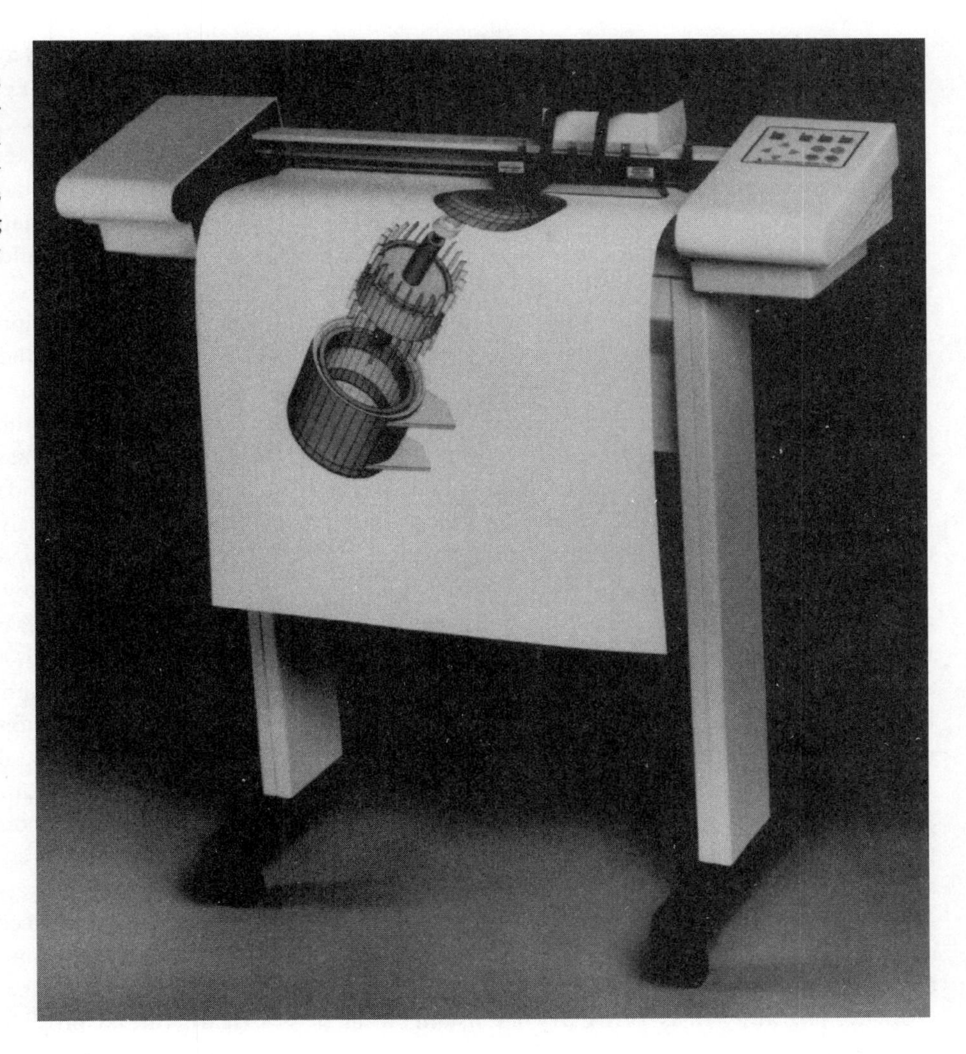

Figure 7. Houston Instruments Plotter. Large plotters like the one shown in this photograph are used to create output for designs generated by computer-aided design (CAD) systems. With CAD, engineers can create and manipulate objects on a computer screen before making physical models or prototypes. *Courtesy: Houston Instrument.*

efficiently. Modern CAM systems are descendants of the numerically controlled machines available in the late 1950s. These machines could be programmed by using paper strips or tapes encoded with instructions for various tasks, such as milling and cutting. Today's CAM programs allow machines to perform a much broader range of tasks.

Robots, which are essentially programmable machines, are also an important part of CAM. Some of the most common robots are stationary arms that rotate along various joints. The arms can use welding devices, hold drills or other tools, or apply paint. Other types of robots may move around a factory on wheels, guided by signals from cables in a factory's corridors and floors. "Intelligent" mail carts and inventory stocking devices are examples of such devices.

Computers in General Business

As computers are beginning to have an enormous impact on manufacturing companies, they are also causing sweeping changes in the way business is done in the office environment. Although mainframe computers have been used in business since the late 1950s for payroll and other accounting functions, the microcomputer revolution of the early 1980s has had a much greater impact on the day-to-day lives of office workers in the U.S., Europe, and Japan. Many of the tasks that were once done manually are now automated through a variety of off-the-shelf software packages. In fact, the availability of truly useful business software for personal computers has been the primary force behind the microcomputer boom.

Microcomputer business software falls into the basic categories described below.

Electronic spreadsheets. The spreadsheet was an important accounting and planning tool long before computers were available. A spreadsheet is a grid with headings that appear at the left of each row and the top of each column. The intersection of a row and column represents a specific item. For example, if the first row in a spreadsheet is labeled "Revenues," and the first column is labeled "January," then the intersection of the first row and column will reflect the revenues received in the month of January. The values in different cells (the intersections of columns and rows in the spreadsheet) may be manipulated mathematically (added, multiplied, and so forth) to obtain useful information, such as total sales or profit margins for a year. In the past, bookkeepers, accountants, and managers typically used adding machines or calculators to compute particular values. If one value changed, an associated section of the spreadsheet would have to be recomputed, making the analysis a laborious and tedious process.

The job of spreadsheet users became much easier in 1982 when VisiCorp introduced VisiCalc, which enabled people to make electronic counterparts of their paper spreadsheets. VisiCalc could instantly add columns or rows and feed the resulting numbers into a grand total, but more importantly, the program could recalculate the entire spreadsheet when the value in a single cell was changed. This allowed users to carry out instant "what-if" analyses, testing various assumptions about budgets, sales, revenues, and so forth. This capability made electronic spreadsheets an invaluable planning tool for managers at all levels in companies.

VisiCalc's successor, Lotus 1-2-3, introduced by Lotus Development Corp. of Cambridge, Massachusetts, added the capabilities of generating business graphs from spreadsheets, and using spreadsheets as simple databases. Lotus went on to become one of the most successful software companies in history and the leader in the spreadsheet sector. In 1989, Lotus announced the introduction of a third-generation version of its popular 1-2-3 program, which enables users to stack spreadsheets in three dimensional "cubes," allowing for easy consolidation of budget and departmental data.

Databases. Like spreadsheets, databases were widely used (although known by other names) long before the first computers. One of the most common everyday databases is the address book. Each entry in an address book may be considered a "record," a discrete unit of information that consists of various information "fields" — name, address, city, state, zip code, and telephone number. The primary disadvantage of an address book is that the records are fixed in place, so you can view them in only one order, usually alphabetically.

An electronic address book database uses the same concept of records, with a record consisting of various pieces of information. The advantage of an electronic address book is that the order of the entries may be quickly sorted in a variety of ways, such as by zip code, state, city, area code, or any other criteria you select. Moreover, with an electronic database, you can select different criteria and extract only those records that meet your specifications. For example, if you had assigned codes to everyone in the database to designate particular characteristics, such as whether they owned a pet and what their income was, you could later extract records of all pet owners with incomes above $25,000 a year. The same principles apply for records containing inventory, membership, or subscription information.

Sophisticated databases were once restricted to mainframe computers and minicomputers, but they became available to microcomputer owners in 1983 when Ashton-Tate, Inc., of Torrance, California, introduced dBASE. With dBASE II, users could develop their own databases through a so-called command language provided with the software. The command language enabled users to create sophisticated database applications without having to learn formal programming languages. Ashton-Tate has since released more advanced versions of its program, and by the late 1980s more than 50 software companies were producing database products for businesses.

Word processing. The first word-processing machines were large pieces of office equipment that allowed users to see lines of text and correct mistakes before anything was printed — a revolutionary concept for the time. They allowed typists to greatly increase their speed at the keyboard, since words "wrapped" automatically from the end of one line to the beginning of the next, eliminating the need to press a carriage return at the end of each line.

The first successful word-processing software for microcomputers, WordStar, was introduced in 1978 by Micropro International of San Jose, California. WordStar enabled microcomputers to perform all the functions of dedicated word processors at a far lower cost. By the mid-1980s, dozens of word-processing software packages had become available, with WordPerfect, from WordPerfect Corp. in Orem, Utah, becoming the market leader. Current word-processing software allows users to "cut and paste" blocks of text, search for and replace words or phrases, check the spelling of a document, and carry out literally hundreds of other functions.

Today, word processing is a fundamental tool in most modern offices, where it is used for everything from daily correspondence to creating reports. Will word processors ever completely replace typewriters? The answer is probably no, unless computer printers are developed that can readily handle the variety of odd-sized labels and paper sheets that are typically fed through a typewriter roller.

Desktop publishing. Until recently, companies that wanted to produce a professional-looking brochure had to give their text to a typesetter, who would re-enter it in a phototypesetting machine, specialized equipment capable of generating various typefaces. The phototypesetter would produce long strips of text called galleys, which would be cut into pieces and pasted onto a board, along with photographs and other images, to create what is called a mechanical. A printing company would then make a negative and a photographic plate of the mechanical, which would be used to print the final product.

Today, more and more businesses are avoiding some of these time-consuming tasks by using microcomputer page layout software instead. This so-called desktop publishing software allows users to import text from a word processor and images from graphics programs and manipulate them to create finished pages.

The first desktop publishing program, PageMaker, was created by Aldus Corp. of Seattle, Washington, for the Apple Macintosh computer. The software was well-suited to Apple's LaserWriter printer, which is capable of generating a number of popular typefaces. With PageMaker and a LaserWriter, business users could create a variety of publications at a fraction of the cost of having the materials typeset or buying conventional phototypesetting equipment.

Once desktop publishing became popular with Macintosh users, Xerox Corp. saw a marketing opportunity to develop desktop publishing software for IBM personal computers and compatibles, which together account for most of the business computing market in the U.S. Xerox purchased a program called Ventura Publisher, and began marketing it as the IBM PC answer to desktop publishing. Aldus Corp. also recognized the opportunity, and developed a version of PageMaker that can be run on IBM machines and compatibles.

Several other software publishers are now producing desktop publishing programs, and more will probably enter the market as the demand for desktop publishing applications grows. Moreover, as the printing quality of advanced laser printers approaches that of phototypesetting machines, companies will probably rely increasingly on desktop publishing to produce their printed materials.

Accounting. Although minicomputers and mainframes continue to handle the payrolls, general ledger, and other accounting functions of large companies, small and medium-sized firms have greatly benefited from a wealth of accounting software. For under $400, a company can purchase accounting software powerful enough to manage organizations with several divisions. Many of the packages automatically print checks, issue invoices and statements, update accounts payable and accounts receivable, and carry out other functions that at one time were done manually in small organizations. Once the programs are set up, they can quickly pay for themselves through reduced bookkeeping and accounting fees.

Graphics. Business charts and graphs have become an essential part of presentations and proposals. This is due in part to the availability of high-quality printers at affordable prices. Laser printers can now be purchased for around $1,000, and color dot-matrix printers are available for under $600. (For more information about these devices see the article "Anatomy of a Computer" in this chapter.) Some of the newer graphics packages allow users to create high-quality photographic slides or generate "slide shows" by using devices that can project computer-generated images.

A relatively new addition to microcomputer graphics software is the free-form drawing program. Some top-tier graphics programs allow users to custom-design graphic elements on the screen and then integrate them into charts or graphs. A number of packages also enable users to create artwork completely on screen and then integrate it into text. These programs are often used together with desktop publishing software, in which the merging of text and images is an essential part of designing a document.

Communications software. Communications programs and modems are used to link computers together, whether they are on the same street or halfway around the world. The software not only enables users to transmit files from computer to computer, but it provides a gateway into electronic mail services, through which letters, reports, and memos can be sent or stored in electronic mailboxes.

Another use of communications software is to "download" or receive information from an electronic data bank, such as Dialog. Dialog, a product of Dialog Information Services, Inc., is a collection of individual databases that contain the abstracts and full texts of articles about a variety of topics ranging from the day's news to science, medicine, and business. Users can search the databases for key words and phrases, and then download the information they want to their personal computers.

For example, suppose you wanted to prepare a report on the use of helicopters in emergency medicine. First, you would select the appropriate business and medical databases, and then you would search for various combinations of key words, such as "helicopter" and "emergency," or "helicopter" and "medical." Dialog would then produce the abstracts for all articles that contain the key words. Other information databases work in a similar way. A typical electronic search may cost about $10 to $50, although the cost can be much higher, particularly for specialized legal or business information.

Personal information managers. One new type of software introduced in the late 1980s is a group of programs designed to help people manage random bits of information. These personal information managers, or PIMs, are essentially databases that allow people to group or categorize information and later structure it as their interests and needs change. One such software package, Lotus Development Corp.'s Agenda, allows users to enter random ideas — such as lists of people to call, appointments to keep, tasks on a project, and so forth — and then assign various categories to each idea.

For example, one category might be a completion date, another might be a priority level, and yet another might be the project's status. At any point, a user can look at the information from a different "view." The user could, for instance, display or print a status view, in which all ideas are organized according to whether they have been completed or not, or instead opt for a priority view, in which ideas are grouped by whether they are considered high, medium, or low priority. At the end of 1988, Agenda was one of 25 different PIMs available, with more products set to enter the market in 1989 and 1990.

Work group software. One of the most promising new categories of software contains what have been called "work group" programs, which were being developed by a number of software publishers in 1989. This software is designed to allow people to work on documents simultaneously from separate computers connected in a network. With work group software, people will be able to include their comments and additions to a growing document without making permanent changes in other people's work. As such, the new software is expected to facilitate team writing programs for people in businesses, universities, and other organizations.

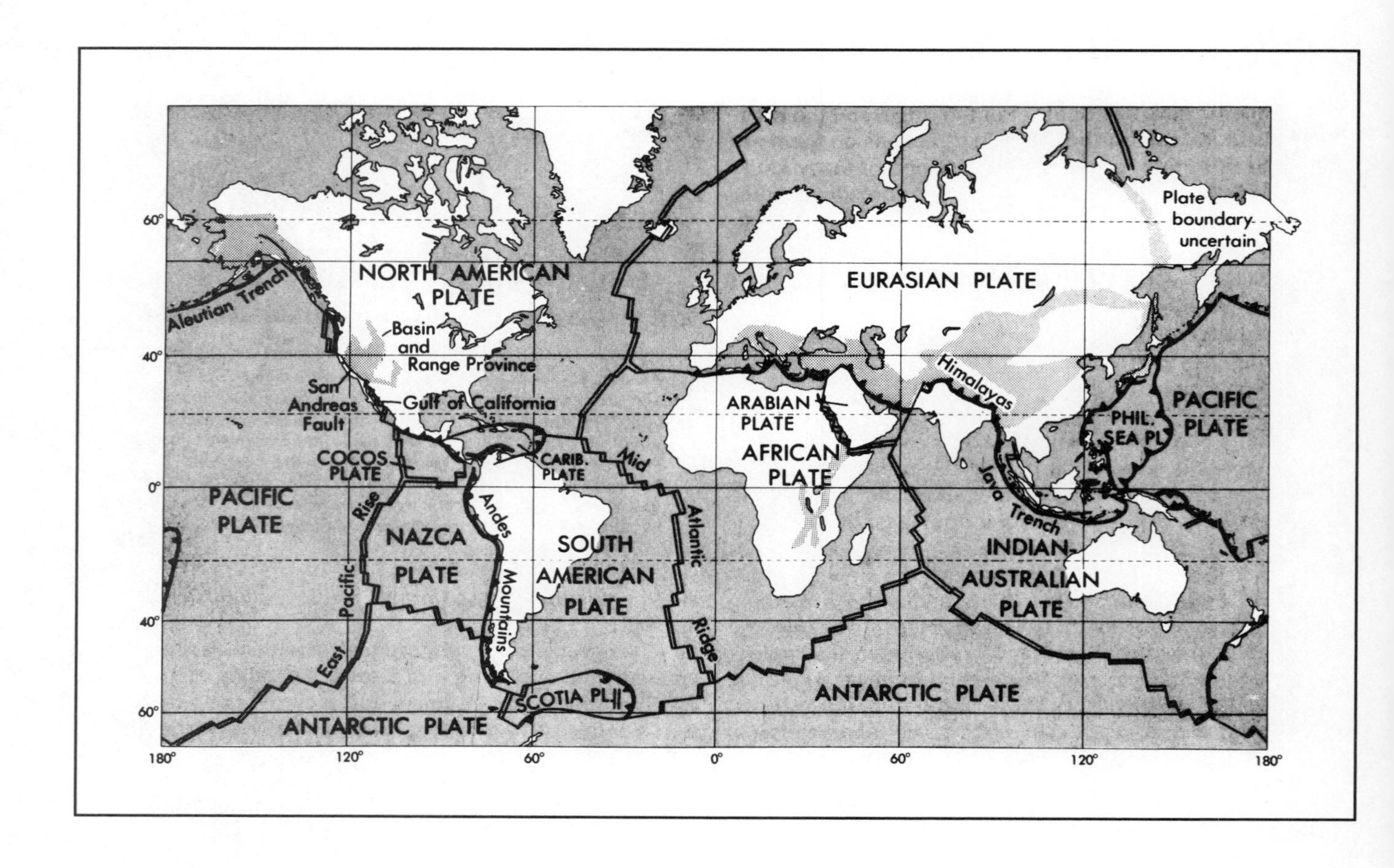

NORTH AMERICAN PLATE
Aleutian Trench
Basin and Range Province
San Andreas Fault
Gulf of California
COCOS PLATE
CARIB. PLATE
PACIFIC PLATE
NAZCA PLATE
East Pacific Rise
Andes Mountains
SOUTH AMERICAN PLATE
Mid Atlantic Ridge
SCOTIA PL
ANTARCTIC PLATE
EURASIAN PLATE
ARABIAN PLATE
AFRICAN PLATE
Himalayas
Java Trench
INDIAN AUSTRALIAN PLATE
PHIL. SEA PL
PACIFIC PLATE
Plate boundary uncertain
ANTARCTIC PLATE
60°
40°
0°
40°
60°
180°
120°
60°
0°
60°
120°
180°

Chapter 6

Earth Sciences

Earthquake Hazards, Monitoring, and Prediction

Earthquakes — the sometimes violent shaking of the ground caused by the sudden displacement of rocks at or below the Earth's surface — are one of the most destructive natural phenomena known. Although earthquakes have caused relatively few deaths in the U.S. in recent years, their toll has been far greater in some nations, including Algeria, Ecuador, Italy, Mexico, the People's Republic of China, the U.S.S.R, and Yemen. The most devastating earthquake of the 1980s, a magnitude-6.9 tremor centered near Spitak in northwestern Armenia, occurred on 7 December 1988. The main shock and a strong aftershock a few minutes later nearly leveled Spitak and heavily damaged many nearby cities and towns, including Leninakan and Kirovakan. Altogether, the earthquakes killed an estimated 25,000 people and left 500,000 homeless.

Even the toll of the Armenian earthquake is relatively small, however, when compared to losses from several catastrophic earthquakes in the past. The most deaths in any earthquake during the twentieth century occurred at T'ang-shan, China, on 28 July 1976, when a magnitude-8.2 shock destroyed nearly all the buildings there, killing 240,000 people and leaving 1 million homeless. Earlier in the century, on 16 December 1920, China's Kansu province was devastated by another major earthquake that took about 200,000 lives.

What Causes Earthquakes?

According to the theory of plate tectonics, also known as "continental drift," the Earth's outermost layer, the lithosphere, is composed of approximately one dozen rigid plates with an average thickness of about 60 miles. Heat-driven currents in the Earth's asthenosphere, a hot layer of rock below the lithosphere, move these plates slowly, at speeds averaging about an inch to a few inches per year. Earthquake and volcanic activity most often occur at the boundaries where plates converge, diverge, or slip past one another. As forces beneath the Earth's surface move two adjacent plates relative to each other, the rocks along the plate boundary may either slide relatively smoothly and steadily past each other or become locked together because of friction between the rocks on each side of the boundary. When rocks along a section of the plate boundary become locked together, the forces pushing the plates cause those rocks to be bent or deformed; geologists refer to this deformation as strain.

Once a strain limit is reached, the rocks at the boundary rupture along a plane, which is called a fault. Fault zones have strong and weak regions. Weak regions within the fault plane slip easily and concentrate strain at strong regions called asperities. The strong regions slip abruptly past each other only during earthquakes, releasing the energy that has accumulated in the strained rock. The seismic event may stop either at a major barrier, such as a kink along the fault, or in a weak region of the fault, where the rocks do not build up a large amount of strain. Seismologists have found that segments that have ruptured in large earthquakes tend to rupture again in large earthquakes with a recurrence period that varies from a few decades to a few centuries.

The sudden release of energy during earthquakes generates seismic waves that emanate in all directions from the area where the fracture occurs. As these waves pass through the ground, they cause it to vibrate or "quake," and these vibrations are known as earthquakes. The magnitude of an earthquake is a measure of the total energy released. The point beneath the surface where seismic waves are first generated is called the focus, and the epicenter is the point on the surface directly above the focus of the earthquake.

Although most earthquakes occur at or near plate boundaries, they may also occur far from plate margins at faults located in areas of weakness in the Earth's crust. The ultimate causes of these intraplate earthquakes are not as well understood as those of earthquakes along plate boundaries. However, the sequence of events that leads to earthquakes — the build-up and sudden release of strain — is the same in both. These earthquakes, which occur along faults, are called tectonic earthquakes.

Small numbers of earthquakes are associated with volcanic activity, but their frequency and size are quite different from those of tectonic earthquakes. Manmade earthquakes may also occur as a result of explosions generated at relatively shallow depths (within a few miles of the surface) by atomic tests. Only a few earthquakes are produced in this way, however, and they are relatively small.

Seismic Risks

More than 80 percent of the world's earthquakes occur along a series of plate boundaries that rim the Pacific Ocean to form what is sometimes called the "Ring of Fire." Many of the world's major cities lie in this seismically active region, which includes the Pacific coasts of North and South America, Japan, the Philippines, Indonesia, and New Zealand. Earthquakes are also common along plate boundaries in the Caribbean and Mediterranean regions, as well as in Afghanistan, Iran, the People's Republic of China, Turkey, and the U.S.S.R.

As the world's population has grown and become concentrated in urban centers during the twentieth century, earthquake-related property losses have increased steadily, and the potential for catastrophic loss of human life has also increased. According to the U.S. Geological Survey (USGS) and the Federal Emergency Management Agency (FEMA), a recurrence today of the magnitude-8.3 earthquake that struck San Francisco in 1906, taking over 2,500 lives and causing about $170 million in damages (in 1978 dollars), could kill about 5,000 people, injure 700,000 others, and cause about $24 billion in damages. A shock of similar size along the Newport-Inglewood Fault near Los Angeles could kill up to 23,000 people and cause property damage of about $45 billion.

Several regions in the U.S. not generally thought to have a high seismic risk have experienced devastating earthquakes during the past few hundred years and could be the sites of major events in the future. For example, a violent earthquake struck Charleston, South Carolina, on 31 August 1886, killing 60 people and damaging nearly every building in the city; the earthquake was felt in over 30 states and parts of eastern Canada.

The single most violent series of earthquakes in U.S. history occurred not in California or Alaska but in the lower Midwest, near the area where the borders of Arkansas, Kentucky, Missouri, and Tennessee meet. During the winter of 1811–1812, three events with Richter magnitudes estimated at 8.6, 8.4, and 8.7 shook an area 20 times larger than that affected by the 1906 San Francisco quake. The earthquakes were centered near New Madrid, Missouri, and were felt as far away as New Hampshire and Quebec. The tremors caused structural damage to buildings in the then-frontier towns of Louisville and St. Louis, and leveled an area about 150 miles by 50 miles. Similar events, which seismologists estimate occur in the region on average about once every 600 to 700 years, would today cause damages, deaths, and injuries comparable to the scenarios described above for earthquakes in San Francisco and Los Angeles.

In recent years, several groups of researchers have reported that other parts of the U.S. might have a much higher seismic risk than was once believed. For decades, geologists have thought that the Meers Fault, which cuts 16 miles across prairies and farmlands in southwestern Oklahoma, was simply a scar left by an ancient fault that had last been active about 200 million years ago. Recent studies have shown, however, that the Meers Fault may have ruptured in a large earthquake as recently as a few hundred to 2,000 years ago and might still be active. D. Burton Slemmons of the University of Nevada in Reno has calculated that if the Meers Fault ruptured again, it could produce a damaging earthquake with a magnitude between 6.5 and 7.5.

New York City, another area not generally considered earthquake-prone, could also be the site of future damaging earthquakes, according to a report by a group of California researchers led by Mark Zoback of Stanford University. In an analysis of National Geodetic Survey records, Zoback's group found that the positions of survey markers in parts of western Long Island and along a 40-mile section of the southern Hudson River had shifted significantly between 1872 and 1973. According to Zoback's group, the movement of the markers shows that surface rocks in southeastern New York are being deformed at a rate comparable to that observed along California's San Andreas Fault. They calculated that, if the deformation in southeastern New York is caused by the steady motion of rocks within the Earth's crust, then the crust must be moving at over 1 inch per year, and a major earthquake might someday occur.

As if to underscore that possibility, a magnitude-4 earthquake occurred a short distance north of New York City on 19 October 1985, just days before the report was published. Nevertheless, seismologists must fully assess the area before they can begin to estimate the risks of damaging tremors there.

Earthquake Precursors

Advances in seismology and plate-tectonic theory have given earth scientists a better understanding of the mechanisms that cause earthquakes and have led to improved techniques for identifying areas with a high seismic risk. No method has yet been developed to predict consistently where and when earthquakes will strike and how strong they will be. However, scientists and laypersons alike have observed that diverse phenomena, ranging from fluctuations in groundwater levels to changes in animal behavior, sometimes precede tremors. In a few instances, these phenomena — known collectively as earthquake precursors — have been used to make successful predictions. By analyzing seismic records for earthquake-prone areas, seismologists have found that certain patterns of seismic activity often occur before major earthquakes. Small events, usually too weak to be felt by people, may take place near active faults, and their occurrence does not generally indicate that a large earthquake is imminent. Seismologists have observed, however, that changes in the location and frequency of these small earthquakes often precede major events.

Several years or more before some major shocks, the number of small earthquakes detected in the region may decrease suddenly and remain unusually low for months or years. This period of low activity, known as quiescence, may end with a swarm of small earthquakes clustered close to the area where a major event later takes place. Before some tremors in Japan, the epicenters of these small earthquakes have formed characteristic doughnut-shaped rings around the epicenter of the upcoming event.

Other geophysical precursors are sometimes also observed. For example, the slope or "tilt" of the ground surface may change rapidly a short time before an event. Since these changes are often very small, seismologists use sensitive instruments known as tiltmeters, which are capable of measuring tilt changes of less than one-ten-thousandth of a degree.

The relative motion of rocks on opposite sides of a fault

may also change before an earthquake. Seismologists use instruments known as geodimeters to detect variations in fault movement. Geodimeters send a pulse of laser light from a point on one side of a fault to a point on the other side. Since light travels at a known velocity, the distance can be calculated easily by timing precisely how long the pulse takes to travel between the points.

Geodimeters have been used to measure the motion of different parts of the San Andreas Fault in California. The San Andreas Fault is not actually a single fault but is instead a fault zone, an area up to thousands of feet wide that consists of numerous interlacing small faults. The San Andreas is the longest active fault zone in the continental U.S. It extends roughly northwest to southeast over 600 miles from Mendocino on the northern California coast to the Imperial Valley on the Mexican border.

Geodimeter measurements near the Salton Sea about 150 miles southeast of Los Angeles showed that the relative movement of rocks on each side of the San Andreas changed in January 1979 from compression (squeezing together) to expansion (pulling apart). In October 1979, 9 months after this change in motion was first observed, a damaging earthquake occurred in California's Imperial Valley, about 20 miles south of where the measurements were made.

Changes in groundwater levels and chemistry may also precede earthquakes. During the 1960s, Chinese seismologists first learned that the concentration of radon, a radioactive gas released by rocks, often increases significantly in groundwater before earthquakes. Since then, researchers in Japan, the People's Republic of China, the U.S., and the U.S.S.R. have been studying the concentration of radon, other gases, and minerals in groundwater, searching for patterns that precede major shocks.

The Chinese have used changes in groundwater radon concentrations to predict several earthquakes, including the magnitude-7.3 earthquake that struck Hai'ch'eng in the province of Liaoning in February 1975. Radon concentrations in groundwater taken from wells near the earthquake's epicenter increased rapidly a few days before the event, and then rose sharply 30 minutes before the it occurred. Groundwater levels in wells near the epicenter also fluctuated significantly in the days preceding the earthquake, although not always in the same direction — levels rose in some wells and fell in others.

Chinese scientists have also pioneered studies of another precursor — changes in animal behavior. In December 1974, less than 2 months before the Hai'ch'eng earthquake, snakes reportedly came out of hibernation and left their holes, only to freeze to death above ground. In January 1975, the number of reports of strange animal behavior increased, particularly reports involving large animals such as cows, dogs, deer, horses, and pigs. Chinese observers reported that some animals were strangely agitated, would not eat, or made odd noises. On the basis of unusual animal behavior and geophysical precursors, Chinese seismologists issued a prediction that an earthquake was imminent, just 9 hours before it happened.

Seismic Gaps

Unfortunately, seismic precursors are often unreliable and misleading. Precursors differ from fault to fault and earthquake to earthquake, and some events apparently have no precursors at all. For example, no such signs were observed before the 1984 Morgan Hill, California, earthquake, which generated very strong ground motions.

Although pinpointing the exact time and place of an earthquake is still beyond science, seismologists are becoming increasingly skilled at identifying segments of faults that are due for a major shock and estimating approximately when an event might occur. This approach is sometimes called earthquake forecasting, as opposed to prediction.

In many parts of the circum-Pacific region, large earthquakes occur fairly regularly along fault segments. For example, if three events with a magnitude of 7 on the Richter scale struck a given fault segment during a 300-year period, that segment would have a recurrence interval of about 100 years for magnitude-7 earthquakes. Seismic research has shown that, as a general rule, major earthquakes, which have magnitudes of 7 or higher, have longer recurrence intervals than smaller events with magnitudes between 4 and 6. This would be expected, since strain along a fault builds up over time. If the strain is released frequently, the resulting earthquakes will be relatively small. If strain accumulates for a longer time, stronger events will generally occur when the fault finally ruptures. Through careful analyses of historic and seismic records, seismologists have constructed maps of earthquake recurrence times in the circum-Pacific region. Such studies have helped identify seismic gaps — segments of a plate boundary that have not ruptured for a relatively long time and that may in a sense be due for a major shock.

For example, Japanese seismologist Kiyoo Mogi reported in 1986 that a magnitude-8 earthquake could be expected to occur in the Tokai gap just southwest of Tokyo, near where the Pacific Plate is sliding beneath Japan. Great earthquakes typically occur in this region every 100 to 150 years, and the fault segment has not ruptured since 1854. Another great earthquake is expected along the coast of Mexico in the area between Tecpan de Galeana and San Marcos in the state of Guerrero. Oaxaca, which is south of Guerrero, was rocked by a magnitude-7.8 shock in 1978, and Michoacan to the north was the epicenter of the 1985 earthquake that caused heavy damage in Mexico City. Several research groups have calculated that magnitude-7.5 earthquakes occur in this region about every 50 years.

Unfortunately, the historic record in many locations extends back only 100 to 200 years — too short a period to calculate earthquake recurrence times of centuries. To overcome this limitation, geologist Kerry Sieh at the California Institute of Technology (Caltech) in Pasadena has supplemented the historic record with data from the geologic record. Sieh has found that the sediments in some creek beds near the San Andreas Fault show evidence of sudden displacements caused by major earthquakes.

Sieh used radioactive carbon dating techniques to measure the age of peat in these displaced beds and thereby determine the dates of prehistoric earthquakes. He found that 12 major earthquakes took place at Pallett Creek in the San Gabriel Mountains during the past 1,700 years. In over a decade of research along the San Andreas, Sieh has also learned that the segment of the fault between San Bernardino and Tejon Pass breaks in a great earthquake of magnitude 8 or more about every 145 years. The segment last broke in 1857, leading to the much-publicized conclusion that there is a high probability of a great earthquake in southern California within the next 50 years.

Steven Wesnousky, a former Caltech researcher now at the Tennessee Earthquake Information Center in Memphis, has created a set of computer-generated earthquake hazard maps by analyzing geological data for hundreds of faults described in the scientific literature. The maps show where the most severe shaking is likely to occur during a 50-year period, how often severe shaking can be expected, and which faults in California might produce ground accelerations that could significantly damage buildings. The maps present seismic-risk information in a form that can be readily understood by the people responsible for planning decisions.

Earthquake Research Programs

The science of predicting earthquakes is still in its infancy. Although seismologists have made some successful long-term predictions (ranging from a few years to a few decades), they have had less success making intermediate-term predictions (ranging a few weeks to a few years) and short-term predictions (ranging from hours up to a few weeks). Researchers do not yet understand, for example, what causes earthquake precursors and why they appear before some events and not others. To reduce the risk of property damage and death, international agencies and governments worldwide are developing programs to monitor and understand earthquakes. The most advanced programs are in the People's Republic of China, Japan, the U.S., and the U.S.S.R.

Chinese program. The earthquake prediction program in China began after the destructive Hsing-t'ai earthquake of 1966. By the late 1970s, the Chinese had established a nationwide network of fully equipped seismic observatories and auxiliary stations collecting data at over 5,000 separate locations. About 10,000 full-time workers and thousands of amateur observers have been involved in the Chinese prediction program.

During 1983, in a joint research project with the USGS, Chinese seismologists finished installing an array of about 40 instruments to monitor ground motions associated with strong events in the Beijing/T'ien-chin region of northeastern China. Six major earthquakes, including the devastating 1976 T'ang-shan earthquake, occurred in this area between 1966 and the mid-1980s. The instruments in the array can be moved to any location in China where a shock is expected.

In the late 1980s, China's State Seismological Bureau and the USGS were cooperating in 13 major projects, including two earthquake prediction experiments, one near the town of Xiaguan in the western Yunnan province and the other in the region surrounding T'ang-shan. A network of sensitive instruments has been installed on the surface and in a series of underground tunnels near Xiaguan. The network monitors the seismic activity, strength of the Earth's gravitational and magnetic fields, and deformation of the Earth's surface. A similar array of instruments was installed near T'ang-shan, together with an advanced computer system to provide seismologists with immediate access to current seismic information for the region.

The biggest project, China's nine-station Digital Seismographic Network, was completed in 1988. Each computer-operated seismograph communicates directly with a data management station in Beijing, and recorded data is shared with scientists around the world. Chinese and American earth scientists are also working together in several field studies. The Chinese have conducted research at faults in the western U.S., and Americans have mapped faults in the Yunnan and Gansu provinces and the Ningsha region, which are all sites of past devastating earthquakes.

Japanese program. Japan is located in an area of high seismic risk, and many cities have suffered heavy casualties and damage during the twentieth century. Japan's greatest recorded earthquake, one of the most devastating in history, struck Tokyo and Yokohama on 1 September 1923. The earthquake and subsequent fires killed 143,000 people and destroyed 575,000 buildings. After another destructive shock in Niigata in June 1964, the Japanese government established an ongoing program to equip seismic observatories throughout the nation and conduct basic research.

As part of the program, the government installed a prototype earthquake prediction network in the Tokai district southwest of Tokyo. The district is located where two plates converge; the Philippine Sea Plate is subducting (sliding beneath) the Eurasian Plate. Researchers are monitoring seismic activity and earthquake precursors, including groundwater temperature and radon gas concentration, magnetic and electrical properties of surface rocks, and changes in the tilt and shape of the Earth's surface.

The Tokai network now contains more than 70 seismic stations covering an area about 220 miles long by 90 miles wide, including the region offshore. The stations collect data on about 40 earthquakes per day, most of which are too weak to be felt. The data are transmitted continuously to the Japan Meteorological Agency in Tokyo, where seismologists analyze the information for signs of future tremors.

Soviet program. During 1982, the U.S.S.R. opened a regional earthquake monitoring center in Dushanbe to study seismic events in the republics of Kazakhstan, Kirgiziya, Tadzhikistan, Turkmeniya, and Uzbekistan. The facility was built as part of an earthquake prediction program that began after damaging shocks struck Ashkhabad and Khait during 1948 and 1949. The Dushanbe center is the first in a network of regional centers that is eventually expected to provide earthquake warnings for the entire country. The center compiles and analyzes data from seismic stations throughout the region, including an advanced research facility monitoring 12,000 square miles in the northern Tien Shan mountains.

Scientists at the Dushanbe station are coordinating an extensive research program in earthquake prediction. Geochemists take samples from deep wells to detect any changes in their gas, mineral, and ion content that might signal a future earthquake. Field geologists gather data on subtle changes in the Earth's magnetic field and measure minute displacements along faults in the region. Biologists monitor the behavior of animals for signs of unusual behavior signaling a seismic event.

Soviet seismologists have reported that their earthquake monitoring and prediction program led to the successful forecast of a strong earthquake near the town of Gazli in Uzbekistan in March 1984. Although the earthquake heavily damaged buildings and injured over 100 people, emergency planning made possible by the prediction allowed relief personnel to rush food, clothing, and tents to the area within a few hours of the shock.

U.S. program. Much earthquake research in the U.S. is supported by the National Earthquake Hazards Reduction Program (NEHRP), which is coordinated by the USGS, FEMA, the National Science Foundation (NSF), and the National Bureau of Standards. Since it began in 1977, NEHRP has sponsored a variety of projects ranging from basic research to programs for improving community planning for earthquakes.

In NEHRP-funded studies, researchers have studied the characteristics of past U.S. earthquakes and refined their estimates about the frequency of events in different regions, leading to better earthquake hazard maps — the basis for establishing regional building codes — as well as response plans to deal with the aftermath of large tremors. Many seismic monitoring networks receive NEHRP funds.

NEHRP has also funded research to develop earthquake-resistant designs for buildings, bridges, roads, and other structures. In 1985, the State University of New York at Buffalo became the headquarters of the new National Earthquake Engineering Research Center, operated by a consortium of institutions and funded initially by the NSF. NEHRP supports engineering research on a shaking table at Buffalo, an earthquake simulator that vibrates scale models of buildings the way an actual earthquake would. Studies of how buildings respond to different types of ground motion could lead to the design of safer, more earthquake-resistant buildings.

In 1985, a consortium of 58 university research groups founded the Incorporated Research Institutions for Seismology (IRIS) to develop a network that includes 500 portable and 60 global, digital, broad-band instruments capable of recording seismic waves over a broad range of frequencies. The recording stations will be the foundation of the new Global Seismic Network (GSN) that will continue operating into the 21st century. A joint IRIS-USGS data collection center has been established in Albuquerque, New Mexico, and the GSN data were expected to become available during 1989. IRIS has already undertaken projects, including the installation of five broad-band stations in the U.S.S.R. for nuclear testing verification under an agreement signed in Moscow in 1988 by IRIS and the Academy of Sciences of the U.S.S.R.

Earthquake Prediction in the U.S.

The National Earthquake Prediction Evaluation Council (NEPEC) was established under NEHRP to review and assess research on U.S. earthquake hazards and predictions. In January 1981, the Council made its first assessment of an earthquake prediction — one that a series of major earthquakes would affect a large region along coastal Peru and northern Chile beginning in 1981. The prediction, which was made by U.S. Bureau of Mines geophysicist Brian Brady and USGS scientist William Spence, was based on their studies of past earthquakes and foreshocks, as well as laboratory research on the ways rocks fracture when subjected to pressure.

According to the prediction, the first severe earthquake would occur in June 1981, and the series would culminate in August or September 1981 with a catastrophic shock several times more powerful than the strongest earthquake recorded anywhere in the world during this century. The prediction was widely publicized and caused great concern in Peru.

Although the earthquakes were not expected to affect the U.S., NEPEC decided to evaluate the prediction in response to a request by the Peruvian government, which was concerned about public fears raised by the forecast. The council sharply criticized the prediction, noting that the researchers had "shown nothing...that lends substance" to their prediction. NEPEC also expressed regret that "an earthquake prediction based on such speculative and vague evidence has received wide credence outside the scientific community." The council's doubts about the prediction proved correct, since none of the predicted tremors took place.

Parkfield earthquake prediction. NEPEC made its first endorsement of an earthquake prediction in April 1985. According to the prediction, an earthquake of magnitude 5.5 to 6.0 will occur along a section of the San Andreas Fault near Parkfield, California, about halfway between San Francisco and Los Angeles, sometime before 1992.

Thomas McEvilly of the University of California at Berkeley and USGS seismologists William Bakun and Allan Lindh first published the prediction of the Parkfield earthquake in 1984, after a detailed study of the area's seismic history. Moderate earthquakes have occurred there with unusual regularity — in 1857, 1881, 1901, 1922, 1934, and 1966. The tremors have taken place an average of once every 22 years, with the only important exception being the 1934 earthquake, which occurred after only 12 years.

Seismic records for the Parkfield area have shown that the last three moderate earthquakes there were nearly identical. Each of the events had a magnitude of about 5.6, began at about the same point on the San Andreas Fault, and ruptured the ground for 12 to 20 miles to the southeast of the origin point. Bakun and McEvilly also noted another astonishing similarity between the 1934 and 1966 events: In both, a foreshock of magnitude 5.1 occurred at the same location about 17 minutes before the main shock.

Researchers have speculated that Parkfield's location, which is in a transition area between two segments of the San Andreas Fault, may explain this recurring seismic pattern. In the fault segment to the northwest of Parkfield, the Pacific Plate slides steadily past the North American Plate without sticking and producing large earthquakes. In the fault segment to the southeast of Parkfield, the Pacific and North American plates tend to lock together, gradually accumulating strain until they periodically break apart in a large earthquake and slide past each other.

If the Parkfield area follows the pattern of activity observed during the past century, then an earthquake of about magnitude 6 would be expected to occur within a few years of 1988. USGS researcher Paul Segall, Caltech's Kerry Sieh and others have raised the possibility that an earthquake in Parkfield might trigger a rupture in the part of the San Andreas Fault immediately to the south. If this adjacent section ruptured, a major tremor of magnitude 7.0 to 7.5 could occur. The Parkfield region of the San Andreas Fault is now the most heavily instrumented seismic zone in the world. The fault is dotted by a network of dozens of seismographs, creepmeters (which measure slow ground slippage across the fault), strainmeters, and water-level meters in boreholes. Many of the instruments relay data to a USGS facility in Menlo Park, California. Computers calculate the epicenters of every detectable tremor almost as soon as it happens and can activate a paging system to alert seismologists about important changes in the area's vital signs.

A series of seismometers has been installed at about

Figure 1. San Andreas Fault. This aerial photo shows the San Andreas Fault in the Carrizo Plains region of central California. The San Andreas is the longest fault zone in the continental U.S. It marks the boundary where the Pacific Plate to the west (left in this photo) and the North American Plate (right in this photo) slide past each other. *Courtesy: Robert Wallace, USGS.*

300-foot intervals in a 3,000-foot-deep abandoned oil well to provide seismologists with subsurface seismic information for the fault segment. Instruments are also in place to measure the strong ground motion that accompanies large earthquakes and that often destroys more sensitive equipment. Other equipment will study the Parkfield earthquake's effects on pipelines and soils and on a model designed to simulate the containment vessel of a nuclear power plant. As of early 1989, the predicted shock had not occurred, although there were minor bursts of seismic activity during February and August 1987 and March 1988.

High-risk areas. In recent years, NEPEC has been evaluating research that identifies areas with high seismic risk, including three sections of the San Andreas Fault in southern California. These areas, a segment extending 25 miles south from Cholame, another between Tejon Pass and San Bernardino, and a third stretching from Palm Springs to the Salton Sea, typically rupture about every 125 to 175 years.

The first two fault segments have not ruptured since the destructive Fort Tejon earthquake in January 1857, which caused extensive damage in San Francisco, Sacramento, and Los Angeles and is believed to have been comparable in magnitude to the San Francisco earthquake of 1906. Although sediments and rock strata show evidence of major earthquakes in the fault segment between Palm Springs and the Salton Sea, the area has not had a major earthquake since at least 1845. On the basis of this evidence, NEPEC concluded that there is a moderate to high probability of a major earthquake in this region during the next few decades.

On 1 October 1987, the Los Angeles metropolitan area suffered California's most damaging earthquake since 1971. Although the magnitude-5.8 event caused several deaths and over $350 million in damages, it released less than one-thirtieth of the total energy that would be unleashed in an earthquake of magnitude 7 or more. Seismologists believe that the recent Los Angeles tremor and two magnitude-6 events in the Imperial Valley during November 1987 have not significantly reduced the risk of larger, catastrophic earthquakes in southern California. The fact that the 1987 Los Angeles earthquake occurred on a formerly unknown, buried fault may mean that the region's seismic risk may actually be greater than previously thought.

NEPEC has also considered estimates of earthquake probability along the Aleutian Islands and the southern coast of Alaska. The Alaska-Aleutian subduction zone, where the Pacific Plate is diving beneath the North American Plate, is one of the most seismically active regions in the world. During the past 90 years, seven earthquakes with magnitudes greater than 7.8 have occurred in this zone. The most destructive of these events was the Prince William Sound earthquake on 28 March 1964, which devastated Anchorage and triggered large ocean waves called tsunamis that damaged property and killed 122 people along the coasts of Alaska, British Columbia, Oregon, and northern California.

Although great earthquakes have historically occurred about once every 12 years in this region, more than 20 years have passed since the last great earthquake shook the remote Rat Islands in the western Aleutians on 4 February 1965. Seismologists have identified three sections of the Alaska-Aleutian subduction zone that they believe are the most likely sites for future earthquakes with a magnitude of 7.8 or greater.

Two of these segments are located on the Shumagin Islands and on Unalaska Island south of the Alaskan Peninsula. The first segment extends from the southeastern end of Kodiak Island to Unimak Island, and the second extends from Unimak Island to Umnak Island. The third segment is centered near Yakataga on the northern coast of the Gulf of Alaska and extends from the town of Cordova to Icy Bay. Seismologists have estimated that both of these regions have a 30 to 90 percent risk of a great earthquake during the next 20 years. Such an event could cause extensive damage and perhaps trigger tsunamis with wave heights of 20 feet to more than 100 feet along the coast. Two major earthquakes with surface magnitudes of 6.9 and 7.6 occurred on the Pacific Plate south of the Yakataga gap during November 1987.

In 1987, several reports suggested that the Pacific Northwest is another area at risk for a great earthquake. Just off the coast of Washington, Oregon, and British Columbia, several small plates are being subducted beneath the North American Plate. The geologic setting in this region, the Cascadia subduction zone, appears similar to that in southern Chile, Colombia, and southwestern Japan, where great earthquakes have struck. In these regions, young oceanic crust is diving beneath a continental plate, and the ocean trench marking the plate boundary is poorly defined. Research by USGS geophysicists James Savage and others indicates that the Cascadia subduction zone is building up strain that could be released suddenly in a great earthquake.

Additional research by Brian Atwater of the University of Washington provides more evidence of seismic risk in the Pacific Northwest. Atwater studied sand and intertidal mud layers along the coast and examined evidence of sea-level change along the coast during the past 7,000 years. If the coastline had been submerged suddenly due to an earthquake, well-vegetated lowland soils would be covered by intertidal muds. Atwater found evidence that rapid submergence has occurred at least six times in the last 7,000 years. Though NEPEC has considered these results, it has concluded that more evidence is needed for a full assessment of the Pacific Northwest's earthquake risk.

Limitations of Earthquake Predictions

Although recent research has made it possible to identify areas with a high long-term seismic risk, seismologists do not yet know enough about the events that precede and trigger earthquakes to make the reliable short- and intermediate-term predictions that would allow government agencies and individuals to prepare for the emergency. NEPEC members Robert Wesson and John Filson have noted that to achieve success in earthquake prediction, researchers must develop new technologies to transform an improved understanding of the Earth into accurate predictions "on an increasingly large scale...over broad regions of seismically active crust and sustained over periods of years to decades."

Researchers also face another important problem: social pressures. "Society is likely to demand increasingly accurate predictions much sooner than the scientific community will be able to provide [them]," Wesson and Filson wrote. "At the same time, society will experience significant problems in coming to grips with the responsible application of reliable earthquake predictions. This will require strenuous efforts on the part of the scientists to educate and communicate. There are certain to be disappointments, but it will be vitally important to reduce the loss of credibility from incorrect predictions and to learn as much as possible from mistakes."

Recent Geological Disasters

The mid- to late 1980s will long be remembered as a period of extremely destructive geological events. Two damaging earthquakes rocked southwestern Mexico in September 1985, and less than 2 months later, Colombia suffered the most catastrophic volcanic eruption to occur in over 80 years. Disaster struck again in August 1986, when a deadly cloud of carbon dioxide burst without warning from Lake Nyos, located in the cone of a dormant volcano in Cameroon. Finally, the most devastating earthquake in more than a decade caused nearly total destruction in Spitak in northwestern Armenia and heavily damaged nearby towns during December 1988.

These events were particularly disastrous because they occurred suddenly, before steps were taken to adequately protect life and property. Thousands of deaths and injuries might have been prevented if scientists had been able to predict the catastrophes and policymakers had taken steps to reduce the risks.

Mexican Earthquake

On 19 September 1985, a violent earthquake of magnitude 8.1 centered near Aguililla in southwestern Mexico destroyed or damaged buildings over a 300,000-square-mile area of Mexico. Geophysicists at Caltech later reported that this shock was actually a double pulse, with two roughly equal tremors lasting 16 seconds each and separated by 25 seconds. A powerful magnitude-7.5 aftershock on September 21 added to the destruction as it toppled buildings weakened by the main earthquake. The events killed at least 9,500 people, injured 20,000, and left more than 100,000 homeless.

Earthquake damages. The 19 September earthquake shook all of southern Mexico and was felt as far away as southern Texas. The Pacific coastal states of Guerrero, Jalisco, and Michoacan, which were nearest to the event's epicenter, suffered severe damage. The earthquake devastated the town of Guzman in Jalisco, where a cathedral and about 400 homes collapsed, killing 36 and injuring 150.

However, nearly all of the deaths and injuries occurred in Mexico City, about 200 miles northeast of the first earthquake's epicenter. More than 400 buildings were destroyed and another 3,100 seriously damaged in Mexico City, with damage estimates ranging from $3 billion to $4 billion. The last major earthquake to affect the city, a magnitude-7.9 tremor on 28 July 1957 centered about 200 miles to the south in the Acapulco area, caused about $60 million in damage and killed 68 people.

In both the 1957 and 1985 earthquakes, the damage in Mexico City was heaviest in the city's central district, which is built atop a drained lake bed covered with a 150-foot-thick layer of soft clay. Such layers are often unstable during earthquakes, and structures built above them are more likely to suffer earthquake damage. The clay layer tended to trap seismic waves, which made the layer oscillate as they traveled through it.

Some of the seismic waves that passed through Mexico City in the September 1985 event had a period of about 2 seconds, which is roughly equal to both the period of the clay layer and the period at which buildings between 5 and 15 stories high sway when shaken in an earthquake. Thus, the seismic waves reinforced both the shaking of the lake bed and the natural swaying motion of some of the multistory buildings constructed on it. This situation is analogous to the way in which a carefully timed push of a moving swing will send it through greater and greater arcs. Although the seismic waves that reached Mexico City were much weaker than those at the event's epicenter, the clay layer amplified the waves sevenfold, so that the ground oscillated 16 inches every 2 seconds. Buildings with a natural swaying period of about 2 seconds were severely damaged, and some poorly built structures were literally shaken apart.

The earthquakes occurred along a boundary where the Orozco fracture zone, on the relatively small Cocos Plate off the coast of Mexico and Central America, is sliding beneath the North American Plate, of which Mexico is a part. Studies of the average frequency and severity of earthquakes along the boundary between the Cocos and North American plates showed that each fault segment ruptures in a major earthquake on average about once every 30 to 60 years. The area affected by the September 1985 tremors was previously stricken by another major earthquake in 1911. Furthermore, seismologists did not observe the precursors that sometimes occur in the weeks to years before a large tremor.

Although casualties and damage in central Mexico City were extensive, the event and its aftershocks destroyed or severely damaged only a very small fraction of the approximately 800,000 buildings in the city as a whole. As a result, the events killed or injured only about 0.2 percent of Mexico City's total population of 18 million people.

Engineering studies. Engineers were nevertheless surprised by the extent of damages and casualties. Many of the deaths and injuries might have been prevented if the buildings had been constructed to withstand the severe ground shaking that occurs in the region during major shocks. The 1957 earthquake had demonstrated that buildings in central Mexico City were particularly vulnerable to earthquakes. Subsequent engineering studies revealed that many buildings had inadequate foundations, unreinforced masonry, and other construction and design problems that contributed to the casualties and damage.

Some of the same design and construction deficiencies contributed to the 1985 disaster. Although Mexico City's construction codes for new buildings such as hospitals were revised after the 1957 earthquake and were considered among the most stringent in the Western Hemisphere, known problems in existing buildings had not been corrected.

Engineers at Atkinson, Johnson & Spurrier, Inc., of San Diego, California, who surveyed damage in Mexico City after the 1985 event, cited a variety of reasons why steps had not been taken to correct known deficiencies. Studies in other earthquake-prone areas around the world had shown that the building practices used in a region usually do not change quickly. Few of the construction firms in Mexico City have extensive experience in building earthquake-resistant structures or reinforcing existing ones. This problem has been compounded by the rapid growth of Mexico City during the past 30 years, a situation that has created social and political pressures to construct new buildings quickly; unfortunately, earthquake-resistant buildings are more expensive and time-consuming to build than unreinforced ones. Correcting structural problems in thousands of existing buildings would be extremely costly, and the money needed to correct these problems has not been available in Mexico's troubled economy.

Colombian Volcanic Eruption

Shortly after the devastating Mexican earthquake, another severe disaster struck Latin America. After nearly a year of increasing seismic and eruptive activity, the 17,700-foot volcano Nevado del Ruiz, located about 80 miles west-northwest of Bogota, erupted violently on 13 November 1985. The volcano ejected large amounts of hot volcanic gases, ash, and rock, which rapidly melted snow and ice at its summit. Torrents of water and runoff from several days of heavy rain combined with volcanic ash, rock, and other debris to form lahars — mudflows of volcanic material — that flowed down 11 river valleys surrounding the volcano.

Advancing down the volcano's steep slopes at speeds averaging 20 miles per hour, the lahars buried the town of Armero, about 30 miles east-northeast of the summit, killing an estimated 20,000 to 23,000 of the town's 25,000 residents. Another lahar inundated low-lying parts of the town of Chinchina about 20 miles west-northwest of the volcano, killing another 2,000 people. The eruption also caused $300 million in property damage and the loss of 60,000 acres of farmland.

In terms of total fatalities, the Colombian eruption was the fourth worst volcanic event ever recorded, exceeded only by the eruptions of Tambora on the island of Sumbawa in Indonesia during April 1815, which killed 92,000; Krakatoa in the Sunda Strait between Java and Sumatra in Indonesia during August 1883, which killed 36,000; and Mt. Pelee in Martinique during May 1902, which killed 28,000.

Historic eruptions. Nevado del Ruiz has a long history of volcanic eruptions and lahars. Its first recorded eruption in March 1595 produced ash and lapilli — volcanic rock fragments up to 1.25 inches in diameter — as well as extensive lahars. A major eruption in February 1845, one of the most devastating in South American history, melted large amounts of snow and ice, creating a lahar that swept down the Lagunillas River Valley over the site where Armero was later built. This lahar buried the town of Ambalema, killing 1,000 people. Volcanic debris deposited by the 1845 eruption later formed the rich topsoil that supported the thriving agricultural community of Armero.

Prelude to eruption. Nevado del Ruiz remained quiet from 1845 until late November 1984, when earthquakes began occurring in the region around the volcano. In December, stronger earthquakes took place, and these were followed by harmonic tremors lasting about 30 minutes. (Harmonic tremors are continuous, rhythmic ground vibrations that are distinct from the discrete, sharp jolts of earthquakes.) From late December until the time of Nevado del Ruiz's first major eruption in September 1985, about 25 to 30 "felt" earthquakes (those strong enough to be perceptible) occurred each month.

Scientists visiting the volcano in January 1985 saw evidence of phreatic activity — explosions caused by the sudden vaporization of groundwater that comes in contact with magma. Small vents called fumaroles, which emit steam and gases, had also become more active. On 6 September, just weeks after an array of seismographs was installed at Nevado del Ruiz, the seismic activity increased and began following a regular pattern in which about 15 minutes of strong, high-frequency tremors took place each hour.

This pattern of seismic activity culminated on 11 September in a moderate eruption of ash and steam, which lasted about 7 hours and deposited a trace of ash on the towns of Manizales, 20 miles to the northwest of the summit, and Chinchina, 20 miles to the west-northwest. Heat from the eruption melted snow and ice, causing a lahar that traveled down the canyon of the Azufrado River northeast of the summit for about 17 miles and reached a maximum depth of 30 to 65 feet. Although activity subsided on 12 September, the volcano emitted steam for the rest of the month and occasionally erupted ash.

Predictions of disaster. As early as 26 September, the Colombian National Institute of Geological and Mining Investigations (INGEOMINAS) reportedly recommended that towns located in river valleys surrounding the volcano be evacuated, but the Colombian government did not follow this advice. In early October, the government invited a team of Italian volcanologists to assess the risk of an eruption. The team concluded that an "extremely dangerous" eruption might take place at any time. They urged that a monitoring system be established to observe the volcano for early signs of activity and mudflows and to issue evacuation warnings if necessary.

During October, Colombian geologists completed preliminary hazard maps that pinpointed Armero and Chinchina as particularly vulnerable to lahars triggered by an eruption. Seismic activity increased on 7 November with a series of earthquake swarms, although the seismic activity was less than that before the 11 September eruption.

13 November eruption. On 10 November, a volcanic tremor began and continued for 3 days until an eruption started at about 3:30 P.M. on 13 November. At around 5:30 P.M., a rain of ash and small pebbles began falling at Armero, and at about this time, INGEOMINAS reportedly recommended that Armero and other populated areas around the volcano be evacuated. At 7:30 P.M., the Red Cross tried to begin the evacuation, but town officials did not respond because they were reluctant to order the nighttime evacuation of thousands of people during a heavy rain.

The eruption intensified at about 8:30 P.M. Seismologist Bernardo Salazar, who was at the El Arbolito seismic station less than 6 miles from the crater, reported that the volcano's seismic activity increased at 9:09 P.M. to a level far exceeding that of the 11 September eruption. At 9:37 P.M., explosions rocked the summit, lighting up the clouds above the crater. Large pieces of hot pumice — a light, porous rock — began falling at El Arbolito, and Salazar left the station for a safer location.

Despite the continued activity, local radio reports during the evening reportedly told Armero residents that the eruption posed no danger and advised them not to evacuate their homes. As the rain of ash intensified, residents began panicking. At around 11:00 P.M., another powerful explosion occurred, causing a strong gust of wind that dropped fiery ash on Armero. The eruption's heat melted ice and snow cover, and this water combined with storm runoff and volcanic debris to form the deadly lahars.

The first mudflow reached Armero with a sound "like a

huge locomotive going at full steam." The lahar was reportedly cold, but each successive wave of mud was warmer and warmer, and the last was described as "smoking hot." The lahars eventually buried over 80 percent of the houses in Armero under as much as 25 feet of mud, killing 20,000 to 23,000. Survivors escaped by climbing to the roofs of their houses or into trees or by running to higher ground.

As news of the cataclysm spread early on 14 November, rescue workers from Colombia's civil defense agencies and the Red Cross were sent to Armero to rescue an estimated 1,000 to 2,000 people trapped alive in the mud. Although the volcano's activity had diminished, rescue workers had to work under the risk that another violent eruption might trigger more lahars. Fortunately, no new mudflows occurred, and relief workers were able to rescue hundreds of people from the mud in the week after the disaster.

Eruption aftermath. Nevado del Ruiz sporadically erupted steam, ash, and volcanic debris throughout 1986 and 1987. In March 1988, the volcano erupted a thick column of ash and steam in its most intense activity since November 1985. Volcanologists believe that the volcano could remain active for years. Researchers from the USGS and the University of Rhode Island (URI) at Kingston estimated that 82 to 95 percent of the snow and ice that capped Nevado del Ruiz before its November 1985 eruption remained unmelted after the event. They warned that this icy accumulation will continue to pose the danger of new mudflows as long as the volcano remains active.

Lake Nyos Disaster

One of the strangest natural catastrophes in recent memory occurred during 1986 in the region surrounding Lake Nyos, a volcanic lake in northwestern Cameroon. At about 9:00 P.M. on 21 August 1986, people in the town of Nyos heard unusual rumbling sounds emanating from nearby Lake Nyos. With no other warning, a massive cloud of carbon dioxide (CO_2) gas burst from the lake. A water fountain produced by the cloud flattened vegetation and eroded soil up to 260 feet above the lake surface. The cloud swept down river valleys adjacent to the lake, blanketing the towns of Nyos, Cha, Subum, and Fang. When the cloud hit, about 5,000 people living in a 24-square-mile region around the lake lost consciousness. Before dispersing, the gas asphyxiated a total of 1,746 people and more than 8,300 livestock at a distance of up to 14 miles from the lake. In Nyos, only 4 of the town's approximately 1,200 residents survived.

The outside world first learned of the disaster from a motorcyclist who saw bodies by the road as he traveled toward Cha. He too was overcome by the cloud but later revived and returned to a nearby town to report the disaster. On 24 August, two Swiss helicopter pilots flew over the area. They found that the lake was calm, although dead vegetation covered its surface and the water had turned a rusty reddish color. International rescue teams and scientists were unable to reach the remote area until 26 August, by which time most of the dead had already been buried. Information on the effects of the gas cloud was gathered through interviews with survivors, as well as animal and a few human autopsies. As of early 1989, Nyos remained a ghost town, because the government of Cameroon had forbidden survivors to return to the area until steps could be taken to ensure that the disaster would not recur.

Studies at Lake Monoun. A similar but smaller event had occurred almost exactly 2 years earlier at Lake Monoun, which is in the same volcanic region of Cameroon. On the morning of 15 August 1984, a cloud of gas emerged from the lake and blanketed a section of road about 650 feet long near the town of Njindoun. Authorities later found 37 people dead along the road. The few survivors of the event said they smelled strange odors, felt nauseous, and then lost consciousness. As at Nyos, villagers had heard rumbling noises like explosions the night before.

Subsequent visits to Lake Monoun revealed that plants by the shore had been flattened and that the lake water had turned a rusty red. The government of Cameroon invited a group of American volcanologists led by Haraldur Sigurdsson of URI to join Cameroonian scientists in studying the region around the lake. The group reported their findings at a meeting of the American Geophysical Union in December 1985.

Lake Monoun and Lake Nyos are located in a region of young volcanoes called the Cameroon volcanic line in northwestern Cameroon. Both lakes fill circular maars, or craters within the cones of volcanoes. Lake Monoun is about 1,200 feet wide and 300 feet deep, and contains water with an extremely high concentration of dissolved CO_2. The volcanologists believe that the CO_2 seeps into the lake from a magma column beneath the crater.

Cameroon's year-round tropical climate allows lakes to

become stratified for long periods, with relatively warm water at the surface and cooler, denser water on the bottom. Over thousands of years, the lower waters in Lake Monoun became saturated with CO_2. A combination of high water pressure near the bottom of the lake and lack of mixing of the lake's upper and lower waters ordinarily keeps the CO_2 in solution. However, events that upset this natural balance can cause the gas to be released quickly.

During the visit to Lake Monoun, Sigurdsson's team found a fresh landslide scar at the lake's eastern rim. They believe that landslide debris falling into the lake disturbed the CO_2-saturated bottom waters. This allowed some of the water to rise to the surface where the gas bubbled out of solution explosively at the lower pressures and formed a deadly cloud. As CO_2 bubbled out of the lake, the dissolved iron in the deep water combined with oxygen dissolved in the surface layer to form the rust-colored precipitate that discolored the surface waters. When the cloud mixed with air, its CO_2 concentration was still much higher than the 0.03 percent normally found in the atmosphere. CO_2 concentrations of 40 percent cause nearly immediate death, and even a 10 percent concentration can cause death quickly.

What happened at Lake Nyos? The catastrophe at Lake Nyos occurred just months after Sigurdsson's group proposed this scenario for Lake Monoun. The U.S. Agency for International Development (AID) sent an American scientific team to Lake Nyos to study the disaster. The Lake Nyos maar is about 4,000 feet in diameter and 680 feet deep. Volcanologists have found evidence of a diatreme—a long vertical conduit filled with ash and fragmented rock—beneath the lake. At a depth of about 2 to 3 kilometers (1.2 to 1.8 miles), the diatreme becomes connected to an accumulation of magma. The AID team concluded that the CO_2 in the lake comes from this deep magma source rather than from the volcano crater at the base of the lake. They believe that if the CO_2 was emitted by volcanic activity just beneath the bottom, the lake's temperature would be higher than what they observed, and the water would be hotter at the bottom than at the top, which was not the case. In addition, the lake waters did not contain unusual concentrations of certain gases, such as hydrogen sulfide, which are typically released during volcanic activity.

Carbon-14 dating of CO_2 in the water showed that much of the gas was more than 35,000 years old. This age is consistent with a magmatic origin, because CO_2 produced by biological activity in the lake would be expected to be much younger. The AID researchers believe that the drop in pressure as magma ascends through the crust allows CO_2 to be released from the magma and then carried by groundwater up through cracks in the crust. The groundwater eventually seeps into the lake bottom and raises the CO_2 concentration of the bottom water. The team calculated that the waters were almost 100 percent saturated with CO_2 before the cloud was released in August 1986, and only 20 to 25 percent saturated afterward. This means that the lake released about 2 million tons or about 0.2 cubic miles of CO_2, enough to lower the lake's water level about 3 feet. As the CO_2 mixed with the atmosphere, it formed a cloud that eventually reached well over 300 feet high. Since CO_2 is denser than air, the cloud hugged the ground as it flowed into valleys near the lake.

The AID team did not determine what triggered the gas release. No earthquakes were recorded on the night of the event, and no evidence of landslides was found at the lake. High winds or torrential rain could have disturbed the lake's layered structure, but no one who might have witnessed such events survived.

In March 1987, more than 200 scientists from 20 countries met in Yaounde, Cameroon, to discuss the gas releases at Lake Nyos and Lake Monoun. Most scientists agreed that the clouds at both lakes were composed of CO_2 derived from a deep magma source rather than from volcanic activity near the surface. However, some French and Italian scientists, including French volcanologist Haroun Tazieff, proposed that the gas emissions may have been triggered by a small volcanic explosion from a vent in the bed of Lake Nyos. They said that a steam explosion could have been triggered as water came in contact with hot magma. Such an event could upset the lake's stratified structure and cause the CO_2 cloud without adding significant amounts of hydrogen sulfide or other volcanic gases to the lake waters. Many scientists rejected this interpretation as more complicated than the facts demanded, but there was insufficient evidence to refute it.

In June 1987, Haraldur Sigurdsson reported that medical specialists at the conference generally agreed that the deaths were caused by CO_2 asphyxiation. Although some survivors spoke of smelling the characteristic rotten egg odor of hydrogen sulfide, this observation could be explained as an olfactory hallucination caused by inhaling air with high CO_2 concentrations.

Geologists have proposed steps to prevent a recurrence of the disaster at Lake Nyos, which still contains significant amounts of CO_2. The AID team has noted that a spillway at the northern end of the lake is built of volcanic ash. Heavy rain or other natural events could breach the spillway, flooding valleys downstream and possibly triggering the release of another CO_2 cloud. Even if the spillway remains intact, the lake's high CO_2 concentration poses a continuing threat to anyone living nearby. Geologists at the meeting in Yaounde proposed that the water level at Lake Nyos be lowered and the spillway removed to prevent catastrophic flooding. Several scientific groups, including the AID team, suggested that the CO_2 be removed from the bottom waters through a pipe that would extend from the lake bottom to the surface. Water would be gradually pumped from the bottom at a rate that would allow a controlled and safe release of the dissolved CO_2.

Aside from Lake Nyos, only Lake Monoun appears to have any potential for releasing another gas cloud. George Kling of the Marine Biological Laboratories in Woods Hole, Massachusetts, has studied all of Cameroon's volcanic lakes and has found high CO_2 levels only in Monoun and Nyos. Although the risks are not fully known, the chance of similar events elsewhere in the world appears to be slight. According to Kling, lakes rarely form permanent density-stratified layers in temperate regions, because seasonal wind and temperature changes cause the lake waters to mix each year. Volcanologists believe that the risk of a similar gas release is also slight even in other tropical regions. Fortunately, few, if any, volcanic lakes outside Cameroon have the unusual combination of a source of highly concentrated CO_2 and the hot, relatively stable climatic conditions that allow immense amounts of the gas to accumulate in stratified lakes.

Armenian Earthquake

The world's most destructive earthquake in over a decade occurred just north of Spitak in northwestern Armenia, U.S.S.R., at about 11:40 A.M. on 7 December 1988. The main shock, with a magnitude of 6.9, was followed a few minutes later by a strong aftershock registering 5.8. The two events nearly leveled Spitak, a city of about 20,000 people, and caused severe damage in other nearby cities and villages, including Leninakan, the second-largest city in Armenia. The earthquakes were felt over the entire Caucasus Mountains region, with shaking reported at Tbilisi in Soviet Georgia to the north and Baku in Azerbaijan to the east. Soviet scientists reported that the event was the most severe to strike Armenia in over 80 years and one of the worst ever recorded in the U.S.S.R.

Armenia is an earthquake-prone region, located near the juncture where the African, Arabian, and Eurasian plates collide, creating the Caucasus Mountains. Nearby areas of eastern Turkey have suffered three highly destructive earthquakes since the mid-1970s, an event in 1975 that killed 2,300 people, one in 1976 that killed 4,000, and another in 1983 that killed 1,300.

The largest fault in the region, the Anatolian Fault, passes through Turkey paralleling the south shore of the Black Sea. Although Armenia has many known faults, the recent earthquake occurred on a relatively small and previously unidentified thrust fault running generally northwest-southeast near Spitak. (A thrust fault is a type of fault that dips into the ground at an angle fairly close to horizontal.) During the Armenian earthquake, the northeastern side of the fault, which included the city of Spitak, rode up over the southwestern side. The event shifted the ground horizontally about 5 feet and vertically about 3 feet. This movement triggered hundreds of landslides, including one that blocked the major water supply to Leninakan, a city of 290,000, and another that blocked one of the main railroads in the region for 10 days.

A team of American seismologists who visited the region concluded that the earthquake was probably the strongest the fault was capable of producing. Although the tremor was strong, it caused far more deaths and damage than most earthquakes of a similar magnitude. Two major factors contributed to the devastation in Armenia: the structure of the underlying rocks and soil and the poor design and construction techniques used in many buildings.

Rock and soil conditions. As in the Mexican earthquake 3 years earlier, damage in the Armenian earthquake was made worse by geological conditions. American scientists who surveyed damage in the region found the most severe damage in cities and towns located atop thick soil. Spitak, for example, was built on a deep soil layer over 600 feet thick. Damage was also heavy in Leninakan, about 30 miles southwest of the epicenter. Much of Leninakan is sited on largely water-logged soil about 1,000 feet thick. The earthquake is thought to have caused the saturated sandy soil to liquefy, undermining building foundations and greatly increasing damage. In contrast, almost no damage occurred in the town of Akhurian, which is actually a few miles closer to the epicenter than Leninakan but built on rocks with only a thin soil cover. In Kirovakan, to the east of the epicenter, the survey team found that damage was most severe in a part of the city built over a filled-in marsh.

Building design and construction. Improper design and construction of buildings played a major role in the catastrophe, according to American engineers and seismologists who visited the region. Part of the problem was that the buildings were not designed for earthquakes as strong as the one that hit the area. According to a report in *Science News*, structural engineer Loring Wyllie, Jr., of H. J. Degenkolb Associates in San Francisco reported that regional building regulations specified that new buildings be able to withstand shocks registering 7 or 8 on the

Figure 2. Armenian Earthquake Damage. **The earthquake that struck northwest Armenia on 7 December 1988 killed an estimated 25,000 people and injured another 15,000. Poor building design and construction — especially the use of loosely connected precast concrete walls — contributed to the high death toll and extensive building damage in the Armenian quake. This apartment building in Leninakan, Armenia's second largest city, was one of hundreds destroyed in the earthquake.** ***Courtesy: Armenpress.***

12-point intensity scale used in the U.S.S.R. The 7 December earthquake measured 10 points on the Soviet scale and was about 10 times more powerful than the strongest tremors recorded in this part of Armenia during the past 60 years.

Buildings with precast concrete walls, a design common in the U.S.S.R., are particularly susceptible to collapse if their component walls and floors are not firmly fastened to each other with supporting structures. When shaken by an earthquake, loosely connected concrete-slab walls can easily fall away from a building's floors, causing total collapse. Likewise, unsupported masonry tends to disintegrate when subjected to a strong earthquake. Materials shortages or improper techniques for producing building materials or prefabricated parts may also contribute to structural failure in some buildings.

In a survey of buildings in the region, engineers found that the majority of the relatively new, high-rise buildings were heavily damaged or destroyed, while many older, low-rise buildings suffered little damage. The severely damaged high-rise buildings were generally built with the loose construction methods described above. The engineers found, however, that high-rise buildings in which the component walls and floors were firmly locked to each other fared much better, often suffering only hairline cracks in the joints.

Like the Mexican earthquake, the Armenian event tragically illustrated the critical importance of proper building design and construction in earthquake-prone regions. After visiting the region, Fred Krimgold of Virginia Polytechnic Institute and State University noted that the Armenian event was particularly deadly. While most strong earthquakes cause several times as many injuries as deaths, this situation was reversed in the Armenian earthquake, with about 25,000 deaths and only 15,000 injuries. Krimgold believes that inadequate building techniques were largely to blame for the unusually high death toll.

Soviet officials reportedly plan to avoid past mistakes in rebuilding the area. New buildings will generally be low-rise — 5 stories or less — and made of stronger concrete that is poured in place rather than precast. If existing buildings in the region are properly reinforced and new buildings are designed and built to withstand severe shaking, then the next major earthquake will almost certainly take fewer lives and cause much less destruction than the tremor of December 1988.

Monitoring Active Volcanoes

In the wake of eruptions during the past few years, Mount St. Helens in Washington and Kilauea in Hawaii have become two of the most intensively studied volcanoes in the world. This research has provided new insights into the processes responsible for eruptions and has given volcanologists a greater understanding of the signs that sometimes precede these events. As a result of these studies, earth scientists have become increasingly skilled in predicting volcanic activity at Mt. St. Helens and Kilauea, raising the hope that it may someday be possible to provide reliable early warnings for damaging eruptions at other volcanoes around the world.

Mount St. Helens

In March 1980, Mount St. Helens in the Cascades Mountains east of Kelso, Washington, became the first active volcano in the continental U.S. in more than 60 years. Less than 2 months later, on 18 May, the mountain erupted suddenly and violently, killing 60 people and causing damages totaling $1.1 billion. After the eruption, the USGS established an observatory in nearby Vancouver, Washington, to monitor Mount St. Helens and more than a dozen other potentially active volcanoes in the Cascades of Washington, Oregon, and northern California. The facility was named the David A. Johnston Cascades Volcano Observatory (CVO), in honor of a USGS volcanologist killed during the May 1980 eruption. Seven years later, in May 1987, the summit of Mount St. Helens was reopened to limited public access for the first time since the eruption, in part because of the success CVO scientists have had in predicting the volcano's activity.

Eruptive history. Research by geologists Dwight Crandell, Donal Mullineaux, and Clifford Hopson beginning in the 1950s showed that Mount St. Helens had a long history of eruptions since its formation 40,000 to 50,000 years ago. They found evidence of nine major periods of activity, each lasting from 100 to 5,000 years and separated by dormant intervals 200 to 15,000 years long. In 1975, Crandell, Mullineaux, and Meyer Rubin forecast that the volcano could erupt "before the end of the century."

The forecast proved true in March 1980, when several minor quakes signaled Mount St. Helens's return to activity. Shortly after noon on 27 March, the volcano erupted for the first time with thunderous explosions that sent ash and steam 6,000 feet above the peak and blasted a hole 250 feet wide in the floor of the existing summit. After this eruption, Mount St. Helens sporadically ejected more ash and steam, and the first harmonic tremors — continuous, rhythmic vibrations — were recorded. The harmonic tremors indicated that magma, volcanic gas, or both were moving within the volcano — signs of a growing risk of a magma eruption. In late April, USGS researchers noticed that a bulge had appeared on the northern side of the volcano. This bulge grew outward at a rate of 5 feet per day, and the part of the summit behind it sank.

At 8:32 A.M. on 18 May, a magnitude-5.1 earthquake rocked the volcano. The bulging northern flank collapsed and created a huge avalanche that roared down the slopes at 155 to 180 miles per hour. Avalanche debris swept across Spirit Lake, several miles northeast of the summit, raising the lake bottom nearly 300 feet. However, most of the debris traveled north and west into the Toutle River, burying the river valley under an average of about 150 feet of rubble. The avalanche debris covered a total of 24 square miles and had a volume of 100 billion cubic feet. The sudden collapse of the northern flank of Mount St. Helens also released a blast of very hot, high-pressure steam and other gases that had been trapped in cracks and voids within the volcano. The force of this blast, which traveled at a speed of about 670 miles per hour, obliterated everything in an extensive area to the north of the summit and toppled large trees up to 19 miles away. Altogether, 230 square miles were devastated by the eruption.

Shortly after the avalanche and lateral blast, a strong explosion rocked the volcano, sending up an ash cloud 12 miles high. By about an hour later, the cloud had traveled east to Yakima, Washington, blocking nearly all the sunlight there, and 2 hours later, it blackened the skies above Spokane. Mount St. Helens continued spewing ash for 9 hours, blanketing some downwind towns with a layer more than 2 inches deep. About 540 million tons of ash were deposited over a 22,000-square-mile area in Washington and Idaho, with ash falls reported as far away as Minnesota, New Mexico, and Oklahoma.

The hot ash, gases, and rocks in the lateral blast and avalanche instantaneously melted snow and ice on the volcano. This water combined with volcanic debris to form volcanic mudflows, or lahars, that traveled up to 90 miles per hour down the steep, narrow stream beds near the volcano. The lahars gradually slowed down in the wider river beds downstream. The mudflows reached depths of 35 to 65 feet on parts of the Toutle River, and they dumped 1.8 billion cubic feet of sediment into the Cowlitz River. This sediment eventually flowed into the Columbia River, reducing the depth of its navigational channel from 39 to 13 feet and severely disrupting shipping traffic.

Monitoring and predicting eruptions. Mount St. Helens is a composite volcano, a type common along the "Ring of Fire" — the Pacific coasts of North and South America, Japan, the Philippines, Indonesia, and New Zealand — the site of over 60 percent of the world's active volcanoes. Composite volcanoes have a symmetrical shape with relatively steep slopes formed of alternating layers of eruption debris, such as ash, rock fragments, and solidified lava. When active, composite volcanoes often erupt explosively, threatening any inhabited areas nearby. They are markedly different from shield volcanoes, including Mauna Loa and Kilauea in Hawaii, which have broad, relatively gently sloping sides composed of solidified lava and which usually erupt nonexplosively.

Since its 1980 eruptions, Mount St. Helens has been monitored continuously. Seismometers placed in the crater and on the mountain's flanks transmit seismic data to the Geophysics Laboratory at the University of Washington in Seattle and the CVO. These data tell geophysicists whether the ground vibrations are caused by earthquakes, harmonic tremors beneath the volcano, or rock falls and gas emissions on the slopes.

Changes in the seismic pattern have been one of the most reliable signs of impending eruptions at Mount St. Helens. The number of shallow earthquakes and the amount of seismic energy released during a given period typically increase several days to 2 weeks before so-called dome-building eruptions. These events are nonexplosive releases of lava that enlarge the volcano's lava dome — a great mass of lava that piles up and solidifies around a volcano's vents. The rate at which seismic energy is released usually rises rapidly just a few hours before a dome-building eruption. Once the eruption begins, the number of shallow earthquakes decreases, and surface events, such as rock falls, become the major cause of seismic activity.

Before large explosive eruptions, the pattern is somewhat different from that preceding dome-building eruptions. Earthquakes are centered more deeply within the volcano, and harmonic tremors, which do not occur in dome-building events, are detected. Seismic activity increases only hours to a few days before explosive eruptions, in contrast to the several days to weeks of activity that precede a dome-building event.

In addition to their seismic observations, researchers monitor changes in the volcano's shape. When conditions permit, they visit the volcano several times a week to observe changes in the shape and appearance of the lava dome and crater floor, look for new cracks and faults, and measure the movement of existing ones. The movement of faults and cracks generally accelerates a few days to 4 weeks before an eruption. A network of tiltmeters measures the slope (tilt) of the crater floor and the flanks and transmits data to the CVO. Measurable changes in tilt generally occur several weeks before an eruption, and the change in tilt accelerates several hours to days before the event. An abrupt reversal in the direction of tilt change is sometimes observed hours to minutes before an eruption and is a strong indication that the event is imminent.

Emissions of sulfur dioxide and other gases from fumaroles — holes or vents — on and around the lava dome have been monitored regularly. Sulfur dioxide emissions peaked at 1,500 tons per day during the summer of 1980, when four explosive eruptions occurred. These emissions then declined to about 100 tons per day in 1983, and to about 20 tons per day in early 1986. Sulfur dioxide emissions have increased temporarily, however, before

Figure 3. Mount St. Helens Eruption. Mount St. Helens in southern Washington erupted violently in May 1980, killing 60 people and causing about $1.1 billion in damages. Since the catastrophe, earth scientists have monitored the volcano continuously and have correctly forecast its subsequent eruptions hours to weeks in advance. *Courtesy: USGS.*

several, but not all, dome-building eruptions. This rise has been attributed to the release of sulfur dioxide from magma rising within the volcano toward the surface.

Aerial photographs taken during the night have proved useful in monitoring Mount St. Helens, particularly when the possibility of an eruption has kept scientists from surveying the dome. Cameras with image intensifiers, which increase the brightness of an image thousands of times, can detect incandescent spots on the dome that are invisible to the unaided eye. The total amount of incandescence increases before dome-building events, and glowing cracks often appear to radiate out from the dome. Once new lava reaches the surface, its movement is easily tracked from the air because it continues glowing until it cools.

By applying what they have learned about Mount St. Helens's behavior, scientists at CVO and the University of Washington successfully predicted all of the eruptions — four explosive and two dome-building — between 12 June 1980 and 5 February 1981 a few hours before each began. They also predicted 12 of 14 dome-building eruptions between April 1981 and October 1986 a few days to weeks in advance. Their success has given the U.S. Forest Service enough confidence to allow the public limited access to the top of the mountain — elevations above 4,800 feet — during the summer and nearly unrestricted access during the winter. Only the lava dome itself is still off-limits to the public.

Even at Mount St. Helens, where the prediction success rate is one of the best in the world, earth scientists still cannot predict the size and duration of eruptions or the occurrence of other events, such as rockslides, which can endanger workers and climbers near the summit. CVO scientists are also unsure about their ability to predict explosive eruptions, which often occur with only a few hours' warning. Former USGS chief of volcanic studies Robert Tilling notes that the ability to predict eruptions has still "not been tested by any sudden, violent, explosive eruption such as the cataclysmic eruption of 18 May 1980, that provided no immediate detectable warning."

Kilauea

Hawaii's shield volcanoes Kilauea and Mauna Loa do not pose the same threat to human life as composite volcanoes such as Mount St. Helens, because their lavas are more fluid and they rarely erupt explosively. These volcanoes are located at the southern end of the Emperor Seamounts-Hawaiian Chain. During typical eruptions, these fluid lavas spout from vents or fissures and then flow downhill and solidify. As lava is deposited layer upon layer, shield volcanoes develop their characteristic cone shape with gently sloping sides. Since shield volcanoes erupt nonexplosively, scientists can study their lava flows close up as they occur, with less risk than at composite volcanoes.

In 1987, the Hawaiian Volcano Observatory (HVO) celebrated its 75th birthday by moving into a new building located less than a mile east of the summit of Kilauea. Many volcano monitoring techniques, such as seismic activity and tilt measurements and sophisticated computer analyses of trends in these data, were originally developed and refined at the HVO.

Kilauea is in a phase of eruptive activity that began in January 1983 and has continued for more than 6 years. The first eruption in 1983 began at Napau Crater on a rift zone 13 miles east of the summit. On the day before the event, HVO instruments recorded changes in tilt, a swarm of small earthquakes, and harmonic tremors — signs that magma was moving up in the volcano. The summit of Kilauea subsided a little over an inch as Napau Crater erupted, and the eruptive activity moved northeast into new fissures later in the first eruption and in the two that followed.

In the fourth eruption, activity stopped migrating at a vent now called Pu'u 'O'o. During 17 subsequent eruptions, each lasting between 9 and 100 hours, the volcano spewed fountains of lava through a lava pond in a newly formed crater. The episodes were separated by periods of no activity lasting 8 to 50 days. The vent at Pu'u 'O'o grew to a cinder cone 400 feet high by episode 20, its growth fed by a local reservoir of magma lying just beneath the cone.

Twenty-six eruptions occurred between July 1984 and June 1986, each lasting an average of 13 hours and separated by pauses averaging about 26 days. At times, lava spurted up in fountains reaching over 1,300 feet high. By March 1986, Pu'u 'O'o was 740 feet high and more than half a mile in diameter. As lava flowed down the south flank of Kilauea, it occasionally engulfed houses, but fortunately no deaths occurred.

HVO scientists were consistently able to predict the onset of the first 47 eruptions hours to days in advance. A computerized network of tiltmeters near the vent at Pu'u 'O'o detected increasing tilt of surface rocks as magma filled the reservoir below, and a network of 50 seismometers across the island monitored earthquake swarms and harmonic tremors that signaled magma movements within the chamber. As eruptions began and lava poured out onto the surface, the tilt decreased. HVO volcanologists found that the volume of lava pouring out of Pu'u 'O'o was about equal to the decrease in the volume of the magma chamber beneath the summit.

The pattern of activity changed again when the 48th eruption began on 19 July 1986. The eruption started at vents near Pu'u 'O'o and then shifted to a new vent about 2 miles east. Lava poured from this vent continuously for months, transforming the vent into a lava lake. This eruption finally ended in late April 1988.

Since 1983, lava flowing south from the vent has destroyed 62 houses and covered a mile of highway along the island's southern coast. In November 1986, the flow reached the Pacific Ocean. As lava spread from the coast, it created about 20 acres of new land. Kilauea's 49th eruption since 1983 began in May 1988 from a new vent, which is called Kupaianaha, and was continuing in early 1989. HVO volcanologists are continuing to track the eruption as part of their long-term mission to study the geology, structure, and dynamics of Hawaiian volcanoes.

The Ocean Bottom: A New Frontier

The ocean bottom — a region nearly 2.5 times greater than the total land area of our planet — is a vast frontier that even today is largely unexplored and uncharted. Until about a century ago, the deep-ocean floor was completely inaccessible, hidden beneath waters averaging over 12,000 feet deep. Totally without light and subjected to intense pressures hundreds of times greater than at the Earth's surface, the deep-ocean bottom is a hostile environment to man, in some ways as forbidding and remote as the void of outer space.

Although researchers have taken samples of deep-ocean rocks and sediments for over a century, the first detailed global investigation of the ocean bottom did not actually start until 1968, with the beginning of the NSF's Deep Sea Drilling Project (DSDP). Using techniques first developed for the offshore oil and gas industry, the DSDP's 400-foot-long drillship, the *Glomar Challenger*, was able to maintain a steady position on the ocean's surface and drill in very deep waters, extracting cores of sediments and rock from the ocean floor.

The *Challenger* completed 96 voyages in a 15-year research program that ended in November 1983. During this time, the vessel logged 375,000 miles and took almost 20,000 core samples of seabed sediments and rocks at 624 drilling sites around the world. The *Challenger*'s cores have allowed geologists to reconstruct what the planet looked like hundreds of millions of years ago and to calculate what it will probably look like millions of years in the future. Today, largely on the strength of evidence gathered during the *Challenger*'s voyages, nearly all earth scientists agree that plate tectonics and continental drift are responsible for many of the geological processes that shape the Earth.

The cores of sediment drilled by the *Challenger* have also yielded information critical to understanding the world's past climates. Deep-ocean sediments provide a climatic record stretching back hundreds of millions of years, because they are largely isolated from the mechanical erosion and the intense chemical and biological activity that rapidly destroy much land-based evidence of past climates. This record has already provided insights into the patterns and causes of past climatic change—information that may be used to predict future climates.

When marine organisms die, parts of their bodies may sink to the sea bottom and be preserved as fossils in the sediments. Since some organisms live only in particular environments — such as waters within a particular temperature range — fossil studies have enabled scientists to deduce the changes in water conditions that accounted for the observed sequence of fossils in core samples. Fossils in deep-ocean cores have been used, for example, to reconstruct and map sea-surface temperatures 18,000 years ago, the peak of the last ice age.

Sediment cores dating back as much as 450,000 years have also provided the first compelling evidence supporting a theory developed by Yugoslavian scientist Milutin Milankovitch that variations in the Earth's orbit are partially responsible for triggering the advance and retreat of glaciers. Researcher James Hayes and his associates at the Lamont-Doherty Geological Observatory in Palisades, New York, have confirmed that the Earth's climate varies in distinct, overlapping cycles of about 100,000, 41,000, and 22,000 years each, reflecting cyclical variations in the shape of the Earth's orbit, the planet's tilt on its axis, and the direction in which the axis points.

Ocean Drilling Program

After the *Challenger*'s retirement and the DSDP's end in 1983, a new scientific drilling project called the Ocean

Drilling Program (ODP) began. Although largely funded by the NSF, the ODP is an international program, with support from Canada, France, Japan, the U.K., West Germany, and the European Science Foundation Ocean Drilling Consortium, a group that includes 12 European nations. Using a more advanced drillship, *JOIDES Resolution*, the ODP is continuing the studies begun by *Challenger*. (The drillship's name is derived from Joint Oceanographic Institutions for Deep Earth Sampling, the organization that runs the program.) The special features of the 470-foot-long *Resolution*, which was originally a commercial drillship, have enabled scientists to conduct seabed research that was impossible with the *Challenger*.

The *Resolution* is larger, stronger, and more powerful than its predecessor — qualities that allow the ship to drill in higher winds and rougher seas. The vessel's hull has been strengthened to withstand the icy conditions characteristic of polar waters. The *Resolution* can also carry up to 30,000 feet of drill pipe, compared to 23,000 feet for the *Challenger*, enabling it to drill deeper into ocean sediments and underlying rock. In addition, the *Resolution*'s computer-controlled positioning system allows the ship to remain in a fixed location while drilling in waters up to 27,000 feet deep.

Unlike the *Challenger*, the *Resolution* is fitted with blowout preventers to avoid accidental oil or gas releases. This equipment enables the vessel to drill safely in regions with potential oil and gas deposits — areas where the *Challenger* did not drill because of the risk of causing an uncontrollable release. The *Resolution* also has 12 laboratories with state-of-the-art computers, a scanning electron microscope, and other scientific equipment that allow researchers to analyze drill cores without going ashore. In addition, the *Resolution* can accommodate up to 50 scientists, compared to the *Challenger*'s capacity of 29.

The *Resolution* completed six scientific cruises in 1985, each about 8 weeks long. On its maiden voyage, the *Resolution* drilled in the Bahama Banks, a large, shallow-water region surrounding the Bahamas. Cores taken in the region indicated that the Banks were above sea level about 100 million years ago, but then a rapid rise in the sea level — possibly caused by climatic change — flooded the region.

During its fifth cruise in 1985, the *Resolution* traveled both to Baffin Bay between northern Canada and Greenland and to the Labrador Sea south of Greenland and took cores from the ocean floor above the Arctic Circle — the first time any scientific drillship had done so. Working in stormy seas and iceberg-laden waters, the *Resolution* collected cores extending up to 3,500 feet beneath the seafloor.

These cores contained evidence indicating that Canada, Greenland, and western Europe were joined together in a single land mass until about 85 million years ago, when they began to split apart. As Greenland and Canada separated, creating Baffin Bay and the Labrador Sea, cold polar waters once trapped in the Arctic Ocean were able to flow into the North Atlantic, causing a major cooling of the region's climate. ODP researchers also discovered evidence that glaciers may have formed in Canada and Greenland 3.5 million to 8 million years ago, much earlier than was formerly believed.

In its last voyage of 1985, the *Resolution* became the first ship ever to drill into the hard, bare rock at the center of a mid-ocean ridge. The advanced drilling equipment penetrated into newly formed ocean crust on a small volcano at the Mid-Atlantic Ridge about 1,400 miles southeast of Bermuda. In the past, drillships had been unable to penetrate into bare rock, because a layer of sediment was needed to support and stabilize the drill when a hole was started. To compensate for the lack of sediment on the volcano, a 20-ton supporting base was lowered from the *Resolution* to the seafloor and used to guide the drill into the volcanic rock. Although only one relatively shallow hole about 110 feet deep was drilled into the volcanic rock during the 1985 voyage, the *Resolution* returned to the site during May and June 1986 to deepen the hole and make additional studies of the new ocean crust.

In 1986, the *Resolution* again completed six research cruises that took the ship from the east coast of Africa to the west coast of South America. Early in the year, ODP scientists studied the seafloor in the Tyrrhenian Sea, a part of the Mediterranean off the west coast of Italy. The core samples showed evidence that the region has had a very active geologic history over the past 5 million years, due to a combination of seafloor spreading in the central Tyrrhenian Sea and subduction along the coast of Sicily and southernmost Italy. In earlier research in the region, the *Challenger* had found evidence that the Mediterranean became closed off from the Atlantic Ocean about 5.5 million years ago and later dried up completely, becoming a vast desert.

In the fall, on her fifth leg of the year, *Resolution* sailed to an old drill hole near the Costa Rica Rift, 200 miles west of Ecuador, deepening it to nearly 1 mile. Data from this hole will help explain the dynamics of seafloor spreading centers, where molten material rises from the mantle and then moves laterally away from the rift to form new seafloor. At the bottom of the hole, marine geologists found sheeted dikes, ribbons of volcanic rock that had penetrated into overlying layers of basalt. These rocks are similar to rocks called ophiolites, which are thought to have formed in deep-ocean crust and which are sometimes thrust onto land by crustal movements. Since geological models of the evolution of the ocean crust had predicted that sheeted dikes would be present at seafloor spreading centers, the discovery of the dikes has provided new support for these models.

The *Resolution*'s last voyage in 1986 was to the continental margin off the western coast of Peru, near where the oceanic Nazca Plate has been subducting beneath the South American Plate since the opening of the Atlantic Ocean. Core samples from this area showed that sediments have been scraped off the oceanic plate onto the continental margin since about 16 million years ago. These sediments contain a rich history of past and present life ranging from ancient whales to bacteria.

The ODP began 1987 with an ambitious voyage to the Weddell Sea off Antarctica, where the *Resolution* made some of its most spectacular discoveries to date. Antarctica is divided by the Transantarctic Mountains into East and West Antarctica, with each region covered by a separate ice sheet. Some scientists have been concerned about the possibility that the smaller West Antarctic ice sheet could become unstable if the Earth's average temperature rises due to increasing atmospheric carbon dioxide from the burning of fossil fuels and deforestation. Cores drilled from Antarctica's margin have shown, however, that the West Antarctic ice sheet has remained stable during the last 4.8 million years, even in relatively warm periods.

Cores from 22 holes confirmed that Antarctica was once warm and ice-free, as evidenced by fern spores and beech tree pollen in the sediment. About 35 million to 40 million years ago, the climate cooled, and the first evidence of glaciation appeared in sediments deposited at that time. By about 13 million years ago, the landmass of East Antarctica became completely covered by a vast ice sheet, and about 5 million years later, West Antarctica was also fully covered.

After completing its Antarctic expedition, the *Resolution* drilled 12 holes in the South Atlantic during March and April 1987. The 1.4 miles of sediment cores recovered contain a nearly complete sedimentary record of the last 90 million years, with information about the region's geological and climatic history. These cores provided evidence supporting the hypothesis that the opening of the Drake Passage between Antarctica and South America 20 million to 25 million years ago allowed the Antarctic Circumpolar Current to become established. This cold current helped create conditions leading to a much colder south polar climate that promoted the growth of the Antarctic ice sheets. The *Resolution*'s research in the Weddell Sea and South Atlantic is giving scientists a much more complete picture of Antarctica's geological and climatic history, an important influence on global climate.

During the rest of 1987, *Resolution* traveled to locations around the Indian Ocean. In July and August, the vessel obtained cores from an accumulation of submerged sediments called the Bengal Fan, south of Sri Lanka. The sediments showed that the collision of the Indian and Eurasian plates, which led to the formation of the Himalayas, began 10 million years earlier than geologists had previously thought. Before the *Resolution*'s voyage, geologists had had an incomplete history of the Himalayas, because all the material that was eroded from the range shortly after the plate collision had been carried down rivers and submarine channels into the Indian Ocean basin.

Later in the year, *Resolution* drilled off the coast of Oman, collecting sediments that are expected to bring new insights into the origin and evolution of the Asian monsoon, the seasonal winds that bring rainy and dry seasons to Asia. Monsoon rains support agriculture and make human settlements possible throughout much of Asia. The *Resolution* continued drilling in the Indian Ocean during 1988. Extensive sampling was conducted on the 1,400-mile-long Kerguelen plateau, the world's largest submerged plateau, which is found in the southern Indian Ocean near Antarctica. In another leg, off northwestern Australia, the *Resolution* drilled into sedimentary rocks up to 220 million years old. These are the oldest sedimentary rocks ever obtained during scientific ocean drilling, dating back to a time when the Americas, Australia, Antarctica, Africa, India, and Europe were joined together in a supercontinent called Pangea.

Manned Exploration of the Ocean Bottom

Travels with Alvin. Just as advanced scientific drillships have made it possible to sample the sediments and crust of the sea bottom, rugged submersibles have allowed researchers to study the ocean floor firsthand. One of the most versatile submersibles, *Alvin*, has been used for deep-sea research since 1964. The 16-ton vessel, which was built by the U.S. Navy, is operated by the Woods Hole Oceanographic Institution in Massachusetts. The submersible is 25 feet long and up to 8 feet wide and contains a 7-foot-diameter spherical compartment that can carry a crew of three — a pilot and two scientists. The sphere is built of a 2-inch-thick titanium alloy to withstand the immense pressures of the deep ocean and dive safely to a maximum depth of about 13,000 feet.

Although *Alvin* contains life-support equipment that could keep a crew alive under water for up to 72 hours, its scientific dives typically last between 6 and 10 hours. During a dive, the submersible can travel at a speed of up to about 2 miles per hour and has a cruising range of about 5 miles. *Alvin*'s crew uses its two remotely controlled mechanical arms to manipulate underwater cameras and lighting systems as well as scientific equipment, such as nets, traps, drilling devices, and vacuum samplers, which suck up fragile marine organisms.

Since 1964, scientists aboard *Alvin* have made some of the most important discoveries in oceanography. By late 1988, the vessel had completed a total of over 2,100 dives. First used for Navy pilot training and search-and-recovery missions, *Alvin* proved its worth in 1966 when it located a missing hydrogen bomb that had fallen into the Mediterranean Sea as a result of a plane collision off the coast of Spain.

In Project FAMOUS — the French-American Mid-Ocean Undersea Study — during 1973 and 1974, *Alvin* and two French submersibles named *Archimede* and *Cyana* dove to a part of the Mid-Atlantic Ridge about 400 miles southwest of Ponta Delgada in the Azores. The Mid-Atlantic Ridge is the world's longest mountain range, stretching 12,000 miles from the Arctic Circle east of Greenland to the Weddell Sea off Antarctica.

The region visited by the submersibles is a rift valley, where two plates — the African Plate and the North American Plate — are moving apart as molten rock rises from the Earth's interior to form new ocean crust. The submersibles found that the steep-walled rift valley is flanked on each side by ranges rising 5,000 feet above the valley floor. In a total of 51 dives to the Ridge, *Alvin*, *Archimede*, and *Cyana* mapped a 60-square-mile area, took more than 100,000 photographs, and collected 3,000 pounds of samples.

New types of life. *Alvin*'s most famous dive took place 3 years later, during a voyage to the Galápagos Ridge about 200 miles northeast of the Galápagos Islands. On 19 February 1977, researchers aboard *Alvin* made one of the most startling biological discoveries of the century. Instead of finding a nearly lifeless environment on the ocean floor as they had expected, *Alvin*'s crew discovered a complex community teeming with a variety of organisms that had never been seen before. These organisms were found in small areas adjacent to hot springs circulating through vents in the ridge.

Oceanographer John Corliss, then of Oregon State University, was aboard *Alvin* on its historic voyage and described the discovery this way: "We sat inside Alvin's titanium pressure sphere amidst all our instrumentation and looked out our ports to see shimmering water streaming up past the submarine with pink fish hovering in the warm water, white crabs scuttling over the rocks, huge white clams and yellow-brown mussels, and the long, white tubes of worms with red plumes, surrounding the hydrothermal [hot-water] vents. It was an experience that those of us who dove will never forget."

Before *Alvin*'s voyage to the Galápagos Ridge, marine scientists had believed that the deep-ocean bottom was nearly barren — a marine "desert" suffering not from a scarcity of water, like deserts on land, but from a lack of light. At the Earth's surface and in the ocean's shallow waters, nearly all life depends either directly or indirectly on light. Photosynthetic plants, which are the base of the world's food chain (more commonly known to biologists as the food web), use sunlight to transform water and carbon dioxide into carbohydrates, which provide life-sustaining energy when consumed by other organisms. These organisms are, in turn, consumed by other creatures and so on throughout the food web. Even for meat-eating animals, the ultimate sources of energy and life are photosynthetic plants and therefore light.

With a few exceptions, photosynthetic plants cannot survive in the darkness of the deep ocean. Even in clear water, bright sunlight will usually dim to the point of being too faint to sustain photosynthesis at depths greater than 500 to 650 feet. Before *Alvin*'s historic voyage, marine biologists had believed that in waters more than a mile below the surface — remote from light and photosynthetic food sources — animal life would be severely limited by a scarcity of energy and food. Most organic matter drifting down from the surface would be consumed by bacteria and other organisms before ever reaching the ocean floor, and bottom dwellers would have little to sustain them.

Consequently, biologists were amazed to find a community of organisms apparently thriving in the dark waters of the Galápagos Ridge 1.5 miles below the ocean surface. Since the discovery of the Galápagos Ridge community in 1977, scientists aboard *Alvin* have returned repeatedly to the vents to continue their studies.

Chemically powered bacteria. Although the marine oasis discovered by *Alvin* did not extend far from the hot-spring vents, the concentration of organisms near the vents rivaled the most productive regions at the ocean surface. Yet even more surprising to biologists than finding organisms living in this remote location was the nature of their food source. Unlike most of the world's organisms, which have photosynthetic organisms at the base of their food web, the vent communities depend on chemoautotrophic — chemically self-feeding — bacteria. Such organisms derive their life-sustaining energy from the oxidation of substances such as sulfur compounds spewing out of the vents at the mid-ocean ridge.

When ordinary seawater circulates through the cracks and porous rock at the vents, it is heated and chemically changed. Sulfate ions, which are common in seawater, lose their oxygen and gain hydrogen to become hydrogen sulfide. Chemoautotrophic bacteria use oxygen in the water to oxidize the hydrogen sulfide, a process that provides energy. By using this chemical energy source instead of light, the bacteria are able to convert carbon dioxide, oxygen, and other inorganic compounds in the water into proteins and other organic compounds. The vent ecosystem is highly productive, with biological activity two to four times higher than in the surface waters and up to 1,000 times greater than on the rest of the deep-ocean floor.

Mollusks and crabs. The chemically powered bacteria are not the only unusual inhabitants of the vents. Many of the organisms there are previously unknown species, and some are so distinct that they have been classified as the first species in new animal families and genera. Unlike typical deep-ocean benthic (bottom-dwelling) organisms, which have low metabolic rates to compensate for the scarcity of food, vent animals have relatively high metabolic rates, similar to those of creatures near the surface. Vent clams may grow 1.5 inches in length per year, far more rapidly than other deep-water clams.

Researcher Kenneth Smith of the Scripps Institution of Oceanography in La Jolla, California, has also found marked differences in growth rates among vent animals of the same species, depending on how close they are to the vents and the sulfide-rich waters. Mussels growing near the vents and bathed in sulfide-rich waters are large and relatively massive, whereas those growing farther away are smaller and lighter, even when the water temperature is nearly identical. When thriving mussels are transplanted to waters farther from the vents, their muscle mass shrinks; conversely, lighter mussels transferred to regions near the vents grow rapidly.

The vent crab provides another striking example of the differences between vent organisms and their counterparts in other environments. Crabs living in shallow environments with plentiful food often have large claws for crushing prey. Such claws are rare in deep-sea crabs, however, since there is usually an inadequate amount of food on the deep-ocean bottom for the crab to maintain the high metabolic rates needed to operate its muscular claw. In contrast, the rich supply of food in the environment of the vents has allowed the vent crab to develop a crushing claw. The crab has also developed very high blood concentrations of the oxygen-carrying compound hemocyanin, enabling the crab to live off stored oxygen when it enters the relatively oxygen-poor waters near the vents to prey on organisms there.

Giant tube worms. The giant tube worms discovered near the vents are even more unusual than the vent clams, crabs, and mussels. The worms, named *Riftia pachyptila Jones* after worm expert Meredith Jones of the Smithsonian Museum of Natural History in Washington, D.C., grow to lengths of 5 feet, while related worms in other environments generally measure only inches at most. Even more peculiar, the tube worm lacks a mouth and a gut and is topped by a feathery plume, which is actually composed of over 200,000 tiny tentacles.

Although marine biologists at first thought that the tube worm somehow absorbs nutrients from the surrounding sea or sediments, they were forced to look for more imaginative solutions when they determined that the worm's environment has insufficient organic material to support its growth. After painstaking dissection and analysis, biologists found that instead of relying on external sources of food, the worms are nourished from within by symbiotic bacteria.

Their research revealed that the "tissue" of the tube worm's trophosome, a large body structure that extends over half the length of the adult worm, is composed of closely packed bacteria — over 100 billion per ounce of tissue. Alissa Arp and James Childress of the University of California at Santa Barbara found that the tube worm's blood, deep red from a rich supply of hemoglobin, absorbs oxygen from the water and transports it, together with carbon dioxide and hydrogen sulfide, to the trophosome.

Ensured a rich supply of carbon dioxide, hydrogen sulfide, and oxygen, the bacteria living inside the worms produce carbohydrates and proteins, which the worm then uses for growth. Biologists have also found evidence that the rapid growth of vent mussels and large vent clams is supported by symbiotic chemoautotrophic bacteria living within them.

Researchers John Corliss, John Baross, and Sara Hoffman of Oregon State University have raised the intriguing possibility that life may have begun at vents billions of years ago. The researchers have suggested that the hot, chemically rich waters of the vents could have provided a unique combination of gases, minerals, water, and thermal energy that might have led to the spontaneous generation of life. With the heat of the vents as an energy source and with clay minerals as catalysts for chemical reactions, substances such as ammonia, hydrogen, metals, and methane found in ocean waters could have reacted to form amino acids and proteins, which might have been building blocks for organisms similar, perhaps, to the bacteria that still survive near the vents.

Other deep-sea communities. Since the late 1970s, scientists studying vents at other mid-ocean ridges have found organisms similar to those at the Galápagos Ridge. Communities of vent organisms have been discovered at the Juan de Fuca Ridge off the coast of Oregon and Washington, the Guaymas Basin in the Gulf of California, and at several locations on the East Pacific Rise, a ridge west of Central and South America.

The first vent communities in the Atlantic Ocean were discovered along the Mid-Atlantic Ridge not by *Alvin*, but by cameras lowered from vessels on the ocean surface. During a cruise in July and August 1985, the National Oceanic and Atmospheric Administration (NOAA) ship *Researcher* observed the first Atlantic vent community on the ridge at a location of about 26°N and 45°W. A second vent community was discovered by the scientific drillship *Resolution* during late 1985 about 250 miles south of the first. These two communities were surprisingly different from those observed at the Pacific vents. Although the clams and mussels that lived in symbiotic association with chemoautotrophic bacteria were not found at the mid-Atlantic sites, other organisms were abundant.

NOAA scientist Peter Rona and his colleagues on the *Researcher* voyage reported that they had observed a few worms living near the vents but did not see any thickets of giant tube worms typical of the vent communities in the Pacific. Anemones and fast-moving shrimp were common near the Atlantic vents. Fish up to about 10 inches long and a few crabs were also found nearby.

At the vent visited by the *Resolution*, researchers found organisms that appeared to be shrimp and anemones, as well as snakelike swimming animals — possibly eels — about 1 to 2 feet long. The researchers named the site the Snake Pit after these creatures. In general, both the Snake Pit and the vent community studied by *Researcher* had smaller, more mobile organisms than those observed at the Pacific vents.

Rich communities much like those found at the Pacific vents, however, have been identified in the Gulf of Mexico, although they are not located near any vents, much to the surprise of scientists. In March 1984, a group of researchers aboard *Alvin*, including Charles Paull of Scripps Institution of Oceanography found one such community on the Florida Escarpment about 200 miles off Florida's west coast, where a shallow submarine platform drops off rapidly into the deep waters of the Gulf of Mexico.

While diving in the *Alvin* at a depth of 10,700 feet, Paull found a site containing an abundance of organisms, including clams, mussels, tube worms, fish, and many other creatures similar to those that Paull had observed earlier along the East Pacific Rise. Paull has suggested that a hydrogen-sulfide-rich brine seeping from rocks in the escarpment supports the organisms there.

In September 1988, NOAA researcher Peter Rona and his colleagues aboard the U.S. Navy submersible *Sea Cliff* discovered the first vent community in U.S. waters on the Gorda Ridge, a mid-ocean spreading center off the Oregon coast. Like other vent communities discovered earlier, the Gorda Ridge community includes a variety of organisms, including tube worms, crabs, fish, and other animals.

Offshore Mineral Resources

Vent sulfide deposits. In addition to biological discoveries at the vents and other locations, *Alvin*'s expeditions have revealed mineral treasures as well. Hot water circulating through the vents leaches copper, iron, manganese, nickel, and other metals from rocks in the Earth's crust. These metals are carried into the water in the form of sulfides. Although heavy metal sulfides will not dissolve in water at normal ocean temperatures, they do in the 650°F waters circulating within vents. When the hot water rises through the vent and is quickly cooled by 36°F bottom water, most of the metal sulfides precipitate out of solution and form massive deposits on the ocean floor nearby. Some of the largest deposits are found around chimneylike structures called "black smokers," so named because dark mineral-rich waters gush from the smokers much like smoke from a chimney.

During a 1981 dive to vents at the Galápagos Ridge, U.S. National Ocean Service scientists aboard *Alvin*

Figure 4. "Black Smoker." Hot sulfide-rich waters gush from a "black smoker" at the East Pacific Rise, a mid-ocean ridge off the west coast of Central and South America. As hot vent waters come in contact with the cold water normally found at the ocean bottom, some of the dissolved sulfide minerals in the hot water precipitate out to form mineral deposits. *Courtesy: Dudley Foster, Woods Hole Oceanographic Institution.*

discovered and mapped a sulfide deposit that is probably as extensive as some of the largest ore deposits on land. *Alvin* recovered several samples containing a high percentage of iron as well as significant amounts of copper, silver, tin, vanadium, and other valuable metals. The copper deposits may prove to be particularly valuable. Unlike copper deposits on land, which tend to be only a few inches thick, those at the Galápagos Ridge are estimated to be about 100 feet thick, 650 feet wide, and 3,000 feet long. The Galápagos deposits may contain several million tons of 10 percent copper ore worth many millions of dollars.

During dives aboard *Alvin* in 1984, University of Hawaii researcher Alexander Malahoff found smaller deposits of zinc-rich sulfides at vents on the Juan de Fuca Ridge about 220 miles west of Newport, Oregon. In 1985, scientists on the NOAA research ship *Surveyor* also reported that the ocean waters over parts of the Gorda Ridge contained high metal concentrations, an indication that metal-rich sulfide deposits might also be found there.

In August 1986, researchers from NOAA's Pacific Marine Environment Laboratory in Seattle detected a so-called megaplume, a massive emission of hydrothermal fluids that clouded water in a region 2,300 feet thick and more than 12 miles in diameter, above vents in the Juan de Fuca Ridge. Water in the plume averaged about one-tenth of a degree warmer than surrounding water and was rich in the minerals characteristic of vent waters. Although the plume took only a few days to form, its energy content equaled the annual heat output of as many as 2,000 typical black smokers. The NOAA oceanographers suggested that megaplumes might be common at hydrothermal vents and said that minerals in these plumes could have a significant effect on creatures living in the oceans.

Scientists on the *Researcher* and *Resolution* had also discovered massive metal-rich sulfide deposits during their cruises to the Mid-Atlantic Ridge in 1985. The *Resolution* drilled 10 holes in the area around one large black smoker and found that sulfide deposits were over 40 feet thick at the base of the smoker and at least 10 to 20 feet thick at a distance of 55 feet from the smoker.

Other marine mineral resources. The metal-rich sulfide deposits at the vents are only one of the many promising mineral resources that may be mined from the ocean bottom in decades to come. Sand and gravel, which are widely used by the construction industry, are the largest and generally most accessible of marine mineral resources, since they are abundant in shallow coastal areas around the world. The U.S. Minerals Management Service (MMS) has estimated that in the continental shelf areas of the contiguous U.S. alone, sand and gravel deposits total about 40 trillion cubic feet.

Mineral resources known as placer deposits are second in abundance to sand and gravel. Placer deposits are formed as waves and currents winnow away relatively light grains of clay and sand from exposed rock, leaving behind denser minerals that are often rich in valuable elements such as chromium, gold, hafnium, platinum, tin, titanium, and zirconium. Researchers have estimated that placer deposits total at least 45 billion cubic feet off the U.S. Atlantic coast and over 70 billion cubic feet off the U.S. Pacific coast.

The USGS has already identified placer deposits rich in titanium ore in shallow waters 5 to 50 miles off the Georgia and Virginia coasts. The U.S. depends heavily on imports for its supply of titanium, a strategic metal that is used, among other things, in the construction of high-strength, lightweight aircraft parts.

Deposits of phosphorite, a mineral widely used as a fertilizer, have been found off the Atlantic coast from the Carolinas to Florida and off the Pacific coast of southern California. Although the full extent of phosphorite deposits in the U.S. is not known, the resources in these two coastal areas alone may total 4 billion tons. The phosphorite deposits off California also appear to be rich in cobalt, which is used in magnetic alloys and for ceramics.

Manganese nodules, another mineral resource, are small, metal-rich lumps that form slowly on the ocean bottom as manganese and other dissolved metals precipitate out of the water and adhere to rock fragments or organic debris, such as fish teeth or pieces of whale bones. Over hundreds of years, the mineral coating builds up until the nodule reaches an inch or more in diameter. Manganese nodules have been found on the Blake Plateau, a wide, submarine plain off Florida's Atlantic coast, and on parts of the deep-ocean floor, often at depths of 10,000 feet or more. In addition to manganese, which is used to increase the strength, hardness, and durability of steel, nodules often contain significant amounts of cobalt, copper, and nickel.

Cobalt-rich manganese crusts — metallic coatings on the sides of some submarine mountains, volcanoes, and cliffs — are formed as metals in seawater are slowly deposited on seafloor rocks. Since manganese crusts are generally found at shallower depths and have higher concentrations of cobalt than manganese nodules, they are probably a more accessible and valuable mineral resource than nodules. During cruises in the Pacific Ocean, USGS researchers have found cobalt-rich manganese crusts an inch or more thick on the sides of submarine mountains and volcanoes off the coasts of Christmas Island, Johnston Island, and the Hawaiian Islands.

Mapping the U.S. Exclusive Economic Zone

Interest in seabed mineral resources has increased in recent years as more than 65 nations around the world have each claimed exclusive rights to living and mineral resources in a so-called Exclusive Economic Zone (EEZ) extending 200 nautical miles (about 230 miles) from their coastlines. In March 1983, U.S. President Ronald Reagan established a 200-nautical-mile EEZ around the U.S. and its territories. This region covers about 4.5 million square miles, an area 1.3 times larger than the entire land area of the United States. NOAA and the USGS have embarked on a decade-long program to map and assess the resources on the seafloor in the EEZ.

NOAA mapping. To conduct its mapping work, NOAA has been using three ships, *Davidson*, *Surveyor*, and *Discoverer*, equipped with high-precision sonar systems that measure the depth of water beneath the ship to the nearest meter (3.3 feet). In all sonar mapping systems, strong pulses of sound are transmitted through the water to the ocean bottom, where the sound is partially reflected back to the instrument that transmitted it. By precisely measuring the time interval between the pulse transmissions and returning echoes, sonar equipment can calculate the depth to the bottom.

For mapping the seafloor, survey ships typically run a series of profiles spaced some distance apart, and cartographers then draw contour lines between points of equal depth. The accuracy of bottom topography maps constructed from this profile data is limited not only by the accuracy of the depth measurements but also by two other problems. Accurate maps cannot be drawn unless a mapmaker knows the ship's precise position when it took the data. Until satellite navigation systems became available, ships had difficulty determining their exact positions because there are relatively few reference points at sea from which to calculate location.

In addition, since conventional sonar systems provide depth measurements only along profile lines, the accuracy of maps made from these lines is limited by their spacing; topographic features smaller than the distance between the lines are missed. Although maps made from many closely spaced profiles are more detailed than those using fewer lines, these maps are also more time-consuming and costly to produce.

To overcome these limitations, the *Davidson* and *Surveyor* are equipped with advanced sonar systems called multibeams that use multiple sonar beams to map a broad swath during one pass over the seafloor. With multibeam systems, there is no gap between profiles, and as a result, a detailed three-dimensional map showing even very small topographic features may be constructed. Satellite positioning systems, which provide ships with a frame of reference, also enable surveyors to determine their profile locations with a high degree of accuracy.

Although NOAA's ships had surveyed roughly 100,000 square miles of the U.S. EEZ by early 1989, NOAA had not yet released completed maps to the public. The U.S. Navy and the Defense Mapping Agency had opposed publication on the grounds that the highly detailed maps produced by the multibeam systems could be used by hostile submarines to navigate through U.S. coastal waters undetected. These maps would also make it easier for an enemy submarine crew to calculate its vessel's precise position—information that would enable the submarine to fire missiles more accurately at a target. Because of these concerns, the NOAA multibeam mapping data were classified as secret and were not available to researchers. In April 1989, however, the Navy announced that it would soon allow NOAA to release most of the multibeam data.

USGS seafloor studies. In contrast to NOAA's EEZ mapping project, the USGS's research program EEZ-SCAN has not been hampered by national security issues, because it relies on a different type of sonar system. This scanning system, called GLORIA, was developed by the Institute of Oceanographic Sciences (IOS) in Wormley, U.K. As in the multibeam systems, GLORIA emits multiple sonar beams into the water and measures the reflected beams from the bottom topography. Unlike the multibeams, however, GLORIA is specifically designed to measure differences in the acoustic reflectivity — rather than the depth — of the seafloor.

Since different types of rocks and sediments reflect different amounts of the incoming sound signals, the strength of the echoes can be used to determine what types of rock or sediments are located on the ocean bottom. The echo data are then processed to produce sonographs — images that in some ways resemble high-altitude aerial photos — of the sediments and rocks on the bottom. According to IOS researchers, GLORIA is able to survey the seafloor about 10 times faster than any competing sonar system at a cost of about $5 per square mile.

During the summer of 1984, the GLORIA system mapped the EEZ along the west coast of the contiguous U.S., an area of about 330,000 square miles. The sonographs revealed previously unknown submarine features, including over 100 mountains and 50 volcanoes, and showed fine details in the structure of submarine canyons, channels, faults, and mid-ocean ridges.

In 1985, GLORIA mapped the EEZ along the Gulf of Mexico from Texas to Florida, an area of 140,000 square miles, and then went on to map the EEZ off Puerto Rico, an area of 58,000 square miles. GLORIA's sonographs of the Gulf EEZ showed an extensive network of submarine channels and thick accumulations of sediment fanning out from the channels like deltas spreading out at river mouths. USGS researchers believe that some of these sediment fans may contain significant amounts of oil and gas. Off Puerto Rico, GLORIA mapped the Puerto Rico Trench which, with a maximum depth of over 28,000 feet, is the deepest part of the Atlantic Ocean.

During 1986, GLORIA mapped the EEZ in the Bering Sea off Alaska from Unimak Pass northwest to the U.S.-U.S.S.R. Convention Line of 1867. Most of this area's geology was unknown. GLORIA imagery showed that parts of the northern flank of the Aleutian Ridge are eroding away, as debris flows and massive slides carry material off the ridge down to the Aleutian Basin along some of the largest submarine canyons in the world. Mapping of the Alaskan EEZ south of the Aleutian Islands and in the Gulf of Alaska continued during 1988, along with surveys of the regions around Hawaii. By the end of the year, GLORIA scans had covered about 1.3 million square nautical miles.

In 1988, while using the GLORIA scanner to map the EEZ around the Hawaiian island chain, USGS scientists detected extensive undersea lava flows surrounding the islands. The finds were surprising, because the undersea flows covered an area equivalent to twice the surface area of the islands. Also, the flows were very young — 50,000 to 200,000 years, far younger than rocks from the eruptions that originally formed the islands. Scientists generally agree that the Hawaiian Islands were formed by the westward passage of the Pacific Plate over a stationary plume of hot magma — called a hot spot — from deep in the Earth's mantle. Thus, islands at the western end of the chain should be older than those toward the east. The existence of young lava fields throughout the Hawaiian chain calls this idea into question.

When the mapping of the U.S. EEZ is completed in the 1990s, scientists will have an unprecedented view of the topography and geology of the coastal seafloor. GLORIA's sonographs of submerged volcanoes off the western U.S. and Hawaii have already led to a greater understanding of submarine volcanic activity and the differences between submarine volcanoes and their counterparts on land. From GLORIA's sonographs of the Gorda and Juan de Fuca ridges off the coast of northern

California, Oregon, and Washington, scientists may gain new information about the formation of crust at mid-ocean ridges and the tectonic forces that move the Earth's continental and oceanic plates.

Studies of undersea fault zones may also lead to better assessments of the risks of earthquakes in coastal areas as well as the dangers of destructive waves called tsunamis, which are generally triggered by offshore earthquakes and volcanic activity. The detailed maps of coastal areas will also be used to estimate the size of undersea resources and identify the most promising locations for oil, gas, and mineral exploration.

El Niño

During 1982 and 1983, unusual and often damaging weather struck many areas around the world. Australia suffered its worst drought in at least a century, with more than $2 billion in losses to crops, livestock, and property. In Indonesia's East Kalimantan province on the island of Borneo, a drought-related fire ravaged 8.5 million acres of rain forest — a region about the size of Taiwan — causing nearly $1.5 billion in timber losses. Abnormally dry conditions were also reported in areas as widely scattered as Bolivia, Mexico, southeastern Africa, India, and the Philippines, causing thousands of deaths and several billion dollars in crop losses.

In sharp contrast, coastal Ecuador and northwestern Peru were deluged by more than 10 feet of rain that swept away crops and homes and broke all previous rainfall records for the region. Flooding brought trade nearly to a halt in Ecuador's five coastal provinces during January 1983 and left 23 people dead and over 100,000 homeless. By the end of June 1983, flooding had killed about 600 people in Ecuador and northern Peru.

In addition to the flooding problems, the people of Ecuador and Peru suffered an abrupt decline in their fish catch, due to abnormally warm water temperatures in the eastern Pacific Ocean during late 1982 and early 1983. The amount of fish meal produced in Peru during the first 7 months of 1983, for example, was barely one-quarter of the amount produced during the same period a year earlier, which was in turn far below the peak levels of the 1960s and early 1970s. In Peru alone, flooding and fishing losses were estimated at $2 billion — more than 10 percent of the nation's gross national product — while in Ecuador, total economic losses reached $640 million.

In the fall of 1982 and the winter of 1982–1983, an unusual number of severe storms also struck parts of the U.S. along the Pacific Ocean and the Gulf of Mexico. Between September 1982 and June 1983, storms along the Pacific coast killed at least 645 people and caused extensive beach erosion, mudslides, and flooding, which resulted in property damage of about $500 million; crop damage from those storms totaled another $600 million. Heavy rain and flooding along the Gulf Coast killed about 50 people and caused losses totaling an estimated $1.1 billion.

The distribution of tropical storms and hurricanes around the world was also severely disrupted during 1982 and 1983. The hurricane seasons for both years were among the quietest of the century in the Atlantic, Gulf, and Caribbean regions, while the eastern North Pacific Ocean had many more storms than average. In the South Pacific, French Polynesia — which had not suffered a hurricane in 75 years — was buffeted by six tropical storms, including five hurricanes, between December 1982 and April 1983. One of the storms, Hurricane Veena, was described as the worst on record in Tahiti. The hurricane destroyed or damaged 7,500 homes, caused property damage estimated at $50 million, killed one person, and left 25,000 people homeless.

Although normal variations in global climate cause droughts and storms in various parts of the world each year, the exceptionally severe climatic conditions covering wide areas during 1982 and 1983 appeared to be part of a consistent global pattern. After extensive study, most researchers agreed that these events were linked to the Southern Oscillation—a periodic change in large-scale Pacific Ocean weather systems — and El Niño — an associated warming of the equatorial Pacific Ocean.

El Niño and the Southern Oscillation

In the normal weather pattern of the tropical Pacific Ocean, the combination of a high-pressure area located to the east of Tahiti and a humid, low-pressure area over Indonesia, northern Australia, and the Indian Ocean helps to produce trade winds blowing toward the west. These winds in turn cause ocean currents in the Pacific that flow westward. As warm surface waters in the eastern Pacific are carried west, relatively cool water rises from the depths to the surface along the western coast of South America. The rise of these cool, nutrient-rich ocean waters sustains one of the world's richest fisheries.

At irregular intervals about every 3 to 4 years, but occasionally as few as 2 years or as many as 10, this normal pattern of ocean temperatures is disrupted in a phenomenon known as El Niño. During El Niño years, the surface waters of the tropical Pacific become unusually warm in a region extending from the South American coast far westward, sometimes beyond the International Date Line (roughly corresponding to 180° longitude) more than 6,000 miles away. (The event is called El Niño, which is Spanish for "the [Christ] child," because the appearance of warm ocean waters off the western South American coast often occurs around Christmas.)

When an El Niño occurs, the presence of a relatively thick layer of warm surface water off western South America prevents the cool, nutrient-rich water from rising. As a result, the productivity of the coastal fishery sharply declines. Studies of the state of the ocean and atmosphere during past El Niño events have shown that El Niño is not an isolated occurrence, but is instead part of a pattern of changes in the global circulation of the oceans and atmosphere. In conjunction with El Niño, the atmospheric pressure rises in the normally low-pressure area over Indonesia, while the pressure falls in the normally high-pressure area in the southeastern Pacific. At about the time that the warming ends, the reverse occurs, with the pressure falling in Indonesia and rising in the southeastern Pacific.

The British meteorologist Sir Gilbert Walker first noted this pattern of changing pressures in the eastern and western tropical Pacific during the early 1920s and named it the Southern Oscillation. Not until the 1960s, however, did researchers associate the occurrence of El Niño with these pressure changes.

To monitor the Southern Oscillation, climatologists use a measure called the Southern Oscillation Index (SO Index), which is based on the atmospheric pressure difference between Tahiti and Darwin, Australia. The pressure at Darwin reflects the strength and location of the low-pressure area usually found near Indonesia. Similarly, although Tahiti is located in the south-central Pacific, the pressure there reflects the strength of the high-pressure area usually found in the southeastern Pacific.

Typically, for a period of at least 18 months before the beginning of a strong El Niño, brisk easterly winds tend to push water from the eastern Pacific toward the west, producing abnormally high sea levels in the west and abnormally low levels in the east. A decline in the SO Index usually marks the start of El Niño, as the atmospheric pressure at Tahiti starts falling. As the difference in atmospheric pressure between the eastern and western Pacific decreases, the easterly winds in the tropical Pacific slacken and often change direction to westerly.

Without easterly winds to provide the force necessary to pile up the warm water in the western Pacific, the warm water that has already accumulated there begins surging eastward at a speed of 3 to 10 feet per second, raising the sea level as it moves. Several weeks later, the surge arrives on the coast of South America, bringing El Niño conditions to Ecuador and Peru. At its maximum extent, the unusually warm waters cover much of the tropical Pacific.

Events of the 1982–1983 El Niño

The 1982–1983 El Niño — one of the most severe climatic events of the twentieth century — was extraordinary both in its geographical extent and in the degree of warming that took place. One of the first signs of the impending event was a fall in the SO Index in early 1982. The index then sharply decreased in May and June and reached record low values later in the year. The easterly trade winds in the western tropical Pacific diminished rapidly in June and reversed direction in July, remaining westerly and unusually strong throughout the remainder of the year.

As the easterlies slackened and were replaced by westerlies, the warm surface water in the western Pacific migrated eastward. Sea level fell in the western Pacific and rose more than 16 inches in the eastern Pacific. During the period of strong westerlies, the large-scale patterns of rainfall across the tropical Pacific shifted eastward. This shift was directly responsible for the drought that affected Australia and Indonesia in 1982 and early 1983, and for the torrential rains that struck parts of the central and eastern equatorial Pacific at about the same time.

In the 1982–1983 El Niño, abnormally warm ocean surface waters first appeared in the western and central tropical Pacific in May 1982; by June they had spread along a line stretching about 7,000 miles from about 170°E eastward to the South American coast. The temperature anomaly (the difference between the observed temperature and normal temperature) increased, and by late 1982, the surface water temperature of a large portion of the eastern tropical Pacific was more than 7°F above normal. Thereafter, the temperature anomaly diminished in the central Pacific but reached a second peak east of 110°W in May and June 1983, when some locations near the South American coast were 14°F above normal. These temperature anomalies were unprecedented, since in most El Niño events water temperatures rise a maximum of only 3°F to 5°F above normal.

During the first half of 1983, while its effects in the eastern equatorial Pacific were intensifying, El Niño began showing some of the first signs of its eventual dissipation. Just as a drop in the SO Index in early 1982 marked the onset of the event, a sharp rise in the index to nearly normal levels between February and May 1983 indicated that the weather conditions in the tropical Pacific were starting to return to normal. The region of above-normal ocean temperatures shrank from its maximum extent of about 8,000 miles in late 1982 to 5,000 miles by the end of July 1983.

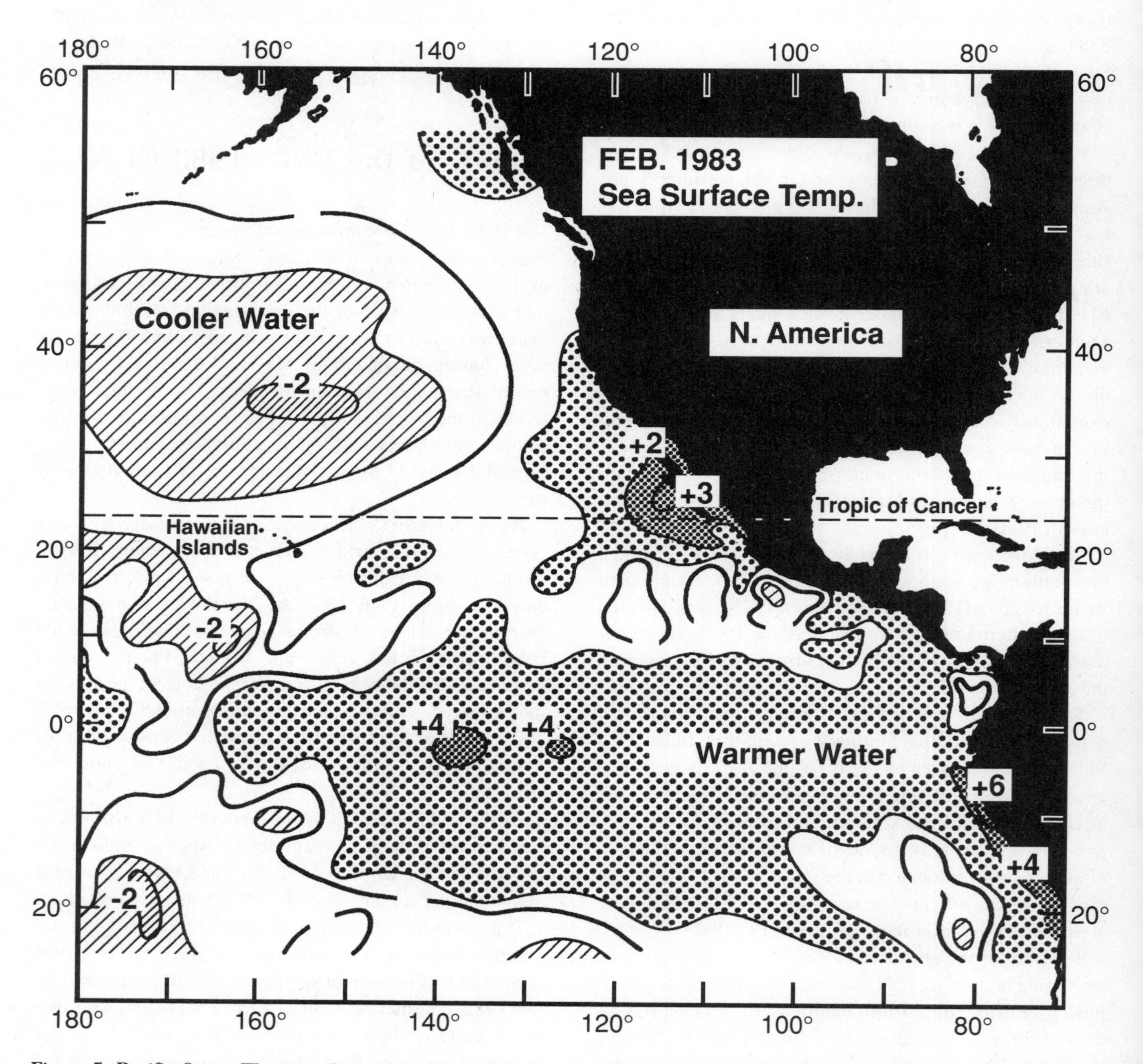

Figure 5. Pacific Ocean Warming. During the peak of the 1982–1983 El Niño, which was probably the most intense El Niño of the 20th century, water temperatures off the western coast of South America reached over 14°F (about 8°C) above normal. The map shows the departure of water temperatures from normal in degrees Celsius (1°C = 1.8°F) during February 1983 as the warming in the eastern Pacific approached its maximum. *Courtesy: Scripps Institution of Oceanography and NOAA.*

According to Eugene Rasmusson, formerly of the National Weather Service (NWS) Climate Analysis Center in Washington, D.C., ocean temperatures in the eastern Pacific declined rapidly in July 1983 and more slowly through September. By the end of 1983, the SO Index had returned to normal, and the precipitation patterns that had been shifted eastward by El Niño had returned to the west again; in some instances, they had moved even farther west than normal. In early 1984, the ocean temperature and precipitation patterns were nearly the reverse of those a year earlier: The trade winds had returned to easterlies and were even stronger than usual, and the water near the South American coast was cool once again.

El Niño Models

Volcano-El Niño Link? In the wake of the 1982–1983 El Niño, researchers developed a variety of models to explain the origin and evolution of El Niño events, as well as to predict their occurrence. Physicist Paul Handler of the University of Illinois at Urbana-Champaign has proposed a controversial hypothesis that volcanic eruptions may trigger El Niño events. After studying about 120 years of climate and volcanic eruption data, Handler noted an apparent correlation between major volcanic eruptions in the low latitudes and the warming of tropical Pacific waters characteristic of El Niño a few months later.

Handler has postulated that clouds of sulfurous gases ejected into the atmosphere by low-latitude volcanic eruptions are ultimately responsible for El Niño events. These gases react with water vapor in the stratosphere to form a haze of minute droplets of sulfuric acid that can persist for many months. (The stratosphere is the atmospheric layer that extends from a height of about 5 miles at the poles and 10 miles at the equator up to about 30 miles.) This cloud of minute droplets, called aerosols, prevents some sunlight from reaching the Earth's surface, and as a result, the surface below cools.

Because land cools more rapidly than the ocean, the decline in surface temperatures in a given period of time would be greater over land than over adjacent areas of ocean. Since large land masses in the Tropics are generally warmer than nearby ocean waters, the presence of an aerosol cloud would cause land temperatures to fall closer to that of the ocean; that is, the temperature difference between the land and the nearby ocean would decrease.

The temperature differences between different parts of the world are, to a large extent, responsible for the normal pattern of winds at the Earth's surface. For example, as the land masses of Australia and Indonesia are heated, the air above them is also warmed. This heated air expands and rises, and more air is sucked in from the Pacific Ocean to replace the rising air. In this way, temperature differences between the Australia-Indonesia region and the adjacent ocean reinforce the normal pattern of easterly winds that blow from the tropical eastern Pacific toward Australia.

According to Handler, when the temperature difference between the Australia-Indonesia land masses and the adjacent ocean decreases due to the presence of a volcanic aerosol cloud, the normal easterlies in the equatorial Pacific weaken and may even reverse, triggering El Niño. Handler has postulated that only low-latitude eruptions can trigger El Niño events, since only these eruptions create the low-latitude aerosol clouds that can reduce heating over Australia and Indonesia and thereby diminish the Pacific easterly winds. Aerosols from high-latitude eruptions never reach the Tropics, because the prevailing winds in the stratosphere tend to carry any aerosol clouds that do form in the high-latitude stratosphere away from the Tropics.

Handler has argued that the 1982–1983 El Niño was triggered by the eruption in April 1982 of the Mexican volcano El Chichón, and that the eruption of the Colombian volcano Nevado del Ruiz in November 1985 triggered the 1986–1987 El Niño. In support of his hypothesis, Handler noted that 3 months after each eruption the waters off the Peruvian coast began warming.

Although Handler's theory has drawn considerable attention, it has not been accepted by most meteorologists and oceanographers who study El Niño. Most of these researchers believe that El Niño events evolve more gradually than the relatively quick response suggested by Handler; some contend that the first signs of the 1982–1983 El Niño were apparent before El Chichón's eruption in 1982 and so could not have been triggered by it. Vernon Kousky of the NWS Climate Analysis Center has also noted that low-latitude volcanic eruptions occur each year, and he has claimed that Handler's model does not adequately explain why El Niño occurs in some years with eruptions and not in others.

Recent studies by Neville Nicholls of Australia's Bureau of Meteorology do not support Handler's theory. Nicholls compared the timing of 10 low-latitude eruptions with variations in the average monthly atmospheric pressure data measured at Darwin, Australia. Relatively high monthly mean pressures at Darwin are widely considered to be indicators of El Niño events. Nicholls found that in years with both eruptions and El Niño events, the average pressure at Darwin sometimes rose well before these eruptions, suggesting that the rise in pressure could not have been caused by the eruptions. Most climatologists now believe that the existing evidence does not support a link between eruptions and El Niño.

Cane-Zebiak hypothesis. Rather than postulating that external events, such as volcanic eruptions, are the cause of El Niño, most researchers have tried to explain the phenomenon completely in terms of interactions between the ocean and the atmosphere. In an attempt to simulate the formation, evolution, and decline of El Niño events, some scientists have developed computer models based on current theories of how the atmosphere and ocean interact over time. With these models, researchers are able to specify a particular set of initial conditions, such as the sea-surface temperature, the wind speed and direction, and the solar radiation intensity in different parts of the world, and then see how changes in one condition, such as wind speed, might trigger changes in the others.

Scientists Mark Cane and Stephen Zebiak of Columbia University's Lamont-Doherty Geological Observatory developed one such model that may explain the evolution of El Niño events. According to the model, El Niño may be triggered by a variety of atmospheric events, some of the most common of which are occasional bursts of westerly winds that occur in the western equatorial Pacific.

Such westerly winds, for example, would push some of the warm surface water in the western Pacific Ocean eastward, resulting in abnormally warm water temperatures in the eastern Pacific. This warm water would, in turn, heat the atmosphere above. The heated air in the eastern Pacific would rise and be replaced by air from adjacent areas. Air flowing into the eastern Pacific from the west would cause westerly surface winds, which would push more water into the eastern Pacific and thereby reduce the upwelling of the relatively cool, deep water there. Once started, the flow of warm water into the eastern Pacific, the resultant atmospheric heating, and the occurrence of westerly winds would tend to be self-perpetuating.

Cane and Zebiak have postulated that, at the peak of El Niño, the accumulation of warm water in the eastern Pacific starts to spread out. Some of the water flows back toward the central and western Pacific, and some moves north and south, leaving the equatorial region entirely. The loss of warm water from the eastern Pacific allows the tropical atmosphere and ocean to return gradually to a normal state, with the easterly trade winds reinforcing the normal Pacific temperature pattern of relatively warm waters in the west and cool waters in the east.

In the meantime, the warm water that has flowed north or south out of the equatorial Pacific gradually circulates through the northern and southern Pacific. This warm water eventually returns toward the equator, making ocean conditions again favorable for El Niño. According to the Cane and Zebiak model, the next El Niño will not occur until the warm waters have returned to the Tropics, a process that usually takes a few years. Once these warm waters return, any sudden change in atmospheric conditions — another burst of westerly winds, for example — could again trigger El Niño.

When Cane and Zebiak ran their computer model to simulate interactions between the atmosphere and the ocean during a hypothetical 90-year period, the model generated El Niño events that occurred at irregular intervals on an average of every 3 to 4 years and that generally peaked around the end of the calendar year. Both of these characteristics are consistent with El Niño events that have actually occurred.

Cane and Zebiak have acknowledged, however, that their model provides only a rough approximation of actual atmospheric and oceanic conditions. For example, in the model's events, the abnormally warm sea-surface temperatures in the eastern Pacific were not as high as some temperatures that have actually been observed, and the length of time during which the water temperatures remained unusually high was somewhat longer than in actual El Niño events.

Nevertheless, Cane and Zebiak believe that the model simulates actual conditions closely enough to provide insights into the nature of El Niño. On the basis of their model, the researchers have hypothesized that El Niño is the result of normal interactions between the tropical ocean and atmosphere, rather than extraordinary events, such as volcanic eruptions. They believe that all of the conditions necessary for triggering and sustaining El Niño exist in the tropical Pacific, and that influences outside the Tropics are not necessary to begin or end an event.

Forecasting El Niño

When Cane and Zebiak tested their model during 1985, using information on actual oceanic and atmospheric conditions for each year since 1970, they found that it often successfully predicted whether El Niño would occur in the following year. In addition, they noted that, based on sea-surface temperature and wind data for 1985, their model predicted that another El Niño might begin in 1986.

By early 1986, other researchers were also predicting that El Niño might occur later in the year. Surface waters off Peru warmed at an unusually rapid rate during December 1985 and January 1986, and in addition, a substantial amount of warm water accumulated in the Pacific west of the International Date Line. In February 1986, the NWS Climate Analysis Center issued an "El Niño Watch." The statement cautioned, however, that all of the conditions usually found at the beginning of El Niño had not yet been observed, and emphasized that the watch did not mean that El Niño was actually forecast to occur.

When the warming off the Peruvian coast continued through February, the center issued a similar statement in March. The NWS meteorologists again noted that some of the signs of El Niño were missing. Compared to the usual conditions at the start of El Niño, the warming of the eastern Pacific waters did not extend over as broad an area along the South American coast, and even in areas where the water temperature was unusually high, the warm water was confined to a relatively shallow depth. The meteorologists also noted that, before El Niño occurs, the sea level is generally higher than normal in the western Pacific and is lower than normal in the eastern Pacific; these conditions had not developed.

However, the prediction of El Niño eventually proved correct. In the late spring of 1986, the easterlies began weakening and continued declining through the summer and fall. By November, all the major features of El Niño were present, including warming of eastern and central Pacific equatorial waters. In January 1987, the sea along the west coast of South America began to warm, leading to above-normal rainfall in Ecuador and Peru, a classic sign of El Niño. The event reached its "mature phase" between December 1986 and February 1987, with the characteristic pattern of above-normal rainfall in the central equatorial Pacific, southern Brazil, and eastern Africa, and below-normal rainfall in southern Africa and northeastern South America. Stronger-than-normal westerlies in the Northern Hemisphere brought an unusually mild winter to Alaska, western Canada, and the Pacific Northwest.

The warm episode peaked over South America during the spring, with continued high rainfall. The Southern Oscillation indicators weakened throughout the summer and fall of 1987 and the event ended by early 1988. Meteorologists now generally agree that there is no "standard" El Niño — every event is somewhat different. The 1982–1983 El Niño produced particularly severe changes in precipitation and weather patterns, but the 1986–1987 event may have been at least partially responsible for record global warmth during 1987 — the warmest year of the approximately 100 years for which global temperature records had been kept. (This record was later broken in 1988.) The winter of 1986–1987 was especially warm over North America, producing temperatures averaging 9°F above normal in western Canada. El Niño caused above-normal rainfall along the U.S. Gulf Coast during the 1986–1987 winter season, and it disrupted the 1987 spring and summer monsoon season in Asia, causing drought in India and Southeast Asia.

Future Ocean and Climate Studies

The 1982–1983 El Niño inspired unprecedented international cooperation in studying its development and far-reaching effects. During the event, scientists from many nations quickly gathered and exchanged biological, meteorological, and oceanographic data, which normally would have taken months or years to compile and distribute. Although information about the 1982–1983 and 1986–1987 El Niño events was still limited in scope, it was far more detailed than that gathered for any previous El Niño. When the 1986–1987 event struck, meteorologists were able to predict its onset.

In the 8 July 1988 *Science*, three research groups assessed the ability of their computer simulations to predict the onset and course of the 1986–1987 El Niño. The simulations evaluated were the Cane-Zebiak model described above and two other models developed by researchers at the Scripps Institution of Oceanography in La Jolla, California, and Florida State University in Tallahassee.

Each of the models used different assumptions about interactions between the atmosphere and ocean that could lead to El Niño. Despite the wide differences between the models, each was able to successfully predict the onset of the 1986–1987 El Niño to within a few months. The models were also able to predict the general course of the El Niño, except that they all forecast that the event would end sooner than it actually did. The researchers concluded that the tropical atmosphere and ocean act as a "coupled" system, and that El Niño is the result of changes in the strength and direction of interacting ocean currents and winds. They noted that while relatively short-lived weather events may slightly alter El Niño, long-term ocean and wind patterns ultimately determine the course of a given event.

Although the partial success of these simulations in modeling the 1986–1987 El Niño is heartening, more detailed information will be needed to make more accurate forecasts of these events. Scientists hope that two programs, one now under way and one scheduled to begin in the 1990s, will provide the data they need to predict the course of El Niño events with greater accuracy in the future.

A 10-year international project called the Tropical Ocean and Global Atmosphere Programme (TOGA), which began in 1985 as part of the World Climate Research Programme, has the objective of monitoring the tropical ocean and atmosphere and studying the ways in which they interact. On the basis of these observations, researchers will be able to develop and refine models of oceanic and atmospheric processes, including El Niño, with the ultimate goal of predicting variations in the tropical ocean and atmosphere.

As part of TOGA, scientists are gathering information on the wind, temperature, and humidity of the atmosphere over the tropical oceans, as well as the temperatures, currents, biological activity, and sea level heights of the oceans themselves. By late 1988, nine countries — Australia, Chile, Colombia, Ecuador, France, New Zealand, the People's Republic of China, Peru, and the U.S. — were actively participating in the research, and several other nations, including Japan, were planning to become involved.

Another major program that will enable scientists to study El Niño in detail — the World Ocean Circulation Experiment (WOCE) — is scheduled to begin in 1990 and to continue for at least 5 years. WOCE will undertake the first comprehensive survey of the physical properties of oceans on a global scale. In addition, it is designed to determine how the oceans influence climate over a period of a decade or more and to develop accurate models of the interaction between the atmosphere and ocean. According to current plans, much of the data needed for the WOCE studies will be collected by instruments carried on advanced oceanographic satellites that the European Space Agency, France, and the U.S. expect to launch beginning in 1990.

The first of these satellites, the European Space Agency's ERS-1, will carry several instruments useful in studying El Niño, including a scatterometer. The scatterometer will beam microwaves to the sea surface and measure the strength of the returning echoes. ERS-1's scatterometer data will enable scientists to measure the roughness of the sea surface, determine the characteristics of small surface waves, and calculate the direction and strength of the wind that generated these waves.

In addition, by taking into account the satellite's speed and movement above the Earth and combining data taken at several points along its flight path, scientists will be able to use ERS-1's scatterometer to produce images of details on the sea surface as small as 100 feet across. These data could be used to study the wavelength and movement of ocean waves. ERS-1 will also carry a radar altimeter, an instrument that emits radar signals and then measures the amount of time each signal takes to reach the sea surface and return to the satellite. The altimeter will be able to measure wave heights to within about 1.5 feet.

The Ocean Topography Experiment satellite (TOPEX/Poseidon) is expected to be launched in the early 1990s as a joint French-U.S. mission. TOPEX/Poseidon will carry a high-precision radar altimeter capable of measuring the elevation of the ocean surface to within about 1 inch. It will provide nearly global coverage of the oceans, with the exception of polar regions north of 63°N or south of 63°S. Using the data collected, TOPEX/Poseidon will generate more than 100 global maps of ocean surface topography over a period of 3 to 5 years. With these detailed maps, oceanographers will be able to follow weekly, seasonal, and annual changes in ocean surface elevation and make deductions about the currents that result from these elevation differences.

Another instrument that may be useful in studying El Niño is the U.S. National Aeronautics and Space Administration's Sea-Viewing Wide-Field Sensor (Sea WiFS), which will map variations in ocean color caused by marine life, sediments, and pollution. The colored pigments in microscopic marine plants called phytoplankton, for example, affect the color of the ocean, generally making it greener. Data on ocean color would enable scientists to study the concentration of phytoplankton, the major food source for fish and other aquatic organisms, and estimate the biological productivity of the ocean. The Sea WiFS's high-resolution coverage of the oceans will enable researchers to observe changes in important fisheries, track marine pollution, and study the movement of sediments from rivers into the oceans.

Altogether, this network of satellites will give scientists an unprecedented global view of the ocean environment and the processes that influence it. Many researchers are confident that as their understanding of the oceans and the atmosphere increases, they will be able to predict the timing and intensity of disruptive events such as El Niño months to a year or more before they occur.

Weather Forecasting

Many of the most damaging and life-threatening types of weather — torrential rains, severe thunderstorms, and tornadoes — begin quickly, strike suddenly, and dissipate rapidly, devastating small regions while leaving neighboring areas untouched. One such event, a tornado, struck the northeastern section of Edmonton, Alberta, in July 1987, killing 27 people and injuring more than 250. Total damages from the tornado exceeded $250 million, the highest ever for any Canadian storm.

Conventional computer models of the atmosphere have limited value in predicting short-lived local storms like the Edmonton tornado, because the available weather data are generally not detailed enough to allow computers to discern the subtle atmospheric changes that precede these storms. In most nations, for example, weather balloon observations are taken just once every 12 hours at locations typically separated by hundreds of miles. With such limited data, conventional forecasting models do a much better job predicting general weather conditions over large regions than they do forecasting specific local events.

Until recently, the observation-intensive approach needed for accurate very short-range forecasts, or "Nowcasts," was not feasible. The cost of equipping and manning many thousands of conventional weather stations was prohibitively high, and the difficulties involved in rapidly collecting and processing the raw weather data from such a network were insurmountable.

Fortunately, scientific and technological advances have overcome most of these problems. Radar systems, automated weather instruments, and satellites are all capable of making detailed, nearly continuous observations over large regions at a relatively low cost. Communications satellites can transmit data around the world cheaply and instantaneously, and modern computers can quickly compile and analyze this large volume of weather information.

In the 1980s, meteorologists and computer scientists have worked together to design computer programs and video equipment capable of transforming raw weather data into words, symbols, and vivid graphic displays that forecasters can interpret easily and quickly. As meteorologists have begun using these new technologies in weather forecasting offices, Nowcasting is becoming a reality.

Advanced Forecasting Equipment

Doppler radar. Doppler radar, which provides more detailed information than conventional weather radar systems, has great potential as a Nowcasting tool. Conventional radars collect precipitation data by transmitting radar pulses that bounce off targets, such as raindrops, snowflakes, and hail, and are then reflected back to the radar receiver. By timing the interval between the pulse transmissions and returning echoes and then measuring the intensity of the echoes, weather radar can show both the location and approximate intensity of precipitation.

Doppler instruments measure not only precipitation location and intensity, but also targets' radial speeds — movements toward or away from the radar. In this way, Doppler radar allows meteorologists to measure the speed of particles carried in a storm's winds. Just as the pitch of a train whistle appears to change slightly from a higher pitch as the train approaches to a lower pitch as it recedes, the motion of precipitation toward or away from the radar antenna causes minute shifts, about 10 parts per billion, in the frequency of the radar echo.

Since the shifts in radar frequency are proportional to the speed of the targets, computers analyzing Doppler radar data can calculate the speed of particles carried in the wind. In addition, by combining data from two Doppler radars, computers can construct images of storm circulation in three dimensions. These images allow meteorologists to pinpoint high winds that threaten life and property on the ground, as well as updrafts, downdrafts, and areas of turbulence that are hazardous to air traffic.

Tracking tornadoes with Doppler radar. Conventional radar is generally ineffective for identifying tornadoes and other dangerous local storms, because the radar images of these storms are usually indistinguishable from harmless weather phenomena. At best, a tornado funnel may appear as a small hook-shaped image on the radar screen, but by the time the hook appears, the tornado may have already touched ground and begun causing damage. Doppler radar, with its ability to measure wind speed and direction, makes it easier to distinguish a tornado from a group of harmless thunderstorms and can even help identify tornado-breeding clouds before the twisters form.

Doppler radar graphically displays the characteristic rotating winds of storm systems called mesocyclones that sometimes produce tornadoes. The tornado itself has an easily recognized signature on the radar screen. Intense winds flow in a tight circle around the center of the tornado funnel, which can vary from tens of feet to a mile or more across. Consequently, the winds on one side of the funnel move in the opposite direction of the winds on the other side.

Different wind speeds and directions appear as different colors on the Doppler radar screen. The color choice is arbitrary, but as a general rule, colors at the red end of the spectrum correspond to wind moving away from the radar; colors such as green and blue on the opposite end of the spectrum correspond to wind blowing toward the radar. Colors close to either end of the spectrum indicate high wind speeds. On the Doppler screen, the pattern of wind motion in a tornado funnel shows up as contrasting colors, such as bright red and deep green, over unusually short distances.

Scientists at the U.S. National Severe Storm Laboratory (NSSL) in Norman, Oklahoma, have been using Doppler radar since the 1970s to measure wind speeds in tornadoes. As of early 1989, the strongest tornado winds ever measured with Doppler radar occurred during a twister in Binger, Oklahoma, on 22 May 1981. Researchers from the NSSL clocked the tornado winds at a peak of 196 miles per hour. Along its 14-mile path, the tornado picked up cattle, oil storage tanks, trucks, and farm machinery, and tossed them aside at speeds of over 40 miles per hour, according to the radar measurements. The Doppler images showed that the tornado was about 0.5 miles wide at its base, with a maximum width of more than 1 mile at an altitude of about 2.5 miles above ground level.

The NSSL team was able to trace the Binger tornado's outline from the ground to a height of 7.5 miles above. They also detected a radar "hole" — an absence of echoes — surrounded by a ring of powerful echoes from precipitation and debris carried by the high winds. This hole, which was nearly free of the precipitation, debris, and dust that reflect radar signals, corresponded to the inside of the tornado funnel.

The discovery of the radar hole was consistent with theories about tornado dynamics. According to these theories, strong updrafts around the periphery of the funnel carry moist surface air high into the atmosphere where the vapor cools and condenses to form precipitation. Within the funnel itself, however, the air is sinking and condensation does not occur. As a result, air in the funnel can be clear and dry.

In tornado-prone Oklahoma, meteorologists have compared the performance of the Doppler radar in identifying tornadoes with that of typical severe weather warning systems, which depend heavily on visual sightings and conventional radar. Doppler radar detected 69 percent of the tornadoes that occurred, compared with 64 percent by the conventional systems. The false alarm rate — the proportion of tornado warnings not followed by storms — was 63 percent for conventional methods, but only 25 percent for Doppler radar.

The most important advantage of Doppler radar, however, is that it detects tornadoes sooner and allows more time to evacuate threatened areas than any other system. Since conventional warning systems rely heavily on visual sightings, which are usually made shortly before the funnel touches down, the warning time averages just 2 minutes. In contrast, with its capacity to identify mesocyclones about 125 miles away, Doppler radar can spot dangerous storm-breeding weather systems over a wide area. This gives meteorologists ample opportunity to issue warnings for specific locations. In the Oklahoma tests, Doppler radar observations allowed an average warning time of more than 20 minutes, long enough for residents to seek safe refuge and even secure some of their belongings.

Wind shear and microbursts. The U.S. Federal Aviation Administration (FAA) has estimated that about 40 percent of aircraft accidents in the U.S. are caused by poor weather conditions during flights. One such hazard known as wind shear — a sudden change in wind speed, wind direction, or both in a short distance — is particularly dangerous during an aircraft takeoff or landing. The most hazardous type of wind shear is caused by a phenomenon known as a microburst, a strong downdraft that begins thousands of feet above ground, descends, and then spreads out horizontally in all directions in a starburst pattern when it reaches the ground. Microbursts usually occur in conjunction with thunderstorms, typically last for less than 20 minutes, and create very strong, damaging winds near the ground. To be classified as a microburst, the downdraft and starburst wind pattern must affect an area on the ground no more than 2.5 miles in diameter; downdrafts and starburst wind patterns that affect larger areas are called downbursts.

Microbursts are dangerous because of the way they

affect an aircraft's flight. Since air flows out in all directions from the center of the microburst, a low-altitude airplane flying into a microburst first encounters a strong head wind. This head wind increases the plane's lift, and the pilot must often reduce engine power to keep the plane from rising. When the plane passes by or through the center of the microburst, the head wind is quickly replaced by a tail wind, which reduces the plane's lift. With reduced lift and engine power, the plane may stall and drop rapidly toward the ground. If the plane is already at a very low altitude, as during a takeoff or landing, the pilot may not be able to increase engine power fast enough to prevent a crash.

The FAA has implicated the abrupt wind shifts and downdrafts associated with microbursts in more than 30 aircraft accidents since 1964, including the Delta Airlines L-1011 crash at the Dallas-Fort Worth Airport on 2 August 1985, which killed 137 persons, and the Pan American World Airways 727 crash in Kenner, Louisiana, on 9 July 1982, which killed 154 persons. In another incident on 1 August 1983, Air Force One, with President Ronald Reagan on board, landed at Andrews Air Force Base near Washington, D.C., just 6 minutes before a very strong microburst struck the base.

Doppler radar and aviation. Scientists involved in the Joint Airport Weather Studies (JAWS) project, which was sponsored by the U.S. departments of commerce, defense, and transportation, studied about 100 microbursts in the area around Denver and Boulder, Colorado, during mid-1982. Using Doppler radar displays, the researchers detected six microbursts near Denver's Stapleton International Airport on 25 May, the eighth day of the study. Research aircraft entered three of the microbursts and measured strong, rapidly shifting winds. Upon detecting one microburst about 6 miles south of a major runway at Stapleton, JAWS meteorologists issued a Nowcast advisory, which warned the NWS and the FAA of the microburst. As a precautionary measure, Stapleton's control tower closed the runway until the danger had passed.

The JAWS data showed that microbursts are a fairly common phenomenon around Denver and Boulder in the late spring and summer, with an average of about one per day within a 20-mile radius of Stapleton Airport. According to John McCarthy, who directed the JAWS project, about 15 flights during a typical summer will have a close encounter with a microburst while taking off or landing at Stapleton.

McCarthy and researcher Robert Serafin of the National Center for Atmospheric Research (NCAR) analyzed the characteristics of 75 microbursts observed in detail during the JAWS project and found that the microbursts were typically about 0.6 to 1.9 miles across. The events lasted between 5 and 15 minutes, with severe wind shear occurring for about 2 to 4 minutes. The average change in wind velocity between the head winds and the tail winds that would have been encountered by an aircraft passing through a microburst was 56 miles per hour, although the maximum change measured was 107 miles per hour.

At the request of the FAA, NCAR meteorologists conducted another wind shear study called CLAWS — for Classify, Locate, and Avoid Wind Shear — at Stapleton during the summer of 1984, after a microburst nearly caused the crash of a United Airlines 727 jet at the airport in May of that year. CLAWS focused on verifying techniques for forecasting microbursts with Doppler radar and on developing systems to give pilots timely advisories about these events.

During the summer of 1987, the FAA used a network of three Doppler radars and a dense array of automated weather instruments at Stapleton Airport to determine whether these instruments could provide airport meteorologists with immediate information on wind shear. At the same time, NCAR meteorologists conducted a project called the Convection Initiation and Downburst Experiment (CINDE) in the Denver area. CINDE used data from the FAA airport experiment, as well as detailed information gathered by a variety of other advanced weather instruments, satellites, and reconnaissance aircraft, to study thunderstorms and microbursts.

The goal of CINDE was to learn why thunderstorms form where they do, how they evolve, and how dangerous storms — including those causing microbursts — can be distinguished from harmless ones. CINDE experimenters focused on the short-term forecasting of storms that produced microbursts, and the conditions in which these storms formed.

Although NCAR scientists are still analyzing data from the CINDE program and earlier tests, some preliminary results have been reported. Meteorologists have found, for example, that some tornadoes develop in a different way than was previously thought. Before these experiments, scientists generally believed that the air circulation responsible for all tornado formation begins in clouds high above the ground. Doppler radar measurements have shown, however, that in some cases the circulation originates relatively near the ground, well before the tornado funnel forms.

CINDE scientists also used Doppler radar to observe the formation of thunderstorms along lines where winds converge from different directions. When these winds meet, air is forced upward; the rising air expands and cools, and water vapor begins to condense, conditions that can promote thunderstorm development. By observing converging winds with Doppler radar, NCAR scientists were able to make accurate short-term predictions of thunderstorms at Stapleton Airport.

Doppler radar installations in the U.S. Beginning in the early 1990s, the NWS plans to replace the approximately 300 conventional weather radars owned and operated worldwide by the U.S. with a new network called NEXRAD, short for Next Generation Weather Radar, consisting of at least 110 Doppler radars. Many of the conventional radars were built in the late 1950s and have been gradually wearing out.

In 1983, the U.S. government selected two contractors to develop NEXRAD prototypes for field testing. These prototypes were later tested, and in December 1987, the government chose a design developed by UNISYS for production. A prototype NEXRAD facility was installed in Norman, Oklahoma, during late 1988 and was expected to undergo tests through the first half of 1989. The first operational radars in the NEXRAD system are expected to be installed in 1990. The total cost of the NEXRAD program, including system design, production, site selection, and program management, is projected to be about $1 billion.

The NEXRAD network will consist of three major components — the Doppler radars themselves, computers to convert raw radar data into a usable form, and a series of computer terminals that will allow meteorologists to gain easy access to the NEXRAD information. The Doppler radars will be designed to operate unmanned, transmitting and receiving radar pulses and then passing the data to computers for processing.

The FAA also plans to install a nationwide radar network called Terminal Doppler Weather Radar (TDWR), which will be specifically designed for airport use. The TDWR network will provide air-traffic controllers and pilots with Nowcasts of severe weather — dangerous wind shear conditions, in particular — thereby reducing the risk of aviation accidents.

The first tests of a prototype TDWR system were conducted in Denver during the summer of 1987, and an operational test took place the following summer. Both tests demonstrated that TDWR can successfully detect and provide timely warnings of wind shear and other weather hazards in the region around Stapleton Airport. When the radar detected abrupt changes in wind speed or direction, air-traffic controllers were alerted, and the controllers radioed the information to pilots of approaching aircraft, who changed their flight patterns to avoid the dangerous winds. The TDWR's wind shear warnings may have helped avert accidents on at least two occasions during July 1988.

Since the first TDWR radars will not be installed until the early 1990s, the FAA has made arrangements with the NWS to temporarily use NEXRAD radars at more than a dozen major airports until the TDWR facilities are available. These NEXRAD radars will be installed at high-traffic airports serving cities such as Atlanta, Chicago, Dallas-Fort Worth, Denver, Miami, New York City, and Washington. In November 1988, the FAA awarded a contract to Raytheon Corp. for the construction of 47 TDWR systems at major airports, and the network may eventually include about 100 radars.

Profilers. Until the early 1980s, the only way that meteorologists could obtain a profile of temperature, wind, humidity, and pressure from the ground to the upper atmosphere was by launching radiosondes — weather balloons equipped with instruments and radios to gather and transmit data back to the Earth. Launching, tracking, and compiling data from radiosondes is time-consuming and costly. Consequently, the balloons are sent up from weather stations typically once every 12 hours. Even for research, radiosondes are rarely launched more often than once every 3 hours. This situation will begin to change during the 1990s, however, when a new system called the profiler becomes more widely available.

The profiler was originally developed by the NOAA Wave Propagation Laboratory in Boulder. It generates continuous profiles of conditions from the ground to the upper boundary of the troposphere without using weather balloons. (The troposphere is the atmosphere's lowest layer, extending from the Earth's surface to a height of about 5 miles at the poles and about 10 miles at the equator.) The troposphere is also known as the "weather layer," because it is the layer in which clouds, rain, and storm systems occur as air mixes both horizontally and vertically.

The first profilers tested by NOAA were Doppler systems that emit radar beams of relatively long wavelengths — about 1 to 10 meters (3.3 to 33 feet) — compared to ordinary radars, which emit pulses with wavelengths of about 10 centimeters (4 inches). The profilers' long-wavelength radars can detect echoes from air alone, unlike ordinary weather radars, which follow winds by tracking the movements of raindrops, snowflakes, dust, or other objects carried by the winds.

With their long-wavelength Doppler radars, the NOAA profilers can provide information on wind speed and direction at altitudes up to 10 miles. This data can be processed rapidly by computers and may be updated as frequently as every 30 minutes — a vast improvement over the twice-daily observations provided by radiosondes. NOAA has selected sites for a demonstration network of 30 profilers that will cover the central third of the U.S. The first profiler in the network was installed in Platteville, Colorado, just north of Denver, during 1988 and was scheduled to be tested through early 1989. Project meteorologists hope the tests will show how profilers can best be used to monitor and predict the weather.

Scientists at NOAA's Wave Propagation Laboratory are also conducting research to improve the profiler's capabilities by adding an instrument called a radiometer, which will measure microwave emissions from the atmosphere at six different wavelengths, or channels. The atmosphere's pattern of microwave emissions depends upon the air's temperature and its water vapor and liquid water content. Data taken at a single wavelength are not sufficient to provide a complete picture of atmospheric conditions; however, it is possible to generate temperature and water-vapor profiles by comparing the emission patterns at different wavelengths. The researchers are trying to improve their radiometer's accuracy, which is still below that of radiosondes.

If profilers prove to be a cost-effective alternative to radiosondes, they could become an important instrument for weather research and day-to-day forecasting. Since profilers could provide frequently updated information on important weather conditions from the Earth's surface to the upper troposphere, researchers could track both the high-altitude jet stream winds, which govern the movement of major weather systems, and observe how local weather phenomena, such as thunderstorms and sea breezes, evolve over very short periods of time. The system would also provide the detailed data needed to refine computer forecasting models, which currently depend heavily on the twice-daily radiosonde observations. In the U.S., for example, the profiler's continuously updated atmospheric profiles would enable the NWS to revise its computer weather predictions much more frequently.

Weather satellites. With the launch of the Tiros 1 weather satellite in April 1960, meteorologists were able for the first time to observe the evolution of weather systems directly and track their movements from birth to death. From its 430-mile-high orbit, Tiros 1 photographed one-quarter of the Earth's surface each day and provided the first sequential photographs of large-scale weather phenomena.

Although Tiros was an important first step, its achievements were small compared to the routine operation of today's weather satellites. Networks of modern weather satellites, some operating in geosynchronous orbits — orbits in which the satellite remains above a fixed point on the Earth — transmit high-resolution photographs of the Earth's clouds and land masses many times each day. These satellites also take measurements of ocean and atmospheric temperatures and the Earth's magnetic field.

The first operational geosynchronous weather satellite, SMS 1, was launched by the U.S. in May 1974. NOAA currently uses its Geostationary Observational Environmental Satellite (GOES) series to collect data over much of North, Central, and South America as well as the western Atlantic and eastern Pacific oceans. In addition, GOES satellites gather environmental data transmitted from up to 10,000 remote instrument stations on land, at sea, and in the air, and then broadcast this information, together with their own weather and magnetic-field observations, to receiving stations on the Earth.

Hurricanes are now routinely tracked by the GOES series and other satellites. When the potential for other types of hazardous weather exists, meteorologists have used GOES satellites to take observations over high-risk areas as often as once every 15 minutes in order to identify severe thunderstorms that might bring high winds and heavy rains or spawn tornadoes.

The most recent satellites in the GOES series contain an instrument called VAS, for visible infrared spin-scan radiometric atmospheric sounder. The VAS has 12 channels capable of mapping water-vapor concentrations and temperature variations both horizontally and vertically through the atmosphere. The instrument's visible light detectors are able to resolve objects, such as clouds, as small as about 0.6 mile across, while its infrared sensors can discern objects as small as about 5 miles across.

High-resolution satellite images have made it possible to study large-scale weather features and to obtain precise measurements of smaller atmospheric processes. For example, by combining data from two geosynchronous satellites viewing the same clouds from different locations, researchers are able to calculate the height of the clouds to an accuracy of about 0.3 mile.

With the launch of GOES-7 on 26 February 1987, NOAA had a full complement of two GOES satellites in orbit, the minimum number needed for complete weather coverage of the entire U.S. The satellites collect weather

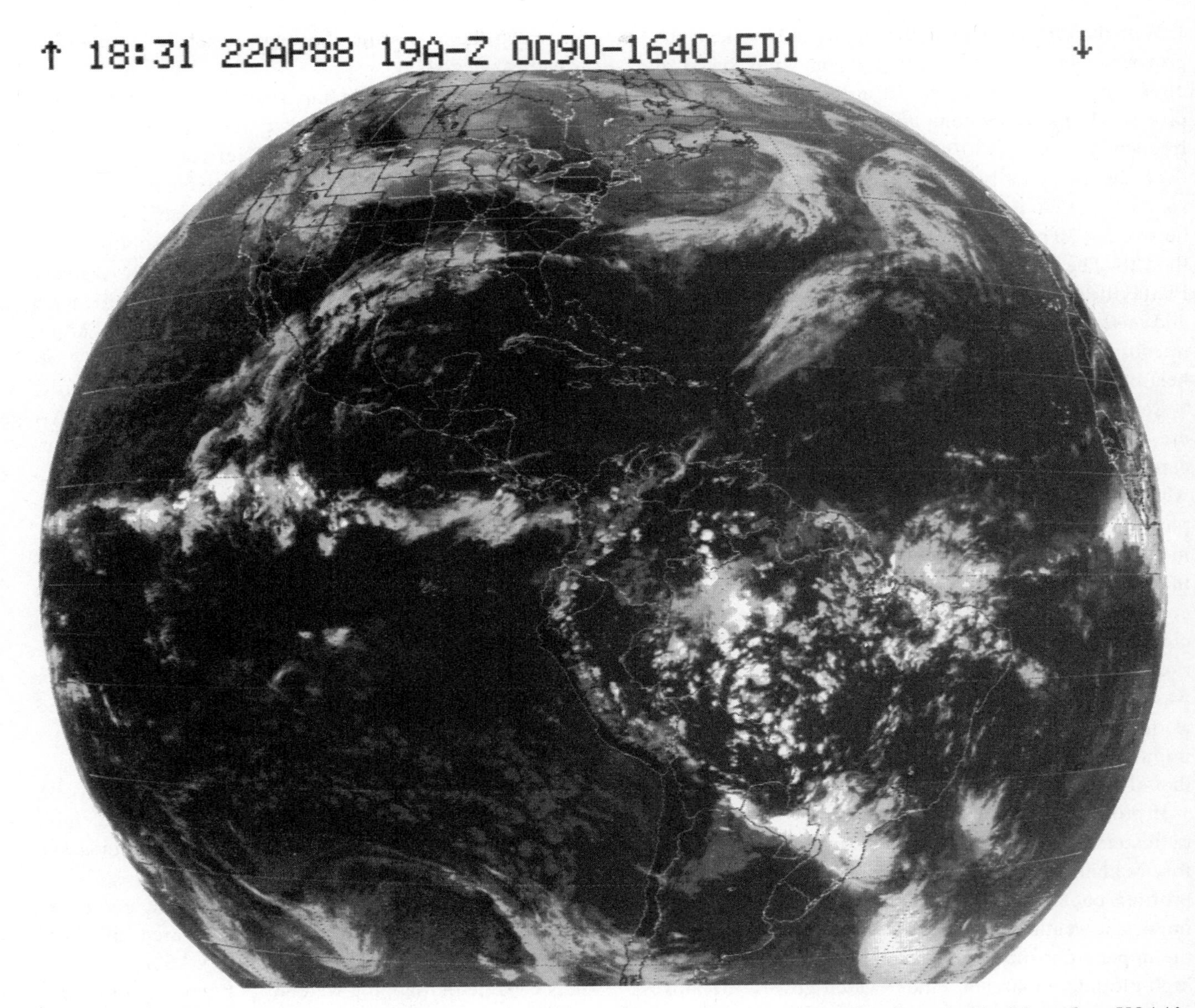

Figure 6. GOES Image of the Western Hemisphere. From their vantage points over 22,000 miles above the Earth's surface, NOAA's GOES weather satellites collect weather data over much of the Western Hemisphere. GOES instruments provide detailed measurements of cloud cover and map variations in the atmosphere's temperature and water-vapor content. *Courtesy: NOAA.*

data from regions as far north as Alaska and northern Canada and as far south as northern South America. GOES-7, which is also called GOES East, provides coverage from western Africa westward to the central U.S. GOES-6 (GOES West) gathered data from the central U.S. westward to the International Date Line in the central Pacific Ocean, but in January 1989 the satellite's navigational instruments failed, rendering it useless. To regain full weather coverage of North America, NOAA has moved GOES-7 westward. A replacement for GOES-6 will not be launched until at least 1990 because of delays in the U.S. space shuttle schedule.

A new series of U.S. weather satellites called GOES NEXT will provide even more detailed weather information when they are launched beginning in the early 1990s. Like the current GOES satellites, the GOES NEXT series will have visible light sensors to obtain cloud images with a resolution of 0.6 mile. GOES NEXT infrared instruments, however, will have a resolution of about 2.5 miles instead of the current 5 miles. These more detailed measurements will enable meteorologists to make better assessments of cloud temperature and water vapor content. In addition, the GOES NEXT satellites will be able to scan a small area repeatedly — perhaps as often as once a minute — allowing scientists to closely monitor the development of severe thunderstorms and tornadoes, which form and evolve rapidly.

Improving Regional Forecasts

In 1980, NOAA embarked on an ongoing program to develop advanced computer and communications systems, as well as software to analyze a variety of regional weather data and to help meteorologists quickly transform the data into accurate local weather forecasts. The planners of the Program for Regional Observing and Forecasting Services (PROFS) in Boulder, Colorado, believe that much of the technology needed to improve local forecasts already exists but has not yet been fully exploited. According to former PROFS scientist Donald Beran, one of the major goals of the program is to provide "a systematic way of transferring this backlog of technology to operational forecasting."

In the program's first major test in the early 1980s, PROFS's computers collected data on spring and summer storms in the Great Plains, including radar information, surface observations, and satellite imagery — the type of information that most forecast offices already receive. What made the test different, however, was the way the computers displayed the data.

The computers combined surface wind data taken from many stations to show regional "streamlines" or wind-flow patterns. By using color in video displays to highlight areas of precipitation tracked on radar and then superimposing this imagery on the wind-flow patterns and on satellite cloud photographs, forecasters were better able to recognize signs of strengthening storms. This computerized system is particularly effective, according to former PROFS scientist Ron Alberty, because "computers are very good at manipulating large amounts of data but not very good at pattern recognition, while humans are the opposite." Using PROFS's computers, forecasters are able to do what they do best — see patterns in the atmosphere and use their forecasting experience to predict what will happen next.

Development of PROFS. Since PROFS is an ongoing program, its weather instruments and the computer hardware and software are being continually modified to improve forecasting accuracy. For example, as computers have evolved rapidly during the 1980s, PROFS programmers have improved software to take advantage of the machines' higher computing speeds and to increase their capacity to process very large amounts of data.

In an early PROFS system developed in 1981, forecasters had to type commands on a keyboard to request a particular type of weather information, such as a satellite photograph. Depending on the complexity of the command and the typing skill of the forecaster, entering a command could take anywhere from around 5 seconds to a minute or more. Once the command was typed, the forecaster typically had to wait about 40 seconds or more for the desired image to appear on the monitor. Since forecasters must analyze many different types of weather data to make predictions, the time meteorologists spent entering data requests and waiting for the information to appear was a serious impediment, particularly when weather conditions were changing rapidly and forecasts were needed quickly.

These problems have now been largely eliminated. Forecasters can request information simply by hitting a single key or using a "mouse" to choose among different types of weather data listed on the computer screen like entrées on a restaurant menu. Once a request is made, forecasters generally wait less than 5 seconds for the information to appear on their screens.

Technological advances have also increased the types of weather information available at a forecaster's fingertips. For example, PROFS scientists have developed software that uses Doppler weather information to display wind speeds and precipitation intensities at up to eight different altitudes in a thunderstorm. This software allows forecasters to monitor thunderstorm development more closely and pinpoint the storms most likely to cause damaging winds, hail, or tornadoes. Other software has given meteorologists immediate access to detailed information on sunshine, precipitation, or barometric pressure at any local weather station in their area. In 1986, PROFS scientists developed software that projects on a computer monitor the future motion of a storm up to about an hour ahead. The program also automatically updates these projections as new data come in.

In addition, PROFS forecasters can zero in on a particular part of a satellite photograph or radar image to study small details in the structure of a storm, or view a time-lapse series of images to see how a storm has moved and evolved over time. Increasingly, new technology is allowing PROFS meteorologists to tailor weather information to particular types of predictions — from Nowcasts to long-range forecasts of the week ahead — and to their own forecasting style. In 1987, a PROFS workstation was placed in service at the Denver Weather Service Office, so that weather forecasters could begin using the new technology in their day-to-day predictions. The workstation, which is called the Denver AWIPS Risk Reduction Requirements Evaluation, consists of a console with several different screens that can display a variety of weather data. The station is a prototype for the AWIPS-90 (Advanced Weather Interactive Processing System for the 1990s), a high-technology system that will be used by forecasters in NWS offices during the years ahead.

New Look at the Earth's Interior

The distance from the Earth's surface to its center is about 4,000 miles, yet the deepest hole ever drilled, at about 8 miles, barely scratches the lithosphere, the Earth's outermost layer. Scientists have had to learn about the Earth's interior indirectly — through rocks brought from the mantle to the surface at mid-ocean ridges and volcanoes and from the study of seismic waves traveling through the interior. Research on the Earth's gravity and magnetic fields as well as laboratory measurements of minerals under high pressure have also provided clues to the conditions deep within our planet. This information has shown that the Earth may be divided into three main layers — crust, mantle, and core — each with a distinct composition and physical properties.

If more were known about the interior, earth scientists might be able to achieve one of the major goals of geology and geophysics — a detailed explanation of the forces responsible for plate tectonics. Most researchers agree that the heat-driven motion of hot rock — thermal convection — within the mantle is responsible for the movement of plates at the Earth's surface. (This motion is driven by the differences in density between the cooling and sinking lithosphere and the hot mantle.) Yet many questions still remain about the mantle's internal structure, patterns of rock flow, and chemical make-up, as well as the effect events in the mantle have on plate motions.

Seismic tomography. A new technique called seismic tomography is beginning to provide fascinating insights into some of these questions. The principle behind seismic tomography is similar to that of the medical technique called computed tomography, which is used in CT scans. In computed tomography, an x-ray generator rotates around a patient's body, exposing it to x-rays from many different angles. X-ray detectors measure the amount of radiation that passes through the body, and a computer combines these data to form a cross-sectional image. In seismic tomography, earthquake-generated seismic waves (analogous to the x-rays) traveling through the Earth (analogous to the patient) are measured by seismometers (analogous to x-ray detectors) at locations around the world. As in computed tomography, a computer combines the data into an image — in this case, an image showing the inner structure of the Earth.

Seismic tomography is possible because the speed at which seismic waves travel is not constant, but depends on the rigidity and compressibility of the rock through which the waves pass, as well as the orientation of mineral crystals in the rock. Seismic waves travel at higher speeds through relatively rigid and incompressible rocks than through less rigid, more compressible rocks. Since hot rock is generally less rigid and more compressible than cooler rock, seismic waves travel more slowly through hot areas of the Earth's interior than through cooler regions. Hot rock is less dense than surrounding cooler rock of the same composition and, as such, will tend to rise; conversely, relatively cool rock is more dense than hot rock and will tend to sink.

The velocity of seismic waves depends not only on the temperature of the rock, but also on the orientation of the crystals in the rock. The crystals of minerals commonly found in the Earth's mantle have three axes, each of which has a characteristic amount of rigidity. Seismic waves will travel faster if their direction of propagation is parallel to the axis of greatest rigidity — the so-called fast axis — than if their direction is parallel to an axis with less rigidity.

Researchers have found that the crystals in rock within the Earth's interior tend to become aligned so that their fast axes are parallel to the direction in which the rock is flowing. Consequently, the velocity of the seismic waves through a particular region will depend on the direction in which the seismic waves are traveling relative to the orientation of the fast axes in the rock crystals. Taking all of these factors into consideration, geophysicists can use data on seismic velocity changes related to rock temperature and crystal orientation to make inferences about the characteristics and motion of materials in the Earth's interior.

Mapping the Earth in three dimensions. Geophysicist Robert Clayton and his associates at Caltech have used earthquake data from 200 Caltech and USGS seismometers

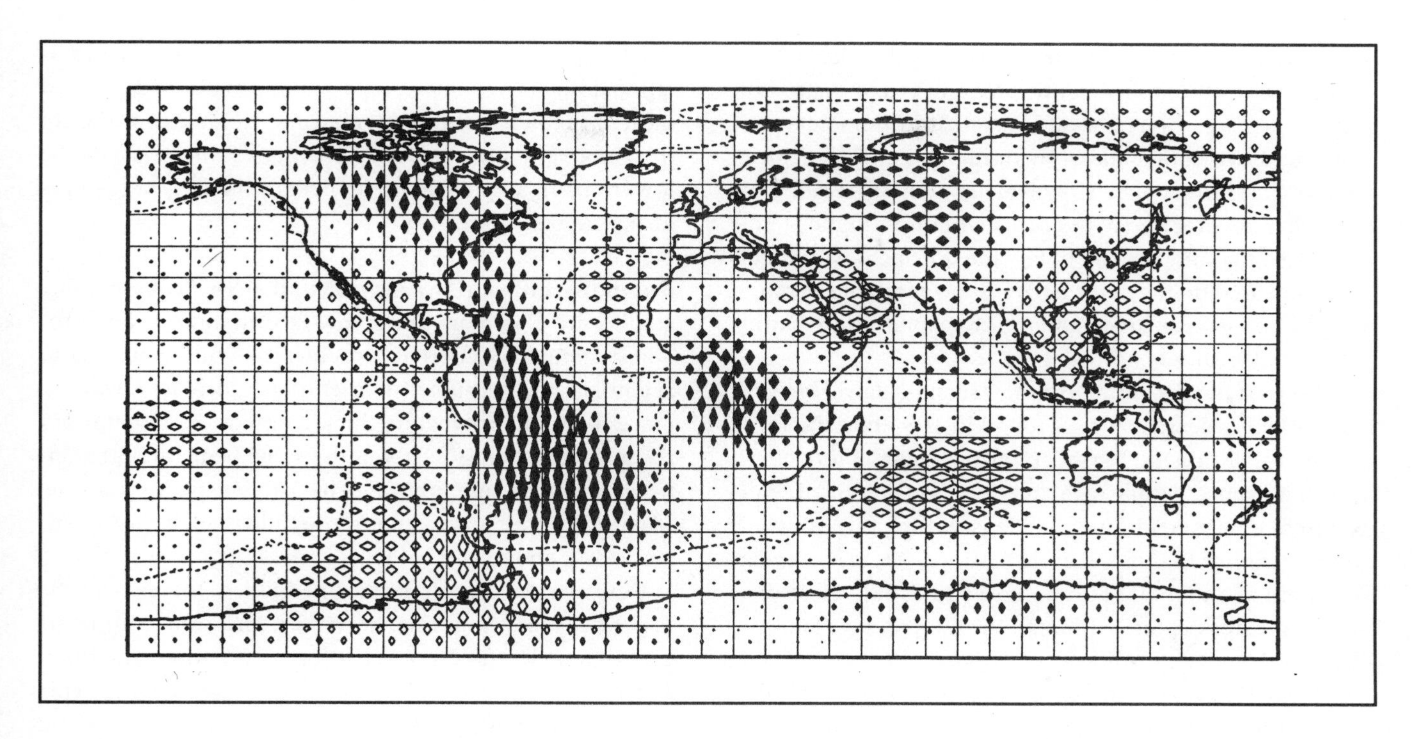

Figure 7. Seismic Tomograph. Geophysicists are able to generate cross-sectional maps, or tomographs, of the Earth's interior at specific depths by using computers to combine seismic data from many observing stations around the world. This tomograph shows variations in the speed at which seismic waves travel through rocks in the Earth's upper mantle at a depth of 100 kilometers (about 60 miles). The size and orientation of the diamonds represent differences in seismic wave speed and direction that provide scientists with information on the temperature and age of rocks and the direction in which the rocks are moving within the Earth. *Courtesy: Seismological Laboratory, California Institute of Technology.*

in California to produce detailed maps of the crust in southern California at two different depths. On one map, which shows the crust at a depth of 10 kilometers (about 6 miles), the San Andreas Fault — the boundary between the Pacific Plate to the west and the North American Plate to the east — is clearly visible as a sharp change in seismic wave velocities across the fault.

As would be expected, the seismic wave velocities measured in rocks west of the fault are characteristic of rocks in the Pacific Plate, while the seismic wave velocities measured to the east of the fault are characteristic of rocks in the North American Plate. At a depth of 30 kilometers (about 19 miles), the San Andreas Fault boundary is less distinct than at the shallower depth, and it does not always closely follow the fault line at the surface.

In 1984, Harvard University geophysicists Adam Dziewonski and John Woodhouse and Caltech geophysicist Don Anderson described their studies of seismic velocities at various depths in the mantle for the Earth as a whole. The researchers reported that the velocity of seismic waves at a depth of 100 kilometers (about 60 miles) in the upper mantle corresponds with what would be expected from knowledge about the locations of plates on the Earth's surface. The mantle beneath the hot, young rock of the mid-ocean ridges has a relatively low seismic velocity, while the relatively cooler, older rock of the continents has a relatively high velocity.

At a depth of 300 kilometers (about 190 miles), however, the patterns of fast and slow velocity no longer closely mirror the patchwork of continents and mid-ocean ridges on the surface. Some areas of pronounced low seismic velocities correspond to known hot spots, including Hawaii in the Pacific Ocean, Iceland and Tristan da Cunha in the Atlantic Ocean, and Kerguelen in the southern Indian Ocean. These hot spots are thought to be regions where hot material is rising toward the Earth's surface from sources deep in the mantle. Another low-velocity area has been found under the Gulf of Aden and the Red Sea.

The pattern of seismic velocities at a depth of about 500 kilometers (about 310 miles) is distinct from that at either 100 or 300 kilometers. The hot, low-velocity area beneath the Red Sea at 300 kilometers is still pronounced, but the data also indicate that many parts of the mid-ocean ridges — such as in the mid-Atlantic region — and some hot spots, which have hot rocks at 100 and 300 kilometers, are situated above cooler rocks (with higher seismic velocities) at 500 kilometers. However, any small regions of low seismic velocity (with correspondingly higher temperatures) that might exist below the hot spots would not have appeared on the maps because the available seismic data did not have a sufficiently high resolution to distinguish small features.

Even with the limitations of their seismic data, Harvard and Caltech researchers first found in 1984 that the flow of material within the Earth's interior is much more complex than most geophysical models had assumed. For example, new temperature and density features, not necessarily similar to those near the surface, begin to appear at a depth of about 300 kilometers and become more distinct at greater depths. Although the temperature anomalies associated with mid-ocean ridges and volcanic regions can often be traced to great depths within the mantle, they do not correspond to simple vertical plumes carrying warm material directly up from the mantle to the surface as most earlier models had assumed. Some researchers have suggested that the hot spots may originate deep in the mantle and the upwelling of the hot material may be influenced by the subduction of colder rocks from the surface. Others have proposed that hot spots may originate at relatively shallow depths in the mantle.

Seismic tomography has helped reconcile maps of the Earth's gravity with structural features deep in the mantle. For features near the surface, the Earth's gravity field is relatively high over areas where there is an excess accumulation of rock or where rock is relatively dense. For example, the excess mass of the Himalayas increases the gravity field in that area. In contrast, the gravity field is relatively low over regions where the total rock mass is unusually low or where the rock is less dense than usual. However, some highs and lows in the Earth's gravity field cannot be explained by mountain ranges and basins on the surface. Work by Bradford Hager and his colleagues at Caltech in 1985 demonstrated that many of these previously unexplained areas of high or low gravity could be accounted for by considering the motion of hot and cold material in the lower mantle.

Hager's group had originally expected that hotter, lighter material rising toward the surface would reduce the gravity field while colder, denser material falling toward the core would raise it. Just the reverse turned out to be true, however. To explain these findings, the researchers proposed that warm, rising mantle material caused an upward warping of the boundaries between layers within the Earth, causing an excess of mass that increased the gravitational field; conversely, cold, sinking material depressed these boundaries and decreased the field. According to this model, large-scale movements of material in the lower mantle explained 80 percent of the largest features of the field.

Further studies of the mantle and core. Tomographic studies are also generating new information and controversy about the internal structure of continents. Earth scientists once assumed that continental plates were no more than 100 kilometers thick and that the structure of the asthenosphere was about the same beneath both continental and oceanic plates. (The asthenosphere is a zone of partially melted rock; it is more fluid than the layers immediately above and below.)

Evidence from tomographic and other studies over the last few years suggests, however, that the continents extend to a depth of at least 150 to 200 kilometers (90 to 125 miles) with "roots" that penetrate even deeper. This evidence includes seismic wave data showing lateral changes in the velocity of waves as they travel through the mantle from areas beneath the continents to those beneath the oceans. To account for these observations, geophysicist Thomas Jordan of the Massachusetts Institute of Technology (MIT) in Cambridge has hypothesized that the continents have roots as deep as 400 kilometers (about 250 miles). Data from tomographic studies have also revealed evidence of differences in the composition and temperature of material underlying oceanic and continental plates. These differences seem related to the large-scale structure and dynamics of the Earth's interior.

In a related controversy, seismologists are debating how far oceanic crust can plunge into the mantle when plates are subducted beneath continents. Seismic tomography may help resolve this question. Scientists have long known that the speed of seismic waves does not always change gradually with depth. Seismic waves accelerate abruptly at depths of about 400 to 670 kilometers (250 to 415 miles) within the mantle. Some theorists have proposed that a change in rock chemistry or crystal structure accounts for these sudden changes in wave speed.

Two competing models have been developed to explain the 670-kilometer discontinuity, which is the boundary between the upper and lower mantle. According to one model, the subducting crust may penetrate through the 670-kilometer discontinuity, with convection extending throughout the mantle. In the other model, the subducting crust does not penetrate through the discontinuity; hence, the upper and lower mantle are distinct layers, with little or no mixing occurring across their boundary.

Some scientists have interpreted the lack of earthquakes at depths below 670 kilometers as evidence that subducting plates cannot penetrate deeper than this discontinuity, a finding that would support the two-layer model. Harvard University researchers John Woodhouse and Domenico Giardini reported in 1984, however, that oceanic crust at some locations, including the Tonga Trench in the South Pacific, thickens and may even penetrate the 670-kilometer discontinuity.

Bradford Hager of Caltech also showed in 1986 that the subducting crust partly penetrates through the discontinuity. However, he believes that the upper mantle is characterized by much higher mixing rates than the lower mantle, and that the discontinuity has to do more with differences in viscosity of the two layers than with differences in their composition.

In 1988, Karen Fischer and Thomas Jordan of MIT and Kenneth Creager, now at the University of Washington, reported evidence that fragments of oceanic crust may penetrate to depths up to 900 to 1,000 kilometers. These viscous fragments apparently encounter strong resistance at a depth of about 650 kilometers and thicken as they are pushed to greater depths. The researchers believe that these findings support the hypothesis that mixing occurs within the mantle and argue against the hypothesis that the mantle is divided into two distinct layers.

Although the evidence of deep subduction is still being debated, geophysicists agree that its implications for the structure and dynamics of the Earth would be profound. The existence of deep subduction would mean that at least some mixing occurs between the upper and lower mantle and that material from the lower mantle may eventually travel up to the continental crust. Models of mantle mixing are already complicated by geochemical data indicating that rocks in the Earth's crust seem to be derived from several chemically distinct sources. This combination of evidence could mean that parts of the upper mantle form discrete zones that do not mix despite the general convecting motion of rock. Future models of the Earth's interior will have to reconcile the apparent inconsistencies in the evidence about the mantle's structure and circulation.

Tomographic studies have been extended to a depth of 2,900 kilometers (1,800 miles), the boundary between the mantle and the outer core. Scientists were surprised to find that the region near the mantle-core boundary is not as smooth and homogeneous as once believed. Instead, it appears to be highly irregular, with many bumps and depressions.

In 1987, Harvard's Andrea Morelli and Adam Dziewonski reported that analyses of waves reflecting off or passing through the boundary showed evidence of relief of up to 6 kilometers (3.7 miles) in that region. The speed of seismic waves along this boundary is not uniform, but instead varies by as much as a few percent. This variation in speed can be interpreted as topography, with some raised areas centered beneath the Atlantic and the eastern Pacific oceans and some depressed regions centered beneath Colombia and Japan. Morelli and Dziewonski speculated that the variations in speed might be caused by a thin layer at the boundary of the outer and inner core. This layer would have thermal and chemical characteristics distinct from the rest of the core and could have shaped the outer core. In 1988, Morelli and Dziewonski also found an apparent association between low-velocity areas in the lower mantle and regions at the core-mantle boundary where the magnetic field changes rapidly. A model of mantle convection developed by W. Richard Peltier and other researchers at the University of Toronto in Ontario supports the existence of boundary-layer topography like that inferred from the seismic studies of Morelli and Dziewonski.

Other tomographic studies by Kenneth Creager, Thomas Jordan, and Bradford Hager have shown similar seismic anomaly patterns. Creager and Jordan believe that the data support the existence of one or more layers of varying thickness at or near the mantle-core boundary instead of at the boundary between the outer and inner cores as in the Morelli-Dziewonski model. According to Creager and Jordan, the layer at the mantle-core boundary would consist of materials with a chemical composition distinct from the surrounding regions. These materials could be thought of as internal continents at the bottom of the lower mantle. If such internal continents exist, they could be linked to convection in the mantle, deep plumes that create volcanic island chains, and possibly even the forces that generate the Earth's magnetic field

Studies of the magnetic field have also brought new information about the nature of the Earth's interior. Most scientists believe that the planet's magnetic field is the result of a self-regenerating dynamo that draws its power from the motion of the electrically conducting fluid in the Earth's outer core.

During 1987, Jeremy Bloxham of Harvard University and David Gubbins of Cambridge University in the U.K. mapped the magnetic field at the boundary between the fluid outer core and the solid mantle. For this purpose, they used data spanning a period between 1715 and 1980. The older information was drawn from records in ships' logs and marine almanacs, including those from the expedition of James Cook in 1768. The researchers identified the existence of a so-called bipolar core spot located beneath the Atlantic Ocean near South Africa. This spot is a magnetic feature in the outer core, and is in a sense similar to sunspots. Other research has shown evidence that lateral temperature variations in the mantle appear to influence the motion of fluid in the outer core. Consequently, the two systems may be coupled. By studying these systems and the ways they interact, researchers hope to learn more about the mechanisms involved in mantle convection.

Several important questions remain unanswered. Geophysicists would like to determine what causes fluctuations in the intensity of the magnetic field and why the magnetic poles gradually drift over time. They would also like to understand what causes the magnetic poles to reverse periodically. Researchers have observed that the intensity of the magnetic field has been declining. Bloxham believes that if the current trend continues, the field may reverse within the next 1,000 years.

By analyzing different types of seismic waves, Harvard researchers found that the Earth's solid inner core — believed to consist of nearly pure iron — is slightly elliptical and seems to have some of the properties of a hexagonal crystal whose axis is aligned with the Earth's rotational axis. Complex questions about the Earth's internal and thermal structure, chemistry, and gravitational and magnetic fields will have to be resolved before the properties of the core-mantle boundary and the inner core are understood.

Simulating high pressure. Although indirect observations of seismic waves passing through the mantle and core, as well as gravitational and magnetic studies, have provided a glimpse of the processes at work inside the Earth, they cannot substitute for direct observations of the rocks within the interior. Since it is impossible to sample rocks from the mantle and core, some scientists have tried instead to simulate the conditions of the Earth's interior in the laboratory. During the late 1960s and early 1970s, diamond-anvil cells replaced hydraulic presses as the instruments for simulating the high temperatures and pressures within the mantle. In a diamond-anvil cell, a steel gasket forms a sealed pressure chamber, which contains two gem-quality diamonds with parallel surfaces facing each other. A mineral sample is placed between the diamonds and then compressed as in a vise. A laser shines through the diamonds to heat the sample to high temperatures. This apparatus can generate pressures more than 1 million times the atmospheric pressure at the Earth's surface.

Minerals can pass through different phases (crystalline structures) as they are subjected to increasingly higher pressures. By studying the characteristics of a variety of mineral samples subjected to extreme pressures, geochemists can test their hypotheses about what materials exist at different depths in the mantle. They can also learn whether the abrupt changes observed in seismic-wave velocity are caused by changes in the materials' crystal structure or chemical composition.

Elise Knittle and Raymond Jeanloz of the University of California in Berkeley have tested the hypothesis that the minerals garnet, olivine, and pyroxene are transformed into the mineral perovskite in the lower mantle below the 670-kilometer discontinuity. In 1987, Knittle and Jeanloz reported that perovskite is stable at 2,000°C (3,600°F) and more than a million times the pressure at Earth's surface, conditions typical of the lower mantle. They concluded that perovskite is probably the most abundant material within the Earth.

A group of scientists from Caltech, the University of California in Berkeley, and the University of Illinois have measured the melting point of iron at pressures comparable to those in the core — 2.5 megabars, more than 2 million times the pressure at the Earth's surface. The melting temperatures of many materials rise as pressure increases. Since geophysicists believe the core is mostly iron, knowledge of the melting point of iron at high pressure is critical to understanding the physical characteristics of the core. By subjecting samples of iron to explosive shocks, the researchers determined that iron has a melting point of about 4,500°C (about 8,100°F) at pressures comparable to those at the core-mantle boundary. After taking into account the fact that elements alloyed with iron in the outer core would lower the melting point of iron, the researchers calculated that the temperature of the core-mantle boundary must be at least 3,500°C (6,300°F).

According to their calculations, the temperature at the boundary between the solid inner and liquid outer core is about 6,300°C (11,300°F), and the temperature at the Earth's center is 6,600°C (11,900°F), which is about 1,100°C (2,000°F) hotter than the surface of the Sun. The researchers believe that the core's high temperatures indicate that it is the ultimate source of energy driving convection in the mantle. They conclude that the apparently large difference between the outer-core and lower-mantle temperatures could indicate that a nonconvective layer exists in the lower mantle and that this layer controls the rate at which heat escapes from the core.

Seismic tomography, gravitational and geomagnetic studies, and high-pressure experiments have brought major advances in geophysical knowledge of the Earth's interior. In the future, higher-resolution tomography and refined laboratory studies promise deeper insights into the inner workings of the planet.

Chapter 7

Environment

Depletion of the Ozone Layer

For perhaps the first time in the long and fractious dispute about the threat posed to the Earth's protective ozone layer by manmade chlorofluorocarbons (CFCs), researchers, government officials, and CFC manufacturers agreed in 1988 on the necessity for speedy corrective action. The agreement was prompted by the discovery of a "hole" in the ozone layer over Antarctica, followed by a definitive report showing that the ozone layer is deteriorating at a much faster rate than was previously believed.

CFCs are molecules consisting of either one or two carbon atoms bonded to chlorine and fluorine atoms. Industry relies on CFCs because they are not flammable or toxic, do not react with other chemicals, and are not degraded by microorganisms. These properties, combined with their light weight, make them ideal for use as propellants in aerosol cans, refrigerants in refrigerators and air conditioners, blowing agents for producing plastic insulating foams, and solvents for cleaning electronic components. As a result, industry used more than 2.1 billion pounds of CFCs worldwide in 1987. In the U.S. alone, more than 5,000 businesses at 375,000 locations rely on CFCs to produce goods and services worth more than $28 billion per year.

The inertness of CFCs at ground level makes them dangerous, however. When CFCs leak into the atmosphere from automobile air conditioners, abandoned refrigerators, or factories, they migrate very slowly — over a 15- to 50-year period — to the stratosphere,

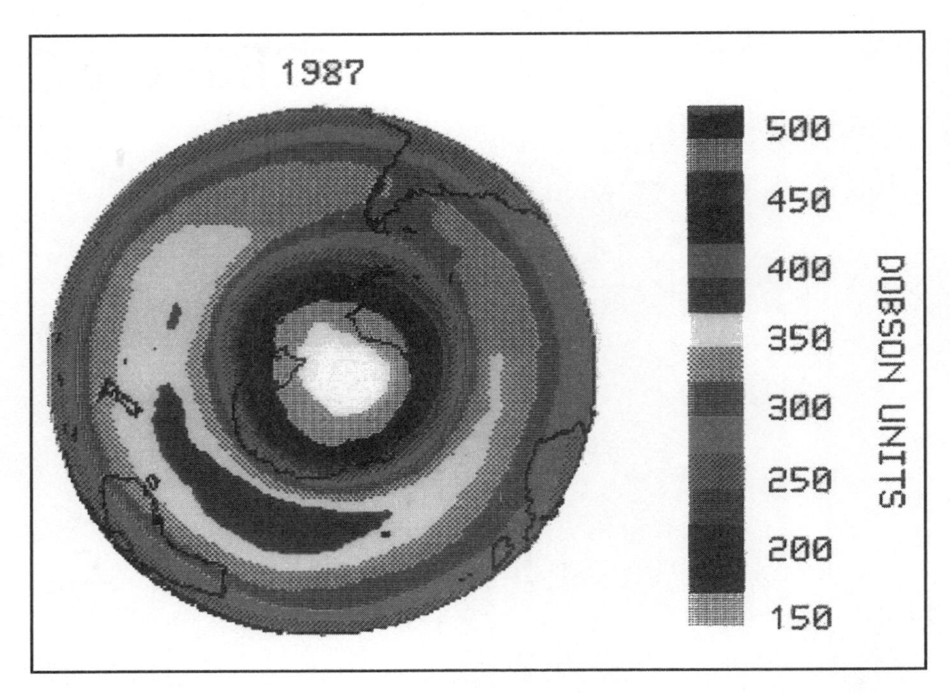

Figure 1. Antarctic Ozone Hole: 1987. Between 1976 and 1987, the concentrations of stratospheric ozone measured each spring at the Halley Bay weather station in Antarctica declined precipitously. In 1985, the concentration was so low that the phenomenon was first described as a "hole" in the ozone layer. In this image, generated by computer from satellite data, the 1987 hole is indicated by the white area near the center of the circle. *Courtesy: Arlin Krueger, NASA.*

the region of the the Earth's atmosphere that extends from 9 to 30 miles above the Earth's surface, and that contains the ozone layer.

Ozone Layer

Ozone, which consists of three oxygen atoms connected in a ring, is a highly reactive molecule that is a severe pollutant when formed near the Earth's surface from automobile exhausts and factory emissions. In the stratosphere, however, ozone is beneficial. Stratospheric ozone forms when oxygen molecules (O_2) are broken down into two oxygen atoms by sunlight; these oxygen atoms combine with oxygen molecules to form ozone (O_3).

Although ozone exists in the stratosphere in only minuscule quantities — about three parts per million — it screens out more than 99 percent of the ultraviolet (UV) radiation in sunlight that would otherwise reach the Earth's surface. In the process, ozone is broken apart into oxygen atoms and oxygen molecules. Ideally, the stratosphere maintains a delicate equilibrium, with incoming sunlight producing just enough ozone to replace that broken down by UV radiation.

Effects of ozone depletion. Because UV radiation can damage nearly all forms of life, most scientists agree that life on Earth did not evolve until after the protective ozone layer had formed. Studies by the U.S. Environmental Protection Agency (EPA) have estimated that, for each 1 percent decrease in stratospheric ozone levels, 2 percent more UV radiation reaches the Earth's surface. A 1987 EPA report concluded that, in the worst case, higher UV radiation from ozone depletion could cause an extra 200 million cases of common skin cancer worldwide in people alive today or born by the year 2075. Cancer researchers have also estimated that each 1 percent increase in UV radiation could increase the incidence of malignant melanoma, the most serious form of skin cancer, by as much as 1 percent.

Other studies have suggested that ozone depletion could cause an additional 50 million cases of cataracts. Another effect of UV radiation — suppression of the immune system — is only poorly understood, but scientists suspect that it might have a devastating impact, particularly on people with immune systems already compromised by viruses, drug treatments, or genetic disorders.

The 1987 EPA report noted that the effect of increased UV radiation on many plant species and on the ecosystem as a whole has not been studied in detail. According to the report, increased UV light falling on the ocean's surface could kill plankton, which serve as a food source for other marine life; increased UV light could also decrease the yield of agricultural crops. Finally, changes in the distribution of ozone within the stratosphere could have an unforeseeable impact on global weather patterns.

In view of such possibilities, the National Science

Foundation (NSF) began funding studies in Antarctica to assess the effect of the ozone hole on the Antarctic environment. The NSF was so concerned about the increased risk to their own researchers as a result of the ozone hole that it issued sun-blocking creams and protective clothing to the scientists working in Antarctica.

Early ozone studies. Concerns about ozone depletion began in the early 1970s when Harold Johnston of the University of California at Berkeley and other scientists warned that nitrogen oxides emitted in the exhaust of the planned Supersonic Transport (SST) could deplete stratospheric ozone. Although the government abandoned the SST program in 1971 for other reasons, including its high cost, the U.S. Congress had already instructed the Department of Transportation to study the SST's possible effects on the atmosphere.

In 1975, the Climatic Impact Assessment Program concluded that nitrogen oxides in the exhaust gases of an SST fleet of 500 commercial aircraft operating in the stratosphere would reduce ozone concentrations by about 10 to 20 percent. (These projections have since been challenged by other research groups.) Although only two commercial SSTs were flying in 1989, both the U.S. and the U.S.S.R. operate a large number of supersonic fighters and bombers in the stratosphere. No one has thoroughly evaluated the potential impact of these planes on ozone depletion.

In 1974, chemists F. Sherwood Rowland and Mario Molina of the University of California at Irvine reasoned that CFCs, which are lighter than oxygen and nitrogen, would float into the stratosphere. Subsequent laboratory experiments indicated that sunlight in the stratosphere could break the CFC molecules apart, releasing highly reactive chlorine atoms that could each destroy millions of ozone molecules. Rowland and Molina estimated that continued growth in the use of CFCs could lead to destruction of as much as 30 percent of the ozone layer by the middle of the twenty-first century. The catastrophic implications of that prediction inspired a movement that, in 1978, led the U.S. to ban the use of CFCs as propellants in aerosol cans.

Although chemists and meteorologists do not fully understand the complex series of reactions by which CFCs destroy ozone, they know that the process involves at least 192 separate chemical reactions — 48 of which are stimulated by light. To predict how severely CFCs will deplete the ozone layer, researchers need to determine the rate at which each of these reactions occurs. Chemists have been trying to measure these rates accurately for two decades. Most of the first accurate measurements seemed to lower the estimates of eventual ozone depletion. By 1984, an NAS report concluded that the eventual ozone depletion would total only 2 to 4 percent if CFC production continued at the then-current rates.

Intensive research also indicated that a variety of other chemicals affect the ozone layer. The industrial solvents carbon tetrachloride and methyl chloroform, for example, deplete ozone, as do halons — bromine-containing compounds used to extinguish fires quickly in confined spaces.

Predictions of relatively minor ozone depletion lulled both the public and government officials into a false sense of security that appeared to justify continued production of CFCs. With no scientific consensus on ozone depletion, further regulatory action to restrict CFCs was generally stalled. In March 1985, however, representatives from 28 nations signed the Vienna Convention for Protection of the Ozone Layer after a period of negotiation sponsored by the United Nations Environment Programme (UNEP). While the Vienna Convention established a protocol for negotiating a treaty to protect the ozone layer, it did not prompt any direct action to reduce CFC emissions. New, disturbing scientific findings in 1985 and 1986, however, created a renewed sense of urgency.

Ozone hole. As researchers continued to refine their understanding of the chemical reactions involved in ozone depletion, their estimates of ozone depletion began to increase. During the same period, they also began to change their method of calculating the estimates. In the past, the estimates had been based on the continued release of CFCs into the environment at current levels. In reality, however, the use of CFCs had risen 25 percent between 1982 and 1986. When researchers factored that growth into their calculations, their predictions changed dramatically.

An April 1985 EPA report predicted the loss of as much as 60 percent of the ozone layer by the middle of the next century if CFC use continued to grow by 4.5 percent annually. Even if that growth rate dropped to 2.5 percent annually, ozone depletion would reach 26 percent by the year 2075.

In 1985, the dramatic discovery of a hole in the ozone layer over Antarctica by meteorologist Joseph Farman and his colleagues at the British Antarctic Survey in Halley Bay, Antarctica, revived public and government concern about the ozone problem. The Farman group said that measurements taken at the Halley Bay station on the Weddell Sea coast at about 75°S indicated a steep drop in ozone concentration during the austral spring of each year since 1976. The ozone levels began falling in late September, as the Sun rose above the horizon at the end of the long Antarctic winter, and reached a minimum in mid-October.

The area of low ozone concentration, which resembled a hole in the ozone layer about the size of the continental U.S., would typically disappear by the end of October as ozone levels returned to their normal values. Measurements made by U.S. satellites subsequently confirmed this discovery. Ironically, the satellites had not initially detected the hole because they had been programmed to dismiss as erroneous any data that showed very low ozone levels.

The seasonal ozone depletion over Antarctica worsened steadily between 1976 and 1987, with ozone levels generally dropping further each year and the hole persisting longer. The depletion followed an unusual pattern of lowest ozone concentrations in odd-numbered years and slightly higher concentrations in even-numbered years — a pattern believed to result from meteorological conditions caused by Earth's orbit. In 1987, total stratospheric ozone above Antarctica fell 50 percent during the austral spring, and the concentration of ozone between 9 and 12 miles high fell a maximum of more than 95 percent. The hole also persisted longer than ever before, lasting from September until early December.

Prompted by the discovery of the ozone hole, the National Aeronautics and Space Administration (NASA) began a 17-month study of ozone levels around the world. In March 1988, the NASA Ozone Trends Panel reported that ozone depletion throughout the world was far greater than anyone had originally predicted. The panel concluded that, since 1969, when satellite measurements had first been taken, CFCs had destroyed about 2.3 percent of the ozone over the U.S. and significantly larger amounts near the North and South Poles. During the winter in some Northern Hemisphere locations, ozone concentrations fell as much as 6 percent below 1969 levels. "The indications are bad, not good," said NASA's Robert Watson, chairman of the panel.

The panel also found that, largely due to variations in the amount of sunlight striking Earth, ozone depletion fluctuates according to season and latitude, and ozone levels drop most sharply near the poles during winter. According to the panel, the average year-round ozone levels in the Northern Hemisphere have decreased an estimated 1.7 percent between latitudes 30° and 39°, 3 percent between 40° and 52°, and 2.3 percent between 53° and 64°. Although the panel did not offer an overall estimate for the entire region from 30° to 64°, Rowland, who was a member of the panel, said that the decline would average about 2.3 percent.

The NASA panel's study, carried out by more than 100 scientists throughout the world, used data from ground-based ozone detectors, most of which were located in the Northern Hemisphere, to calibrate the satellite's ozone detectors. While data from the satellites had previously suggested that the ozone layer was being depleted at a rate of about 1 percent per year, the new study indicated that the satellite instruments deteriorated over time and thus were exaggerating the amount of depletion. Nonetheless, the data did establish a much greater level of ozone depletion than had been previously expected.

Although the panel did not study the Southern Hemisphere except for the region near the South Pole, experts predicted that they would find similar ozone depletion levels there. Such a depletion would raise health concerns for residents of Australia and New Zealand, which lie in an area with an estimated depletion of 4 percent. That depletion level is large enough to cause a significant increase in skin cancer in those countries if residents do not take special precautions to protect themselves from UV light.

Explaining the Ozone Hole

Antarctica's climate differs from any other on Earth. The continent, surrounded on all sides by water, is circled by the world's coldest ocean current, the Antarctic Circumpolar Current. As the continent enters its long winter each year, the cold temperatures bring a weather pattern called the polar vortex, which isolates the continental land mass by preventing warmer weather systems from the lower latitudes from reaching it. As a result, stratospheric temperatures over the continent fall below –130°F, much colder than elsewhere in the stratosphere and cold enough to form ice clouds even in the very dry air there. Elsewhere around the world, the stratosphere is not cold enough for such ice clouds to form.

Many researchers initially believed that these unusual climatic conditions in the Antarctic atmosphere caused the ozone hole. Intensive studies showed, however, that while climate certainly contributes to the development of the ozone hole, the primary cause of ozone depletion, without question, is CFCs. Evidence to support that conclusion came from two intensive studies of the Antarctic stratosphere during the austral springs of 1986 and 1987.

Linking ozone depletion to CFCs. In 1986, NOAA and NSF organized the National Ozone Expedition (NOZE) to coincide with the austral spring and the appearance of the ozone hole in late September. During August 1986, four scientific teams flew to McMurdo Station, Antarctica, bringing a variety of instruments to measure concentrations of ozone and hundreds of chlorine and nitrogen compounds at various levels in the stratosphere. In a series of reports published during 1987, these teams documented the depletion of ozone in the hole and the presence of sufficient quantities of chlorine compounds to cause that depletion. In retrospect, the evidence obtained during this mission was probably adequate to prove that CFCs were the cause of the hole, but many skeptics remained unconvinced until they saw evidence from a second mission the following year.

More than 150 scientists went to Antarctica during August and September 1987 for the second study. As part of the project, NASA repeatedly flew a heavily instrumented Douglas DC-8 through the ozone-depleted region at an altitude of 7.5 miles and an ER-2 (a U-2 spy plane modified for scientific studies) at an altitude of 11 miles. The scientific flights took place over a period of 6 weeks, with the aircraft logging more than 175,000 miles. The airborne studies were complemented by ground-based measurements, but it was the flights that provided definitive evidence of the role of CFCs in ozone depletion.

The ER-2 carried a fluorescence spectrometer designed by chemists James Anderson and William Brune of Harvard University in Cambridge, Massachusetts. They designed the spectrometer to measure the concentrations of chlorine monoxide, a key component of the chain reactions that scientists had proposed for the destruction of ozone. The plane also carried ozone-measuring spectrophotometers developed by atmospheric scientists Michael Proffit of the National Oceanic and Atmospheric Administration (NOAA) and Walter Starr of NASA.

Flights in September before the Sun rose over Antarctica showed that the ozone layer there was intact, but that levels of chlorine monoxide were 500 times higher than the levels measured at comparable altitudes in the mid-latitudes. After the Sun rose above the horizon later in the season, ozone levels began to drop dramatically, but only in those areas where Anderson and Brune had measured high chlorine monoxide levels. "There was a very clear relationship between ozone and chlorine monoxide, a dramatic anti-correlation," Anderson said.

The ultimate explanation for this destruction was provided by Molina, now at NASA's Jet Propulsion Laboratory in Pasadena, California. Earlier, various researchers had shown that chlorine monoxide is trapped in the stratosphere above Antarctica by adsorption onto the surface of the frozen water droplets that make up the polar stratospheric clouds. Through laboratory experiments, Molina demonstrated that chlorine monoxide on the surface of ice crystals breaks down ozone much faster than it does when it is floating freely as a gas. Researchers now believe that the reaction pathway identified by Molina accounts for 80 percent of the ozone destruction over Antarctica.

Some chemical models suggest that ozone depletion tends to perpetuate itself: The less ozone there is in the hole, the less solar radiation that is absorbed in the upper atmosphere, and the colder the hole becomes. Cooling promotes the formation of ice crystals and polar stratospheric clouds, leading to even more ozone depletion. The greater the cooling and the ozone depletion, the longer it takes for the spring sunlight to warm the ozone layer and for the polar vortex to give way to another weather pattern.

An Arctic hole? Despite gross similarities, atmospheric conditions in the Arctic differ substantially from those in the Antarctic. Although a polar vortex does form during the Arctic winter, it is much weaker and shorter-lived than the one that forms over Antarctica. Consequently, air within the north polar vortex is not as isolated from weather systems in the lower latitudes. Warmer air from the mid-latitudes frequently reaches the Arctic, making the average winter stratospheric temperature higher than that over the South Pole. Meteorologists have thus questioned whether the polar stratospheric clouds that promote ozone depletion at the South Pole can form over the North Pole. If the clouds can form, researchers fear that ozone depletion in the Arctic would cause far more damage than the Antarctic hole because more people live there.

In May 1988, meteorologist Wayne Evans of Environment Canada reported evidence that the Arctic may undergo an annual ozone depletion cycle similar to that observed in the Antarctic hole but much smaller in magnitude. Evans found that measurements made with balloon-borne instruments indicated that Arctic ozone levels had declined from January through April in each year since 1986. NOAA chemist George Mount and his associates later announced that they had detected chlorine monoxide concentrations as much as five times higher than expected in the stratosphere above Thule, Greenland, during February 1988.

In January 1989, an international research team began studies in the Arctic similar to those carried out during the preceding 2 years in Antarctica. Researchers used the same two NASA planes, this time based in Stavanger,

Norway, to measure atmospheric concentrations of the suspect chemicals. In February 1989, Evans announced that his balloon-borne instruments had detected the stratospheric clouds, and the international team reported that it had detected the presence of large quantities of chlorine monoxide in the region. Before the Sun rose over the Arctic, however, winds had dispersed the chlorine compounds, and no significant ozone hole developed.

Containing the CFC Threat

Most scientists now agree that CFCs released into the atmosphere since the 1930s will continue damaging the ozone layer in the coming decades. They also agree that, although nothing can be done about the chemicals already released, the Earth's inhabitants would be wise to reduce future CFC production as much as possible.

Ozone treaty. In September 1987, representatives of 46 nations agreed to the landmark Montreal Protocol on Substances that Deplete the Ozone Layer — a treaty designed to protect the layer by cutting world CFC production in half. Signatories to the treaty, which took effect on 1 January 1989, agreed to hold domestic CFC consumption at 1986 levels and limit production to 110 percent of that year's levels by July 1990. The production freeze will be followed by a 20 percent reduction in CFC usage by 30 June 1994 and another 30 percent decrease by 30 June 1999. Consumption of halons is also limited to 1986 levels.

The Montreal Protocol also included a special provision for developing nations, whose current use of CFCs is relatively low and whose economic growth might suffer as a result of tight restrictions on their future CFC use. These nations would be allowed to increase their consumption to 0.66 pound per capita, compared with the current average consumption of 1.76 pounds per capita in the developed world. The U.S.S.R. will also be permitted to complete CFC production plants under construction when the treaty was signed. As a result of these exceptions, production of CFCs might decline only about 35 percent by 1999.

Many scientists and some government officials believe that the projected reduction is not enough. In September 1988, then-EPA Administrator Lee Thomas called for a complete phaseout of CFC production as soon as possible. Thomas also called for reductions in the use of halons and a freeze in the production of methyl chloroform. Scientists and government leaders in other countries have made similar recommendations.

At a March 1989 meeting in London convened to explore ways to preserve the ozone layer, the 12 member nations of the European Economic Community pledged to ban CFC production and use by the year 2000. The day after that pledge, President George Bush, who had not attended the meeting, made a similar vow. At least 20 countries that had not previously signed the Montreal Protocol have agreed to do so; however, the People's Republic of China, India, and some other developing countries would not agree to the protocol, arguing that their economic growth would severely suffer if they did. Representatives of these countries said that they could adhere to the protocol only if industrialized nations helped them finance the development and use of alternatives to CFCs.

In March 1988, E. I. du Pont de Nemours & Co. of Wilmington, Delaware, which produces 25 percent of the world's CFCs, announced that it would phase out CFC production. Although Du Pont did not set a timetable, Joseph Steed, an environmental manager in the company's CFC production division, said that the company could reduce its CFC production 95 percent by the year 2000. Pennwalt Corp. of Philadelphia, Pennsylvania, and Imperial Chemical Industries, Ltd., of London, U.K. — both large producers of CFCs — also said they would phase out CFC production.

The week after Du Pont's announcement, the Foodservice and Packaging Institute, a trade association, announced that American producers of styrofoam fast-food containers would voluntarily phase out the manufacture of containers made with CFCs. Although this use accounts for only 2 percent of the CFC market in the U.S., environmentalists hope that the food-service industry's decision to stop using CFCs signals the beginning of a trend away from the chemicals.

CFC substitutes. A 1986 report by Alan Miller and Irving Mintzer of the World Resources Institute in Washington, D.C., concluded that CFC releases could be reduced immediately if equipment were maintained better, CFCs in old refrigerators and air conditioners were recycled, and safe substitutes for CFCs were used whenever possible. For example, manufacturers can use hydrocarbons, instead of CFCs, to form air bubbles in plastic foam for fast-food containers and other applications, even though the hydrocarbons do not insulate as well as the CFCs.

The electronics industry could use new equipment to capture and recycle the CFCs that are used to remove solder resin from electronic components, and it might even be able to replace CFCs in many cases. American Telephone and Telegraph Co. (AT&T), for example, uses 3 million pounds of CFCs every year for cleaning equipment. In January 1988, AT&T and Petrofirm, Inc., of Fernandina Beach, Florida, announced that they had developed a new organic solvent, called Bioact EC-7, which would replace 20 to 30 percent of the CFCs that AT&T uses within 2 years, and more later. Although several smaller companies have also bought the organic solvent, many electronics companies have not purchased it because they would need to redesign their cleaning equipment before using Bioact EC-7.

Most companies, however, are committed to modifying CFCs, for example, by using less chlorine and more fluorine in them, or replacing at least one chlorine or fluorine atom with a hydrogen atom to make the compounds break down before they reach the stratosphere. Because these materials have different physical properties than the CFCs now in use, however, their use will force the redesign of air conditioners and refrigerators. Because the substitutes do not work as efficiently as existing CFCs, automobile air conditioners that use the new compounds, for instance, would be heavier and bulkier, thereby reducing the cars' gas mileage.

The new materials are also more expensive. CFCs typically cost between 50 and 70 cents per pound, while the alternatives would sell for $1.50 or more. In a $700 refrigerator or a $500 automobile air conditioner, the increased cost would be only about $15; however, since CFCs represent about 25 to 40 percent of the cost of insulating foams, the price of foams would increase substantially.

In any case, the major impediment to the chemical alternatives is the time required to test their potential toxicity. In December 1987, Du Pont organized a consortium of CFC producers to cooperate in the safety testing of the two most promising alternatives. That testing will cost at least $6 million and take 5 years to complete. Nonetheless, the company is so confident that the chemicals are safe that it plans to complete a production plant by 1992 — a year before the safety testing is scheduled to be finished. Imperial Chemical Industries will complete its own plant at about the same time.

For the first time in many years, environmentalists are growing confident that the problem of ozone depletion will eventually be solved — even though it seems certain that further depletion will occur before the situation begins to improve. Ecologists now hope the spirit of international cooperation that has brought about quick action on the ozone problem will also lead to new steps to reduce other global problems, including greenhouse warming and acid rain.

Greenhouse Effect

The Earth's climate is closely linked to the amount of carbon dioxide and other trace gases in its atmosphere through a phenomenon known as the greenhouse effect, in which these gases trap heat from the Sun and prevent it from re-radiating into space. During periods in the Earth's history when atmospheric levels of carbon dioxide have been low, the planet experienced ice ages during which a large number of species became extinct. When carbon dioxide reached its lowest level 800 million years ago, glaciers expanded from the poles to the equator.

In contrast, when levels of carbon dioxide were above historic norms, the Earth's temperature was unusually high. During the Cretaceous Period (65 million to 135 million years ago) when dinosaurs roamed the planet, carbon dioxide levels may have been twice current levels, and global temperatures averaged 8°F warmer than today. Although the link between past carbon dioxide levels and global temperature is strong, scientists are not yet sure whether changes in carbon dioxide levels caused the climate changes.

Nevertheless, many scientists now fear that the Earth is entering another period of unusual warmth caused by the accumulation of carbon dioxide and other greenhouse gases in the atmosphere. Since the mid-1800s, the concentration of carbon dioxide in the atmosphere has increased from 270 parts per million (ppm) to 350 ppm, and the average global surface temperature has risen by 0.5° to 1.25°F. A 1983 U.S. National Academy of Sciences (NAS) report predicted that the carbon dioxide level could reach 600 ppm within 50 to 100 years if fossil fuels continued to be burned at then-current rates.

Such an increase could eventually raise global temperatures by as much as 8°F, the report concluded, producing a climate comparable to that during the Cretaceous Period. Global warming might cause a partial melting of the polar ice caps, which would, in turn, cause flooding in many coastal cities. Oceanographers have calculated that the sea level has already risen 1 to 3 inches since the 1850s. Global warming would also increase the rate of evaporation of the oceans, which could increase rainfall an average of over 10 percent worldwide. At the same time, the patterns of rainfall would change dramatically. Some climate models project that prime agricultural areas in the U.S. and Europe could grow hotter and drier, decreasing productivity and increasing the need for irrigation. Marginal croplands in Canada and the U.S.S.R., in contrast, would become much more productive.

While most scientists believe that the effects of global warming could be dramatic, there are many uncertainties in existing models of the greenhouse effect. For example, an indeterminate amount of the carbon dioxide released by burning fossil fuels will be absorbed in the ocean, which would help moderate global warming. Furthermore, increased warming of the oceans would lead to more evaporation and increased cloud formation. These clouds would reflect some of the sunlight back into space, and some researchers believe that the cloud cover might largely cancel out the greenhouse warming.

Most climatologists, however, fear that greenhouse warming is inevitable, and question only when the Earth will first suffer its effects. Some believe that it has already arrived. During the 1980s, the Earth experienced the hottest years in recorded history. During the hot summer of 1988, climatologist James Hansen of NASA's Goddard Institute for Space Studies in New York City startled the U.S. Senate Energy and Natural Resources Committee by stating that the warming has already begun. According to Hansen, "It is time to stop waffling so much....The evidence is pretty strong that the greenhouse effect is here." Other researchers believe that they have found evidence of greenhouse warming in changing rainfall patterns and other climatic characteristics.

Most researchers have resisted attributing the intense heat and drought in the U.S. during 1988 to greenhouse warming. "Climate is, by definition, the average over many years," said physicist John Firor of the National Center for Atmospheric Research (NCAR) in Boulder, Colorado, "so there is no way you can tell if 1 year is part of a climate change or not." Climatologist Alan Hecht of NOAA's Environmental Research Laboratories in Boulder noted, however, that "What the drought does underscore...is the real possibility that, in the future, we may see droughts more frequently and more intensively than we have in the past. Given that possibility, this drought is important as a case study of how we respond to it and how we would anticipate and respond to any future drought."

Greenhouse Gases

Meteorologists have debated the climatological effects of carbon dioxide since the late 1800s. While carbon dioxide allows the Sun's visible light to reach the Earth's surface and warm it, it also absorbs heat (infrared radiation) that would otherwise be radiated back into space — thus trapping heat somewhat like the glass panes of a greenhouse. "We've always had a greenhouse effect — that's the least controversial theory in meteorology," according to climatologist Steven Schneider of NCAR. "If it didn't work, Mars wouldn't be a deep freeze, Venus wouldn't be a hothouse, and the Earth wouldn't be just right." In fact, without the greenhouse effect, the Earth's average surface temperature would be near 0°F, rather than the current 59°F. The controversy now focuses on how much temperatures will rise if levels of carbon dioxide and the other greenhouse gases continue to increase.

Carbon dioxide forms during the combustion or decay of organic materials, such as trees, plants, and fossil fuels. Volcanoes also release it from the Earth's interior. Animals and plants produce it when their cells oxidize sugars for energy, and plants extract it from the air during the process of photosynthesis, converting it back into sugars. Compared to nitrogen, oxygen, water vapor, and argon, which together account for over 99.9 percent of the Earth's atmosphere, carbon dioxide occurs in only minute concentrations.

Most of the carbon dioxide increase in the atmosphere since the mid-1800s can be attributed directly to two human activities: the burning of fossil fuels and the permanent clearing of trees and other vegetation for agriculture. For the world as a whole, fossil fuel combustion produces about four to five times as much carbon dioxide as deforestation. In the U.S., about 35 percent of the carbon dioxide results from electricity production, 30 percent from cars and trucks, 24 percent from industry, and 11 percent from heating residential buildings. In developing countries, deforestation is a major source of carbon dioxide. When trees and other vegetation are cut down and burned or left in the forests to rot, the carbon they contain is released into the atmosphere in the form of carbon dioxide. If the trees and other vegetation are not replanted, this carbon dioxide will remain in the atmosphere and will not be consumed through photosynthesis.

Some of the strongest evidence to date linking atmospheric carbon dioxide levels and global temperatures was reported in three papers in the 31 August 1987 *Nature* by a team of researchers headed by Claude Lorius and his colleagues at the Laboratory of Glaciology and Environ-

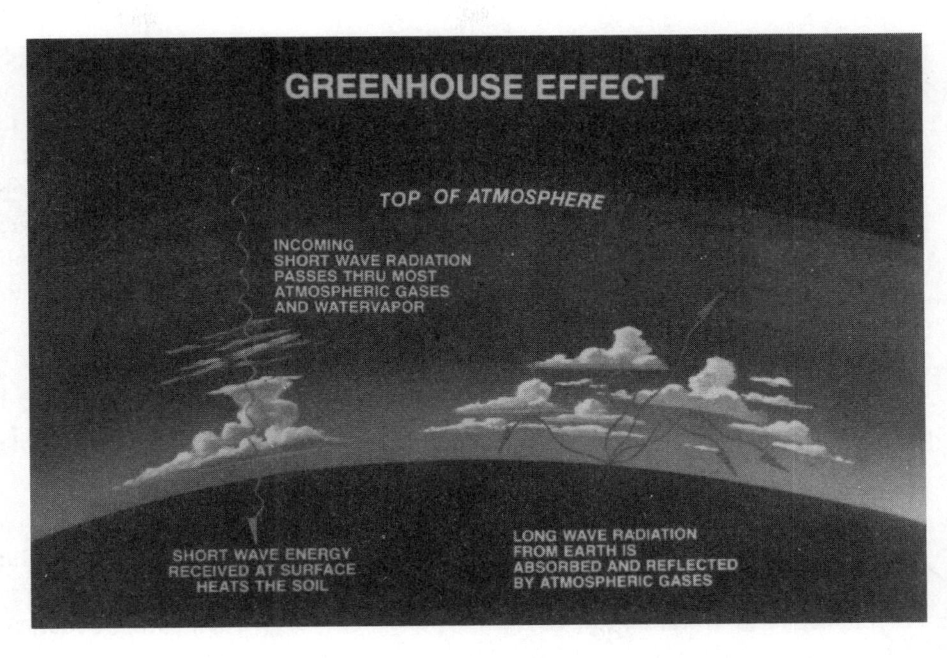

Figure 2. Greenhouse Effect. Carbon dioxide and the other so-called greenhouse gases in the atmosphere allow sunlight to pass through the atmosphere to the Earth's surface. Much of this solar energy is re-radiated from the surface as heat, and the greenhouse gases trap some of this heat in the atmosphere. Most climatologists believe that rising concentrations of greenhouse gases will lead to a warmer climate during the next century. *Courtesy: NOAA.*

mental Geophysics in St. Martin d'Heres, France. Lorius's team studied ice samples that were retrieved from a 6,784-foot-deep hole drilled by Soviet scientists at their Vostok Station in eastern Antarctica. The oldest ice samples date back 160,000 years.

As snow in the Antarctic region was compressed into ice by the weight of the fresh snow above it, the ice trapped small bubbles of air and locked them in place. Lorius and his colleagues measured the carbon dioxide concentrations in the bubbles at 3-foot intervals down the entire length of the recovered ice core. By measuring the proportion of deuterium (an isotope of hydrogen containing one proton and one neutron) in the ice, they could also estimate air temperatures over the centuries. (The amount of deuterium in rain or snow increases with increasing temperature.)

In the *Nature* papers, Lorius and his colleagues reported that carbon dioxide levels in the atmosphere dropped 40 percent and average global temperatures fell about 18°F on the last two occasions when glaciers swept as far south as St. Louis, Missouri, and New York City. Both temperatures and carbon dioxide levels returned to normal as the glaciers receded. "We found a very strong correlation between carbon dioxide fluctuations and temperature fluctuations over the entire 160,000 years," according to Lorius. He hypothesized that "changes in carbon dioxide magnified changes in temperature that resulted from variations in the Earth's orbit around the Sun. These results... suggest that carbon dioxide was definitely responsible for the warming and cooling."

More recently, an international team of investigators suggested that depletion of carbon dioxide from the atmosphere may have caused the most severe ice age in world history. The glaciation occurred over a period beginning about 1 billion years ago and ending 600 million years ago. At its height, about 800 million years ago, glaciers blanketed the equator at sea level — a phenomenon that apparently has not happened before or since. The massive glaciation — documented by researchers in the multinational Precambrian Paleobiology Research Group –Proterozoic (PPRG–P), headed by geophysicist J. William Schopf of the University of California at Los Angeles — apparently occurred when both the weathering of the soil and the rapid growth of bacteria, algae, and other microorganisms in the ocean depleted most of the carbon dioxide in the atmosphere.

The resulting cold caused the extinction of as many as two-thirds of all species then alive. The PPRG–P researchers speculated that the world was saved from entering a perpetually frozen state only by the release of carbon dioxide from volcanoes, which allowed the planet to warm up again. The massive die-off of microorganisms, according to the researchers, opened new biological niches and set the stage for the sudden appearance of larger, multicellular organisms once the climate returned to normal.

Methane. In recent years, climatologists have discovered that other gases — most notably methane, nitrous oxide, ground-level ozone, and chlorofluorocarbons — are responsible for a large portion of the greenhouse effect. Methane, which contains one carbon atom linked to four hydrogen atoms, is a colorless, odorless, highly flammable gas. It is the principal component of natural gas, and forms when bacteria decompose organic matter in the absence of significant amounts of oxygen. It thus occurs in relatively large quantities wherever matter decomposes underwater, such as in rice paddies and swamps, as well as in the rumens of cattle. Termites also produce it when they digest wood. According to chemist Patrick Zimmerman of NCAR, termites alone — and particularly the 2,400 species that live near the equator — may produce up to 165 million tons of methane per year, roughly half the worldwide total. In comparison, according to chemist Paul W. Crutzen and his colleagues at the Max Planck Institute for Chemistry in Mainz, West Germany, the world's 1.2 billion head of cattle produce some 60 million tons of methane annually, while rice paddies give off roughly 115 million tons.

Atmospheric methane levels have doubled in the last 200 years. Chemist F. Sherwood Rowland of the University of California at Irvine reported in late 1988 that the current concentration of methane in the atmosphere is about 1.7 ppm, and it has been rising at an average rate of 0.018 ppm per year for the last decade.

Nitrous oxide. The atmospheric concentration of nitrous oxide, which consists of two nitrogen atoms bonded to one oxygen atom, was 304 parts per billion (ppb) in 1987, and has been increasing at a rate of about 0.25 percent per year, according to the 1987 UNEP report *The Greenhouse Gases.* Nitrous oxide is formed as a byproduct of combustion and as a result of microbial action in the soil, especially in the presence of nitrogen-rich fertilizers. The increasing use of fertilizers in agriculture and the burning of fossil fuels, timber, and crop residues are thought to account for most of the nitrous oxide entering the atmosphere.

Chlorofluorocarbons and ground-level ozone. CFCs are the only synthetic gases contributing to the greenhouse warming. They are widely used as refrigerants in refrigerators and air conditioners, as cleaning solvents in the electronics industry, and as foaming agents for the production of insulating plastic foams. They have also been used as propellants in spray cans, although the U.S. banned that use in 1978. CFCs are powerful greenhouse gases: one CFC molecule can absorb as much solar radiation as 10,000 molecules of carbon dioxide. In 1987, the lower atmosphere contained 0.63 ppb of CFCs, according to *The Greenhouse Gases.* (For further information about the effects of CFCs, see the article "Depletion of the Ozone Layer" in this chapter.)

The Earth's stratospheric ozone layer shields life at the surface from the Sun's harmful ultraviolet rays. At ground level, however, ozone is an important air pollutant that can reduce crop yields, damage property, and affect human health. Ground-level ozone is also a greenhouse gas that could contribute significantly to global warming.

Predicting the Effects of Climatic Change

Scientists hoping to forecast the effects of increasing levels of carbon dioxide and other greenhouse gases are unable to draw on past experiences because the projected temperature increases are higher than those encountered in recorded history. They must therefore base their predictions on either geological records or computer models simulating global climate.

Modeling the atmosphere. Computer models must consider a large number of variables that influence the Earth's climate, including the behavior of clouds, the mechanisms of ocean circulation, and the various ways in which the ocean affects the atmosphere. As a result, climate prediction models have become enormously complex and generate different conclusions depending on the scientific assumptions used. Nevertheless, most of the major computer models forecast warming in the range of 2.7° to 8.1°F by the year 2030, with wide variations in the regional distribution of those increases.

One of the largest uncertainties involves the response of the oceans to atmospheric warming. Because ocean waters will absorb some of the increased heat from the atmosphere, thereby slowing the global temperature rise, the oceans could delay the warming trend by 10 to 40 years, depending on the extent to which the ocean surface waters mix with deep waters. The U.N. World Climate Programme is sponsoring a study called the World Ocean Circulation Experiment (WOCE), which should help scientists better understand how ocean circulation will influence possible warming. WOCE is expected to provide new insights into the circulation of deep waters, which sink near the poles and travel along the ocean floor for 500 to 2,000 years before surfacing again near the equator.

Preliminary results of one study released in June 1987 indicated that the oceans may absorb more carbon dioxide than scientists had previously thought. Taro Takahashi of Columbia University's Lamont-Doherty Geological Observatory in Palisades, New York, measured the carbon dioxide content of surface water in the North Pacific over the course of a year. Because carbon dioxide dissolves more easily in cold water, Takahashi had expected that relatively cold winter waters would be a stronger "sink" for carbon dioxide than warmer summer waters. His measurements, however, showed exactly the opposite — the waters absorbed the most carbon dioxide between July and September.

Takahashi's discovery implies that, as the atmosphere warms, the oceans may be able to absorb an increasing amount of carbon dioxide. Takahashi speculated that increased photosynthetic activity during the summer allowed plankton to remove more carbon dioxide from the water. As the plankton remove carbon dioxide from the water, more of the gas will be removed from the atmosphere.

Another major uncertainty in modeling the atmosphere centers on clouds, which can either warm or cool the Earth's surface. Much of the sunlight is able to pass through relatively thin high-level clouds and reach the Earth's surface, where it is partially absorbed. When this absorbed solar energy is later radiated from the surface as heat, the high clouds trap much of the heat in the atmosphere. In contrast, low-level clouds are generally thicker than high clouds, and they reflect most incoming sunlight back into space before it can reach the surface, thereby cooling the ground below. Researchers are not sure whether the warming or cooling effects of clouds would predominate if the Earth began to warm up, but they have received some clues from new satellite measurements.

In early 1989, climatologist Veerabhadran Ramanathan of the University of Chicago in Illinois and his colleagues reported preliminary results obtained with a satellite called the Earth Radiation Budget Experiment, launched by the space shuttle in 1984, and two polar-orbiting weather satellites. The group used the satellites to measure the amount of sunlight reaching the Earth and the amount reflected back into space (the Earth's radiation budget). The researchers found that clouds generate a net cooling effect; without them, the Earth would grow significantly warmer. But the scientists cautioned that their finding might not necessarily hold true in the future.

Climatologists have also become concerned about other effects that may play a bigger role than was previously anticipated. These effects include:

- Breakdown of methane hydrates, complexes of methane and water found in arctic permafrost and in mud on the continental shelf. As the planet warms, these hydrates would become unstable and release their methane, thereby accelerating the warming trend.
- Vegetation decay, which will accelerate as the Earth warms. That decay would release more methane and carbon dioxide into the atmosphere.
- Swamps and rice paddies, where increasing temperatures will boost the production of methane more than that of carbon dioxide.
- Increased absorption of the Sun's radiation by land uncovered as the polar caps shrink and by vegetation growing on that land.

A new model prepared by EPA climatologist Daniel Lashof incorporates these effects into the greenhouse scenario and concludes that actual global warming could range between 6.3° and 11.3°F during the next century. "But given the uncertainties, a range as high as...18° cannot be ruled out," Lashof said.

Climate changes. Regardless of the degree of global warming, the greenhouse effect will not cause a uniform rise in temperatures around the world. Instead, a warming will change the Earth's complex climate system in ways that will result in regional changes in temperature, precipitation, and evaporation rates.

Studies of past climates have shown, for example, that during the period about 4,000 to 8,000 years ago, when Earth's climate was about 1.8° to 4.9°F warmer than it is now, world precipitation patterns looked much different than they do today. Large regions of Canada, Scandinavia, and the U.S. received less moisture than they do now, while eastern China, southern Europe, and southern India received more. The Middle East and North Africa also had a wetter climate than they have today. That climate probably benefited agriculture and helped nurture the early civilizations that developed in the region.

Future global warming will probably produce further changes in rainfall patterns, and some evidence even suggests that those changes may have already begun. In mid-1987, a team of researchers headed by climatologists John Eischeid and Henry Diaz of the Environmental Research Laboratories in Boulder, Colorado, reported that rainfall over much of Europe, the U.S., and the U.S.S.R. has increased by more than 10 percent over the last 30 to 40 years, while rainfall in regions closer to the equator, including North Africa and the Middle East, has declined by a similar amount. Those changes match the predictions of most computer models for the early stages of greenhouse warming.

In late 1988, NASA's Hansen and his colleagues predicted, based on their model, that the Midwest farm belt and the southeastern U.S. would feel the effects of greenhouse warming before many other regions of the world. Other regions that will experience early warming include the East Indies, the Bay of Bengal in India, the Mongolian-Manchurian region of Asia, and both polar regions. Interestingly, according to Hansen's model, California and the U.S. Pacific coast will be among the last regions in the world to experience the warming, lagging behind the rest of the country by as much as 25 years.

Hansen's model identified regions of potential early warming by focusing on interactions between the oceans, which warm very slowly, and the land, which warms much more rapidly. In the U.S., for example, this differential warming produces large regions of high atmospheric pressure off the Atlantic and Pacific coasts. Because winds flow clockwise around high-pressure regions in the Northern Hemisphere, high-pressure areas in the Atlantic would bring warm winds blowing from the south into the Southeast and Midwest. A similar clockwise flow around the high-pressure region in the Pacific, however, would carry cool air to California.

Hansen and his colleagues warned that more frequent extreme weather events resulting from greenhouse warming could threaten the biosphere more seriously than the increase in average temperatures. In the past, for example, Omaha, Nebraska, has experienced runs of at least five consecutive days with temperatures above 95°F only three times per decade, on average. The model predicted that such runs will occur five times per decade in the 1990s and seven times per decade by 2020. Such runs of high temperature would be damaging to crops and hazardous to humans, Hansen warned.

Greenhouse warming could also increase the strength of tropical storms and hurricanes, according to meteorologist Stephen Leatherman of the University of Maryland in College Park. In late 1988, Leatherman said that increased water temperatures in the ocean could increase the power of hurricanes. While hurricanes now achieve maximum wind speeds of about 175 miles per hour, according to Leatherman, they could reach a maximum of 220 miles per hour by the middle of the next century.

Agricultural impact. Because most crops grow best within a narrow range of temperatures, precipitation levels, and other climatic factors, any climatic change would greatly affect agricultural production. Models such as Hansen's predict that greenhouse warming will tend to extend growing seasons at high-latitude locations and shift the most productive agricultural areas toward the poles. In North America, for example, the regions with optimal temperatures for corn and soybean growth would tend to shift northward from the U.S. into Canada, while across the Atlantic, the optimal climate for agriculture would shift from central Europe into the U.S.S.R.

In the major wheat-growing areas around the world, such as the U.S. Midwest, a 1.8°F warming could decrease yields by 1 to 9 percent, while a 3.6°F increase could decrease yields by 3 to 17 percent, according to Martin Parry of the Atmospheric Impact Research Group at the University of Birmingham in the U.K. Perry also predicted that the wheat-growing area of Canada would be substantially expanded due to higher winter temperatures and increased rainfall, while yields in Mexico would plummet.

Paul Waggoner of the Connecticut Agricultural Experiment Station in New Haven has found that, if all other factors such as temperature and precipitation remain constant, a higher level of carbon dioxide can increase the photosynthetic activity and growth rate of 20 of the world's major food crops, including rice, soybeans, and wheat. Unfortunately, however, it can also increase the growth rate of many of the world's most damaging weeds. Some types of crops, including corn, sorghum, and millet, do not respond to increased carbon dioxide.

The increased productivity of some crops might, however, be offset by changes in the activity of pests. Biologists David Lincoln of the University of South Carolina in Columbia and Nasser Sionit and Boyd Strain of Duke University in Durham, North Carolina, reported in late 1984 that, with increased concentrations of carbon dioxide, plant pests consume more plant tissue, resulting in a net loss of vegetation. They found that caterpillar larvae eat 80 percent more leaves when the carbon dioxide concentration is 650 ppm than they do when it is 350 ppm.

The biologists attributed the increased voraciousness of the larvae to changes in the nutrient composition of the leaves: When the plants are exposed to higher concentrations of carbon dioxide, their leaves contain higher concentrations of sugars and lower concentrations of essential nitrogen-containing compounds, such as proteins. The leaves thus have less nutritive value than normal, so that the larvae have to consume greater quantities to derive the same nourishment.

Ecological effects. Given the complexity of ecosystems and the many unknown factors — including the rate and magnitude of climatic change and the adaptive capacity of different organisms — scientists cannot reliably predict the ecological effects of greenhouse warming. Some biologists speculate that global warming will greatly influence ecosystems around the world, many of which are already highly stressed by desertification and urban expansion. As the climate changes, the location of such habitats as deserts, forests, prairies, and tundra will shift; species unable to move to new areas or adapt to the changing climate could become extinct.

World Wildlife Fund biologists Robert Peters and Joan Darling reported in late 1985 that climatic change could cause local extinctions in two interrelated ways: the climate of a species' habitat could change until it no longer meets the species' physical needs, or climate change could alter the interactions among species — by allowing a new predator or competitor to thrive, for example. Such a change in species interactions could eliminate a formerly successful species from an area even if it could still tolerate the climate.

If the climate in the U.S. warms, the boreal forests in the upper Midwest (which consist primarily of fir, pine, and spruce) will retreat slowly into Canada, according to biologist Daniel B. Botkin of the University of California at Santa Barbara. They will gradually be replaced by a northern hardwood forest consisting of beech, sugar maple, and yellow birch. At its southern limit, the northern hardwood forest would die, eventually giving way to hickory and oak.

One casualty of these forest changes may be the endangered Kirtland's warbler, Botkin said. This songbird nests only in a specific part of southern Michigan — at the southern edge of the jack pine forest's geographic range — and only in young stands of trees that grow after a fire. According to a computer model developed by Botkin, the jack pine forests in this area will die out as the climate becomes warmer. They will be replaced by white pine and red maple, which will in turn give way to treeless shrubland. When the jack pine dies out, Botkin said, the Kirtland's warbler will disappear.

Researchers believe a variety of other species, especially those restricted to parks or protected areas, will meet the same fate for similar reasons. Species confined to an area because of surrounding development would not be able to migrate to another area with a more favorable climate. Geographic barriers might also hinder the migration of species. Peters and Darling have pointed out that, after the Pleistocene ice ages, a number of plant species found in northern latitudes around the world died out in Europe while surviving in North America, presumably because east-west barriers such as the Pyrenees and Alps blocked the southward spread of the plants' seeds in Europe.

In addition, global warming might create changes in prevailing winds, ocean currents, and water temperatures that could dramatically affect marine ecosystems. As atmospheric and oceanic circulation patterns change, the location of biologically important areas of upwelling (the rise of cold, nutrient-rich waters to the ocean surface) would also shift, perhaps to less accessible regions. Areas of upwelling support some of the world's major fisheries, and reduced accessibility could reduce the world's food supply.

Rising sea levels. Scientists agree that global warming would cause sea levels to rise, both because sea water expands as it warms and because polar sea ice and glaciers would melt. In early 1989, oceanographer Richard Peltier of the University of Toronto in Ontario reported that the level of the ocean is rising 1/12 of an inch per year, about twice as fast as researchers had previously believed. According to Robert Thomas of the Space Department of the Royal Aircraft Establishment in the U.K., global sea level could rise an average of 3 to 5.5 feet by the year 2100, with a rise of 3.5 feet most likely if current projections of greenhouse warming prove correct.

A relatively small rise in sea level would cause coastal flooding, destruction of coastal marshes, shoreline erosion, and an increase in the salinity of bays and rivers, according to James Titus of the EPA. Some low-lying islands, such as the Marshall Islands and the Maldives, could virtually disappear, forcing a relocation of their entire populations to other countries. According to Titus, 16 percent of the people in Egypt currently live in areas that would be lost to flooding if the sea level rose 3.5 feet, and 9 percent of the people in Bangladesh live in similar areas.

While richer nations may be able to build sea walls and dikes to protect their coastal cities, as the Dutch have done for centuries, poorer nations may simply lose some of their land area to the sea. Likewise, popular tourist beaches may be replenished with imported sand, but mangrove areas and marshlands may simply disappear. The EPA has calculated that as much as 80 percent of the coastal wetlands in the U.S. could be destroyed by a 7-foot rise in sea level. Wetlands in Louisiana, which account for about 40 percent of the nation's total, are especially vulnerable to such flooding. Although some wetland areas may become reestablished further inland along the new shoreline, existing coastal development will prevent that in many cases. Because saltwater wetlands serve as a nursery to many of the world's fish species, the loss of those wetlands could lead to a sharp decline in the size of commercial fish harvests.

Forestalling Global Warming

The greenhouse effect may be difficult to prevent because the greenhouse gases are by-products of essential human activities. Many researchers believe, however, that the extent of greenhouse warming can be reduced by adopting strong measures such as increasing the efficiency of energy use, changing fuel sources, reducing CFC production, and perhaps even planting more trees around the world.

Using less energy. An obvious starting point for delaying the greenhouse effect is to cut energy use by increasing energy efficiency, particularly in the U.S., where much energy is simply wasted. According to Scott Denman of the Safe Energy Communication Council in Washington, D.C., the U.S. spends about 10 percent of its gross national product on energy, while both Japan and West Germany, two of the U.S.'s foremost international competitors, spend only about 5 percent. Increased efficiency in the U.S. would thus both mitigate greenhouse warming by reducing the output of greenhouse gases and improve the nation's international competitiveness by lowering energy expenditures.

Energy use could be reduced, for example, by improving the efficiency of electrical motors and appliances, insulating homes better, and increasing the efficiency of power generation through the use of technologies such as cogeneration. Improvements in energy efficiency could readily be achieved in the automobile industry. Even though the U.S. government now requires that new cars in an automobile manufacturer's fleet average 26 miles per gallon, environmentalists urge that the standard be raised to as much as 50 miles per gallon over the next two decades.

Changes in the pattern of fuel consumption could also reduce the greenhouse effect. For a given energy output, burning coal, for example, releases about one-third more carbon dioxide than burning petroleum and twice as much as burning natural gas. The largest users of coal are U.S. utilities, which account for 7.5 percent of the world's carbon dioxide output, according to the Electric Power Research Institute in Palo Alto, California. The utilities have resisted switching to either petroleum or natural gas, however, because both cost more than coal, their prices undergo wide fluctuations, and their availability is subject to disruption — such as during cold spells, when more natural gas is required for heating homes.

Removing carbon dioxide from the exhaust gases of power plants is probably not a realistic option, since the required equipment would at least double the capital cost of the power plants and increase the cost of electricity by 75 percent. Trapping carbon dioxide would also create storage problems for the wastes produced in the process.

Some researchers argue that the increased use of nuclear power plants offers the best solution now available for the problem of carbon dioxide-free generation of electricity. Most argue, however, that the problems of plant reliability and nuclear waste disposal must be solved before the number of nuclear facilities is increased. Others propose more investment in solar energy and related technologies, which also produce no carbon dioxide. Solar energy is still in the early stages of development and is not expected to be widely used at any time in the near future. Nevertheless, the cost of solar energy has been falling, and it is already competitive with fossil fuels in some locations. Interest in solar energy will probably continue to grow as governments and the public become increasingly aware of the environmental costs of fossil fuels.

One idealistic — but perhaps impractical — proposal for mitigating greenhouse warming involves planting large numbers of trees to absorb carbon dioxide. For example, governments might require energy companies that build new power plants to plant a certain number of trees to offset the increased carbon dioxide production. One company — Applied Energy Services of Arlington, Virginia — has already done so, arranging with World Resources Institute for a large forestry project in Guatemala to counter the carbon dioxide emissions from a coal-fired power plant that it built in Connecticut.

A significant impact on the greenhouse effect, however, would require forest plantings on an unprecedented scale. Gregg Marland of the Oak Ridge National Laboratory in Tennessee told the Senate Energy Committee in 1988 that 2.2 acres of sycamores on a plantation in Georgia can absorb about 7.5 tons of carbon every year. About 3 million square miles of trees, covering an area roughly the size of Australia, would thus be necessary to absorb the roughly 5 billion tons of carbon released every year by burning fossil fuels. Marland noted that this area would roughly equal all the tropical forest that has been cleared in the last 10,000 years.

Changing energy and development policies. Strong measures such as increased energy efficiency, protection of tropical forests, and a reduction of CFC emissions could postpone the greenhouse effect and limit its impact, according to a 1987 study by Irving Mintzer of the World Resources Institute. While most computer models have focused on the extent of warming or the possible effects of climate change, Mintzer's model considers the impact of different energy use policies on greenhouse warming.

Mintzer's model analyzed four policy options, ranging from the "High Emissions Scenario" in which government policies encourage coal use, tropical deforestation continues, and CFC production proceeds unabated, to the "Slow Buildup Scenario," in which strong government policies increase energy efficiency, promote solar energy alternatives, control deforestation, sponsor massive reforestation projects, and freeze CFC production at its 1985 level. The "Base Case Scenario" projected a continuation of present policies, while the "Modest Policies Scenario" projected moderate energy conservation and tropical reforestation policies. All scenarios assumed the same level of global economic growth.

The report concluded that, "depending on which policies are adopted, the year when we are irreversibly committed to a warming of [2.7° to 8°F] above the pre-industrial temperature varies by approximately six decades." In the High Emissions Scenario, the Earth reaches this point in 2015. In the Slow Buildup Scenario, an equivalent warming is postponed until after 2075. According to the High Emissions Scenario, the Earth would undergo a buildup of greenhouse gases that would cause a warming of 9.5° to 29°F by 2075. By contrast, the Slow Buildup Scenario would commit the planet to a warming of only 2.5° to 7.5°F by the same date. According to Mintzer, "Clearly the magnitude and timing of planetary warming can be substantially affected by policy choices made now and implemented over the next several decades."

Acid Rain

In the isolated Vermillion Bay region of Ontario, 5 hours' drive from Winnipeg, researchers from Canada's Department of Fisheries and Oceans have been deliberately pouring sulfuric and nitric acids into a few once-pristine lakes in an unprecedented attempt to chart the biological effects of acid rain — a pollution problem that is destroying lakes, streams, and forests around the world. This Canadian study has shown how a complex ecosystem can be rapidly damaged by increased acidity.

During the first 3 years of the study in the early 1980s, Lake 302, for instance, changed very little, according to biologist David Schindler, who has led the research project at Vermillion Bay for more than 20 years. By the fourth year, however, Lake 302's natural ability to neutralize acid had been exhausted, and the water was about five times more acidic than normal. Freshwater shrimp had disappeared, and fathead minnows failed to reproduce. While phytoplankton — the microscopic plant life at the base of the food chain — still survived, the distribution of species had changed.

By years 5 and 6, Lake 302's acidity had risen to about 10 to 12 times normal, and mats of slimy algae blanketed its surface during the summer. Crayfish became vulnerable to parasites, suffered skeletal damage, and developed brittle egg-holding stalks. Lake trout had become emaciated, looking more like eels than trout. By year 7, when the lake's acidity was about 30 times normal, the trout had stopped reproducing altogether. Within another 3 years, all fish species had disappeared from Lake 302, as well as all crayfish, leeches, and mayflies. The water appeared clear, but in fact the lake had become sterile and lifeless.

This type of biological devastation has been repeated thousands of times worldwide every year — not in scientific experiments, but as a by-product of industrial activities, such as smelting, and the widespread use of fossil fuels for producing electricity and powering automobiles and trucks. In eastern Canada alone, sulfuric and nitric acid by-products of fossil-fuel combustion have so acidified 15,000 lakes that they no longer contain fish, according to Canada's Prime Minister Brian Mulroney.

Tens of thousands of other lakes in the region could suffer the same fate in the early 1990s if acid rain continues unabated. Acidic pollutants are killing fish and other aquatic organisms in lakes and streams, promoting the undesirable growth of algae, accelerating the loss of forests already stressed by other types of air pollution, and endangering the health of humans who must breathe contaminated air.

Unlike many other forms of pollution, which affect primarily the locations where they originate, acid rain can fall hundreds of miles from the pollution source, often crossing national boundaries. As much as half the acid rain descending on Canada, for example, results from pollutants produced in the U.S., while Canadian sources themselves account for 10 to 15 percent of the U.S. problem. Similarly, according to Sweden's Ministry of Agriculture, Norway receives three times as much sulfuric acid as it produces, and Sweden receives 1.7 times as much. Most of that pollution originates in central Europe.

The acid rain problem is not confined to wealthy, industrialized countries. A study by UNEP and the International Council of Scientific Unions demonstrated a significant acid rain problem in the People's Republic of China, which has the highest level of sulfur emissions in the world. Large forested areas in southwestern China, which are particularly vulnerable to acid rain because the region's soils and water have a naturally high acid content, have been damaged by sulfur emissions from nearby towns and industry.

What Is Acid Rain?

Although the term "acid rain" is commonly used to describe acid precipitation falling as rain, scientists generally prefer the term "acid deposition," because snow, fog, sleet, and hail can also contain acids. The phenomenon does not even require moisture; in a process called dry deposition, gaseous nitric and sulfuric acids — either as free-floating gas or attached to tiny particles — can land on leaves, soil, water, or building surfaces. According to the 1988 annual report of the U.S. National Acid Precipitation Assessment Program (NAPAP), dry deposition accounts for at least half the total acid deposition in the U.S., and for as much as 90 percent in some arid areas, such as the Southwest.

In the atmosphere, a series of reactions converts gaseous or dissolved sulfur dioxide (SO_2) into sulfuric acid (H_2SO_4) and nitrogen dioxide (NO_2) into nitric acid (HNO_3). The gaseous reaction takes place so slowly that, in the absence of precipitation, sulfuric acid can form far downwind from the sulfur dioxide emission source, according to the 1988 World Resources Institute report entitled *Ill Winds: Airborne Pollutants' Toll on Trees and Crops.*

Sulfur oxides are produced by the burning of fossil fuels containing sulfur; nitrogen oxides are created when nitrogen and oxygen in the air are exposed to the high temperatures of fossil-fuel combustion. The 1987 NAPAP report, *The Causes and Effects of Acid Deposition*, estimated that sulfur oxides come primarily from power plants (65 percent), but also from industrial processes (13 percent), transportation (3 percent), and other combustion sources (18 percent). Nitrogen oxides, in contrast, are produced mainly by transportation activities (41 percent), and power plants (29 percent), with the rest resulting from industrial processes (3 percent) and other combustion sources (24 percent).

Natural events, such as volcanic eruptions and forest fires, release substantial amounts of sulfur and nitrogen compounds into the atmosphere. In addition, natural microbial processes in the oceans, salt marshes, and coastal mud flats generate gaseous sulfur compounds by chemically reducing the sulfates in seawater. Lightning naturally generates nitrogen oxides when its heat causes atmospheric nitrogen and oxygen to combine. Soil bacteria also release nitrogen oxides.

In its 1987 *State of the Environment* report, UNEP concluded that, for the world as a whole, natural emissions of sulfur and nitrogen roughly equal manmade emissions. According to UNEP, however, the relative contributions of natural and manmade sources differ dramatically in areas such as Europe and North America. In 1983, Sweden's Ministry of Agriculture estimated that about 90 percent of sulfur deposition in northern Europe is produced by human activities. Similarly, the 1987 NAPAP report estimated that natural sources account for only 6 percent of North America's total sulfur dioxide emissions and 12 percent of its nitrogen oxide emissions.

Measuring acidity. Scientists use the pH scale to describe the acidity or alkalinity of a substance. The pH is the negative logarithm of the concentration of positively charged hydrogen ions in a substance. (Ions are atoms or molecules that have acquired a net positive or negative electric charge by the loss or gain of one or more electrons.) Substances with a pH value of 7.0 are neutral, those below 7.0 are acidic, and those above 7.0 are alkaline. Because the pH scale is logarithmic, every decrease or increase of one unit on the scale corresponds to a 10-fold increase or decrease in acidity. Consequently, a drop in pH from 6 to 5 corresponds to a 10-fold increase in acidity, while a drop in pH from 6 to 4 corresponds to a 100-fold increase.

Scientists have calculated that the pH of natural, unpolluted rainwater should be about 5.6 (slightly acidic), because atmospheric carbon dioxide dissolves in rainwater to form a weak solution of carbonic acid. The general term "acid rain" thus refers to precipitation with a pH below 5.6.

History of acid rain. People have recognized the problem of air pollution from human activities for at least 700 years. In England, coal smoke darkened the sky at least as early as the thirteenth century, and the problem reached serious proportions by the seventeenth century. In the book *Fumifugium* published in 1661, John Evelyn reported that coal-related pollution in London was causing corrosion in structures and materials and harming plants and animals. English demographer John Graunt noted in 1662 that the rapid increase in coal use after 1600 was partially responsible for the increasingly unhealthy conditions in the London area.

English chemist Robert Angus Smith probably recognized the effects of acid rain before anyone else. In his article "On the Air and Rain of Manchester," published in 1852, Smith noted that sulfuric acid pollution had caused metals to rust and colors in prints and dyed goods to fade. In 1872, Smith wrote *Air and Rain*, in which he coined the

Figure 3. World's Tallest Smokestack. This 1,250-foot-high smokestack at Inco Ltd.'s nickel smelter complex in Sudbury, Ontario, was until recently one of the largest point sources of sulfur dioxide emissions in North America, releasing about 2,000 tons of sulfur dioxide per day into the air. Although very tall smokestacks are generally built to reduce air pollution close to the ground, they may be responsible for delivering sulfur dioxide and nitrous oxides to altitudes where the emissions can be transported long distances before being deposited as acid rain, acid snow, or dry particles on the ground. Inco is implementing pollution abatement measures, and aims to contain 90 percent of its sulfur dioxide emissions by 1994. *Courtesy: Inco Ltd.*

term "acid rain" to describe precipitation with a high sulfuric acid content. He noted that acid rain often fell on urban areas, particularly those whose industries burned coal with a high sulfur content.

Modern-day sulfur emissions. East German scientist Detlev Moller estimated in a 1984 article in *Atmospheric Environment* that manmade global emissions of sulfur in the early 1980s had reached about 90 million tons per year. Worldwide, according to Moller, 53 percent of sulfur dioxide emissions were derived from burning coal, 28 percent from oil, 11 percent from lignite (a type of coal), and the remaining 8 percent from smelting and other activities. He noted, however, that the portion of sulfur dioxide emissions from different sources varies widely among countries.

Sulfur dioxide emissions in the U.S. peaked with heavy coal use during World War II and then again in the early 1970s when emissions reached 30 million metric tons per year, according to the 1987 NAPAP report. Since the Clean Air Act was passed in 1970, however, manmade emissions of sulfur dioxide have decreased, dropping to 22 million metric tons by 1985. The decrease in sulfur dioxide emissions resulted from the installation of pollution-control equipment on new power plants, and increased use of the low-sulfur fuels. It occurred despite the fact that U.S. coal consumption actually rose during this period.

Although the total release of sulfur dioxide has decreased, the damage caused by acid deposition has increased, largely due to the introduction of tall stacks on power plants as a method of dispersing air pollution. The tall stacks inject sulfur dioxide and nitrogen oxides high into the atmosphere, where air currents carry them long distances before depositing them on lakes, forests, and fields far from their sources, according to William Moomaw, a senior researcher at the World Resources Institute.

Ecological and Health Effects

The effects of acid deposition depend on the geology and biology of the affected region. Alkaline soils can withstand acid deposition without ill effects for a longer time than less alkaline soils, because the alkaline substances neutralize the acid. Similarly, limestone and other carbonate rocks neutralize acid, while rocks such as quartz do not.

Because the topography of a region can influence local weather patterns, it can also determine how seriously acid rain affects the flora and fauna. For example, a watershed with rivers running down steep, rocky slopes or extending over a large area might concentrate a large amount of acid rain in a lake. In contrast, deep soils give trees a reserve of nutrients, better enabling them to withstand the effects of acid deposition.

A combination of climatic and geologic factors have made many parts of the U.S., eastern Canada, Scandinavia, and various European countries particularly vulnerable to acid deposition. Each of these areas is located near major sources of pollution. The prevailing winds carry heavy concentrations of pollutants into the regions, and precipitation helps deposit pollutants there. The soils and bedrock in these regions also have only a limited natural capacity to neutralize the acid.

Lakes, rivers, and streams. Some of the best-known effects of acidification were observed in lakes in New York, Ontario, and Scandinavia in the 1960s and 1970s. When acidified snow melts during the spring, large volumes of acidic water enter such lakes, causing their pH to drop rapidly for a period of a few days to a few weeks. During the spring snow melt, the waters in Panther Lake in New York's Adirondack Mountains, for example, can become 100 times as acidic as normal, with the pH dropping from 7 to 5. Such sudden and short-lived increases in acidity can produce severe and potentially lethal chemical shocks to aquatic life.

In regions with chronic acid deposition, the pH of many lakes declines and their water chemistry changes, as the Vermillion Bay project demonstrated. When phytoplankton and zooplankton, which give lake water some of its color, die, their absence increases the water's transparency. Also, tannins and other organic compounds from plants tend to lose color as acidity increases. Ironically, as the water becomes clearer and looks cleaner, the lake is actually dying.

When the pH of a lake or river falls below 6.5, females of some fish species may produce eggs that fail to hatch. Fish fry that hatch in waters with a low pH have a high mortality rate, and many are born with severe deformities. Most adult fish can survive in waters with a pH of 6.0, but the number of species that can do so decreases rapidly as the pH drops below this value. Increases in acidity also disrupt the internal chemistry of adult fish, depleting calcium from their bones and tissues. As calcium is lost, a fish's skeleton is less able to retain its form. In the worst cases, a fish's movements can actually pull its skeleton out of shape, leading to such deformities as humped backs.

Falling pH levels also cause aluminum, nickel, zinc, and other metals to be released into the lake water from the surrounding acidified soils and lake sediments. The aluminum can irritate fish membranes and cause mucus discharges that clog gills and interfere with respiration. In addition, at a pH of 5.0 to 6.0, aluminum hydroxides precipitate out of the water as a white solid that also clogs fish gills.

As the pH of a lake falls below 6.0, the lake ecosystem is seriously disturbed; clams, snails, and crayfish die, followed by many important aquatic insects, including the mayfly, damselfly, and dragonfly, and, finally, some species of fish. At pH values below 5.0, nearly all the commercial and sport fish species disappear, and as the pH drops below 4.5, only acid-insensitive algae, fungi, and moss growing on the lake bottom and acid-insensitive animal and plant plankton remain in the lake.

A growing number of fragile lakes have suffered severe damage due to acid deposition during the past few decades. In a 1984 report entitled *Acid Rain and Transported Air Pollutants*, the U.S. Office of Technology Assessment (OTA) concluded that about 17,000 lakes and 112,000 miles of streams in the 31 states that border or lie east of the Mississippi River are at risk. Of these, about 20 percent — 3,000 lakes and 23,000 miles of streams — are extremely vulnerable to any additional acid pollution or have already become acidic.

While thousands of lakes in Quebec and Ontario have become so acidified that they no longer contain fish, several times that number may soon suffer a similar fate. In Nova Scotia, some 13 rivers that once supported large salmon populations have become barren because of increasing acidity, and several other rivers are threatened with a similar fate.

The Swedish government reported that, as of 1986, more than 40 percent of the country's 85,000 lakes had a pH lower than 5.9. The extent of acidification has been increasing rapidly in northern Sweden, which has in the past been less severely affected than southern areas.

The EPA has estimated that about 7,000 lakes in the western U.S. have bedrock, soil, terrain, and alkalinity characteristics that put them at risk of rapid acidification. A 1985 report by the World Resources Institute indicated that mountains in the western U.S., with generally steeper slopes and thinner soil bases than the mountains in the East, allow acid rain to flow quickly into lakes and streams before it can be neutralized.

These conclusions about the West were supported in a 1985 report by the Natural Resources Defense Council in New York City and the National Clean Air Fund in Washington, D.C. "The most sensitive areas of the West coincide with our 'crown jewel' national parks and dozens of other resource areas," according to Deborah Sheiman, the report's author. She noted that the water in 55 U.S. national forests, 23 wilderness areas, and 11 national parks and recreation areas does not contain minerals that would neutralize acid rain. In addition, according to Sheiman, most of these areas are receiving abnormally acidic rainfall. The most vulnerable sites include the Glacier Peak Wilderness area in Washington, Mount Hood in Oregon, Yosemite National Park in California, and Rocky Mountain National Park in Colorado.

Forests. The rapid and widespread death of forests in Central Europe and the decline of forests in eastern North America have focused attention on the impact of acid rain and other pollutants on woodlands. Over the past 50 years, 19 forest "die-backs" — the death of large numbers of trees — have occurred. Scientists have implicated air pollutants in seven of these die-backs, according to *World Resources 1986*. All but one of the pollution-related die-backs took place after 1978. The Earthscan news service in London, U.K., has reported that about 17 million acres of forests have been damaged in 15 European countries.

Scientists generally agree that forest decline is the result of many interacting factors, and that air pollution is sometimes a major contributor to the problem. Acidity, along with high levels of manmade ozone, damages tree foliage directly by oxidizing certain compounds inside leaf cells. Acid deposition also leaches essential nutrients, such as calcium, magnesium, phosphorus, and potassium, from the soil, and it releases naturally present toxic metals such as aluminum and manganese, which are then absorbed by tree roots. The aluminum destroys the fine feeder roots, thus reducing a tree's total root mass. As the trees weaken, their resistance to fungi, viruses, and other pathogens also decreases.

The gradual weakening and death of forests has been linked to acid rain and other pollutants in many parts of Europe and North America. West Germany's Federal Ministry of Food, Agriculture, and Forestry reported in 1984 that about 50 percent of West Germany's forests had suffered damage, and by 1986, that percentage had risen to 54 percent — nearly 10 million acres. The worst-affected areas are located in Bavaria and the regions of Baden and Württemberg, where the Black Forest is located. Natural stresses, including insect pests and abnormal weather conditions, have aggravated the damage caused by acid rain and other air pollutants.

According to *World Resources 1988-89*, a 1987 study of forest damage in Europe from air pollutants and other stresses found that almost 15 percent of the timber volume in 17 countries had suffered moderate or severe damage. Damage appeared particularly severe in Switzerland (25 percent of the national timber volume affected), West Germany (21 percent), the Netherlands (20 percent), and Sweden (20 percent).

In Czechoslovakia, the most seriously affected country in Eastern Europe, pollution from high-sulfur coal burned at power stations has destroyed a significant percentage of the nation's forests. In 1983, the Czechoslovak Academy of Sciences estimated that 500,000 to 750,000 acres of forests had already died, with up to 2.5 million acres at risk of further damage. The scientific group projected that, if the trend continued, over half of Czechoslovakia's forests would be dying by the end of the century.

Although evidence from the U.S. and Canada is still inconclusive, many researchers believe that symptoms of forest decline in eastern North America may signal the onset of problems similar to those in Europe. For instance, botanists have reported that sugar maple trees are suffering from die-back in parts of Quebec. When the decline was first observed in 1982, an aerial survey revealed that 32 percent of the trees were damaged or dead in some locations. By 1986, 82 percent of the trees had been affected. An Ontario government committee concluded that "serious and widespread soil and forest effects are

Figure 4. Acid Rain Damage in Vermont Forest. These red spruce trees near Montpelier, Vermont, are suffering from a condition known as die-back, which may be caused in part by high soil acidity due to acid deposition. *Courtesy: David Like.*

expected over the next 25 to 100 years if the acidity of precipitation remains at a constant level."

Crops. Tests conducted by EPA on the sensitivity of 35 crop varieties to simulated acid rain indicated that five varieties suffered foliage damage at pH 4.0; a total of 28 varieties suffered damage at pH 3.5; and 31 suffered damage at pH 3.0. The tests revealed that, under mildly acidic conditions, the yields of beets, broccoli, carrots, mustard greens, and radishes would decrease, while the yields of alfalfa, green peppers, orchard grass, strawberries, and tomatoes would actually increase, probably because of the fertilizing effect of the nitrogen and sulfur compounds in the rain. The results for the other crops under mildly acidic conditions were inconclusive.

Damage to materials. Acid deposition increases the rate at which such materials as aluminum, copper, nickel, steel, limestone, marble, leather, paper, plastics, and textiles corrode or deteriorate. The acids in acid rain literally dissolve some stones and metals.

Acid rain and other types of air pollution ha /e taken a particularly heavy toll on works of art and architecture. The Caryatids — marble statues of six maidens carved by the sculptor Phidias in the fifth century B.C. — were recently moved from the Acropolis in Athens, Greece, to a museum and replaced with replicas because acid rain had eroded them so severely.

The surface of Cleopatra's Needle, an Egyptian granite obelisk carved in about 1500 B.C., began flaking within a few years after the monument was relocated to heavily polluted New York City; it had survived virtually unchanged in Egypt for about 35 centuries. Acid rain has also been rapidly defacing the Parthenon in Athens, the Colosseum in Rome, the Taj Mahal near Agra, India, and many other historic buildings, monuments, and statues around the world.

Visibility problems. According to the 1984 OTA report *Acid Rain and Transported Air Pollutants*, "episodic regional haze over large segments of the East tends to curtail some segments of general aviation aircraft and slow commercial, military, and other instrument flight operations on the order of 2 to 12 percent of the time during the summer." Although the effect of atmospheric sulfates on visibility varies from region to region, visibility reductions correspond roughly with increases in sulfate concentrations. The greatest deterioration in visibility has occurred in the Southeast — one of the few regions in the U.S. where total sulfur emissions have risen in recent years.

In 1986, the National Park Service reported that sulfates are the single most important contributor to reduced visibility in national parks. Sulfates in the eastern U.S. account for about 70 percent of the visibility impairment in the summer and 50 percent for the entire year, according to OTA.

Human health effects. Although scientists have developed little firm evidence linking acid rain to human illness, they fear that breathing acidic pollutants can harm people who already suffer from asthma, bronchitis, and heart disease. In fact, researchers believe that high acid levels were responsible for the deaths associated with episodes of high pollution in the Meuse Valley of Belgium in 1930; in Donora, Pennsylvania, in 1948; and in London during the early 1950s.

Japanese researcher Tetsuzo Kitagawa reported in 1984 that acid pollution caused 600 cases of severe lung disease over a period of 8 years in a small sector of the city of Yokkaichi in central Japan. All the victims lived close to a titanium dioxide pigment plant that emitted 100 to 300 tons of sulfuric acid into the atmosphere each month. The average acid concentrations in the air and the incidence of lung disease declined with increasing distance from the plant. Most convincingly, new lung disease cases fell sharply when the plant installed emission control equipment to trap the sulfuric acid.

Epidemiologist David Bates of the University of British Columbia in Vancouver has also established evidence suggesting a link between acid rain and health. Bates studied hospital admissions in 79 hospitals in southern Ontario and compared them with pollutant concentrations at 15 monitoring stations in the area. He found that admissions for respiratory diseases in the summer correlated with sulfur dioxide and ozone concentrations and with temperature. The strongest correlation was between sulfuric acid concentrations and asthma admissions, particularly among young people. Bates did not, however, find any correlations for nitrogen oxides or other pollutants. Drawing on Bates's study and others, OTA has estimated that as many as 50,000 premature deaths per year in the U.S. and Canada might be attributed to sulfates and other particulates, although this estimate has a high degree of uncertainty.

Acidification of water supplies is also a potential health hazard, because the solubility of many toxic metals, such as aluminum, cadmium, lead, and mercury increases sharply with decreasing pH. Acidic water can thus leach metals from soil, lake sediments, metal pipes used in water systems, and the soldering materials used to join pipes together. It can also free asbestos fibers from the cement-asbestos pipes used in some water systems.

Lead pollution is a widespread problem with well-documented adverse health effects. While researchers have not yet found evidence of increased lead concentrations in large public water systems in the U.S., they have detected high lead levels in some smaller systems. William Sharpe of Pennsylvania State University in University Park concluded that about 1 percent of homes in the eastern U.S. with private water supplies use cisterns that hold water collected with roof catchment systems. A substantial percentage of those holding tanks are lined with lead, and not surprisingly, Sharpe found high concentrations of lead in the water. Many older homes in the U.S. also use lead or copper pipes joined together with lead solder, which can present problems if the pipes carry acidic water.

Michael Moore of the University of Glasgow in Scotland has found high lead levels in the water supplies of Glasgow and Ayr, both of which are located in an area with the worst acid rain in the U.K. The pH of water in the Glasgow reservoir is about 4.3, while that in the Ayr reservoir is between 4.5 and 5.0. In both places, authorities must add lime to the water to reduce its acidity and lead content. The water's high lead content has not yet been firmly linked to any health problems in either area, however.

Economic costs. Economists cannot easily estimate the actual and potential costs of acid deposition because scientists still do not agree about the effects of acid deposition on ecosystems, people, and materials. UNEP's 1987 *State of the Environment* report, however, estimated that the annual materials damage from acid rain in industrialized countries is between $2 and $10 per capita (between 0.10 percent and 0.23 percent of the gross national product).

Combating Acid Rain

In its first major review of national strategies to curb air pollution, the Economic Commission for Europe (ECE) reported in 1987 that 23 of the 31 ECE countries reduced their sulfur dioxide emissions between 1980 and 1985, with 10 countries achieving reductions of 30 percent or more. Eleven countries in Western Europe have said that they planned to reduce their sulfur dioxide emissions 50 percent or more by 1995. Countries outside the ECE are also reducing their emissions. According to UNEP's Global Environmental Monitoring System, which has collected data at 52 sites around the world, the average annual concentrations of sulfur dioxide at those sites fell from 73 micrograms per cubic meter of air in 1973 to 40 micrograms per cubic meter in 1980.

Nitrogen oxide emissions remained steady or declined only slightly in most European and North American countries between 1980 and 1985, according to the ECE review. Trends in nitrogen oxide emissions vary widely among countries. Only Japan has shown a marked decrease in nitrogen oxide emissions between 1960 and 1980, primarily because of strong emission controls on both vehicles and power plants there. Japan achieved the reduction despite a dramatic increase in the number of vehicles in use, according to the Japan Environment Agency's 1982 report *Quality of the Environment in Japan*. In the U.S., nitrogen oxide emissions have tripled since 1940 because of the increased use of automobiles but are now declining slightly, according to EPA data.

Uncertainties about how sulfur and nitrogen compounds are transformed into acid rain have slowed or prevented the adoption of acid-rain legislation in some nations. Opponents of controls on sulfur and nitrogen oxide emissions have claimed that reduction of emissions may not necessarily lead to a proportional decline in acid rain. By the late 1980s, however, scientists had marshaled evidence supporting a direct link between changes in sulfur dioxide and nitrogen oxide emissions and changes in acid deposition rates. One study showed that, between 1965 and the early 1980s, sulfur dioxide emissions in the northeastern U.S. fell about 40 percent. During that same period, sulfur deposition at the Hubbard Brook Experimental Forest in New Hampshire dropped 33 percent.

This study led a National Academy of Sciences (NAS) panel to conclude in 1983 that a 50 percent reduction in sulfur dioxide would lead to about a 50 percent drop in sulfuric acid deposition. When the NAS evaluated more recent evidence in 1986, it again found a strong association linking sulfur dioxide emissions, atmospheric concentrations of sulfate particles, visibility ranges, sulfate concentrations in precipitation, and sulfate levels in streams monitored by the U.S. Geological Survey.

The NAS report also noted that the highest sulfur emission and sulfate deposition rates are found in the Northeast and Midwest. Members of the U.S. Congress drew heavily on this report when drafting proposed U.S. acid-rain legislation that would control the emissions of sulfur dioxide and nitrogen oxides. As of mid-1989, however, the legislation had not been passed.

The most effective way to control acid deposition is to reduce overall energy use. This can be achieved by harnessing more of the energy in fuels, by using energy-efficient machinery, vehicles, and consumer goods, and by adopting a variety of energy conservation measures. (For more information about the effects of reducing energy use, see the "Greenhouse Effect" article in this chapter.) The substitution of natural gas, which is virtually sulfur-free, for coal and oil could also reduce acid deposition.

Low-sulfur fuels and clean coal. Conversion to low-sulfur fuels in plants that now rely on high-sulfur fuels is another direct way to reduce sulfur emissions. The sulfur content of coal varies from about 0.4 to 5 percent. Hard coals, such as anthracite, are usually low in sulfur and are most desirable for use in power plants. Soft coals, such as bituminous, sub-bituminous, and lignite (brown coal), generally contain more sulfur, but cost less than anthracite and are more widely available.

Switching to low-sulfur coal could reduce sulfur emissions at coal-burning facilities by 30 to 90 percent. As part of an effort to reduce sulfur emissions 65 percent by 1995 and 80 percent by the year 2000, the Swedish government has prohibited the burning of any coal or oil that causes the emission of more than 0.05 gram of sulfur (for new coal-burning plants or existing noncoal plants) or 0.10 gram of sulfur (for existing coal-burning plants) per one million joules of energy produced.

The use of low-sulfur fuels, however, can add significantly to the cost of power generation, because new transportation lines might have to be built to carry low-sulfur coal from distant sources to local power plants. Engineers would also need to modify the equipment in some power plants in order for them to burn low-sulfur coal — a measure that would increase the cost of generating power.

Another approach to reducing emissions involves cleaning the coal before it is burned. Coal contains two kinds of sulfur: pyritic sulfur, which is contained in small particles of the mineral iron pyrite dispersed throughout the coal, and organic sulfur, which is chemically bonded to the carbon in coal. Current commercial coal-cleaning techniques involve physically separating the relatively heavy mineral particles from the coal. Because this method deals only with pyritic sulfur, it can remove only 10 to 30 percent of the total sulfur in the coal, according to the 1987 NAPAP report. Organic sulfur can be removed only by chemical treatment under high temperature and pressure. This process, still in the experimental stage, may eventually remove 90 to 95 percent of the total sulfur, but at a high cost.

Improved combustion technologies. Many plants that burn coal reduce emissions by flue-gas desulfurization, a process in which scrubbers remove 90 to 95 percent of the sulfur dioxide from the exhaust gas before it leaves the smokestack. The equipment for this process can be built into new plants or added to existing plants. In general, flue-gas desulfurization involves spraying an absorbent substance, typically lime or limestone, into emission gases in the stack. The absorbent removes the sulfur dioxide from the emission mixture.

Several new technologies are now providing further reductions in sulfur emissions at comparable costs. Both integrated gasification combined cycle (IGCC) and atmospheric fluidized bed combustion (AFBC) are now being tested on a commercial scale. A third new technology — pressurized fluidized bed combustion (PFBC) — which removes more pollutants than AFBC, is just entering commercial-scale testing.

Unlike scrubbers, IGCC equipment cleans the coal before combustion. A coal/water slurry is injected into a pressurized gasifier, where the coal is partially oxidized at high temperature and pressure to produce "synthesis gas," consisting of hydrogen, carbon monoxide, carbon dioxide, hydrogen sulfide, carbonyl sulfide, and nitrogen. After cooling and processing, the gas is passed through a solvent that absorbs the gaseous sulfur compounds. The clean fuel gas can then be burned in a gas turbine to produce electricity.

PFBC uses a special technique for burning coal to prevent sulfur emissions: the bottom of the firebox is filled with inert granular particles of sand and limestone suspended by air blown in through holes in the bottom of the firebox. Sulfur from the coal burned in a fluidized bed reacts with the limestone particles to form a heavy, dry sulfate, which is trapped in the firebox. PFBC takes place under 8 to 10 atmospheres of pressure and requires a smaller firebox than that used in AFBC.

Costs of sulfur reduction. In its 1987 report *America's Clean Coal Commitment,* the U.S. Department of Energy (DOE) estimated that, between 1975 and 1985, utility companies spent $62 billion to reduce sulfur dioxide emissions. Of this total, they spent $11 billion on coal cleaning, $17 billion on flue gas desulfurization, and $34 billion on switching to cleaner fuels, according to the report.

The 1987 NAPAP report estimated the added costs of reducing sulfur dioxide and nitrogen oxide emissions with AFBC, PFBC, and IGCC systems installed at coal-powered plants. The cost of removing 1 metric ton of sulfur dioxide or nitrogen oxide by the systems varied substantially, depending on both the technology used and the type of coal burned. The cost of removal at a 500-megawatt plant burning coal with a 3.5-percent sulfur content was estimated to be about $305 per metric ton for AFBC, $560 per metric ton for PFBC, and $380 per metric ton for IGCC. If coal with a sulfur content of 2.0 percent was used, the cost of removing 1 ton of sulfur dioxide or nitrogen oxide rose to about $380 for AFBC, $795 for PFBC, and $630 for IGCC.

Removal of nitrogen oxides. In the U.S., the emission of nitrogen oxides from new vehicles has fallen from an average of 3.0 grams per mile in 1973 to 1.0 gram per mile in 1987. According to the 1987 NAPAP report, additional significant reductions are unlikely without fundamentally new technology or alternative fuels. The use of alternative fuels, such as methanol or compressed natural gas, is now being considered as a way to reduce pollution, although the overall effect that these fuels will have on air quality is still unclear.

An interesting new technology, developed at Sandia National Laboratories in Albuquerque, New Mexico, could reduce nitrogen oxide emissions from diesel trucks and cars by 99 percent. It involves injecting cyanuric acid (a chemical commonly used in chlorinating swimming pools) into a vehicle's diesel exhaust system, where the heat vaporizes it. The gasified cyanuric acid then reacts with the nitrogen oxides in the exhaust system and converts them to harmless nitrogen. A modified version of this approach might also benefit power plants and other stationary sources of nitrogen oxides.

Fuel Tech, Inc., of Stamford, Connecticut, is developing a similar process to reduce nitrogen oxide emissions from stationary sources. This new process reduces emissions 75 percent at a fraction of the cost of the current best available technology.

Liming. Liming, which entails dumping limestone or other alkaline materials into water to neutralize acids, is now being used to lower the acidity of lakes and rivers. Studies by EPA have shown that liming can help prevent acid rain-related damage to aquatic plants and animals. It can also cause aluminum and other metals to precipitate from the water onto the lake bottom or river bed, thereby reducing the risk for organisms living in the water. The results of liming are not permanent, however, and the process must be repeated every few years. It also costs an estimated $50 for every acre of water surface treated.

Sweden began liming its lakes in 1976, and by 1986 had treated about 3,000 lakes at a total cost of about $25 million. Swedish officials acknowledge that liming provides only a stopgap measure for acidification. It does not restore lakes to a pristine state, and is not practical for use in the thousands of Swedish lakes in remote and inaccessible areas.

International Agreements to Limit Emissions

In recent years, government leaders have become increasingly aware that air pollution control requires international cooperation. According to the 1972 Stockholm Declaration, which was developed at a U.N.-sponsored conference, nations have "the responsibility to ensure that activities within their jurisdiction or control do not cause damage to the environment of other states or areas beyond the limits of national jurisdiction."

A decade after the Stockholm Declaration, the Swedish government sponsored the Stockholm Conference on the Acidification of the Environment, at which government officials and scientists from 22 developed countries, principally in Europe and North America, met to discuss acid deposition. Both the government officials and the scientists issued recommendations that, if implemented, would significantly reduce the problem.

The scientists agreed on the following points: (1) existing, commercially available technology could dramatically reduce the amount of sulfur dioxide and nitrogen oxide emissions; (2) the prospect of improved emissions-control technology in the future should not justify delaying actions to reduce current emissions; (3) any decrease in deposition will benefit the environment; and (4) sulfur deposition should not exceed 0.5 gram of sulfur per square meter (about 4.5 pounds per acre) per year in sensitive aquatic areas.

This recommended maximum level of sulfur deposition is significantly lower than the annual deposition levels of 2 to 3 grams per square meter in Nova Scotia and parts of Sweden, the 4.3 grams per square meter in sections of New York's Adirondack Mountains, the more than 5 grams per square meter in large areas of central Europe, and the 10 grams or more per square meter in some forests of West Germany. According to deposition figures published in 1983 by the Swedish Ministry of Agriculture, the annual sulfur deposition per square meter averages about 5 grams in West Germany, 7 grams in East Germany, and 10 grams in Czechoslovakia.

At the Stockholm Conference, government officials also agreed to eliminate the use of tall smokestacks, because these stacks reduce local pollution at the expense of neighboring areas or nations, and to support programs for using the best available technology to reduce both sulfur dioxide and nitrogen oxide emissions.

European agreements. Although 34 countries signed the ECE Convention on Long-Range Transboundary Air Pollution in 1979, agreeing in principle to reduce their sulfur dioxide emissions, they set no firm timetable for such reductions. This situation changed with the formation of the so-called 30 Percent Club at the 1984 International Conference of Ministers on Acid Rain in Ottawa, Ontario. Representatives from 10 Western European countries pledged to reduce sulfur dioxide emissions 30 percent below their 1980 levels by 1993. A total of 21 countries had joined the 30 Percent Club by 1989. Unfortunately, three of the largest sulfur-dioxide polluters — Poland, the U.K., and the U.S. — have refused to agree to

30 percent reductions on the grounds of incomplete scientific data.

ECE nations have forged a similar accord on nitrogen oxide emissions. The agreement, in essence, calls on member nations to freeze nitrogen oxide emissions at 1987 levels by 1994. Representatives from the U.S. and a number of ECE nations signed the agreement in late 1988.

In 1985, the governments of Canada's seven easternmost provinces announced that they would reduce their sulfur dioxide emissions 50 percent by 1994. In the U.S. Congress, legislators introduced several bills aimed at revising the Clean Air Act of 1970 with new requirements for reduced sulfur dioxide emissions, but the 100th Congress adjourned in October 1988 without taking any action.

The 1987 NAPAP report concluded that "available observations and current theory suggest that there will not be an abrupt change in aquatic systems, crops, or forests at current levels of air pollution." The report was sharply criticized by U.S. environmental groups and by Canada's Environment Minister Tom McMillan, who dismissed it as "voodoo science." The report made a number of controversial assumptions, among them defining an acidified lake as one with a pH of 5.0 — a level at which most fish die — rather than 6.0 — a level that seriously disrupts the ecosystem.

James Gibson, the chairman of the NAS Committee on Acid Rain, noted in 1986 that "the connection between acid rain and environmental damage is real, but it is more variable and complex than many people have supposed." In an earlier study, OTA concluded that "it is unlikely that a definitive model linking particular sources to specific receptors [areas receiving pollution] will be developed within the next decade. Policy decisions to control or not to control emissions will have to be made without the benefit of such precise information." In 1989, Canada's Prime Minister Brian Mulroney, urging the U.S. to agree to emission controls, observed that, "in this matter, time is not our ally but our enemy. For what would be said of a generation that sought the stars but permitted its lakes and streams to languish and die?"

Biological Diversity

The durian, an unusual fruit much prized by Indonesians for its exotic flavor of strawberries and garlic, mysteriously became scarce in the mid-1970s. The reason for its decline was discovered serendipitously by American biologist Anthony Start, who had been studying a seemingly unrelated subject. Start was monitoring the behavior of a species of Malaysian bat that inhabited caves about 60 miles inland, far from the durian groves. The bat, which feeds exclusively on pollen and nectar, has a digestive system so specialized, according to Start, that an accidentally swallowed spider will ultimately appear in the bat's guano "perfect, undigested."

By analyzing pollen grains in the bats' droppings, Start determined that the bats fed mainly on a genus of continuously flowering trees found only in the mangroves along the coast. Occasionally, however, on their way home from the mangroves, the bats would make "gourmet flower" stops in the rain forest to sample trees, such as the durian, with very short blooming cycles. The bats, it turned out, were the tree's only known pollinator: without bats, there would be no durians.

Unfortunately, in the mid-1970s, a local cement factory had begun removing limestone from the bat caves, disturbing the creatures' homes. At the same time, shrimp farming was becoming established in the mangroves, interfering with the bats' foraging grounds. As the bats' life cycle was disrupted, the durian was not pollinated and fruit production declined. The problem was ultimately resolved by closing the cement factory and making the bat caves part of a nature sanctuary.

Such intricate interrelationships among species abound in nature. Increasingly, humanity's interference with the complex web of life, by means of forest-clearing or pollution, has destroyed entire species, precipitating a chain reaction of extinctions throughout the web. Such species extinctions pose "a threat to civilization second only to the threat of thermonuclear war," according to nine prominent American scientists who call themselves the Club of Earth. The group's joint statement, issued in late 1986 at the National Forum on Biodiversity in Washington, D.C., said: "While a majority of the species threatened with extinction are completely unknown, the results of their loss could be an unprecedented human tragedy."

In his keynote address at the forum, entomologist Edward O. Wilson of Harvard University in Cambridge, Massachusetts, noted: "We are locked in a race. We must hurry to acquire the knowledge on which a wide policy of conservation and development can be based for centuries to come, before opportunities of unimaginable magnitude are closed out forever." Wilson and other members of the Club of Earth likened the ongoing extinctions to those that destroyed the dinosaurs and many other species 65 million years ago. Some believe that the magnitude of the extinctions in the coming decades will be even greater.

Species at Risk

Writing in the December 1985 *BioScience*, Wilson asked, "How many species of organisms are there on Earth? We don't know, not even to the nearest order of magnitude." Since Carolus Linnaeus developed the binomial system for naming organisms in 1753, Wilson noted, scientists have identified about 1.7 million species: about 440,000 plants, 47,000 vertebrates, 750,000 insects, and 470,000 assorted invertebrates and microorganisms. That number may represent only a small fraction of the true total. While most of the classified organisms live in the temperate zone, far more species may exist in the biologically rich but less thoroughly studied Tropics. Estimates of the total number of species worldwide range from 5 million to 30 million.

The enormous biological diversity of tropical rain forests was confirmed in a 1988 census of a 0.2-square-mile plot of lowland Malaysian rain forest about 100 miles southeast of the capital city of Kuala Lumpur. Botanist Ira Rubinoff of the Smithsonian Institution Tropical Research Institute in Panama City, Panama, and his colleagues reported that the plot contained 835 species of trees — more species than are found in all of Canada and the United States. Other researchers have begun similar surveys in India, China, Africa, and Southeast Asia.

Because scientists do not know the precise number of species on the planet, they can only guess at the number at risk of extinction. At the Forum on Biodiversity, Wilson himself postulated that "at least half the fauna and flora, comprising hundreds of thousands or even millions of species, will eventually be lost."

Many other biologists, however, regard such estimates as grossly inflated. Ariel Lugo of the U.S. Forest Service in Rio Piedras, Puerto Rico, has argued that a slight shift in assumptions would yield a loss of only 9 percent of species by the year 2000. Even so, a 9 percent loss from 5 million species would represent 450,000 extinct species.

In the hope of identifying and preserving species that might be valuable to humans, Wilson and other biologists have proposed a massive effort to catalog the world's species. The prospect is daunting, however. In 1985, Wilson calculated that cataloging 10 million species would require 25,000 taxonomists (specialists in species classification) working for 40 years. Unfortunately, Wilson observed, no more than 1,500 taxonomists worldwide are skilled in identifying tropical species.

Researchers have actually identified only a small fraction of the immense number of organisms threatened with extinction. By the end of 1988, the Endangered Species Office of the U.S. Fish and Wildlife Service, for example, had officially designated about 1,000 species around the world as endangered — on the brink of extinction throughout all significant portions of their habitat. The list included 313 mammals, 223 birds, 159 plants, 114 reptiles, 83 fishes, 30 clams, 16 amphibians, 13 insects, 9 snails, and 6 crustaceans.

In another listing of endangered species, the Conservation Monitoring Unit of the International Union for the Conservation of Nature and Natural Resources (IUCN) cited 4,589 animal species as endangered in 1988, including 2,125 invertebrates, 1,073 birds, 596 fishes, 555 mammals, 186 reptiles, and 54 amphibians. The group also estimated that 60,000 plant species could become extinct by the year 2050 if current trends continue.

Benefits of Biological Diversity

The term "biological diversity" most often refers to the number of species in a particular region, although some biologists also use it to describe genetic variability within a species, the number of species that make up a community, the number of communities in an ecosystem (the combination of flora, fauna, and their environment), or the number of ecosystems in a region.

Genetic variability allows a species to adapt to changing environmental conditions through natural selection — the process that drives evolution. As environmental conditions change, individuals with genetic traits most suitable to the new environment will be more likely to thrive than those lacking the desirable traits. A species with a large population and wide genetic variability among individuals thus stands a better chance of surviving environmental change than a species with a small population and low genetic variability.

Biological diversity is of vital importance to humans for many reasons. A diverse ecosystem sustains and improves agriculture, provides opportunities for medical discoveries and industrial innovations, maintains "ecosystem services," such as watershed protection and soil fertility, and offers recreational, cultural, and spiritual benefits.

Agricultural benefits. The corn blight that destroyed 15 percent of the U.S. corn crop in 1970 vividly illustrates the damage that may occur when farmers rely too heavily on a single crop variety. Fortunately, in that particular case, farmers recovered quickly because disease-resistant corn varieties had been preserved by the U.S. Department of Agriculture and were quickly cross-bred with commercial varieties. Biologists fear, however, that the loss of biological diversity caused by species extinction will impair humanity's ability to respond as successfully to future agricutural crises of a similar nature.

The genetic diversity of plants and animals remains a largely untapped resource, but one with great potential for helping to solve environmental problems. Researchers have, for example, cross-bred the inedible but salt-tolerant plant *Lycopersicon cheesmanii* from the Galápagos Islands with the domestic tomato, *L. esculentum*, to yield a salt-tolerant tomato plant that has an edible fruit and that can be irrigated with 70 percent seawater. Researchers hope that farmers can grow the new variety in areas that have become desertified because of too much salt in the soil. Botanists have identified other plants that produce higher yields in regions with arid climates or poor soils and are cross-breeding them with conventional varieties to develop hardier and more productive crops.

One of these newly identified plants — *Zea diploperennis* — a type of tall Mexican grass very closely related to corn — is a perennial that does not need to be replanted after each harvest as corn does. If the two species could be successfully cross-bred to yield perennial corn, farmers could save the millions of dollars spent plowing corn stubble under each year and sowing new seed. *Z. diploperennis* was discovered at only three small sites in a remote Mexican mountain range and could easily have been destroyed before anyone recognized its value.

Genetic diversity also plays an important role in livestock. In its 1987 report *Technologies to Maintain Biological Diversity*, OTA concluded that the principal future threat to livestock production is the replacement of native breeds in developing countries with imported breeds from developed countries. Farms with only a few cows, such as those in developing countries, find artificial insemination cheaper than keeping a bull, but those countries lack facilities to collect, freeze, and distribute semen from local cattle breeds. They therefore import most semen from commercial studs in developed countries, thereby replacing their native cows with foreign-bred species and reducing genetic diversity.

Medical and industrial uses. Pharmacologist Norman Farnsworth of the University of Illinois in Chicago has found that at least 40 percent of the prescriptions filled in the U.S. contain drugs whose primary active ingredients come from natural sources — 25 percent from plants, about 13 percent from microorganisms, and about 3 percent from animals.

Drugs derived from plants include belladonna, a substance that relieves asthma and some stomach-ulcer symptoms; digitalis, a heart stimulant obtained from the foxglove plant; morphine, a powerful pain killer extracted from the poppy; and vinblastine and vincristine, two chemicals derived from the periwinkle plant, both of which are now widely used in chemotherapy for certain types of cancer. *World Resources 1986*, published by the World Resources Institute in Washington, D.C., reported that sales of plant-derived medicines total $20 billion per year in the U.S. alone and twice that amount in the entire industrial world. Farnsworth estimated that as many as 5,000 of the world's approximately 500,000 plant species might eventually yield valuable drugs.

Many seemingly unlikely species are now used in medical research. In its 1987 report, OTA listed some intriguing contributors to biomedical research:

- Sea urchin eggs, which are used extensively in experimental embryology, in studies of cell structure and fertilization, and in tests to gauge the risk of birth defects from experimental drugs;
- Horseshoe crabs, which provide an extract used in a fast, sensitive test to determine whether vaccines are contaminated with bacterial toxins;
- Certain butterfly species, which are used in research on cancer, anemia, and viral disease;
- Sponges, the study of which has contributed to the fields of structural chemistry, pharmaceutical chemistry, and developmental biology, and has led to the discovery of new chemical compounds. For example, *D*-arabinosylcytosine, a synthetic anticancer agent, was developed as a result of studies of a chemical substance in a Jamaican sponge.

Many species are also promising sources of raw materials. For example, the fast-growing kenaf plant, native to parts of tropical Africa but also grown in higher latitudes, produces long fibers suitable for making many types of paper, including newsprint. Kenaf offers an important advantage over the conifer trees typically used for paper pulp because it can be harvested annually, whereas conifers can be harvested only about once every 15 years. Another promising plant, the jojoba bush, grows in arid regions and produces seeds rich in a liquid wax that effectively replaces oil from the sperm whale as a lubricant for use at high pressures and temperatures.

Ecosystem services. Natural ecosystems provide humans with a number of essential "services" whose monetary value is difficult, if not impossible, to measure. According to Walter Westman of the University of California at Davis, these ecosystem services include the absorption and breakdown of pollutants, the cycling of nutrients, the degradation of organic waste, the maintenance of a balance of gases in the air, and the regulation of radiation balance and climate — the functions, in short, that are crucial to producing clean air, pure water, and a green Earth.

The loss of a single species can cause a ripple effect in an ecosystem, leading to further extinctions, as well as other problems. For example, 22 species of bird feed on fruit from Cost Rica's canopy tree, *Casearia corymboso*, and several of them are entirely dependent on it during part of the year. According to biologist Paul Ehrlich of Stanford University in California, if the tree became extinct, "these birds would go also, with ramifications for other trees that depend on the birds for seed dispersal, and thus on the herbivores dependent on those trees, and so on." Similarly, biologist Peter Raven of the Missouri Botanical Gardens in St. Louis has estimated that, for every plant species that becomes extinct, another 10 to 30 species also die out.

Recreational, cultural, and spiritual benefits. Many people cite moral, religious, or spiritual reasons for the preservation of species. In the 1978 book *The Arrogance of Humanism*, David Ehrenfeld insisted that species "should be conserved because they exist and because this existence itself is but the present expression of a continuing historical process of immense antiquity and majesty."

In the report *Environmental Quality — 1980,* the U.S. Council on Environmental Quality noted that experiencing the beauty and diversity of nature enriches people emotionally and spiritually, fulfilling important human needs. According to the report: "Perhaps this ancient kinship [between humans and all other living organisms] explains why it is essential to the psychological well-being of so many people to collect, hunt, read about, watch, photograph, or be surrounded by diverse living things. There would be little excitement in only one kind of flower, bird, or butterfly; their fascination lies in their different sizes, shapes, colors, sounds, behaviors and abundance, their complexity, and their freedom."

The U.S. Fish and Wildlife Service has estimated that at least 83 million people per year in the U.S. engage in fishing, hunting, and nonconsumptive uses of wildlife, such as hiking, birdwatching, and nature photography. The trend toward nonconsumptive uses has brought increased tourism income to developing countries with spectacular wildlife resources. According to *World Resources 1986*, safari-related tourism in Kenya now ranks second only to coffee exports as a source of foreign exchange.

Causes of Extinction

Habitat destruction. Many ecologists believe that the destruction or disruption of an organism's natural habitat contributes more than anything else to the growing number of extinctions. In 1987, the world's human population passed 5 billion, and it will probably increase to 6 billion by the year 2000. The 1980 *Global 2000 Report* observed that the projected growth in human population during the next several decades should create "enormous economic and political pressure to convert the planet's remaining wild lands to other uses."

Obviously, as population grows, so does the need for agricultural land. According to *World Resources 1987*, the world's crop land expanded 170 percent between 1850 and 1980, usually at the expense of forests and grasslands. The developing regions underwent much larger changes. Southeast Asia, for example, lost 7 percent of its forests and 25 percent of its grasslands as crop acreage rose 670 percent. South Asia lost 43 percent of its forests and 1 percent of its grasslands, while crop land increased 196 percent. In North Africa and the Middle East, 60 percent of the forests and 5 percent of the grasslands were lost, as crop acreage expanded 294 percent.

Clearing land kills plants directly and harms animals indirectly by disturbing their breeding sites and food sources. Clearing can also strongly affect species remaining in undisturbed parcels of land, such as national parks or refuges, particularly when the undisturbed area is smaller than an organism's original natural habitat. Such effects have been demonstrated in an ongoing study by biologist Thomas Lovejoy of the World Wildlife Fund. In 1976, Lovejoy and his colleagues began a 20-year project by carving out seven isolated reserves in Brazil's Amazon forest, three of about 2.5 acres each, three of 25 acres each, and one of 250 acres.

By 1988, preliminary results from the studies of these reserves had shown that species requiring a large range from which to obtain food were the first to disappear from the smaller reserves. When fruit-eating monkeys, for example, could not find enough fruit on trees in their

25-acre reserve, they abandoned it. In contrast, a band of red howler monkeys, which live on leaves, have apparently been sustaining themselves.

Habitat destruction is taking the greatest toll on species in tropical forests, where humans have been clearing broad areas for agriculture, logging, and development. According to the U.N., tropical forests have been disappearing at the rate of 27.9 million acres per year, resulting in the loss of habitats for untold numbers of species. The rate of habitat loss is lower in many developed countries, where farmers have already cleared much of the forested land for agriculture. The total losses have been substantial, however.

In its 1987 study "Sliding Toward Extinction," for example, the California Nature Conservancy noted that, since 1848, nearly 30 million acres in California have been converted to crop land or urban settlements. In the process, the state has lost 80 percent of its coastal wetlands, 89 percent of its riparian woodlands (those along a river or stream), 94 percent of its interior woodlands, and 99 percent of its grassy woodlands. Altogether, 30 native California animals either have become extinct or no longer exist in the state.

Pollution. Environmental pollution has seriously threatened many species. Researchers have linked the pesticide DDT to the overall decline in the population of the endangered brown pelican, the peregrine falcon, the bald eagle, and a variety of other birds. Fortunately, many affected species have recovered since the U.S. banned DDT in 1972.

In the early 1980s, however, scientists found other toxic chemicals contaminating a number of wildlife refuges. The most widely publicized problem was at Kesterson National Wildlife Refuge in California, where researchers traced bird deaths and deformities to high levels of selenium-tainted irrigation water draining into the refuge. Reports of serious contamination in water at other refuges prompted the U.S. Fish and Wildlife Service in 1986 to undertake a systematic survey of its 430 refuges. The study revealed pollution problems severe enough to require corrective action at 9 refuges and in-depth monitoring at 26. Another 43 refuges were identified as having potential pollution problems. The sources of contamination at the refuges included hazardous waste sites, industrial and municipal waste-water discharges, and irrigation water containing agricultural chemicals.

Wildlife is also at risk from acid rain, which has killed the fish in thousands of lakes in Scandinavia, the northeastern U.S., and eastern Canada. According to the Ontario Ministry of the Environment, the Canadian aurora trout may have already become extinct because of the growing acidification of lakes throughout its natural habitat. (For more information, see the article "Acid Rain" in this chapter.)

Diversion of water. On 19 January 1982, the Tecopa pupfish became the first species whose extinction removed it from the U.S. endangered species list. The pupfish died out because developers diverted water from its only known habitats — two hot springs near Tecopa, California. In the 1983 book *Vanishing Fishes of North America*, biologists R. Dana Ono, James Williams, and Anne Wagner reported that problems for the pupfish began around 1940 when entrepreneurs built bathhouses around the springs. Development of the area continued, and in 1965, the outflows of the two springs were joined and the remaining channel straightened, causing both the temperature and the flow rate of the waters downstream to increase. These changing water conditions at least partially caused the pupfish's demise.

The dusky seaside sparrow also became extinct when its habitat was disrupted. The last surviving member of the species died in 1987 after 7 years in a Florida zoo. The gradual extinction of the duskies began in the 1960s, when the federal government flooded the swampy marshlands of Florida's Merritt Island, the duskies' prime habitat, in order to reduce the number of mosquitoes at the nearby Kennedy Space Center. In late 1979, the U.S. Fish and Wildlife Service authorized the capture of the remaining dusky sparrows to prevent their extinction. Researchers could find only 6 of the original population of more than 2,000 birds, however, and they captured 5, all of them males.

Overexploitation. Fossil records indicate that overhunting by early humans may have been the primary cause of the extinction of many large animals throughout the world. Jared Diamond has noted that the arrival of humans in Australia over 30,000 years ago roughly coincided with the disappearance of nearly all the indigenous mammals with weights greater than 100 pounds. At that time, 17 genera and 46 species of marsupials became extinct, including *Diprotodon*, a herbivore as large as a rhinoceros; *Genyornis*, a flightless bird; *Megalania*, a giant lizard; *Meiolania*, a giant tortoise; and *Wonambi*, a giant snake. According to Diamond, the animal species most vulnerable to extinction are generally flightless, large, or both. Such species typically produce relatively few young, exist in low population densities, and generally make easy targets for hunters. All but four of the 41 largest marsupial species in Australia became extinct after humans first reached the continent.

In North and South America, about three-quarters of the genera of large mammals disappeared during a wave of extinctions that reached its peak about 10,000 to 12,000 years ago. Some scientists believe these extinctions were caused by the Clovis people, a band of hunters who used a distinctively shaped arrowhead first found near Clovis, New Mexico. Clovis arrowheads have been found with fossils of many of the large herbivores that died out. A similar situation may have arisen with the arrival of Polynesians in New Zealand about 1000 A.D., when at least 30 bird species became extinct. Another eight New Zealand bird species became extinct during the two centuries after the arrival of Europeans.

Poaching and unregulated hunting continue to threaten many species in developing countries. Endangered African and Asian rhinoceros species, for example, are hunted illegally for their valuable horn despite international agreements banning trade in rhinoceros products. Asians use rhinoceros horns in fever-reducing medicines and will pay over $10,000 a pound for them. Hunting has decimated the rhinoceros population, which declined from 70,000 in 1970 to 11,500 in 1987, according to the World Wildlife Fund, which monitors the illegal trade in endangered and threatened species.

Ivory poachers are also destroying the elephant population in the Central African Republic. A June 1985 survey conducted by the World Wildlife Fund found a "catastrophic reduction" in the elephant population in that country during the preceding 4 years. Although authorities believe that most of the poachers are horsemen from the Sudan and Chad who kill the elephants for their ivory tusks, people living in the area also kill the animals for their ivory and meat.

An aerial survey of a 25,600-square-mile region of the Central African Republic identified 7,800 elephant carcasses and only 4,300 live elephants. Overall, the World Wildlife Fund estimated that the country's total elephant population had fallen from 80,000 to 15,000 in a decade. In Africa as a whole, according to IUCN, the number of elephants dropped from 1 million in 1981 to 764,000 in 1987.

Illegal trade in raw African ivory totaled about $55 million in 1986, an amount of ivory equivalent to that in the tusks of about 89,000 elephants, according to *World Resources 1988-89*. In an attempt to stem the seemingly uncontrollable poaching, the Convention on International Trade in Endangered Species set quotas for legal ivory shipments for each exporting country. In 1986, the year that the quotas went into effect, 19 African nations legally exported $28 million in ivory. Although poaching has continued, authorities believe that the quota system will eventually succeed in restricting the amount of ivory on the world market.

Overfishing. The blue pike, a species that once supported a large sport and commercial fishery in Lake Erie and Lake Ontario, is now believed to be extinct due to overfishing. In *Vanishing Fishes of North America*, biologist R. Dana Ono noted that, between the beginning of fisheries records in 1885 and the last report of a blue-pike catch in 1962, fishermen took about 1 billion pounds of marketable blue pike from the two lakes.

Although the major fish species in commercial ocean fisheries are not currently threatened with extinction, the U.N. Food and Agriculture Organization (FAO) reported that, by the late 1970s, more than 25 of the world's most important ocean fisheries had declined significantly from historic levels. The anticipated increase in worldwide demand for fish and fish products from 93.5 million tons in 1985 to 110 million tons by the year 2000 will likely outpace any expansion of traditional fisheries.

The FAO's 1987 report *Agriculture: Toward 2000* noted that "all important stocks of demersal (bottom dwelling) species are either fully exploited or overfished.... Crustacean species generally are heavily exploited and many, if not most, stocks are depleted." To meet the projected demand, fishermen will need to increase their catches of small shoaling species such as herring, begin marketing so-called trash fish that would normally be discarded, and expand fish-farming operations.

Whaling. Environmentalists around the world hailed the July 1982 International Whaling Commission decision to impose a global ban on the commercial harvesting of all whales beginning in 1986. Decades of overhunting have severely reduced the world's whale population and threatened many species with extinction. In the year before the ban took effect, 14,000 whales were harvested worldwide, a third of them by Japan. Although all whaling countries eventually agreed to abide by the ban, Japan announced that it would continue to harvest about 300 whales per year, ostensibly for research purposes. A number of other marine mammals, including some species of dolphins, manatees, and dugongs, continue to be threatened by hunting, coastal pollution, and habitat destruction.

Preserving Endangered Species

Environmentalists often divide programs to protect endangered species into two basic groups: species-oriented approaches and ecosystem approaches. Species-oriented programs generally try to conserve either a single species or a group of closely associated species. Ecosystem approaches, in contrast, strive to preserve an entire ecosystem so that all the species in it will endure.

Species-oriented approaches. Species-oriented programs designate particular species for protection. The national or international agencies responsible for these programs generally select species on the basis of their utility and the degree of threat. Some are selected because their beauty, majesty, or unusualness appeal to humans. Such species include exceptional flowering plants, birds of prey, and carnivorous plants.

Species-oriented programs often reflect the biases of the people who run them. Mammals and birds together account for over half of the 950 endangered species on the U.S. Fish and Wildlife Service's list. In contrast, only 6 crustacean and 13 insect species appear on the list, even though mollusk and insect species far outnumber mammal and bird species.

Individual species may be protected through legal mechanisms that restrict or ban their killing, collection, or trading. In addition, living organisms may be placed in aquariums, botanical gardens, or zoos, while eggs, embryos, seeds, sperm, or tissues may be preserved in so-called gene banks. For example, the California condor, Pere David's deer from China, and Cooke's kokio (a small tree native to Hawaii) no longer exist in the wild, but still live in zoos and botanical gardens.

The preservation of living organisms seldom succeeds outside their natural environments, however. Unless endangered organisms produce viable offspring, their removal from natural environments will only increase the risk of extinction. Because cages often lack the rich stimulation of the natural environment, zoo animals sometimes behave abnormally and may not be able to mate or adequately care for their young. Only one California condor, for example, has so far been bred in captivity.

Confinement in a small zoo community also makes a population vulnerable to disease. Biologists believe, for example, that some black-footed ferrets first caught for a captive-breeding program contracted canine distemper — a fatal disease — after initially being confined with infected animals. When only a few individuals of a species reside in zoos, inbreeding may also occur, reducing offsprings' prospects for long-term survival. Fortunately, animal exchanges among zoos can partially alleviate this problem.

Seed banks around the world have been used by government agencies for decades to preserve the genetic material from thousands of cultivated and wild plant varieties, particularly food crops such as barley, beans, rice, and wheat. The Consultative Group on International Agricultural Research (CGIAR), established by the U.N. and the World Bank in 1971, has been coordinating the efforts of major agricultural research centers worldwide to develop and preserve plant resources. By 1988, CGIAR germplasm banks contained more than 350,000 genetically distinct samples of crop seeds, including 76,500 types of rice, 67,000 types of wheat, 29,000 types of beans, 24,000 types of sorghum, and 23,000 types of barley.

The U.S. operates over 40 separate germplasm storage facilities. According to Henry Shands, director of the Agricultural Research Service's National Plant Germplasm System, such storage is particularly important because "the U.S. is a have-not nation in terms of germplasm resources." The U.S. has no naturally occurring wild varieties of its major food crops. Wheat and barley species came originally from the Middle East, tomatoes from Central and South America, and soybeans from Asia. Virtually every major food crop is native to some other region of the world. Shand said, "We've got to have a certain amount of gene-bank defense to protect our nation's food supply in the future."

Ecosystem approaches. An ecosystem approach aims to preserve the diversity of an entire ecosystem rather than just a single species or a group of species. In this way, the ecosystem approach protects not only a relatively small number of designated species, but also many "neglected" organisms, such as insects, mollusks, and plants, which are an integral part of the overall ecosystem. Unlike organisms confined in aquariums, zoos, and other captive environments, organisms in the ecosystem can move about freely, find food, mate, and raise their young with little or no disruption.

Still, conservators must take steps to ensure that the protected area is large enough to preserve all the organisms living in it. Studies at U.S. and Canadian parks have demonstrated that the preservation of large animals requires extensive parcels of land. In general, the smaller

Figure 5. Courting Condors. Thousands of condors once ranged throughout the U.S., but by 1900 hunting had reduced the species to a small population restricted to southern California. Today, only 28 adult California condors are known to exist, and all of them are in protective captivity in California. A captive breeding program has been established to try to rebuild the population of this critically endangered species. In 1987, the condor pair pictured here began "courting," and on 29 April 1988, they produced the first chick conceived and hatched in captivity. Scientists hope to begin releasing captive-bred condors into the wild by 1992. *Courtesy: Ron Garrison, Copyright © Zoological Society of San Diego, 1988.*

the preserve, the more limited the number of species that can be protected.

Ecosystems cannot always escape events that occur outside their boundaries. For example, the upstream diversion of water for irrigation or drinking water supplies can seriously threaten a protected ecosystem downstream, and pollutants such as acid rain or spilled oil may be carried hundreds of miles by winds or water currents into a protected area. To prevent severe damage to an ecosystem, planners must consider both the types of events that may occur and the most appropriate responses to them. Even with the best planning, however, unforeseen events, such as climatic change, can threaten ecosystems and their unique organisms.

International conventions. The boundaries of important ecosystems, such as the Amazon rain forest, do not follow national borders. Some species, such as migratory birds, may pass through many nations during their seasonal travels. The destruction of any part of the migratory route may threaten their survival. In addition, winds and water currents may carry pollutants from one country to another, threatening organisms over a large area. Many animals also spend part or all of their life cycles in regions outside national jurisdictions, such as in the open ocean and Antarctica, and can only be protected by international agreements preventing habitat destruction and overhunting in those areas.

Four major global treaties have been established to preserve species and ecosystems: (1) the Convention on Wetlands of International Importance, Especially as Waterfowl Habitat (Wetlands Convention); (2) the Convention Concerning the Protection of the World Cultural and Natural Heritage (World Heritage Convention); (3) the Convention on Conservation of Migratory Species of Wild Animals (Migratory Species Convention); and (4) the Convention on International Trade in Endangered Species of Wild Fauna and Flora (CITES).

The Wetlands Convention, which was adopted in Ramsar, Iran, in February 1971 and took effect in December 1975, requires signatory nations to set aside at least one wetland for conservation. By 1989, 47 nations had ratified the convention. Wetlands support natural fisheries by providing spawning grounds or other vital habitats for two-thirds of all food fish. They also help maintain water tables, stabilize shorelines, and provide water storage and purification, flood control, and recreation. As of 1988, 404 sites covering more than 50 million acres worldwide had been designated for the Wetlands Convention's protected list.

The World Heritage Convention states that nations must accept the obligation to protect unique cultural and natural areas that are the heritage of all people. It also holds that the international community should provide financial assistance to nations that cannot afford protective measures on their own. The convention was adopted in Paris, France, in November 1972, and took effect in December 1975; 92 nations had ratified it by 1989. A fund established by the convention helps nations finance their protection schemes. As of 1988, 78 natural areas worldwide had been designated as World Heritage sites under the convention. In addition to setting up this roster of sites, the convention provides training and technical assistance to help nations preserve the protected areas.

The Migratory Species Convention, which was adopted in Bonn, West Germany, in June 1979, attempts to safeguard migratory species in danger of extinction. Signatory nations to the convention, which took effect in November 1983, have pledged to protect endangered migratory species within their jurisdictions and also to reach bilateral and multilateral agreements to ensure the protection of those species throughout all parts of the birds' ranges. As of 1989, 22 nations, most of them European and African, had ratified the convention.

CITES was adopted in Washington, D.C., in March 1973 and took effect in July 1975; by 1989, 101 nations had ratified it. CITES established a system for either prohibiting or controlling trade in certain endangered species. Specifically, CITES set up a network of national management and scientific authorities that cooperate with one another and the CITES secretariat in overseeing international trade in endangered species. Each nation's management authority has accepted responsibility for supervising the mechanics of the endangered-species trade, such as issuing permits, and also for implementing and enforcing trade controls within the nation. The scientific authority has shouldered the responsibility for ensuring that an import or export permit will not harm the endangered species traded.

Finding the money for conservation. Most debt-ridden Third World countries have little money available for conservation programs. Environmentalists refer to many of the designated reserves in those areas as "paper parks" because they exist on paper but have no staff or budget. Some conservationists have argued that developed countries should help fund projects to preserve tropical species diversity in developing countries because species preservation involves the global community, not just individual countries acting on their own. The governments of most developed countries, however, have responded with meager sums, at best.

Nongovernmental organizations, in contrast, have experimented with novel arrangements to finance conservation. Perhaps the most active of these groups has been the Nature Conservancy of Arlington, Virginia, which had bought over 2.6 million acres of fragile environments by 1989. Its properties, purchased with donated funds, include habitats for more than 1,000 species of endangered plants and animals. Among its 900 tracts, there are 64,000 acres of wetlands along the Gulf Coast of Florida, 13,000 acres of threatened desert near Palm Springs, California, and 60 acres along the Mianus River Gorge in Westchester County, New York. The Nature Conservancy has also entered into cooperative agreements to preserve land and collect environmental data in 10 Central and South American countries.

In July 1987, Conservation International of Washington, D.C., announced an unusual agreement that may serve as a model for future wildlife preservation. The group arranged a "debt-for-nature" exchange, in which Conservation International purchased $650,000 worth of Bolivia's $4-billion debt to U.S. banks for $97,500. It retired the debt in exchange for an agreement by the Bolivian government to set aside 4 million acres of Amazon rain forest. (The banks agreed to settle for fifteen cents on every dollar owed because they assumed Bolivia would otherwise default on the entire debt.)

The 4 million acres of Amazon rain forest became part of the buffer zone surrounding the Beni Biosphere Reserve, a tropical forest near the headwaters of the Amazon. Conservationists hailed the arrangement as a creative approach to conservation financing, and expressed the hope that similar agreements can be used to save many other endangered areas. Given the failure of many governments to take the initiative in preserving biological diversity, such private efforts may well represent the best hope for preserving the world's genetic heritage.

Deforestation and Desertification

In Malaysia, Central America, and Brazil, subsistence farmers are slashing and burning tropical forests to clear land and plant crops that will keep them alive for another 2 or 3 years. In the drought-stricken Sahel region of

Africa, nomadic shepherds are gathering their herds around scarce watering holes, where the livestock trample and kill vegetation, converting the already depleted soil into wind-blown sand. In short, the world's forests are shrinking, and its deserts are growing. These changes portend a radically altered planet.

Deforestation and desertification are leading to the disappearance of thousands of known and perhaps millions of as-yet-uncatalogued species of plants and animals — living resources that might have provided a new heart or cancer drug, better agricultural crops, or faster-growing species of livestock. The problem even extends to humanity's own genetic heritage: In the year 1500, an estimated 9 million aboriginal Indians lived in the rain forests of the Amazon basin; today, more than 100 tribes have become extinct, and only 200,000 Indians remain. They are among humanity's last links with its most distant ancestors.

But the effects go much further. Forests release abundant moisture into the atmosphere, and this moisture condenses into clouds that provide rain to surrounding areas. The loss of forests can thus greatly increase the aridity of nearby regions. Deserts, in contrast, are islands of heat that can, by their mere presence, discourage rainfall and produce climatic change in nearby areas. The loss of forests in the Amazon could produce drought in North America, while desertification in Africa could lead to higher temperatures and lower agricultural productivity in Europe.

The loss of forests also allows rainwater to erode the soil rather than soak in and replenish the groundwater. The rain that rolls off the deforested slopes of the Himalayas may gather into life-threatening floods in India and Pakistan. Finally, when the wood harvested from forests is burned or allowed to decompose, it releases vast quantities of carbon dioxide — the major contributor to the greenhouse effect that scientists believe is warming the entire Earth.

Part of the blame for the situation lies with the industrialized countries, which are consuming more than their share of the resources of developing countries to sustain a lifestyle that subsistence farmers or nomadic herders in the Third World can scarcely imagine. According to the U.N., the industrialized countries have only 25 percent of the world's population, but they consume 75 percent of its energy and 85 percent of the lumber that developing countries produce.

Blame also lies with the leaders of developing countries, who have not enacted and enforced measures that would halt the loss of forests and limit the spread of deserts. And ultimately, some of the blame also lies with the farmers and the herders — people fighting a daily battle for survival.

World's Shrinking Forests

Forests have sustained humanity since the dawn of time. They clear the air, moderate the climate, protect soil from erosion, and keep water clean. To early humans, forests offered a place of refuge from enemies, fuel for fires, vegetation and game for sustenance, and medicines for their ills. As civilization developed, the forests also provided building materials for homes and other structures, paper for recording ideas, and thousands of other useful products. Forests have always symbolized health, wealth, and power. As civilizations destroyed the forests that sustained them, they grew dependent on others for essential raw materials and became vulnerable to disruptions in supply.

Sicily, for example, was once a heavily wooded, fertile island — a flower of ancient Greek culture and prosperity. But Sicily was so thoroughly deforested by a succession of conquerors that it became almost desertlike by the thirteenth century and has never regained its former wealth. Or consider the region of North Africa from Tunisia to Morocco, which was colonized by Rome after the fall of Carthage in 146 B.C. From that time until the sixth century A.D., the region's fertile fields and bountiful forests made it the "granary of Rome." Today, Rome's former granary is little more than barren desert.

In 5,000 years, the ever-expanding human population has cleared three-fifths of the world's forests. Where dense stands of trees once covered half of the Earth's land area, they now cast their shade on only 20 percent, and the rate of loss is rapidly accelerating. According to the U.N., Africa has lost 23 percent of its forests since 1950, Central America has lost 38 percent, and the Himalayan watershed 40 percent.

The World Wildlife Fund has estimated that humans are now slashing and burning tropical forests at a rate of 50 acres per minute — 27 million acres per year. And the U.S. government's *Global 2000 Report*, published in 1980, predicted that the world's forests would shrink from covering 25 percent of the globe's ice-free land surface in 1956 to about 17 percent by the year 2000 and only 14 percent by 2020.

Forests and their problems. The U.N. has estimated that the world's forests cover about 14.7 million square miles — of which 42 percent lie in developed countries and the remainder in developing regions. More than 90 percent of the forests in developing nations, encompassing 7.75 million square miles, are tropical forests, which are home to more than half the world's known species and millions more that are still unclassified. Much of the remaining forest lies in the far northern areas of Canada, Scandinavia, and the U.S.S.R., where trees are too small, too slow-growing, or too distant from markets for profitable harvesting.

Ecologists often classify the wide variety of forest types into two general categories — closed-canopy forests and open woodlands. The dense vegetation of closed-canopy forests, which account for about 80 percent of the total, allows relatively little direct sunlight to reach the forest floor. In contrast, the wider spacing of trees in open woodlands allows direct sunlight to reach much of the ground. Open woodlands may include agricultural land or rangeland.

Ecologists also distinguish between moist and dry tropical forests, each of which has distinct characteristics. The world's moist tropical forests include the tropical rain forests and other closed-canopy forests found mainly in Latin America, Southeast Asia, and Zaire. The world's dry tropical forests include the open woodlands common in India, Mexico, much of Africa, and parts of eastern Brazil. Moist tropical forests have suffered the loss of plant and animal species as a result of rapid population growth and development pressures. In addition, their water quality has been declining due to slash-and-burn agriculture, the clearing of upland forests, and destructive logging practices. People living in dry tropical forests are plagued by equally difficult problems: shortages of fuelwood, lack of fodder (often taken from the lower branches of trees) for pastoral herds, and erosion of agricultural land and rangelands.

The total forested area of developed countries in temperate regions has remained relatively stable in recent years. Between 1950 and 1980, forest cover actually increased in Europe, Australia, and New Zealand, while it decreased slightly in Canada and the U.S. Of course, the original forests in most developed countries had already been extensively cleared for agriculture and development, and much of the remainder has been altered by logging and timber replanting.

Even developed regions face problems, however. Tropical lowland rain forests in the state of Hawaii, for example, covered half the islands' surface area only 200 years ago. Today, they cover less than 5 percent. Hawaii's forests have been cleared to make way for plantations, ranches, and urban development. Experts fear that much of what remains is too fragmented to survive. The loss is particularly disturbing to ecologists because 95 percent of the state's flowering plants and 99 percent of its animals and birds are unique to the islands.

Hawaii's problems are shared by many developing countries throughout the world. According to U.N. estimates, the world loses 42,500 square miles of tropical forests — an area the size of Ohio — each year because of agricultural clearing and fuelwood harvesting. Logging destroys an additional 17,000 square miles of tropical forests each year — an area the size of Maryland and New Jersey combined. Although much of this logging involves selective cutting of only the most valuable trees, it severely disturbs forest ecosystems by removing the most valuable species and damaging many more trees than are removed. Logging roads also open the forest to settlement, which can lead to further clearing for agriculture, building materials, and fuelwood.

Of the developing countries, only the People's Republic of China and South Korea have begun reforesting land at a faster rate than the forests are being cleared. According to the U.N., other developing countries replant an average of only one-tenth of the deforested area. The Worldwatch Institute has estimated that reforestation projects would have to expand fivefold worldwide — fifteenfold in Africa — to prevent forest cover from shrinking further. Other potential remedies for deforestation include managing forests more aggressively to minimize losses to insects, disease, and fire, curbing wasteful and illegal logging practices, and enacting social reforms that would give local residents a greater economic incentive to preserve their forests.

Causes of Deforestation

Fuelwood. At least 2.25 billion people worldwide rely on wood as their primary source of energy for cooking and heating. In developing nations, more than 80 percent of the wood harvested or gathered each year is used for cooking and heating. Still, according to the 1983 U.N. report *Fuelwood Supplies in Developing Countries*, more than 100 million people lack sufficient wood supplies to meet the "minimum requirements" for those purposes. In addition, nearly 1.2 billion people, primarily in Asia, meet their fuel needs by using fuelwood faster than it is being regrown.

The most acute fuelwood shortages have occurred in the arid regions of Africa, the mountainous areas of Asia (especially the Himalayas), and the Andean plateau of South America. The U.N. predicted that, by the year 2000, 3 billion people either will be unable to obtain enough fuelwood to meet their minimum daily needs or will use fuelwood faster than it is being regrown.

In the early 1980s, the U.N. analyzed the supply and demand for fuelwood in seven countries in the impoverished Sahel region of Africa and in 12 countries in eastern and southern Africa. It found that, in 1980, 78 percent of the 139 million people in these 19 countries lived in areas with overall fuelwood deficits. If the population of these countries increases 74 percent between 1980 and 2000, as is projected, then 97 percent of their residents would live in regions with fuelwood deficits. The areas with the most severe current deficits are the heavily populated highland regions of Ethiopia, Kenya, and Malawi in eastern Africa, where industry also uses wood for fuel. In Malawi, for example, industry accounts for 45 percent of the total fuelwood consumption.

In the early 1970s, many development specialists believed that the fuelwood problem in developing countries would take care of itself as economic conditions in these nations improved and people shifted from wood to other energy sources, such as kerosene. Rising energy prices and global economic problems during the 1970s and early 1980s prevented this shift from happening, and fuelwood use remained high. More recently, however, governments and international agencies have begun establishing programs to plant more trees and develop ways to use fuelwood more efficiently.

Many groups, for example, have designed more efficient cookstoves. Traditional methods of cooking over an open fire use only about 10 percent of the fire's energy. Although a number of more efficient stoves have been field tested, none have become widely used, because the stoves are relatively complicated and few parts are available to repair them. Furthermore, none of the prototype stoves has offered the other benefit of an open fire — providing heat and light.

Planting trees to replace lost forests and provide new sources of fuelwood requires an enormous investment of money and effort. The Club du Sahel, an international organization established to provide assistance to the drought-stricken Sahel, has estimated that 370,000 acres of tree plantations would have to be planted in the region each year to meet fuelwood needs in the year 2000. This planting rate would be about 50 times greater than the reforestation rate in the Sahel during the late 1970s.

Logging. The value of forest products worldwide totaled $300 billion in 1985, according to the U.N. Exports of forest products, including logs, lumber, wood panels, and paper, reached $50 billion that year, most of it from developed countries. Forestry exports in developing countries amounted to only $6.6 billion in 1985, or about 13 percent of the total.

The harvesting of forest products in developing countries parallels the course taken in the U.S. and other developed countries during their early years. To generate export income, governments in Third World countries have granted lumber concessions in public forests. Concessionaires often take the most valuable timber from a large area of forest, damaging the remaining trees through careless harvesting and rarely making any attempt to replant the area. In their desperate efforts to obtain foreign currencies, most of the heavily forested developing countries have set timber prices too low to enable them to replace their lost resources, according to Robert Repetto of the World Resources Institute.

Repetto described a particularly disturbing example of poor forest management in the Ivory Coast, where low concession fees have produced the highest deforestation rates in the world. As a result of the concession policy, the annual rate of deforestation has accelerated from 3.9 percent of the total forest area in the late 1960s to 5 percent in the late 1970s and an estimated 7 percent in 1989. Timber contractors have virtually exhausted the more valuable species, allowing the less valuable species to move in and replace them. By 1988, the country had lost 90 percent of its former 37 million acres of forest. Although hardwood timber is still the country's third largest export, Ivory Coast officials admit that they may have to start importing wood within 10 years.

Similarly, unless immediate steps are taken to halt deforestation, the once rich forests in the eastern states of Sabah and Sarawak in Malaysia will completely disappear by the end of the century, according to Friends of the Earth, an international environmental group. Friends of the Earth has also predicted that, in Costa Rica, the last remaining forests outside national parks will probably disappear by 1995.

A computer model developed by Alan Grainger of the Oxford Forestry Institute in Oxford, England, has predicted that hardwood timber exports from Asia, which

Figure 6. Fuelwood Scarcity. In developing countries, the majority of people depend on fuelwood to meet their household energy needs. Approximately two-thirds of those nations are experiencing acute fuelwood shortages, and many of the most severely affected countries are in sub-Saharan Africa. Each day these women must walk several miles from their village in Mali to gather the firewood they need. *Courtesy: F. Mattioli, World Food Programme.*

now account for 80 percent of all tropical hardwood exports, will decline in the next 10 years as resources dwindle. According to the model, tropical timber exports worldwide will peak before the year 2010, with commercial timber reserves declining to only one-quarter of current levels by 2020.

Even some developed nations are having difficulty harvesting their forests in a sustainable fashion. The U.S.S.R. has virtually ignored timber in its far north, while overharvesting in accessible regions near rivers, roads, and rail lines. Because the gap between harvesting and reforestation continues to grow, officials have estimated that Soviet timber resources will last no more than 50 years. Canada also harvests primarily in accessible regions, which has led some forestry experts to predict shortages of high-quality accessible timber within 20 years.

Slash-and-burn agriculture. Perhaps the most wasteful of all forestry practices is slash-and-burn agriculture, practiced by farmers throughout the Third World. In slash-and-burn agriculture, farmers cut several acres of trees, burn them to enrich the soil, and then plant crops on the cleared area. Unfortunately, this practice produces virtually no new agricultural land in the long term, while causing massive erosion of the soil, clogging dams and river basins with sediment, and converting productive land into desert.

Despite their lush profusion of plant life, most tropical forests grow on relatively poor soil, with most of the available nutrients tied up in the biomass. As plants die and decay on the forest floor, they contribute their nutrients directly back into the root systems of the forest for recycling. In this way, the forest continually recycles a limited amount of nutrients to support generation after generation of life. When farmers burn the forest, some of these nutrients still enter the soil, but much is washed away by rain and lost forever. The remainder is gradually consumed by the agricultural crops.

In the first year after a forest is cleared, the land yields a good crop. In the second year, the yield typically drops by half or more. By the third year, the land produces only 10 to 25 percent of the first year's yield, hardly enough to justify planting. After the third year, the farmers must abandon the cleared area and clear more forest, starting the cycle anew. Meanwhile, rainstorms wash soil from the abandoned land, leaving behind barren earth unsuitable for most uses.

Thailand is an example of how a country can incur extensive ecosystem damage as a result of slash-and-burn agriculture combined with heavy commercial exploita tion of forest resources. A 1987 report by the Thailand Develop-ment Research Institute noted that such activities reduced the area of the country's forests by 45 percent between 1961 and 1986. According to the report, the 500,000 people in the hills of Thailand who practice slash-and-burn agriculture have caused "extensive denudation of critical watershed areas," which has led to widespread flooding.

The Thai government has launched reforestation programs in some of the most severely affected areas, but by early 1989, those programs had made little progress in restoring the forest cover. In December 1985, Thailand adopted a National Forest Policy that set a goal of maintaining forests over 40 percent of the country's total area, with 15 percent devoted to conservation and 25 percent dedicated to sustainable timber production. Unfortunately, the country has not yet implemented the plan.

Combating Deforestation

Properly designed and managed tree plantations can produce a continuous supply of fuelwood and commercial timber. Such plantations now provide a significant amount of the wood supply in some industrial nations. Although some tropical countries have established tree plantations, current planting rates are generally not high enough to meet the growing demand for wood. According to Grainger's computer model of tropical forestry, tree plantations will provide only 1.7 percent of the wood harvested in the Tropics in the year 2000. As more plantations are established and more trees reach maturity, however, plantations may satisfy a much larger portion of the world's wood needs later in the twenty-first century.

Grainger and many other tropical forestry analysts believe that only better management of natural tropical forests, improved land-use planning, and the adoption of sustainable agricultural practices can prevent widespread deforestation. As of 1988, however, only about 3 percent of the world's tropical forests were properly managed, according to the U.N.

Tropical Forestry Action Plan. The Tropical Forestry Action Plan is an ambitious framework designed to conserve tropical forests while at the same time increasing their productivity. The plan originated with the U.N., the World Bank, and the World Resources Institute in 1985 and has gained considerable international support. It calls for independent and government-supported development-assistance agencies to coordinate their forestry-related programs and for those agencies, as well as governments, small farmers, and private industry, to spend $8 billion over a 5-year period to improve forestry management.

The plan stresses fuelwood production, watershed rehabilitation, sustainable production of wood for industry, conservation of forest ecosystems, and projects in social forestry, in which the residents of a region are taught to manage their forest resources properly. At a 1987 conference in Bellagio, Italy, representatives of major assistance agencies and government officials from 60 nations recommended 10 steps to advance the goals of the plan. These steps included expanding efforts to monitor deforestation, removing economic subsidies that encourage overharvesting, and increasing support for forestry research.

The World Bank, which funds many development projects in the Third World, has promised to make available a total of $350 million for forestry projects in 1989, more than twice its 1987 financial commitment. In the past, many environmental groups had criticized the World Bank for supporting projects that favored development at the expense of the environment. Shortly after Barber Conable became World Bank president in 1987, however, he promised to redirect the bank's efforts to develop environmentally sustainable projects. The bank has subsequently rejected several development projects that would have resulted in the destruction of forested areas.

International Tropical Timber Agreement. In 1985, representatives of 41 nations that together account for 95 percent of the tropical timber trade completed the International Tropical Timber Agreement (ITTA). Unlike most trade agreements, ITTA officially endorses conservation, committing parties to "encourage the development of national policies aimed at sustainable utilization and conservation of tropical forests and their genetic resources, and at maintaining the ecological balance of the regions concerned."

Desertification

Desertification, the degradation of fragile ecosystems in arid regions, may be the world's oldest environmental problem. UNEP's 1987 *State of the Environment* report estimated that 7.7 million square miles of fertile land have been lost to desertification throughout human history — an amount considerably larger than the 5.8 million square miles of land now cultivated for crops. In his 1982 book *Desertification*, Alan Grainger noted that early civilizations developed in regions with fertile soils suitable for agriculture. When soil quality deteriorated, either the civilizations declined or the people moved to other areas. "It is no coincidence that many ruins of great temples and palaces are today found amid sandy wastelands," he noted.

Figure 7. Restoring Soils and Reforesting with *Leucaena*. In tropical countries, the fast-growing tree *Leucaena leucocephala* is being widely used to halt erosion, restore soils, and reforest areas denuded by slash-and-burn agriculture. *Leucaena's* deep-growing leguminous root system enables it to thrive even in drought-stricken environments and to fertilize worn-out soils by enriching them with nitrogenous compounds. On the island of Cebu in the Philippines, farmers no longer have to cut down forest to open up new agricultural land. Even on steep hillsides such as this one, they are able to rotate their crops among *Leucaena*-enriched, terraced plots. *Courtesy: World Neighbors.*

For example, Mesopotamia — the fertile crescent formed by the flood plains of the Tigris and Euphrates rivers — supported a civilization of 17 million to 25 million people in the year 2000 B.C. The irrigation systems installed by the ancient Sumerians to grow wheat and barley, however, eventually left the soil waterlogged and contaminated with salt. Today the region supports only 10 million people.

Timbuktu, the oasis at the edge of the Sahara Desert, was known for centuries as the "Pearl of the Desert" because of its rich production of dates, oranges, and gum arabic. Overgrazing of nearby grasslands, however, reduced the land's productivity, and overpopulation depleted its water sources. In addition, climatic changes caused in part by desertification shifted the bed of the Niger River — once the main thoroughfare to the city — to a location 15 miles away, further isolating Timbuktu, which has no roads leading to it. Today, Timbuktu is a desert town that supports only a small population.

Expanding deserts. As defined by the U.N., desertification is the "process of degradation of fragile ecosystems in arid, semiarid, and subarid lands. Desertification is manmade, the result of misuse or overuse of the land by human beings and, in pastoral economies, by their livestock." Other types of misuse include salinization by improper irrigation systems and deforestation. Population pressure provides a common backdrop to all these problems, as a growing number of people try to subsist on less and less arable land.

A 1984 assessment by the U.N. indicated that about 17.5 million square miles of land had been affected by desertification, with 30 percent severely or very severely desertified. Moderate desertification affected about 470 million people, and severe desertification affected another 190 million. According to the U.N. assessment, about 81,000 square miles of land were rendered nearly or completely useless each year.

In regions where people are barely able to survive in good years, drought can have tragic consequences. For example, drought in Ethiopia and other parts of sub-Saharan Africa during the mid-1980s caused widespread famine, malnutrition, and deaths, and accelerated ongoing desertification in the region. At the peak of the crisis in 1984 and 1985, 30 million to 35 million people in 21 countries suffered serious malnutrition and about 10 million became homeless.

Many experts are particularly concerned about the possibility that desertification may be self-perpetuating, which could make it extremely difficult to reverse. According to this hypothesis, the loss of vegetation in desertified areas actually increases the likelihood of future droughts through a complex series of biological and atmospheric interactions. As vegetation declines, the Earth's surface reflects more sunlight back into the atmosphere, which becomes hotter and drier.

Overgrazing. In dry areas, with irregular rain and unevenly distributed vegetation, herders usually adopt a nomadic system, moving their herds with the shifting rainfall patterns. Nomadic herders were blamed for causing a famine in the Sahel in the early 1970s because they allowed their herds to grow too large and overgraze the land. According to *Desertification*, the number of cattle in Niger increased 450 percent between 1938 and 1961, when it reached 3.5 million. It rose another 29 percent by 1970. Similar increases occurred in other sub-Saharan countries.

Overgrazing leads to "xerification," or drying out, a phenomenon that precedes desertification. If a range is overstocked, or if livestock cluster in one area, such as around a watering hole, they eat most of the plant cover. In response, the plants draw nutrients up from their roots to grow new leaves, causing the root area to shrink and absorb less water during rains. As the more palatable plants are killed off by overgrazing, less palatable species become dominant. And as even these are overgrazed, the plant cover becomes scarce, the soil is trampled and compacted, and severe erosion may occur. When the infrequent but occasionally heavy rains characteristic of many arid zones occur, the water simply runs off rather than penetrating into the soil and recharging the groundwater.

To support the rapidly growing population of sub-Saharan Africa, more and more livestock are grazing on an ever-shrinking range. As a safeguard against an uncertain future, nomads have increased their herd sizes, building the herds in good years to forestall the effects of bad years. During the 1960s and 1970s, development-assistance agencies misguidedly instituted programs in the Sahel to settle the nomads around newly dug watering holes. Unfortunately, these settlement programs led to severe degradation of the surrounding land, especially as herd sizes increased.

A study of two grazing regions in Niger prepared for the U.N. Conference on Desertification in 1977 found that grazing around watering holes could not be successfully restricted. According to the study, "Grazing loads were two to three times those envisioned.... The cumulative effects of trampling and of grazing during the wet season resulted in devastated areas extending [6 to 7.5 miles] around watering points. It was a failure of pastures, not of water, that led to the death of stock."

Salinization from improper irrigation. In 1985, about 1 million square miles of agricultural land, producing 30 percent of the world's harvest, was under irrigation. Irrigation reduces the risk of crop failure during droughts and increases the yields of cereal crops up to sixfold in arid regions. However, improper irrigation can leave behind large quantities of salt, which can actually destroy the fertility of crop land. In the world's arid regions, over 1,900 square miles of irrigated land become desertified every year — roughly the same area that is freshly irrigated each year. Despite their benefits, irrigation systems can be expensive to build and maintain and can sometimes lead to problems, such as outbreaks of disease caused by waterborne parasites.

In addition, if irrigated land does not have adequate drainage, the soil becomes waterlogged, preventing salts from both the soil and irrigation water from being leached away. Water evaporates from the soil during the dry season, leaving behind the salts. Over time, the soil becomes increasingly salinized to the point that it can no longer support conventional agricultural crops.

Although farmers can reclaim salinized soil by constructing proper drainage systems, these systems are often expensive. Most governments will not provide financial assistance because they dislike investing heavily in dry lands. Lacking assistance, farmers usually abandon salinized lands. While biologists and genetic engineers are attempting to develop new agricultural crops that can grow and even thrive on saline soil, such crops may not be widely available for many years.

U.N. Conference on Desertification. During the early 1970s, the countries in the Sahel suffered through a catastropic drought that left 600,000 people dead and reduced much of the region to a barren wasteland. In an effort to prevent similar catastrophies in the future, the U.N. organized the 1977 Conference on Desertification (UNCOD) in Nairobi, Kenya. After studying the available scientific data, the conference delegates concluded that desertification stemmed primarily from improper land use rather than from climatic problems such as droughts.

The delegates also formulated the Plan of Action to Combat Desertification, which consists of 28 recommendations that call for all nations to end desertification by the year 2000 by improving range, soil, and water management and by implementing reforestation programs. UNEP, the agency charged with carrying out the plan, created a special desertification unit to coordinate projects with national governments and development agencies around the world. The delegates proposed expenditures of $4.5 billion each year during the following two decades to fight desertification. Although this proposed investment would be substantial, it represents a mere fraction of the cost of damage caused by desertification. In 1980, the U.N. estimated that desertification reduced world agricultural production by $26 billion per year.

In 1984, in its first assessment of the desertification

problem since the conference, UNEP noted little progress by any nation. In the introduction to the assessment report, the agency's director, Mostafa K. Tolba, concluded that nations had simply refused to make the fight against desertification a political priority. "We need to look no further than the absurdly inadequate level of contributions to the special account set up in 1979 to finance the [plan of action] for an illustration of the low priority nations attach to tackling the problem. By the end of 1983, it had received less than $50,000 — all from developing countries. The special machinery the General Assembly set up to mobilize funds to tackle desertification has raised in its 6 years of existence only $26 million."

In UNEP's 1987 report, *Rolling Back the Desert*, Tolba again criticized the continuing lack of progress against desertification: "Where are we 10 years after UNCOD? It grieves me to say it, but more land and tragically more people are affected by desertification today than in 1977." Tolba added, however, that "UNEP has not lost hope. I personally believe that desertification is stoppable.... Our goal is to roll back the desert."

Unable to mobilize national governments or development agencies into supporting projects to control desertification, Tolba's agency has redirected its efforts to local nongovernmental organizations that have been conducting small-scale but effective projects to halt desertification. UNEP has formed regional networks to improve communication and cooperation among these local organizations, and has published handbooks and pamphlets to help them tackle the problem.

Some of these efforts have attained moderate success. In a project funded by CARE in the Majjia valley in Niger, trees were planted as windbreaks, and in an Oxfam-sponsored project in Burkina Faso, stones were placed together along the land's contour lines to trap silt and moisture. Millet yields in both regions increased substantially after these simple steps were taken. According to UNEP's 1985 report, a Tanzanian project to reclaim land by planting trees and controlling grazing in the Kondoa district resulted in an "almost unbelievable ecological transformation from a desolate landscape of bare ground" to usable farmland.

Ironically, however, some well-intentioned efforts to reduce desertification have accomplished the opposite result. The drilling of wells, for example, has led some nomads to increase the size of their herds beyond the carrying capacity of the land. The herds eat and trample the vegetation to the point of destruction. When drought comes, as it always does, the animals die not from thirst but from lack of forage. Clearly, humanity still has much to learn about preserving the land.

Hazardous Waste Management

More than a decade has passed since the federal government declared the neighborhood around the Love Canal landfill site in Niagara Falls, New York, an emergency area. Tons of hazardous wastes buried there during the 1940s and 1950s had severely contaminated the groundwater, soil, and surface water and posed a serious threat to human health. Noxious chemicals seeping up through the ground forced hundreds of families to evacuate the area and abandon formerly valuable property, which was then surrounded by chainlink fences and dotted with skull and crossbones signs to warn of potential dangers. The government will spend tens of millions of dollars to try to remedy a tragedy that could have been avoided in the first place.

For many Americans, the Love Canal site has come to symbolize the consequences of indiscriminate hazardous waste disposal. Yet, despite all the headlines prompted by Love Canal, the citizens of the U.S. and most other countries have barely begun to comprehend the extent to which the traditional practices of dumping and burying industrial chemicals and other hazardous wastes have jeopardized human health and the environment.

The safe and permanent disposal of hazardous wastes is a critical issue in industrialized and developing countries alike. According to a Roper Organization Poll conducted in 1986 for the EPA, 93 percent of the U.S. citizens surveyed were concerned about the disposal of hazardous wastes. A 1986 Louis Harris survey on the country's environmental problems found that 68 percent of the respondents cited cleaning up hazardous waste sites as a very serious problem, while another 24 percent considered it a somewhat serious problem. Only 6 percent termed it little or no problem. In addition, 67 percent of the Harris poll's respondents favored the passage of a bill in Congress to reauthorize the 1980 Comprehensive Environmental Response, Compensation, and Liability Act (CERCLA), the law that created the so-called Superfund and authorized EPA to undertake investigations and cleanups at the nation's most serious hazardous waste sites.

National concern over the improper management of hazardous wastes, and the need for improved environmental protection in general, also was highly evident in a 1986 *New York Times*/CBS poll, which showed that 66 percent of the respondents believed that "protecting the environment is so important that requirements and standards cannot be too high, and continuing environmental improvements must be made regardless of cost."

People throughout the world fear for the future of the environment. The Organization for Economic Cooperation and Development (OECD) found in a 1987 survey that industrial waste disposal worried Italian, Danish, Dutch, and British citizens more than water pollution, air pollution, and nuclear waste disposal. A poll conducted in West Germany during 1986 revealed that 61 percent of the respondents thought that the "poisoning of the environment" by chemical wastes could "most decisively endanger mankind." Only 59 percent gave nuclear war that distinction. In a survey by the European Economic Community of its 12 member nations in 1986, 72 percent of all the people surveyed agreed that protection of the environment is an "urgent and immediate problem."

Scope of the Problem

Experts cannot reliably estimate the annual amounts of hazardous waste that different countries generate and discharge because nations have different definitions of what constitutes a "hazardous waste." As the Worldwatch Institute noted in its 1987 report, *Defusing the Toxics Threat: Controlling Pesticides and Industrial Waste*, a country's official figures on its hazardous or "special" or "industrial" waste generation and disposal may be more the result of that country's legal definitions of waste than a realistic assessment of the volume of contaminants actually discharged into the environment.

Most countries, however, define the term "hazardous waste" in a way that resembles the definition established in the Resource Conservation and Recovery Act (RCRA), a major federal environmental law that the U.S. Congress passed in 1976 to regulate, among other things, the generation, transportation, treatment, storage, and disposal of hazardous wastes. RCRA defined hazardous waste as waste that, "because of its quantity, concentration, or physical, chemical, or infectious characteristics," may cause serious illness or death or pose a "substantial present or potential hazard to human health or the environment" when improperly managed.

Many different substances can pose a hazard. Nearly 60,000 substances appear on the National Institute of Occupational Health and Safety's Registry of Toxic Substances. All of these display mutagenic or carcinogenic effects or cause birth defects at some level of exposure. Under RCRA, EPA defines as hazardous any wastes with at least one of four characteristics — ignitability, corrosiveness, reactivity, and toxicity. The agency has also listed as hazardous almost 500 specific industrial process wastes and chemicals. In addition, under the Superfund law, EPA requires that a company report an accidental release of any of more than 700 "hazardous substances" to the federal government whenever the spill exceeds a specified maximum amount.

Hazardous waste and its generators. According to OECD, the U.S. leads the world in annual hazardous waste generation. In its 1987 *Environmental Compendium*, OECD reported that, between 1980 and 1986, the U.S. generated about 250 million tons of hazardous waste per year — by far the greatest amount in any noncommunist country. West Germany ranked second, generating about 4.9 million tons per year, followed by Canada, with 3.29 million tons, France, with 2.02 million, and the U.K., with 1.5 million tons. Japan produced about 768,000 tons per year. In *World Resources 1988-89*, the World Resources Institute noted that figures on annual hazardous waste generation in the U.S.S.R. and other communist countries, such as Bulgaria and Czechoslovakia, were unavailable.

Other recent estimates of the amount of hazardous waste generated annually in the U.S. closely match the OECD figure. A 1985 Congressional Budget Office report — *Hazardous Waste Management: Recent Changes and Policy Alternatives* — estimated the total amount at 265.6 million metric tons during 1983. According to the Congressional Budget Office, while the chemical industry was responsible for about 127.3 million metric tons, or 48 percent, of this total amount, nearly all manufacturing and many nonmanufacturing industries contributed a portion. The primary metals industry accounted for about 47.7 million metric tons, or 28 percent of the total amount; the petroleum and coal products industry accounted for about 31.4 million metric tons, or 12 percent; and all other major industries, such as motor freight transportation, accounted for 59.2 million metric tons, or 13 percent.

Not all hazardous wastes fall under the jurisdiction of RCRA, which Congress passed in 1976 and amended in 1984 to regulate hazardous wastes more strictly. In a 1988 survey of its member companies, the Chemical Manufacturers Association (CMA) found that a representative number of U.S. chemical plants generated a total of 220.5 million tons of hazardous waste during 1986. Of this total, CMA noted, only 4.3 million tons, or 2 percent, came

Figure 8. Evacuated Homes at Love Canal. This residential area in Niagara Falls, New York, was built atop an abandoned canal into which more than 21,000 tons of hazardous chemicals had been dumped during the 1940s and early 1950s. When it became evident that chemicals were leaching from the top and sides of the site, 239 families in the houses immediately surrounding the landfill were relocated in 1978, and an additional 750 residents were relocated in 1980. *Courtesy: Copyright 1979, David Rinehart.*

under RCRA regulation. The other 216.2 million tons, or 98 percent, represented wastewater, which is covered by the Clean Water Act (CWA), not RCRA. According to CMA, most of these wastewaters found their way into publicly owned wastewater treatment facilities. The Congressional Budget Office report confirmed that, during 1983, about 22 percent of the hazardous wastes generated in the U.S. went into either surface waters or CWA-regulated wastewater treatment facilities.

According to a 1986 EPA survey of 3,000 U.S. hazardous waste treatment, storage, disposal (TSD), and recycling facilities, facilities not regulated under RCRA handle over half of the hazardous waste generated in the U.S. each year. Manufacturing plants' own wastewater treatment and solvent-recovery facilities, as well as their facilities that use hazardous waste as fuel, managed most of the estimated 322 million metric tons of hazardous wastes that went to non-RCRA facilities in 1985. RCRA facilities, on the other hand, managed 243 million metric tons of hazardous wastes that year.

The Conservation Foundation, which discussed the EPA survey in its *State of the Environment: A View toward the Nineties*, explained the divergence of these figures from other accepted estimates of the total amount of U.S. hazardous waste generation and management by pointing out that the survey "included waste regulated as hazardous under either state or federal laws," and that the "definition of hazardous waste varies from state to state."

Not all hazardous waste is generated by large industries. J. Winston Porter, then-EPA Assistant Administrator for Solid Waste and Emergency Response, reported at a 1986 hearing held by the Environment, Energy, and

Natural Resources Subcommittee of the U.S. House of Representatives' Government Operations Committee that about 77,000 RCRA-regulated facilities in the U.S. generate what the agency defines as large quantities of hazardous waste — more than 1,000 kilograms per month.

In 1984, Congress had amended RCRA to incorporate "small-quantity generators" (SQGs) — facilities that generate between 100 and 1,000 kilograms of hazardous waste per month. About 125,000 SQGs entered the RCRA regulatory system at that time; however, after completing a national SQG survey in 1985, EPA concluded that these generators produced only about 940,000 metric tons per year, or less than 1 percent of the amount of hazardous waste generated. According to the SQG survey, nonmanufacturing firms, such as those engaged in construction, laundry, and photographic processing, account for about 85 percent of SQGs; the other 15 percent includes printers and enterprises that manufacture metals, chemicals, and textiles.

Hazardous waste sites. Although there are certainly hundreds of thousands of uncontrolled hazardous waste sites around the world, no one can estimate the true number. A 1988 report by the U.S. General Accounting Office (GAO) — *Superfund: Extent of Nation's Potential Hazardous Waste Problem Still Unknown* — concluded that, because EPA has failed to maintain a proper inventory of hazardous waste sites in the U.S., no one knows the actual number. Under CERCLA, the GAO report noted, EPA was supposed to compile and annually update a list of all TSD facilities and hazardous waste sites. This list was to provide candidates for inclusion on the National Priorities List (NPL), which identifies all the sites that require urgent attention and qualify for Superfund-financed cleanups.

The GAO report said that—although the list, known as the Comprehensive Environmental Response, Compensation, and Liability Information System (CERCLIS) inventory, contained 27,200 sites as of August 1987—other information from EPA and various federal agencies indicates that it does not include between 130,340 and 425,480 potential sites. As of November 1988, the CERCLIS inventory listed only 30,013 sites. Moreover, as of June 1988, the NPL contained only 799 final and 378 proposed sites. New Jersey ranks first in the number of final and proposed NPL sites, with 110, followed by Pennsylvania with 97, California with 88, Michigan with 81, and New York with 76.

The U.S. Office of Technology Assessment (OTA) estimated in its 1985 *Superfund Strategy* that one day the NPL could contain as many as 10,000 sites, and that the total cleanup costs for these sites could run as high as $100 billion. According to OTA, EPA's 1985 projections of a 2,500-site NPL and total cleanup costs of about $23 billion dramatically underestimated the scope of the problem. At least 1,000 of the hazardous waste management facilities and 5,000 of the solid waste management facilities operating in 1985 would make the list of potential NPL sites, OTA said. Revising the process by which sites make the list could add another 2,000 sites.

OTA recently revised its cleanup cost projections, stating in a 1989 report entitled *Assessing Contractor Use in Superfund* that, "today, with new information on how many sites require cleanup...a more realistic estimate is perhaps $500 billion...over at least 50 years."

More than just NPL sites require cleanup in the United States. After completing an analysis of RCRA facility assessments undertaken at 550 RCRA hazardous waste management facilities during 1987, EPA projected that 77 percent of all operating RCRA land disposal facilities, 70 percent of all closing RCRA land disposal facilities, and 56 percent of all operating and closing RCRA incinerator facilities may be releasing contaminants that will require corrective action. In a 1988 report on RCRA facility closures, GAO concurred that the government will need to clean up as many as 2,500 RCRA facilities.

Other countries also face huge expenditures to clean up hazardous waste sites. For example, according to the Worldwatch Institute's 1987 paper, West Germany, which depends on groundwater as its primary source of drinking water, may have to contend with as many as 35,000 uncontrolled hazardous waste sites and spend at least 18 billion German marks, or about $10 billion, during the next decade to control them. Denmark projects that it will spend one billion Danish kroner, or about $140 million, to address contamination problems at 2,000 sites. The Netherlands has estimated that it will need to spend about $5.6 billion during the next 15 to 20 years to clean up at least 5,000 sites that threaten drinking water supplies or abut residential areas. The World Resources Institute and the International Institute for Environment and Development reported in *World Resources 1987* that up to 95 percent of the hazardous wastes generated in Eastern Europe go into landfills, and that few of these facilities have "adequate safeguards to protect the environment from contamination." Some Eastern European nations, including Hungary and Poland, have begun regulating hazardous waste only during the past few years.

Resource Conservation and Recovery Act

In October 1976, Congress took the first major step toward regulating hazardous waste management in the U.S. by enacting RCRA, which established a national "cradle-to-grave" system for tracking hazardous wastes from their generation, through their transport, storage, and treatment, and finally to their destruction or disposal.

EPA issued the first RCRA regulations in May 1980. These regulations required waste generators to determine whether their wastes are hazardous, to securely package and properly label any hazardous wastes that must be transported to off-site TSD facilities, to prepare a "Uniform Hazardous Waste Manifest" that identifies the type and quantity of hazardous waste being transported and that must accompany the waste wherever it goes, and to comply with recordkeeping and reporting requirements.

Under these RCRA regulations, the owner/operator of an existing TSD facility received "interim status," which would permit the facility to continue operating until EPA had finally determined whether to issue a final operating permit. If EPA decided not to do so, it would require the facility to close. Issuance of a final RCRA permit depended on whether a given owner/operator could meet certain design, siting, and performance standards, as well as requirements for closure and post-closure procedures and emergency response plans in the event of an accident.

In addition, a TSD facility owner/operator would have to comply with requirements regarding facility inspection, groundwater monitoring, recordkeeping, and financial capability. RCRA rules also mandated that, if a land disposal facility did not have groundwater-monitoring systems, leachate-collection systems, and double liners made of impermeable synthetic material, its owners/operators would have to sample and analyze the groundwater around the facility's TSD units at least twice a year to detect any leakage that might contaminate underlying aquifers.

Groundwater contamination. It soon became clear, however, that the contamination of the nation's groundwater by substances migrating from hazardous waste TSD units — even those regulated under RCRA — had escalated into a much more serious problem than experts had first estimated. In a 1981 assessment, EPA found that over 70 percent of the 80,000 hazardous waste surface impoundments in the U.S. lacked leak-proof liner systems to prevent contaminants from permeating the surrounding soil and groundwater. About 30 percent of those impoundments were built above usable groundwater aquifers, the agency said.

The Congressional Budget Office concluded in its 1981 report that about 72,000, or 90 percent, of the surface impoundments in the U.S. threatened the integrity of the nation's groundwater supplies in one way or another. According to the U.S. Council on Environmental Quality's 1981 report *Contamination of Groundwater by Toxic Chemicals*, the presence of organic hazardous substances forced the closure of hundreds of drinking water wells between 1978 and 1981. In its 1984 report *Groundwater Protection Strategy,* EPA announced that contamination had rendered more than 8,000 private, public, and industrial wells unusable during that year. Confronted with this growing body of evidence of increasing groundwater contamination, Congress moved to strengthen the federal hazardous waste management law.

1984 RCRA Amendments. In November 1984, Congress enacted the RCRA Amendments, which broadened the scope of the existing RCRA regulatory program and imposed much stricter controls on hazardous waste generators and TSD facilities. Almost every major provision of the amended law reflected Congress's concern about the need to further protect U.S. groundwater resources.

For example, the amended RCRA regulations required that, by late November 1988, the owners/operators of existing hazardous waste surface impoundments meet new "minimum technology requirements" by installing double liners and groundwater-monitoring and leachate-collection systems. If any owner/operator failed to retrofit its surface impoundments with the mandated equipment by the deadline, it would have to submit a closure plan to EPA. The regulations also required installation of double liners and leachate-collection and groundwater-monitoring systems in all new and remodeled landfills and surface impoundments.

Congress did not extend these minimum technology requirements to existing hazardous waste landfills, because EPA had determined that the risk incurred by workers to retrofit these units would far outweigh any possible benefits. Congress did require, however, that owners/operators of all existing and new land disposal facilities conduct groundwater monitoring, regardless of whether their facilities met the minimum technology requirements.

In addition, the 1984 Amendments increased EPA's authority to compel an owner/operator to clean up any contamination that resulted from a release or leak. The original RCRA legislation required EPA either to prove that an imminent hazard existed before requiring a facility owner/operator to correct contamination, or to establish a schedule for corrective action when it issued or renewed a facility's final permit — a process that could take years. Now EPA can order an owner/operator to clean up the contamination resulting from a release or leak even if it occurred before RCRA or its Amendments were enacted.

Problems under RCRA. Ironically, while the 1984 RCRA Amendments may have raised public awareness of the fragility of the U.S.'s groundwater resources, the revisions to the law have caused some problems. For instance, the new regulations forced the closure of hazardous waste land disposal facilities that could not meet the stringent design, performance, groundwater-monitoring, and financial responsibility requirements by November 1985.

EPA's J. Winston Porter testified before the Environment, Energy, and Natural Resources Subcommittee of the House of Representatives' Government Operations Committee in December 1987 that the number of land disposal facilities that had closed or planned to close "had increased dramatically" since the passage of the 1984 RCRA Amendments. According to Porter, between November 1985 and December 1987, 995 — or about 65 percent — of the 1,535 land disposal facilities that had operated in the U.S. before the Amendments took effect either decided to close or were required to do so because they could not comply with the new RCRA standards.

The sudden and simultaneous closure of so many RCRA facilities nearly overwhelmed EPA, which lacked both the financial resources and the personnel to handle all of the complex paperwork that the closures generated. As Representative Michael Synar pointed out at the December 1987 House Environment, Energy, and Natural Resources Subcommittee hearing, the agency's failure to close the inactive facilities promptly and properly could transform them into Superfund sites or, in his words, "ticking toxic time bombs."

The stricter rules created another double-edged sword. While the increased cost of disposing of hazardous wastes has encouraged economically attractive alternative waste management methods — including incineration and chemical and biological treatment, as well as waste reduction and recycling techniques — the soaring costs have also prompted some hazardous waste producers and TSD facility owners/operators to dispose of hazardous waste illegally.

Land disposal ban. The 1984 RCRA Amendments strongly encouraged treatment over land disposal by banning the disposal of all untreated hazardous wastes in landfills, surface impoundments, underground injection wells, and other land-based units after May 1990. The Amendments specifically stated that "reliance on land disposal should be minimized or eliminated, and land disposal...should be the least favored method for managing hazardous wastes."

Congress designed the land disposal ban to take effect through a series of deadlines called "hammers." Under RCRA's hammer provisions, EPA's failure to establish a treatment method or level for any prohibited waste will result in the automatic banning of that waste altogether. The agency has already issued several sets of regulations to implement the land disposal ban. For instance, it prohibited the landfilling of containerized hazardous liquids in February 1986, the land disposal of wastes containing solvents and dioxins in November 1986, and the "California List" — untreated liquid and solid wastes containing cyanides, halogenated organic compounds, metals, and polychlorinated biphenyls (PCBs) — in July 1987.

In addition, the 1984 RCRA Amendments incorporated into existing RCRA regulations a list of wastes that EPA had identified as hazardous, and required that the agency ban the land disposal of the first one-third of these wastes by August 1988, the second one-third by June 1989, and the final one-third by May 1990.

Underground storage tanks. The RCRA Amendments also established a comprehensive regulatory program — enforced by state and local agencies with EPA oversight and assistance — for the 1.4 million underground storage tanks (USTs) containing petroleum products, such as gasoline and crude oil, and substances that are considered hazardous under CERCLA. Of these 1.4 million USTs — which RCRA defines as tanks with 10 percent or more of their volume, including associated piping, located underground — about 95 percent hold petroleum products, while 50 percent store gasoline at service stations. Only about 54,000 USTs, or 4 percent, contain CERCLA hazardous substances.

EPA has estimated that 80 percent of all current USTs consist of unprotected bare-steel tanks, which will ultimately corrode and leak. In 1986, the agency reported that about 75,000 of the 1.4 million USTs had already leaked, and it projected that 350,000 more would begin leaking by 1991.

Under RCRA, EPA established regulations for tank systems containing RCRA hazardous wastes. These regulations set out standards for tank design, installation, operation, inspection, and maintenance, as well as requirements for primary and secondary containment systems with leak-detection capabilities, release response plans, and closure and post-closure procedures. The regulations, issued in September 1988, require that operators protect all new petroleum USTs against corrosion and that manufacturers include equipment to prevent and detect leaks and overflows. Owners/operators of existing petroleum USTs must retrofit their tanks with corrosion-protection, leak-detection, and spill/overflow detection equipment by December 1998. The standards also required new chemical USTs to include secondary containment systems such as double walls, concrete vaults, or impenetrable liners, as well as leak detection and spill/overflow systems.

The RCRA UST technical standards also mandate that UST owners/operators clean up any contamination caused by a leaking UST. In addition, upon final closure of a chemical or petroleum UST, the owner/operator must empty the tank, determine whether it has caused any environmental damage, and either remove the tank or fill it with inert materials.

In October 1988, EPA issued a final rule requiring that the owners/operators of petroleum-containing USTs demonstrate the financial capability to take prompt corrective action and compensate third parties for injuries or damages that might be caused by releases from such tanks.

Infectious wastes. During July 1988, hospital debris began washing ashore on the beaches of Massachusetts, New Jersey, New York, and Rhode Island. Only 1 month earlier, EPA had solicited public comments on the risks posed to human health and the environment by so-called infectious wastes, as well as on the agency's potential role in regulating such wastes under RCRA. Infectious wastes, which can cause infectious disease, include the cultures of infectious agents, human blood and blood products, pathological wastes, contaminated hypodermic needles and scalpels, and contaminated body parts and bedding.

Although EPA had not established regulations for these wastes by early 1989, it had considered regulating them as hazardous wastes under RCRA Subtitle C, or as nonhazardous solid wastes under RCRA Subtitle D. In May 1986, the agency issued nonbinding guidelines for infectious waste handling, storage, and disposal, but in November 1987, an agency task force determined that states and hospitals already had adequate infectious waste regulatory policies.

According to EPA, as of July 1988, 39 states had adopted infectious waste regulations, with five more soon to follow suit. A 1988 Council of State Governments report entitled *State Infectious Waste Regulatory Programs* concluded that, while 28 states had not developed any regulatory structure to address infectious wastes when EPA issued its nonbinding guidelines in 1986, since then, 25 of the states had begun to write and issue regulations to control infectious waste management.

Most states focus their infectious waste management programs on hospitals, which routinely generate the highest volume of infectious waste, according to the Council of State Governments report. In most states, hospital licensing bureaus monitor the on-site generation, treatment, and disposal of infectious wastes, while solid waste management authorities regulate their off-site disposal. The report noted that infectious waste falls into the category of "special" waste in 26 states, hazardous waste in 13 states, and nonhazardous waste in 11 states.

In November 1988, President Ronald Reagan signed into law a bill requiring EPA to develop a "cradle-to-grave" medical waste tracking system that would monitor the transport, treatment, and disposal of medical wastes in 10 states: Connecticut, Illinois, Indiana, Michigan, Minnesota, New Jersey, New York, Ohio, Pennsylvania, and Wisconsin. The program, which took effect in June 1989, requires medical waste generators, transporters, and treatment and disposal facilities to maintain manifest forms in order to monitor the movements of medical wastes.

Illinois, Indiana, Michigan, Minnesota, Ohio, Pennsylvania, and Wisconsin chose to withdraw from the program, as allowed under the law, because they either have established or are now developing their own tracking programs. Louisiana, Rhode Island, and the District of Columbia, however, have decided to participate, and EPA will report to Congress on the program's success at the end of a 2-year demonstration period.

Superfund

When Congress passed RCRA in 1976, it took a step toward preventing the future contamination of the U.S.'s soil, groundwater, and surface water by hazardous wastes. The law did not, however, provide EPA with either the authority or the funding to address actual or potential releases of substances at the nation's uncontrolled hazardous waste sites. Nor did RCRA enable EPA to compel the responsible parties to clean up or at least pay for cleaning up any resulting environmental damage. In December

1980, at the height of the national controversy over Love Canal, Congress enacted CERCLA, which created the $1.6-billion Hazardous Substances Response Trust Fund, (also called Superfund), to cover the costs of cleaning up abandoned hazardous waste sites.

To finance the fund, Congress authorized EPA to collect a tax on 42 specified feedstock chemicals, as well as on crude oil and imported petroleum. This tax generated about 86 percent of the fund's revenues. Although about 600 companies paid the CERCLA-mandated tax, the 10 largest petroleum and chemical companies accounted for almost 50 percent of the annual tax revenues.

CERCLA expanded the National Oil and Hazardous Substances Pollution Contingency Plan, or National Contingency Plan (NCP), to provide a regulatory framework for EPA's Superfund program. NCP outlines the procedures and sets the standards for federal, state, and private party responses to releases of oil and hazardous substances, and provides EPA with two options for cleaning up uncontrolled hazardous waste sites — short-term emergency responses designed to mitigate an immediate threat, and remedial actions aimed at long-term removal of a threat.

Under the first option, EPA could initiate action before determining responsibility for the problem. To recover its investigative and remedial costs, as well as damages, the agency could sue any potentially responsible parties (PRPs) that it identified: the past and present owners and operators of the site, every generator that had ever disposed of hazardous waste there, even the transporters that had carried wastes to the site. If EPA could not locate any PRPs at a site, then the Superfund would absorb the cleanup costs.

The second option enabled EPA to force PRPs to clean up a site, either through a mutual agreement between the agency and the parties or through an enforcement order requiring the PRPs to take action. If the PRPs refused to comply with the order, EPA could seek punitive damages of up to three times the cleanup costs. When CERCLA's authority expired in 1985, however, the agency had completed remedial actions at only 10 sites.

Superfund overhauled. When Congress saw that CERCLA had failed to address the growing number of uncontrolled hazardous waste sites in the U.S., it passed, in October 1986, the Superfund Amendments and Reauthorization Act (SARA), which not only extended the life of the Superfund program for another 5 years but also increased its funding more than fivefold, to $8.5 billion. SARA maintained CERCLA's basic approach of authorizing EPA to initiate cleanups at Superfund sites, compel PRPs to perform cleanups, and recover costs. The new law, however, established much stricter standards for cleaning up those sites, expanded the already extensive role of the states in the cleanup process, and required EPA to encourage more public participation in that process.

To guarantee that the U.S. properly managed its RCRA hazardous wastes and the wastes excavated from Superfund-funded and NPL sites, SARA required that every state assure EPA no later than October 1989 that the state can adequately treat or dispose of all hazardous wastes generated within its borders over the next 20 years. Any state that could not make this assurance would receive no cleanup money from the Superfund.

Chemical accidents such as the December 1984 release of the poisonous gas methyl isocyanate from a pesticide plant in Bhopal, India — an incident that killed over 2,600 Bhopal residents and injured at least 186,000 others — motivated Congress to take action to ensure that state and local governments respond quickly and effectively to environmental emergencies. As a result, SARA established the Emergency Planning and Community Right-to-Know Act, or SARA Title III, which requires states and communities to collect and make publicly available information concerning toxic chemical use in their jurisdictions. Under Title III, facilities producing or handling more than a specified amount of any of about 400 hazardous chemicals must submit to state and local emergency committees and local fire departments detailed information about the toxic chemicals used at their plants.

In the case of USTs, SARA added to RCRA new authorities for federal and state investigation and cleanup of contamination from USTs leaking petroleum products. To clean up such contamination, it established a $500 million trust fund, which would be raised from a 0.1-cent-per-gallon tax on motor fuels. Under SARA, EPA can use the trust fund to pay for remedial actions at leaking UST sites where it cannot identify the PRPs or where the PRPs have gone bankrupt; the agency may also undertake emergency cleanups and sue the PRPs to recover its costs.

Effectiveness of the Superfund program. Just as the 1984 RCRA Amendments emphasized treatment over land disposal, SARA requires EPA "to the maximum extent practicable" to reduce permanently the volume, toxicity, or mobility of hazardous substances with alternative treatment technologies rather than with containment techniques. Off-site transport and disposal without treatment under SARA became a "least favored" cleanup method.

In July 1987, EPA issued a document outlining the key criteria that the agency's regional offices should use when selecting Superfund site remedies: overall protection of human health and the environment; compliance with other federal and state laws; permanence; short-term effectivness; reduction of toxicity, mobility, or volume through treatment; implementability; cost-effectiveness; state concerns and acceptance; and community concerns and acceptance.

Some environmentalists worried that these guidelines allowed cost-effectiveness to outweigh other factors in the remedy selection process. For example, environmental groups such as the Natural Resources Defense Council charged that, in placing considerations about the cost of a site remedy on the same level as or above considerations regarding the remedy's safety and permanence, EPA had illegally subverted the priorities of the Superfund statute.

In the end, EPA's revised draft of the guidelines — published for comments in November 1988 — included a three-tiered framework for ranking the nine criteria. According to EPA, the proposed framework clearly reflects SARA's mandates that remedial actions must, in descending order of importance, fully protect human health and the environment, achieve permanent cleanups via treatment technologies whenever possible, and be accomplished at an acceptable cost.

Critics of EPA have cited flaws in the agency's implementation of the Superfund program. From the enactment of CERCLA in 1980 through December 1988, the agency had completed cleanup activities at only 48 NPL sites, and had managed to remove only 18 of them from the NPL. In a June 1988 report entitled *Are We Cleaning Up? Ten Superfund Case Studies*, OTA concluded that "Superfund remains largely ineffective and inefficient" because of a "lack of centralized management control" and the inexperience, poor training, and high turnover rate of the Superfund work force.

Calling the selection of a cleanup technology at a Superfund site a "crucial step," the OTA report found EPA's decisions "questionable" in the ten cases studied: "the range of cleanup alternatives was too narrow," thus precluding the use of "many good, permanently effective waste treatment technologies"; "the analysis was not comprehensive and was not fair to different technologies; the study work was not internally consistent; mistakes were made in calculations and [cost] estimates; critical assumptions were false"; and "conclusions were stated without analysis or documentation." In addition, according to the report, EPA too often chose temporary remedies, such as containment and capping on-site, which it found "cheaper in the short run."

According to an April 1988 report, *Status of EPA's Superfund Program,* by the House Appropriations Committee, EPA "needs to intensify its enforcement efforts under SARA." In its report, the committee criticized the agency's "propensity to place far greater reliance on the use of the Superfund to finance cleanups rather than to vigorously pursue an enforcement program designed to have PRPs take full responsibility for the funding and management of cleanup actions." The report suggested that EPA's failure to use its enforcement authority aggressively to force PRPs to undertake and fund remedial work at Superfund sites could undermine the overall effectiveness of the Superfund program.

Using EPA's estimated average cleanup costs of about $21 million per site, the House Appropriations Committee report concluded that, if the agency continued to pay for NPL site cleanups itself, EPA would be able to complete cleanups at only about 400 of the 951 final NPL sites then listed. Moreover, if average cleanup costs for an NPL site increased to between $30 million and $40 million, as expected under SARA requirements, then the agency would be able, on its own, to fund the cleanup of fewer than 300 of those 951 sites.

International Waste Transport

In September 1988, the Karin B, a German-registered ship carrying about 2,100 tons of hazardous waste generated in Italy and West Germany, returned to Italy from Nigeria to unload her cargo after being denied permission to dock in England, France, Spain, West Germany, and the Netherlands. The Italian government had originally sent the Karin B from Italy to Koko, Nigeria, with a cargo of more than 6,000 drums of chlorinated solvents, waste resins, and PCBs for disposal there. The Nigerian government, however, charging that industrial nations were using its country as an inexpensive dumping ground, forced the Italian government to order the Karin B to return to Italy.

The odyssey of the Karin B has illustrated the growing controversy over international hazardous waste transport and the sometimes exploitative relationship between waste-exporting countries — usually the U.S. and the highly industrialized nations of Western Europe — and waste-importing countries — most often developing African nations with few, if any, regulations controlling hazardous waste disposal.

In African countries, dumping wastes in a landfill can cost as little as $20 per ton. As a result, according to a 1988 audit of EPA's waste-export regulation program by the agency's Inspector General's Office, waste brokers in the U.S. may have exported thousands of tons of wastes without notifying EPA of the shipments, in clear violation of RCRA. The Inspector General's audit found that, because EPA and the U.S. Customs Service have not jointly developed a comprehensive system for monitoring international hazardous waste shipments from the U.S., these brokers have been able to avoid meeting the RCRA requirements.

The current RCRA hazardous waste export regulations, which EPA issued in August 1986, require waste generators to notify EPA that they intend to export hazardous wastes at least 60 days before actually beginning shipments. A generator must also provide EPA with the following information: the foreign TSD facility receiving the waste; the type and volume of waste exported; the frequency of shipments; and the mode of transportation used to ship the waste. Before the hazardous waste leaves the U.S., EPA requires a written acknowledgement of consent from the importing country; a manifest from the transporters identifying both the date and location of a shipment's departure from the U.S.; and an annual report, filed with EPA every year, identifying the quantities of waste exported to foreign TSD facilities and the number of shipments undertaken during the preceding calendar year.

Throughout 1988 and 1989, both importing and exporting countries took steps to control the transboundary movement of hazardous wastes. In May 1988, the Council of Ministers for the Organization of African Unity, which is a coalition of African nations, passed a resolution denouncing the unauthorized dumping of hazardous waste in Africa as a crime against Africa and the African people. The resolution, which did not identify any countries or hazardous waste exporters by name, condemned all "transnational corporations involved in the introduction, in any form, of nuclear or industrial wastes in Africa," and demanded that the responsible corporations clean up contaminated areas.

In June 1988, a bill was introduced into the U.S. Senate that would prohibit the export of hazardous and solid waste from the U.S. unless the waste exporter obtained an EPA permit certifying that the management of the waste in the receiving country will meet both the U.S.'s and the importing country's requirements for hazardous waste handling, treatment, storage, and disposal. In addition, the bill, entitled the Waste Export Control Act, would require prospective hazardous and solid waste exporters to provide EPA with information about the type, toxicity, and volume of the hazardous or solid waste being transported.

International efforts to regulate waste. Representatives from 44 countries gathered in Geneva, Switzerland, at a November 1988 meeting sponsored by UNEP to draft a treaty on regulating transboundary shipments of hazardous waste and promoting the international exchange of information and technologies to help control that waste. Under the draft treaty, which the countries had been negotiating since 1985, contracting nations would adopt a program — similar to the U.S.'s notification program — whereby hazardous waste exporters would be required to provide the maximum amount of information possible to an importing country before delivering a waste shipment. In addition, each contracting nation would be required to obtain from the waste exporter a "hazardous waste movement" document and ensure that any waste it imports is packaged, labeled, and transported "in conformity with generally accepted and recognized international rules and standards."

The draft UNEP treaty also proposes the establishment of an international agency that would accept information from countries on transboundary waste movements, accidents that occurred during waste shipments, and steps taken by contracting countries to implement the treaty's provisions. The treaty also encourages the establishment of an international fund to create centers, especially in developing countries, for research into hazardous waste management and disposal.

One provision in the draft UNEP treaty has sparked considerable debate. That provision would prevent a contracting nation from negotiating a trade agreement with a noncontracting nation unless the agreement reflected the aims of the treaty. During the November 1988 meeting, representatives from Norway proposed including a limited waste trade ban in the treaty to prohibit a contracting nation from exporting hazardous waste to a noncontracting nation unless the two countries had reached an agreement to provide for the satisfactory disposal of that waste. Supporters of the limited waste trade ban have argued that, without such a ban, developing countries lacking the capacity to dispose of hazardous waste adequately could choose "poison over poverty."

Western European countries have also undertaken efforts to halt pollution in their own "backyards." In December 1987, Belgium, Denmark, France, the Netherlands, Norway, Sweden, the U.K., and West Germany agreed to the following measures aimed at preventing the further contamination of the North Sea, the Mediterranean Sea, the Baltic Sea, and portions of the Atlantic Ocean along the European coastline: (1) phase out the ocean dumping of most industrial wastes and sewage sludge by the end of 1989; (2) reduce the amount of hazardous waste incinerated at sea by not less than 65 percent by 1991 and completely by the end of 1994; and (3) cut in half by 1995 the amounts of pollutants, such as nitrates and phosphorus, that the eight countries currently discharge into the marine environment. Each year, about 74 million metric tons of industrial wastes and sewage reach the North Sea.

In October 1988, the 65 countries participating in the London Dumping Convention resolved that they would "minimize or substantially reduce" the use of ocean incineration to dispose of liquid hazardous wastes by 1991, and would re-evaluate "as early in 1992 as possible" the use of ocean incineration as a means of disposing of hazardous wastes.

Hazardous Waste Management Technologies

In the wake of the 1984 RCRA Amendments and SARA, the hazardous waste management industry in the U.S. has entered a period of growth and change. The traditional approach of placing untreated hazardous wastes in land-based disposal units — whether landfills, surface impoundments, or underground injection wells — has given way to consideration of innovative treatment technologies, such as incineration and bioremediation.

Incineration. Incineration, an effective and economical way to destroy hazardous wastes, is perhaps the most rapidly expanding sector of the hazardous waste management industry. According to testimony by EPA's J. Winston Porter before a September 1986 hearing of the Environment, Energy, and Natural Resources Subcommittee of the House of Representatives' Government Operations Committee, about 340 incinerators operating at private industrial facilities and 13 commercial incinerators were burning about 2 million metric tons of hazardous wastes per year, with the 13 commercial incinerators burning about half the total amount.

A survey published by the consulting firm ICF, Inc., in 1986 found that the use of incineration and thermal treatment in hazardous waste management was growing at such a pace that the incinerators operated by the 18 responding companies were running at almost 90 percent of their full capacity. These companies predicted severe and chronic shortages in incineration and thermal treatment capacity by the late 1980s.

According to its 1988 survey — which focused on the 1986 and 1987 activities of the 14 companies believed to operate at least 70 percent of the commercial hazardous waste management facilities in the U.S. and control at least 40 percent of the revenues — ICF found, however, that the incineration capacity of these companies "grew at a faster pace than the increase in waste volumes received for incineration," increasing by 98 percent from 318,000 wet metric tons in 1985 to 631,000 wet metric tons in 1987.

In addition, according to the survey, even greater incineration capacity increases — anywhere from 200 to 300 percent — were expected to occur between 1988 and 1991. The companies attributed "much of the turnaround" to three trends: the decline in PCB-contaminated liquid incineration and the subsequent "freeing up" of incinerator capacity for RCRA hazardous wastes; the entry of cement kilns, light aggregate kilns, and industrial boilers — known as recycle kilns — into the thermal treatment services market; and the expected issuance, by 1991, of final RCRA permits to a large number of hazardous waste incinerators.

Several other factors will determine whether the commercial hazardous waste incineration industry will continue to grow, according to the survey: whether EPA can quickly process new and final RCRA operating permit applications for the construction or expansion of incinerators; whether the agency will stringently regulate recycle kilns; and whether it will select incineration as the best available technology for treating certain wastes affected by the RCRA land disposal ban.

The most common type of incinerator — the liquid injection system — pumps liquid organic waste through atomizing nozzles into a combustion chamber, where the temperatures reach as high as 3,000°F. The waste typically stays in the combustion chamber for 0.5 to 2 seconds, during which time high temperatures decompose the waste's hazardous constituents into substances nonhazardous enough to meet RCRA standards (elimination of at least 99.99 percent of the hazardous constituents). Liquid injection systems, like other types of incinerators, do not destroy heavy metals or other inorganic hazardous wastes; instead, these substances end up in the incinerator ash, which itself may be hazardous and require treatment and/or disposal in a secure landfill.

Another common type of incinerator is the rotary kiln. The rotation of its cylindrical combustion chamber swirls the waste inside, thereby exposing more of the waste material for burning and improving the efficiency of the process. Unlike the liquid injection system, the rotary system can burn both liquid and solid organic and inorganic hazardous wastes.

In January 1989, EPA issued the first-ever permit for a new type of system: a mobile hazardous waste incinerator that uses infrared thermal treatment technology. During the infrared incineration process, a rotating belt transports hazardous waste through the incinerator's main combustion chamber, where an infrared heating element creates temperatures capable of vaporizing the waste's hazardous constituents. The vaporized waste then moves into a secondary combustion chamber, where another infrared heating element uses even higher temperatures to further oxidize the hazardous constituents. Unlike rotary kiln incinerators, which require combustion beds made of heavy material such as refractory brick, the infrared incinerator uses a belt made of lightweight fiber to support the waste material in the combustion chamber. Compared to other mobile incinerators, infrared incinerators can handle larger volumes of waste and are more easily transportable.

Under its Superfund Innovative Technology Evaluation (SITE) Program, EPA has begun testing a circulating bed combustion system that achieves high efficiency by entraining the wastes in a high-speed, circulating air-flow, which creates very turbulent conditions in the combustion chamber. This approach so completely destroys the hazardous constituents that it makes an afterburner unnecessary. EPA initiated the SITE program in February 1986 to accelerate the development, demonstration, monitoring, and use of new or innovative treatment technologies.

Stabilization and solidification. Stabilization and solidification technologies are used to reduce the possibility that hazardous components will leach from a waste after it enters a landfill. Substances are chemically and physically immobilized in a solid, impermeable mix of cement, incinerator ash, kiln dust, or plastic. Stabilization renders the hazardous components less toxic and more leach-resistant through chemical reactions, whereas solidification involves encapsulating the components in a solidifying agent or a sorbent, or chemically fixing them in a solid chunk. These technologies work especially well for wastes containing heavy metals and inorganic salts.

Recently, several companies have begun developing "in-situ," or in-place, stabilization and solidification processes designed to isolate and treat contaminated soils without excavating them. For example, Battelle Pacific Northwest Laboratories in Richland, Washington, has developed what it calls an "in-situ vitrification" process for treating contaminated soils at hazardous waste sites.

Originally designed for application to soils contaminated with radioactive materials, in-situ vitrification uses electrodes implanted in the ground to generate an electric current that heats and destroys organic hazardous constituents and converts contaminated soils and sludges into a chemically inert, leach-resistant, glasslike product. Battelle has determined that the process, which EPA plans to demonstrate under its SITE Program, could provide a practical alternative for treating soils and sludges containing heavy metals and organic chemical wastes.

Chemical and physical treatment. Other chemical and physical treatment processes degrade a waste's hazardous constituents into less toxic forms or remove the constituents from the waste altogether. These treatment processes include neutralizing acid and alkaline wastes; oxidizing cyanides; recovering solvents and other organic hazardous wastes through adsorption, distillation, extraction, evaporation, and fractionation; and recovering metals chemically through precipitation and reduction and physically through electrodialysis, freeze crystallization, ion exchange, reverse osmosis, and ultrafiltration.

Bioremediation. Biotreatment, or bioremediation — the use of microorganisms to degrade the hazardous constituents in a waste into nonhazardous substances, such as water and carbon dioxide — has rapidly emerged as a way to treat wastes and soil contaminated with crude oil, chlorinated hydrocarbons, PCBs, and pesticides. Although still in its infancy, the technology promises to provide more effective and less expensive on-site treatment than existing treatment methods, and EPA has already begun encouraging its use in site remediation projects.

For example, in October 1987, the PRPs at the French Ltd. NPL site near Houston, Texas, completed a bioremediation demonstration project using naturally occurring bacteria to degrade the organic hazardous waste contained in part of a 7.3-acre waste lagoon. The lagoon had been used from 1966 to 1971 to dispose of toxic petrochemical wastes and contained about 80,000 tons of contaminated sludge. The experiment isolated a 0.3-acre area that contained some of the lagoon's most highly contaminated sections. Scientists pumped about 1,600 cubic feet of air per minute into the lagoon water and into the sludge on the lagoon bottom. Liquid nitrogen-phosphorous fertilizer was added to the lagoon to stimulate the growth of the bacteria, which then "ate" the contaminants. Through the degrading action of the bacteria, the contaminants (except for the heavy metals) were converted into water, carbon dioxide, and other harmless by-products.

Researchers at the University of California at Riverside have developed a biological process that uses naturally occurring fungi to clean up soil contaminated by selenium, a highly toxic chemical found in agricultural drainage water. The scientists cultivated the fungi in field tests at 46 selenium-contaminated plots in the Kesterson Wildlife Refuge near San Jose, California, where selenium killed hundreds of ducks and other waterfowl between 1983 and 1985.

The University of California process, activated by aerating and irrigating the contaminated soil and adding fertilizers such as orange peels, cow manure, and straw, stimulates three types of fungi — *Penicillium, Acremonium,* and *Uloclabium* — to consume selenium. A "detoxification mechanism" in the fungi converts the selenium into the gases dimethylselenide and dimethyldiselenide, which disperse into the atmosphere and are 500 to 700 times less toxic than selenium itself. Initial results have shown that the process has removed substantial amounts of selenium from the soil. In laboratory tests conducted during 1986, the fungi's natural conversion of selenium to a harmless gas was accelerated tenfold.

Scientists at Michigan State University (MSU) in East Lansing, Michigan, have used anaerobic bacteria to degrade PCBs. Through a process known as "bacterial dechlorination," the anaerobic bacteria remove the chlorine atoms from the PCB molecules, rendering the PCBs less toxic and more susceptible to further biodegradation by more common aerobic bacteria. In the experiments, conducted over a period of 16 weeks, the bacteria successfully removed over half of the chlorine present in sediment samples containing at least 700 parts per million of PCBs. According to John Tiedje of MSU's Department of Crop and Soil Sciences, the experiment marked "the first time that anyone has shown that the most persistent of PCBs can be degraded by bacteria."

Emerging treatment technologies. Other innovative hazardous waste treatment technologies are still in the research and development stage. For example, an engineering firm in Vancouver, British Columbia, plans to begin large-scale testing of a sonic hazardous waste treatment process, which uses sound waves to destroy organic chemicals contained in hazardous wastes. During the process, a sonic generator emits vibrations through a slurry of the hazardous waste undergoing treatment. The vibrations form a pressure wave that, if sufficiently intense, causes vapor bubbles to form and then violently collapse. This action in turn produces localized high temperatures and high pressures that ultimately degrade hazardous constituents into nonhazardous substances such as water and carbon dioxide.

In October 1988, EPA selected seven developers of hazardous waste cleanup technologies to receive a total of $1.88 million under the agency's Emerging Technologies Program (ETP). ETP promotes the development of treatment technologies not yet ready for full-scale demonstration and evaluation by EPA's SITE program.

These seven technologies include processes that remove heavy metals from groundwater through ultrafiltration and chemical treatment; electroacoustically decontaminate soil containing organic compounds and metals; recover metals from groundwater by using algae immobilized in a silica gel polymer; remove metals from degraded waters by using the natural geochemical and biological processes inherent in a wetlands ecosystem; and oxidize organic compounds in water through a laser-stimulated photochemical process.

Use of hazardous waste management technologies. In its 1985 report *Hazardous Waste Management: Recent Changes and Policy Alternatives*, the Congressional Budget Office examined the methods that 70 different U.S. industries used to manage the 265.6 million metric tons of hazardous waste that they generated in 1983. It found land disposal methods to be the most popular waste management techniques, with underground injection wells accounting for 66.8 million metric tons, surface impoundments accounting for 49.5 million, waste landfills for 34.2 million, sanitary landfills for 26.7 million, and land treatment facilities for 2.9 million.

According to the Congressional Budget Office report, the industries discharged about 58.9 million metric tons of hazardous waste either directly into sewer systems or into wastewater treatment plants and then into sewer systems. Only about 2.7 million metric tons of hazardous wastes were incinerated, and an additional 9.5 million metric tons were burned in industrial boilers. Finally, two alternative treatment technologies, oxidation and ion exchange, accounted for 3 million and 500,000 metric tons, respectively.

In 1988, CMA issued the results of its "1986 CMA Hazardous Waste Survey," which covered 529 plants owned and operated by 33 of the 50 largest U.S. chemical companies. According to the survey, the total amount of hazardous waste generated by the plants in 1986 exceeded 220 million tons (216 million tons of wastewater and 4 million tons of solid waste). Of the total amount of hazardous wastewater generated, the companies treated 199 million tons before discharging it to surface waters or publicly owned wastewater treatment facilities, and they disposed of 17 million tons via underground injection. According to the CMA survey, 3 million of the 4 million tons of hazardous solid waste generated was recycled.

The CMA survey also analyzed the hazardous waste management practices of 221 chemical plants that have responded each year since 1981. It found that the use of incineration to treat hazardous wastes increased 28 percent between 1981 and 1986, with the biggest jump occurring

Figure 9. Sampling at a Missouri Dioxin Site. Scientists at the Denney farm site near Aurora, Missouri, wore protective "moon" suits with self-contained breathing apparatus as they took samples to determine the concentration of the highly toxic chemical dioxin there. The Denney farm is one of many dioxin-contaminated sites in Missouri that have been linked to chemical wastes generated between 1969 and 1971 at the defunct Northeastern Pharmaceutical and Chemical Co. facility in Verona, Missouri. *Courtesy: EPA.*

between 1985 and 1986; treatment by means other than incineration dropped 73 percent over that time period. The amount of hazardous waste landfilled decreased 65 percent during that period, while the amount disposed of by underground injection dropped 20 percent, and the amount disposed of by other means decreased 80 percent. These findings underscored the movement away from land disposal and toward treatment.

According to the 1988 ICF survey of the 14 major hazardous waste management companies, the treatment and disposal facilities operated by these companies received a total of at least 4.2 million wet metric tons of hazardous wastes in 1986 — a decrease of 8 percent from the 4.6 million wet metric tons received in 1985. The companies attributed this decline to temporary and permanent facility closures due to regulatory and enforcement actions; regulatory restrictions on the amount of wastes received, as well as self-imposed restrictions to conserve treatment and disposal capacity; and "slowdowns in certain service markets," such as wastewater treatment. In particular, the survey found that deep-well injection decreased 34 percent, from 412,000 wet metric tons in 1985 to 271,000 wet metric tons in 1986, while incineration rose 31 percent, from 269,000 wet metric tons in 1985 to 351,000 tons in 1986.

In 1987, according to the ICF survey, the total volume of waste received by the 14 companies surveyed grew 6.5 percent, to at least 4.6 million wet metric tons. The companies attributed the increase to improved marketing, the expansion of their treatment and disposal capacity through construction or acquisition; and the "rapidly growing market" for handling wastes from Superfund cleanups and RCRA corrective actions.

ICF reported that incineration use increased 36 percent between 1986 and 1987, to 476,000 wet metric tons, as a result of the RCRA land disposal restrictions and "generators' growing preference for total destruction" of their wastes. The survey also noted, however, a 5 percent increase in the volume of waste landfilled, from 2.37 million wet metric tons in 1986 to 2.47 million wet metric tons in 1987. According to some of the surveyed companies, greater use of stabilization and solidification technologies before disposal may have made landfilling a more attractive option.

The disposal of hazardous wastes by deep-well injection and the treatment and disposal of hazardous wastes in landfills and surface impoundments should decrease in the early 1990s, according to most respondents to the ICF survey. In particular, they predicted a bleak future for surface impoundments, which will probably be strictly regulated or phased out and replaced by treatment in tanks.

Some companies dismissed the negative effects of both the RCRA land disposal restrictions and waste minimization strategies. Instead, they predicted the expanded use of secure landfills as "residual repositories" for hazardous wastes that have been stabilized or solidified. Large generators are relying increasingly on on-site pretreatment, as well as on-site incineration, in-house solvent recovery, and waste minimization to reduce the volume of waste sent off-site and gain more control over the destination of the waste and avoid future liability. Smaller generators, faced with costly regulatory requirements under RCRA, will probably make greater use of off-site treatment and disposal services, the survey concluded.

Hazardous Waste Reduction

As the problems of hazardous waste management become more complex, reducing the amount of hazardous waste that industry generates becomes a more and more compelling option. Hazardous waste reduction can involve any practice that reduces the volume of substances released from a facility or entering the waste stream, such as modifying equipment, altering processes, redesigning products, using different raw materials, or improving management, training, and inventory control.

In a 1987 report entitled *From Pollution to Prevention: A Progress Report on Waste Reduction*, OTA asserted that "human, organizational, and institutional obstacles in industry and government" have hindered the advance of waste reduction as an alternative to treatment and disposal. OTA recommended that, to remove these obstacles, Congress should introduce legislation that would create a nonregulatory technical assistance program to help industry reduce the amount of waste generated. Such legislation should focus on source reduction, rather than "end-of-the-pipe" pollution controls, and should address all hazardous wastes, rather than just those specifically regulated under RCRA. OTA also recommended that such legislation concentrate on providing assistance to encourage waste reduction, rather than requiring compliance with existing and new regulations, and emphasize that waste reduction directly benefits industry through increased efficiency and profits, as well as lower costs for pollution control and waste disposal.

During 1988, the House of Representatives considered the Hazardous Waste Reduction Act, a bill that was intended to reduce the generation and promote the recycling of hazardous wastes. Introduced in June 1987, the bill would require facility owners/operators to prepare 2-year estimates of the quantities of toxic chemicals that their facilities recycle and to report to EPA information about the following: all source-reduction and recycling efforts; the amounts of toxic chemicals that enter their waste streams; and the amounts of chemicals released into air, water, and land.

In addition, the bill would require EPA to create an Office of Waste Reduction to serve as a clearinghouse for the compilation and distribution of technical data on waste reduction techniques. The office would also distribute matching funds annually to states for hazardous waste reduction research programs.

In January 1989, EPA issued on its own initiative a proposed strategy that would establish pollution prevention as an official agency policy for the first time. According to the proposed strategy, although waste management activities focusing on the treatment, control, and disposal of pollutants have led to "significant progress" toward improving environmental quality, further improvements demand multimedia source reduction and recycling programs to reduce or eliminate the discharge of pollutants into the environment.

To emphasize its new emphasis on multimedia pollutant reduction, EPA replaced the term "waste minimization" with the term "pollution prevention." With this change, the agency hoped to eliminate a term perceived by some as closely tied to single-medium hazardous waste management strategies under RCRA and emphasize pollution prevention as applicable "beyond the RCRA hazardous waste context."

EPA's Pollution Prevention Office, created within the agency's Office of Policy, Planning, and Evaluation in August 1988, would shoulder primary responsibility for implementing the new strategy, while state and local governments would play a major role in encouraging industry to develop pollution prevention programs. EPA intends to help states implement those programs by creating a clearinghouse that will provide state and local officials with technical information on pollution prevention.

Exxon Valdez Oil Spill

On 24 March 1989, the Exxon Valdez, a supertanker equipped with modern navigation and operations systems, ran aground on a well-known and well-marked reef and released about 10.9 million gallons of oil into Alaska's pristine and ecologically sensitive Prince William Sound. What caused this incident, which ranks as the largest oil spill in U.S. waters and the most expensive oil spill cleanup ever? What will be the environmental effects, both short-term and long-term, of the incident?

At 9:21 P.M. on 23 March 1989, the 987-foot-long tanker Exxon Valdez, which was laden with more than 53 million gallons of crude oil from Alaska's North Slope, left Valdez, Alaska, for an oil refinery in California. The vessel traveled without difficulty to the entrance of Prince William Sound under the command of a local harbor pilot. At 11:25 P.M., the tanker's captain, Joseph Hazelwood, routinely reported to the U.S. Coast Guard station in Valdez that the local pilot had disembarked, and requested permission to leave the Sound's outbound vessel traffic lane to avoid ice chunks that were breaking off the nearby Columbia Glacier—a common request by tanker captains during the spring months.

The Coast Guard station granted permission to Hazelwood to take the Exxon Valdez on a southerly course, so that the tanker would travel into the empty inbound vessel traffic lane before beginning its southwesterly voyage out of the Sound and into the Gulf of Alaska. At 11:55 P.M., however, the Coast Guard's radar system located the Exxon Valdez about a half-mile to the east of the inbound vessel traffic lane and still moving directly south. Apparently, shortly after receiving permission to steer around the ice in the outbound vessel traffic lane, Hazelwood had left the bridge to do paperwork in his cabin and ordered the ship's third mate, who was not certified to pilot a tanker through Prince William Sound, to stand command on the bridge.

Before leaving the bridge, Hazelwood had placed the Exxon Valdez on autopilot — against regulations in confined waters such as Prince William Sound — and told the mate to turn the tanker back toward the west, onto its proper course, when the tanker moved past Busby Island. In the meantime, the Exxon Valdez autopilot had activated a program that gradually accelerated the vessel to ocean traveling speed. When the third mate attempted to turn the tanker back toward the west, he did not realize until too late that the Exxon Valdez was on autopilot. He also did not realize that the tanker was traveling at such a high speed that it would have had to start its turn several minutes earlier to avoid hitting Bligh Reef, which the tanker was rapidly approaching. The mate finally turned off the autopilot and called for "hard right rudder," in order to turn the tanker sharply to the west and away from the reef. However, the Exxon Valdez was still a quarter of a mile to the east of the inbound vessel traffic lane and moving quickly when it first struck Bligh Reef, about 25 miles south of Valdez.

The Exxon Valdez was still navigable after this first impact, but it was so far east of deep water that, as the third mate was trying to direct the tanker away from the shoreline, the vessel struck a second part of the shallow reef. It was this second impact, at 12:04 A.M. on 24 March, that impaled the tanker on a pinnacle of Bligh Reef. At 12:28 A.M. on 24 March, the Exxon Valdez radioed the Coast Guard station in Valdez to report that it had run "hard aground" on the reef. The reef had ripped out a portion of the tanker's hull on the starboard side and torn open all 11 of its cargo tanks. Within 5 hours, about 10.9 million gallons of oil had been released into Prince William Sound.

Spill Response

The next day, Exxon announced that "the full resources of the company, including people with special expertise and additional equipment, have been brought to bear" to manage the ensuing vessel salvage and spill cleanup operations. By the end of the first week of April, the more than 40 million gallons of oil that remained in the tanker's cargo tanks after the grounding had been offloaded, and the tanker had been carefully floated off Bligh Reef and towed a distance of 25 miles to Naked Island in Prince William Sound for temporary repairs. Captain Hazelwood — who had been immediately fired by Exxon — was arrested on criminal charges of operating a vessel while intoxicated, reckless endangerment, and negligent discharge of oil.

In July, the Exxon Valdez sailed into the port of San Diego, California, for extensive repairs after a long trip that was not just marked by fears that the tanker might capsize and break up in the rough seas of the Pacific Ocean, but was also delayed by the discovery that oil was still escaping from the tanker.

While the salvage efforts proceeded fairly smoothly, the cleanup operation was another matter. Within a week of the spill, state and local officials and area fishermen's groups became increasingly dismayed at what they considered slow, poorly coordinated, and largely ineffective efforts to control and clean up the massive oil slick. Without waiting for federal assistance, these parties began to intensify their ongoing efforts to prevent the slick from contaminating the state's fish hatcheries.

Meanwhile, federal and state officials began hearings into both the causes of the incident and the problems in the response and cleanup. It became apparent that Exxon and Alyeska Pipeline Service Co. did not respond to the spill in the way described in Alyeska's state-approved spill contingency plan. (Alyeska is the manager of the Trans-Alaska Pipeline and Valdez Marine Terminal and the designated first-responder to any spills involving oil from the pipeline.) Exxon offered a public apology for the spill even as it and the Coast Guard traded charges of responsibility for the confusion and delay in the cleanup.

The oil spill contingency plan for Prince William Sound, which had been prepared by Alyeska and approved by the Alaska Department of Environmental Conservation

Figure 10. Cleaning Up the Largest Oil Spill in U.S. History. On 24 March 1989, the Exxon Valdez tanker grounded in Prince William Sound about 25 miles south of Valdez, Alaska, and spilled more than 10.9 million gallons of crude oil. The incident ranks as the largest oil spill in U.S. history and the costliest oil spill cleanup ever, involving expenditures of more than $1 billion. The spilled oil eventually contaminated 1,100 miles of shoreline in Prince William Sound and along the Alaskan coast. At the peak of the cleanup activities, Exxon Corp. deployed a work force of 11,000 people. *Courtesy: Al Grillo, Anchorage Times Publishing Co., Inc.*

(ADEC), presented two possible scenarios for responding to tanker spills — one for a 168,000-gallon spill and the other for an 8.4-million gallon spill. The plan's description of the latter scenario is brief because, according to the plan, "Alyeska believes it is highly unlikely that a spill of this magnitude would occur." However, on 24 March 1989, a spill worse than the "highly unlikely" scenario happened, and both Alyeska and Exxon appear to have been unprepared.

Investigations by the Coast Guard, the National Transportation Safety Board, and various congressional committees have since revealed the extent to which Exxon and Alyeska's lack of readiness contributed to the impact of the Exxon Valdez spill. Although Alyeska's spill contingency plan required a damaged tanker to be completely encircled with booms — floating barriers that contain or deflect spilled oil — within 5 hours of the release, Exxon and Alyeska took 35 hours to encircle the Exxon Valdez. By the time the tanker was surrounded by containment booms, the released oil had floated miles away and formed a virtually solid 50-square-mile slick.

In addition, although the plan required Alyeska to notify Alaska when a significant piece of cleanup equipment was taken out of service, the company failed to inform the state that the booms needed to encircle the Exxon Valdez had been removed from its emergency response barge in Valdez so that the barge could be repaired. Moreover, when Alyeska employees finally loaded the barge, they placed on the vessel equipment to remove the oil remaining in the Exxon Valdez rather than containment booms to control the spill. Alyeska then needed additional time to unload and reload the barge with the proper equipment.

Dispersant use. From the outset, dispersant use in Prince William Sound was a controversial issue among the principal participants in the Exxon Valdez oil spill response. Dispersants are chemicals designed to break up an oil slick into small droplets. These droplets, which become suspended in the water below the surface, are then subject to natural processes such as biodegradation and weathering.

As the National Response Team (NRT) report on the Exxon Valdez spill noted, the use of dispersants as a response option involves "a choice between leaving the spill untreated and floating on the surface of the water, where it may threaten surface resources, or treated and distributed in the water column, where it may threaten subsurface resources." Noting that dispersant use has been controversial because, in the past, some dispersants were found to be highly toxic, the report said that "it must be concluded that the harm from dispersant use on the subsurface environment is likely to be less than the potential harm of untreated oil."

Exxon favored using dispersants on the Exxon Valdez oil slick, but environmentalists and local fishermen opposed their use, fearing potential toxic effects on fish and wildlife during the migration season. Congressional investigations found, however, that neither Alyeska nor Exxon had a sufficient supply of chemical dispersants on hand to manage the spill anyway. And even if the companies had had sufficient quantities of chemical dispersant to begin breaking up the slick, they were not ready to begin test dispersant applications until almost a day after the spill began. According to one witness, "they then did so by ineffectually tossing buckets of chemicals out of the door of a helicopter."

The small-scale applications of chemical dispersants that were conducted on the Exxon Valdez oil slick for 2 days after the spill did not produce favorable results, according to ADEC and the Coast Guard. Three conditions are necessary to effectively disperse oil with chemicals: proper timing of application; sufficient wave energy to mix the dispersants into the oil; and an adequate supply of dispersants. Even under good conditions, dispersants are generally most effective when applied within 12 hours after the spill—before the slick "weathers," becoming increasingly viscous and resistant to chemical action.

Several factors worked against Exxon in its efforts to break up the slick with chemicals. During the 2-day period following the spill when dispersant use would have been most effective, the waters of Prince William Sound were calm and flat and the winds were light, so that there was not sufficient wave energy for the dispersants to work. Regardless of whether dispersants would have worked to break up the rapidly spreading oil slick, Exxon had neither enough dispersants for a massive application nor enough personnel or equipment to apply the dispersants to the entire spill.

After it became clear that Exxon and Alyeska's efforts to halt the movement of the oil were ineffective, Exxon claimed that federal and state delays in approving both the application of chemical dispersants and the use of in-situ burning on the slick were a major cause of the oil's rapid spread across Prince William Sound and beyond.

Local action and federal response. When Exxon assumed control of the cleanup operation from Alyeska on 25 March, it turned down the offers of local fishermen and other residents to mobilize an effort to protect Prince William Sound's ecologically sensitive areas, especially the five major fish hatcheries located in Eshamy, Esther, Main, and Sawmill Bays. After the oil slick spread into the Gulf of Alaska, the local fishermen took action and launched a massive hatchery-protection effort with administrative and logistical support from ADEC.

The fishermen's efforts, which reminded one resident of the rescue of Allied troops from Dunkirk by a flotilla of fishing boats during World War II, involved deploying containment booms to corral oil and then skimming, or removing the oil from the water, with industrial vacuum-loader trucks — commonly known as "super suckers" — that were mounted on barges. According to Prince William Sound Aquaculture Corp., the strong currents and typically high tides in the Sound broke many of the containment booms laid down by the fishermen, so that boat crews had to check each of the booms every 2 hours to ensure that they were still in place to collect the floating oil for eventual removal.

On 7 April, after insisting for 2 weeks that Exxon had sufficient resources to properly manage the response to the Exxon Valdez oil spill, President George Bush ordered the expansion of the federal government's participation in the ongoing cleanup efforts in Prince William Sound. Calling the incident a "tragic environmental disaster," Bush announced that he had authorized the Coast Guard to coordinate the efforts of federal agencies involved in the cleanup and accelerate the pace of the cleanup efforts. Bush also said that he had directed the U.S. Department of Defense to provide all necessary equipment, personnel, and technical and logistical support to assist in the cleanup.

Acknowledging that "the job of cleaning up the oil from both the sea and the affected land areas will be massive, prolonged, and frustrating," Bush conceded that Exxon's efforts, standing alone, were not enough to enable cleanup crews to achieve what he said should be the ultimate goal of the cleanup operation: the "complete restoration of the ecology and the economy of Prince William Sound."

EPA Administrator William Reilly was assigned by Bush to oversee long-range planning for activities to restore the environment of Prince William Sound and other areas affected by the spill. According to Bush, EPA will draw on the expertise of leading scientists and oil spill experts in this work, as well as consult with other federal agencies that are assessing the environmental impacts of the spill. These agencies include the National Oceanic and Atmospheric Administration, the U.S. Department of Agriculture, and the U.S. Department of the Interior.

Plan for Long-Term Cleanup

In April, Exxon began the long and difficult job of removing the spilled oil from the coastline in Prince William Sound and the Gulf of Alaska by flushing the heavily and moderately oiled beaches and rocky shorelines with sea water and then collecting the oil from the water with skimmers. Workers also began using steam cleaners and high-pressure hot-water washers to remove oil from rocks and from areas where it had concentrated or hardened. When Exxon's original beach cleanup plan — which proposed that 364 miles of shoreline be completely restored by 15 September 1989 — was announced, it was met with severe criticism from almost all sides as setting near-impossible goals for the cleanup. For example, Coast

Guard Commandant Paul Yost told a Senate subcommittee that, in the most favorable scenario, cleanup crews could hope to accomplish only "a rough cleanup of the gross amount of oil" on the beaches. Even this goal, he said, "is an object we will strive for with no guarantee of success." Exxon was also criticized for suggesting that lightly contaminated sections of shoreline, especially in the Gulf of Alaska, "might not require cleaning if environmental considerations indicate that allowing natural forces to operate produces a better result" than mechanical flushing and recovery.

After publishing an expanded beach cleanup plan in May, Exxon asserted that it "fully expects to meet the milestones established in the revised plan and finish the shoreline cleanup job by 15 September." Other participants in the cleanup, including ADEC and the Coast Guard, were not as optimistic about the cleanup's progress. ADEC said that even the revised plan grossly underestimated the amount of contaminated shoreline that needed cleanup, and failed to address the restoration of sensitive natural areas such as Kenai Peninsula, Cook Inlet, and Kodiak Island.

In addition, according to ADEC, Exxon's cleanup plan did not provide information about the number, size, and allocation of cleanup crews in Prince William Sound and along the Alaskan south-central coastline. It also lacked "clear and detailed" cleanup schedules with definite dates by which cleanup activities would be completed in specific areas.

ADEC sharply criticized the cold-water flush-and-float method as a "relatively inefficient cleaning process" that might be not only ineffective — even with three passes on each section of beach — but also inappropriate in many of the areas that it was intended to restore. According to ADEC, Exxon should have considered the use of other cleanup techniques, such as removing tarballs from the beaches by hand, and should have allowed experimental cleanup techniques to be tested and used.

Environmental Effects

The initial impact of the Exxon Valdez oil on the fish and wildlife in Prince William Sound and along the Alaskan coast as far southwest as Kodiak Island was severe. The Sound, among the largest tidal estuarine systems in North America, is one of the most ecologically sensitive areas in the United States. The spilled oil has also threatened several national parks; Prince William Sound is within the boundaries of the Chugach National Forest, the Kenai Fjords National Park and the Katmai National Park are both located on the oiled Kenai Peninsula, and the Kodiak National Wildlife Refuge is on Kodiak Island.

The U.S. Forest Service estimated that, after the spill, the Exxon Valdez oil drifted southwest — out of Prince William Sound, into the Gulf of Alaska, and down Alaska's south-central coast — in sheen, "mousse," and tarball form at an average of 17.5 miles per day, eventually contaminating more than 1,000 miles of shoreline. To illustrate the magnitude of the spill at a House subcommittee hearing in April, ADEC Commissioner Dennis Kelso noted that, "if this had happened on America's Atlantic seaboard, filthy foam and tarballs would be washing up on beaches from Boston, around Cape Cod and down Long Island Sound, along the sandy beaches of New Jersey, past Cape May, to the mouth of the Chesapeake Bay, and perhaps up the Potomac itself."

According to the NRT report on the spill, "many of the conditions present during the Exxon Valdez spill increased, rather than diminished, the severity of its impacts relative to other large spills." Several factors determine the effects of an oil spill on a given ecosystem. These include the type and amount of oil spilled and the degree to which the oil has weathered and emulsified — mixed with water to form a mousselike substance — as well as the location of the spill, the season in which the spill occurred, and the types of habitats and organisms affected. Another major factor is the adequacy of the spill response.

In the Exxon Valdez spill, the type of oil lost was Alaska North Slope crude oil, which can be particularly destructive to a marine environment. Compared to oil from the Middle East or Mexico, North Slope crude contains higher amounts of toxic hydrocarbons, such as benzene, toluene, and xylene. These volatile compounds readily dissolve into seawater, efficiently spreading the oil's poisonous effects. In addition, even after weathering, North Slope crude remains resistant to natural degradation for years.

The Exxon Valdez spill occurred in a subarctic region at the beginning of spring, and affected over 1,000 miles of vulnerable shorelines. Oil percolated rapidly and deeply into the south-central Alaskan coast's gravel and cobble beaches, where cleanup crews found hardened oil from the spill embedded at depths of 3.5 feet. In addition, the narrow inlets and coves of Prince William Sound and the south-central coast trapped much of the spilled oil, which was then deposited onshore by each high tide. Oil may remain for years in the area's partially enclosed bays, marshes, and tidal flats, where the lack of strong wave and wind action leaves only slow-working natural biological and chemical processes to break up and disperse the stranded oil.

The low average temperature of the Prince William Sound region will allow the oil to persist longer there than it might persist in a more temperate climate by slowing the rate at which the Exxon Valdez oil naturally weathers and degrades and retarding the release of the hydrocarbons' toxic components. The heavier hydrocarbon components will sink to the ocean floor, where they may remain in the soft sediments for over a decade. Such persistence "provides the potential for long-term exposures and sublethal chronic effects, as well as short-term exposures and acute effects," according to the NRT report.

Effects on the food chain. What effects will the long-term presence of petroleum hydrocarbons have on the wildlife and fish of Prince William Sound and the Gulf of Alaska? Before the spill, the National Marine Fisheries Service's Auks Bay Laboratory in Juneau, Alaska, had performed extensive research on the effects of petroleum hydrocarbons on various species of Alaskan aquatic life. The laboratory concluded that long-term exposure to sublethal concentrations of petroleum hydrocarbons reduces the reproduction and growth rates of most of these species, alters their behavior by inhibiting their nervous system responses, and increases their vulnerability to disease.

During the spring, the populations of phytoplankton and zooplankton — the microscopic plants and fish and shellfish larvae and eggs that form the bottom layer of the marine food chain — increase dramatically. "In the open waters of the Sound and Gulf," the NRT report said, the impact of the Exxon Valdez oil "probably will be short-lived and local because of the quick replacement of plankton by the same organisms from unaffected areas." According to the report, however, "for some species...the mortality of planktonic eggs and larvae may be reflected in long-term population effects."

Unlike most oil slicks, which consist of oil patches that cover between 20 and 30 percent of the water surface in a spill area, the Exxon Valdez slick initially covered virtually all of the water surface in the area that it affected. In addition, the spilled oil mixed with the water and emulsified into a mousselike substance whose volume was four times that of the unmixed oil. The amount of oil in the water and the toxicity of its various components killed or inhibited the growth of much plankton, reducing the Sound's food supply.

Effects on birds. Birds were perhaps the most visibly affected wildlife along the Alaskan coastline that was impacted by the Exxon Valdez spill. As of September 1989, the U.S. Fish and Wildlife Service reported that the corpses of more than 33,000 birds killed by the spill had been recovered. The Fish and Wildlife Service noted that this figure represents only a small percentage of the birds actually killed.

About 350,000 birds of 18 different species breed and nest in Prince William Sound in the spring, and at least 100,000 migrate there each summer. An additional 20 million to 30 million shorebirds and waterfowl of more than 200 species stop to feed in the Copper River delta on the Sound's eastern shore during their annual migration to breeding grounds in western Alaska. One Cordova resident said that, in April, the air is normally filled with the noisy cries of huge flocks of birds on the water. "This year," she said, "since the spill, there has been an eerie silence. The birds just aren't out there."

Because the spring migration and breeding seasons were just beginning when the tanker hit Bligh Reef, a large number of shorebirds and waterfowl died from the direct effects of swimming in the oil-covered water and nesting on the oiled beaches. Oiled birds usually die from hypothermia and shock; they also become weighed down by the oil and drown.

Immediately after the Exxon Valdez spill, the U.S. Fish and Wildlife Service counted over 91,000 waterbirds — primarily diving ducks, grebes, and loons — in and around Prince William Sound. About half of these birds were either in or near areas impacted by the spill. Another U.S. Fish and Wildlife survey of the Sound at the time found 80 oiled birds per 100 meters of beach. The International Bird Rescue Research Center, which is based in Berkeley, California, set up a rescue center in Valdez following the spill and collected oiled birds for cleaning. After being bathed in a detergent solution and thoroughly rinsed with a Water-Pik, most of the birds were held at the center for about 8 hours and then released if it appeared they would be able to survive in the wild.

Scavenging birds, like bald eagles and golden eagles, ingested the oil's toxic components by consuming oil-killed birds and fish. When these birds eat oil-contaminated food, their stomachs and intestines become coated with oil and can no longer absorb nutrients or water. The birds eventually die of starvation and dehydration. This is especially troubling because bald eagles are an endangered species, and the 5,000 bald eagles that live in the Prince William Sound region are the largest remaining group of these birds anywhere in the world.

Of Alaska's birds that did not die from the acute, short-term effects of the spill, many have died and will die from indirect, long-term effects such as habitat loss, loss of food resources, decreased reproductive capacity, and a high mortality rate among chicks and eggs.

Effects on mammals. Other marine species that feed on the populations of phytoplankton and zooplankton were also just beginning their annual migration through the Sound when the Exxon Valdez spill occurred. With the exception of sea otters, marine mammals appear to have fared better after the spill than birds. About 23 different species of marine mammals — including harbor seals, sea lions, dolphins and porpoises, and gray, humpback, and killer whales — live at least part of the year in Prince William Sound or the Gulf of Alaska.

These mammals are not as vulnerable to damage from oiling as sea otters because they depend on fat or blubber, rather than fur, to maintain their normal body temperature. When oil fouls the insulating fur of a sea otter, however, the otter not only becomes vulnerable to death from hypothermia, but also may drown, because it loses its buoyancy when just 10 percent of its body becomes covered with oil. Moreover, as scientists examined the otters killed by the Exxon Valdez oil, they learned that the oil caused acute damage to organs such as the kidneys, liver, and lungs, and destroyed the animals' immune systems. The spilled oil also apparently caused premature delivery of otter pups and spontaneous abortion among pregnant female otters.

Of the estimated 10,000 to 12,000 sea otters that live in Prince William Sound — the largest concentration of sea otters anywhere in North America — as many as 2,500 were located in the most heavily oiled areas of the sound. The Exxon Valdez oil also threatened the 2,500 to 3,000 otters living in the Gulf of Alaska along the Kenai Peninsula and the 4,000 to 6,000 around Kodiak Island.

By the end of April, only 135 out of the hundreds of otters that had been collected from the waters and beaches of Prince William Sound had made it alive to an otter rescue center set up in Valdez. Of these, 80 died despite efforts to carefully clean and feed them and give them oxygen, antibiotics, and antitoxins. The U.S. Fish and Wildlife Service reported that, by September 1989, almost 1,000 otters had died as a result of the Exxon Valdez spill.

Effects on fish. Prince William Sound is one of the nation's most productive commercial fishing grounds. Five varieties of salmon — as well as many other commercially valuable fish and shellfish, including cod, halibut, Pacific herring, sablefish, Tanner and Dungeness crabs, and shrimp — are abundant there. The Exxon Valdez spill, however, forced the Alaska Department of Fish and Game to cancel the Sound's 1989 black cod season and cut short the pot shrimp season. In addition, ADEC banned fishing for Pacific herring in the Sound for 1989 after determining that at least 87 percent of the herring spawning grounds there were "heavily oiled." In 1988, herring fishermen had earned about $14 million by catching more than 10,000 tons of the fish and their eggs — called roe — in Prince William Sound.

Herring typically spawn, or lay their eggs, during April and May in the floating eel grass and kelp beds that grow in the Sound's intertidal and subtidal areas. They do not spawn until they are at least 3 years old, but once they begin, they return to their birthplace annually for spawning. In the spring of 1989, both the adult herring returning to Prince William Sound to lay their roe and the roe itself were exposed to the Exxon Valdez oil that contaminated the Sound's tidal areas. While the spilled oil smothered or poisoned some of the adult herring and reduced their food supply, its more insidious effect may have been to cause what the NRT report called "developmental abnormalities" in the growing embryos of the herring roe, which were probably deposited in eel grass and kelp contaminated by the spill.

Because herring do not spawn until the age of three, the full impact of the Exxon Valdez spill on herring in Alaskan waters may not be known until scientists can estimate the sizes of the spawning adult populations returning to affected areas of Prince William Sound in 1992, 1993, and 1994, and then determine how many of these spawning adults hatched from eggs that survived the spill.

The Exxon Valdez spill also affected Alaska's salmon industry. The salmon hatcheries in Prince William Sound were about to release more than 650 million fry — recently hatched fish—from their indoor incubators into net holding pens in the Sound's estuaries. The fry are usually released during April and May, at the height of the plankton bloom, and they spend up to 12 weeks feeding and growing in these shallow waters before migrating to the Gulf of Alaska.

After the Exxon Valdez spill, hatchery operators feared that the oil would directly kill the fry, which are extremely sensitive to low concentrations of hydrocarbons, or slow their growth by reducing the amount of plankton available to them for food. The 1988 salmon catch of about 14.9 million fish was valued at $76 million, and the Alaska Department of Fish and Game had projected that the 1989 catch would be four times larger because of recent milder-than-average winters and improvement of salmon stocks by aquaculture. The size of the future catches is now, in the words of one fisherman, "anybody's guess."

In addition, immobile intertidal organisms, such as barnacles and mussels, which live in the highly stressful environment of rising and falling tides, may suffer high mortality rates as a result of the spill, and their populations may take years to recover.

Future Oil Spill Research

William Evans, the National Oceanic and Atmospheric Administration's undersecretary for oceans and atmosphere, testified before a Senate subcommittee in early April that "this is without a doubt the worst oil spill that has occurred in the history of the United States." Several hundred scientists from more than 25 federal and state agencies, industry groups, and academic institutions arrived in Valdez almost immediately after the spill to begin assessing the immediate and long-term damages to the natural resources of Prince William Sound and the Gulf of Alaska.

The spill has prompted the environmental, scientific, and oil industry communities to increase their efforts to identify and understand the impact that major oil spills have on the world's waters and coastlines. Demands for action have been made, proposals have been set forth, and many new efforts will soon be under way.

For example, a coalition of six groups — the Environmental Policy Institute, Friends of the Earth, the Oceanic Society, Citizen Action, Clean Water Action, and the U.S. Public Interest Research Group — in May 1989 issued a list of "imperatives for action" related to the Exxon Valdez oil spill. On the list, accompanying the demand that Exxon be held liable for the costs of cleaning up the spill and compensating the appropriate parties for spill-related damages, were calls for intensive research into the short-term and long-term environmental effects of oil spills and for the implementation of species restoration programs for sea otters and other affected fish and wildlife.

The coalition of environmental groups also urged the federal government to ensure that "ecologically sound and effective cleanup initiatives...[are] carried out now and in the months and years ahead," especially in light of the fact that Exxon and Alyeska "failed to move quickly enough in the hours and days immediately following the spill" and then responded with "far too little prepositioned equipment and personnel." The oil industry, the coalition said, should create a $5 billion trust fund to finance activities to "improve and advance environmental protection and education in Alaska and all other states where the oil industry maintains...facilities."

The executive director of the Oceanic Society told two congressional subcommittees in April that the federal government should use money provided by Exxon to initiate a comprehensive study of the long-term environmental effects of oil spills, as well as coordinate a study of all relevant federal research and development activities to evaluate the existing "state of the art" in oil and hazardous substances response equipment.

In June, an oil industry group called the Oil Spill Task Force announced a new 5-year, $250-million program to prevent, contain, and clean up major oil spills in the U.S. The task force plans to undertake a research program, funded with $30 million to $35 million, to both determine the biological and chemical effects of oil spills on the environment and improve oil containment and recovery techniques. The overall program will be implemented by a specially created industry association—the Petroleum Industry Response Organization—whose primary goal is to establish five regional response centers with around-the-clock response teams and sufficient equipment to handle massive oil spills.

While some scientists believe that the ecosystems along Alaska's southern coastline may recover in as few as 3 to 5 years, others believe that, despite nature's power to regenerate itself, human and natural efforts to clean up the Exxon Valdez spill will take far longer.

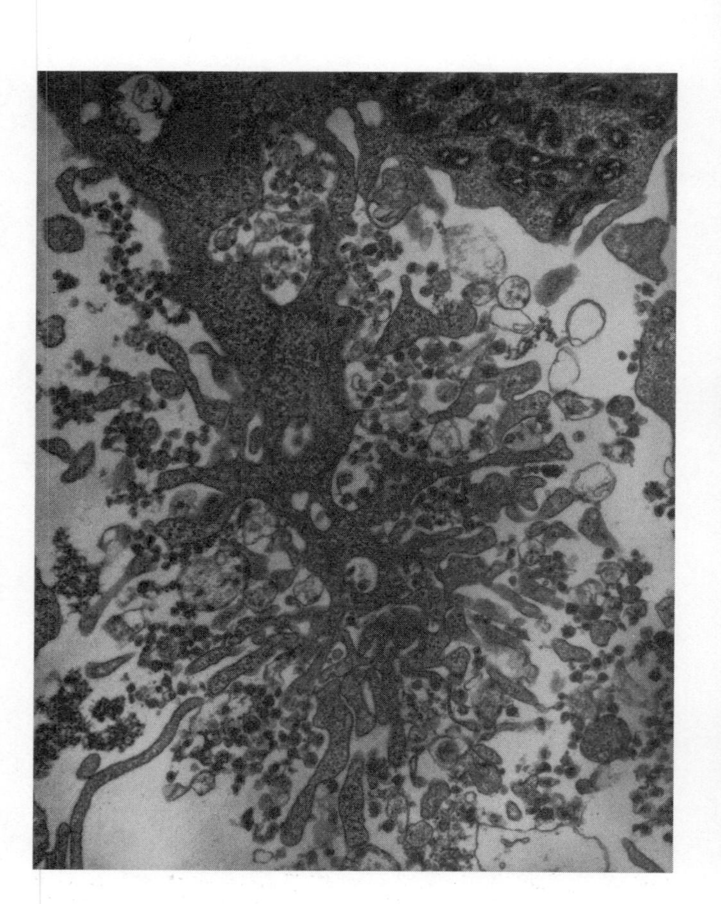

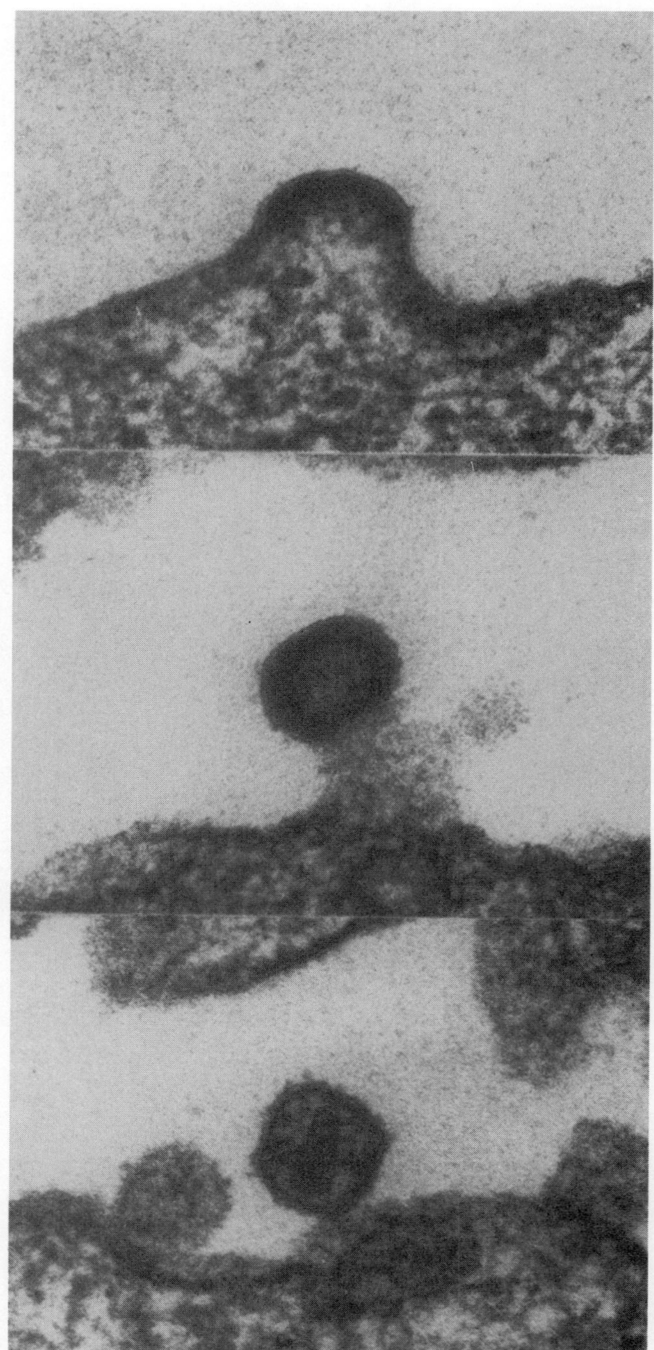

Chapter 8

Medicine

Acquired Immunodeficiency Syndrome

Acquired immunodeficiency syndrome, known more commonly as AIDS, is a rare entity in the annals of modern medicine — a truly new disease. It is even more unusual in that, unlike most diseases affecting humans, it is caused by a retrovirus, which operates within its target cells in a way opposite to that in which usual human viruses work.

Like the nuclei of cells they infect, most viruses contain cores of deoxyribonucleic acid (DNA). Such DNA viruses use the machinery of the cells they invade to make copies of themselves, causing prompt cell death and disease symptoms. Retroviruses contain cores of ribonucleic acid (RNA), a molecule that cells use as a template or messenger for genetic information. A retrovirus uses its RNA to code for genes that make target cells translate its RNA into DNA. The DNA derived from the virus can then lie dormant within the target cell's DNA for years without causing any symptoms of disease. This latency period ends when the viral DNA is somehow activated, causing the infected cell to begin manufacturing new retroviruses, which kill the host cell as they break through the cell membrane. The human immunodeficiency virus (HIV), which is thought to cause AIDS, is particularly deadly because it attacks key white blood cells that orchestrate the normal immune response.

HIV's ability to "hide" in its target cells explains how the retrovirus was able to spread from its still-unknown origins for years — perhaps decades — before the world medical community became aware of its presence. The proliferation of HIV was accelerated by cultural and socioeconomic changes in developed and Third World countries. For instance, the decolonialization of Africa brought many people from rural regions, where HIV may have existed unnoticed for decades, into densely populated cities. In the developed world, the so-called sexual revolution, coupled with the explosion of illicit intravenous (IV) drug use and an increase in intercontinental jet travel, increased the rate at which a virus like HIV could spread. Because the disease has spread so widely, medical researchers now describe AIDS as pandemic.

Despite rapid increases in rates of HIV infection worldwide, the virus is not transmitted through casual contact. In its fourth progress report on AIDS, issued in late 1988, the World Health Organization (WHO) stated, "There is no evidence to suggest that HIV transmission involves insects, food, water, toilets, swimming pools, sweat, tears, shared eating and drinking utensils, or other items such as secondhand clothing or telephones."

WHO and leading AIDS researchers have concluded that there are only three modes through which HIV is transmitted. One is sexual intercourse, either heterosexual or homosexual. Another is exposure to blood, blood products, donated organs, or semen contaminated with HIV. Transfusion of HIV-contaminated blood is known to be the most efficient means of transmission. In developed countries, however, where all blood has been screened for the presence of HIV antibodies since 1985, the principal route of exposure to HIV-contaminated blood is through the sharing of unsterilized needles and syringes by users of illicit IV drugs. Finally, HIV can be transmitted from an infected mother to her child before, during, or shortly after birth.

Given the modes of transmission, it is not surprising that AIDS is overwhelmingly a disease of the young — those who are sexually most active and most likely to engage in risky drug-using behavior, as well as babies born to infected parents. In the U.S., two-thirds of AIDS patients are under the age of 40, while in developing nations the proportion of young people with AIDS is even higher.

Emergence of a Pandemic

By early 1989, with the pandemic still in its early stages, more than 300,000 cases of AIDS had occurred in over 140 nations. Jonathan M. Mann, director of the WHO Global Program on AIDS, estimated in 1988 that "between 5 and 10 million people are already thought to be infected, although only a tiny fraction of them know it yet," and that several hundred million people around the world "behave in ways that make them vulnerable to HIV infection." Once infected, carriers of HIV are thought to be capable of spreading the virus to others for the rest of their lives. This assures that AIDS will remain a global health problem into the twenty-first century.

As of mid-1988, the U.S. accounted for about two-thirds of the world's officially reported AIDS cases, and the U.S. Public Health Service (PHS) estimated that between 1 million and 1.5 million Americans were infected with HIV. Official reports are misleading, however, because there is wide variation among nations' disease-surveillance systems and their willingness to report AIDS. Most evidence suggests that the part of the world most heavily affected by HIV infection and AIDS is a belt across Central Africa from Zaire and the Congo in the west to the countries clustered around Lake Victoria — Kenya, Uganda, Rwanda, Burundi, and Tanzania — in the east. Within these nations, AIDS is most highly concentrated in cities such as Kinshasha, Zaire; Nairobi, Kenya; and Butare, Rwanda.

In the October 1988 *Scientific American*, Mann and James Chin of WHO, Peter Piot of the Institute of Tropical Medicine in Antwerp, Belgium, and Thomas Quinn of the U.S. National Institutes of Health (NIH) wrote, "Close to half of all patients in the medical wards of hospitals in those [African] cities are currently infected with HIV." They noted that 10 to 25 percent of young women are also infected in those cities, "and that will mean an increase in child mortality of at least 25 percent," a catastrophe that may well nullify the hard-won gains in child survival achieved in the past two decades.

Although AIDS was first recognized as a distinct entity in 1981, there is evidence that HIV first began to be transmitted regularly among homosexual men in America's largest coastal cities during the early to mid-1970s. The long latency period between HIV infection and the onset of AIDS explains why cases of AIDS did not begin showing up in large numbers until the early 1980s. The incubation period of AIDS could not be established until the late 1980s, after scientists had followed a large number of cases over a long period of time. Several reports have concluded that the average incubation period between infection with HIV and the appearance of AIDS is about 8 years. This incubation period varies widely among individuals, however. Drawing from a number of studies, officials at the U.S. Centers for Disease Control (CDC) believe that only about half of those infected with HIV will develop AIDS within 10 years; some cases do not show up until even later.

After the first cases of AIDS were recognized in mid-1981, new cases began coming to light at a rapid pace. At first, the number of cases diagnosed doubled every few months. By early 1983, the number of AIDS cases in the U.S. had surpassed 1,000. By August 1989, the cumulative total of U.S. AIDS cases exceeded 100,000, of whom over 59,000 were known to have died. Using two different forecasting models, PHS estimates that the number of Americans ever diagnosed with AIDS will reach 365,000 by the end of 1992.

Risk Factors

For a short period in early 1982, some researchers proposed calling the emerging immune disease "Gay-Related Immune Deficiency," or GRID, because it had been seen only in homosexual men. And, in fact, by the end of 1988, gay men who claimed to have no history of intravenous drug use accounted for about 62 percent of all people diagnosed with AIDS in the U.S. In Canada, Europe, Australia, New Zealand, and parts of Latin America, the majority of AIDS cases occur among gay men. But this fact belies the complexity and the evolution of the disease in these regions, which WHO terms Pattern I nations.

In many parts of Pattern I nations, AIDS cases are now being diagnosed more frequently among IV drug users, their sexual partners, and their children. While the male to female ratio of AIDS cases now ranges from 10:1 to 15:1, heterosexual transmission is increasing and with it, transmission from mothers to infants. In the late 1980s, this trend was still concentrated in cities with large homosexual communities and a high rate of illicit drug abuse, but the mobility of urban populations has enabled the virus to spread outside of narrowly defined "risk groups."

Transfusion-associated AIDS. Although the first cases of AIDS were among gay men, it became apparent within the first 6 months of 1982 that AIDS was not a so-called gay disease. Cases began appearing among heterosexual hemophiliacs and blood-transfusion recipients, an indication that whatever was causing the illness was capable of being transmitted through the blood. Hemophiliacs, who number about 15,500 in the U.S.,were especially at risk because they are dependent upon injections of a blood-clotting protein extracted from plasma obtained from thousands of donors. If one of those donors was infected with HIV, the clotting factor was able to transmit the virus.

In developed countries, the risk of HIV infection from transfusion of blood and blood products, as well as from transplantation of donor organs, was nearly eliminated in 1985, when a sensitive and rapid test was developed to reveal the presence of antibodies to the newly discovered AIDS virus. Beginning in 1985, hemophiliacs also achieved additional (though not total) protection from new heat-treatment methods that kill the virus in preparations of blood-clotting factor.

Before blood-screening and heat-treatment methods became available, however, approximately 70 percent of people with hemophilia-A (deficiency of a clotting protein called Factor VIII) and 35 percent of those with hemophilia-B (deficiency of Factor IX) acquired HIV infection. Hemophiliacs and people with other coagulation disorders who were infected through clotting factors accounted for 1 percent of all U.S. AIDS cases as of early 1989.

When U.S. blood banks began screening all blood for HIV antibodies in May 1985, about 35 out of every 100,000 units tested positive. By mid-1987, that rate had declined to 12 per 100,000 as a result of the exclusion of previously identified infected donors and high-risk individuals from the donor pool. Studies have shown, however, that it is possible for a recently infected person to donate blood before his or her body has had time to develop antibodies, which means that there remains a very small risk of acquiring HIV from blood transfusion.

As a result of transfusion-associated HIV infections acquired before mass screening of donated blood, over 2,000 Americans had developed AIDS by early 1989, and about 1,500 of them had died of the disease. In March 1987, CDC officials estimated that about 12,000 Americans had undetected HIV infections acquired through transfusions that occurred between 1978 and 1985. The PHS advised Americans who had transfusions between 1977 and 1985 to be tested for HIV antibodies, so that people who are infected may take precautions to avoid infecting their sexual partners.

Intravenous drug use. Like transfusions, the sharing of contaminated needles and syringes by IV drug users, often in urban "shooting galleries" where unsterilized injection equipment may be rented, has proved to be a highly efficient way to transmit HIV. As addicts prepare to inject themselves with drugs, a small amount of blood is drawn into the syringe; if the blood contains HIV, the virus will be injected directly into the next user's bloodstream. Heterosexual IV drug users account for one out of every five U.S. cases of AIDS among adults and adolescents. Another 7 percent are homosexual or bisexual IV drug users. Thus, IV drug use may account for more than one-quarter of all AIDS cases.

The 1980s have seen an explosion in illicit drug-using behavior. With an estimated 500,000 IV drug users in the U.S., drug-use-related HIV infection opens avenues for the disease to spread widely and to penetrate beyond the relatively self-contained population represented by urban gay men. Female drug users often engage in prostitution to support their habit, and they may infect their non-drug-using male clients. Because HIV can be transmitted across the placenta to the unborn fetus, IV drug use is a major factor in the rapidly growing epidemic of pediatric AIDS.

Heterosexual cases of AIDS are largely the result of HIV transmission from IV drug users to their sexual partners. Overall, heterosexual cases have accounted for over 4 percent of all adult cases of AIDS since 1981. A CDC analysis in mid-1988 showed that about three-fifths of these cases involved sexual contact with someone in another risk category, either an IV drug user or a bisexual man. The remaining heterosexual AIDS cases involved people who came originally from countries such as Haiti, where heterosexual transmission is the major mode of HIV transmission.

Because IV drug use is most concentrated in poor, inner-city Black and Hispanic communities, those communities are disproportionately burdened by AIDS. Available data indicate that future AIDS cases — as extrapolated from HIV infection rates among apparently healthy people — will continue to affect minority groups disproportionately. The CDC reported in 1988 that blood tests of nearly 1.8 million military recruits showed that Blacks were 3.6 times more likely than whites to test positive for HIV, and Hispanics were 2.5 times more likely to be infected. CDC statistics from June 1988 showed that the proportion of new AIDS cases among homosexual men and all white adults began falling in mid-1987, while the proportion of cases involving Black and Hispanic adults and all adult women was rising steadily. AIDS researcher Andrew Moss of San Francisco General Hospital noted in early 1989 that, in the U.S., heterosexual AIDS is generally a poor people's disease. For example, the vast majority of pediatric AIDS cases — the fastest-growing component of the U.S. epidemic — involve the children of the poor.

Pediatric AIDS. More than three-quarters of the 1,400 pediatric cases reported to the CDC by early 1989 involved children born to a parent with AIDS or at risk for the disease, usually a mother who used IV drugs or was the sex partner of an IV drug user. Another one-fifth of pediatric cases were children who were hemophilic or had received blood transfusions.

Elaine J. Abrams of the New York City Collaborative Group for the Study of Maternal Transmission of HIV reported in 1988 that among 48 New York City babies born with HIV antibodies in their blood, 43 percent developed symptoms of immune deficiency in their first year of life. Charles Lawrence of the New York State Department of Health and his coworkers have suggested that many HIV-infected children will probably survive for years before succumbing to AIDS. A multinational team, led by Pratibha Datta of the University of Nairobi, has reported that infected infants born to mothers in later stages of the disease may sicken and die earlier than those born to healthier HIV-infected women.

Health care workers. Although health care and biomedical laboratory workers are at risk of HIV infection, they are not listed separately in the CDC's weekly statistics on the disease. The CDC has reported that among the over 2,800 cases listed as "undetermined" in early 1989 was one health care worker "who seroconverted to HIV [developed antibodies indicating infection] and developed AIDS after documented needlestick to blood." Of the approximately 1,200 other health care workers monitored by the CDC following accidental exposure to the blood or body fluids of AIDS patients, only four had developed HIV antibodies by the end of 1987. The CDC estimated in 1988 that a health care worker's risk of infection after a needlestick injury is less than 1 percent and the risk entailed in incidents involving exposure of workers' mucous membranes or nonintact skin to fluids from AIDS patients is probably far less than that. Nevertheless, the CDC has advised health care workers to assume that any patient could be infected with HIV and to take "universal precautions," which include using latex gloves and protective equipment when performing procedures likely to expose them to patients' blood or body fluids.

AIDS Outside Pattern I Nations

A central question of the AIDS pandemic, and one that has caused much confusion among both the public and policymakers, is why the pattern of the disease differs so sharply between most industrialized nations (the Pattern I countries) and other countries, where males and females are equally affected.

Pattern II countries. This second pattern, as defined by WHO, occurs in sub-Saharan Africa, some parts of Latin America, and some Caribbean countries, including Haiti. In these nations, more than 1 percent of the entire population is infected with HIV, although AIDS cases are mostly concentrated in urban areas, where typically only 10 to 20 percent of the population live. In the fourth progress report on the WHO Global Program on AIDS in October 1988, Jonathan Mann noted that up to 25 percent of the young and middle-aged urban population, those between 15 and 49 years of age, are infected. Because males and females are equally affected, mother-to-infant transmission of HIV is common throughout all strata of society. Since IV drug use and homosexuality are thought to be very uncommon in these countries, these behaviors do not contribute significantly to infection. Transmission of HIV through contaminated blood remains a significant problem because health budgets and logistical problems have not permitted universal blood-donor screening. The widespread recycling of unsterilized needles and syringes for both medical injections and traditional skin-piercing rituals may also contribute to HIV infection.

Scientists agree that the strikingly different epidemiology of AIDS in Pattern II countries does not stem from any biological difference in HIV but from behavioral and social factors that favor a different pattern of spread. For instance, the relatively high frequency of female prostitution and the existence of polygamy or "unions libres" appears to be a leading factor in the infection of young urban males in many Central African cities. High rates of sexually transmitted genital ulcer diseases, such as syphilis, gonorrhea, chancroid, chlamydia, and herpes, also promote the spread of HIV.

HIV-2 disease. Another hallmark of West African AIDS is the emergence of a second human immunodeficiency disease caused by a distinct virus, now dubbed HIV-2. The first evidence for such a virus was reported in late 1985 by Francis Barin of the University of Tours in France and an international research team. The virus was first found in the blood of prostitutes in Dakar, Senegal. At about the same time, Luc Montagnier and Francois Clavel of the Pasteur Institute in Paris, France, found the organism in the blood of two men from Guinea-Bissau.

Genetic analysis revealed that HIV-2 is more closely related to a virus that causes AIDS in monkeys, called simian immunodeficiency virus (SIV), than it is to HIV-1. Yet SIV cannot be the immediate "ancestor" of HIV-2, as was first suggested, because the viruses share only about 50 percent of their genetic material. In 1988, a University of Tokyo team lead by Masanori Hayami analyzed the entire genetic sequence of SIV isolated from the African green monkey (which frequently harbors SIV but does not suffer immunodeficiency disease) and concluded that the monkey virus is equally and distantly related to both HIV-1 and HIV-2. That conclusion has been disputed, however, by Temple F. Smith and coworkers at Dana-Farber Cancer Institute in Boston, Massachusetts, who argue that SIV resembles HIV-2 more closely.

The similarities and differences among SIV, HIV-1, and HIV-2 are of more than academic interest, since researchers hope that discovering the reasons why one variety may be less virulent than another may offer clues to effective vaccine and drug strategies. Some researchers at first thought that HIV-2 might not actually cause illness, since many infected West Africans appeared to be disease-free, and even that HIV-2 infection might confer immunity, acting as a sort of natural vaccination against HIV. However, Montagnier's group has shown that this is not the case. Instead, the relative virulence (disease-causing capacity) of HIV-2 may be less or the latency period may be years longer.

In the late 1980s, HIV-2 was almost entirely confined to the West African nations of Senegal, Cape Verde, Guinea-Bissau, Burkina Faso, Ivory Coast, and Benin. However, some health officials are concerned that HIV-2 infection may begin to spread among U.S. Portuguese-speaking minorities, such as the large Cape Verdean population in Massachusetts and other northeastern states.

Pattern III countries. The third pattern of HIV disease, according to WHO's classification, occurs in northern Africa, Eastern Europe, the Middle East, the U.S.S.R., and all of Asia. The introduction of HIV into these regions apparently occurred in the early to mid-1980s, and few deaths have yet been reported. Most of the relatively few AIDS cases have involved people who have traveled to AIDS-endemic areas or have had sexual contact with natives of those areas. A small number of cases have also been traced to imported blood products. In its October 1988 report, WHO noted, "How extensively HIV will spread in Pattern III countries is difficult to predict. However, where IV drug use is prevalent, HIV and AIDS will be a major potential problem."

How HIV Works

The worldwide effort to understand the causes of AIDS has been aided by a remarkable confluence of new scientific knowledge and technology that has focused the powers of molecular biology, virology, cellular biology, and genetic engineering on this scientific and human challenge. For example, scientists' recognition that retroviruses could play a role in human disease dates only from 1980 (the year before AIDS was recognized), when Dr. Robert C. Gallo of the National Cancer Institute (NCI) isolated the first human retrovirus, human T-lymphotropic virus type I (HTLV-I), which causes an unusual form of adult leukemia.

Without modern scientific tools, the history of AIDS to date would have been significantly darker. For example, far more people would have become infected through contaminated blood transfusions while scientists groped to understand the nature of the infectious agent. At the opening plenary session of the Fourth International Conference on AIDS in Stockholm in June 1988, Gallo stated, "There is no doubt in my mind that if the AIDS epidemic had come upon us, say, in the 1960s, we would still be stumbling in the dark." Instead, Gallo and his former scientific rival, Luc Montagnier of the Pasteur Institute, concluded in a jointly written article in the October 1988 *Scientific American*: "In spite of the startling nature of the epidemic, science responded quickly."

By the end of 1983, Montagnier had isolated the AIDS virus from the lymph glands of a young homosexual man with chronically swollen glands and determined that it was a retrovirus. He named it lymphadenopathy-associated virus (LAV) and developed a reasonably sensitive antibody test for it. The following year, Gallo and his colleague Mikulas Popovic were able to grow the virus in quantity (a difficult feat, since the virus tended to kill most cells it infected). They named it HTLV-III, in an effort to place it in the same family as Gallo's earlier retroviral discoveries. By 1985, they had developed a quick and sensitive antibody test that could be widely applied in blood screening and in hundreds of epidemiological and clinical studies of AIDS.

A bitter struggle between Gallo and Montagnier ensued over credit for the discoveries, the name of the virus, and the patent rights to the resulting antibody tests. The feud was eventually resolved and, in a legal settlement, the two agreed to set up a Franco-American AIDS foundation to receive royalties from the AIDS blood test. The foundation uses the royalties to fund AIDS-related projects such as the antibody testing of blood in Africa, intended to reduce transfusion-associated infections. In 1986, an international committee of virologists proposed the neutral term "human immunodeficiency virus," by which the virus is now known.

Attacking the immune system. Although its genetic blueprint is 100,000 times smaller than any human cell's, HIV can cause tremendous damage once it enters a host. HIV's destructive capacity stems from its ability to infect cells that play a critical role in monitoring the human immune system. The most common feature of the various cells targeted by HIV is a marker on their surfaces called the CD4 antigen, a sort of docking site that allows certain specialized cells to recognize each other and interact appropriately as they carry out their physiologic functions. HIV bears a protein on its surface called gp120 (gp stands for glycoprotein, a complex of sugar and protein, while 120 signifies the molecular weight) that recognizes and latches onto the CD4 antigen. CD4 is most abundant on immune-system cells called T4 helper cells.

Although the name suggests that they are merely servants, T4 helper cells are more appropriately thought of as master cells of the immune system, orchestrators of the complex symphony of immune response that must be ready to defend the body against a wide range of invading viruses, parasites, bacteria, and fungi. T4 cells recognize when human cells have been infected and recruit other immune cells called cytotoxic lymphocytes and natural killer cells to attack and destroy the infected cells. By secreting hormonelike molecules called lymphokines, T4 cells also activate highly mobile scavenger cells called monocytes and macrophages to seek out and engulf infected cells and the debris left over from their destruction.

All of these functions are part of what is called cell-mediated immunity, because they involve direct attacks on the infection by various immune cells. But T4 cells are also vital components in the second arm of immune response, called humoral or antibody-dependent immunity, because they stimulate immune cells called B cells to produce antibodies so that free-floating infectious microbes can be inactivated before they infect cells.

By gradually depleting the body's stores of T4 cells, HIV can cripple the entire immune system. The retrovirus can also infect any cell that bears CD4 receptors, including the bone-marrow progenitors of all white and red blood cells and the Langerhans cells of the skin, some antibody-producing B cells, as well as monocytes and macrophages. The latter cells are not killed outright by the virus but become reservoirs of infection that can carry HIV directly into the brain, causing central nervous system infection that leads to mental confusion, personality changes, memory loss, and eventually profound dementia and loss of muscle control. Finally, HIV can apparently gain entry into some cells that do not bear detectable

CD4 receptors, including certain cells in the intestine and brain.

Charting the course of the disease. Most HIV disease is characterized by a slow, irreversible progression from latent infection to total immune collapse with a succession of "opportunistic" infections — illnesses caused by microbes that are harmless to people whose immune systems are intact but devastating to people with AIDS. At any time during the course of HIV disease, patients can develop a set of so-called constitutional symptoms (generalized symptoms not related to a particular organ system) known as AIDS-related complex or ARC. These symptoms typically involve persistent night sweats, chronic diarrhea, and progressive weight loss.

The course of HIV disease has been organized into a six-stage classification scheme by Robert A. Redfield and his colleagues at the Walter Reed Army Institute of Research in Washington, D.C.

In the first stage, many people suffer an acute, mononucleosis-like illness weeks to months after exposure to HIV. In addition to swollen glands, fatigue, and fever, some recently infected people suffer symptoms of neurological impairment, such as headaches or even encephalitis, an inflammation of the brain. After recovering from this brief illness, an infected person typically feels well for 6 months to a year. Then, according to Redfield and his colleagues, most people develop chronical lymph node disease, a condition called lymphadenopathy, which may last several years without any other signs of disease.

The third and fourth stages show a decline in immune function as measured by T4 cell counts, which fall from the normal level of about 800 cells per cubic millimeter of blood to fewer than 400 cells. Stage four is also characterized by an abnormal reaction to a skin test called delayed hypersensitivity response, which measures T-cell activation following an injection of foreign proteins.

As T4 levels continue to decline, patients enter stage five, when they develop symptoms of immune collapse. These symptoms most often involve chronic infections of the skin and mucous membranes, including thrush, an infection of the mouth caused by the fungus *Candida albicans* and marked by white spots and ulcers; ulcers of the mouth, genital area, or skin surrounding the anus caused by herpes simplex virus; and oral hairy leukoplakia, an infection of unknown origin that produces fuzzy white patches on the tongue.

The sixth and final stage is AIDS itself, when once-exotic and often life-threatening infections may strike a wide variety of organs. These illnesses often begin with persistent coughing and shortness of breath, symptoms of PCP, a virulent form of pneumonia caused by a protozoan called *Pneumocystis carinii* that is harmless to people whose immune system is sound. According to the CDC, PCP is the primary cause of death for three out of five AIDS patients. Studies have shown that about half of HIV-infected people with T4 cell levels below 200 develop PCP within a year.

The prognosis is grim once patients have entered stage six, with most dying within 2 years. Many researchers hoped that not all HIV-infected people would progress to the immune collapse represented by stages four through six, but long-term studies of AIDS have shown that the course of illness appears inevitable. A Walter Reed study found that 90 percent of a group of HIV-infected patients progressed one or more stages over an average of 3 years.

A long-term study of 200 HIV-infected homosexual men in San Francisco, by Robin Edison and colleagues at San Francisco General Hospital, found a relentless progression toward AIDS and death. Among people followed for 7 years, 61 percent had developed AIDS and 47 percent had died. A CDC team led by Kung-Jong Lui estimated in 1988 that an infected person has a 99 percent chance of developing AIDS, based on an extrapolation of the experience of 84 San Francisco men.

Opportunistic infections. Aside from PCP, people with end-stage AIDS often develop other strange, difficult-to-treat parasitic infections caused by ubiquitous, normally harmless organisms. Toxoplasmosis typically afflicts the brain and can cause seizures and coma. A fungus called *Cryptococcus*, often harbored by pigeons, can cause meningitis (inflammation of the lining of the brain and spinal cord) as well as damage to tissues of the liver, bone, skin, and other parts of the body. Cryptosporidiosis often causes the chronic diarrhea that makes life miserable for AIDS patients and can cause life-threatening dehydration. Histoplasmosis is the source of chronic fevers and disseminated infections of the liver and bone marrow.

Viruses that often lie dormant in the tissue of HIV-infected people are unleashed as T4 levels plummet. Hepatitis B, a viral infection transmitted in the same ways as HIV, can flare up, causing liver damage that disturbs metabolism and disrupts blood clotting. Herpesviruses are especially troublesome for AIDS patients, since by nature they establish occult, lifelong infections. Epstein-Barr virus, one member of the herpes family that infects perhaps 80 percent of the world's population, may give rise to B-cell lymphomas, described in greater detail below. Cytomegalovirus (CMV), which is common among gay men, can cause runaway infections of the lungs, colon, liver, adrenal glands, and the retina of the eye, rapidly causing blindness.

The bacteria *Streptococcus pneumoniae*, *Pneumococcus*, and *Hemophilus influenzae*, all can cause pneumonia in people with AIDS. One of the gravest and most exotic bacterial infections associated with AIDS is caused by *Mycobacterium avium-intracellulare*, a complex of organisms whose effects can mimic tuberculosis not only of the lungs but of the liver, bone marrow, and gastrointestinal tract. Common tuberculosis (TB) itself is on the rise, especially among IV drug users. Extrapulmonary TB (infections outside the lung, a relatively rare condition before the advent of AIDS) is becoming common in inner-city hospitals.

Latency and replication. One of the key questions AIDS researchers confront is how the virus can remain dormant for years in which its victims are asymptomatic but infectious, and then suddenly replicate, causing a relatively rapid decline. In an attempt to understand the highly variable course of HIV infections, scientists have sought to establish cofactors, which are environmental agents that might influence how the disease progresses. The search has generally not been fruitful. One exception is a finding reported in early 1989 by Paolo Lusso of Bionetics Research, Inc., in Rockville, Maryland, and Robert C. Gallo. The researchers discovered that T cells infected by both HIV and a newly discovered virus called herpesvirus VI die considerably faster than T cells infected by HIV alone. The herpesvirus appears to trigger HIV replication, according to Lusso and Gallo, who are investigating the relevance of the test-tube study for humans. Most HIV-positive people show evidence of exposure to herpesvirus VI, since most people are infected with the virus in childhood. The herpesvirus appears harmless except in the presence of HIV.

Once HIV attaches itself to a CD4 receptor on a target cell's membrane, it punches a small hole through the membrane, into which it injects its RNA core along with a package of enzymes. In one scenario, the enzymes allow HIV to direct the cell's machinery to manufacture more HIV particles, using the viral RNA as a template. But in other cases, the virus uses a set of enzymes known as reverse transcriptase to translate its RNA nucleus into DNA, which becomes integrated with the cell's own DNA. The virus can then remain latent in the host cell for an indefinite amount of time. Scientists do not yet know what determines which of these two courses the HIV viruses follow, but they hope to find clues by analyzing the genetic structure of the virus.

By mid-1989, scientists had identified 11 HIV genes. Only three of these genes, called gag, pol, and env, are known to code for structural components of the virus. The others encode a repertoire of regulatory proteins that, in their interaction, can speed up, slow down, or totally suppress viral replication. One of the genes, called vif, appears to make the virus more efficient at infecting cells. The function of another two, called vpu and vpr, is not yet understood. In 1988, Klaus Strebel and his colleagues at the National Institute of Allergy and Infectious Diseases speculated that vpr may coordinate the assembly of viral components before HIV emerges from infected cells.

One of the most remarkable features of HIV is the gene called tat, short for "trans-activating," a term signifying that the gene codes for a protein that exerts its effect at a distance rather than on adjacent genes or their protein products. Discovered in 1985 by William A. Haseltine of the Dana–Farber Cancer Institute and Gallo's NCI colleague Flossie Wong-Staal, tat can accelerate HIV's replication to a rate 1,000 times faster than that of mutant viruses lacking the gene. This gene enables HIV to reproduce much faster than any other known virus.

Haseltine and Wong-Staal have speculated that such a frenzied replication rate might prove self-defeating in long-lived hosts such as humans, because the hosts would often die before transmitting the virus to other people. This, they believe, may explain why HIV contains other regulatory genes that oppose or dampen the stimulating effects of tat. For instance, a gene called nef (for "negative-regulatory factor") represses all HIV genes, perhaps accounting for the virus's ability to remain dormant for years until external stimuli activate the T4 cell in which the HIV gene resides. Another, called rev (for "regulator of virion-protein expression"), favors viral growth by repressing regulatory genes and activating those that code for HIV structural proteins.

Further understanding of the genes of the two HIVs and their cousin SIV may give researchers important clues for developing new drug treatments or vaccine designs. The University of Tokyo team that analyzed the entire genetic sequence of SIV found that the African green monkey virus lacks the vpr gene found in both HIVs. They and other researchers speculate that vpr may increase pathogenicity or virulence, and its lack may permit SIV to exist in its monkey hosts without causing illness.

Role of gp120. Though researchers do not yet fully understand the mechanisms governing infection, they have gained some insight into how the virus functions. Scientists were at first puzzled that HIV could cause such devastating immunosuppressive effects when so few T4 cells seem to be infected by the virus — no more than about 1 in 1,000 at a given time. This mystery was at least partially solved by Haseltine's discovery that HIV was responsible for killing even uninfected T4 cells through the formation of syncytia. These are giant, multinucleated, nonfunctioning masses that form when up to 500 T4 cells fuse together. Syncytia occur because the surfaces of infected T4 cells are studded with gp120, the HIV surface glycoprotein. In lock-and-key fashion, the CD4 surface receptors of uninfected T4 cells latch onto the gp120, which leads to fusion of the infected and uninfected cells. The ability of HIV to induce syncytia formation greatly amplifies the cell-killing capacity of a single virus.

Syncytia formation also enables HIV to pass from one T4 cell to another without emerging into the bloodstream, where antibodies against the virus might recognize and destroy it. Some researchers fear this may thwart the effectiveness of an HIV vaccine, since vaccines rely upon antibodies' access to free viruses circulating in the blood. The expression of gp120 on the surface of infected T4 cells may also mark them for destruction by natural killer cells and cytotoxic T cells, and possibly by antibodies. Thus, the immune system may kill its own T4 cells, even though the retrovirus inside them is not actively replicating.

The gp120 protein itself may also produce some symptoms of AIDS. NIH researcher Douglas E. Brennan and his coworkers have reported that gp120 detached from HIV or infected cells can bind to and kill uninfected mouse brain cells, which bear CD4 receptors on their surfaces. Since gp120 often falls off the fragile HIV and may be shed from infected cells as well, this might explain some of the neurological effects of HIV infection.

Figure 1. AIDS Education. The AIDS virus is most often spread when people engage in certain high-risk behaviors, such as unprotected sexual activity or needle-sharing during intravenous drug use. In the past few years, many government and private organizations worldwide have launched campaigns to teach people how to protect themselves against AIDS. *Courtesy: Carlos Gaggero, Pan-American Health Organization.*

HIV and cancer. AIDS scientists also do not understand how HIV infection sometimes leads to the appearance of rare cancers such as Kaposi's sarcoma, which causes blood vessel overgrowth and later invades vital organs, and B-cell lymphoma, a rare cancer of antibody-producing white blood cells that has become more common in cities heavily affected by AIDS. HIV-infected people occasionally also suffer rare malignancies of the rectum and tongue.

B-cell lymphomas may result from several different mechanisms. HIV infection leads to chronic stimulation of antibody-producing B cells, perhaps through the abnormal release of lymphokines by other immune cells infected with HIV. This chronic B-cell activation could lead to malignancy. Alternatively, about 5 percent of B cells bear CD4 antigens on their surface, permitting them to become infected by HIV, which may somehow directly trigger cancerous changes. Another possibility is that latent Epstein-Barr infections may be activated as immune function declines. Epstein-Barr virus has been implicated in a B-cell cancer called Burkitt's lymphoma that is endemic to Africa.

For unknown reasons, the prevalence of Kaposi's sarcoma among homosexual men has fallen sharply during the course of the AIDS pandemic. Alan R. Lifson of the San Francisco Health Department and his colleagues reported in 1988 that the proportion of San Francisco AIDS cases involving Kaposi's sarcoma declined from 77 percent in 1981 to 26 percent in 1987. The decline of Kaposi's sarcoma has been paralleled by a fall in the transmission of CMV, a virus that is often sexually transmitted among gay men. Some researchers believe that this coincidence suggests that CMV may be a cofactor for Kaposi's sarcoma in people infected with HIV and that the decline in Kaposi's sarcoma in San Francisco may be due in part to changes in the sexual behavior of gay men in that city.

Another possible clue was suggested by NCI researcher Jonathan Vogel and his colleagues, They created Kaposi's-like lesions in mice, which are normally free of such tumors, by splicing the tat gene into the animals' genetic material while they were in the embryo stage. The researchers concluded that the HIV tat gene's capacity to induce tumors in mice suggests that HIV can cause cancer. At the 1988 Stockholm meeting, Gallo and his NCI colleagues proposed that Kaposi's sarcoma is not caused by HIV itself but by the virus's ability to stimulate T4 cells to secrete a substance that makes blood vessels grow, which Gallo called the Kaposi's sarcoma growth factor. The NCI group said they had partially purified the growth factor and found that it caused Kaposi's-like lesions even in uninfected mice.

Treating AIDS and HIV Infection

Since the 1940s, many infectious diseases have been brought under control with antibiotics. The successes of the antibiotic era, however, have mostly been in the treatment of diseases caused by bacteria and some other nonviral pathogens. The treatment of viral infections has been far less successful. Since human retroviral infections were discovered only recently, the prospects of developing effective treatments to combat them are still unclear.

Bacteria and many other human pathogens are cellular organisms whose characteristics are very different from those of human cells. Drugs that exploit these differences can attack a microbe while leaving its host's cells largely unscathed (the so-called magic bullet approach). Viruses, on the other hand, are noncellular parasites, nothing more than tiny packets of genetic information that use their host's intracellular mechanisms for their reproduction. Drugs aimed at interrupting the viral life cycle are likely to interrupt the host cell's too, including healthy, uninfected cells.

Retroviruses may be even more difficult to treat than ordinary viruses, because they can remain concealed within the host cell's DNA for extended periods of time. Drugs against retroviruses are more likely to resemble cancer chemotherapy agents, which achieve only partial success and cause a variety of adverse side effects, than the highly effective antibiotics.

During the time immediately after the retroviral origin of AIDS was discovered, some scientists doubted that an effective therapy could ever be found. Their pessimism has given way to a mood of cautious optimism, however. By late 1988, a large group of companies was working on nearly 40 different antiretroviral drugs and immune-stimulating agents.

Zidovudine. This optimism is largely due to the success of azidothymidine (AZT) (more recently known as zidovudine), one of the first 15 drugs tried against HIV. Zidovudine was first synthesized as an anticancer drug by Jerome P. Horwitz of the Michigan Cancer Foundation in 1964. The drug was ineffective against cancer, but its ability to interfere with DNA synthesis prompted NCI researcher Samuel Broder and his colleagues to test whether it could stop the replication of HIV in the test tube. It did, and the treatment was then given to a person with AIDS in July 1985. After several weeks on zidovudine, the patient gained weight, his T4-cell count increased, and he regained the ability to mount an immune response to the delayed hypersensitivity skin test. Eighteen other AIDS patients were then given zidovudine, and 14 improved measurably.

The Burroughs-Wellcome Co., a British pharmaceutical firm, quickly collaborated with 12 U.S. medical centers to organize a randomized clinical trial, involving 280 AIDS patients, comparing zidovudine to a placebo (a harmless pill containing no medication). Sixteen weeks into the trial, the biostatisticians overseeing the study discovered that 19 patients in the placebo group had died, compared to only one in the group receiving zidovudine. The trial was halted, since it was no longer considered ethical to withhold the drug from patients receiving the placebo. The U.S. Food and Drug Administration (FDA) permitted the drug to be released on a "compassionate use" basis, a decision that led to its preapproval use by nearly 5,000 AIDS patients.

Eventually, information from the multicenter trial showed that among patients receiving zidovudine, mortality decreased sixfold by the end of 36 weeks. The FDA approved zidovudine in March 1987, and by December of that year, about 10,000 AIDS patients were thought to be taking the drug.

There is no doubt that zidovudine extends the lives of people with AIDS. Burroughs-Wellcome epidemiologist Terri Creagh-Kirk and her colleagues reported in late 1988 that in a followup study of over 4,800 AIDS patients who started the drug after surviving a first bout of PCP, 73 percent of those on zidovudine were still alive 44 weeks after they began taking the drug. Before zidovudine became available, AIDS patients had only a 50 percent survival rate during a similar period after their first episode of PCP. Furthermore, the drug has the important advantage of being able to penetrate the blood-brain barrier, which prevents many substances from reaching the brain. There is evidence that neurological symptoms of AIDS are less severe among both children and adults taking zidovudine.

Based on these and similar results, researchers in 1988 began to investigate whether zidovudine might be beneficial for HIV-infected people without symptoms of AIDS. A study of 18 healthy HIV-infected people found that two-thirds had lower levels of HIV and higher T4 cell counts after taking zidovudine over a period of weeks. In August 1989, federal officials announced that zidovudine significantly delayed the progression of HIV disease in two large-scale trials — one involving patients with early ARC and the other involving patients without symptoms but with abnormal T4 cell counts. The officials urged that patients at risk of HIV infection be tested and, if the results are positive, be considered candidates for zidovudine treatment if their T4 cell counts fall below 500 per cubic millimeter.

Although these results are encouraging, zidovudine does not cure AIDS or eradicate HIV infection. Because zidovudine interferes with DNA synthesis, it is highly toxic to blood-cell-producing bone marrow. Up to 15 percent of AIDS patients cannot tolerate the drug at all, and a substantial fraction must discontinue it after some months, at least temporarily, because they become severely anemic or develop other blood cell problems. There is also some evidence that HIV synthesis rebounds sharply when zidovudine treatment is stopped.

ddC. Researchers are testing other agents that, like zidovudine, block reverse transcriptase, the enzyme that transcribes the retrovirus's RNA into DNA, but that do not cause the same side effects. One promising alternative is ddC, or dideoxycytidine. A clinical trial of ddC led by Thomas Merigan of Stanford University revealed that ddC inhibits HIV replication even at low doses. At higher, more effective doses, however, the drug causes a painful nerve disorder affecting the hands, feet, and legs after 8 to 12 weeks of treatment. Nevertheless, since ddC is not toxic to bone marrow, trials are under way to alternate ddC with zidovudine in the hope of gaining the antiviral benefit of both while minimizing toxicity. Two related antiviral drugs, ddI (dideoxyinosine) and ddA (dideoxyadenosine), also showed promise in studies reported in mid-1989.

Barring HIV's entry. Blocking reverse transcriptase is an attractive approach because human cells have no need of the enzyme, but it is not the only way that HIV can be fought. Another avenue of attack is to block the virus at an early stage, before it enters the cell. One agent that uses this approach is dextran sulfate. Test-tube studies have shown that this drug interferes with HIV's ability to bind to cell membrane receptors, and it also inhibits the formation of syncytia, the large masses of T4 cells.

Dextran sulfate has been used in some countries to lower cholesterol and reduce blood clotting, and it appears to be nontoxic to humans. For these reasons, dextran sulfate has attracted a wide underground following among people with AIDS, some of whom have imported it from Japan, where its medical use is approved. In early 1989, Donald I. Abrams and his coworkers at San Francisco General Hospital reported, however, that preliminary trials indicate that the drug appears to have no measurable effect.

Another method of blocking HIV's entrance into host cells is to produce quantities of decoy CD4 molecules — the receptors the virus uses to gain entry to cells. These decoys are intended to attract and bind free-floating HIV before it can infect cells. Animal studies have shown that daily injections of CD4 for 50 days sharply decreased signs of SIV infection in four macaque monkeys and caused no ill effects in uninfected monkeys. Safety testing of soluble CD4 began at several institutions in August 1988.

Scientists have already begun working on second-generation versions of the decoys. One goal is to increase the synthetic decoys' power to kill cells infected with HIV. NIH researchers Bernard Moss and his colleagues engineered a hybrid molecule containing both soluble CD4 and *Pseudomonas exotoxin-A*, a highly lethal poison. The CD4 will carry the poison to cells infected with HIV. In test-tube experiments, the hybrid molecules left non-infected cells unharmed.

Many AIDS researchers believe that the most effective approach will be to administer combinations of drugs simultaneously or in succession, just as combination chemotherapy is used to combat cancer. In early 1989, researcher Martin S. Hirsch of the Massachusetts General Hospital reported that some promising combinations now being tested include zidovudine plus interferon-alpha and zidovudine plus acyclovir. Patients may also be given immune-modulating agents such as interleukin-2 (a T-cell growth factor), antibody-containing immune globulin, interferons, and granulocyte-macrophage colony stimulating factor, a hormone that stimulates the production of immune cells.

Fighting opportunistic infections. Other therapies under development attempt to cope with the opportunistic infections and cancers that plague HIV-infected patients in the latter stages of AIDS. For example, the natural antiviral substance alpha-interferon (IF-a), now produced through genetic engineering, can cause the remission of Kaposi's sarcoma, the most common AIDS-associated cancer.

In one large study of IF-a, Paul A. Voldberding and his colleagues at San Francisco General Hospital found that 35 percent of 114 Kaposi's sarcoma patients given the drug achieved complete or partial remission of the blood-vessel tumor. People who received the highest doses had 45 percent remission rates, and those in earlier stages of disease or without systemic symptoms had remission rates up to 60 percent. Only 6 percent of the patients dropped out of the study because of adverse reactions to the drug.

Probably the most important success to date in treating the opportunistic infections of AIDS is the use of pentamidine in aerosol form to prevent episodes of PCP. This type of pneumonia kills more people with AIDS than any other single infection, accounting for at least 60 percent of deaths. Patients usually survive a first bout through the use of intravenous pentamidine or sulfa drugs. Many patients cannot tolerate sulfa drugs; even those who can are still at high risk of dying within a year after their second or third episode of PCP.

Several studies have strongly suggested that inhaling pentamidine in an aerosol form once a month can prevent recurrences of PCP. In 1988, many doctors began to offer the treatment, and some major insurance companies began to pay for it, even though the drug was not approved for aerosol administration. In early 1989, the FDA improved access to the drug by classifying it as one of a group of potentially life-saving experimental drugs, and final approval occurred later in the year.

Two other experimental drugs for the treatment of CMV retinitis — a rapidly progressing blindness — illustrate the unprecedented pressure on researchers and regulators to find, study, and approve AIDS drugs as quickly as possible. A number of uncontrolled studies (in which subjects taking a drug are not carefully matched with others not taking it) convinced many AIDS specialists that a drug called DHPG or ganciclovir arrested the progress of CMV retinitis in approximately 80 percent of patients. When the drug was stopped, the infection invariably came back. On the basis of this evidence, the FDA permitted the drug's manufacturer, Syntex, to distribute ganciclovir to approximately 4,000 patients on the basis of a "compassionate use" provision in its drug approval process.

When Syntex applied for full market approval, however, the FDA denied it on the basis that no controlled studies had been conducted. Some researchers protested that they could not ethically conduct controlled studies because they believed patients with CMV retinitis should receive the drug. A compromise was struck in late 1988, in which the manufacturer agreed to complete a limited study, designed to determine at what point in the disease the treatment should be started.

Patients have also begun to pressure researchers for access to a second experimental drug for CMV retinitis, phosphonoformate, which also has antiretroviral activity. Patients taking ganciclovir to prevent blindness cannot simultaneously take zidovudine to suppress HIV because both drugs are toxic to bone marrow. Some patients may also go blind despite their use of ganciclovir. In this case, however, both the FDA and the manufacturers of phosphonoformate, Astra Pharmaceutical Products, Inc., have decided not to give patients access to the drug under the "compassionate use" provision because of the difficulties encountered with ganciclovir. This decision has effectively restricted the use of phosphonoformate to patients who qualify for clinical trials.

Although the success of therapies such as zidovudine and aerosolized pentamidine is heartening, the reduction of some opportunistic conditions may simply mean that other conditions will arise and take their place. Another disturbing possibility is that the pathogens that plague people with AIDS may develop resistance to the available drugs. For example, reports published in early 1989 demonstrated that the herpes simplex virus has become resistant to the widely used antiviral drug acyclovir in a few cases, and that CMV has likewise become resistant to ganciclovir. These findings are ominous to many virologists, who fear that HIV itself might also become resistant to the drugs used to treat it.

Heart Disease

Heart disease is the leading killer of Americans, accounting for more than one of every three U.S. deaths — 20 percent more than from all forms of cancer combined. When deaths from stroke and other circulatory-system diseases that share many of the same underlying causes are also considered, cardiovascular diseases are responsible for nearly half of all deaths among Americans.

In many countries, however, heart disease is relatively rare; the annual heart disease death rate in Japan, for example, is only 15 percent of the U.S. rate. Such large differences suggest — and other evidence confirms — that heart disease is not an inevitable consequence of aging in most people. Although genetic factors increase the heart disease risk in a small percentage of the population, heart disease now appears to be mainly the result of specific living habits that differ from country to country and person to person.

Guided by this insight, researchers have concentrated on identifying the risk factors that lead to heart disease. Risk-factor studies have provided strong evidence linking fatal heart disease with cigarette smoking, high blood pressure, high blood cholesterol, stress, and obesity. These findings led in 1984 to the first official U.S. government recommendations for changes in the American diet to prevent heart disease and to more urgent warnings on the dangers of cigarette smoking and uncontrolled high blood pressure.

Heart Disorders

Life depends on the uninterrupted, effective functioning of the heart; any interference with the heart's normal pumping action can produce disability or death. The heart, which is made mostly of muscle tissue and is about the size of a pair of fists, is composed of four chambers. These chambers are connected to each other and to the body's circulatory system by valves that allow the blood to flow through in one direction only. Blood enters the heart's right atrium from the body through two veins, the inferior vena cava from the lower part of the body and superior vena cava from the upper part of the body. (Vessels carrying blood to the heart are called veins; those carrying blood from the heart are called arteries.) The atrium then pumps the blood into the right ventricle, the chamber below the right atrium.

Each time the heart beats, the right ventricle contracts, squeezing blood into a short, wide vessel called the pulmonary trunk. This artery divides into two branches called the pulmonary arteries that carry blood to the lungs, where waste carbon dioxide picked up from the body's cells is exchanged for fresh oxygen. This freshly oxygenated blood then flows from each lung through two pulmonary veins to the left atrium and, from there, into the left ventricle. In each heartbeat, the left ventricle — the most powerful of the heart's four chambers — also contracts, squeezing blood into the wide artery called the aorta. From the aorta, arteries branch into narrower and narrower vessels and finally into the capillaries, which carry blood to every living cell in the body. Capillaries then join to form wider and wider vessels that feed into the superior and inferior vena cava. These two large veins return the blood to the right atrium of the heart, thereby completing the circuit of the blood through the body.

Heart attack. The heart does not nourish itself directly from the large quantity of blood — several quarts per minute — flowing through its chambers. Instead, short blood vessels called coronary arteries branch off from the aorta and carry a small portion of the total blood flow back to the heart tissue. Interruption of this vital flow, usually caused by a blood clot in a coronary artery, deprives the heart muscle of the fresh blood and oxygen it needs to survive. This deficiency of blood supply is called ischemia; if it continues for more than a short time, a myocardial infarction, or MI — death of heart muscle — occurs.

Myocardial infarctions, which are commonly called heart attacks, can be so slight as to go unrecognized, or they can be massive, painful, and rapidly lethal. MIs large enough to be noticed are called heart attacks. Heart attacks strike as many as 1.5 million Americans a year and kill about 524,000, according to the American Heart Association (AHA). The death rate from all forms of cardiovascular disease was 417 per 100,000 Americans in 1985; MIs were responsible for 234 of these 417, or 56 percent, according to the National Center for Health Statistics. Among men between the ages of 35 and 50, MIs were responsible for about one-third of deaths from all causes in the U.S. For reasons not clear, men at any given age under 70 were more vulnerable to heart attacks than women of the same age.

About half of all heart attack victims die before receiving medical treatment. Because half of all heart attack deaths occur within 2.5 hours after pain is first noticed, doctors have encouraged people to learn to recognize the

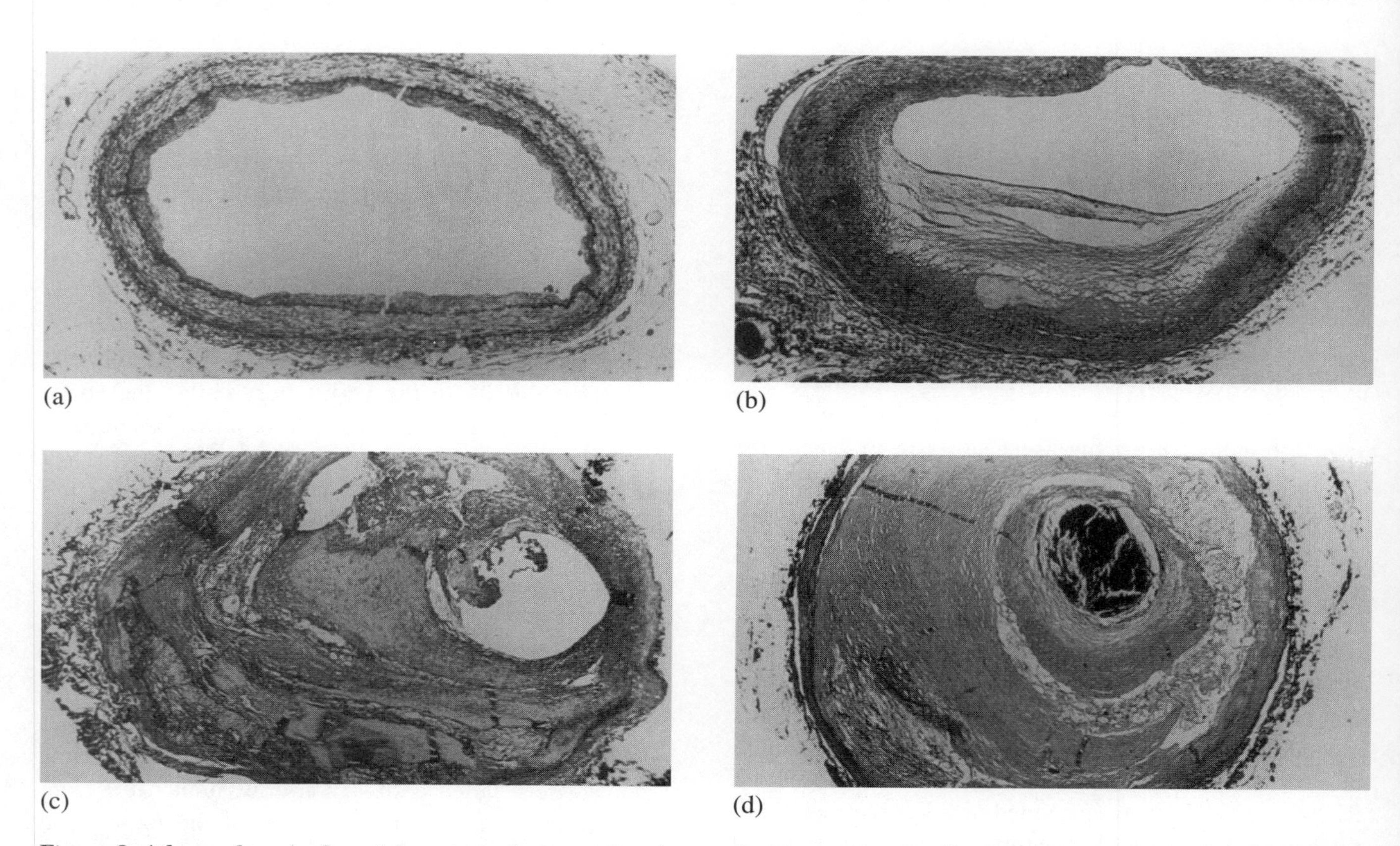

Figure 2. Atherosclerosis. One of the principal causes of cardiovascular disease is atherosclerosis — the accumulation of plaque on the lining of arteries. Plaque, which typically consists of deposits of cholesterol, calcium, and other substances, may partially or entirely block the flow of blood and cause a heart attack or stroke. These enlarged sequential photographs (a–d) of very thin cross-sections of arteries illustrate the gradual buildup of plaque. *Courtesy: American Heart Association.*

symptoms of a heart attack and seek immediate help. These symptoms include prolonged squeezing pain or pressure in the center of the chest — victims often clench their fists over their hearts in describing their condition. The pain may spread to the shoulder, neck, arm, hand, and sometimes the jaw. The pain, which may come and go, is sometimes accompanied by shortness of breath, nausea, vomiting, and sweating.

The most common cause of a heart attack — and the leading cause of death in the U.S. — is coronary heart disease (CHD), damage to the heart due to insufficient blood supply. This damage is caused by the narrowing of the coronary arteries by the gradual accumulation of a waxy, cholesterol-containing deposit called atherosclerotic plaque on the arteries' inside walls. Several other forms of heart disease are also common in the U.S., including congenital heart defects, rheumatic heart disease, and congestive heart failure.

Although lifestyle clearly plays a major role in heart disease, many scientists believe that a tendency toward heart disease is inherited. If the genes controlling that inheritance can be found, scientists may be able to identify individuals who have the greatest risk of developing heart disease — and who may be able to take steps to prevent the disease.

Genetic basis of heart disease. One inherited condition that greatly increases the risk of CHD is familial hypercholesterolemia, or FH, which is inherited by about one in 500 Americans. Physicians Joseph Goldstein and Michael Brown of the University of Texas won the Nobel Prize in 1985 for their discovery that this disease is the result of a defect in the biochemical mechanism for removing a particular kind of lipoprotein — a substance consisting of protein linked with fat or fatlike substances called lipids — from the bloodstream. Normally each person carries two genes coding for the cell-surface receptors designed to receive low-density lipoprotein, or LDL, from the bloodstream and take it into the cell, where the cholesterol can be used for cell growth and repair or the manufacture of necessary hormones. People with defects in both LDL receptor genes have a rapid buildup of plaque that causes massive heart attacks in childhood. Those who inherit a single defective LDL receptor gene have a milder form of FH; nonetheless, the excess LDL in their bloodstream causes atherosclerotic plaque to build up in the coronary arteries, resulting in heart attacks as early as the thirties and forties.

However, LDL receptor defects account for less than 5 percent of premature heart attacks (those affecting people under age 50). Recently, researchers have focused on the other half of the interaction between cells and LDL, the protein component of LDL called apolipoprotein B-100 (apo-B), which latches onto the cell receptor. In 1986, research teams from the U.S., Canada, and the U.K. detailed the amino acid structure of apo-B — the largest protein sequenced up to that time. Since then, many researchers have identified mutant forms of apo-B both in animals and in human families that appear to have an unusually high risk of coronary artery disease. However, as of 1989, it was still too early to predict whether screening people for certain types of apo-B would be useful for targeting risk-reduction efforts, or if drugs could be developed to counteract the effect of harmful apo-B mutants.

Congestive heart failure. In congestive heart failure, either the right or left chamber of the heart — and sometimes both — cannot pump sufficient blood to satisfy the body's needs. When the left ventricle fails to pump into the aorta all the blood that arrives from the lungs, blood backs up in the pulmonary veins, and fluid accumulates in the lungs. When congestive heart failure occurs in the right ventricle, blood backs up into the vena cava, and fluid tends to accumulate in the legs and feet. Congestive heart failure may result from a variety of heart-weakening afflictions, such as congenital defects and rheumatic heart disease, or from damage sustained during heart attacks.

The most common early symptoms of congestive heart failure are fatigue and shortness of breath upon mild exertion. Other symptoms include tachycardia — literally "swift heart" — and accumulation of excess fluid in the legs and/or the lungs. Congestive heart failure can also cause damage in the kidneys, liver, lungs, and other organs, as the blood flow to those organs is reduced and the fluid balance is disturbed. Treatment for congestive heart failure most often involves the drug digoxin, which slows and deepens the heart's contractions. Diuretic drugs are also used to reduce the total amount of fluid in the body by increasing the amount of urine excreted.

Risk Factors of Heart Disease

Framingham Study. While scientists had long suspected that smoking, high levels of blood cholesterol, and high blood pressure (hypertension) increase the risk of heart disease, large-scale studies were required to actually establish how much these factors increase the risk. Of the many studies on the causes of heart disease, the Framingham Study — begun in 1950 and still under way in 1989 — is the longest-running and most comprehensive.

In prospective studies such as the Framingham Study, a group of initially healthy people is examined periodically over many years. Information is gathered about a variety of health-related factors such as height, weight, blood pressure, diet, and smoking and drinking habits. A significant number of these people eventually get heart disease; this group of victims is then compared with the rest of the participants to discover what, if anything, is different about them and their lives. Prospective studies often follow thousands of people for decades and are considered among the most reliable indicators of risk factors for disease.

The Framingham Study initially included 5,209 residents of Framingham, Massachusetts, between the ages of 30 and 62 — about half the middle-aged population of the town. The study, which has continued to receive support mainly from the National Heart, Lung, and Blood Institute (NHLBI) in Bethesda, Maryland, has established that high blood pressure, high levels of blood cholesterol, cigarette smoking, high blood sugar, obesity, lack of exercise, and high levels of stress are all more common among heart disease victims than among healthy people.

In general, the study has found that a few indicators, such as blood pressure and blood cholesterol level, are the most powerful predictors of who is likely to get heart disease. Among the 4,000 children of the original Framingham subjects, heredity has been found to be far less important than environment in determining which people will die of heart disease. The implication of these findings is that heart disease should be largely preventable through changes in lifestyle, such as quitting smoking, altering diet, and controlling high blood pressure.

Like several other studies, the Framingham Study has shown that high cholesterol levels in the blood are strongly associated with a high risk of heart attack. For example, people with 300 milligrams of cholesterol per deciliter (about 0.2 pint) of blood had a heart attack risk four times higher than that of people with only 180 milligrams per deciliter (mg/dl) of blood. In 1986, the average cholesterol level for Americans was about 220 mg/dl, while Chinese — who get heart disease relatively rarely — averaged 170 mg/dl. The ideal cholesterol level is probably between 130 and 190 mg/dl, according to physician William B. Kannel of the Framingham Study.

Cholesterol is a waxy substance produced naturally in the liver but also consumed in such animal and dairy products as eggs, beef, kidney, liver, shellfish, and cheese. While cholesterol is an essential component of human cells, excess concentrations of cholesterol in the blood contribute to the formation of atherosclerotic plaque, which narrows the channel for blood flow in arteries.

In 1978, the Framingham Study was among the first to find that an even better predictor of heart disease risk than the total cholesterol level is the ratio of the total amount of blood cholesterol to the amount of cholesterol bound to proteins called high-density lipoproteins (HDLs). Because cholesterol does not dissolve in the blood, it is transported through the circulatory system by two types of carrier proteins: HDLs and LDLs. LDLs transport cholesterol to body tissues, while HDLs carry cholesterol to the liver for destruction and disposal.

The higher the ratio of cholesterol in all forms to HDL cholesterol, the higher the risk of heart disease. American men have an average ratio of 4.4 to 1; men whose ratio is 7.1 to 1 have twice the average risk of heart disease, whereas those whose ratio is 3.4 to 1 have only half the risk. White women follow the same pattern as white men, except that white women always have a lower risk of heart disease than white men with the same blood cholesterol ratios and similar risk factors. This sex difference is found only among white people; for reasons still unknown, black men and women with the same cholesterol ratios have about the same risk.

In populations as diverse as Framingham (U.S.), Tel Aviv (Israel), and Kuopio (Finland), researchers have recently found that half of all heart attacks and coronary deaths occur in people defined as low- to medium-risk by conventional total-cholesterol guidelines. The Framingham researchers and other scientists currently believe that measuring HDL, either its ratio with total cholesterol or alone, can help discriminate which people are at greater risk for a heart attack among those with equal total cholesterol levels.

One 1988 study of 15,000 U.S. physicians, conducted by Meir J. Stampfer and his colleagues at Harvard Medical School, found that those who had heart attacks and those who did not had nearly equal total cholesterol levels. However, heart attack victims had significantly lower levels of HDL. "We found a 6 percent decline in heart attack risk for every 1 milligram increase in HDL," Stampfer noted.

The Framingham Study has also shown that heart disease risk doubles with moderate smoking and increases steadily with increases in blood pressure. Combinations of risk factors can dramatically increase a person's chances of dying from heart disease. For example, cigarette smokers with high blood pressure and high levels of blood cholesterol and blood sugar are 10 times more likely to develop heart disease than smokers who have only high blood pressure.

Smoking. In 1964, the U.S. Surgeon General first identified cigarette smoking as an important contributor to heart disease and stroke deaths in men. The 1980 Surgeon General's Report on Smoking and Health extended that finding to women, especially those who both smoke and use oral contraceptives.

The 1989 Surgeon General's Report, which summarized 25 years of research on the health consequences of smoking, concluded: "The findings from several prospective studies involving more than 20 million person-years of observation in North America, Northern Europe, and Japan have been remarkably similar: Cigarette smokers are at increased risk for fatal and nonfatal myocardial infarction and sudden death." Smokers have a 70 percent higher coronary death rate and a two- to four-fold greater risk of sudden-death heart attacks, the report said.

During the 1970s, the multicenter Pooling Project Research Group, which analyzed data from dozens of studies, found smoking to be equal to either high cholesterol or high blood pressure as a coronary risk factor for middle-aged men. In a report published in 1978, the Pooling Project concluded that smoking in combination with either high cholesterol or hypertension doubled the risk of major heart attacks, while men with all three risk factors had more than triple the risk.

Further evidence of the danger to the heart of smoking was obtained by researchers in the Coronary Artery Surgery Study group, a team of investigators at many U.S. hospitals who were studying the effectiveness of bypass surgery. While screening patients for surgery, the researchers identified over 4,100 smokers with proven coronary artery disease and studied these patients' health on a prospective basis. All of the patients in that group were asked to quit smoking, but only 1,500 did. In 1986, after following the smokers for 5 years, the researchers reported that 22 percent of the smokers had died of heart disease, compared to only 15 percent of those who had quit.

The Surgeon General's Report concluded that about 390,000 U.S. deaths were attributable to smoking in 1985, far more than any other preventable cause of death. Smoking accounted for 21 percent of heart disease deaths, 18 percent of stroke deaths, and 87 percent of lung cancer deaths, the report concluded. A large body of research demonstrates that smoking increases the risk of heart disease and stroke in several ways. The 1983 Surgeon General's report, for instance, noted that autopsy studies have demonstrated a strong correlation between smoking and the buildup of plaque in both the aorta and the coronary arteries. A 1988 study by K. M. Galan of the St. Louis University Medical School underscored this finding by showing that cigarette smokers who continued to smoke after a coronary angioplasty — a procedure to widen clogged coronary arteries — were more likely to experience reclogging of the arteries severe enough to require repeating the procedure. The mechanism by which smoking contributes to atherosclerosis, however, is not fully understood, though recent evidence suggests that it adversely affects the HDL/LDL ratio and causes repetitive injury to the lining of the arteries by disturbing normal enzyme activity.

In addition to this chronic effect of smoking on arteries, however, scientists have reasoned that smoking must also exert acute or short-term effects, because the risk of cardiovascular events such as heart attack and stroke declines dramatically within the first year after someone stops smoking. Smoking can contribute to myocardial ischemia (inadequate blood flow to the heart muscle) by causing spasms of coronary arteries already narrowed by atherosclerosis. Smoking also appears to increase the irritability of heart muscle, lowering the threshold for ventricular fibrillation, an irregular, ineffective fluttering of the heart's main pumping chambers that underlies most episodes of sudden cardiac death.

Smoking is also thought to increase the risk of blood clots that can suddenly block a coronary artery or a cerebral artery, leading to heart attack or stroke. Many researchers are currently investigating the mechanisms behind this state of hypercoagulability — abnormally high clotting activity. Garret FitzGerald and coworkers at Vanderbilt University in Nashville reported in 1988 that smokers have elevated levels of plasma fibrinogen, a substance that initiates clotting, and they also seem to have more activated platelets, the disklike blood components that adhere to one another in the earliest stage of clot formation.

The risk of stroke among smokers has recently been clarified by major prospective studies. The largest of these, conducted by Graham A. Colditz and his colleagues at Harvard Medical School, followed nearly 120,000 women for 8 years. In 1988, the Harvard group found that light smokers (people who smoked 1 to 14 cigarettes daily) had 2.2 times more risk of strokes than women who never smoked, while heavy smokers (25 or more cigarettes daily) had 3.7 times more stroke risk. This strong association persisted even when other cardiovascular risk factors were taken into account, and women who stopped smoking reduced their stroke risk nearly to that of people who never smoked within 2 years. Virtually the same result was found by P. A. Wolf and colleagues at the Framingham Study, who reported in 1988 that smoking was independently related to the risk of stroke among over 4,200 middle-aged men and women followed for 26 years.

Prevention of Heart Disease

Blood pressure. Several major studies published during the mid-1980s demonstrated that the incidence of heart disease can be reduced by reducing hypertension (high blood pressure). The benefit of lowering blood pressure was greatest in people whose blood pressure was initially the highest and in those who were at the greatest risk of heart disease because of age or other factors.

Blood pressure provides a measure of how hard the heart is working. When the heart contracts (systole), blood is squeezed into the circulatory system, raising the blood pressure to a maximum value, called the systolic pressure. As the heart muscles relax (diastole), the blood pressure falls to a minimum value, called the diastolic pressure. Both systolic and diastolic pressures have medical significance, and both are usually written together, separated by a slash. Thus, a typical blood pressure is given as 120/80, signifying systolic and diastolic pressures of 120 and 80 millimeters of mercury, respectively. (A millimeter of mercury is a unit of pressure.)

Studies conducted during the 1960s established that severe hypertension — for example, a diastolic blood pressure of 115 to 129 — could be dramatically reduced by drug treatment and that this reduction lowered the risk of cardiovascular disease. The majority of hypertensive people in the U.S., however, have diastolic pressures in the range of 90 to 114 and are otherwise completely healthy. These people, mostly men, represent 70 to 80 percent of all hypertensive Americans and about 15 percent of the entire U.S. population. The use of drug therapy for this group of more than 30 million people has been evaluated in several studies reported in recent years.

The largest of these studies was the Multiple Risk Factor Intervention Trial (MRFIT), which was sponsored by the NHLBI and involved over 8,000 subjects and 250 researchers in 28 different institutions. MRFIT was designed to test, among other things, whether offering drug treatment and/or diet and health advice to people with mild hypertension would decrease their chances of dying from CHD. The participants in the study were randomly assigned either to the "usual care" group, whose members received routine medical examinations, or to the "special intervention" group, whose members were encouraged to stop smoking if they smoked and to improve their diets. About two-thirds of the special intervention group members were also given drugs to lower their blood pressure.

To the surprise of most observers, the results of MRFIT, which were reported in various journals between 1982 and 1985, revealed no significant difference in heart disease deaths between the two groups. Although the death rate in the special intervention group fell as expected, the death rate in the usual care group also fell. The reason for the improved health of the usual care group, researchers now agree, was that these subjects changed their risk-related behavior and reduced their blood pressure even without receiving specific medical advice.

As a result of being asked to participate in the study, members of the usual care group realized that they had a high risk of suffering a heart attack. This prompted many of them to change their diets, stop smoking, and seek help for their high blood pressure. For example, within 6 years after the start of the study, 50 percent of the smokers in the special intervention group had quit smoking; without special encouragement, 29 percent of the smokers in the usual care group had also quit. In addition, MRFIT found that both groups changed their other risk-related habits in similar ways.

The first of four newer studies — each of which lasted for more than a decade — was conducted by the Medical Research Council (MRC) Working Party in the U.K. Physicians in this study screened about 500,000 healthy people between the ages of 35 and 64, and identified over 17,000 patients with a diastolic blood pressure between 90 and 109. The subjects were divided into three groups; one group received placebos, one received diuretics, and one received drugs called beta-blockers, which reduce blood pressure by blocking the effect of hormones such as adrenaline. The MRC group reported in 1985 that the overall incidence of cardiovascular events such as strokes and heart attacks was about 20 percent lower in the treated groups. The rate of nonfatal strokes in the treated groups was 45 percent lower than the rate in the placebo group. The rates of fatal strokes and MIs were the same in the treated and untreated groups.

Similar results were obtained in a study conducted by the European Working Party on Hypertension in the Elderly (EWPHE), which studied 840 subjects over the age of 60 with diastolic pressures between 90 and 119. The researchers reported in 1985 that the overall incidence of cardiovascular events was reduced by about 25 percent in the groups treated with diuretics or beta-blockers, and that the incidence of nonfatal strokes was reduced by 52 percent. The main difference between the results of the MRC and the EWPHE studies was that the latter study, for unknown reasons, showed a significant reduction in fatal heart attacks in the treated group.

A third study, the International Prospective Primary Prevention Study in Hypertension (IPPPSH), monitored over 6,300 patients who had never had a heart attack; half received beta-blockers and half did not. About one-third of the patients in each group also received diuretics. The researchers reported in 1985 that all the patients had a similar incidence of MI, stroke, and sudden death, indicating that beta-blockers offered no benefit for patients who have not had a heart attack.

A much larger study in which the effects of smoking and cholesterol were considered along with hypertension was conducted in Belgium, Italy, Poland, and the U.K. by the World Health Organization. Researchers studied nearly 61,000 men with moderate hypertension who worked in 80 factories in the four countries. Half of the men received advice on lowering their cholesterol level through diet, controlling smoking, reducing weight and high blood pressure, and exercising regularly. Drugs were also given to reduce the men's hypertension, but not their cholesterol levels. The other half of the men received no advice and served as controls. The researchers reported in 1986 that the treated group had about a 10 percent lower incidence of CHD, a 7 percent lower incidence of fatal CHD, a 15 percent lower incidence of nonfatal MIs, and 5 percent fewer total deaths.

In summary, the conclusion of all these studies is that lowering blood pressure reduced the risk of heart disease even for people with mild hypertension.

Salt, calcium, and high blood pressure. For many years, doctors have recommended reducing salt consumption as a way to lower blood pressure without drugs. This approach is, however, effective in only about one-third of people with mild hypertension. To study the effect of salt and other minerals in the diet, epidemiologist David McCarron and his colleagues at Oregon Health Sciences University in Portland analyzed blood pressure and diet information for more than 10,000 American adults. They reported in 1984 that low calcium intake — not high salt intake — was the dietary factor most closely correlated with hypertension. McCarron's results were consistent with earlier studies by other researchers showing that the incidence of hypertension and heart disease is lowest in areas with a high concentration of calcium in the water supply.

McCarron and his colleague Cynthia Morris have suggested that supplementing the diet with calcium can reduce the incidence of hypertension and heart disease. They reported in 1985 that daily 1 gram doses of calcium lowered blood pressure in individuals with normal blood pressure as well as those with hypertension.

However, the hypothesis that low calcium is a cause of essential hypertension and dietary supplements can be used to treat it remains controversial. In a critique published in the December 1986 *Annals of Internal Medicine*, Norman Kaplan of the University of Texas in Dallas and Roderick Meese of Duke University in Durham, North Carolina, concluded that the calcium hypothesis "is based on the use of only a portion of the available experimental data and the clinical evidence remains inconclusive." A year later in the same journal, McCarron and Morris reasserted that epidemiologic and clinical studies "support a protective role for calcium in regulating arterial blood pressure."

By 1989, no major health organization had deviated from the conventional wisdom that hypertension should be controlled primarily by salt avoidance, weight reduction, and drug therapy when necessary. Several other ways to reduce hypertension are also recommended. Exercise — preferably in sustained sessions of about a half-hour three or four times a week — lowers blood pressure and also helps to control weight. People who quit smoking benefit from lowered blood pressure as well as from dramatically reduced risks of heart disease. Blood pressure can also be lowered by reducing consumption of caffeine, which is found in coffee, tea, chocolate, and many over-the-counter drugs. In addition, stress-reducing techniques such as meditation have produced decreases in blood pressure.

New drug for hypertension. In 1981, biochemist Adolfo de Bold of Queen's University and Hospital in Toronto, Ontario, discovered that the heart secretes at least two hormones — the first indication that the heart has functions other than pumping blood. By 1986, one of the hormones was being used experimentally to treat hypertension. The hormone is called atrial natriuretic factor (ANF) because it is secreted by the heart's atria. It lowers blood pressure by increasing the rate at which kidneys excrete sodium in the urine.

Researchers at California Biotechnology, Inc., in Palo Alto have used genetic engineering techniques to produce ANF in substantial quantities. Since 1985, a number of researchers have begun clinical trials of the hormone. Recent work by John H. Laragh of Cornell University Medical School and others has shown that people with severe hypertension have high levels of ANF in their blood, but their response to the hormone is blunted. For that reason, Laragh and his colleagues have found that patients with congestive heart failure — a frequent end result of chronic severe hypertension — did not respond to ANF the way people with normal blood pressure do.

ANF is being intensely investigated as a possible treatment for hypertension and congestive heart failure. It appears to have few side effects, but its chief disadvantage is that it must be injected. Researchers are working to develop a form that can be administered in a pill or as a nasal spray.

Cholesterol. In October 1987, the NHLBI, in association with the American Heart Association (AHA) and 21 other health-oriented organizations, issued guidelines telling physicians how to evalute and treat adults with high blood cholesterol. The guidelines were part of a National Cholesterol Education Program launched in February 1986 to educate both physicians and the public about the benefits to be gained by reducing the levels of cholesterol in the blood. The new program follows on the heels of several major studies demonstrating that, at least for people who start with cholesterol levels above 250 mg/dl (and probably for those below this level as well), every 1 percent reduction in blood cholesterol levels is accompanied by a 2 percent reduction in the incidence of CHD.

Because numerous studies have shown that reducing the amount of cholesterol or saturated fats in the diet can reduce the cholesterol level in the blood, most scientists believe that people who follow a cholesterol-lowering diet will have a lower risk of heart disease than those who do not. Saturated fats are a major contributor to cholesterol formation, while polyunsaturated fats are relatively harmless. (Saturated fats do not have any double bonds in their molecular structure, because all of their carbon atoms are fully saturated with hydrogen atoms. In contrast, polyunsaturated fats have two or more double bonds.) Animal fat, butter, and palm and coconut oils are high in saturated fats, whereas safflower, sunflower, and cotton oils consist mainly of polyunsaturated fats. While vegetable and animal products contain both saturated and polyunsaturated fats, only animal products contain cholesterol.

In 1964, the AHA recommended a diet emphasizing low-fat, low-cholesterol foods such as skim milk, low-fat cheeses, lean meat, vegetables, and fruit. In the 25 years since these recommendations were released, Americans have in fact changed their eating habits. According to the U.S. Department of Agriculture, per capita consumption of animal fats, including butter and oils with a high saturated fat content, declined by 39 percent between 1963 and 1980. Americans' switch toward polyunsaturated fats between the 1950s and the mid-1980s was even more dramatic, according to Canadian nutritionist Alison M. Stephen. In the 1950s, Americans ate about five times more saturated fat than polyunsaturated fat, but by the mid-1980s, the ratio was one to one, Stephen reported at an AHA meeting in November 1988. This change in diet is thought to be partially responsible for the 36 percent decrease in cardiovascular disease that occurred between 1963 and 1983 — although a 27 percent decline in smoking and a decrease in the incidence of hypertension also contributed.

MRFIT has provided strong evidence demonstrating the dangers of high cholesterol. Summarizing data from the 325,000 men screened for MRFIT, William B. Kannel and other study researchers reported in 1986 that men whose cholesterol levels were in the top 10 percent had four times the normal risk of contracting CHD. High cholesterol was especially harmful among those who also had other risk factors, such as smoking and high blood pressure.

The benefits of reducing cholesterol have been documented in the Lipid Research Clinics–Coronary Primary Prevention Trial (LRC–CPPT), the largest cholesterol and heart disease study ever undertaken. In the LRC–CPPT, which was begun in the early 1970s under the sponsorship of the NHLBI, nearly 500,000 men were screened for possible inclusion in the study, and about 3,800 men between the ages of 35 and 59 were selected. Each participant was advised to follow a moderate cholesterol-lowering diet; in addition, half of the men were given the cholesterol-lowering drug cholestyramine. In studies published during 1984 and 1985, the researchers reported that the blood cholesterol levels of people who received the drug decreased an average of 13 percent, compared to a decline of only 5 percent in the diet-only group. Furthermore, the diet-and-drug group suffered 19 percent fewer CHD deaths and nonfatal heart attacks than the diet-only group.

According to the study's authors, the value of the cholesterol-lowering drug must have been even greater than these numbers suggest because not all of the subjects consistently took the prescribed medicine, which has an unpleasant taste and often produces side effects such as constipation. The researchers estimated, in fact, that strict adherence to the drug program would probably have produced a 25 percent reduction in blood cholesterol levels and a corresponding 50 percent decline in CHD deaths and nonfatal heart attacks. A significant implication of the study was that the cholesterol-related risk of heart disease can be reduced in middle age, even though arterial deposits of cholesterol have been accumulating for decades.

Largely on the basis of the LRC–CPPT report, a consensus panel at the National Institutes of Health concluded that, "beyond a reasonable doubt," lowering blood cholesterol levels reduces the risk of heart attack. The panel concluded that 25 percent of Americans have a dangerously high risk of heart disease due to their high cholesterol levels and advised that all people in this group lower their cholesterol levels through changes in diet and, if necessary, through the use of drugs.

This recommendation was strengthened by data from the 4,100-man Helsinki Heart Study, which showed that the drug gemfibrozil achieved a 37 percent reduction in nonfatal heart attacks and a 26 percent reduction in coronary deaths compared with a placebo pill. Perhaps because gemfibrozil raises levels of the "good cholesterol" HDL, as well as lowering the "bad" LDL, the men taking it had twice the protective benefit observed in the LRC–CPPT subjects — that is, a 4 percent decline in heart attack risk for each 1 percent drop in cholesterol, instead of a 2 percent risk reduction for each 1 percent that cholesterol was lowered.

In light of these and other, more recent, studies suggesting that atherosclerosis can actually be reversed by diet and drugs, a consensus was developing in the late 1980s concerning the benefits of aggressive cholesterol-lowering therapy for the one in four Americans deemed at high risk of heart attack. To put this strategy into effect, the National Cholesterol Education Program (NCEP) urges that all adults aged 20 years or over be screened for high cholesterol at least once every 5 years. Those with initial total cholesterol readings of 200 mg/dl or higher should be retested to be sure that the test results are accurate, according to the NCEP guidelines. Those with borderline-high cholesterols (200 to 239 mg/dl) should receive dietary education and annual followup unless they also have symptoms of heart disease or two other risk factors, which should trigger further evaluation of their LDL levels. All those with high cholesterol (240 mg/dl or more) should be evaluated further for possible diet and drug therapy.

There was still controversy in 1989 over the relative importance of LDL and HDL levels as indicators for aggressive treatment. While the NCEP relies largely on LDL, a number of studies presented at the November 1988 AHA meeting suggest that less-than-ideal levels of HDL are an important risk factor even among people whose total cholesterol falls in the desirable range.

Fish oil. In recent years, a number of studies have suggested that regular consumption of cold-water fish such as salmon, tuna, and cod can lower the levels of LDL and triglycerides (fatty acids present in very low-density lipoprotein, or VLDL), reduce the stickiness of platelets and thus blood clot formation, lower blood pressure, and perhaps reduce the risk of heart attacks and coronary deaths. For example, in a dietary comparison among over 850 middle-aged men in the Netherlands, epidemiologist Daan Kromhout and colleagues at the University of Leiden in the Netherlands reported that the death rate from heart disease was over 50 percent lower among men who ate fish regularly, compared to those who never ate fish. Other studies have found that two omega-3 fatty acids, which are plentiful in cold-water fish, have beneficial effects on blood clotting and blood lipids (fats and related substances such as cholesterol).

Such reports have led some to advocate taking capsules of fish oil regularly to lower the risk of heart disease, or at least consuming more fish. While experts on diet and heart disease agree that substituting fish for red meat on a regular basis is probably beneficial, the verdict is still out on the wisdom of taking fish-oil supplements. "It is far too early to advocate widespread prophylactic or therapeutic use of dietary fish oils to prevent or reduce atherosclerotic cardiovascular disease," S. H. Goodnight, Jr., of the Oregon Health Sciences University wrote in the July 1988 issue of *Seminars in Thrombosis and Hemostasis*. "Additional studies are needed, especially in respect to the prevention or regression of atherosclerosis, inhibition of arterial thrombosis, and long-term toxicity."

Heart Disease Treatment

The 1980s have been an active period for development of new therapies for heart disease across a broad front. New drugs called beta-blockers emerged to reduce the pain of angina pectoris — squeezing, oppressive chest pain — and were shown to extend the lives of heart attack patients, including many who might otherwise require coronary artery bypass surgery. Another class of new agents, called calcium channel blockers, became widely used for angina, though they are not as useful as once hoped in limiting the size of MIs or reducing the long-term risks of death from heart disease.

In the late 1980s, the immediate treatment of heart attacks was revolutionized by the use of the clot-breaking drugs streptokinase, urokinase, and tissue plasminogen activator (tPA) to restore blood flow to threatened areas of heart muscle and thus limit the size of the infarction. The artery-widening technique called angioplasty also became widely used, especially among heart attack survivors.

Meanwhile, progress in microelectronics led to the production of both programmable cardiac pacemakers for patients with a variety of heart-rhythm disturbances and, more dramatically, implantable defibrillators that use electric shock to stop ventricular fibrillation and make the heart beat more normally.

Thanks largely to the availability of the antirejection drug cyclosporine, transplantation of human hearts became routine. For the first time in medical history, a human being — Utah dentist Barney Clark — received an artificial heart, not as a temporary measure to support him while awaiting a human heart transplant, but as a permanent substitute for his own failing heart. By June 1986, five other patients had received similar artificial hearts. The survival of two implant patients for at least several months established that artificial hearts can sustain human life. However, all eventually died, and the occurrence of strokes in five of the six recipients raised serious doubts about the future of artificial heart programs. However, artificial hearts have proved useful as temporary "bridges" to transplants when an appropriate human donor heart is not immediately available for a dying patient.

These many advances were part of a continuing redirection of heart disease treatment that began in the early 1960s. The theme of this change has been the increasingly aggressive management of heart disorders with drugs and surgery. Before the 1960s, victims of cardiac arrest — people whose hearts spontaneously stopped beating — were simply pronounced dead because there was no practical way to revive them. By the late 1970s, however, most larger US hospitals had established elaborate cardiac emergency care units, equipped to resuscitate and treat heart-attack victims. Today, most patients who suffer cardiac arrest in hospitals are revived.

One indication of the increasingly aggressive management of heart disorders is that coronary bypass surgery had, by the early to mid-1980s, become one of the most frequently performed major surgical operations in the U.S. In this operation, short segments of a blood vessel, generally either a vein from the patient's leg or a dispensable artery in the chest, are used to circumvent obstructions in the coronary arteries so that an adequate supply of blood can once again reach the heart muscle "downstream" from the narrowed artery. In 1980, about 137,000 Americans had coronary bypass operations annually, according to the AHA. By 1989, the estimated total was well over 300,000 per year. The numbers have grown partly because of a rapid increase in the numbers of hospitals and surgeons capable of performing open-heart surgery and partly because doctors are now recommending the operation to more older patients than in the past. According to Mortimer Buckley of the Massachusetts General Hospital, the average age of coronary bypass patients in 1988 at that hospital was 66, and the average number of coronary vessels bypassed was four, indicating that the great majority of surgical patients had advanced coronary disease.

Controlling angina. Angina pectoris is usually brought on by exertion or stress, a condition known as stable angina. During exertion or times of stress, the heart needs more blood and oxygen, and coronary arteries narrowed by plaque may be unable to supply the amount needed. Changes in diet and exercise patterns are often recommended to help control angina. Angina patients who smoke cigarettes often get relief from their symptoms if they quit smoking. However, drugs such as nitroglycerine or beta-blockers may still be needed. When drugs fail to quell angina or when the pain comes more frequently, bypass surgery or angioplasty is often recommended.

Drug treatment. To alleviate the pain of angina, drug treatments are designed to decrease the heart's need for blood, increase its blood supply, or both. The nitrate-containing drug trinitroglycerol — often called nitroglycerin — causes the coronary arteries to relax, increasing blood flow to the heart. Nitroglycerine has long been used in the form of small tablets held under the tongue (called sublingual nitroglycerin) until they dissolve, providing immediate but transient relief. During the 1970s, longer-acting nitrates were developed for preventive use — not to provide temporary relief from angina but to prevent heart attacks and diminish their severity if they occur. However, in January 1989, the U.S. Food and Drug Administration (FDA) released a major study, the largest ever conducted on angina patients, which showed that when nitroglycerin doses are delivered to the body continuously, 24 hours a day, the drug quickly loses its effectiveness. An FDA advisory committee recommended that angina patients be given an overnight break from the drug each day to preserve its effectiveness, and the agency was expected to incorporate that advice into labeling changes for long-acting nitrate drugs.

A major new class of drugs called beta-blockers proved effective in treating angina during the late 1970s. Another new class, called calcium channel blockers, is widely used to treat angina, either alone or in combination with beta-blockers, but it has not lived up to its early promise in limiting MI damage.

Beta-blockers reduce the heart's demand for blood by blocking beta-adrenergic receptors, the targets of heart-stimulating hormones such as adrenaline. The effect is to calm the heart, reducing its oxygen demand, and to dilate diseased blood vessels, supplying more oxygen to the heart. Beta-blockers have other benefits as well. They prevent arrhythmias, potentially fatal disturbances of heart rhythm. Some reduce blood pressure throughout the body and thus the workload of the heart. Dozens of controlled studies involving 20,000 heart attack patients showed that long-term use of beta-blockers reduced the risk of death by 21 percent, on average, and the risk of subsequent heart attacks by 24 percent, according to a 1989 report by Ake Hjalmarson of Sahlgren's Hospital in Göteborg, Sweden.

The long-term protective effect of calcium-channel blockers, also called calcium antagonists, is still controversial. A 1987 report on a multicenter trial of one such drug, diltiazem hydrochloride, showed that it strongly benefited one high-risk group of heart attack patients — those with chest pain and electrocardiogram (ECG) abnormalities in the early days after their MI. People who took diltiazem for 2 weeks after their first heart attack had about a 5 percent risk of a second heart attack, compared with a 20 percent risk among patients who received a placebo. However, other studies have failed to demonstrate this immediate post-MI benefit. "Of four major clinical studies," cardiologists Robert Kloner and Eugene Braunwald wrote in the 30 January 1987 *American Journal of Cardiology*, "only one has shown that calcium antagonists are capable of reducing infarct size in man. However, in most of these studies, drug therapy commenced relatively late — 4 or more hours after symptoms." Kloner and Braunwald suggested that, for diltiazem to make a difference, the zone of threatened heart muscle may have to be small, the drug may have be administered very early, and clot-dissolving drugs may also have to be used soon after heart attack symptoms appear.

A longer followup study of the patients in the multicenter diltiazem trial, presented at the November 1988 AHA sessions, found that patients whose infarction had not penetrated the full thickness of the heart wall benefited from long-term diltiazem treatment. After a year on the drug, these patients had a 9 percent rate of subsequent heart attacks and coronary deaths compared to 15 percent in placebo patients. However, an analysis of 21 trials of calcium channel blockers, carried out by Peter Held and his colleagues at the NHLBI, found that long-term use of these drugs does not prevent either first heart attacks or later ones, nor does it reduce MI mortality.

Bypass surgery. When drug treatment fails to help a person with severe angina, surgery is considered. Coronary artery bypass surgery, which came into use in 1967, is now by far the most common surgical remedy for angina. During bypass surgery, one or several short lengths of blood vessels are first removed from the patient's leg, or another site served by more than one blood vessel. Each segment is then used to create an alternate blood pathway — a bypass — around an obstruction in an artery near the heart. Surgery to create two such shunts is called a double bypass, three shunts constitute a triple bypass, and four a quadruple bypass.

Bypass surgery usually produces good results in angina patients with no other apparent heart disorders. About 85 percent of angina patients who underwent bypass surgery in the 1980s enjoyed dramatic relief from their symptoms, and no more than 1 to 3 percent of the patients died during the operation. Epidemiologist Babette Stanton and her colleagues at the Boston University School of Medicine reported in 1985 that only about one-quarter of bypass recipients were rehospitalized in the 6 months after their operation, and fully one-quarter of those hospitalizations were for problems unrelated to the heart.

Although grafted blood vessels tend to close up with time, typically 80 to 85 percent of the grafts remain open 1 year after the operation, and multiple bypass operations often remain effective for 5 years or more. In recent years, surgeons have learned that the source of the blood vessel used for the bypass can make a big difference in the long-term outcome of the surgery. In 1986, surgeon Floyd Loop of the Cleveland Clinic Foundation in Ohio compared the results of over 3,600 bypasses in which a saphenous vein from the thigh had been used and about 2,300 bypasses in which the internal mammary artery from the chest had been used. Loop found that the veins were much more likely than the arteries to become reclogged after surgery. Patients with a vein bypass had a 60 percent higher risk of death over the 10 years covered by the study than patients with artery bypasses. Those with vein bypasses were also 40 percent more likely to suffer heart attacks. This evidence has convinced many American surgeons to use internal mammary artery grafts whenever possible.

Since the advent of bypass surgery, there has been debate over who actually benefits from it. The operation is almost certainly beneficial for patients with severe damage to heart muscle and for those with severe angina that does not respond to drug therapy. Researchers were not sure, however, whether the procedure was equally beneficial for patients with a milder form of angina. New evidence that bypass surgery is, in fact, not beneficial for such patients was produced by the Coronary Artery Surgery Study (CASS) sponsored by NHLBI.

In CASS, conducted between August 1975 and April 1983, about 16,600 patients with suspected or confirmed heart disease were screened. The investigators chose 780 patients who had mild, so-called stable angina — in which chest pains occurred only after exertion — or who had suffered heart attacks but did not have subsequent heart disease symptoms. Half of the 780 were randomly assigned to bypass surgery, and the other half received drug therapy.

After 5 years in CASS, 8 percent of the drug-only group and 5 percent of the bypass group had died — a difference judged not to be statistically meaningful. The incidence of nonfatal heart attacks in the two groups was also similar. Because there was no clear improvement in survival from the bypass procedure, the CASS authors recommended that patients with stable angina can safely delay bypass surgery until their symptoms worsen to the point that surgery is needed.

Advocates of bypass surgery have argued that, while the operation may not extend the lives of many patients, it can improve the quality of life after surgery. One part of CASS focused on lifestyle. Overall, once they had recovered from the surgery itself, the bypass patients suffered less chest pain, had fewer limitations on their activities, and required less drug therapy than patients in the drug-only group. However, there were no differences in the employment status and recreational interests between the groups.

After reviewing the findings of CASS and other recent bypass studies, Harvard cardiologist Eugene Braunwald concluded in 1983 that bypass surgery offers little or no advantage over drug therapy to the large number of patients whose condition corresponds to that of the patients who participated in CASS. Braunwald noted that CASS is likely to be the basis of decisions about the use of bypass surgery for years to come because no other large, controlled studies of bypass surgery were under way and such studies require about 10 years to complete.

More recently, as more CASS followup data have become available, there has been continued debate over how to identify the patients most likely to benefit from surgery.

"The interpretation that surgery does not produce benefit over medical therapy is exactly right in the gross numbers," Massachusetts General Hospital researcher Mortimer J. Buckley said at an AHA seminar in February 1989. "But we must look at where the benefit lies, not just the gross numbers. We need to be sure we're not throwing into the bundle the patient who can really be helped."

Buckley and others have noted that 40 percent of patients with no symptoms or only mild angina who were originally assigned to medical therapy in CASS eventually underwent surgery, according to a 10-year followup study. University of Washington researcher Edwin L. Alderman and his CASS colleagues have noted that surgery produced a 16 percent 10-year survival advantage among patients who had three-vessel coronary disease and impaired heart-pumping function at the beginning of the study.

Balloon angioplasty. Opening clogged coronary arteries in angina victims without the risk, pain, and expense of bypass surgery is the aim of a procedure that was first described in 1978 by cardiologist Andreas Gruentzig and his colleagues at University Hospital in Zurich, Switzerland. The procedure involves inserting a cardiac catheter tipped with a strong, inflatable balloon into a partially blocked coronary artery to compress the atherosclerotic plaque clogging the artery. By 1988, according to cardiologist David Faxon of Boston University Medical Center, the procedure was being used to treat more than 200,000 patients annually in the U.S. — approximately two-thirds as many as were receiving bypass grafts.

As in other cardiac catheterization procedures, the catheter tube is typically inserted through the thigh and guided through a blood vessel to the heart with the aid of a live x-ray image. Once the catheter tip has been maneuvered to the narrowed part of the blocked coronary artery, the balloon is inflated and deflated — several times if necessary — to reopen the blood vessel and clear a passage for the blood. Percutaneous transluminal coronary angioplasty (PTCA), as the procedure is called, can be used only if at least a small passage remains open in the artery, so that the balloon-tipped catheter can fit inside. The coronary artery must also not be contorted too much, and the plaque must not have become so hard that it cannot be compressed. Angioplasty could be an appropriate treatment for about two-thirds of patients with single-vessel and multivessel coronary disease, according to Faxon.

Given the high cost of a bypass operation (about $15,000 to $25,000 in 1989), the need for hospitalization, and the pain and risk of open-heart surgery, PTCA is relatively simple, inexpensive, and safe. Local anesthesia is used for the procedure, only a 2-day stay in the hospital is usually required, and the cost is only about $5,000. However, within 6 months of PCTA, the treated arteries in about 40 percent of patients become partially clogged again, and by 1 year about two-thirds have some obstructions. While many such patients have a second angioplasty, others go on to have bypass surgery.

To reach patients who cannot be treated with balloon angioplasty and to address the problem of recurring obstructions, many investigators are working on advanced devices that use laser beams or heat to clear out the blockage. As of 1989, however, those approaches still had problems of their own, including a risk of damage to the arteries, larger and more cumbersome equipment, and high rates of recurring blockage. Nevertheless, many cardiologists expect that improved versions of some of these second-generation devices will have advantages over balloon angioplasty in the future.

Treating heart attacks. Because half of heart attack deaths occur within 2.5 hours after the attack begins, quick diagnosis and action are required to help the patient survive. The first step is often emergency electrical defibrillation, a technique that has been widely used since the 1970s. Most hospital emergency rooms are now equipped with electrical defibrillators, which supply a controlled burst of electricity to a pair of hand-held paddles placed against the chest of a heart attack victim. This electrical shock frequently halts the fibrillation — rapid, irregular twitching of the heart in which no blood is pumped — and restores normal heart action. A 1984 study by cardiologist Kenneth Stults and his colleages at the University of Iowa Hospital in Iowa City showed that equipping ambulances with portable defibrillators and teaching ambulance technicians how to use them could significantly improve survival rates for patients needing defibrillation.

Speed in treating heart attack victims is vitally important not only to reduce the risk of immediate death but to protect the heart from further damage. When a heart muscle's blood supply has been blocked during a heart attack, the muscle does not die immediately, but dies gradually over a period of hours. In the late 1970s, doctors began to develop drug strategies to limit the damage done to heart tissue after an MI by restoring blood and oxygen flow to the heart as soon as possible and by reducing the heart's oxygen demands.

By 1989, it appeared that some of these strategies were paying off. In a 1988 review of 28 clinical trials involving 27,500 heart attack patients, Salim Yusuf of the NHLBI concluded that prompt treatment with beta-blockers achieved a 15 percent reduction in early death, subsequent heart attack, or ventricular fibrillation, and also may have reduced the size of the area where muscle tissue dies, as indicated by enzyme measurements that reflect the amount of muscle death.

Dissolving blood clots. Heart researchers have recently developed another tool to help save threatened heart tissue and improve the outlook for patients with advanced coronary disease: the means to restore blood flow to the heart and keep tissue damage to a minimum.

Most heart attacks are caused when a blood clot blocks the flow of blood to the heart, causing the death of heart muscle tissue. By the mid-1980s, studies had shown that damage to the heart could be minimized by thrombolysis, in which enzymes are used to dissolve blood clots by triggering the destruction of fibrin, a key constituent of clots. The three most important clot-dissolving enzymes are streptokinase, which is obtained from streptococcal bacteria; urokinase, originally a derivative of urine; and tissue plasminogen activator, or tPA, which was originally isolated from tissues of the uterus but is now produced by genetically engineered bacteria. A fourth clot-dissolving agent called APSAC, a modified form of streptokinase, has shown preliminary promise in increasing survival after a heart attack.

The use of thrombolytic therapy has soared in the past decade as study after study has shown that clot-dissolving drugs not only restore blood flow to heart tissue but also increase survival after MI if treatment begins early enough — within 4 hours after the symptoms start and preferably sooner than that. By 1988, four out of six large studies from Italy, the Netherlands, West Germany, New Zealand, the U.K., and the U.S. had found that intravenous streptokinase treatment significantly lowered MI mortality. Even the two studies that failed to show an improvement in survival rates found that the streptokinase-treated patients had better heart function.

Streptokinase appears to be even more effective in combination with simple aspirin, which inhibits clotting by making blood platelets less sticky. The Second International Study of Infarct Survival (ISIS-2), which compared various combinations of streptokinase, aspirin, and placebo in over 17,000 heart attack patients in 400 hospitals across Europe, North America, and Australia, showed that only 8 percent of patients who received both streptokinase and aspirin died within 5 weeks, compared to about 13 percent among patients given placebos. In 1988, Peter Sleight of Oxford University, the ISIS-2 chairman, said the results could have a huge impact on heart attack survival.

The principal drawback of streptokinase is that it can produce internal bleeding throughout the body, especially at the doses needed to dissolve blood clots when it is given intravenously. The advantage of tPA is that it is more selective, that is, its effects are targeted on the surface of clots, and it does not interfere as much with the body's clotting system. Its primary disadvantage is that it is considerably more expensive than streptokinase and it has a half-life in the bloodstream of only a few minutes, so it must be given continuously over several hours, or new clots will form.

In 1984, the NHLBI organized a large trial of tPA called the Thrombolysis in Myocardial Infarction (TIMI) trial. The $31 million study was originally designed to compare tPA with streptokinase in 4,000 heart attack victims at 25 U.S. medical centers. However, preliminary results from the study showed that tPA cleared coronary clots in 66 percent of 118 patients, while streptokinase removed blockages in only 36 percent of 122 patients. These results were so impressive that the study design was changed to give all patients tPA at varying dosages and time schedules.

The U.S. FDA touched off a storm of protest among cardiologists in May 1987, when it refused to license tPA on the basis of available evidence. However, in November 1987, the agency did approve both streptokinase and tPA. At a press conference, FDA Commissioner Frank Young said studies found that tPA dissolved coronary clots in 71 percent of patients treated within 6 hours of the onset of symptoms. Harvard's Braunwald noted at the time that TIMI results showed that brain hemorrhages occurred in only 7 of the 1,200 tPA patients, a rate of just 0.6 percent.

Post-MI treatments are still evolving rapidly, marked by the controversy that often occurs in a fast-moving field with such high stakes in both human lives and financial rewards. (The annual market for thrombolytic agents is estimated to exceed $1 billion.) For example, the results of TIMI II, one part of the ongoing TIMI trial, sparked heated debate among cardiologists. TIMI II assessed early tPA treatment in combination with angiography and early angioplasty in patients for whom it was appropriate, compared to tPA and "watchful waiting," with angioplasty deferred until symptoms of angina developed.

Contrary to the expectations (and clinical practice) of many U.S. cardiologists, the TIMI II results, published in the 9 March 1989 *New England Journal of Medicine (NEJM),* showed no advantage in angioplasty performed within 48 hours of MI symptoms compared to the watchful waiting strategy. Both groups had comparable death rates and heart attack rates, and their heart function was similar.

The controversy may be settled (or further inflamed) by the results of an ongoing study of 600 MI patients called the Thrombolysis and Angioplasty in MI (TAMI) study, which is comparing urokinase, tPA, or both, with and without immediate angioplasty. The implications for the health care system are immense, according to physicians Alan D. Geurci and Richard S. Ross of Johns Hopkins University, who wrote a *NEJM* editorial accompanying the TIMI II report. If immediate angiography and angioplasty had proved superior in the TIMI II study, it would have implied "a vast increase in the number of urgent interhospital transfers [of MI patients] and a substantial proliferation of facilities for angiography, angioplasty, and cardiac surgery. The costs of transporting patients, training staff, and expanding facilities, together with the associated complexities of quality control, can only be imagined."

Preventing repeat heart attacks. Heart attack survivors are monitored closely during the first few days after the event and are treated in a variety of ways to limit damage to their hearts. Once a patient has survived a week or more, however, the risk of immediate death tends to be relatively small. Doctors then focus their attention on ways to relieve the symptoms of heart disease and prevent a second heart attack.

In 1977, researchers discovered that ordinary aspirin helps to prevent clots from forming in the blood vessels of men recovering from hip replacement operations. (For unknown reasons, women showed no similar benefit from aspirin.) Since then, doctors have speculated that aspirin might also be useful in preventing heart attacks by inhibiting the formation of clots in coronary arteries.

This hypothesis was dramatically confirmed with the publication in 1988 of results from the Physicians' Health Study, a project involving over 22,000 male U.S. doctors. Doctors who were randomly assigned to take a 325-milligram tablet of buffered aspirin every other day had a 47 percent lower risk of heart attacks than those who took a placebo pill. However, a smaller study of 5,000 British physicians, led by Sir Richard Doll and Richard Peto, both of Oxford University, and published at about the same time as the American study, failed to find a difference between those who took 500 milligrams of aspirin a day and a placebo group.

Harvard researcher Charles Hennekens, the principal author of the U.S. study, said the discrepancy between the two outcomes might be explained by the fact that physicians are an unusually healthy group, so the beneficial effect might not be apparent except in studies involving very large numbers of people, as in the U.S. study. The Oxford researchers, after evaluating 31 aspirin trials involving 29,000 patients, concluded that aspirin carries a clear benefit for people with a known risk of heart attack or stroke, but they were cautious about advising all men over age 40 to take aspirin to reduce the risk of heart attacks, because aspirin may increase the risk of strokes and other bleeding problems.

Pacemakers and cardioverters. In many heart disorders, the heart's chambers are fully capable of pumping blood, but something disturbs the rhythmical electrical impulses that signal the heart's chambers to alternately contract and relax. These impulses originate in a group of so-called pacemaker cells located on the right atrium and spread along specific pathways throughout the heart, signaling muscles to contract and relax in a well-defined sequence. Disturbances in this rhythm can often be controlled by drugs such as quinidine and certain beta-blockers; in many cases, however, damage to the heart leads to uncoordinated or irregular heartbeats — arrhythmias — which cannot be treated with drugs.

When drug treatment fails, an artificial cardiac pacemaker is often used to provide a steady, rhythmical signal that overrides the signal from the pacemaker cells or retransmits the signal around a part of the heart that is not conducting the signals properly. Conventional electrical pacemakers, which usually weigh a few ounces and are powered by batteries lasting 3 years or more, are generally implanted under the skin of the chest and connected to the heart by wires in a procedure that does not require major surgery. Over 215,000 Americans each year have pacemakers inserted or undergo operations to maintain existing pacemakers, according to the AHA.

The first pacemakers, which were introduced in the early 1960s, simply amplified and retransmitted the heart's own pacemaking signals. Other early pacemakers generated a steady output of pulses, whose strength and frequency were set at the factory. The newest generation of pacemakers, however, allow the strength and frequency of the pacing signals to be changed after the pacemaker has been implanted. The programmed responses of one such pacemaker, developed by Anthony Richards at the National Heart Hospital in London, U.K., can be changed with a small radio transmitter held close to the patient's chest. Signals from the external transmitter alter the rhythm and intensity of the pacemaker's signals by resetting a microprocessor inside the device. The microprocessor also records information on episodes of arrhythmia for later analysis.

A pacemaker that changes the frequency of the pacing signals automatically in response to changes in the level of exertion was approved by the FDA in 1986. The 1.5-ounce device, which is slightly larger than a silver dollar, is called the Activitrax and is manufactured by Medtronic,

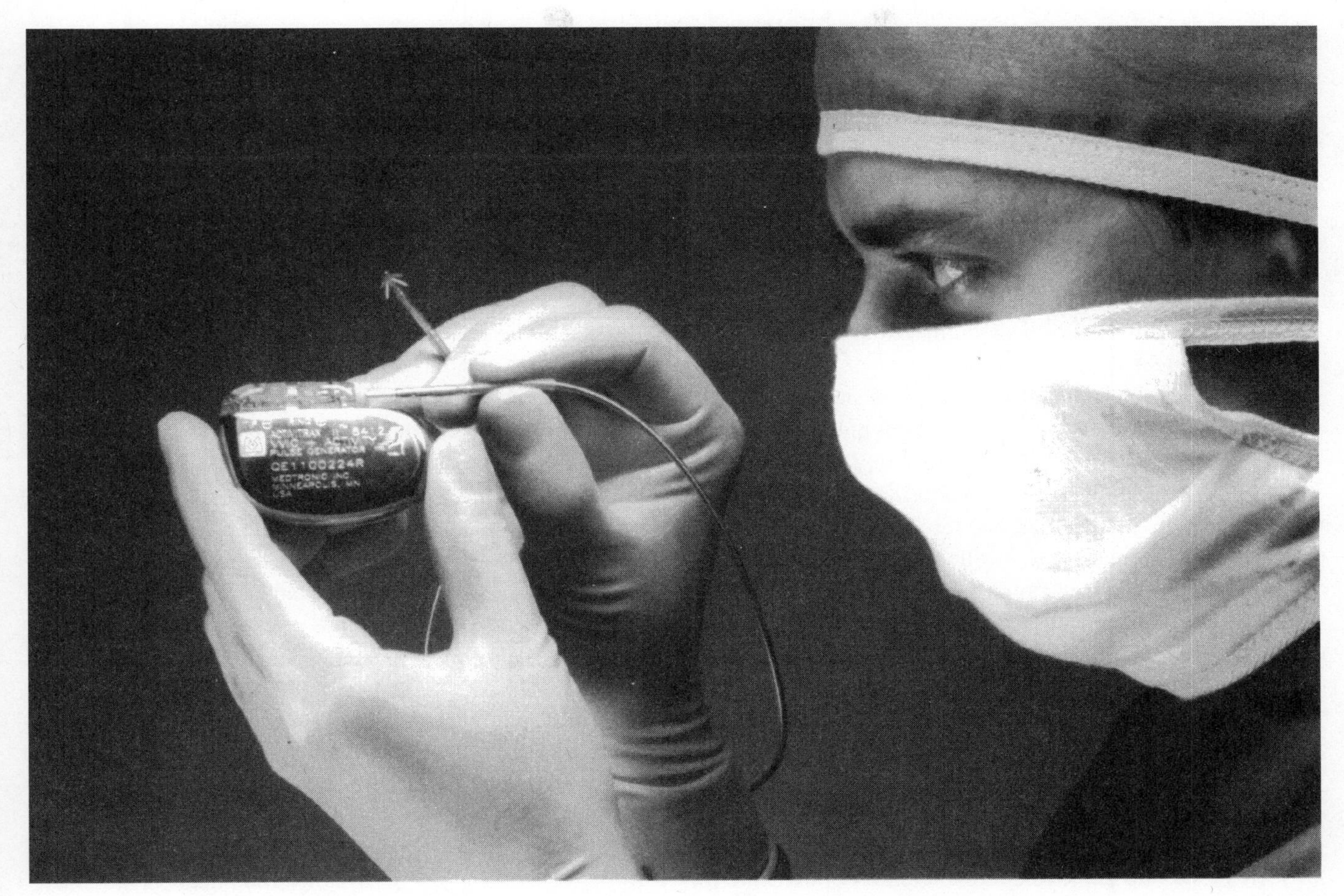

Figure 3. Responsive Pacemaker. Unlike pacemakers that are programmed to pace the heartbeat at a fixed rate, the Activitrax II, manufactured by Medtronic, Inc., can regulate the heartbeat in response to the heart's changing requirements. Such pacemakers have been implanted in patients between 1 day and 107 years old. *Courtesy: Medtronic, Inc.*

Inc., of Minneapolis, Minnesota. The key to the pacemaker's flexibility is a minute crystal that — much like the needle on a phonograph — converts pressures resulting from body movements into electrical signals that adjust the pacing rate. The pacing rate can vary between 60 and 150 beats per minute.

At the time the Activitrax was approved, the pacemaker had been implanted in 300 patients in the U.S. and more than 4,000 throughout the world. The battery in the device is expected to last 8 to 10 years if a person is sedentary, or 6 to 7 years if a person is active. The pacemaker's rate can also be adjusted from outside the body by passing a programmed magnetic device over the implantation site. The cost of the Activitrax is about $5,000, somewhat more expensive than a conventional pacemaker.

Until about 1982, the most sophisticated pacemakers were able to control only bradycardia — an abnormally slow heartbeat. Since then, several new devices have been developed to manage a much wider variety of heart disorders. Unlike earlier pacemakers, these devices do not continuously emit signals to pace the heart; instead, they emit signals only when an abnormal heart rhythm is detected.

The most dramatic such device is actually a miniature version of the defibrillators used in ambulance and emergency rooms to jolt hearts out of the otherwise-fatal rhythm disorder known as ventricular fibrillation. Marketed by several manufacturers for around $6,000, the device senses when fibrillation begins and then, within 20 seconds, sends a strong electric shock through wires permanently connected to the heart. This device, invented by Mieczyslaw Mirowski at Johns Hopkins University School of Medicine, also responds to ventricular tachycardia (abnormally rapid heartbeat) by delivering an electric shock to convert the abnormal rhythm back to a normal heartbeat, a process called cardioversion.

Defibrillation and cardioversion require much stronger bursts of electricity than the tiny signals used in ordinary pacemakers. To supply the needed energy, Mirowski's device — called a defibrillator/cardioverter — requires a power pack that is too large to fit easily within the chest; instead, the half-pound unit is implanted in the abdomen. The lithium batteries in the unit supply enough energy for as many as 100 shocks during their 3-year lifetime. In early applications of the device, electrical connections from the device to the heart had to be made by open-heart surgery. More recently, however, researchers have been able to make these connections by means of cardiac catheters.

Some new, experimental cardioverters have been developed that are no larger than conventional pacemakers but considerably "smarter." For instance, a device called the Intertach was developed by Intermedics, Inc., of Angleton, Texas, and was first implanted in a human in 1985. The microprocessor-controlled device is programmed to discriminate between rapid heart rates resulting from exercise or emotion and those caused by tachycardia. When tachycardia is detected, the Intertach responds with a preprogrammed sequence of mild electrical pulses to momentarily "jam" the electrical pathways. If the first sequence does not work, the device tries several others until it finds one that does work.

A record of all of the device's activity is stored in its internal memory, which can be read in the physician's office by using a simple telephone telemetry device held over the implant site. The same communications device can be used to reprogram the Intertach without surgery.

Manufacturers are developing second-generation implantable cardioverters that incorporate defibrillators. A successful combination defibrillator/cardioverter could be widely used. Each year, 700,000 people survive heart attacks in the U.S. alone and, according to Roger Winkle of the Stanford University Medical School, all of these people are at risk of sudden death from arrhythmias and therefore are candidates for implantable defibrillators/cardioverters.

Heart transplants. The first human heart transplants, performed by heart surgeon Christiaan Barnard in Cape Town, South Africa, in 1967, demonstrated that people could survive the operation and function normally with a donor heart for at least a limited time. In the first year after Barnard's initial success, about 100 people received new hearts. The prospects for long-term survival of these transplant recipients were poor, however, mostly because the patients' bodies rejected the donor hearts. As a result, by the mid-1970s, only 20 to 40 heart transplant operations were being performed worldwide each year.

By the early 1980s, better methods of controlling rejection had been developed, and 60 to 75 percent of heart transplant patients were surviving at least 1 year after their operations; on average, about half of the transplant recipients were alive 3 years after their surgery. The single most important factor in this improvement was the introduction of the rejection-suppressing drug cyclosporine. As the prospects for heart transplant patients have improved, the number of centers performing these procedures and the patients receiving them has grown rapidly. By 1988, more than 1,000 such operations were being performed annually in over 100 centers in the U.S. alone. Cyclosporine has also made heart-lung transplants feasible; 41 such operations were performed in 1987 and 45 North American hospitals were offering the procedure in 1988.

Whether the extension of life for a relatively few patients justifies the massive hospital resources required by heart transplants was, however, an issue actively debated in the 1980s. Many physicians questioned whether heart transplants provide "the greatest good for the greatest number." Critics have pointed out, for example, that each heart transplant consumes resources comparable to those used in six to eight open-heart operations. As the operation became more established, however, much of this debate faded. One reason it is not more prominent, some speculate, is that the number of operations is effectively limited by the number of suitable donor hearts available to transplant. About 20 to 30 percent of heart transplant candidates die waiting for a donor organ, a statistic that is leading many transplant centers to accept hearts from older donors — up to age 55 or 60 in some places, compared to the age limit of 35 (for male donors) or 40 (for females) prevalent until the late 1980s.

Because of this chronic shortage of human donor hearts, some experts still hope that artificial hearts will prove to be the ultimate cure for irreparable heart damage. One California surgeon, Leonard L. Bailey of Loma Linda University, briefly turned to nonhuman primates as a source for organs. In an operation that attracted worldwide attention, Bailey in 1984 attempted unsuccessfully to save the life of a newborn baby by replacing her deformed heart with that of a young baboon.

Several surgeons, including Bailey, have attempted transplants in newborns with hypoplastic left-heart syndrome — a birth defect involving a severely deformed left ventricle — in a handful of infants. Obtaining viable donor hearts is an almost insurmountable problem in these cases, but Bailey reported at the November 1988 AHA sessions on 5 children who have survived an average of 17 months with transplanted hearts. Surgeons at the

Figure 4. Jarvik-7 Artificial Heart. The Jarvik-7 artificial heart, pictured here with its inventor Robert Jarvik, was first implanted in 61-year-old dentist Barney Clark during late 1982. Although Jarvik-7 hearts were permanently implanted in six patients, all suffered serious complications and none survived longer than 21 months. Nevertheless, Jarvik-7 hearts have been used successfully as "bridges to transplant" in over 100 patients waiting for human hearts. *Courtesy: Brad Nelson, University of Utah School of Medicine.*

University of Pittsburgh, reporting at the same meeting on 26 heart transplant recipients between 3 weeks and 16 years old, noted a "disappointing 42 percent" survival rate at 6 years, due largely to the prevalence of accelerated atherosclerosis in the donor heart.

Artificial hearts. The first successful implantation of an artificial heart into a human captured worldwide attention in late 1982 and early 1983, when 61-year-old dentist Barney Clark lived for 112 days after receiving an artificial heart called the Jarvik-7. Clark's surgeon, William DeVries, reestablished his program at Humana Hospital in Louisville, Kentucky, where on 25 November 1984 he implanted a second Jarvik-7 artificial heart in 53-year-old William Schroeder, a retired postal worker from Jasper, Indiana. The device sustained Schroeder for 620 days — about 21 months — making him the longest-surviving implant recipient. But Schroeder's recovery was repeatedly set back by a series of incapacitating strokes. During his struggle, artificial hearts were permanently implanted in four other patients, but all four died. DeVries attempted to find other suitable candidates for permanent Jarvik-7 implants, but the use of human heart transplants in older and sicker patients, combined with the unfavorable course of the first six recipients, effectively ended the program with Schroeder's death.

By the end of 1988, however, Jarvik hearts had been used as temporary "bridges to transplant" in well over 100 patients in 22 U.S. centers. In a report on the first 100 Jarvik "bridge" patients presented at the AHA's 1988 scientific sessions, Lyle D. Joyce of the Minneapolis Heart Institute and coworkers said 70 patients underwent cardiac transplantation after spending an average of 23 days on the artificial heart. (Among the entire group, the range of time spent on the device was 1 day to 243 days.) Survival following transplant was 77 percent at 30 days and 54 percent at 6 months. Three Jarvik heart recipients had strokes and six suffered transient ischemic attacks — "little strokes," which are typically caused by small blood clots in the brain.

If artificial hearts are to become widely used, they must be made unobtrusive and reliable enough to allow their recipients freedom of movement and the chance to resume a reasonably normal existence. To accomplish this, most researchers agree, will require the development of completely implantable artificial hearts that can operate without being connected to equipment outside the body. Such hearts will have to be electrically powered, rather than air-powered like the Jarvik-7.

Two different designs of fully implantable, electrically powered hearts were being tested in animals at the Hershey Medical Center and the Cleveland Clinic in recent years, but these devices will not be ready for human tests before the early 1990s. Totally implantable hearts were also under development in Japan. A research group at the Tokyo University Institute of Medical Electronics, under the direction of Kazuhiku Atsumi, has reported that they expected to have a totally implantable heart ready for animal testing by 1989 and a device ready for human implantation by the mid-1990s.

Critics of artificial hearts, including bioethicist Daniel Callahan of the Hastings Center in Hastings on Hudson, New York, have argued against their use on the basis that they do not provide an adequate quality of life and that their cost is excessive. "We're not going to create healthy people with long life expectancies," Callahan said in a February 1986 debate with the heart's inventor, Robert Jarvik, held in Washington, D.C. "We're going to create people who are going to be chronically ill."

Callahan also noted that, according to 1985 NHLBI estimates, there are 17,000 to 35,000 potential candidates for artificial hearts each year. At a cost of $150,000 per implant, the procedure could add as much as $5 billion to the U.S. medical bill each year. "One hates to argue against something that has value for some individuals," Callahan said, "but that money could be better spent on health education aimed at promoting behavioral and dietary changes to prevent the heart disease that necessitates artificial hearts."

Heart-assist devices. Completely implantable devices that assist the heart but do not entirely replace it are favored by many researchers as a more practical alternative to total artificial hearts. The heart-assist devices that would benefit the largest number of heart disease patients are the electrically powered left-ventricle assist devices (LVADs). LVADs are installed in the chest near the heart with one tube carrying blood to the device from the left ventricle and another tube returning the blood from the device to the aorta.

Once installed, the LVAD can completely take over the pumping function of the left ventricle — the heart's main pumping chamber — either temporarily while the heart heals after surgery, or permanently. Left-ventricle failures are the most frequent cause of heart disability and death; among all patients who eventually die from heart failure, as many as 80 percent could be sustained by an LVAD. These statistics imply that as many as 40,000 Americans each year could be candidates for LVADs.

Because the potential U.S. market for implantable LVADs and artificial hearts is so large, as much as $2 billion annually, strong competition has developed among several medical technology firms to produce the first practical devices. Many researchers believe that this technology, once developed, will return to active life thousands of people who would otherwise die or be invalids. But whether the early commercial devices will permit implant patients to lead full lives or commit them to years of marginal health in hospitals remains an open question.

Cancer

Since 1937, the U.S. government has conducted a much-publicized war on cancer. In 1989, cancer research accounted for $1.5 billion of the National Institutes of Health (NIH) $7.5 billion budget. Yet there is widespread debate about whether scientists are winning the war against cancer or simply holding their ground. Based on figures from the National Cancer Institute (NCI) Surveillance, Epidemiology, and End Results (SEER) program, the American Cancer Society (ACS) estimated that 985,000 people found out they had cancer for the first time in 1988. (This total does not include nonmelanoma skin cancer, the most common type of cancer in the U.S., which accounts for more than 500,000 new cancer cases each year but causes only about 2,000 deaths.) Cancer ranks as the second leading killer of Americans and now claims more than 500,000 lives each year.

Many researchers have questioned whether the war on cancer might not be better fought by focusing on prevention rather than treatment. The overall cancer death rate has actually increased by 8 percent since 1950, but that increase is largely due to a nearly 250 percent increase in the rate of lung cancer deaths, which occur primarily among smokers. When lung cancer deaths are removed from the statistics, the overall cancer death rate has decreased 13 percent. Though lifestyle factors — such as cigarette smoking and diet — as well as environmental factors are widely believed to play a role in cancer,

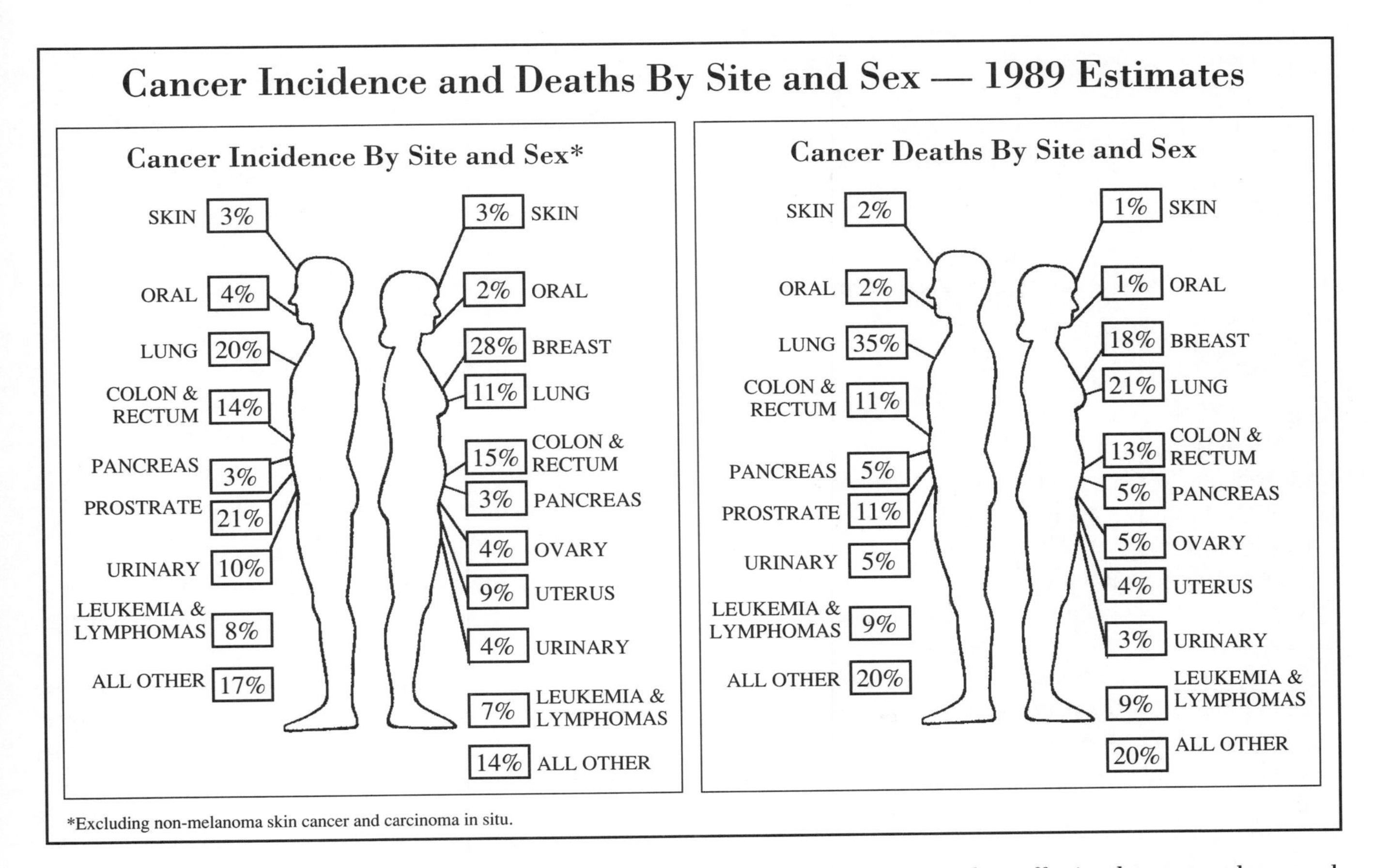

Figure 5. U.S. Cancer Cases and Deaths. During 1989, the most common cancers in men were those affecting the prostate, lungs, and colon or rectum, while in women, the most common were those affecting the breasts, colon or rectum, and lungs. Lung cancer accounted for the highest percentage of cancer deaths among both men and women. *Courtesy: American Cancer Society.*

scientists have found it difficult to determine how much exposure to specific agents will cause a given number of cancer cases in the general population. Industrial chemicals and radiation are known to increase the cancer risk in certain groups of heavily exposed individuals, but the risk to people with only occasional exposure is not known. Asbestos, for example, is an important cause of cancer among insulation workers, but scientists do not yet know what effect everyday exposure to asbestos in office buildings has on the people who work there.

Because cancer is actually a group of more than 100 different illnesses, scientists have had to develop a battery of diagnostic tests and treatments for the different forms of the disease. Widely used diagnostic techniques include cell-structure analysis (cytologic studies), endoscopy, radioisotope scanning, x-ray imaging, ultrasound imaging, and a variety of experimental biochemical and immunological techniques. Exploratory surgery is also needed for diagnosis in many cases, but the advent of noninvasive techniques, including computed tomography (also known as CT scanning) and magnetic resonance imaging, has greatly reduced the need for such surgery. Many cancer researchers are particularly excited about the possible use of monoclonal antibodies to detect and treat cancers.

For many years, cancer treatment has centered on surgery, radiation, chemotherapy, and hormone therapy, but another method — immunotherapy — is now emerging. Compounds such as interleukin-2 are being used experimentally to bolster the body's immune system, which may be weakened by the fight against tumor cells. At least in a few cases, these experimental techniques have been successful.

Cancer Formation and Growth

Cancer begins when a single normal cell is subtly altered or damaged, but not so much that it cannot continue to function and reproduce. The alteration often occurs in the cell's machinery regulating growth and maturation. As the new cell and its daughter cells reproduce, often at a very rapid pace, they form a growing mass of tissue that is different from the surrounding tissue, creating a tumor or a neoplasm — from the Greek roots "neos," meaning new, and "plasma," meaning formation.

Benign tumors consist of nearly normal cells that, although altered or damaged, have not acquired the ability to invade surrounding tissue or spread beyond the site where they originate. Such tumors may grow rapidly or slowly or stop growing altogether. In any case, benign tumors remain localized and can generally be completely removed by surgery if necessary.

Although malignant tumors likewise develop from a single cell, the cell's growth control machinery is more severely deranged. The growing tumor may invade neighboring tissues and organs, ultimately spreading to other parts of the body. Clumps of malignant cells or even single malignant cells may travel through the blood or lymph system to be deposited at a distant site in the body, giving rise to new, secondary tumors. As discussed below, several distinct changes appear to be necessary to convert a normal cell into a cancerous one. Cells in benign tumors, which have already undergone one or more of these changes, can sometimes undergo additional changes and develop into malignant growths.

A tumor one centimeter (0.4 inch) in diameter — about the smallest that can be detected by standard imaging equipment — contains about one billion cells. A treatment that kills 99.999 percent of the tumor will still leave 10,000 cells alive — more than enough to seed the growth of new tumors.

Gene damage and cancer formation. Evidence continues to accumulate that damage to genes — the blueprints that cells use to make proteins for a wide variety of functions — is an essential step in the formation of some cancers. Indeed, there is growing agreement among scientists that most, and perhaps all, cancers arise from alterations in cells' normal genetic instructions.

Support for the gene-damage theory comes from direct observation of genetic changes in laboratory experiments, as well as extensive evidence showing that most of the industrial chemicals that cause cancer in humans do so by damaging genes. Moreover, most of the chemicals that have produced cancers in laboratory animals are known to cause damage to the animal's genes, strongly suggesting that they are also potentially damaging to people.

RNA tumor viruses and oncogenes. Viruses are increasingly being linked to gene damage and cancer formation. Indeed, the clearest evidence that gene damage is ultimately responsible for the development of cancer has come from experiments studying how RNA viruses "turn on" the cancer process in certain cells. RNA is one of the molecules involved in carrying out a cell's DNA instructions for reproduction, growth, and maturation. RNA viruses — also known as retroviruses because of the unusual way they function in the cells they infect — attach themselves to specific cell targets, such as white blood cells. Once attached, they are then able to inject their own genes into a cell's nucleus and disrupt the chain of command between DNA and normal cell operations. Retroviruses may carry cancer-causing genes called oncogenes, from the Greek root "onkos," which means "mass." Genetic engineering techniques have allowed researchers to isolate specific genes involved in the normal day-to-day operation of cells, as well as oncogenes that have been inserted into cells by retroviruses.

After discovering viral oncogenes, researchers learned that oncogenes are also present in apparently normal cells uninfected by a virus. In cells, the genes lie dormant until being activated by a triggering event, beginning the cancer process. These forerunners of cancer are called proto-oncogenes. Proto-oncogenes may be altered in a process known as a rearrangement, in which normal genetic material from one arm of a chromosome crosses over to the chromosome's other arm or to another chromosome entirely, switching places with different genetic material.

Genes altered by rearrangement contain new instructions for making chemicals in the cell, but the genes often lie dormant for a long period of time before any change begins. Exposure to chemicals in tobacco smoke or other cancer-causing agents may be needed to turn on these dormant genes. Once the genes are activated, they apparently begin a multistep process that leads to the disruption of the normal pattern of cell growth and reproduction, ultimately causing a tumor to form.

In 1988, University of California researcher Abraham de Vos and a group of American and Japanese researchers reported that they had determined the complete three-dimensional structure of the product of one of these oncogenes, called c-H-*ras*, a member of the most common family of oncogenes, the *ras* family. These oncogenes have been found in cancers of the breast, colon, bladder, lung, and other sites. By studying the detailed structure of the chain of amino acids that make up the protein coded for by the c-H-*ras* oncogene, the researchers have discovered important clues about the process that leads to the rearrangement of genes and how an oncogene is created. This information may someday be useful in the design of drugs to prevent proto-oncogenes from being transformed to oncogenes.

It now appears that all normal cells may contain proto-oncogenes, meaning that all the cells in the body have the potential to become cancer cells. Several dozen proto-oncogenes and oncogenes have been discovered so far, but the functions of most of these genes and their roles in the development of cancers are still largely unknown.

In 1988, Michael Karin and his colleagues at the University of California at San Diego reported finding what they believe is an important link in the final stages of the initiation of cancer, a protein called activator protein 1, or AP-1. This protein appears to carry the message to turn on the process that causes abnormal cell growth and multiplication.

Lifestyle Factors Contributing to Cancer

Smoking. Since 1964, when a U.S. Surgeon General's report publicly identified cigarette smoking as the cause of most lung cancer deaths in the U.S., scientists have linked smoking to a wide variety of other cancers. In addition to causing about 85 percent of new lung cancer cases, smoking appears to double the risk of getting bladder cancer, and also to significantly increase the risk of cancer of the esophagus, larynx, mouth, or pancreas.

On the twenty-fifth anniversary of the landmark 1964 report, Surgeon General C. Everett Koop called on Americans to create a smoke-free society by the year 2000 to prevent many of the 200,000 lung cancer deaths predicted for that year. The U.S. Office of Technology Assessment estimated that in the mid-1980s, the total cost of smoking-related health problems and lost productivity was approximately $65 billion each year — or about $2.20 per cigarette pack sold.

The question of whether breathing other people's cigarette smoke causes cancer became an important health issue in the 1980s, when the U.S. Surgeon General and the National Academy of Sciences in Washington, D.C., issued reports concluding that the number of people injured by involuntary, or passive, smoking was much higher than the number injured by other environmental agents already regulated by the Clean Air Act. The reports contributed, in part, to legislation that imposed a ban on smoking on U.S. airline flights lasting less than 2 hours. By late 1989, there was growing support in the U.S. Congress for a ban on smoking in all domestic airline flights.

In a review article published in late 1988, Jonathan E. Fielding and Kenneth J. Phenow of the University of California at Los Angeles analyzed 18 major studies of passive smoking and found that the risk of lung cancer was about 34 percent higher for nonsmokers regularly exposed to other people's cigarette smoke. James Repace of the U.S. Environmental Protection Agency (EPA) and Alfred Lowrey at the Naval Research Laboratory in Washington, D.C., had estimated in 1987 that 4,700 of the approximately 150,000 annual lung cancer deaths in the late 1980s could be attributed to tobacco-smoke exposure in nonsmokers.

Smokeless tobacco. The use of smokeless tobacco in the form of snuff or chewing tobacco is thought to account for 4 percent of all cancers in men and 2 percent of all cancers in women. The majority of cancers associated with smokeless tobacco are oral and throat cancers. The oral cancer generally occurs at the site where the tobacco is most commonly chewed, usually in the cheek or on the gums. The ACS has estimated that about 30,000 Americans develop oral cancer and over 9,000 die each year. One study sponsored by NCI and the University of North Carolina has found that snuff users have at least 13 times the risk of suffering cheek and gum cancers as nonusers.

Public health officials have been alarmed by a sharp increase in the number of young people using smokeless tobacco products, particularly in the Southeast and Southwest. At a recent NIH consensus development conference on the health implications of smokeless tobacco, researchers reported that 13 percent of third-grade boys in Oklahoma dip snuff or chew tobacco, and that the number rises to 39 percent by eleventh grade. Other expert groups, including an advisory panel to the Surgeon General, have estimated that sales of smokeless tobacco rose 11 percent each year during the 1970s and early 1980s, as the number of users grew to between 7 million and 12 million. Although the production of snuff and chewing tobacco has since leveled off, nationwide surveys indicate that, on average, 17 percent of boys and 2 percent of girls between third and twelfth grade use smokeless tobacco more than once a week.

Alcohol consumption. Scientists have long debated the existence of a link between breast cancer and alcohol consumption. In May 1987, two large studies by researchers at NCI and Harvard University concluded that women who consumed even moderate amounts of alcohol appeared to have a significantly higher risk of developing breast cancer as nondrinkers. In the Harvard study, which involved about 90,000 women, women who had an average of three to nine drinks a week had about a 30 percent higher risk of developing breast cancer than nondrinkers, while women who drank more than that amount had about a 60 percent higher risk. One year later, however, a major study by the American Health Foundation in New York City concluded that there was "no consistent trend [of increased risk of breast cancer] with increasing alcohol consumption."

In August 1988, researchers Matthew P. Longnecker and his associates at the Harvard School of Public Health reported on their analysis of all the major studies of alcohol and breast cancer. They concluded that there is strong evidence of an association between alcohol consumption and breast cancer, but cautioned that there is, as yet, no proof that alcohol actually causes breast cancer.

Other cancers have also been linked to alcohol. For example, heavy beer drinking has been linked to cancers of the colon, rectum, and liver. Excessive consumption of alcoholic beverages, when combined with cigarette smoking, was found to be a probable cause of cancers of the upper respiratory tract in a 1988 study involving over 4,000 men in six European countries.

Dietary factors. Cancer researchers believe that diet plays a role in some cancers, because particular types of cancer are much more common in some countries than in others, and eating habits vary widely among countries. Researchers have suggested that some foods might contain cancer-causing chemicals, while others might contain substances that offer protection against cancer.

The diet-cancer link is hotly debated, mainly because so many studies have been inconclusive. Even when there is evidence of an association, researchers cannot tell with certainty which foods cause or protect against cancer. Proving a cause-and-effect relationship is exceedingly difficult, because people typically eat hundreds of different types of food prepared in a variety of ways, and their diet changes from day to day and year to year as different foods become available. One experimental approach has been to measure the incidence of cancers in animals placed on carefully monitored diets. Such experiments are difficult to interpret and sometimes unreliable because of large differences in the effects of foods on different species, and because very large amounts of the suspect foods must be used to produce a significant difference in disease rates in the relatively small number of animals typically used for the tests.

Increasing the intake of fiber by eating bran and whole grain foods has been widely advocated as a way to reduce the risk of cancer of the colon and rectum. Because fiber speeds the flow of digested food through the bowel, it is presumed that the risk of cancer is lowered by shortening the time tissues are exposed to possible cancer-causing agents in food and drink. In November 1988, however, epidemiologist Martha L. Slattery of the University of Utah in Salt Lake City reported that a study comparing colon cancer patients and healthy individuals suggested that only fruit and vegetable fiber are beneficial and that grains offer no protective effect. NCI is currently analyzing the fiber content of 400 foods in hopes of determining which fibers protect against cancer.

Other dietary factors being investigated by scientists include beta-carotene (a substance found in carrots and certain other vegetables and fruits), folic acid, selenium, and vitamins A, C, E, and B_{12}. These nutrients have been linked to a reduced risk of cancers of the bladder, breast, cervix, colon, esophagus, lung, skin, and stomach. But existing evidence suggests that these nutrients may play a role only in preventing cancer rather than in reducing existing tumors.

Another approach to studying the connection between diet and cancer has been to search for specific substances in foods that may cause cancer. Since many mutagenic, or gene-damaging, chemicals can cause cancer, researchers have analyzed foods to determine which ones contain such chemicals.

Dietary recommendations. In the early 1980s, the National Academy of Sciences issued its first official recommendations for dietary changes to reduce the risk of cancer, and these guidelines have since been supported by dozens of medical organizations, including the ACS and the NCI. The recommendations, which were based on a review of more than 1,000 studies, advised Americans to eat fruits, vegetables, and whole-grain cereal products daily; minimize their intake of smoked, cured, and pickled foods; and drink alcoholic beverages in moderation, if at all. The report also urged Americans to reduce their total fat intake from about 40 percent of the calories consumed to about 30 percent.

Many researchers have found that the high fat consumption common in North America and many European nations appears to be associated with an increased risk of several types of cancer, especially cancer of the breast, colon, and prostate. Human studies have not yet shown whether changing to a lower-fat diet reduces a person's cancer risk. However, animal experiments and research comparing cancer rates in different regions suggests that reducing total fat intake is likely to reduce the risk of these types of cancer.

Cancer researchers and nutritionists have also emphasized the importance of eating more dark-green and yellow or orange vegetables, such as broccoli and carrots. Many studies over the years have shown that people who eat these vegetables have a lower incidence of bladder, larynx, and lung cancers. Scientists have proposed that the apparent protective effect of these vegetables is due to beta-carotene, a chemical converted in the body into vitamin A, as well as vitamin A itself. Studies have also suggested that the consumption of fruits and vegetables rich in vitamin C, especially citrus fruits, provides some protection against cancer of the esophagus, lungs, and larynx.

The possible role chemical additives and preservatives play in cancer development continues to be investigated. Nitrites, nitrates, and nitrosamines found in bacon, bologna, frankfurters, ham, sausages, and smoked fish have been linked to stomach and esophageal cancer, but the final results from large studies are not yet available. Studies are also under way to determine whether benzo(a)pyrene, a known cancer-causing chemical produced during charcoal-broiling, increases the risk of cancer in people who often eat foods cooked in this way.

Caffeine. Caffeine has been suspected of causing several types of cancer, including pancreatic cancer and breast cancer. By 1989, however, researchers had failed to document a conclusive link. Some scientists have suggested that caffeine aggravates fibrocystic breast disease, a benign but often painful malady that appears to be a forerunner to cancer in some women. Some have suggested that caffeine may worsen the condition to such an extent that breast cancer develops.

A 1988 review of breast cancer studies in 44 countries concluded, however, that caffeine plays no role in breast cancer development, even in women with fibrocystic disease. The study, led by Hugh Phelps of the Maine Medical Center in Portland, attributed the nation-to-nation variation in breast cancer rates to differences in fat intake, with higher-fat diets being associated with higher rates of breast cancer.

Birth-control pills. Since their introduction 30 years ago, birth-control pills have been linked with several health problems, including a possible increased risk of certain cancers. Surveys published in 1988 by the Alan Guttmacher Institute in New York City indicated that more than 13 million women in the U.S. used birth-control pills each year. Because birth-control pills contain estrogen — a hormone that plays a critical role in the female reproductive system — researchers have closely monitored their effects to determine whether they raise the risk of cancers of the breast, cervix, endometrium (the lining of the uterus), and the ovaries.

The dose of estrogen in birth-control pills was substantially reduced in the 1970s in an effort to reduce the side effects, including elevated cancer risk, associated with earlier, relatively high estrogen-level pills. To assess the current risk of birth-control pills, Samuel Shapiro and his colleagues at the Boston University School of Medicine compared the incidence of breast cancer in more than 400 women who took birth-control pills and 400 who did not. Their findings, which were presented before an FDA panel in early 1989, indicated that a woman's risk of breast cancer apparently increases with the length of time she takes birth-control pills.

At the same meeting, the results of two larger studies were also presented. Bruce Stadel, an FDA staff scientist, presented a reanalysis of an earlier CDC study involving about 5,600 women. Stadel found that not all women who used the pill had a higher risk of breast cancer. However, birth-control pill users who started menstruating before they were 13 years old and who did not have children had a higher risk of developing breast cancer. Women who used birth-control pills for 4 to 7 years had a 30 percent greater risk, while those who used birth-control pills for 12 years or more had a risk 12 times greater than that of the women who had never taken the pills.

The other study, by Clifford Kay and associates of the Manchester Research Unit of the Royal College of General Practitioners in the U.K., followed 46,000 women. They found that women who had taken the pill had three times the usual risk of developing breast cancer between the ages of 30 and 34, but there was apparently no added risk at younger or older ages.

Pending further studies, the FDA panel concluded that no changes in the use of birth-control pills appeared necessary and that warning labels were not warranted. At the same time, the NCI announced that it would conduct another study of 2,000 women to provide a more definitive answer to the questions raised by the new studies, and to examine whether nutrition and alcohol consumption might influence the risk of breast cancer among pill users. The first findings of that study are expected in 1993.

Viruses

Some strains of the sexually transmitted papillomavirus are known to cause cervical cancer. The virus is also responsible for the nonmalignant condition called genital warts, which has been increasing rapidly in the U.S. during recent years. Although this rise has prompted some researchers to warn of an impending epidemic of cervical cancer, University of Minnesota researcher Peter J. Lynch believes that this may not necessarily occur, since only a few of the more than 45 types of human papillomavirus have been strongly linked to cancer. Another type of virus, Epstein-Barr virus, best known as the cause of infectious mononucleosis, was definitively linked to Burkitt's lymphoma, a cancer of the lymph nodes, in the early 1980s. As described in the article on AIDS, people infected with the human immunodeficiency virus sometimes develop certain rare cancers, such as Kaposi's sarcoma and B-cell lymphoma.

Environmental Factors

Radon. Exposure to radon gas in buildings causes between 5,000 and 20,000 cases of lung cancer each year, ranking second only to smoking as a cause of lung cancer, according to a warning issued in 1988 by the EPA and the U.S. Surgeon General's office. At least 8 million homes in the U.S. are estimated to have high radon levels. The government has recommended that Americans check their homes for radon, which can leak into houses through cracks in basement floors and foundations and through openings around pipes and wiring. According to the EPA report, the buildup of high concentrations of radon is easily prevented by sealing foundation cracks and maintaining adequate ventilation under the house or in the basement to allow the gas to escape.

The EPA recommends that people living in homes with radon levels between 20 and 200 picocuries per liter take remedial action within a few years. Immediate action should be taken if levels are above 200 picocuries per liter. The risk of exposure to radon is considered so serious that public health officials in some states have begun special programs to test houses and help people take action to reduce radon levels in their homes. Some scientists have estimated that for every 100 people spending most of their time in a house with radon levels of 20 picocuries per liter, between 6 and 21 would die of lung cancer over a 70-year period.

Other radiation sources. Over the years, researchers have investigated the possibility of a connection between cancer and exposure to the electromagnetic radiation of radio waves, microwaves, and low-frequency waves. In a review article published in late 1988, Roy Shore of the New York University Medical Center in New York City concluded that the association between these forms of radiation and cancer is, at most, weak. Previous studies have shown little, if any, evidence that the microwaves emitted by ovens and transmitting towers have any cancer-causing potential. Studies of workers with high exposures to the low-frequency electromagnetic fields found around high-voltage power transmission lines have shown evidence that they may have a somewhat elevated risk for leukemia, although the results of the studies are generally inconclusive. However, a 1987 report issued by David Savitz of the University of North Carolina in Chapel Hill showed that the risk of cancer among children living near high-power lines is 1.7 times higher than that of other children.

Several reports in 1988 indicated that nuclear weapons plants in the U.S. are a major source of air and groundwater contamination with both radioactive materials and cancer-causing chemicals. Some people living near these plants have claimed that rates of cancer and other illnesses are unusually high in their communities, but these claims have not yet been fully evaluated. The U.S. Department of Energy has estimated that the cleanup of 16 weapons plants could take decades and cost between $66 billion and $110 billion.

Asbestos. The EPA estimated in a 1988 report that one in five commercial buildings in the U.S. has building materials containing easily broken asbestos fibers, which are associated with a high risk of lung cancer. The study concluded that asbestos-containing materials were significantly damaged in 43 percent of the more than 700,000 affected buildings, increasing the likelihood that the fibers would become airborne. Airborne asbestos has repeatedly been shown to be an important cause of lung cancer and other respiratory diseases at manufacturing plants where asbestos is used.

Sunlight. Skin cancers are very common in the U.S., with more than 500,000 new cases expected in 1990, according to the Skin Cancer Foundation in New York City. Fortunately, most skin cancers are completely curable. During the past decade, however, the incidence of malignant melanoma, the most dangerous type of skin cancer, has nearly doubled, a situation some dermatologists have called an epidemic. The lifetime risk of developing melanoma has risen from 1 chance in 1,500 in 1930 to 1 in 150 today. New York University researchers Alfred Kopf, Darrell Rigel, and Robert Friedman have predicted that, by the year 2000, the risk will have increased to 1 in 90. In 1989, the ACS estimated that 28,000 new melanoma cases are diagnosed annually and that approximately 5,800 cancer deaths are due to malignant melanoma each year. Early detection and removal of melanomas could greatly reduce the death rate, since early treatment generally leads to a complete cure.

There is overwhelming evidence that ultraviolet radiation from the sun is the major cause of melanoma and that taking protective measures, such as applying sunscreen while outdoors, could prevent a large number of these tumors. Scientists have long known that skin cancers are most common in parts of the world where sunlight is intense. Light-skinned people, who have little protective skin pigment, also have a relatively high risk of getting skin cancer if they sunbathe regularly, particularly if they have suffered severe sunburns as children or during their teenage years.

Scientists are now studying what implications the depletion of the ozone layer may have for skin cancer rates. The ozone layer in the stratosphere protects life on Earth by preventing most of the Sun's harmful ultraviolet radiation from reaching the Earth's surface. The depletion of the ozone layer that has recently been observed near the North and South poles allows more ultraviolet radiation to reach Earth and could theoretically cause more skin cancer. EPA consultant Janice D. Longstreth has predicted that an additional 154 million skin cancer cases and 3.2 million deaths worldwide will occur among people born before 2075 unless steps are taken to stop ozone depletion.

Many dermatologists believe that tanning parlors may also contribute to the rising skin cancer rate. In the late 1980s, some states began regulating tanning booths to help prevent burn injuries.

Diagnosing Cancer

Radiology. Ordinary x-ray pictures continue to be the most widely used technique for diagnosing cancer. To increase the detail and clarity of x-rays, contrast materials such as barium are sometimes used to highlight particular organs or blood vessels on x-rays. Contrast methods include barium swallow and barium enema to detect cancers in the stomach and colon, arteriography for imaging blood vessels in many organs in the body, and lymphangiography for visualizing the lymph nodes. These methods all use radiopaque "dyes" that show up as darkened or opaque images on the x-ray film. A variation of this technique uses low-level radioactive elements called radioisotopes to track the movement of fluids through different parts of the body. The elements emit radiation, which exposes special types of film.

Marvin H. Chasen of the University of Texas Cancer Center in Houston believes that a new refinement of arteriography called digital radiography may be especially promising. By using computers to enhance radiographic images, the technique can reveal tumors that were previously undetected.

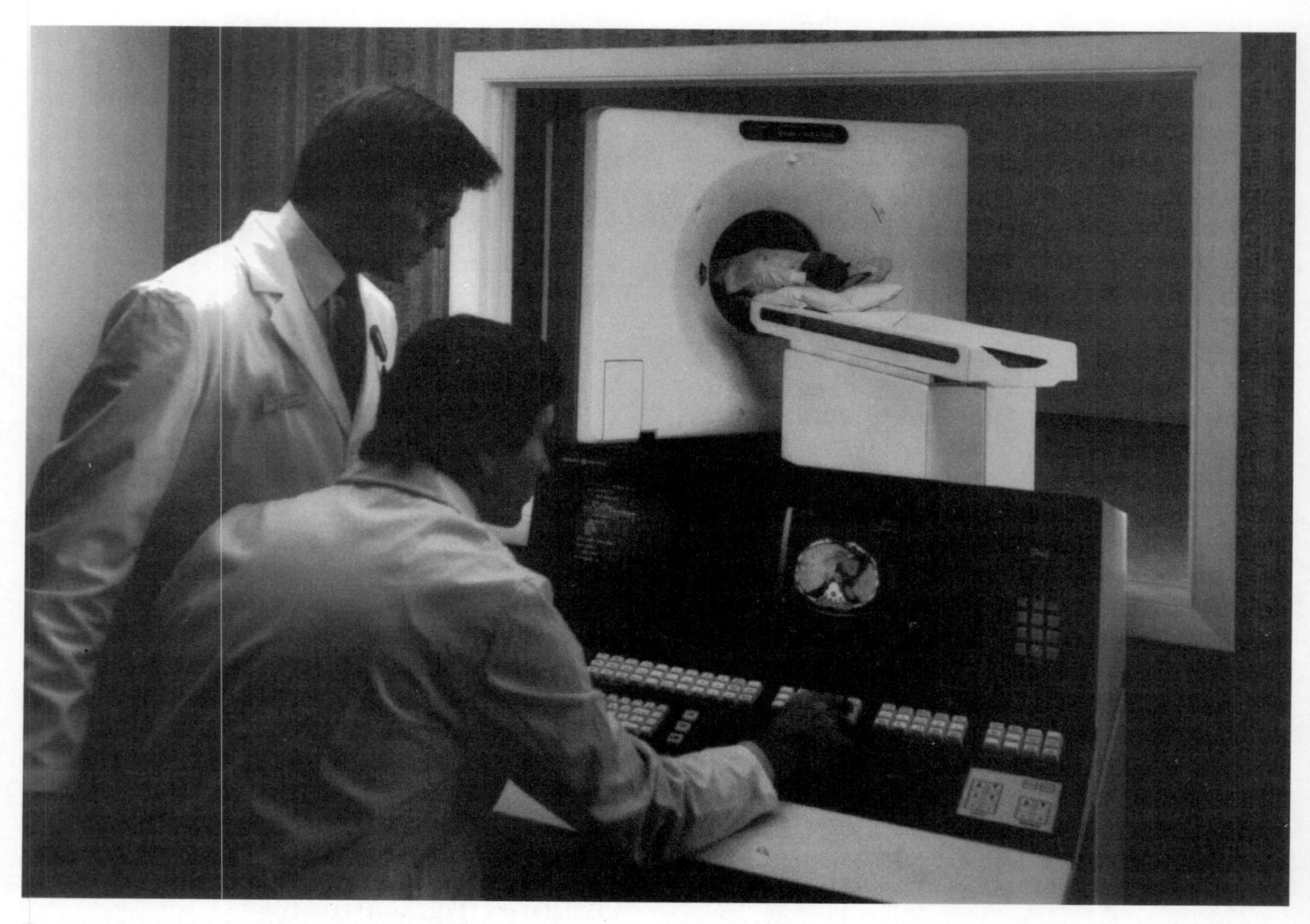

Figure 6. CT Scanner. Computed tomography (CT) scanners measure the amount of x-radiation absorbed by the body when it is exposed to x-rays from many different angles. A computer combines these data to form an image of a cross-section through the body. This technique is often used to detect tumors. *Courtesy: General Electric Co.*

In another technique called computed tomography (CT), an x-ray source is rotated around a patient's body, and a computer is used to analyze the patterns of x-rays after they have passed through the body. The computer produces cross-sectional images showing the structure of the body's internal organs. Abnormal structures, such as tumors, sometimes appear on the scan. Magnetic resonance imaging (MRI), a technique known to chemists for many years, has recently proved its usefulness in medical applications. Chasen believes that MRI "has the potential for surpassing CT in the evaluation of lung cancer." Unlike x-rays or CT scans, MRI allows physicians to readily distinguish blood vessels from possibly malignant tissue. Researcher Ian C. P. Smith of the Canadian National Research Council reported in June 1988 that he had used MRI techniques to correctly identify 27 cancers with no potential to metastasize (spread beyond the original site) in 200 patients with colon or breast cancer.

Ultrasound imaging. Ultrasound — sound at a frequency higher than the human ear can hear — can be used to produce images of tissues within the body. In this technique, an instrument called a transducer is used both to direct ultrasound toward an organ or body cavity and to detect returning echoes. These echoes are processed into an image of the structures within the body. Ultrasound is now used to diagnose cancers in the abdominal lymph nodes, breast, liver, pelvis, and several other sites. The technique can also help physicians direct needles during biopsies of the kidney, liver, and other organs.

Endoscopy. An endoscope is an instrument that generally consists of a fiber-optic tube inserted into the body (either through a natural orifice, such as the nose, mouth, or rectum, or through a small incision) and optical equipment that allows physicians to view organs and body cavities. Endoscopy is now often used to examine the esophagus, stomach, intestines, abdomen, chest, lymph nodes, and reproductive organs. Endoscopes may include equipment to suck or cut cell and tissue samples for analysis. In addition, endoscopes can be used to deliver high-intensity laser light to burn away cancerous cells.

Analysis of cells and chromosomes. Cancers can be diagnosed by analyzing cell samples under a microscope. The Pap test, in which cells from the uterine cervix are removed and examined, continues to be the principal technique for diagnosing cervical cancer. Cell studies are also used to diagnose cancers of the bladder, mouth, respiratory tract, and uterine wall.

Karyotyping (the analysis of a cell's chromosomes for genetic defects) has shown promise as a technique for diagnosing certain leukemias and lymphomas since the early 1980s, but it has not yet become a standard clinical procedure. Certain chromosome abnormalities, such as the exchange of a particular segment of one chromosome with that of another, can be detected through karyotyping.

As recombinant DNA techniques continue to be refined, scientists have found them increasingly useful for detecting genetic aberrations that may be too subtle to be discerned by karyotype analysis. This method has also been used to identify oncogenes and proto-oncogenes. Recombinant DNA techniques are particularly useful for detecting duplications of individual genes — a phenomenon called amplification — and deletions of genes from chromosomes.

Enzyme and chemical levels in the blood. Doctors monitor enzyme levels in the blood to aid in the diagnosis of pancreatic and prostate cancers, and they check calcium levels to detect the presence of bone cancer. Likewise, the levels of two enzymes called lactic dehydrogenase and 5'-nucleotidase are often much higher than normal in patients with liver cancer and can be detected with simple blood tests. Abnormal chemical and enzyme levels in the blood do not provide a foolproof diagnosis, however. Instead, they supplement other diagnostic information and may be useful in monitoring a person's response to treatment.

Markers. Tumor markers are substances that are carried on or produced by cancer cells and are therefore signs that a person has cancer. One such tumor marker, carcinoembryonic antigen (CEA), is present in very high levels in advanced breast, colon/rectal, lung, pancreas, and stomach cancers, as well as in other cancers that have spread to the liver. The discovery of this marker has allowed scientists to monitor colon/rectum and breast cancer patients for signs of recurrence after they have been treated.

Scientists have found that the presence or absence of molecules called receptors that bind the hormone estrogen can provide valuable information in breast cancer cases. In general, breast cancer cells carrying the receptors respond better to treatment, and such cancers recur less rapidly after surgery. Tests to detect these receptors may have valuable diagnostic uses, since they are a good indicator of a breast cancer's probable response to therapy with cytotoxic (cell-damaging) drugs and x-rays, and the likelihood of recurrence after surgery.

Monoclonal antibodies. Biotechnology has enabled researchers to produce substances called monoclonal antibodies that bind specifically to proteins on the surface of cancer cells. Monoclonal antibodies are now being used to detect cancers of the ovary, colon, pancreas, and colon, as well as leukemias and lymphomas. These antibodies detect cancer cells with far greater accuracy than CEA. Monoclonal antibodies specific to molecules associated with tumor cells have also been used to produce images of tumors and to determine the extent to which they have spread. Typically, this has been accomplished by adding radioactive substances to the antibodies, which are then injected into the patient, where they bind to cancer cells. Radiation emissions from the antibodies can be detected to reveal the location of cancer cells.

Because some monoclonal antibodies can distinguish cancer cells from normal cells, researchers have sought to use these substances not only to diagnose cancers but to destroy them. In principle, this can be achieved by attaching a radioactive substance or chemical poison to the antibody, which then delivers the fatal blow directly to cancer cells, without killing healthy cells. By the late 1980s, however, relatively few patients had been treated with this type of therapy.

Treating Cancer

The effectiveness of antibiotics, such as penicillin, in curing infectious diseases such as pneumonia, syphilis, and tuberculosis, is probably responsible for the popular notion that a powerful, general cure for cancer might someday be found. The curative power of these antibiotics, however, is based on the vast biological differences between the disease-causing microbes, such as bacteria, and their human hosts. These differences have made it possible to find a variety of both naturally occurring and synthetic chemicals — the antibiotics — that interfere with the functioning of microbes while causing little or no harm to human cells. The differences between normal human cells and human cancer cells are, however, far smaller and subtler. As a result, chemicals that kill cancer cells generally tend to kill normal cells as well.

Unfortunately, the impact of new cancer therapies in the 1980s has not been large. Consistently effective cures have been found for certain relatively rare cancers, such as certain childhood leukemias and testicular cancer. But the death rates for the most common types of cancer in the U.S. have not changed substantially during the past decade. In 1988, lung cancer, the biggest killer, had only a 13 percent 5-year survival rate after the diagnosis was made. (Five-year survival is generally used as a benchmark for measuring the effectiveness of treatment.) The approximate 5-year survival rates for other common cancers in whites were 53 percent for colon/rectum cancer, 75 percent for breast cancer, and 72 percent for prostate cancer. In general, 5-year survival rates for blacks were significantly lower than for whites.

Surgery. Cancers that have not spread to distant sites through the blood or lymph system can often be cured by surgical removal. Frequently, however, cancers are not diagnosed until after they have begun to spread, making a surgical cure nearly impossible. For example, more than 70 percent of lung cancer cases are detected after the cancer has spread beyond the lung. Only 7 percent of Americans diagnosed with lung cancer that has spread will survive 5 years after diagnosis, compared to 33 percent of people whose cancers are detected before spreading.

Not all cancers spread rapidly. For example, two-thirds of all women diagnosed with cervical cancer survive 5 years or more after the cancer is diagnosed, because this type of cancer usually grows at its original site for many years before spreading. The chance of survival for cervical cancer patients is more than 85 percent if the cancer is detected before spreading.

However, even when cancer does not appear to have spread at the time of surgery, microscopic clumps of cells may have already traveled to distant sites. In fact, more than half of all patients with apparently localized tumors that are surgically removed eventually develop cancers at distant sites, an indication that the tumors had already begun spreading at the time of diagnosis. For this reason, the majority of cancer patients are treated not only with surgery but with radiation, anticancer drugs, hormones, or other techniques. Radiation, usually from an x-ray machine, is used to destroy cancer cells in specific parts of the body. Chemotherapy with cytotoxic chemicals is often used to destroy cancer cells that may have metastasized to distant organs and tissues.

Several ongoing studies are attempting to determine what type of treatment is most effective for breast cancer. Until the 1970s, U.S. physicians had long believed that extensive surgery provided a better chance for long-term survival than more limited procedures. Surgeons generally favored an operation called radical mastectomy, in which not only the breast but also nearby muscle and many underarm lymph nodes are removed. Although the operation was painful and disfiguring, it was thought to offer the only effective means of preventing a recurrence of the disease. In recent years, however, researchers have found that less extreme treatments are sometimes as effective as the radical mastectomy. In the modified radical mastectomy, more of the tissue around the breast is preserved, and in the simple mastectomy only the breast itself is removed.

When breast cancer is detected at an early stage, breast-conserving surgery (also called lumpectomy), may be performed. Many women choose the lumpectomy because it is less disfiguring than the other types of breast cancer surgery. In a lumpectomy, only the tumor and a margin of the surrounding breast tissue are removed. The overlying skin is left intact, and a silicone implant may be inserted to replace the lost tissue. A pathologist carefully evaluates the excised tissue to ensure that the tumor has been completely removed. In addition, samples of lymph node tissue are assessed to establish the stage of the cancer — important information for determining the appropriate treatment. The breast tissue that remains after a lumpectomy is treated with radiation (described below) to destroy any cancerous cells that may remain.

Followup studies have shown that when breast cancer has not spread to the underarm lymph nodes, the prospects for long-term survival are essentially the same after either simple mastectomy or radical mastectomy. Researchers have also found that for women treated in the early stages of breast cancer, the survival rate following lumpectomy with radiation therapy is comparable to that after more extensive surgery.

Radiation therapy. Soon after the German physicist Wilhelm Konrad Roentgen discovered x-rays in 1895, they were found to have the ability to damage and kill human tissues. By the 1930s, scientists had learned to use radiation to treat cancers by repeatedly focusing a powerful beam of x-rays on cancerous cells. Cells exposed to lethal doses of x-rays — hundreds to thousands of times higher than the doses used for diagnostic x-rays — are damaged so extensively that they cannot divide and grow.

In cases of head, larynx, and neck cancers, in which surgery is especially difficult, x-ray treatment may be used to replace surgery altogether; more often, however, it is combined with surgery, chemotherapy, or both. Radiation is also used to treat cancers of the bladder, cervix, prostate, and skin, as well as Hodgkin's disease and certain testicular cancers. After a tumor is removed by surgery, doctors often subject the area surrounding the tumor site to radiation in an attempt to kill any cancer cells that may remain.

A major disadvantage of radiation therapy is that the x-ray beam also damages healthy tissues near the site being treated. X-rays that deliver enough radiation to destroy a tumor could cause so much damage to nearby healthy tissues that they could kill the patient. For this reason, scientists have developed sophisticated ways to target high doses of radiation onto a tumor. In addition to conventional x-ray machines, three other methods of radiation delivery were being used at the end of the 1980s: radioisotopes in the form of radiopharmaceuticals, implanted radiation sources, and particle beams.

In one form of particle-beam therapy, scientists use neutrons — electrically neutral subatomic particles — to bombard tumors that have not responded to conventional radiation therapy. The advantage of neutron beams and other subatomic-particle beams lies in their ability to deliver a large dose of radiation to a deep-lying tumor with only limited harm to overlying tissues. In a review of neutron-beam therapy published in 1988, Frank R. Hendrickson of the Midwest Institute for Neutron Therapy in Batavia, Illinois, reported that about 9,000 patients had been treated with neutron-beam therapy worldwide. Studies have shown that the treatment produced long-term remission in about half of the treated patients with bladder cancer and over 70 percent of the patients with salivary-gland tumors resistant to traditional therapy. Difficult-to-treat prostate cancer has proved to be very sensitive to neutron-beam therapy, according to Hendrickson.

Chemotherapy. Cancers that are not localized in a particular site cannot be treated effectively by surgery or localized radiation. Instead, they are often treated with cytotoxic drugs that are circulated throughout the entire body or injected into a particular organ or part of the body. These cytotoxic chemicals are selected for their ability to kill cancer cells without killing the patient. Most new chemotherapy drugs have been discovered by testing thousands of chemicals to identify promising compounds that fight cancer and can be safely used on patients. By the late 1980s, approximately 40 anticancer drugs had been approved by the FDA.

Though cancer cells are often only subtly different from normal cells, it should be possible, in principle, to develop drugs that exploit just these differences and kill cancer cells while leaving normal cells unscathed. So little is understood about the complex biochemical changes in cancer cells, however, that few researchers anticipate that such drugs will be developed at any time in the near future. Unfortunately, existing anticancer drugs have only limited selectivity and are often highly toxic to healthy cells. To maximize the chance of destroying the cancer, these drugs must be used in the largest tolerable doses — that is, amounts only slightly smaller than what would severely poison the patient. Consequently, chemotherapy has a reputation for causing distressing side effects, such as nausea, vomiting, and hair loss.

A particularly serious side effect of chemotherapy is reduced resistance to infection due to supression of the bone marrow's blood-cell producing function. To counter this problem, a growth factor may be administered to stimulate the production of white blood cells. Scientists use genetic engineering techniques to manufacture large amounts of one such growth factor, called granulocyte-macrophage colony stimulating factor (GM-CSF). Saroj Vadhan-Raj and associates at the University of Texas reported in 1988 that GM-CSF can significantly improve the blood cell counts of patients with advanced malignant cancers. In other studies, the therapy has shown promise for boosting the immune systems of patients with AIDS.

Several anticancer drugs may even cause new cancers in a small percentage of patients, but the immediate benefits of chemotherapy outweigh the much lower risk that a new cancer might develop later.

To improve the effectiveness of chemotherapy and reduce its harmful side effects, drugs are often given in combination. For example, one drug might have general anticancer activity but also cause harm to the liver; another might generally inhibit cancer cells but damage the bone marrow. By combining moderate doses of these two drugs, it may be possible to treat a cancer without causing severe damage to either the liver or bone marrow.

By the late 1980s, chemotherapy had significantly increased the survival of patients with many kinds of cancers, including acute lymphoblastic leukemia, Burkitt's lymphoma, embryonal cell cancer, Hodgkin's disease, neuroblastoma, testicular cancer, and Wilms' tumor. Even when a complete cure is not possible, chemotherapy can be used to shrink cancers to relieve symptoms and prolong life in some cases of breast and prostate cancers, certain lymphomas and leukemias, osteogenic sarcoma, and small-cell lung cancer. Chemotherapy has had the biggest impact on childhood cancers. Over the past two decades, according to NCI statistics, chemotherapy in children younger than 15 years old has reduced cancer mortality by at least 50 percent in leukemia and 30 percent in non-Hodgkin's lymphoma.

Hormone therapy. Hormones released by endocrine cells control the growth of certain tissues in the body. Cancer cells arising in these hormone-responsive tissues sometimes retain the hormone dependence of normal cells. When this occurs, cancer growth can sometimes be restrained by eliminating the normal source of the hormone or by administering chemicals that interfere with its action.

Prostate cancer, for example, is the most common malignancy and the third leading cause of cancer deaths in American men. Since the male hormone testosterone stimulates the growth of prostate cancer cells, most treatments for this cancer attempt to reduce testosterone levels. This is accomplished either by removing the testicles, which produce testosterone, or by using hormones or other drugs to stop testosterone production. Monthly injections of one promising long-acting drug called goserelin acetate have been found to reduce testosterone levels, relieve symptoms, and bring about remission in some prostate cancer patients. One study led by physician Ian Holdaway of Auckland Hospital in New Zealand found that about half of a group of patients treated with monthly goserelin acetate injections had either stabilized or were in remission 1 year after beginning treatment. The researchers also found that the drug caused few side effects. However, further research is still needed to determine whether goserelin acetate used alone or in combination with other drugs will be more effective than existing treatments for prostate cancer.

A hormonal treatment that has proved useful for treating breast cancer is the drug tamoxifen, which binds to estrogen receptors. Tamoxifen has been approved as a preventive followup treatment for women with breast cancer. In a 1988 review article assessing progress in tamoxifen therapy, Sewa S. Legha of the M. D. Anderson Hospital and Tumor Institute in Houston, Texas, concluded that the drug is mainly useful in controlling cancers whose cells bear estrogen receptors. For reasons not yet clear, the drug is also most effective in postmenopausal women whose breast cancers have spread to the lymph nodes. Its use in younger women remains controversial, Legha said.

Boosting the immune system. For many decades, immunologists have hoped to enlist the body's natural defense mechanisms to fight cancers. Ideally, the immune system would learn to recognize cancer cells and attack them specifically, unlike toxic chemicals, x-rays, or other agents, which kill many healthy cells along with the targeted cancer cells.

Although early attempts to boost the immune system failed, several research groups, including Ralph Freedman and his colleagues at M. D. Anderson Hospital and Tumor Institute, are now working on this approach. In 1988, they reported that an extract made from an influenza virus appeared to enhance immunity to cancer, much as a vaccine boosts a person's immunity to mumps or measles. After injecting a mixture of virally infected cancer cells grown in the laboratory, Freedman found that tumor size was reduced in a small number of women with advanced ovarian cancer.

Another approach of vaccination to boost immunity has been used in the treatment of advanced melanoma. Jean-Claude Bystryn of New York University, one of the pioneers of this technique, reported positive results in early 1988. The vaccine was derived from four cell lines of melanoma cells incubated in the laboratory. The vaccine significantly boosted the immune systems of nearly two-thirds of 55 patients with advanced malignant melanoma, but the researchers are not yet sure whether this immune response will actually improve the patients' prospects for long-term survival.

One immune-boosting treatment that has received a great deal of attention in the scientific and lay press has been interleukin-2, a molecule secreted by white blood cells that can have powerful effects on other cells in the immune system. In 1988, NCI researcher Steven A. Rosenberg reported on a study of 221 patients with advanced cancers of various types treated in a way designed to reduce side effects. The treatment was successful in only a small number of patients with relatively few types of cancer. Sixteen patients went into complete remission and 26 experienced partial regressions (at least a 50 percent reduction of the tumor) with this treatment.

Rosenberg and his team are investigating the combined use of interleukin-2 and tumor-infiltrating lymphocytes, which are potent cancer-fighting cells, to improve the efficacy of this treatment. Tumor-infiltrating lymphocytes are taken from a cancer patient and then grown in a laboratory in larger numbers. The enhanced cells are then reinjected into the patient, where the cells search out and destroy cancer cells. Rosenberg believes that researchers may someday be able to make tumor-infiltrating lymphocytes even deadlier by splicing a gene for alpha-interferon or interleukin-2 into the lymphocytes so that they will produce and deliver these cancer-fighting substances directly to tumor cells.

Alzheimer's Disease

Only in the last 15 years have physicians and scientists recognized that memory loss and gradually worsening speech difficulties in older Americans are not part of the natural aging process, but instead can be key symptoms of an illness called Alzheimer's disease. This disease robs hundreds of thousands of people of their memories and minds each year, and it causes more than 100,000 deaths annually — making it one of the leading causes of death in the U.S. Although researchers have not yet developed a cure or even an effective treatment for this condition, their knowledge about the underlying biology of the disease has expanded rapidly during the past few years.

The medical community once thought Alzheimer's disease was a relatively uncommon form of dementia — the loss of intellectual capacity. By the late 1980s, however, doctors accepted Alzheimer's disease as the single most important cause of dementia, accounting for about two-thirds of all cases. The disease afflicts 4 percent of people between 65 and 74 years old, and the proportion may increase to as much as 30 percent among people 85 or older. Altogether, between 2.5 and 3 million Americans suffer from the condition, according to the National Institute of Aging (NIA) in Bethesda, Maryland.

Physicians still cannot easily diagnose Alzheimer's disease because its symptoms are like those of many other conditions and because there is no one test to confirm the disease, facts that explain why it took so long for the medical community to recognize the scope of the disease. Even when a doctor notes the typical symptoms of Alzheimer's disease, he or she must also perform an extensive neurologic evaluation to rule out other causes of the symptoms, including depression, adverse drug reactions, nutritional deficiencies, head injury, stroke, liver disease, and vitamin B_{12} deficiency, among others.

Alzheimer's disease patients live an average of about 8 to 10 years after the first symptoms of the disease appear, but individual patients may live much longer — up to 25 years. Many of these people cannot take care of themselves and must be institutionalized at an annual cost, according to NIA, of more than $13 billion in 1985; nearly half of all persons admitted to nursing homes are there because of Alzheimer's disease. With an additional $31 billion spent in 1985 by families on home care and $43 billion lost in the reduced economic activity of family caregivers, Alzheimer's disease clearly imposes a heavy burden on the U.S. economy.

Since other advances in modern medicine will continue to lengthen the lifespans of people living in the U.S. and other industrialized countries, the number of elderly people suffering from Alzheimer's disease is expected to skyrocket by the turn of the century. Gene H. Stollerman, editor of the *Journal of the American Geriatrics Society*, observed this cruel twist of fate in the foreword to the book, *Clinical Management of Alzheimer's Disease*: "Having conquered many of the [medical] disasters of our lifetime ... we now face the blighting of the promise of a healthy life that can now extend with increasing frequency into the seventies, eighties, and even nineties by a condition that robs us of our very personhood in old age, just when we should be enjoying its full glory." If scientists do not find a cure or effective treatment, the number of Americans with Alzheimer's disease and related disorders could increase to 7.4 million by the year 2040, according to a recent report issued by the U.S. Office of Technology Assessment.

On the positive side, these and other facts have stimulated a substantially increased level of funding for research into the disease, with funding jumping from about a few million dollars per year in the mid-1970s to $120 million in 1989. The new research projects have uncovered many potentially effective approaches for treating Alzheimer's disease.

Symptoms and Diagnosis of Alzheimer's Disease

The symptoms of Alzheimer's disease follow a pattern of progressively worsening impairment. At first, a person may occasionally forget or have trouble learning new information such as phone numbers or names. Finding words during normal speech becomes increasingly difficult, and families may seek medical attention for the person during this phase, perhaps after he has become lost while traveling a familiar route.

Later, the person suffers even greater confusion and more serious amnesia, and may begin having delusions and become agitated. A phenomenon called "sundowning" may occur, with the person showing little confusion during the daytime, but becoming quite confused, agitated, and difficult to manage during the night when visual cues diminish. As the disease progresses, the person becomes increasingly childlike and self-centered and cannot remember recent events, although memories from childhood and early life may still be clear.

People with advanced Alzheimer's disease cannot read, write, calculate, or care for themselves. They become irritable, suffer rapid swings of emotion, and often experience hallucinations and delusions of persecution. The patient loses almost all of the ability to speak and to control urinary and fecal functions. Perhaps the most traumatic time for family members is when the patient becomes bedridden, and the family must endure not only emotional strain but also the enormous financial cost of institutionalization — typically over $20,000 per year. Although the number of private insurance policies that cover long-term care for Alzheimer's disease and other chronic illnesses has increased in recent years, the cost of institutional care remains prohibitively expensive for most American families.

Despite the lack of a specific test for Alzheimer's disease, researchers have shown that expert neurologists make the correct diagnosis about 90 percent of the time, particularly when they obtain brain scans with computed tomography or magnetic resonance imaging. Such scans can help physicians rule out brain tumors and other causes of memory loss and dementia.

A completely conclusive diagnosis of the disease, however, requires examination of a brain tissue sample, taken during an autopsy or during brain surgery for other reasons. Brain tissue affected by Alzheimer's disease will reveal a characteristic pattern of damage: Many neurons — nerve cells — will have died, particularly in parts of the brain responsible for memory and thought processes, and large numbers of neurofibrillary tangles and amyloid plaques will be present in scattered areas of the brain, particularly in regions responsible for memory and thought processes.

Neurofibrillary tangles consist of two protein filaments, called paired helical filaments, which wrap around each other. Masses of disorganized tangles indicate that the normal microscopic architecture of a nerve cell has been altered or destroyed. Amyloid plaques are composed of a core of amyloid protein and debris from dead nerve cells. Amyloid may also accumulate inside blood vessels, and it is thought to cause some strokes in the elderly.

Although plaques, tangles, and other structural abnormalities can also be found in the brains of elderly people who are not suffering from dementia, they are usually present in larger numbers and are more widely distributed throughout the brain in Alzheimer's disease. Since doctors usually cannot take a sample of brain tissue from a living patient, it is not possible to document these changes in a patient who otherwise appears to have Alzheimer's disease. Thus, doctors are never totally certain that they have made the right diagnosis, although brain scans have greatly increased their confidence.

Causes of Alzheimer's Disease

During most of the 80 years since the German neurologist Alois Alzheimer first identified the disease bearing his name, physicians attributed its cause to the effects of aging or to hardening of the arteries, a condition in which brain cells die because they receive insufficient blood. Both of these theories were discredited during the 1960s, when researchers who performed autopsies on a large number of elderly institutionalized patients discovered that patients who had suffered from dementia often had many plaques and tangles in their brains and that people without dementia did not usually have a large number of plaques and tangles. These findings made researchers realize that many, if not most, of the elderly patients who had been labeled as simply senile or as suffering from hardening of the arteries actually had the same disorder that Alzheimer had described in 1907.

But finding the relationship between dementia, plaques, and tangles did not explain what caused Alzheimer's disease. Scientists began searching intensively for potential causes and eventually proposed several major hypotheses about the disease. Despite some striking differences among these models, each is supported by a growing amount of experimental evidence, and each suggests different avenues for therapy or cure.

Genetic susceptibility. Once scientists accepted evidence that the dementia associated with Alzheimer's disease was not a normal part of aging, they began looking for patterns of the disease within families. In one of the first such studies, Leonard L. Heston and coworkers at the University of Minnesota in Minneapolis reported in 1981 that as many as ten members in four or five generations of one family could succumb to Alzheimer's disease. Researchers have found that about 5 to 10 percent of Alzheimer's disease patients have a familial form of the disease that is dominantly inherited; that is, a person who

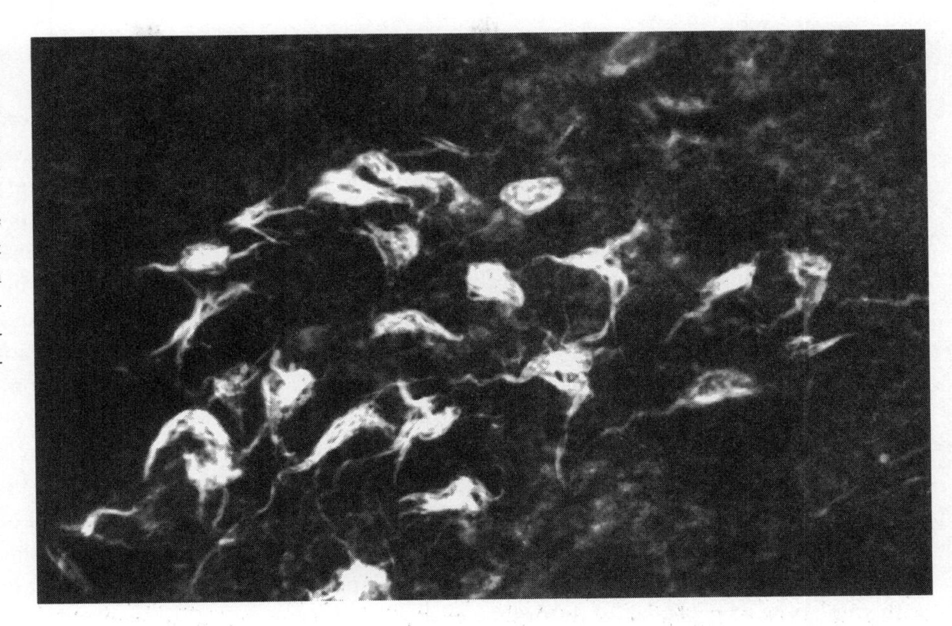

Figure 7. Neurofibrillary Tangles. The white objects in this photomicrograph are neurofibrillary tangles in the portion of the brain that serves as an input system to the hippocampal formation — an important memory-related structure. Such tangles disrupt the flow of nerve impulses. Some researchers believe that this structural damage may play a role in the memory impairment that characterizes Alzheimer's disease. *Courtesy: Bradley Hyman, Gary Van Hoesen, and Antonio Damasio, University of Iowa.*

has one parent with familial disease has a 50 percent chance of inheriting the condition. In patients with familial Alzheimer's disease, the symptoms often appear earlier than usual — typically by age 45 to 50, rather than in the sixties, seventies, and eighties for people with nonfamilial disease. The inheritance of early-onset Alzheimer's disease appears to be linked to a gene on chromosome 21. Researchers have also identified another gene on chromosome 21 that governs amyloid production in all people. It is not yet clear, however, whether this second gene plays any role in Alzheimer's disease.

Some researchers are intrigued by the possible connections between chromosome 21 and Alzheimer's disease, because chromosome 21 abnormalities are also present in another neurological condition, Down's syndrome, which is a major cause of mental retardation. About 95 percent of people with Down's syndrome have an extra copy of chromosome 21, and most of the remaining 5 percent have genetic defects involving the chromosome. Neuroscientists have long known that, like Alzheimer's disease patients, people with Down's syndrome have large numbers of neurofibrillary tangles in their brains. Future studies of the similarities and differences in the genetic makeup of people with Alzheimer's disease and Down's syndrome may lead to new insights into both conditions.

Protein abnormalities. The neurofibrillary tangles and the neuronal plaques characteristic of the brains of people with Alzheimer's disease contain unusual proteins absent from the brains of young healthy people. Researchers have found that the filaments that make up the tangles contain many different proteins, but most notably one called A68, which occurs in large amounts in the brains of Alzheimer's disease patients. When researchers first isolated A68, they hoped that it would provide a marker that could easily be detected by a blood test or spinal tap. But further studies identified the same protein in the neurons of people afflicted by other diseases, as well as in normal brains of infants under 2 years old.

Neuronal plaques were also studied intensively during the late 1980s, with scientists discovering that they contain deposits of minute fibers composed of amyloid. Although researchers have linked amyloid production to genes on chromosome 21, they still do not understand the mechanism that controls the accumulation of amyloid deposits.

Nevertheless, scientists speculate that amyloid deposits form when the normal protein-processing machinery within the tissue malfunctions, perhaps as a result of a genetic mutation. Other researchers have also suggested that key enzymes called proteases, which break proteins into smaller parts, may be a missing cog in this machinery. Thus, plaque formation may stem from the failure of some subtle "house-cleaning" apparatus in the brain that normally clears away the protein debris that accumulates during a cell's life.

Some recent research has focused on the protein parent of amyloid, amyloid precursor protein (APP). William M. Pardridge and associates at the University of California at Los Angeles (UCLA) School of Medicine, have studied APP to locate the precise area where the protein-processing machinery breaks it apart to form the amyloid found in plaques. Analyses of blood samples and spinal taps from Alzheimer's disease patients and normal subjects have found similar levels of APP. This finding suggests that the cellular machinery controlling APP breakdown somehow malfunctions in Alzheimer's disease patients and causes amyloid deposits. Pardridge and other researchers speculate that one or more proteases are responsible for this malfunction. The presence of a protease inhibitor — a substance that would stop a protease from breaking up a protein — might lead to inadequate breakdown of APP, causing an amyloid deposit to accumulate.

Pardridge views the amyloid deposits inside and outside nerve cells as "the primary event" in Alzheimer's disease. At the 1988 annual meeting of the American Association for the Advancement of Science, Alzheimer's disease researcher Donald Price of the Johns Hopkins School of Medicine said that the role of the proteases in amyloid deposits "will clearly be a major area of research in the future." Finding a way to control amyloid deposits with some type of protease inhibitor may eventually lead to an effective treatment for Alzheimer's disease. Pardridge believes that a yet-to-be-identified gene defect may cause a protease-inhibiting enzyme to malfunction, leading to an overproduction of amyloid.

Environmental toxins. As with cancer, respiratory diseases, skin diseases, and other conditions, scientists have long suspected that an environmental toxin may contribute to Alzheimer's disease. In 1973, Donald R. Crapper-McLachlan of the University of Toronto in Ontario, Canada, reported that brain tissue from people with Alzheimer's disease contained two to three times the normal amount of aluminum, with the highest concentrations being in tissue near the neurofibrillary tangles. When other researchers could not produce similar results, however, the idea of a link between aluminum and the disease fell out of favor until the 1980s, when it once again attracted serious interest. During a 1988 conference on Alzheimer's disease and related disorders in Las Vegas, Nevada, several research groups presented evidence that aluminum is found along with amyloid at the core of neuronal plaques as well as in the neurofibrillary tangles. Toronto's Crapper-McLachlan and coworkers reported that high amounts of aluminum also collect on the genetic material in some brain cells of people with Alzheimer's disease. It is still unclear whether the accumulation of aluminum is a contributing cause of the disease or is perhaps a result of it.

Some investigators have conducted population studies that appear to show a link between the amount of aluminum in drinking water and the prevalence of Alzheimer's disease. One British study, for example, determined that the rate of Alzheimer's disease was 1.7 times higher in regions with high levels of aluminum compared to areas with low levels.

It would be practically impossible for people in industrial countries to avoid contact with aluminum, because it is used in a wide variety of consumer goods, ranging from deodorant products and skin creams to instant chocolate mixes. Nevertheless, scientists know that the human body absorbs some forms of aluminum more readily than others. Aluminum cans and pots, once suspected of being a major aluminum source, are no longer considered such because they add only about 10 to 15 percent to the natural levels of aluminum found in food.

Aluminum has been linked to dementia among Japanese living on the Kii Peninsula in Japan, the Chamorro tribe on the island of Guam, and residents of a small part of New Guinea. The soil and water in each of these areas have a very high concentration of the metal.

Crapper-McLachlan's group and others have attempted to treat Alzheimer's disease with chelating agents — chemicals that bind to aluminum and other metals and remove them from the body. Theo Kruck of the University of Toronto reported in 1988 that the chelator deferoxamine selectively removes aluminum from the brain in Alzheimer's disease patients. Early findings from a study of the drug's therapeutic use have shown improvement in some people exposed to high amounts of aluminum, but final judgment awaits completion of the study in 1990.

Infectious agents. Although Alzheimer's disease patients do not have overt signs of viral infection, such as fever, at least a dozen researchers have postulated that Alzheimer's disease may be linked to an infectious agent. Although many scientists remain skeptical because no such agent has yet been isolated, that does not necessarily rule out the possibility that an unidentified virus may cause the disease.

The infectious agent hypothesis gained support when researchers found an amyloid protein called the prion protein in three Alzheimer's-like dementias, Creutzfeldt-Jakob disease, scrapie, and kuru. Each of these conditions is caused by an infectious agent, but none produces the conventional signs of an infection. In all three diseases, there is a long incubation period — usually at least one-quarter of the victim's lifespan — between exposure to the agent and the appearance of symptoms. The agents responsible for these conditions are called slow viruses, because of their long incubation periods.

In 1988, a group of researchers at the Yale University School of Medicine reported new evidence to suggest that at least some cases of Alzheimer's disease may be caused by a slow virus. The Yale group took samples of white blood cells from 11 volunteers who had close relatives with Alzheimer's disease and then injected the cells into the brains of young hamsters. The researchers found that cells from 5 of the 11 volunteers appeared to cause Alzheimer's-like symptoms in the hamsters after incubation periods averaging about 1 year. When brain tissue was taken from the affected animals and injected into the brains of healthy animals, those animals also developed brain degeneration, thereby providing further evidence that a virus or other disease agent was responsible for the condition.

However, even if a virus is involved in some cases of Alzheimer's disease, this does not mean that the disease is spread by everyday contact or through blood transfusions. On the contrary, disease experts have found no evidence that Alzheimer's disease can be transmitted in these ways. The Yale team has speculated that the virus might be common to many or all people, and that other factors, such as genetic susceptibility or some life experience, might be required for the disease to develop.

Reduced blood flow. Scientists using noninvasive imaging techniques to study the brains of Alzheimer's disease patients have found not only a reduced blood supply to various parts of the brain, but also a correlation between the drop in blood flow and the severity of a patient's symptoms.

Researchers have found amyloid deposits similar to the plaques seen on the surface along the walls of certain blood vessels. Much as fatty deposits can block blood flow to the heart, these amyloid deposits appear to reduce blood flow within the brain. Pardridge and colleagues at UCLA reported in 1987 that the composition of the amyloid protein found in these blood vessels is similar to the amyloid proteins found in the neuronal plaques and neurofibrillary tangles. If researchers find a way to block amyloid accumulation in nerves, perhaps with a protease inhibitor, they may also be able to prevent amyloid deposits in blood vessels.

Reduced acetylcholine production. The first clear biochemical abnormality associated with Alzheimer's disease was reported independently in 1976 by Peter Davies, now at the Albert Einstein College of Medicine, and David Bowen of the Institute of Neurology in London. Both men found unusually low amounts of the enzyme choline acetyltransferase in two areas of the brains of Alzheimer's disease patients, the hippocampus and the cerebral cortex. Choline acetyltransferase is involved in the synthesis of the neurotransmitter acetylcholine, which transmits messages between neurons. The reduction in the enzyme level was proportional to both the extent of memory loss and the number of amyloid plaques.

Interestingly, the hippocampus and the cerebral cortex play major roles in the learning of new information, memory, mood, and many other aspects of behavior. Since Alzheimer's disease patients invariably suffer memory loss, researchers have extensively tested the acetylcholine-deficiency hypothesis. In the late 1980s, studies continued to confirm and extend Davies and Bowen's original findings, and they suggested that the cognitive impairment might be reduced by increasing the concentration of acetylcholine in the brain. Because acetylcholine itself breaks down too rapidly in the bloodstream to be used as a drug, researchers have looked for other compounds that will stimulate acetylcholine production in the brain.

A large trial of one such drug called tetrahydroaminoacridine (THA) is being conducted at a number of research centers across the U.S. Preliminary results suggest that THA may temporarily improve some symptoms in Alzheimer's disease patients. The trial was suspended for a time in 1987 when researchers became concerned about possible liver damage from the drug, but with a reduction in dosage, the trial soon resumed.

Other researchers believe that a protein called nerve growth factor may help keep nerve cells from degenerating and thus slow the progression of at least some Alzheimer's disease symptoms. Found at low levels in most animals, nerve growth factor stimulates new growth of acetylcholine-producing nerves.

Many researchers think that future treatment may involve transplanting healthy or genetically altered cells that make the growth factor directly into the brains of Alzheimer's disease patients. Scientists have already successfully transplanted fetal cells that make another neurotransmitter, dopamine, into the brains of patients with Parkinson's disease, although the results have been mixed. Using a variation of this technique, Lars Olson of the Karolinska Institute in Stockholm, Sweden, found that when he transplanted into mice genetically altered cells that produce nerve growth factor, the cells did stimulate new growth of acetylcholine-producing neurons.

Although other therapies designed to increase acetylcholine production have also been studied, two prominent candidates, physostigmine and lecithin, have failed to live up to expectations. At one time, physostigmine seemed to hold great promise as a treatment for memory loss, but while its use improved memory somewhat in about one-quarter to one-half of the patients tested, other symptoms of dementia progressed unabated. Numerous trials using lecithin — a fatty substance found in the blood and bile, as well as brain, nerves, and other animal tissues — showed that it could not consistently halt memory loss.

A successful treatment for Alzheimer's disease will probably not be developed at any time in the near future. None of more than 50 drugs now being studied in clinical trials worldwide appears to be the breakthrough cure for which scientists and patients hope. Until more effective therapies become available, physicians will continue treating the individual symptoms of the disease in an effort to improve their patients' quality of life.

Human Genetics

DNA probes — short segments of DNA used to study the genetic composition of organisms — provide a powerful way of identifying genetic defects that cause human disease. Since the late 1970s, researchers have identified DNA sequences, called genetic markers, that have been linked to a variety of diseases. The markers make possible the prenatal diagnosis of certain inherited disorders, and they can also help researchers hone in on the defective genes that cause inherited diseases. Identifying the cause of particular diseases may ultimately lead to new forms of treatment.

Finding Genetic Links to Disease

To use genetic probes, researchers break the DNA from cells into small fragments with substances called restriction enzymes. (For more information on restriction enzymes, see the article "Biotechnology Techniques" in the "Biology" chapter.) The fragments are separated on a gel by electrophoresis, a process in which fragments of different sizes migrate across the gel at different rates when subjected to an electric field. The fragments are then immobilized on a membrane and incubated with radioactively labeled genetic probes to reveal characteristic patterns. By comparing the electrophoretic patterns of the individual members of a family with a high incidence of a specific genetic disorder, it is possible to identify DNA fragments found only in the family members with the disorder. These fragments are markers for the disorder.

In many cases, a genetic disease will involve the substitution of one base for another in the DNA sequence of the defective gene. If that substitution has occurred in a sequence recognized by a restriction enzyme, the enzyme will not be able to cleave the gene at that point. The researcher might then observe two bands in the electrophoretic pattern for healthy people but only one band in the pattern for diseased individuals. Once again, the probes enable the researcher to tie the genetic defect to a specific region on a chromosome.

Since the DNA sequence recognized by a particular restriction enzyme represents such a small fraction of a gene's total DNA, it might at first seem highly unlikely that one could manage to cleave the gene precisely at the site of the defect. Fortunately, many different types of restriction enzymes are available, each capable of recognizing a specific nucleotide sequence. By using large numbers of restriction enzymes, researchers can usually find at least one enzyme that will produce a difference in the electrophoretic patterns of healthy and diseased people. These restriction enzymes can then be used to screen cells for genetic defects. This technique is especially useful because the researcher does not need to know what the specific defect is.

Muscular dystrophy. Genetic screening techniques have revealed the biochemical cause of Duchenne muscular dystrophy. A team headed by human geneticist Louis Kunkel of Children's Hospital in Boston, Massachusetts, reported in late 1987 that victims of this hereditary disease lack a key protein that helps trigger muscle contraction.

Duchenne muscular dystrophy, the most serious form of muscular dystrophy, strikes one in every 3,500 males in the U.S. The disease, which rarely affects females, is usually diagnosed when muscle weakness develops in children between ages 3 and 5. The continuing muscle weakness almost always leaves victims confined to wheelchairs by age 11, and people usually die before age 30 from respiratory failure. There is no effective therapy for the disease.

The discovery of dystrophin, the missing protein in Duchenne muscular dystrophy, was made possible in 1986 when Kunkel's team identified part of the defective gene responsible for the disease. The researchers found that a DNA probe was accurate 98 percent of the time in predicting which fetuses would develop the disease. The discovery gave new hope to the sisters of people with Duchenne muscular dystrophy, since these women often carry the gene for the disease and are therefore at risk for having a child with the condition. Until recently, many sisters of victims routinely aborted male fetuses rather than take the chance that they would have sons with the disease. With the new test, it is now often possible to tell whether a woman carries the gene.

To search for the actual protein that is disrupted in Duchenne muscular dystrophy, Kunkel's group determined the DNA sequence for a part of the gene, determined what protein fragment was coded for by that part, and then synthesized that portion of the protein. Next, they injected the synthesized protein fragment into sheep and mice, which made antibodies against it. The group then used the antibodies as a probe on cell samples to search for a protein coded for by the gene. Kunkel found that the antibody bound to the protein from muscle cells of healthy individuals, but not to the corresponding cells of two boys with Duchenne muscular dystrophy. The antibody also did not bind to muscle cell protein from mice with a muscular dystrophy-like disease. Kunkel concluded that the antibody binds to a protein that is not produced in muscular dystrophy victims because the gene that should code for it is defective.

Kunkel and his colleagues then began looking for the protein's precise location within the muscle cells and found that the protein is located in the muscle cell membrane. When this protein (dystrophin) is absent, muscle cells malfunction and eventually die. Once dystrophin's role is known more precisely, researchers may be able to identify other proteins that can take its place. Kunkel noted that dystrophin is present in extremely small concentrations, 0.002 percent of total muscle protein, which explains why it had not been observed before.

Kunkel's team also studied muscle cells from patients with Becker's muscular dystrophy, a less severe and rarer form of the disease that occurs in about 1 in 35,000 male births. Most victims of Becker's muscular dystrophy also become wheelchair-bound, but usually not until later in life. They also generally live longer than people with Duchenne muscular dystrophy. Kunkel's group found that patients with Becker's muscular dystrophy have some dystrophin in their muscle cells, but not as much as in healthy individuals. This discovery may help distinguish between patients with these two forms of the disease, a distinction that is now often quite difficult to make. As with Duchenne muscular dystrophy, there is no cure for Becker's muscular dystrophy, but patients may be comforted to know they will not be as severely disabled. Some way might also eventually be found to cause their cells to produce more dystrophin.

Cystic fibrosis. Scientists have been conducting an intensive search for the gene that causes cystic fibrosis, one of the most common fatal inherited diseases. Cystic fibrosis is characterized by excessive secretion of mucus that clogs the lungs and impairs breathing and by an insufficiency of digestive enzymes. About one in every 20 people in the U.S. carries the defective gene for cystic fibrosis. Only people who have inherited the defective gene from both of their parents develop the disease, which affects about 30,000 people in the U.S. Geneticist Raymond White of the University of Utah in Salt Lake City reported in late 1985 that a probe for a gene located on chromosome 7 and associated with the onset of leukemia was about 95 percent accurate in identifying cystic fibrosis prenatally in about 80 percent of families in which one child had already been born with the disease. Several laboratories later began experimental prenatal testing for cystic fibrosis in fetuses carried by women who had already given birth to a child with the disease.

In 1987, a team headed by Robert Williamson of St. Mary's Hospital Medical School in London reported that they had identified either the gene responsible for cystic fibrosis or a DNA segment very close to it on chromosome 7. The discovery, Williamson says, extends the accuracy of prenatal testing to nearly 99 percent in about 90 percent of families in which a child has had cystic fibrosis. In August 1989, a research team led by Francis Collins of the University of Michigan in Ann Arbor and Lap-Chee Tsui of the Hospital for Sick Children in Toronto reported that it had identified and cloned a gene for cystic fibrosis on chromosome 7.

Tourette syndrome. A wide range of behavioral disorders may be caused by the genetic defect responsible for Tourette syndrome, according to geneticist David Comings and psychologist Brenda G. Comings of the City of Hope National Medical Center in Los Angeles, California. In a series of papers published in late 1987, the researchers said that the defect, for which they discovered a marker, may also be involved in hyperactivity, conduct problems, stuttering, dyslexia, panic attacks, phobias, depression, sleep problems, compulsive behavior, and mania.

The Comingses believe that the genetic defect responsible for Tourette syndrome is surprisingly common, occurring in one of every 83 people, although not everyone who has the defect has Tourette symptoms. The disorder's best-known symptoms are tics — repetitive movements such as eye-blinking or finger-tapping — and compulsive shouting of obscenities. But the new results indicate that many people who do not have these symptoms may also suffer from the syndrome.

The most important implication of the Comingses'

findings may be the recognition that people with many of the disorders linked to the gene cannot be treated with psychotherapy alone. The Comingses concluded that because Tourette syndrome is caused by a chemical imbalance in the forebrain, psychotherapy will be ineffective unless it is accompanied by drug therapy to relieve the imbalance. One of the most common symptoms of Tourette syndrome in children is hyperactivity, in which students are disruptive in class and cannot concentrate, sit still, or complete an assignment. Hyperactivity is often treated with low doses of stimulants, particularly methylphenidate, but such drugs tend to worsen the tics and other symptoms of Tourette syndrome.

The Comingses' research is controversial, however. Geneticist David Pauls of Yale University reported results of studies similar to those performed by the Comingses at the November 1988 meeting of the American Society of Human Genetics in New Orleans. Pauls's conclusions were far different from those of the Comingses: His group found that the conditions identified by the Comingses were no more common among Tourette victims than among the population at large. David Comings presented evidence at the same meeting supporting his original findings, sparking a heated discussion among geneticists at the session. Researchers expect that the dispute will be resolved in the near future, since six different research groups are now seeking the gene responsible for Tourette syndrome. Once the gene is found, researchers will be able to test individuals with the other conditions identified by the Comingses to determine whether they also have the defective gene.

Gaucher's disease. Researchers have discovered the defective gene that causes Gaucher's disease, an inherited disorder affecting about 1 in every 600 Jews of Eastern European descent. Victims of Gaucher's disease have a deficiency of an enzyme called glucocerebrosidase, which breaks down a specific fat in certain body tissues. In the absence of the enzyme, the fat collects in the cells, causing complications that can include an enlarged spleen and liver, bone deterioration, and neurological problems. Victims of the most severe form of this disease die before reaching the age of 3.

In early 1987, geneticist Edward I. Ginns and his colleagues at the National Institute of Mental Health reported that they had identified a gene defect that apparently plays a role in the disease. The researchers found that the gene was defective in about 15 percent of people with the mildest form of Gaucher's disease and in more than 85 percent of those with the most severe form. On the basis of his studies, Ginns believes that it should be possible to identify people who will develop the most severe form of the disease. By late 1988, Ginns's group had used genetic engineering techniques to insert the healthy form of the gene into cultured cells taken from people with Gaucher's disease and showed that it corrected the biochemical deficiency in the laboratory. They were also attempting to achieve the same feat in the bone marrow of mice with the disease. If the engineered bone marrow cells can be successfully reimplanted in the mice, they may produce blood cells that secrete the missing chemical.

Other conditions. Two research teams, one at Harvard Medical School and the Massachusetts General Hospital and the other at the University of Utah, independently reported in 1987 that they had identified the location of the gene that causes von Recklinghausen's neurofibromatosis, commonly known as "Elephant Man's disease." The condition, which is characterized by benign lumps on the skin and nerve-cell tumors called neurofibromas, affects about 100,000 Americans. The two groups reported that the gene is located in the middle of chromosome 17. The researchers speculated that the identification of the gene could lead to the first treatment for the disease and perhaps eventually to a cure.

In the past few years, researchers have also made progress toward isolating genes responsible for a variety of other genetic diseases. These include bilateral acoustic neurofibromatosis, a rare disease affecting the nervous system; Friedreich's ataxia, which causes muscle weakness and loss of coordination and affects speech and the heart; and several mental disorders, including schizophrenia, Huntington's disease, and manic-depressive illness.

Human Genome Initiative

When geneticist Raymond White of the University of Utah announced in 1985 that he had narrowed the search for the defective gene that causes cystic fibrosis to a small area on chromosome 7, he predicted that the gene would be found within 2 years. But the task proved much more difficult than he imagined, and it was not until August 1989 that the discovery was finally announced. According to Robert K. Dressing, president of the Cystic Fibrosis Foundation, if White had possessed the complete genetic blueprint of humans, then the gene would have been found much sooner.

Dressing and scientists around the world would like to have that blueprint — the sequence of the 3 billion bases that comprise the genetic material of humans, known as the human genome. To obtain the sequence, many of the world's leading geneticists are proposing the largest single project in the history of biology, the Human Genome Initiative. The project could involve hundreds of scientists and technicians working for 10 to 20 years and, by some estimates, could cost as much as $3 billion.

Scientists believe that the knowledge gained by having the complete sequence of the genome would yield new insights into the causes of an estimated 3,500 inherited diseases, as well as cancer, diabetes, and heart disease, which are all thought to have a genetic component. With these insights may come new therapies, and perhaps even cures. Obtaining the sequence of the human genome is the Holy Grail of biology, according to Nobel laureate Walter Gilbert of Harvard University, the "ultimate answer to the commandment 'Know thyself.'"

But research in biology has traditionally been conducted in groups consisting of fewer than 20 people, often with minimal oversight from funding agencies. Consequently, the prospect of a project as large as the Human Genome Initiative, which would require unusual coordination among the participants to ensure efficiency and minimize duplication, has generated many questions. Who will run the project? Who will fund it? How might the resulting information be used — or misused? For example, what would prevent an insurance company from discriminating against a person identified as having a genetic predisposition for cancer?

Mapping the genome. Ten years ago, no rational scientist could have argued convincingly for the sequencing of the 3 billion bases in the genome, because the process was far beyond the reach of existing technology. But the quest for the genome has now been made possible by the advent of genetic engineering, the development of new techniques for separating large DNA fragments, and great increases in the speed and power of computers. Computers would be especially important for identifying the order of fragments and for handling the massive amounts of information that would be generated. "Sequencing the genome is a supercomputer project as much as a biology project," says biophysicist Charles P. DeLisi of the Mount Sinai School of Medicine in New York City.

Humans have about 100,000 individual genes in 22 pairs of chromosomes and the sex chromosomes. If the DNA in one human cell could be uncoiled and stretched out in a straight line, it would be about 8 feet long. But if it were enlarged so that each base was the size of a small letter of type, the DNA would form a huge sentence about 5,000 miles long, stretching across the U.S. and partway back again. So far, researchers have sequenced about 12 million bases — about 20 miles' worth.

Researchers have two goals in the genome initiative: to create a "physical map" and to sequence the entire genome. A physical map would be a collection of perhaps 100,000 DNA fragments that span the entire genome and whose positions on the chromosomes are known. A sequence would give the order in which each of the 3 billion bases occur in the genome. Mapping might be compared to the process of locating every city and town in a country, while sequencing would be the equivalent of producing a detailed street map of the entire country.

Some critics of sequencing, including geneticist Francisco J. Ayala of the University of California at Irvine, argue that mapping will provide so much information that sequencing will not be necessary. But supporters of sequencing claim that, while mapping may lead to the identification of most genes associated with disease, sequencing would provide much more information about how genetic information is expressed and how cells work.

In essence, sequencing involves chemically removing one base at a time from the end of a DNA fragment and identifying each base as it is removed. An accomplished technician can sequence perhaps 1,000 bases per day. In 1986, biologist Leroy Hood and his colleagues at the California Institute of Technology in Pasadena announced the development of an automated DNA analyzer that can sequence 10,000 bases per day. He expects that improvements to the instrument will enable it to handle 200,000 bases per day by 1990. Researchers at the Institute of Physical and Chemical Research at Wako, Japan, have built a prototype of an analyzer they claim will eventually be able to sequence 1 million bases per day, which would make it possible to sequence the entire genome in 10 years — provided financial support is available.

But scientists cannot simply jump in and start sequencing each chromosome at one end, according to molecular biologist Cassandra Smith of Columbia University in New York City. "To study chromosomes, you have to break them down into small pieces that can be manipulated in the laboratory." Researchers must thus begin by making a physical map of the genome. Smith and Columbia molecular biologist Charles Cantor are mapping chromosome 21, the smallest human chromosome and the easiest to work with. This chromosome contains, among other things, the genes for several interferons — proteins that help fight disease.

To map a chromosome, researchers take perhaps 100 million cells — such as white blood cells, cultured skin

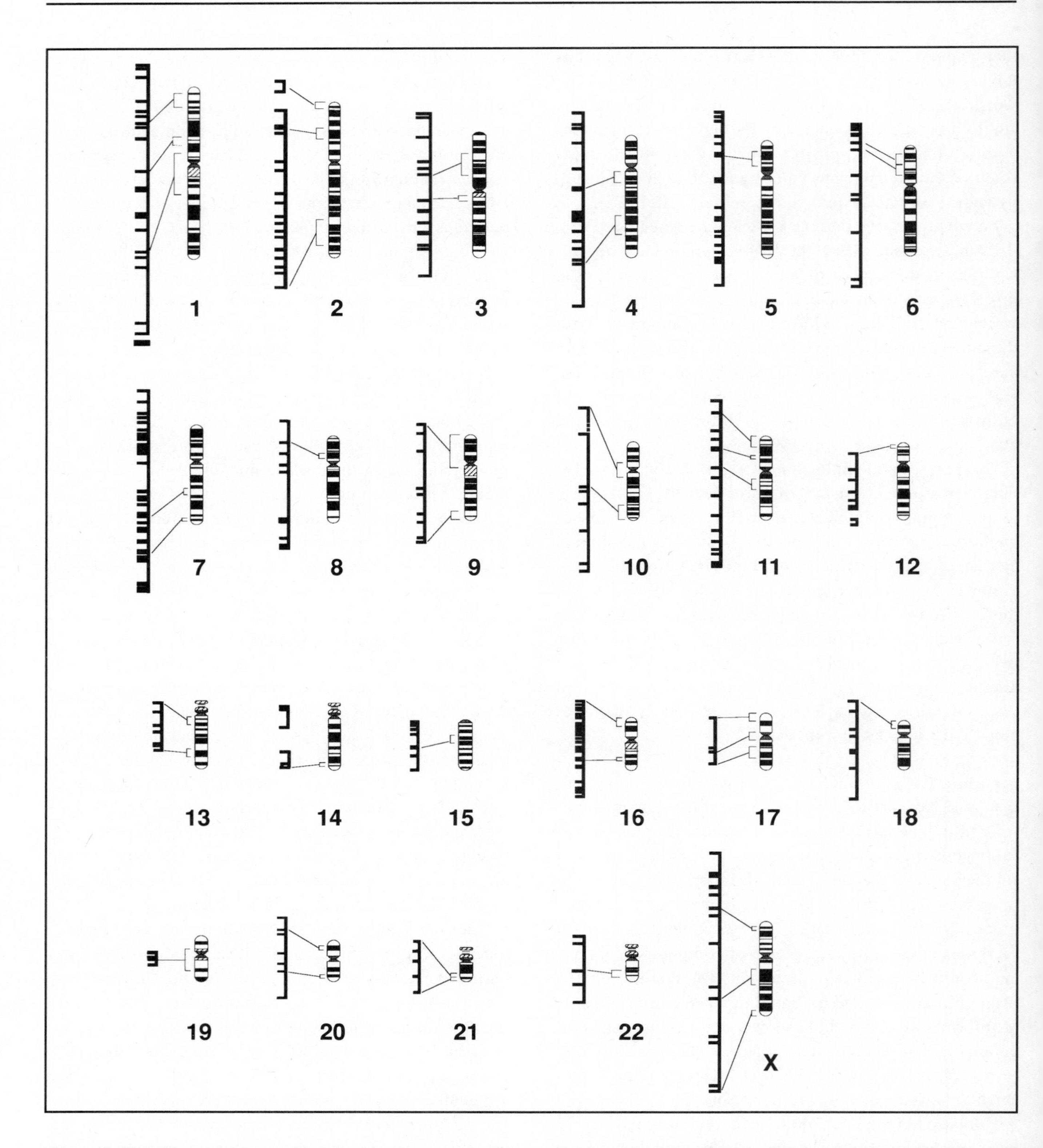

Figure 8. First Human Genome Map. In this diagram — which was developed by researchers at Collaborative Research, Inc., and the Massachusetts Institute of Technology — the numbered capsule-shaped objects are 23 human chromosomes. The ladder-like vertical diagram to the left of each chromosome is a "map" of the entire chromosome, showing the relative distance between DNA probes that bind to specific sequences of DNA on the chromosome itself. Diagonal lines between the maps and bracketed regions on the chromosomes indicate the actual physical locations on the chromosomes of the DNA sequences to which the probes bind. Maps like this one help scientists locate the specific genes or portions of genes that cause heritable disorders such as cystic fibrosis and muscular dystrophy. *Courtesy: Copyright 1987 by Cell Press. Reproduced from Helen Donis-Keller et al., "A Genetic Linkage Map of the Human Genome,"* Cell, *Vol. 51, 23 October 1987, pp. 319–337.*

cells, or placental cells — and rupture the membranes with detergents and enzymes to free the chromosomes, which are then separated with a laser-based sorter. The researchers next use restriction enzymes to break the chromosome into a number of pieces, perhaps 50. Using genetic engineering techniques, the scientists insert each of these pieces into the DNA of a yeast cell, producing 50 separate cell lines. When the yeast cells reproduce, they generate large amounts of the fragments. Next, the researchers use sophisticated matching techniques to determine where each fragment is located on the chromosome.

This entire process must be repeated at least once on each of the fragments, beginning with the use of restriction enzymes to break the original fragments into a smaller set of fragments. "The idea is to divide and conquer," Smith says. The map is thus a collection of yeast cultures containing DNA fragments of characteristic sizes, as well as a guide to fragments' locations. Scientists at Lawrence Berkeley Laboratory in California have already used this process to map chromosome 16, which carries, among other genes, those that influence some forms of kidney disease and leukemia. Lawrence Berkeley scientists are also working on chromosome 19, which carries genes that are involved in the repair of DNA damaged by radiation.

First map. In late 1987, researchers from MIT and Collaborative Research, Inc., of Bedford, Massachusetts, announced that they had made a map composed of over 400 DNA fragments, called genetic markers, scattered more or less evenly across the human genome. While the map will be useful for identifying genes linked to diseases, the researchers conceded that it would not be as useful as a physical map because it is not as detailed. Some researchers at other institutions have objected to the group's claim that they created a true genome map. The University of Utah's White, for example, has argued that the MIT-Collaborative map has too little detail and too many gaps to be considered the first genome map.

Also in 1987, the U.S. Department of Energy (DOE) directed its laboratories in Berkeley and Livermore to expand their efforts to map the genome, and asked its Los Alamos National Laboratory in Albuquerque, New Mexico, to develop improved sequencing technology and data handling techniques. DOE's immediate goal is to produce a physical map of the entire genome by 1993, according to David Smith, head of the department's office of health and environmental research. That will require nearly doubling its mapping budget to about $20 million annually. "We are not proposing that we [sequence the genome]," Smith said, "but that we create the resources and technologies that would allow this to be done."

Some scientists would prefer that the National Institutes of Health (NIH), with its long history of funding biological and medical research, take the lead in the Human Genome Initiative. But others disagree. "We don't want NIH . . . running this," says Harvard's Gilbert, because "their funds for other research would be cut back." Some biologists would like Congress to appropriate more money specifically for a sequencing project, but by early 1989 Congress had not yet done so.

In May 1988, a committee of the National Academy of Sciences issued a report favoring the Human Genome Initiative. The report, *Mapping Our Genes. The Genome Project: How Big, How Fast?,* urged that NIH, DOE, and the National Science Foundation cooperate more in their genetic research. The report also recommended that one agency be placed in charge of the project, but did not specify which one. The panel predicted that the cost of the genome project would rise from $47 million in the first year to $228 million in the fifth year, then remain steady at $200 million per year "until at least the year 2000."

By the summer of 1988, the three agencies had agreed that NIH would take the lead role in the initiative. In September, NIH announced that James D. Watson would become an associate director to coordinate planning. Watson will work 2 days a week at NIH, retaining his position as director of the Cold Spring Harbor Laboratory in New York.

Despite this progress, some scientists, such as Ayala, argue that a sequencing project is unnecessary, and that mapping will provide all the benefits promised by sequencing enthusiasts at a fraction of the cost. Such mapping has already led to the discovery of genetic markers associated with a number of diseases, and more are continually being found. Even if it would be useful to have the sequence of every gene, the genes themselves account for only 10 percent of the genome, according to Ayala. He claims that sequencing the genome's other 90 percent, which now appears to be indecipherable "junk" DNA with no known function, "is not likely to provide meaningful insights."

But Nobel laureate Paul Berg of Stanford University in California disagrees. "I call it arrogance to assume we already know what is junk and what is real. My premise is . . . that information which now appears undecipherable will, in fact, have substantial meaning in terms of understanding how the human genome works."

Some scientists fear that sequencing information developed as a result of the initiative will be used inappropriately. "If an insurance company finds out that you have a predisposition to a heart attack or cancer at age 50, will they insure you?" asks Stanford computer analyst Douglas

Brutlag, who is developing computer systems for handling genetic information and expresses a widely shared concern. "Or will they increase your rate? Who has a right to know your genetic makeup? That's a tough question." Nonetheless, he concludes, "We will sequence the human genome. Society should start thinking now about how to handle the information."

Genetic Screening

The growing number of probes that can be used to test for inherited diseases and the availability of increasingly sophisticated techniques for identifying disease susceptibility in fetuses and adults raises many ethical questions. For example, when genetic screening is used to identify birth defects prenatally, is it ethical to end the pregnancy if the fetus has a specific defect? What if the screening simply shows that the fetus is not the desired sex or that it has some harmless characteristic the parents don't like? Should a 20-year-old man be told that he has Huntington's disease and will develop severely disabling symptoms before dying at a relatively young age, or should he be allowed to live with uncertainty? Is the risk that he will pass the disease to his offspring a sufficient reason to inform him of the disease? Will a person with an identified genetic disorder such as Huntington's disease be able to obtain health insurance?

These questions are troubling, and they have sweeping implications, because every human is thought to carry an average of four to eight harmful genes. The prognosis is not entirely gloomy, however, according to physician Charles Whitten, director of the National Sickle Cell Anemia Foundation in Washington, D.C. Whitten noted that similar questions arose in the early 1970s when it became possible to test for sickle cell disease. Many militant groups of blacks called sickle cell screening a step toward genocide and urged people not to undergo such screening. But by 1985, most major black organizations advocated screening, and Whitten predicted that sickle cell testing will become so widespread that failure to recommend it to couples at risk could become grounds for a malpractice suit.

Mandatory genetic screening of a company's employees is another highly controversial issue. Such screening might be conducted to identify individuals with genetic defects that could make them more susceptible to the effects of hazardous chemicals or other agents encountered in the workplace. For example, people with a deficiency of the enzyme glucose-6-phosphate dehydrogenase have an unusually high risk of developing anemia when exposed to certain chemicals such as naphthalene. Similarly, a deficiency of the enzyme alpha-1-antitrypsin can predispose individuals to developing lung disorders when they are exposed to lung irritants. Companies now involved in screening generally justify it on the basis of protecting the health of their workers and reducing their potential liability for illnesses contracted while on the job.

In spite of these potential benefits, many people fear that employee genetic screening could lead to job discrimination, since individuals might be denied jobs simply because of their genetic heritage. Moreover, since many genetic defects are largely confined to particular ethnic or racial groups, screening of prospective employees could give rise to charges of hiring discrimination based on racial or ethnic background. By 1988, four states — New Jersey, Florida, Louisiana, and North Carolina — had begun to regulate genetic screening by employers. For example, New Jersey has made it illegal to use screening to determine whether prospective employees carry the gene for Tay-Sachs disease, because of concerns that an employer might not hire someone who may have a child with this chronic, costly ailment.

Although some supporters of genetic screening claim the technology is morally neutral, many ethicists disagree. University of Minnesota ethicist Arthur Caplan has said, "prenatal screening today is done with the intent to terminate pregnancies that are defective. The practice is defensible on certain moral grounds. My fear is that there will be an inevitable slide down the slope from definition of genetic disease to promulgation of eugenic goals. I don't think we have to make that slide. But we shouldn't try to fool ourselves [about the choices]."

Chapter 9

Physics

Theoretical Particle Physics

Rapid advances since the 1960s in understanding the theoretical basis for particle physics have culminated in the "standard model" of elementary particles. This model ties together and explains many intricate and baffling observations. It classifies all the known subatomic particles as either leptons or combinations of elementary particles called quarks. The standard model also requires that forces be transmitted between particles by another set of particles called gauge bosons.

In the early 1980s, physicists finally worked out the details of the part of the standard model that deals with the so-called weak force and electromagnetism. They now believe that these two forces may be considered aspects of a single force called the electroweak force. Although the standard model also describes the strong force, which binds an atomic nucleus together, it is more difficult to make specific predictions from what is now known about this force. Physicists have also not yet been able to weave the fourth principal force in nature, gravity, into a comprehensive theory that unites it with the other three forces.

A crucial test of the standard model is its ability to predict accurately the existence and properties of previously undiscovered particles. As one after another of the particles predicted by the model has been discovered in high-energy accelerators, confidence in the model has grown. The standard model has also contributed greatly to an understanding of how the universe emerged in the first instants of the Big Bang about 15 billion years ago. The model does not, however, provide final answers to a number of fundamental questions in physics. Although it specifies the different possible combinations of quarks that make up particles, such as protons and neutrons, it does not explain how the quarks are bound together to make stable particles.

Elementary Particles

In the 1920s, physicists knew of only three elementary particles: the positively charged proton, the negatively charged electron, and the electrically neutral, massless photon — the quantum or "particle" of light. At that time, physicists believed these elementary entities were the constituents of all the matter and energy in the universe. Atoms consisted of nuclei surrounded by electrons, with the nuclei themselves made up of protons. They believed photons carried electromagnetic energy, such as light, between atoms.

In 1932, British physicist James Chadwick discovered that atomic nuclei also contain electrically neutral particles, which were named neutrons. That same year, American physicist Carl Anderson discovered the positron, the antiparticle of the electron. Over the next two decades, researchers found still other particles, mesons, that seemed to play a role in binding nuclei together.

By the early 1960s, high-energy experiments had revealed so many different particles that physicists had to abandon the simple model that matter consists of only a few basic constituents. The dozens of newly discovered particles seemed related to each other in intriguing and complex ways that no theory at the time seemed to explain. For example, physicists found that many particles were related, in that they could decay into others, and they understood a few basic rules governing these decays. One such rule, charge conservation, required that a particle with a particular charge always decays into lighter particles with charges adding up to the charge of the original particle. Many other characteristics of particle decays, however, remained a mystery.

Quarks

In 1963, physicist Murray Gell-Mann of the California Institute of Technology (Caltech) in Pasadena proposed the quark theory, which succeeded in explaining many of the mysteries related to particle decays. A similar theory was developed at about the same time by Yuval Ne'eman, then of Tel Aviv University in Israel, and independently by George Zweig, also of Caltech. Using the mathematical concept of symmetry groups, Gell-Mann, Ne'eman, and Zweig classified many of the known subatomic particles into groups. They postulated that hadrons — particles that interact with each other through the strong and electroweak forces — are combinations of a few truly elementary particles, which became known as quarks. All three believed that hadrons could be classified according to the number and types of quarks they contained.

The quark theory proposed three types, or "flavors," of quarks — up, down, and strange (u, d, and s) — along with their antiparticles $\bar{u}$, $\bar{d}$, and $\bar{s}$. Since then, several other types of quarks have been discovered. The names of the quarks do not signify that they point up or down, nor is the strange quark particularly odd; they are simply convenient names physicists use to distinguish the particles. In contrast to all other subatomic particles, which carry integer electrical charges of +1, 0, or –1, quarks are thought to carry fractional charges of –1/3 or +2/3.

According to the quark theory, hadrons fall into two general classes — mesons and baryons. Each meson consists of a quark and an antiquark, while each baryon is made up of either three quarks or three antiquarks. Neutrons, for example, are baryons that consist of two down quarks and one up quark, which is written ddu. Protons — another type of baryon — consist of the quark combination uud. Among the mesons, the negative pi meson is $\bar{u}d$, and the positive pi meson is $u\bar{d}$. (For more information about the bestiary of different subatomic particles, see the "Guide to Subatomic Particles" following this article.)

Four Forces of Nature

Physicists recognize four forces that act within and upon particles: the weak force and the strong force, which both act within an atom's nucleus, and the more familiar forces, electromagnetism and gravity, whose influence extends to unlimited distances. Within their extremely short ranges of effect, the weak and strong forces are extremely powerful and account for most subatomic phenomena. Because of their limited range, however, people are not familiar with them in everyday life. The relative strengths of the four forces, their approximate ranges of influence, and their roles in nature are summarized in Table 1.

Not every particle is subject to all four principal forces. Depending on the types of particles involved, two particles may interact through two, three, or even all four of the principal forces. Two electrons, for example, may interact through the gravitational, electromagnetic, and weak forces, but not through the strong force, whereas two protons may interact by means of all four principal forces.

The force easiest to understand is the electromagnetic force. By the late 1940s, physicists had completed the theory of the electromagnetic force, called quantum electrodynamics (QED). Attempts to develop a similar theory for the other forces did not succeed until the late 1960s and early 1970s, when a theory of the weak force was developed. Combined with QED, that theory, in turn, produced the so-called electroweak theory.

Both the electroweak theory and the more complicated theory of the strong force, called quantum chromodynamics, incorporate the concept of gauge fields, in which forces are conveyed from one particle to another by other particles called gauge bosons. This concept was introduced into particle physics in 1954 by C. N. Yang and Robert Mills, then at Brookhaven National Laboratory in Upton, New York, in what has become known as the Yang-Mills theory of gauge fields.

The gauge particle of electromagnetism is the photon — the quantum of light. According to QED, two nearby electrons interact with one another by exchanging photons. For example, one electron will emit a photon, which travels through space to the second electron, which absorbs the photon and then itself emits another photon, which travels back to the first electron, at which point the whole process repeats itself. As a result of this kind of infinite exchange, the two electrons repel each other. Physicists call these exchange photons virtual particles, because even though theory insists they must exist, it also dictates that they cannot be "seen" directly.

Fifth Force?

Physicists have long assumed that there are only four fundamental physical forces in nature — gravity, and the electromagnetic, weak, and strong forces. There have been doubters, however, and a number of experiments have been performed to try to prove the existence of a fifth force. One possible explanation for why this force has gone undetected would be if its strength was less than 1 percent of gravity's strength and its range of influence was between about 1 meter and 100 kilometers.

The fifth force would presumably arise from the existence of a new exchange particle or boson, but as a practical matter the force might be detectable in macroscopic experiments, that is, observable without the aid of magnifying instruments. After reexamining the old records of early twentieth century experiments comparing the gravitational accelerations of objects made of differ-

Table 1. Four Forces of Nature

Force	Relative Strength (Strong Force = 1)	Range of Influence	Role in Nature
Strong	1	10^{-13} centimeter or less	holds the nucleus together
Electromagnetic	10^{-2}	unlimited	responsible for all electrical phenomena and for all chemical properties of molecules
Weak	10^{-5}	10^{-15} centimeter or less	partly responsible for energy release in stars and certain kinds of radioactivity
Gravitational	10^{-39}	unlimited	holds planets, stars, galaxies, and clusters of galaxies together

ent materials, some physicists claim to have found evidence of a nongravitational component related to the different materials' nuclear composition.

Groups at Brookhaven National Laboratory and at the University of Washington in Seattle, who have developed sensitive gravity detectors (accelerometers or torsion balances) and tested them near hillsides, have recently found what appears to be a nongravitational force component consistent with ideas of a fifth force. While these findings and the results of similar experiments are intriguing, there is still no conclusive proof of a fifth force, and the scientific community is skeptical about its existence.

Electroweak Theory

The electroweak theory, which unites both the electromagnetic and weak forces in a single framework, postulates the existence of four gauge particles: one to transmit the electromagnetic force and three others to carry the weak force. As in QED, the photon carries the electromagnetic force. Three other gauge particles, which carry the weak force, are the electrically charged W^+ and W^- particles and the neutral Z^0 particle, collectively referred to as "intermediate vector bosons." Unlike the massless photon, they appear to be relatively heavy, more than 80 times as massive as the proton.

Credit for formulating the electroweak theory generally goes to Steven Weinberg, then at the Massachusetts Institute of Technology (MIT), Abdus Salam of the International Center for Theoretical Physics in Trieste, Italy, John Ward of Johns Hopkins University in Baltimore, Maryland, and Sheldon Glashow of Harvard University in Cambridge, Massachusetts. Three of the theorists — Weinberg, Glashow, and Salam — shared the 1979 Nobel Prize in Physics for their accomplishment.

To formulate the electroweak theory, the theorists had to surmount two major obstacles. One of these obstacles arose because the originators of the electroweak theory chose to model the electroweak force in the manner of the Yang-Mills theory of gauge fields. In general, gauge theories that postulate the existence of massless gauge particles describe only long-range forces, while those that postulate the existence of nonzero-mass gauge particles describe only short-range forces. Since the original Yang-Mills theory described only massless gauge particles, it therefore applied only to long-range forces.

To modify the Yang-Mills theory to include the short-range weak force, Weinberg and the others turned to an idea proposed earlier by Peter Higgs of the University of Edinburgh in Scotland and, independently, by Thomas Kibble of Imperial College in London, U.K. The "Higgs-Kibble mechanism," as it became known, proposed the existence of a particle that binds to the intermediate vector bosons and gives them their mass. Since gauge particles with nonzero masses act only over short distances, their inclusion in the new electroweak theory made it possible to describe the short-range weak force, as well as the long-range electromagnetic force.

The second major difficulty encountered by the physicists developing the electroweak theory was a mathematical problem that had arisen earlier when the QED theory was being developed. According to QED, the force between two electrons, for example, is given by a sum of mathematical terms. When theorists in the 1940s finished calculating the actual numerical values of these terms, they discovered that many of them had infinite values. That turned out not to matter, because some of the infinite terms in the series were added and others subtracted in the computations, so that the infinities canceled one another; the calculated value of the force agreed with experimental evidence. However, years of additional theoretical work were needed to prove that the cancellation was exact.

The same problem involving infinities arose in the electroweak theory. Because the mathematical solution developed for QED did not apply to the electroweak theory, the theorists had to solve the problem of infinite terms all over again. When, in 1971, Gerard t'Hooft at the University of Utrecht in the Netherlands succeeded in solving the problem, the electroweak theory finally stood complete.

The electroweak theory successfully predicted a new phenomenon called neutral weak currents — interactions through the weak force but not involving the exchange of electric charge. Before the electroweak theory was proposed, physicists had thought that two particles could interact through the weak force only by exchanging an electrically charged particle. The proposed existence of the uncharged Z^0 meant that particles could interact through the weak force without the involvement of any charged particles. During 1973, the first neutral weak currents were observed at the European Center for Nuclear Research (CERN) near Geneva, Switzerland, and at the Fermi National Accelerator Laboratory (Fermilab) in Batavia, Illinois. That observation dramatically increased confidence in the electroweak theory.

Theory of the Strong Force

While the theory of quarks successfully explained the existence of a large number of hadrons, it did not provide a way to understand how quarks interact with one another, nor how the hadrons form other hadrons. To explain how quarks interact, physicists proposed a theory of the strong force, called quantum chromodynamics (QCD), which postulates that in addition to electric charge, quarks possess a "color charge." Each color charge combines with other color charges according to rules analogous to the rules for adding complementary colors. (Quarks are not truly colored; color simply provides a convenient way of thinking about how the charges combine.)

As with the electroweak theory, theorists patterned QCD on the earlier QED theory, and the problem of infinite terms arose once again. Although physicists could use QCD to predict the existence of certain particles, dealing with the infinite terms has made it difficult to produce precise estimates of particle masses. Like the theory of the electroweak force, QCD includes the concept of gauge particles. While the electroweak theory holds that the electromagnetic and weak forces are transmitted by four gauge particles — the photon, W^-, W^+, and Z^0— QCD postulates the existence of eight different types of massless gauge particles called gluons, so named because they "glue" the quarks together to form hadrons. According to QCD, gluons function first and foremost as carriers of "color force" between quarks and between gluons themselves.

Color is a generalization of the usual concept of charge. Whereas electric charge exists in two varieties — positive and negative — color charge exists in a total of six colors. According to QCD, each quark possesses a color charge of red, green, or blue, and each antiquark a color charge of mauve, yellow, or turquoise. In the theory of color charge, white equals neutral, with the color charges of the quarks in any hadron adding up to white, as the standard rules for combining colored light would dictate.

Thus, the charges of the quark and antiquark that make up a meson must add up to white; a meson, for example, might consist of a red quark and a turquoise antiquark. In addition, the charges of the three quarks that make up a baryon must be red, green, and blue in order to add up to white, while the charges of the three antiquarks in a baryon antiparticle must be mauve, yellow, and turquoise. Unlike the hadrons, each of the eight types of gluons display a characteristic pair of color charges, which in general do not combine to make white — for example, green and yellow.

Since hadrons are color-neutral, one might assume that the color force does not act between them. This is not entirely true, however, because the color charges do not cancel one another exactly. This imperfect cancellation results in a small residual color force between hadrons — the strong force. This works much the same way as the imperfect cancellation of electrical forces in atoms. Atoms are electrically neutral because they contain an equal number of positively charged protons and negatively charged electrons, but because the charges are spread out in space, they do not exactly cancel at small distances. The residual electrical force between atoms accounts for certain types of chemical binding.

Because gluons — the carriers of the color force — possess color charge of their own, they can interact with each other by exchanging other gluons. In contrast, the gauge particles of QED — the photons — do not attract or repel each other. They carry electromagnetic forces between charged particles, but not from one photon to another, since they themselves are uncharged.

Theoretically, gluons can bind to other gluons to form particles called glueballs. Because of the enormous energy of each gluon, physicists expect that a particle formed from gluons would have a large mass. (Energy can be converted to mass in an amount given by Einstein's famous formula $E = mc^2$, where c is the speed of light, E stands for the total energy available, and m is the mass of the particles.) Glueballs are predicted to have masses about 150 times the mass of a proton. This means that glueballs could be created in existing accelerators, but they would be difficult to observe if produced.

Nonetheless, City University of New York physicist Kenzo Ishikawa suggested in 1981 that evidence for such gluon-gluon particles had already appeared during electron-positron collision experiments at Stanford University's Positron Electron Project (PEP) accelerator. With this possible exception, however, researchers have established no evidence for the existence of glueballs.

Physicists think that another feature of gluons — their lack of mass — should make it impossible for anyone to observe isolated, individual quarks. University of Chicago theorist Yoichiro Nambu and other physicists have proposed that the color force carried by gluons not only extends to large distances (massless gauge particles produce forces with long-range influence) but, unlike every other known force, remains undiminished as quarks are pulled apart. If the color force possesses this property, free quarks can never be created, because the gluon bonds that hold quarks together in hadrons could never be broken.

Discovery of New Quarks

Although the original quark theory proposed three types or flavors of quarks, nothing in the theory prohibited the existence of additional flavors. In fact, soon after the quark theory was introduced in 1963, Glashow and James Bjorken of the Stanford Linear Accelerator Center (SLAC) in California proposed that a fourth quark, called "charm," must exist along with the three quarks — up, down, and strange — initially hypothesized by the authors of the quark theory.

The existence of the charm quark implied that additional combinations of quarks should exist in the form of mesons and baryons. For example, a meson that is a combination of a charm and an anticharm quark not only should exist in nature, but it should also be possible to create it with particle collisions of sufficiently high energy. Physicists at Brookhaven National Laboratory and at SLAC (who initially named it, respectively, the J and the psi) discovered this "J/psi" meson in 1974, thus confirming the existence of the "charmed" quark. (Samuel C. C. Ting of MIT and Burton Richter of Stanford University shared the 1976 Nobel Prize in Physics for this discovery.) Likewise, the discovery in 1977 of the upsilon meson confirmed the existence of another quark called the "bottom" or "beauty" quark. That brought the total number of quark flavors thought to exist up to six: up, down, strange, charm, bottom (beauty), and top. Although accelerator experiments have yet to show firm evidence for the existence of the top quark, the heaviest of the quarks, virtually all particle physicists believe that it will be found.

Grand Unified Theory

A grand unified theory (GUT) would combine the electroweak theory and QCD into a single unified theory describing the electromagnetic, weak, and strong forces. Eventually, physicists might add the more difficult-to-integrate gravitational force, creating a truly sweeping unification of the forces. In order to describe the three forces, a GUT would require the existence of 24 types of gauge particles: the photon, the three intermediate vector bosons (W^+, W^-, and Z^0), the eight gluons, and 12 superheavy bosons, which have no counterparts in the other force theories. According to GUT, these superheavy bosons, also called X bosons, can transform leptons into quarks and quarks into leptons.

GUT hypothesizes that during the extreme high-temperature conditions of the universe's early moments, quarks and leptons were interconvertible. As the universe cooled and the enormous energies at these temperatures became unavailable, the quarks and leptons began to acquire distinct identities. According to GUT, however, the transformation of quarks into leptons — and vice versa — through superheavy bosons still remains a theoretical possibility. Theorists estimate the masses of the superheavy bosons at about 10^{15} billion electron-volts, about one million billion times greater than the mass of a proton. (For an explanation of the use of the electron-volt as a unit of mass, see the article "Experimental Particle Physics.")

About 10^{-40} second after the Big Bang, according to GUT, virtually unlimited amounts of energy were available to create gauge bosons. The relatively massive superheavy bosons could be created as readily as the much lighter intermediate vector bosons, and these two types of gauge particles could be created as readily as the massless photons and gluons. At this point, the theory holds, the forces associated with these four types of gauge particles were of comparable strength. Only later, as the universe cooled and individual particles coalesced, did the forces acquire their distinct characteristics and very different strengths.

After about 10^{-35} second, the universe had cooled to the extent that the massive superheavy bosons could no longer be created readily, and they almost immediately decayed into a variety of lighter particles. Once the superheavy bosons disappeared, leptons and quarks were no longer interconvertible. From that time forward, a distinction appeared between the electroweak force — through which leptons interact — and the strong force, which affects only quarks.

After about 10^{-12} second, intermediate vector bosons could no longer be created as readily as photons; since then, the weak force has been distinguishable from the electromagnetic force. Thus, an initially simple universe, in which only one type of particle and one type of force were distinguishable, became increasingly complex as it cooled.

Without GUT, physicists could not explain how the present variety of forces and subatomic particles could have evolved from an initially homogeneous universe. Now, the theory has provided a natural way to address that question, although it will take time for physicists to develop the theory to the point that they are satisfied with it. One version of the theory, proposed by Harvard physicists Sheldon Glashow and Howard Georgi in 1974, predicted a lifetime for the proton that ultimately disagreed with experimental measurements. While this conflict has led some physicists to doubt the validity of at least that version of GUT, interest in a unified theory remains high. Work will no doubt continue until one version is widely accepted, or until some observation clearly proves that the idea of a grand unified theory is unworkable.

Guide to Subatomic Particles

By the late 1970s, physicists had consolidated much of what is known about particles into the so-called standard model of elementary particles. According to the model, three types of truly elementary particles exist in nature — leptons, quarks, and gauge bosons. Leptons, such as the electron and muon, are considered to be simple, structureless particles. Quarks, in combinations of twos and threes, make up the hadrons, which include the proton, neutron, and more than a hundred other particles. Gauge bosons, according to the standard model, are carrier or exchange particles that convey forces between other particles. For example, two electrons repel one another by exchanging photons — the gauge bosons of the electromagnetic force.

The standard model classifies all forces between particles into four principal ones: the strong, weak, electromagnetic, and gravitational forces. Each force is described by its own theory, although physicists have had considerable success in unifying the theories governing some of the forces. In the late 1940s, the theory of the electromagnetic force — quantum electrodynamics or QED — was the first to be completed, and continues to serve as a model for the other three more complicated force theories. The weak force, which acts only at very short range and is responsible for energy release in stars and certain kinds of radioactivity, is described by a theory that was essentially completed by the early 1970s. The theory of the weak force was soon integrated with QED into the electroweak theory.

The strong force, which binds atomic nuclei together, is described by quantum chromodynamics, or QCD. QCD is now sufficiently complete to provide a framework for understanding the strong force and the binding of quarks to form hadrons. In the early 1980s, progress was also made toward a combined theory of the electroweak and strong forces, called the grand unified theory or GUT; there are different versions of this theory, but all under the rubric, GUT. Relatively little progress has been made, however, toward combining the theory of gravity with the theories of the electroweak and strong forces. Some physicists are hopeful that two recent theoretical developments — supersymmetry and string theory — might lead to such a unification.

Leptons and the Electroweak Force

Leptons, by definition, are particles that interact with each other only through the electromagnetic and weak forces (and gravity). Their behavior is completely described by the electroweak theory. A total of six leptons have been discovered, along with their six associated antiparticles. (Each charged particle in nature has a corresponding antiparticle, with identical mass but electric charge of the opposite sign.) The three charged leptons are the electron, muon, and tau, and the three corresponding antiparticles are the positron, antimuon, and antitau, respectively.

The three other leptons — the electron neutrino, muon neutrino, and tau neutrino — are neutral and are generally considered to have zero mass, although some experiments, as described in the "Experimental Particle Physics" article, have suggested that at least one of the neutrinos may have a small mass. The antiparticles corresponding to the three neutrinos are called the electron antineutrino, muon antineutrino, and tau antineutrino.

The electroweak force is carried between particles by four gauge bosons: the photon, which is familiar as the quantum of light, and the three intermediate vector bosons, which were observed for the first time in 1983. According to the electroweak theory, the intermediate vector bosons appear to have nonzero masses because a set of neutral particles called Higgs particles is bound to each of them. One of these, H^0, is thought to be observable as an independent particle, whereas the other three Higgs particles are permanently bound to the intermediate vector bosons and theoretically cannot be observed in isolation.

Table 2. Leptons, Gauge Bosons, and Quarks

Particle	Symbol	Electric Charge	Mass†
Leptons			
electron	e	−1	0.511 MeV
muon	μ^-	−1	105.7 MeV
tau	τ^-	−1	1,784 MeV
electron neutrino	ν_ϵ	0	0*
muon neutrino	ν_μ	0	0*
tau neutrino	ν_τ	0	0*
Gauge Bosons			
photon	γ	0	0
intermediate	W^-	−1	81 GeV
vector bosons	W^+	+1	81 GeV
	Z^0	0	~ 93 GeV
Quarks			
up	u	+2/3	several hundred MeV
down	d	−1/3	several hundred MeV
strange	s	−1/3	several hundred MeV
charm	c	+2/3	1.55 GeV
bottom (beauty)	b	−1/3	5.2 GeV
top (not yet discovered)	t	+2/3	more than 30 GeV

†Physicists use the energy units of million electron-volts (MeV) and billion or giga-electron-volts (GeV) to represent the masses of subatomic particles. These units reflect the energy equivalent of each subatomic particle's mass as described by Einstein's formula $E = mc^2$.

*Not known to be strictly zero. The possible range of each mass is limited by the experiments that have been conducted. For example, experiments indicate that the electron neutrino's mass is below about 10 eV.

Hadrons and the Strong Force

By definition, hadrons are particles that interact through the strong force as well as the electroweak force (and gravity). More than 100 different hadron particles are known, and with the possible exception of the proton, all of them are unstable, that is, they decay into other particles. According to QCD, all hadrons are combinations of quarks, bound together by the color force, which is carried by gauge bosons called gluons. Hadrons are classified as either baryons, which are combinations of three quarks, or mesons, which are combinations of a quark and an antiquark. The strong force acts between hadrons and, according to theory, is the residue of the color force between quarks that does not cancel out when quarks combine to form hadrons.

There are six types, or "flavors," of quarks. However, none have been observed as free particles, nor will they ever be; quarks are permanently bound within hadrons. In addition to electrical charge, each quark flavor possesses a "color charge" of either red, green, or blue, and each of the oppositely charged antiquarks ($\bar{u}$, $\bar{d}$, $\bar{s}$, $\bar{c}$, $\bar{b}$, and $\bar{t}$) possesses an "anticolor" of either turquoise, mauve, or yellow. These color charges are the source of the color force between quark particles in the same sense that positive and negative charges are the source of the electrostatic force between protons and electrons.

Gluons, the gauge particles of QCD, are massless and electrically neutral. Each of the eight different types of gluons has a different combination of a color charge and an anticolor charge; for example, red-mauve, red-yellow, green-yellow, and blue-mauve. Gluons carry the color force that binds the quarks into hadrons.

Table 3. Selected Baryons and Mesons

Particle	Symbol	Electric Charge	Mass	Quark Constituents
Baryons				
proton	p	+1	938 MeV	uud
neutron	n	0	940 MeV	ddu
neutral sigma	Σ^0	0	1,192 MeV	uds
omega minus	Ω^-	–1	1,672 MeV	sss
Mesons				
positive pion	π^+	+1	140 MeV	$u\bar{d}$
negative pion	π^-	–1	140 MeV	$\bar{u}d$
J/psi	J/Ψ	0	3,100 MeV	$c\bar{c}$
neutral kaon	K^0	0	498 MeV	$d\bar{s}$

More than 50 different baryons have been observed, but how many different baryons there actually are in nature is not known. The existence of a charged baryon composed of three quarks implies the existence of an antiparticle made up of three corresponding antiquarks. For example, the proton p, with charge +1, consists of the three quarks uud; the antiproton $\bar{p}$, with charge –1, consists of the combination $\bar{u}\bar{u}\bar{d}$. The color charges of the quarks within a baryon can be "added" according to the laws for combining colored light. The three colors must always add up to white, just as the electrical charges of the quarks within an electrically neutral particle must add up to zero. A baryon must consist of one red, one blue, and one green quark, while its antiparticle must consist of one turquoise, one yellow, and one mauve quark.

Mesons, the other type of hadron, consist of one quark and one antiquark. Each charged meson has a corresponding antiparticle; for example, the negative pion is the antiparticle of the positive pion. The quark composition of the neutral J/psi — $c\bar{c}$ — reveals why it has no distinct antiparticle: the anti-J/psi is the same as the J/psi. Like baryons, the mesons are always color-neutral, or white. For example, a meson might consist of a yellow antiquark and a blue quark.

To date, the standard model has proved adequate to describe all particles that have been discovered. But the standard model does not in itself provide a complete theory of elementary particles; it cannot, for example, predict the magnitude of the strong force nor can it predict the masses of the individual quarks. The ambitious new theories — supersymmetry and string theory — are attempts to extend the standard model to provide more information on the fundamental forces and the masses of particles.

Accelerators

Almost everything that scientists have learned in recent years about the properties of subatomic particles has come from high-energy collision experiments. In an experiment of this kind, a beam of electrons, protons, or other subatomic particles collides with a stationary target; alternatively, a beam of subatomic particles can strike a second beam heading toward it. Physicists use elaborate monitoring equipment to measure the speed and direction of the particles emerging from the collisions and thus determine how the collisions have affected the incident particles and any particles created during the collisions. For example, in so-called elastic collisions between two protons, the particles bounce off one another like billiard balls. Physicists can measure both the angle and energy with which each proton leaves the collision, and thus calculate the forces between the colliding particles. At high energies, two colliding particles can also penetrate each other for an instant, revealing information about their internal composition.

At very high collision energies, some of the energy of motion carried into the collision by the particles transforms into matter — that is, other particles. Since the 1930s, experiments involving such high-energy collisions have led to the discovery of dozens of new particles, adding more and more detail to the physicist's picture of the fabric of nature. Because higher collision energies can bring into existence particles that cannot be created at lower energies, particle physicists have constantly sought to increase the collision energies obtained with their accelerators.

Figure 1. Search for Elementary Particles. Collision experiments at high-energy accelerators have helped to confirm the standard model of elementary particles. In these experiments, a beam of electrons, protons, or other subatomic particles is collided with a fixed target, or else with a second, oncoming beam of subatomic particles. The standard model has accurately predicted the existence and properties of particles created in these collisions. At Fermilab near Chicago, Illinois, the main accelerator is a fixed-target proton accelerator, although it has operated as a proton-antiproton colliding beam machine since 1987. The outline of the 3.8-mile-circumference accelerator ring is clearly visible in this aerial view of Fermilab. *Courtesy: Fermi National Accelerator Laboratory.*

The cyclotron invented by physicist Ernest Lawrence at the University of California at Berkeley in the 1930s accelerated charged particles to energies of several MeV. Since then, new technology, improved accelerator designs, and increased government support have made possible a succession of ever more powerful machines. The most powerful of these high-energy accelerators at Fermilab and CERN have accelerated protons to energies over a trillion electron-volts (TeV), more than 100,000 times greater than that of the original Lawrence cyclotron. Some research groups are planning still larger machines capable of accelerating protons to energies of 20 TeV or more. In head-on collisions between protons and antiprotons the combined energy will be double these figures.

Accelerator Principles

The simplest types of particle accelerators, including the everyday television picture tube, use electrostatic fields to speed up charged particles; electrostatic force arises from the simple attraction or repulsion between two charged entities. Simple electrostatic attraction draws charged particles, such as the electrons boiled off the red-hot filament at the back of a TV picture tube, to an oppositely charged metal plate, or electrode, located between the filament and the viewing screen. The larger the electrode's charge, which is proportional to the voltage between it and the red-hot filament, the stronger the electrostatic attraction, and therefore the greater the particle's final energy when it passes to the electrode.

Although the particle's final energy increases in direct proportion to the voltage between the electrode and the filament, it does not depend on the distance between them. An electron drawn to a positively charged plate by a voltage of 1 volt attains, by definition, a final kinetic energy of 1 electron-volt or 1 eV. If, as in a typical television picture tube, the voltage is 20,000 volts, the electron's final energy will reach 20,000 eV. Researchers in subatomic physics use electron-volts as standard units whether the particles are electrons or some other type, and whether their energies come from electrostatic acceleration or some other source (such as the thermal motion of neutrons in a nuclear reactor), as must always be the case with electrically neutral particles. As a particle travels faster, its energy increases.

The voltages that existing electrostatic accelerators can achieve are constrained by the limits of electrical insulators, so this type of accelerator cannot produce energies beyond a few MeV. Today's giant accelerators circumvent the problems inherent in producing very high voltages by subjecting groups of particles to a succession of relatively small voltages whose cumulative effect accelerates them to extremely high energies.

Linear Accelerators

In a linear accelerator or "linac," a series of particle accelerations occurs as the charged particles proceed down a long, narrow tube containing a number of metal compartments called resonant cavities. These cavities replace the electrodes used in an electrostatic system. Instead of being connected to a constant source of extremely high voltage, the cavities are attached to high-frequency, relatively low-voltage electromagnetic generators similar to radio transmitters. These generators produce an alternating electric field in the cavities, which reverses direction millions of times a second, alternately pointing forward and backward along the direction of the linac tube.

Linacs, like all accelerators using alternating electric fields instead of electrostatic fields, can produce only pulsed bursts of high-energy particles. If a continuous beam tried to pass through a linac, the system would decelerate the particles as often as it speeded them up because they would not arrive at the resonant cavities at the proper moments in the cycles of the alternating electric fields. However, by timing the arrivals of bunched groups of particles at each resonant cavity to coincide with the correct moment in the cycle of the alternating electric field, the system can feed energy to each particle group just as it reaches the entrance of each cavity. Since the particles' energy is boosted thousands of times as they pass down the length of the accelerator tube, linacs can create very high energies without extremely high voltages.

SLAC

The world's largest and most powerful linac operates at the Stanford Linear Accelerator Center (SLAC) at California's Stanford University. It has accelerated electrons to energies as high as 22 GeV, and it has produced as many as 360 bursts of high-energy electrons per second. The accelerator tube is 2 miles long and 4 inches in diameter, with nearly 100,000 resonant cavities to boost the speed of electrons.

Most experiments at SLAC have involved colliding a high-energy beam of electrons with stationary, solid targets such as metal blocks or with liquid hydrogen within a particle detector called a hydrogen bubble chamber. Instruments monitor the energy and direction of each particle emerging from the collisions.

Recently, SLAC's accelerator has been modified into a novel "colliding beam" configuration. In the first tests of the $120-million Stanford Linear Collider (SLC) — the new incarnation of the SLAC linac — during April 1989, American physicists detected the decay products of the Z^0 particle. SLC's objective is to produce thousands of Z^0 particles to further test the electroweak theory.

The SLC is the first linear accelerator in the world designed to collide electrons with positrons. After the two beams of oppositely charged particles emerge from two parallel, 2-mile, straight acceleration paths, magnets bend them onto two separate curved paths that intersect, allowing the particles to collide at energies up to 100 GeV as of mid-1989. Physicists believe that the SLC design will prevent the severe energy losses from particles in the form of so-called synchrotron radiation that commonly occur in circular accelerators.

Cyclotrons and Other Circular Accelerators

Like linacs, circular accelerators accelerate charged particles to high energies through a succession of relatively

Figure 2. Magnet at World's Largest Cyclotron. The major component in the giant TRIUMF cyclotron at the University of British Columbia is this 55-foot-diameter, 4,000-ton magnet, shown during the machine's construction. The "pinwheel" magnet design is a modern refinement of the simple circular magnet used in the Lawrence cyclotron to deflect particles into spiral orbits; otherwise the TRIUMF, which has a peak energy of 500 MeV, operates on the same principles as Ernest Lawrence's original table-top cyclotron. *Courtesy: TRIUMF.*

modest energy boosts. Instead of assembling many thousands of resonant cavities in a row as linacs do, circular accelerators use powerful magnets to deflect the particles into a circular path so that they pass through the same energy-boosting cavity repeatedly, building up speed each time.

In a circular accelerator, the particle beam must follow its curved path very precisely, since even a small loss of particles in each pass around the accelerator would quickly reduce the beam intensity to zero. Designing magnets to guide the particles was a technological challenge, because the faster a charged particle moves through a magnetic field, the less the field deflects it. Thus, a magnet strong enough to deflect the particles into a circular path at the outset of acceleration will become inadequate after the particles have made a few trips around the machine and have greatly increased their speeds.

Instead of increasing the magnetic strength to contain the particles in a fixed orbit as they moved faster, the original 1930s Lawrence cyclotron solved this problem by adopting a design that used a fixed magnetic field that allowed the particles to travel on a spiral path with an ever-expanding radius. In this arrangement, particles are first injected into the center of the space between the parallel circular faces of the two poles of a large electromagnet.

The circular chamber inside which the particles move is divided into two D-shaped compartments, called "dees," separated by a small gap. The system uses a high-frequency, high-voltage power source connected to the dees, and the particle orbits are arranged such that each time a particle crosses the gap it receives an energy boost. As the particles gather energy on each successive orbit, they travel in an ever-widening spiral until they reach the outer edge of the magnet. At that point, a small steering magnet deflects them into the desired target.

The largest cyclotron, called TRIUMF, is located at the University of British Columbia in Vancouver and can accelerate protons to energies of 500 MeV. The more powerful the magnetic field, the smaller are the orbits of particles with a given energy. The TRIUMF magnet had to be over 55 feet in diameter, because even the most powerful magnets cannot deflect protons with tremendously high energies into very tight circular paths. Since a larger magnet would not be practical, scientists have designed machines called synchrotrons that can accelerate particles to still higher energies by combining the principles of both the linac and the cyclotron.

Fixed-Target Synchrotrons

The world's two largest accelerators are synchrotrons: the Tevatron at Fermilab and the Super Proton Synchrotron (SPS) at CERN, which is operated by a consortium of Western European countries. A synchrotron consists of an accelerating tube bent into a ring up to several miles in circumference. Powerful electromagnets, which deflect high-speed protons into fixed, roughly circular orbits, surround the tube. At intervals along the tube, resonant cavities fed by powerful radio frequency power sources boost the particles' energy each time they pass through.

Because higher magnetic fields are required to deflect the particles as they gain energy, scientists have designed the magnets around the ring to adjust their field strengths rapidly and thus confine the particles to circular orbits. This arrangement is different from that of cyclotrons, in which the magnetic fields remain constant and the particles move in ever-widening spiral orbits.

As in a linac, particles enter into a synchrotron in groups rather than in a continuous beam. When a newly injected group of particles begins its trip around the synchrotron's accelerator tube, the magnets are operating at low power. In just 1 second, the group of particles may complete 50,000 energy-boosting circuits around the ring, and the electric current that powers the magnets will increase automatically to raise the magnetic field intensity 10-fold.

After 1 more second, the particles will have reached the highest energy attainable in the machine, and the magnets, which by then will be operating at peak strength, will be just strong enough to deflect the particles into the desired circular orbit. At that time, the machine ejects the particles from a specific point in the ring to collide with their target, while it injects a second group of particles into the ring to begin another 3-second cycle. Since the increase in the guide magnets' strength is precisely synchronized with the particles' increasing energy, scientists call this machine a synchrotron.

The Fermilab's large proton synchrotron, the Tevatron, with a main accelerator ring 3.8 miles in circumference, could initially accelerate protons to energies of 400 GeV. Fermilab later undertook a project called Energy Saver/Doubler to more than double the accelerator's peak energy. The Tevatron routinely accelerated protons to 800 GeV, and modifications to the machine eventually increased the maximum achievable beam energy to about 1 TeV.

The Energy Saver/Doubler project involved installing a new acceleration ring a few inches beneath the original ring. Protons were accelerated first in the original ring and then, at a specific point in the acceleration cycle, switched into the new ring for acceleration up to their final energy. To produce more powerful magnetic fields, scientists used superconducting wire in the guide magnets around the new ring. The Tevatron was, in fact, the first large-scale accelerator to use these powerful superconducting magnets. With them, a machine can accelerate particles to higher energies in the newly added ring than would ever have been possible with the conventional magnets in the original ring. By avoiding the wasted electrical energy (dissipated as heat) associated with conventional magnets, the superconducting magnets have also reduced Fermilab's energy consumption by 75 percent.

In scale and design, the world's other large synchrotron, the 450-GeV SPS at CERN, was until 1981, very similar to the original Fermilab accelerator. Then, at a cost of about 100 million Swiss francs, CERN modified the SPS from its original design as a fixed-target proton synchrotron into a proton-antiproton colliding-beam accelerator. Fermilab also undertook a similar conversion project at the Tevatron to allow the accelerator to operate as a colliding-beam machine. In 1988, Tevatron began accelerating counter-rotating beams of protons and antiprotons with combined collision energies of 2 TeV, making it the world's most powerful particle accelerator.

Colliding-Beam Accelerators

When a high-speed particle strikes a stationary one, much of the moving particle's energy is transferred to the second particle, which speeds off in the same general direction as the incoming particle was traveling. From the particle physicist's point of view, this energy is wasted because it does not create new particles or help disrupt the forces that hold the particles together. To probe the inner details of particle structure and behavior, physicists prefer collisions in which most of the energy goes into disrupting or creating particles.

The ideal type of collision for this purpose involves two particles of equal energy traveling toward one another from opposite directions and colliding head-on. This type of collision "wastes" no energy. In fact, collisions between two 25-GeV proton beams approaching from opposite directions will make available as much energy for disrupting subatomic forces as will collisions between particles in a single 1,300-GeV beam and those in a fixed target.

If a particle and its antiparticle collide, the two will often annihilate one another. The energy created in the annihilation may appear in the form of two high-energy gamma rays emitted in opposite directions from the collision point. Some of the energy released, however, may appear as mass in the form of newly created particles; for example, high-energy proton-antiproton collisions produce pi mesons. With head-on collisions, physicists can create many different types of particles, subject to the general restriction that the mass of the created particles may not total more than the combined mass equivalent energy of the colliding particle and antiparticle.

Physicists have long recognized that, in principle, colliding-beam accelerators could produce high energies, but until the mid-1970s, the disadvantages and technical difficulties of such machines outweighed their attractiveness. For one thing, colliding-beam machines produced only small numbers of collisions. In a conventional fixed-target accelerator, even when a beam containing a billion particles strikes a solid target containing 10^{23} particles, many of the incoming particles do not pass close enough to a target nucleus to interact with it. In early colliding-beam accelerators, the target was a second beam that contained only 1 billion particles and was millions of times less dense than a fixed target. Collisions were so rare that researchers could not obtain enough useful data in a reasonable length of time.

By the mid-1970s, however, technical improvements had made it possible to produce particle beams 0.2 centimeters wide. This greatly increased the density of particles within the beams, as well as the number of collisions that two colliding beams could produce. At the same time, the urgency of testing new particle theories, which can only be done by attaining the higher energies possible with colliding-beam accelerators, made scientists place a high priority on building such machines. Not only are most new and planned machines colliding-beam accelerators, but several older machines have been converted into that type.

CERN. Researchers use the ring at CERN's SPS to accelerate protons, and also antiprotons traveling in the opposite direction along a slightly different path, to energies of 320 GeV. At several points around the ring, small magnets can be activated to deflect the protons and antiprotons so that they collide nearly head-on, with an effective collision energy of 640 GeV. Monitoring devices detect the particles emerging from the collisions and record their energies and other properties. By 1990, the effective collision energy available at the SPS may reach 900 GeV.

The Large Electron-Positron accelerator (LEP) scheduled for completion at CERN on the French-Swiss border in 1989, will produce counter-rotating beams of electrons and positrons. LEP requires a much larger accelerator ring than the SPS — 17 miles in circumference — because small rings cannot efficiently accelerate low-mass charged particles such as electrons. When forced to travel in an orbit of small diameter, an electron loses energy through electromagnetic radiation much faster than does a considerably more massive proton.

In fact, when confined in a small ring, electrons with energies above a few GeV lose energy through radiation as fast as an accelerator can supply it. Higher energies and reduced rates of radiation loss can be achieved only by increasing the diameter of the orbit. When experiments at LEP begin in 1989, the operators of the accelerator will strive toward the goal of producing 100 GeV electrons and positrons.

Scientists using LEP will begin their work with a detailed study of the Z^0 particle, which was discovered at the CERN proton-antiproton collider in 1983. Although the Z^0 plays a key role in the electroweak theory of elementary particles, researchers at the SPS at CERN have been able to record no more than several hundred Z^0 particles. LEP should produce 100 million Z^0 particles per year, however, thus permitting physicists to observe rare Z^0 decays, which are predicted to occur perhaps only once for every 1 million Z^0 particles produced.

Fermilab. The Fermilab Tevatron, which is 4 miles in diameter, operates as a proton-antiproton colliding beam machine. Physicists first create the necessary antiprotons, which do not exist naturally on Earth, by means of high-energy particle collisions inside a nearby smaller accelerator, which also brings them to moderately high energies. They are then directed into a storage ring, where they remain until they are transferred to the main ring of Tevatron for final acceleration to 1 TeV. The antiprotons are collided with protons of equal energy so that the total energy of the collisions is 2 TeV. The converted Tevatron generates ten times more collisions per second than the SPS at CERN because its beams carry more particles.

Electron-Positron Colliding-Beam Accelerators

Other large colliding-beam accelerators planned or under construction will, like the LEP machine at CERN, accelerate and collide electrons and positrons. The SLC uses the existing 2-mile-long linear accelerator to accelerate both electrons and positrons. Although the 200 GeV collision energy ultimately achievable at SLC will allow as many as 3 million Z^0 particles to be produced each year, SLC will produce far fewer collisions in a given time period than CERN's LEP.

At SLC, the particle beams can collide at only one point where the semicircular tunnels meet. In contrast, the storage-ring design at LEP gives the particles many opportunities to collide. In LEP, particles remain within their original rings until they collide with an oncoming particle at one of the several points where the electron and positron beams cross over. Particles not scattered out of the beam by a collision simply continue to orbit around the ring until they eventually collide at another cross-over point. Although the SLC produces fewer collisions than LEP, it cost only about one-quarter as much to build. It also allowed U.S. physicists to conduct high-energy electron-positron collision experiments sooner than their European counterparts at LEP.

Future Accelerators

In July 1983, the High-Energy Physics Advisory Panel, which was established to advise the U.S. Department of Energy on accelerator projects, recommended the construction of a Superconducting Super Collider (SSC), a colossal synchrotron machine many times larger than the LEP at CERN. The machine received the support of the Reagan administration, and in 1988, Waxahachie, Texas, was selected as its site, subject to the approval of the U.S. Congress. With a circumference of 53 miles and an estimated cost of $5 billion to $8 billion, the SSC would be by far the largest and most expensive research instrument on Earth. The completed SSC would produce proton and antiproton beams, each with an energy per particle of 10 to 20 TeV and combined collision energies of 20 to 40 TeV. Like existing proton synchrotrons, the SSC would consist of a ring of magnets to guide bunched groups of particles in a circular path inside an underground tunnel, as well as a set of high-frequency power sources to accelerate bunched groups of protons around the ring.

A rival plan for yet another proton–proton collider, the Large Hadron Collider, with collision energies greater than several TeV was presented by Herwig Schopper, director of CERN, at the 1984 meeting of the International Committee on Future Accelerators held in Tokyo. According to the plan, the magnet ring of this colliding-beam machine would be built in the same tunnel as the LEP accelerator. According to some estimates, the use of the LEP tunnel might limit the machine's construction cost to under $1 billion.

The Large Hadron Collider would produce proton-proton collisions with an effective energy of 10 TeV. Its peak collision energy could reach 18 TeV if new, more powerful superconducting magnets can be built. To avoid duplication of effort, Schopper and others have urged U.S. physicists to consider a collaborative project, but a joint U.S.-European facility seems unlikely at present.

Machines such as SSC may approach the limit of what science can achieve at an acceptable cost with existing accelerator technology. In the words of University of Texas physicist Steven Weinberg, quoted in the 27 June 1985 *Science*, "Perhaps the world economy will be able to afford even larger accelerators, but surely not much larger." Even the more modest LEP, for example, will require as much electrical power for routine operation as a city of 150,000 people.

New accelerator technologies now under development, however, might make it possible to achieve still higher energies at an acceptable cost. Some accelerator designers and particle physicists predict that accelerators based on laser technology will be developed in the next 20 to 30 years. With these machines, researchers would use powerful bursts of laser light to speed particles to energies perhaps as high as 100 TeV. Other physicists doubt that laser accelerators will be operating before the middle of the twenty-first century.

One type of laser device now under development is the collective ion accelerator, which accelerates a dense cloud

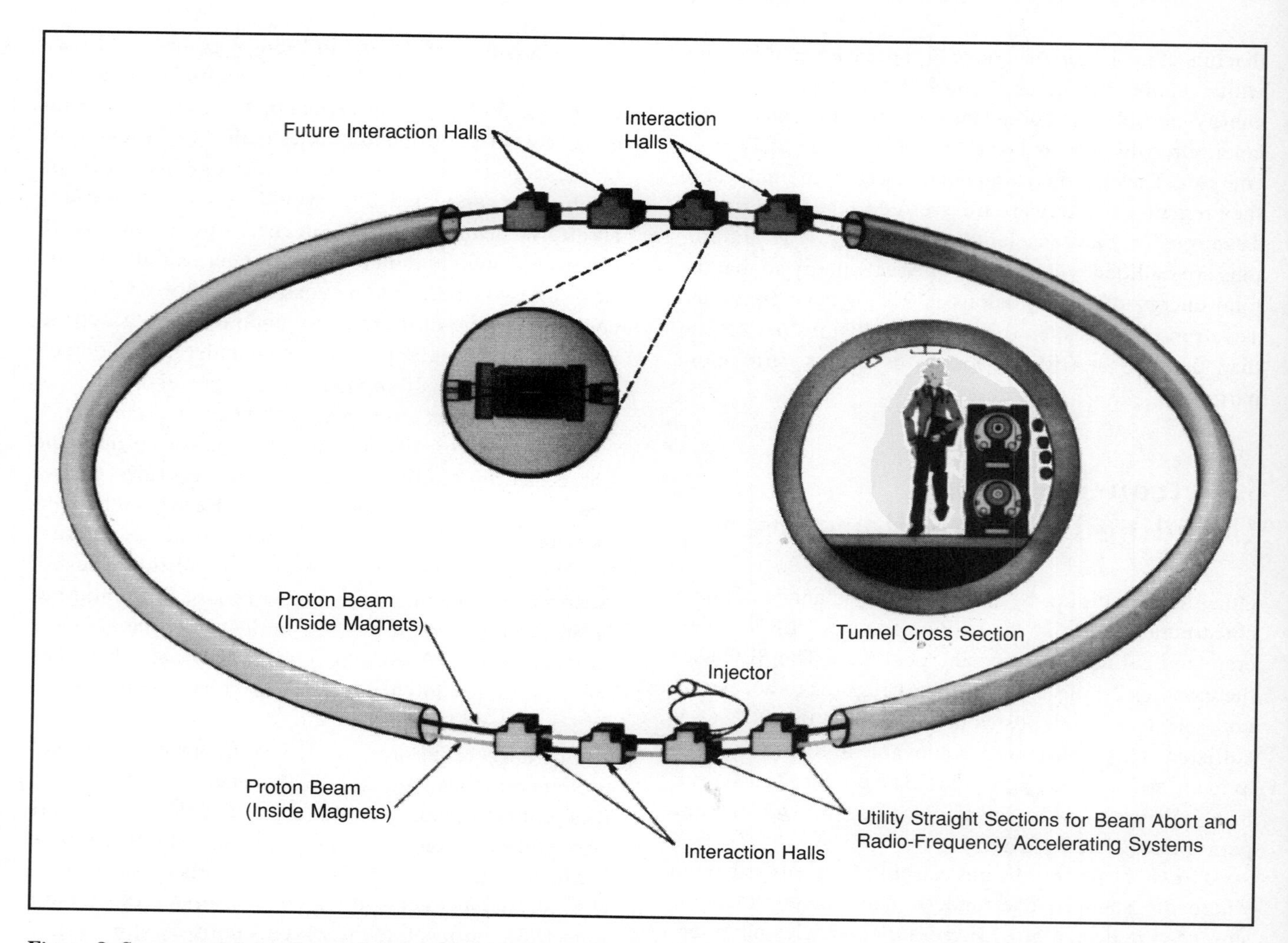

Figure 3. Superconducting Super Collider. With a diameter as large as 32 miles, the main ring of the proposed Superconducting Super Collider (SSC) would accelerate beams of protons and antiprotons in opposite directions. Both types of particles would be first accelerated to billion electron-volt energies in a linear accelerator, or linac, then transferred into a series of two booster accelerators, and finally injected into the main SSC ring. At several points called collision halls, the two beams would collide head-on with energies of 10 to 20 TeV. *Courtesy: SSC Central Design Group, Lawrence Berkeley Laboratory.*

of negatively charged electrons to high speeds. This fast-moving pulse of electrons, in turn, pushes forward a smaller number of positively charged ions, such as protons, accelerating them to high speeds. An experimental model of such a device, the Ionization Front Accelerator (IFA), was successfully tested at Sandia National Laboratories in Albuquerque, New Mexico, in 1985.

A separate device preaccelerates each cloud of electrons to moderately high speeds, at which point they are injected into one end of a linear accelerator tube. The number of electrons in each cloud always totals fewer than the so-called space-charge limit, the point at which electrical repulsion between nearby electrons would cause them to fly apart instead of remaining together in a group.

In the IFA, the laser beam enters the acceleration tube from the side, and as it progresses down the tube, it creates an advancing wave, or front, of positively charged ions in the cesium gas that fills the acceleration tube. This traveling ionization front draws the electron cloud forward. Unlike other accelerators, the IFA does not use either resonant cavities or high-frequency power supplies to accelerate particles to high energies. Instead, it uses a process called photoionization, in which a photon of an appropriate wavelength provides enough energy to break the electrical bonds holding one or more electrons to an atom. With photoionization, the IFA's laser beam, which is focused just ahead of the electron cloud in the acceleration tube, strips from the cesium atoms many electrons that then fly to the wall of the tube. The positively charged cesium ions left behind also respond to electrical repulsive forces, which tend to drive the ions outward, but their much greater mass causes the ions to move much more slowly than the lighter electrons. As a result, the positively charged cesium ions remain in place long enough to attract the cloud of electrons toward them.

As the laser beam used for acceleration maintains its position just in front of the electron cloud, it constantly draws the electrons forward, steadily increasing their speed and energy. With this elaborate arrangement, the IFA can accelerate pulses of small positive ions, such as hydrogen nuclei (protons), to high energies.

The main advantage of the IFA design is that it does not require complex equipment, such as precisely machined resonant cavities or powerful, computer-controlled magnets, to achieve extremely high particle energies. In principle, designers could simply lengthen the acceleration tube and modify the laser system to accelerate electrons and trapped positive ions to ever-higher energies. In fact, some accelerator designers believe that laser accelerators similar to the IFA might rival the CERN and Fermilab accelerators by the early twenty-first century.

Experimental Particle Physics

Since the early 1970s, physicists have established the existence of two groups of particles predicted by the standard model of elementary particles. One group includes combinations of quarks; the other consists of gauge bosons, which carry forces between particles. As researchers using high-energy accelerators have found one after another of the predicted particles, their confidence in the standard model has risen. While most physicists accept the model as essentially correct, they welcome experimental results that show unexpected departures from it — because such findings can provide critical information to further refine the theory.

Not every particle physics experiment, however, has involved the use of accelerators. One such class of experiments has tried to determine whether protons are stable, or whether they eventually decay into lighter particles. Other experiments have attempted to determine whether neutrinos possess mass. Some researchers have also tried to detect magnetic monopoles, hypothetical particles with a single magnetic pole.

Accelerator Experiments

Since the 1950s, most particle physics experiments have relied on high-energy accelerators. In a typical experiment, a beam of high-velocity particles emerging from an accelerator strikes a solid metal target, causing particles to fly off in many directions. By using an array of specialized detectors, physicists can then evaluate the pattern in which the particles scatter. These patterns allow researchers to determine what forces are acting between and within particles, and to derive particle properties such as charge and angular momentum.

At sufficiently high collision energies, some energy may be transformed into matter and new particles will form. Experiments with high-energy accelerators provide the only way for physicists to observe such short-lived particles as the intermediate vector bosons predicted by the standard model. The minimum collision energy needed to create a particle may be derived from Albert Einstein's formula, $E = mc^2$. A particle of mass m can be created only if an amount of energy equal to mc^2 is supplied, where c stands for the speed of light. The same formula applies to the opposite process, the transformation of mass into energy through the mutual annihilation of a particle and an antiparticle.

For example, Einstein's formula determines that a positive pi meson with a mass of approximately 140 MeV/c^2 can be created only in particle collisions with energies of at least 140 MeV. For convenience, physicists generally refer to particle mass in terms of the energy units MeV (million electron-volts) and GeV (giga-electron-volts, meaning billion electron-volts), without including the symbol c^2. Thus, they would say that the positive pi meson has a mass of 140 MeV, instead of 140 MeV/c^2. In practice, the number of positive pi mesons produced in a given number of collisions involving two protons increases as collision energies rise above the minimum value. Particle collisions with several times the minimum amount of energy are needed to ensure that many particles will be produced.

The physicists responsible for recommending which types of accelerators should be built have recently based their recommendations on the amount of energy needed to create particles predicted by the standard model. For example, in 1976 Carlo Rubbia of Harvard University and CERN along with David Cline of the University of Wisconsin in Madison and Peter McIntyre of Texas A&M University in College Station, proposed that an existing proton accelerator using a fixed target be converted into a proton-antiproton colliding-beam accelerator to create sufficient energy to produce the relatively heavy W^+,

W^-, and Z^0 particles (responsible for transmitting the weak force).

Many physicists criticized the Rubbia proposal, because few believed that existing technology could achieve the extremely precise accelerator beam control required to make the beams of particles collide. However, engineers and physicists did solve the technical problems involved, and by the summer of 1981, CERN had converted its fixed target accelerator — the SPS — into a colliding-beam machine capable of colliding protons and antiprotons, each with energies of several hundred GeVs.

Intermediate Vector Bosons

Two protons interact with one another by exchanging gauge bosons. When two protons are close to one another, the force-carrying bosons come into existence, travel between the two protons, and then disappear in as little as a trillionth of a second. This process occurs repeatedly when protons are in relatively close proximity.

The standard model predicts the existence of three intermediate vector bosons: W^+, W^-, and Z^0, and it suggests that the W particles should each have a mass of approximately 83 GeV (about 88 times heavier than a proton) and that the Z^0 particle should be slightly heavier (about 94 GeV). Before the SPS at CERN was converted into a colliding-beam machine, no accelerator could actually produce the high collision energies needed to create such massive particles. Once the conversion was completed in 1981, two groups of physicists at CERN began to search for the W^+, W^-, and Z^0 particles in the particle debris of 540-GeV collisions at the new accelerator. By December 1982, both the UA1 (Underground Area 1) group headed by Carlo Rubbia and the UA2 (Underground Area 2) group led by Pierre Durriulat had observed several events that they interpreted as evidence of the existence of W^+ and W^- particles. By July 1983 they also had reported evidence of the first Z^0 particles.

Both the UA1 and UA2 groups found masses of 81 GeV for the W^+ and W^- particles, fairly close to the predicted 83 GeV. For the Z^0, the experimental value was also close to the theoretical prediction of 93.8 GeV, although the two research groups disagreed slightly: UA1 reported a Z^0 mass of 95.2 GeV, while UA2 estimated 91.2 GeV. The standard model's prediction of the existence and approximate mass of these particles is some of the best evidence yet for the model.

UA1 Experiment

The UA1 group at CERN tried to detect and identify the few collisions (among the billions of proton-antiproton collisions at the SPS) that produced particles with the characteristics predicted for the W^+, W^-, and Z^0 particles. To accomplish this, the group monitored one of the points at the SPS where the beams of protons and antiprotons intersect, collide, and annihilate one another. The UA2 group monitored a different beam intersection point. Although each of the beams contained about 600 billion particles, only a tiny fraction collided with oncoming particles. Thus, in a typical month of operation, the UA1 detectors record only about a billion collisions.

From some proton-antiproton collisions, dozens of newly created particles emerge, while from other encounters, only the two colliding particles emerge. Among the billion collision events recorded in a given month, only about five involved the creation of W^+, W^-, or Z^0 particles, all of which immediately decayed.

The UA1 particle detector is a complex apparatus about as big as a medium-sized house. Enormous electromagnets surround the central detection chamber, which in turn surrounds the UA1 proton-antiproton collision point. The magnets deflect the charged particles that emerge from the collisions along a curved path; the curvature and direction of a particle's path reveal the particle's ratio of charge to mass.

Within the central detection chamber resides a system of concentric cylinders, each containing a closely spaced grid of fine wires connected to an elaborate system of electronic monitors that sense when a charged particle passes nearby and then compute the particle's location. These monitors enable the detector to track the path of any charged particle emerging from a collision.

A huge instrument called a calorimeter also surrounds the central detection chamber. The calorimeter, an array of pieces of clear plastic surrounded by sensitive electronic light detectors or phototubes, measures the energy carried by particles emerging from proton–antiproton collisions. A particle passing through one of the plastic pieces produces a characteristic flash of light, which the light detectors convert into an electric signal, the magnitude of which reveals the particle's energy. Although some particles, such as neutrinos, always pass through the calorimeter without being detected because they interact so weakly with matter, other particles, such as protons and electrons, are registered by it.

Figure 4. Discovery of W and Z Particles. New particles created by proton-antiproton collisions at CERN's Super Proton Synchrotron are detected and analyzed by elaborate detectors, which surround the points where proton and antiproton beams collide. Collision data recorded by the 2,000-ton UA1 detector pictured here provided the first evidence for the existence of W and Z particles. *Courtesy: CERN.*

The signals from the phototubes and the central detection chamber are carried to a short-term information storage device called a buffer. Then high-speed computers analyze the collision data — several million bytes of information per event — in an effort to isolate data about particles of interest. If the computers determine that an interesting event has taken place, the contents of the buffer are transferred immediately to a powerful computer, which stores the data on magnetic tape for more detailed analysis later.

While the computer processes the data gleaned from one event, it cannot perform any other tasks. Thus, the UA1 detector system can store data on only about 30 events per second, or about 140,000 of the billion events registered in the buffer each month. In the weeks and months following an experiment, researchers analyze the captured data to reconstruct the sequence of events that took place during each collision. By studying data from a multitude of collisions, they identify the few events that involve the creation and decay of W or Z particles.

The design and execution of the UA1 experiments involves hundreds of physicists and engineers. A similar sort of collaborative effort, but even greater in scope than UA1, was planned for 1989 at CERN's LEP (Large

Electron–Positron accelerator) facility to study the decay patterns of W and Z particles. This experiment, called L3, will involve nearly 400 physicists and more than 1,000 engineers. Nine countries, including the U.S., U.S.S.R., and People's Republic of China, will provide funding to cover the $200 million cost of the huge L3 detector.

Engineers built the L3 detector in a mammoth chamber 150 feet beneath the Earth's surface in the part of the LEP ring that is in Switzerland. MIT physicist Samuel C. C. Ting, who shared the 1976 Nobel Prize in Physics with Stanford's Burton Richter for the independent discovery of the fourth or "charm" quark, has served as the L3 project's scientific director since its inception in 1982. He describes the effort as "the largest experiment of its kind in particle physics." The L3 experiment is designed to provide evidence for the elusive Higgs boson — a hypothetical particle whose properties may provide critical information for determining the masses of many subatomic particles.

Top Quark

In 1983, Vernon Barger of the University of Wisconsin, and British physicists A. D. Martin of the University of Durham and R. J. N. Phillips of the Rutherford Appleton Laboratory, noted that several events recorded by the UA1 group appeared to establish the existence of yet another elementary particle — the suspected sixth type of quark known as the top quark.

The UA1 detectors recorded six events during 1984 that were once considered convincing evidence for the existence of the top quark. Recorded as part of the W particle search, the events involved the creation of a W^+ particle, which was thought to decay into a top quark and an antibottom quark. The suspected top quark itself seemed to decay into a bottom quark, a neutrino, and either a positron or a heavier antimuon particle. Although too few events had been observed by the UA1 group to pinpoint the top quark's mass, physicists Barger, Martin, and Phillips did make a rough estimate of 35 GeV based on the early UA1 data. By late 1987, the UA1 group had obtained experimental evidence that the top quark's mass must be greater than about 45 GeV, but the paucity of observations makes even this uncertain.

The top quark would be the heaviest of the six types of quarks predicted by theorists. Its relatively large mass — dozens of times that of the proton — explains why it has been so difficult to confirm.

Sparticles and Gauginos

When CERN physicists were analyzing data collected by the UA1 experiment during 1983, they found a few peculiar events among the hundreds of thousands of proton-antiproton collisions analyzed in detail. At first, they attributed these peculiarities to experimental error or chance cosmic-ray interference, but they changed their minds when small, but statistically significant, numbers of the events were also recorded during experiments in 1984.

A few events stood out because they involved either a monojet — a single burst of newly created particles, all moving at high speed in approximately the same direction away from a proton-antiproton collision point — or a bijet — two bursts of particles flying off in two different directions.

Jets commonly occur in high-energy particle collisions. Massive particles with very high energies often decay not into two or three particles, but into ten or more particles that form a jet. Usually, however, jets occur in oppositely directed pairs, with the momentum carried by one jet balancing the equal but opposite momentum carried by the second jet, thus fulfilling the requirement that momentum be conserved. In monojet or bijet events where the jets do not behave this way, a momentum balance can occur only by adding the momentum of at least one other particle created in the collision.

Analysis of the monojet and bijet events recorded by the UA1 experiment led physicists to conclude that an additional high-energy particle had been created but had escaped detection. Some physicists have suggested that these undetected entities may be so-called superparticles, a new class of elementary particles predicted by the theory of supersymmetry.

Supersymmetry, which was first proposed in the 1970s, provides a framework to explain why some particles play the role of force-carrying gauge bosons while others behave as interacting elementary particles. According to supersymmetry, each particle in nature exists in one form as a gauge boson but also has a counterpart that is an interacting elementary particle.

Physicists classify all known elementary particles as either fermions or bosons. Fermions display a physical characteristic called intrinsic angular momentum or "spin" equal to an odd multiple of 1/2. In contrast, the spin of bosons equals an even multiple of 1/2. According to supersymmetry, each fermion has a counterpart — a superparticle or "sparticle" — with the properties of a gauge boson. Likewise, supersymmetry suggests that each gauge boson has a corresponding fermion particle called a gaugino. For example, the existence of electrons and quarks, both fermions, implies the existence of selectrons and squarks, which have the properties of gauge bosons. The theory also predicts the existence of fermion counterparts for the gluon and photon bosons, called gluinos and photinos, respectively.

Until 1983, experimental searches had found no evidence for the existence of superparticles. However, since the theory made no firm predictions of superparticle masses, some physicists thought that the particles might simply be so massive that existing accelerators could not supply enough energy to create them. In 1985, CERN physicist John Ellis outlined one way in which sparticles and gauginos could explain the peculiar monojet and bijet events. By comparing the particle creation and decay sequences observed in the UA1 detector with the large number of sequences that might have occurred, Ellis eventually eliminated all but a few possible explanations. While no sequence involving only known particles could explain the monojet and bijet patterns, one involving the theoretical squarks and photinos might explain them.

Ellis proposed that each proton–antiproton annihilation produced a squark and an antisquark. He further postulated that the squark almost immediately decayed into a quark and a photino, while the antisquark decayed into an antiquark and a photino. The quark and antiquark then decayed into either a monojet or a bijet. If only one of the particles possessed enough energy to produce a jet, a monojet resulted; if both particles did, a bijet occurred. In theory, the photinos produced in the squark and antisquark decay events carried off large amounts of momentum, but they passed through the detectors unnoticed because they lacked electric charge and did not interact through the strong force.

Because supersymmetry theory makes a multitude of other predictions, particle physicists hope that further experiments will eventually confirm or discredit the supersymmetric interpretation of monojets and bijets. While most physicists consider supersymmetry an intriguing idea, there is still little firm experimental evidence to support it. The idea of supersymmetry appeals to many physicists because it offers a way to understand the apparently arbitrary division of particles into force-carrying gauge bosons and the group of quarks and leptons. Supersymmetry suggests that each type of particle actually exists in both forms — ordinary and supersymmetric — although only one member of each pair has so far been detected.

Superstring Theory

Superstring theory, another speculative idea that has gained many followers in the 1980s, proposes that fundamental elementary particles exist not as dimensionless points, as the standard model assumes, but as unimaginably short, one-dimensional "strings." Unlike supersymmetry, superstring theory offers no predictions — such as the existence of squarks and photinos — that physicists could test experimentally within the foreseeable future. Superstring theory appeals to some physicists, however, because it seems to point naturally to a unified theory of all four forces of nature, including gravity. Despite decades of effort, theorists had been unable to develop a theory that provides a unified explanation of gravity and the other fundamental forces.

Superstring theory may also provide explanations for several otherwise seemingly arbitrary properties of elementary particles. Particle physicist Yoshiro Nambu of the University of Chicago first proposed the basic superstring theory in 1970, but his ideas attracted relatively little attention until the early 1980s, when physicists John Schwarz at Caltech and Michael Green at the University of London described how the mathematical difficulties in the theory could be resolved.

According to the standard model, the elementary building blocks of matter — quarks and leptons — are dimensionless points. Hadrons appear to have dimensions — about one ten-trillionth (10^{-13}) of a centimeter in the case of the proton — only because their component quarks can move about in space. Superstring theory, by contrast, suggests that all elementary particles extend about 10^{-33} centimeter in one spatial dimension. Some versions of the theory propose that only closed loops of this length occur; other versions propose both open strings and closed loops.

Although this length is fantastically smaller than the size of a proton, it is still not zero, and this assumption profoundly changes the mathematical description of elementary particles. Superstring theory may, in fact, provide a way to eliminate some of the computational difficulties in the quantitative theory of the strong and electroweak force. Attempts to make numerical predictions based on the theory of quantum chromodynamics (QCD), which describes hadrons and the strong force, are simplified when physicists substitute the new finite values for the size of elementary particles into their equations.

QCD and the standard model also do not seem to provide any way to include gravity with the three other principal forces in a single unified theory. In contrast, superstring theory not only can explain gravity, but actually requires it — the existence of gravity follows from the way the theory describes particle interactions. Just as two strings can join end-to-end to form a longer string, a single string can form a closed loop when its ends join together. The theory predicts that in their lowest energy states all such loops should behave as massless particles with two units of spin (so-called spin-2 particles). These properties — a zero mass and spin of 2 — match those predicted for gravitons, the gauge bosons that theoretically carry the gravitational force.

Superstring theory also explains the origin of the gauge bosons that carry the forces described by the electroweak theory and QCD. In superstring theory, a particle in its lowest energy state corresponds to a straight, nonvibrating string. When two such particles collide, the ends of their strings momentarily fuse and then may separate. This process is considered equivalent to the exchange of a massless, spin-1 particle. Gauge theories assume that particles act upon each other through the exchange of such spin-1 particles, which are called vector bosons. Whereas gauge theories must arbitrarily introduce the existence of spin-1 particles, their existence would be expected according to superstring theory.

Nevertheless, superstring theory creates some vexing problems. For example, the theory can describe fermions — particles whose spin is an odd multiple of 1/2, such as quarks and leptons — only if it assumes that space has nine dimensions, not the three assumed in conventional physics. To describe bosons — particles whose spins are even multiples of 1/2 — superstring theory requires that space has 25 dimensions.

Some physicists have argued that the world could have undetected spatial dimensions that may be curled up or "compacted," thus defying detection. A two-dimensional surface, rolled up tightly like a paper scroll, for example, could be essentially indistinguishable from a line. Superstring theory does not explain, however, why nature should have 6 or 22 permanently compacted spatial dimensions.

Proton Stability

Most of the known subnuclear particles are unstable and decay into lighter particles in times ranging from as little as 10^{-16} second for the neutral pi meson to as much as 15 minutes for a free neutron, that is, a neutron outside an atomic nucleus. Neutrons found within atomic nuclei, however, are stable; otherwise all atoms would have disappeared long ago. Before the 1970s, most particle physicists also considered the proton to be absolutely stable. According to GUT, however, which describes not only the electromagnetic and weak forces but also the strong force, the proton should be very slightly unstable.

A widely known version of GUT, formulated in 1974 by Sheldon Glashow and Howard Georgi of Harvard University and known as the minimal SU(5) theory, predicted that protons should, on average, decay into lighter particles (usually a neutral pi meson and a positron) within 2.5×10^{31} years. This time span, billions of billions of times longer than the estimated age of the universe, defies the imagination; yet physicists have begun to search for proton decays by monitoring very large numbers of protons for occasional isolated occurrences.

Measuring the lifetime of the proton has proved far more difficult than measuring the lifetimes of other particles. For example, researchers have determined the lifetimes of the neutron, pi mesons, and other particles either by counting the number of decays in a small sample of particles, or by measuring the distance a beam containing a known number of the particles travels before dissipating into decay products. Of course, monitoring small numbers of particles would all but eliminate any chance of detecting the extraordinarily infrequent proton decays predicted by GUT.

Instead, physicists have attempted to monitor huge numbers of protons at once by surrounding large masses of matter (plates of iron weighing hundreds of tons, or tanks of liquid weighing thousands of tons) with detectors capable of registering a single proton decay. The first experimental search for proton decay began in 1954 when Frederick Reines and Clyde Cowan, then at Los Alamos National Laboratory in New Mexico, and Maurice Goldhaber of Brookhaven National Laboratory filled a large underground chamber with tons of a special scintillation fluid and surrounded the chamber with an array of 2,000 phototubes. They buried the tank far underground to protect it from cosmic rays, which can produce false signals.

When charged particles or gamma rays interact with a scintillation fluid such as carbon tetrachloride, they produce tiny flashes of light (so-called Cerenkov radiation), bright enough to stimulate the phototubes and produce electrical signals. Proton decays, if they occur, would most often produce a neutral pi meson and a positron. Although the neutral pi meson itself would not produce a flash of light in the scintillation fluid, the particle decays instantly into a pair of gamma rays, which do produce characteristic flashes of light.

By recording signals from the phototubes, the physicists could monitor vast numbers of protons continuously (a single cubic inch of scintillation fluid contains about

Figure 5. Search for Proton Decay. An experiment to detect proton decay is being conducted in a tank in an abandoned salt mine in Painesville, Ohio. Here, a physicist inspects experimental equipment placed on a special "mattress" that floats atop 8,000 tons of water in the tank. Sensitive phototubes monitor the water to detect the tiny flashes of light that would signify the decay of a proton inside a water molecule into a positron and a neutral pi meson. According to one proposed version of the grand unified theory of fundamental forces, about 20 proton decays should occur in this mass of water each year. *Courtesy: Robert Kalmbach, University of Michigan.*

5×10^{24} protons). During the experiment, the researchers observed no proton decays. Based on the calculated number of protons in the scintillation fluid, the length of the observation period, and the known efficiency of the detection system, Reines and Cowan calculated that the average decay time of the proton, if it decays at all, must be more than 10^{22} years. This impressively high lower bound on a proton lifetime was still far too short to test the predictions of different versions of GUT, which predict proton lifetimes of 10^{31} years or longer.

To increase the likelihood of observing proton decays, researchers have since built more sensitive detectors at sites around the world. These have been either liquid chamber detectors like the one used by Reines and Cowan, or layered-tracking devices. Layered-tracking detectors consist of iron plates sandwiched between layers of particle detectors. The detection of gamma rays or other products of proton decay would be evidence that a proton decay had occurred within the iron plates. The layered arrangement makes it possible for physicists to reconstruct the tracks of any subnuclear particle that passes through the sandwich. This feature allows them to discriminate between true proton decays originating in the iron plates and other events, such as those caused by cosmic rays from space.

Experiments using layered-tracking detectors have begun in the Kolar gold fields in southern India, in both the Mont Blanc tunnel and the Frejus tunnel along the French-Italian border, and in an iron mine in Soudan, Minnesota. Research based on liquid chamber detectors has been conducted in a mine in Kamioka, Japan, in the Silver King mine in Park City, Utah, and in a Morton Thiokol salt mine in Painesville, Ohio. The Painesville chamber is the largest and most sensitive of all the proton-decay detectors. Sponsored primarily by the University of California at Irvine, the University of Michigan at Ann Arbor, and

Brookhaven National Laboratory, this project is called "IMB," an acronym that combines the first letters of words in the names of the participating institutions.

The IMB detector incorporates the same principles as those in the 1954 Reines-Cowan detector, except that it uses water instead of scintillation fluid. The chamber is approximately the size of a six-story building (60 feet by 80 feet by 70 feet), and is located about 2,000 feet underground in an abandoned salt mine. Although the chamber holds 8,000 tons of water, the detector has been programmed to ignore events occurring within a few feet of the chamber's walls, which are often the result of cosmic rays or other outside interference. The researchers hope that the decay of any one of the 10^{33} protons within the 3,300-ton effective mass of the water will illuminate some of the approximately 2,000 phototubes inside the chamber.

Despite the 2,000 feet of soil and rock resting above the IMB detector, cosmic rays still penetrate to the chamber. Over 160 muons from cosmic rays and other sources light the phototubes each minute, but it is generally possible for physicists to distinguish these events from proton decays by analyzing the sequence in which the phototubes fire. For example, muons or other particles entering the chamber from outside trigger the outermost phototubes first, and these are discounted as external events rather than proton decay.

When members of the IMB team published a review of their experiment in 1985, they had analyzed data from 204 days of operation. Among the 169 events they classified as possibly originating within the chamber, none displayed the characteristics expected for the decay of a proton into its most likely products, a pi meson and positron. Although the proton can theoretically decay into other products, the IMB researchers had decided to concentrate their search on the easily distinguishable pi meson-positron decay. Considering that they failed to observe any proton decay event, the IMB researchers calculated that the average lifetime of the proton must be longer than 1.7×10^{32} years — nearly seven times longer than the maximum lifetime predicted by the minimal SU(5) version of GUT. Other versions of GUT predict even longer proton lifetimes, but to test these theories, physicists need to know whether protons are stable over these longer periods. As of early 1989, experimental evidence indicated that the average lifetime of a proton is at least 10^{34} years.

Detectors even larger and more sensitive than the IMB detector can be built, but their increased sensitivity also increases the frequency of ambiguous events. Certain cosmic ray events are almost indistinguishable from hypothetical proton decays, and these ambiguous events would be registered more frequently by larger and more sensitive detectors. Some physicists have therefore suggested that science may never be able to detect enough proton decay events to provide a statistically significant signal above the background of cosmic ray events. Nevertheless, groups around the world are continuing their efforts to observe proton decay.

Neutrino Oscillations and Neutrino Mass

Since the late 1970s, physicists have attempted to determine whether neutrinos are massless, as had generally been thought, or whether they have small nonzero masses. Three kinds of neutrinos exist, each of which is related to either the electron, muon, or tau particle. When certain particle decays produce muons, they are accompanied by muon neutrinos; likewise electrons are associated with electron neutrinos, and tau particles are associated with tau neutrinos. The negatively charged electron, muon, and tau, together with the three associated neutral neutrinos, belong to the lepton family. All leptons interact with other particles through the weak force. Charged leptons also interact through electromagnetism, but not through the strong force.

Since neutrinos interact with other particles only through the weak force, which is much weaker than the strong and the electromagnetic forces, neutrinos hardly interact with matter at all. For example, a neutrino could pass through literally a light-year (about 6 trillion miles) of lead with little chance of interacting with either electrons or particles in atomic nuclei. As a result, physicists have found it extremely difficult to perform neutrino experiments because the particle's presence can be detected only if it interacts with matter. This problem had until recently discouraged physicists from even attempting to determine whether the neutrino has zero mass or some very small nonzero mass.

In 1980, Frederick Reines of the University of California at Irvine reported experimental observations that he interpreted as evidence for "neutrino oscillations," in which one type of neutrino, the electron neutrino, changes briefly into another type, the muon neutrino, and back again. The neutrino oscillation observations imply that at least one of these two types of neutrinos must possess a nonzero mass, although the mass may be as low as a few electron-volts (eV).

In his experiment, Reines measured the number of electron neutrinos emitted from a nuclear reactor to see whether their numbers matched or fell short of the theoretical prediction. The latter result could indicate that neutrino oscillations were occurring. Since neutrinos are so difficult to detect, Reines could observe only a small fraction of the total number of electron neutrinos passing through his detectors. However, when he extrapolated the total number of electron neutrinos from the number detected, he found that the number was smaller than theoretically predicted. Reines attributed the discrepancy to neutrino oscillations, in which electron neutrinos emitted by the reactor changed briefly into muon neutrinos before reaching his detector.

Such temporary changes in particle identity, as strange as they may seem, are an established phenomenon in particle physics. However, the theory of quantum mechanics predicts that the phenomenon can take place only if the particles involved have different masses. Thus, for neutrino oscillations to occur, at least one of the two neutrinos involved must have a nonzero mass. Reines concluded that the muon neutrinos and the electron neutrinos differ in mass by approximately 1 eV, and that at least one of the two neutrinos has a mass of no less than 1 eV. However, other experimenters, including Felix Boehm of Caltech and his European collaborators, have not found similar evidence of neutrino oscillations.

Another way to determine whether neutrinos have nonzero masses is by studying the radioactive decay of atomic nuclei. In a process called beta decay, certain nuclei simultaneously emit an electron and an electron antineutrino. Although observers cannot detect the neutrino directly in this kind of experiment, they can deduce its mass by precisely measuring the energies of the emitted electrons. Since the combined energy of all the emitted particles remains fixed for the beta decay of a given type of nucleus, any energy not carried off by the electrons must be carried off by the neutrino. Thus, by measuring particle energies in beta decays, physicists could calculate neutrino mass.

In 1984, V. Ljubinov of the Institute of Theoretical and Experimental Physics in Moscow announced that his studies of beta decays involving the radioactive hydrogen isotope tritium implied that the electron neutrino mass must be between 20 and 45 eV. Another tritium-decay experiment, performed by J. J. Simpson at the University of Guelph in Ontario, Canada, showed evidence of a much higher mass, about 17,000 eV. However, when physicists at Princeton University in New Jersey attempted to confirm Simpson's results by measuring the energies of electrons from the beta decay of the isotope sulfur-35, they ruled out a neutrino mass that large.

Physicists conducting another kind of experiment in the quest for neutrino mass used the high-energy accelerator at Fermilab to create neutrinos. Like the Reines and Boehm experiments, these researchers attempted to observe neutrino oscillations. Although the Fermilab experiment in the early 1980s did not detect neutrino oscillations, it could detect such oscillations only if the mass difference was in the range of 5 to 33 eV. Thus, the Fermilab finding, which may conflict with the Soviet result, does not necessarily contradict Reines's finding of a neutrino oscillation involving neutrinos with a mass difference of approximately 1 eV, or with Simpson's finding of a very high neutrino mass.

The neutrino mass issue is of interest not only to particle physicists, but also to astrophysicists and cosmologists. As discussed in the "Cosmology" article in this chapter, the density of mass in the universe will determine whether it will continue expanding forever or reach a maximum size and then collapse on itself. The amount of visible or luminous mass detected by astronomers falls significantly short of the amount needed to "close" the universe (make it collapse on itself). However, since neutrinos are thought to outnumber protons and neutrons in the universe by a factor of 10 billion to one, even a small neutrino mass would mean that the universe has substantially more mass than astronomers have observed. In fact, the additional mass from neutrinos of finite mass could conceivably be sufficient to produce a closed universe.

In February 1987, the dramatic appearance of Supernova 1987A in the relatively nearby galaxy called the Large Magellanic Cloud provided physicists with more data with which to estimate what mass, if any, neutrinos have. A huge number of neutrinos were produced by that supernova explosion. Even though Supernova 1987A is approximately 160,000 light-years from Earth, astronomers estimated that 60 billion neutrinos passed through each square inch of Earth's surface when the neutrino pulse arrived in 1987. Within a span of less than 20 seconds, Japanese and American researchers (at the Kamiokande II neutrino detector and the IMB detector, respectively) observed a total of about 20 neutrino encounters with the atoms in their detectors — more than enough to tie these neutrinos unambiguously to Supernova 1987A.

Based on the timing of the arrival of neutrinos from the supernova, John N. Bahcall of the Institute for Advanced Study in Princeton, Sheldon Glashow of Harvard University, and many others calculated that the electron neutrino's rest mass is no greater than 11 eV and is probably zero.

Magnetic Monopole

Some versions of GUT predict that magnetic monopoles — particles with a single pole of magnetism (north or south) — formed during the Big Bang. Although several versions of GUT indicate that some of these monopoles should have survived to the present day, numerous experiments have failed to detect them. In 1982, physicist Blas Cabrera of Stanford University in California did report that he had detected a single monopole, but neither Cabrera nor any of the other physicists conducting similar experiments have confirmed that finding.

According to the classical theory of electromagnetism, electric current requires the existence of individual charged particles such as electrons. Magnetism, however, does not need individual magnetic "charges," but instead derives from the motion of particles with electric charge (for example, from the motion of electrons in atoms).

In an ordinary bar magnet, individual iron atoms act like tiny magnets not because they contain magnetic charges, but because their electrons create a magnetic field as they move around the atomic nucleus. Thus, cutting a bar magnet in half between its north and south ends (poles) does not produce a "north" and a "south" magnet, but instead two smaller ordinary bar magnets, each with two opposite poles. The same thing happens no matter how many times the bars are cut. Because magnetic materials had been found only in this two-pole form, by the mid-nineteenth century, most physicists had accepted as a rule of nature that individual magnetic charges, or monopoles, did not exist.

The possibility of magnetic monopoles, however, remained a tantalizing concept to theoretical physicists. As early as 1931, British physicist Paul Dirac at Cambridge University demonstrated mathematically that magnetic monopoles could explain a remarkable fact of nature, that electrical charge occurs only in multiples of a fundamental unit and not in fractions of that unit or in continuously variable amounts, as does mass. Dirac even predicted the amount of so-called magnetic charge that an individual magnetic monopole would carry, although he did not specify its mass.

Interest in magnetic monopoles, which had declined after numerous searches failed to find them, revived in the early 1980s. In part, the renewed interest sprang from theoretical studies, including one by Gerard t'Hooft of the University of Utrecht in the Netherlands, in which magnetic monopoles play an important role. According to t'Hooft's calculations, extremely massive monopoles, each with a mass 10^{16} times greater than that of a proton, must have been created in the early moments of the Big Bang and might continue to exist in significant numbers. Interest in monopoles also increased with the development of sensitive detectors, such as Cabrera's, that could possibly reveal the passage of such rare particles.

The monopole detector developed by Cabrera made use of a principle used in electrical generators, that a moving magnet passing near a wire produces an electric current in the wire. To achieve the high level of sensitivity needed to detect individual monopoles, Cabrera used superconducting wire connected to an extremely sensitive current-measuring device. Because superconductors have no electrical resistance whatsoever, a pulse of current induced in a loop of superconducting wire will continue to flow indefinitely, provided the wire remains at the very low temperatures needed for superconductivity.

Cabrera's experiment used a 1-inch-diameter loop of niobium wire cooled to the temperature of liquid helium. The loop was surrounded by a superconducting metal shield to protect it from the influence of nearby electrical equipment and other interference, such as disturbances in the Earth's magnetic field. A magnetic monopole, which would be expected to pass easily through the shield, would, in the process, cause an abrupt change in the current circulating in the loop.

During February 1982, in the midst of a 151-day observation period, the apparatus recorded a single event, the apparent passage of a magnetic monopole with a magnetic charge close to the value predicted by Dirac. The apparatus garnered no additional information on the particle's properties because the equipment had not been designed to measure the particle's mass, movement, or electric charge.

To confirm that he had detected a monopole and to observe enough additional monopoles to estimate their frequency of occurrence, Cabrera began using a more sensitive detector. The detector had a loop area almost 50 times larger than that of the first detector, and therefore would be much more likely to detect the passage of a monopole. Despite this increased sensitivity, however, the Stanford researcher recorded no additional events, which led Cabrera and other physicists to question whether the 1982 event was really the result of a monopole.

Scientists at other institutions have nevertheless continued the search with similar followup experiments. A research group at Imperial College, for example, has been searching for monopoles with a detector consisting of three loops surrounded by a superconducting shield. In the first 6 months of investigation, the group observed no magnetic monopole events. Taking into account the larger size of their detector, the Imperial College physicists calculated its total exposure (measured in terms of detector area times observation time) to be 230 times greater than Cabrera's.

Physicist Henry Frisch at the University of Chicago and his colleagues at Fermilab and the University of Michigan have operated an even more elaborate detector but have also failed to observe any monopoles. Nevertheless, monopoles may exist, but in such small numbers that longer observation periods or larger detectors will be needed to observe them. Scientist Claudia Tesche, a member of a team at the IBM Research Center in Yorktown Heights, New York, which also operates a monopole detector, has suggested that so few monopoles may exist that observers will need a detector as large as a football field to ensure that they detect a single one.

The lack of evidence for magnetic monopoles is consistent with the inflationary theory of cosmology, which is explained in the "Cosmology" article. Among other things, the inflationary theory postulates that the universe expanded very rapidly and dramatically during its early moments. Any monopoles created during the Big Bang would have since been distributed over the immense volume of the universe at such a low density that it is extremely unlikely that one would ever be detected.

Nuclear Physics

In the 1980s, high-energy collision experiments with beams of elementary particles and ionized atoms revealed some surprising properties of atomic nuclei. Nuclear physicists also created three and possibly a fourth new unstable heavy element.

The atoms that make up ordinary matter are typically about one ten-billionth of a meter in diameter. Most of an atom's volume is occupied by a diffuse cloud of orbiting electrons. Although the nucleus is only about one ten-thousandth the diameter of the atom, it is so extraordinarily dense that it contains more than 99.9 percent of an atom's mass and most of its energy.

The relatively low-mass orbiting electrons are responsible for such properties of matter as color, stiffness, strength, and chemical reactivity. The compact nucleus, on the other hand, has no influence on these properties and remains unchanged even through such violent chemical reactions as the explosion of dynamite.

Chemical reactions involve only the rearrangement of electrons. At the extreme temperatures and pressures in the cores of stars, however, the nuclei of hydrogen and other atoms can fuse together. In this process of thermonuclear fusion, the energy responsible for starlight and sunlight is released. In fission nuclear reactions, on the other hand, certain heavy nuclei are broken apart and release energy. Fission reactions occur naturally in radioactive decay as well as in manmade nuclear reactors.

The field of nuclear physics can be traced back to 1911, when English physicist Ernest Rutherford proposed the modern view of the atom as a dense central nucleus embedded in a cloud of electrons. The nucleus of each element is now known to contain a characteristic number of particles called protons and neutrons, which are bound together by powerful forces. Neutrons are electrically neutral, while protons carry one unit of positive electrical charge. As a result, the number of protons in a nucleus determines its total electrical charge.

In an electrically neutral atom, the number of negatively charged electrons held around the nucleus by electrostatic attraction equals the number of protons. The number of electrons, in turn, establishes the chemical nature of the atom — the ways in which the atom can combine with other atoms to form molecules. For example, the single electron in the hydrogen atom determines that atom's chemical properties, including its flammability and gaseousness at room temperature. Likewise, the six electrons in the carbon atom and the 92 electrons in the uranium atom determine their distinctive chemical behavior.

Although the neutrons in an atomic nucleus have almost no influence on an atom's chemical nature, they do affect the stability of the nucleus. A carbon nucleus containing six neutrons is permanently stable, whereas one with eight neutrons — called carbon-14 because the total number of neutrons and protons in the nucleus is 14 — will spontaneously decay and release energy in the process.

Unstable nuclei such as carbon-14 are called radioactive, because when they spontaneously decay, some of the energy they release appears in the form of radiation, such as gamma rays or x-rays. Carbon-14 nuclei decay, on average, after an interval of 5,700 years. An individual carbon-14 nucleus may exist for many times longer than this period or for only a few minutes, but on average, half the carbon-14 nuclei in a sample of carbon will have decayed when a period of 5,700 years — the "half-life" of carbon-14 — has elapsed.

Other nuclei, such as those of the newly made heavy elements, decay not over thousands of years but within a few thousandths of a second after the nuclei are created in a particle accelerator or nuclear reactor. By studying the relative stability of different nuclei and the ways in which they decay, scientists have been able to determine how particles in a nucleus are bound together. They have discovered what combinations of neutrons and protons will bind to form a stable nucleus and how strong the binding energy will be.

Nuclear physicists have found that nuclei occur in a variety of shapes, from simple spheres to elongated football-like shapes, all of which may undergo complex vibrations and rotations. Protons or neutrons may move individually or collectively, depending on the structure of a particular nucleus. It is even possible for all the protons in a nucleus to move in a group relative to the neutrons. Despite these motions, the density of nuclear matter has been found to remain relatively constant.

To obtain more detailed information on nuclear structure, physicists have used beams of high-energy particles in three different types of collision experiments. In one type, a high-speed proton or other particle colliding with a nucleus is deflected (scattered) in a way that depends on the shape of the nucleus and its structure. In another type of experiment, the incoming particle causes the target nucleus to split (fission) into two or more smaller fragments. In the third type, an incoming particle is absorbed into the target nucleus, fusing with it to form a heavier nucleus.

Scattering experiments using high-energy particles, which can penetrate a target nucleus more deeply than particles with lower energy, have helped to refine models of nuclear structure. Fission experiments have revealed unexpected qualities of nuclear matter in the phenomenon of so-called anomalons, and fusion studies have created at least three new elements — unnilseptium, unniloctium, and unnilennium — whose nuclei contain 107, 108, and 109 protons, respectively. Fusion experiments have also provided possible evidence for another new element with 110 protons.

Compressed Nuclear Matter

In the early 1970s, researchers were able for the first time to accelerate heavy nuclei, such as argon nuclei made up of 18 protons and 22 neutrons, to high speeds. (Accelerators that propel heavy nuclei are called heavy-ion accelerators.) Before that time, only single protons and small nuclei, such as those of helium with two protons and two neutrons (alpha particles) had been used as projectiles in scattering experiments.

Heavy ions can produce more energetic and more revealing collisions in scattering experiments than can single protons accelerated to the same speed. Even the heaviest naturally occurring nuclei — uranium nuclei containing more than 230 neutrons and protons — can now be accelerated to at least 95 percent of the speed of light.

Among the several accelerators where heavy-ion research is under way are the Bevalac at Lawrence Berkeley Laboratory (LBL) in California, the Synchrophasotron at the Joint Institute for Nuclear Research in Dubna, U.S.S.R., and the Laboratory for Heavy Ion Research (GSI) in Darmstadt, West Germany. In an experiment conducted at the Bevalac, a team of scientists from LBL and GSI used a beam of niobium-93 ions (nuclei of the element niobium, containing a total of 93 protons and neutrons) to bombard a thin metal target also of niobium. In some of the resulting nuclear collisions, several protons and neutrons emerged not in the usual nearly random pattern but instead bunched together in a group and perpendicular to the incident beam.

In the early 1980s, theoretical physicist Walter Greiner of the University of Frankfurt in West Germany had predicted that such "sidesplash," as the perpendicular emission of protons and neutrons is called, would be observed from heavy-ion collisions. At these high energies, nuclei are momentarily compressed to densities greater than normal. These compressed nuclei, Greiner suggested, would be likely to eject particles in the sidesplash pattern. Under ordinary conditions, typical atomic nuclei have a density of about 2.5×10^{14} grams per cubic centimeter — about 250 trillion times greater than that of liquid water. According to theory, the nuclear density in sidesplash may reach values several times higher than normal. Such dense nuclear matter is thought to exist naturally in neutron stars — stars that have collapsed inward under their own gravity — but neutron star conditions had not been observed in laboratory experiments.

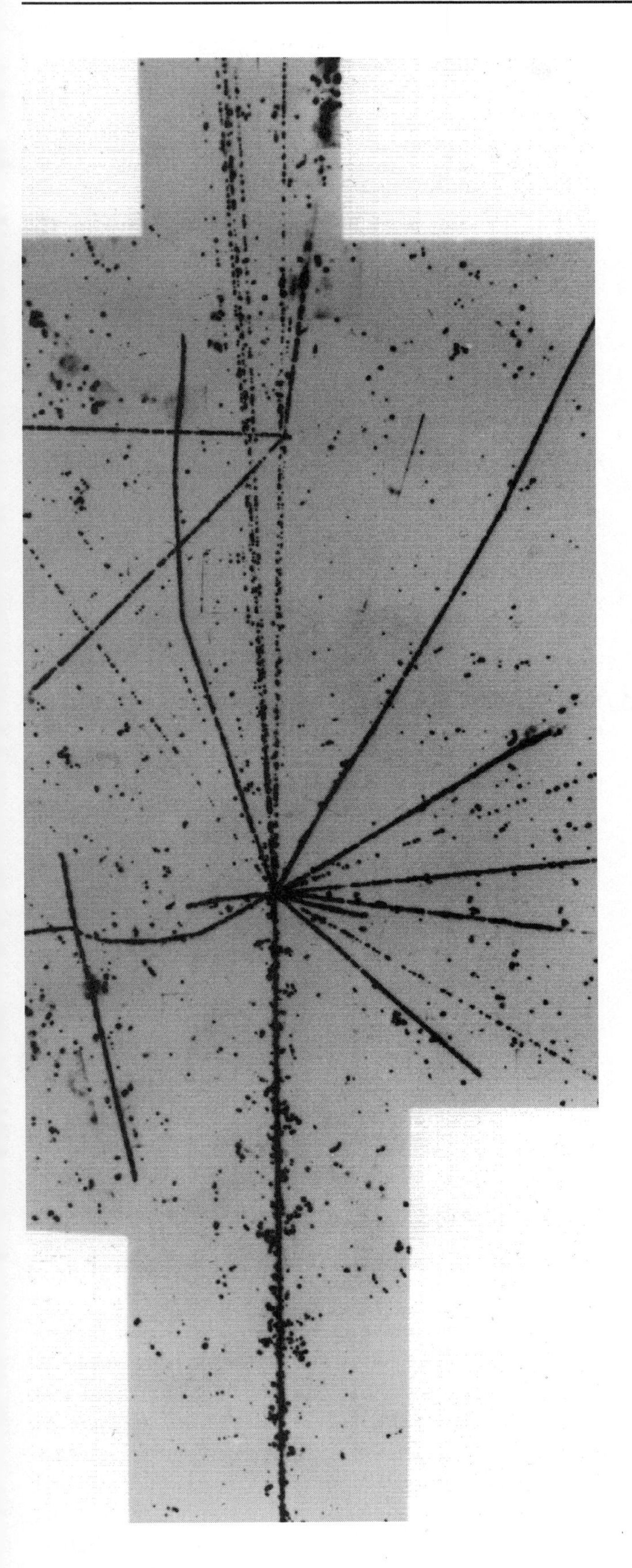

The discovery of evidence for compressed nuclear matter at Bevalac has encouraged physicists to consider whether nuclear matter might undergo further changes under even more extreme conditions. Some nuclear physicists have suggested that, just as water can change phase from gas to liquid to solid under increasing pressure, nuclear matter may likewise change phase in the fleeting instant of a high-energy collision.

Anomalons

Evidence for yet another new form of nuclear matter may have been discovered in a series of puzzling fission experiments performed at the Bevalac. The first of these studies, conducted in 1980 by LBL researcher Erwin Friedlander and his colleagues, involved colliding beams of oxygen, iron, or other nuclei with moderately heavy nuclei in a stationary target.

The target consisted of a stack of glass plates coated with a thick layer of light-sensitive emulsion. Long used in nuclear physics experiments, the light-sensitive emulsions contain silver or other moderately heavy nuclei and are sensitive to the passage of charged particles. Charged particles traveling through such a "nuclear emulsion" leave a trail that becomes visible when the emulsion is developed in much the same manner as ordinary photographic film. By developing all of the plates in the stack and then superimposing them on one another, physicists can reconstruct the path of any charged particles that pass through the stack.

Under the conditions of Friedlander's experiments, many particle collisions caused either the nuclei in the beam, the stationary nuclei in the target, or both to fission

Figure 6. Anomalous Nuclear Decays. When a beam of high-energy heavy ions strikes a photographic plate, subatomic particles and nuclear fragments are produced within the emulsion of the plate. These particles and fragments leave tracks that become visible when the plate is developed. A series of experiments first reported in 1980 revealed occasional tracks only an inch long — about one-third the expected length. In this image, two such short tracks are visible originating from a central point. As of 1989, nuclear physicists had not agreed upon an explanation for these "anomalon" tracks, which may represent super-heavy nuclei or unusual combinations of quark particles. *Courtesy: Lawrence Berkeley Laboratory, University of California.*

into smaller fragments. These fragments normally travel a characteristic distance through the emulsion before colliding with another stationary nucleus — a distance determined by the charge and energy of the fragment.

In one of Friedlander's experiments at the Bevalac, iron nuclei were collided with the target. Most of the tracks observed were approximately 3 inches long, but about 5 percent of the tracks were much shorter — about 1 inch long. Since current nuclear theory provides no ready explanation for the anomaly of these short tracks, the scientists used the term "anomalons" to describe the fragments that must have produced them.

One possible explanation for the anomalons' unexpectedly frequent interactions with the target nuclei is that they are not fragments of nuclei at all, but are instead very large nuclei that are created by fusion of the beam nuclei and stationary target nuclei. Such giant nuclei would interact more frequently with the target nuclei and would produce shorter tracks than small nuclear fragments.

Another explanation proposes that the collisions produce fragments of the expected size, but that at least one of the neutrons or protons within the fragments has been changed into a new particle. The proponents of this idea have suggested that, although neutrons and protons ordinarily consist of a bound group of three particles called quarks, high-energy collisions might sometimes reshuffle these quarks, leaving some in groups of two or six, rather than three. So-called demon nuclei — nuclear fragments containing such reshuffled quark groups — could perhaps interact with nuclei in the emulsion more readily than ordinary nuclear fragments and produce the observed short tracks.

Other studies, however, have raised doubts about whether anomalons exist at all. A second LBL team, consisting of John Stevenson, J. Musser, and S. Barwick, reported that experiments in 1984 did not produce anomalons under conditions in which they should have been produced. Unlike Friedlander, Stevenson and his colleagues used a transparent target made of lucite, in which the passage of collision fragments could be detected by recording the characteristic light, called Cerenkov radiation, emitted when high-speed charged particles pass through a transparent material at a speed greater than the speed of light in that material.

However, a third group of physicists from the State University of New York at Buffalo reported that they had, in fact, detected anomalons during their experiments at the Bevalac. Another group of physicists from Jadavpur University in Calcutta, India, also recorded anomalons in emulsion experiments. In these experiments, conducted at the U.S.S.R.'s Joint Institute for Nuclear Research, nuclear emulsions were bombarded with high-energy carbon-12 nuclei, containing six neutrons and six protons. Among the nuclear fragments emerging from these collisions were nuclei that appeared to be alpha particles, and some of these nuclei left typical anomalon tracks. Before this experiment, only much larger nuclear fragments had been found to leave anomalon tracks. Conflicting results were also obtained with an experimental technique known as track etching, in which the path of a nuclear fragment through a piece of clear plastic is analyzed.

In 1987, a group of American and European physicists working at Fermilab reported new limits on the detectability of anomalons. The team passed a high-energy beam of protons and other particles through a liquid hydrogen-filled bubble chamber. When a charged particle passes through such a chamber, it leaves a trail of bubbles that can be recorded photographically. After analyzing thousands of interactions, the researchers concluded that if anomalons exist at all, they must occur in only a minute fraction of interactions. The actual proportion would depend on how readily anomalons interact with normal matter. Because of the mixed results of various experiments, physicists remain uncertain about whether anomalons actually exist.

New Heavy Elements

Three new elements were created between 1981 and 1985 by bombarding targets made of moderately heavy atoms, such as bismuth, with beams of high-energy heavy ions. In a small number of collisions, the two colliding nuclei fuse together, and the nucleus of a new heavier element forms. Since such a nucleus placed among electrons can easily acquire the electrons needed to make an atom, physicists consider that producing the nucleus of a new element is equivalent to producing the element itself. The three new elements have atomic numbers 107, 108, and 109, representing the number of protons contained in a nucleus of the element.

Elements with atomic numbers 1 (hydrogen) through 92 (uranium) occur naturally on Earth, and before 1940 no element with an atomic number above that of uranium had been detected. By 1974, however, an additional number of heavier, so-called transuranic elements with atomic numbers 93 (neptunium) through 106 (unnilhexium) had been created by bombarding the nuclei of moderately heavy elements with various particles, including the nuclei of lighter elements. These artificially created elements are all unstable, in some cases having a half-life of a few thousandths of a second, which explains why they are not seen in nature and, with a few exceptions, have been produced only in minute quantities. They are interesting primarily for what they reveal about how protons and neutrons bind together to form nuclei.

Elements 107 and 109 were created in February 1981 and August 1982, respectively, using the heavy-ion accelerator known as UNILAC at GSI. The researchers at UNILAC bombarded a bismuth (atomic number 83) target with a beam of chromium (atomic number 24) ions to produce element 107. To make element 109, they used a bismuth target once again, but this time bombarded the sample with a beam of iron (atomic number 26) ions. Only a single isolated nucleus of element 109 was detected, but the evidence of that observation was considered sufficient to establish the existence of the new element. In 1976, a group at the Soviet Joint Institute for Nuclear Research had reported creating element 107 in another form containing fewer neutrons than the form of element 107 produced at UNILAC. Many scientists in the West, however, considered that the supporting evidence of the Soviet report was inconclusive.

In March 1984, a group of 14 physicists at UNILAC succeeded in creating element 108. A beam of iron ions was used to bombard a target of lead (atomic number 82). Like all the other new elements, element 108 is unstable, with a half-life of 0.002 second. This brief half-life is, however, more than ten times longer than nuclear theorists had predicted. This surprise, in addition to posing an interesting theoretical problem, encouraged experimental physicists to try to make still heavier elements, since these will be easier to detect if they exist longer than originally predicted.

In 1987, Soviet physicist Yuri Oganessian and an international group at the Joint Institute for Nuclear Research claimed to have created element 110 in collisions between argon and uranium and between calcium and thorium. They claimed that element 110's half-life is about 0.009 second, but the existence of the new element has not yet been confirmed by other research groups.

Physicists continue to look for a theoretically predicted "island of stability" in nuclei beyond element 109 and concentrated around element 114 — specific nuclei with the requisite number of protons and neutrons to hold together long enough to be detected. In particular, nuclear physicists have predicted that it should be possible to produce element 116 in bombardment experiments. In 1985, however, Peter Armbruster and his colleagues reported that their experiments at both GSI and LBL had failed to produce element 116. A beam of calcium-48 ions was used to bombard a curium-248 target, but no superheavy products were detected. Further experiments will be needed to determine whether some form of element 116 can be produced.

Cosmology

Cosmology is the study of the overall structure of the universe — its origin, history, and destiny. In their efforts to understand the universe, cosmologists must rely on observations of the present-day cosmos in deducing the properties of the past and future universe. Since light from the most distant visible galaxies reaches Earth after traveling through space for about 15 billion years, the radiation now reaching Earth from those galaxies provides important information about what the universe was like relatively soon after it was formed. Many of the concepts of modern cosmology are highly abstract, defying intuition and everyday experience. Nevertheless, modern cosmological theories, which rely heavily on mathematics, are progressing toward unified explanations of the largest phenomena in the universe and phenomena on a subatomic scale.

Cosmologists have increased their knowledge of the universe and its evolution in many ways during the past 20 years. Observations of nearby stars and galaxies have brought new insights into the distribution of galaxies in space and the relative abundance of different kinds of matter in the cosmos. New studies of the faint microwave background radiation found throughout the universe have confirmed that the universe began in an extremely high-temperature state billions of years ago and has been expanding and cooling ever since.

Advances in elementary particle physics have also allowed cosmologists to reconstruct the sequence of events that occurred in the first second of the primordial Big Bang explosion in far greater detail than previously possible. The term Big Bang refers to the explosive origin and expansion of Einstein's four-dimensional space-time. The consequences of the Big Bang continue even today, as evidenced by the recession of galaxies away from one another. As a general rule, galaxies appear to be receding in all directions from our own, and more distant galaxies appear to be moving away faster than nearby ones.

Perhaps the most impressive development in contemporary cosmology began in 1980 with the development of the inflationary theory of the universe. This theory suggests that an expansion of previously unimagined magnitude and suddenness occurred during the first moments of the Big Bang. Intimately linked to theories of particle physics, inflation seems to explain many mysteries that previously eluded cosmologists, including why the expansion of the universe appears to be extremely uniform in all directions. Inflationary theory has led to a renaissance in cosmology the likes of which has not been seen since the 1920s, when astronomers first discovered that the universe was expanding.

Chronology of the Early Universe

The current theory of how the universe evolved during the first few minutes of the Big Bang has been so widely accepted that cosmologists often refer to it as the "standard model." (Although it shares the same name, the standard model of cosmology is not the same as the standard model of elementary particles.) According to the standard model of cosmology, the Big Bang did not begin at a specific point in the universe and "explode into space." Instead, matter and space-time itself expanded everywhere at once, so that distances increased between clumps of matter in the initially highly compressed cosmos. While this strange idea can stretch the imagination to its limits, the concept lies at the core of modern cosmological thinking.

After an inflationary instant — the first 10^{-30} second — the universe was a maelstrom of radiation. As the universe cooled and expanded, particles formed; during the first full second, all of the types of particles that now exist, such as electrons, neutrons, and protons, came into existence.

According to the standard model, the universe began at fantastically high pressures and temperatures. At 10^{-30} second after the Big Bang, when the average temperature of the universe was about one thousand trillion trillion Kelvin (10^{27} K), the elementary particles called leptons and quarks began to form. (Physicists generally use the Kelvin temperature scale rather than the Fahrenheit scale. One Kelvin is equal to 1.8 Fahrenheit degrees. Zero Kelvin — 0 K — which is equivalent to about −459.7°F, is the lowest temperature possible and is referred to as absolute zero.)

By the time the universe was one-trillionth (10^{-12}) of a second old, the average temperature had declined to one quadrillion Kelvin (10^{15} K). At this temperature, the particles possessed a thermal energy (the energy of their motion) of several hundred billion electron-volts — which is about equal to the highest energies that can now be reached in high-energy particle accelerators.

A millionth (10^{-6}) of a second into the Big Bang, quarks began to coalesce into neutrons, protons, and other particles, collectively called hadrons. However, the 10 trillion Kelvin temperature (10^{13} K) was still far too high to allow nuclei to form from neutrons and protons. Nuclei cannot form as long as free neutrons and protons possess energies much greater than the binding energies that hold them together in a nucleus. Such energetic neutrons and protons immediately escape any nucleus they temporarily form.

Atomic nuclei began forming at about 1 second, when the temperature had fallen to 10 billion Kelvin (10^{10} K). During the first 3 minutes of the universe's existence, only the lightest nuclei, such as those of hydrogen and helium, could form. Even today, these light nuclei far outnumber the heavier nuclei that formed much later through fusion reactions within stars and in explosions of supernovae.

During the time when the universe was between 3 minutes old and about 100,000 years old, atoms did not form because the temperatures were still too high to allow electrons to settle into stable configurations around nuclei and form neutral atoms. After about 100,000 years, when the temperature had fallen to 10,000 K, light atoms finally started forming, eventually gathering into massive celestial objects, from which stars and galaxies ultimately evolved.

Evidence for the Standard Model

Cosmologists cite three kinds of observations to support the standard model. First, as described below, observations of galaxies throughout the sky support Hubble's Law, which states that, with the exception of some nearby galaxies, all observable galaxies are receding from one another with velocities proportional to the distances they are apart.

Second, sensitive radio receivers have detected faint microwave radiation, sometimes called the three-degree (3 K) cosmic background radiation, which is uniform in all directions. In other words, the universe, on average, emits radiation as if it were a black body at a temperature of about 3 K. Arno Penzias and Robert Wilson of Bell Laboratories in Holmdel, New Jersey, received the 1978 Nobel Prize in Physics for detecting this radiation in 1964.

According to the standard model, the 3 K radiation originated when the universe was still very hot. Any kind of matter heated to high temperatures glows with a spectrum of colors whose intensities are determined by the specific temperature. As an object's temperature rises to thousands of degrees, the wavelengths of the emitted light decrease, shifting gradually from dull red to bright red and eventually to blue-white. Conversely, at very low temperatures, matter radiates faint microwaves, with wavelengths much longer than visible light.

The photons that make up the 3 K radiation travel, of course, at the speed of light. To observers on Earth, this radiation appears drastically shifted from the short wavelengths of blue-white radiation emitted in the remote past by the extremely hot early universe to the longer wavelengths of the microwave radiation characteristically emitted by objects at 3 K. This change comes about as a result of a phenomenon called the red shift, described below. Measurements made with sensitive microwave detectors during the 1970s confirmed that the cosmic background radiation displays both the wavelength distribution and uniformity predicted by the standard model.

The third line of evidence supporting the standard model involves the proportions of the various low-mass atoms in the universe. The amounts of hydrogen, helium, and lithium observed are consistent with the standard model's predictions for what should have formed during the first 3 minutes. Advances in elementary particle physics in the late 1970s and 1980s have also broadened cosmologists' understanding of how — and, to some extent, in what proportions — subatomic particles such as protons and neutrons came into being in the aftermath of the initial Big Bang. Theory suggests, and measurements confirm, that there are about 1 billion photons for every proton or neutron.

Red-Shift Measurements and the Expanding Universe

The observation that light from all known distant galaxies appears to have large red shifts provides strong evidence supporting the theory of an expanding universe. By using a device called a spectrograph, astronomers can measure the pattern of light wavelengths (called a spectrum) coming from stars in distant galaxies. Astronomers can compare the wavelengths of light received from the stars' hydrogen atoms with those emitted by hot hydrogen atoms in the laboratory. If a galaxy is moving toward or away from Earth, its spectrum will appear shifted. Just as the pitch or frequency of a train locomotive's whistle appears to drop in frequency (with a corresponding increase in wavelength) as the train moves away, so the wavelengths of light from a galaxy moving away from Earth appear longer than if the galaxy remained stationary. Scientists refer to such light as "red-shifted," because the lines in its spectrum have shifted toward the longer-wavelength (red) end of the light spectrum. Conversely, light from a galaxy moving toward us would appear to have shorter wavelengths and would be termed blue-shifted, since the lines in its spectrum would be shifted toward the blue end of the light spectrum.

Because distant galaxies are moving rapidly away from Earth, their light appears red-shifted. This red shift can be used in a simple formula to compute the velocity at which the galaxy is receding. Red-shift measurements of the light from distant galaxies are critically important in modern cosmology, because they allow astronomers to calculate the large-scale expansion of the universe and to estimate the distance to faraway galaxies.

In 1929, astronomer Edwin Hubble at the Mount Wilson Observatory near Pasadena, California, discovered that the galaxies farthest from Earth had the greatest red shifts and must therefore be moving away from Earth at the highest speeds. Hubble also found that the velocity of a galaxy receding from Earth is directly proportional to its distance from Earth. This ratio of velocity to distance was called the Hubble constant, and the observed relationship between velocity and distance became known as Hubble's Law. The value of the Hubble constant is still imprecisely known, but it is on the order of 15 to 30 kilometers per second per million light-years. (A light-year is the distance that light travels in 1 year, about 6 trillion miles.)

To calculate the distance to a faraway galaxy, astronomers first measure the galaxy's red shift and compute the galaxy's recessional velocity. Using the Hubble constant, astronomers can then compute the galaxy's distance from Earth. This method assumes that the expansion of the universe has remained uniform over time. This is thought not to be true, however, because the gravitational effect of all the mass in the universe should gradually slow its expansion. Thus, the value of the Hubble constant should slowly decrease over time.

Since measured red shifts can provide estimates of the distance from Earth to observed galaxies, astronomers have been able to extend their two-dimensional photographic images of the distribution of galaxies in the sky into a three-dimensional portrait. Sensitive red-shift measurements made during the 1980s have revealed what appeared to be an unexpected unevenness in the distribution of galaxies. In the past, most scientists had assumed that galaxies would be evenly distributed throughout space. Now, cosmologists must try to explain the various scales of galactic clustering that have been observed: "local" clusters of galaxies consisting of a few dozen members, clusters of such clusters, and even "superclusters" extending hundreds of millions of light-years.

In 1981, astronomers at the Kitt Peak National Observatory near Tucson, Arizona, conducted a survey of the red shifts of a large number of galaxies. The survey revealed a nearly complete absence of galaxies in an immense region of space almost 300 million light-years across. (For comparison, the Milky Way Galaxy is only about 100,000 light-years across, and the distance between the Milky Way and the Andromeda Galaxy, one of its nearest neighbors, is over 2 million light-years.) The 1981 survey, and a number of others since, seemed to show that many galaxies congregate into either filament-shaped or pancake-shaped clusters, with vast, nearly empty "void" regions between them.

A theory first proposed in 1972 by the late Yakov Zel'dovich of the Institute of Applied Mathematics in Moscow, U.S.S.R., might account for the uneven distribution of galaxies. According to the theory, which is often called the "pancake model," matter in the early universe coalesced into large clusters and superclusters, which later collapsed into flat planes or "pancakes" of matter. These, in turn, eventually broke up somewhat unevenly into clusters and individual galaxies. To date, cosmologists have not yet reached a consensus about whether this model is true, nor do they agree on an explanation for the large-scale structures observed. Some cosmologists believe that chance events in the first instants of the universe are responsible for at least some of the distribution of matter now observed.

At a meeting of the American Astronomical Society during early 1989, astronomers from the University of Michigan in Ann Arbor reported evidence that galaxies do apparently exist in one large void region in the constellation Boötes. Using data from the Infrared Astronomy Satellite (IRAS) launched in 1983, the researchers detected infrared radiation coming from what seemed to be exceptionally dusty galaxies in the void. Ground-based measurements by Harvard University astronomer Robert P. Kirshner also appeared to confirm this finding. It may turn out that large intergalactic voids do not exist after all.

Evidence for Dark Matter

For Zel'dovich's pancake model of the universe to be correct, the universe must contain a certain minimum average density of mass. While the density of matter in known stars and galaxies falls short of this critical density, astronomers believe that much of the mass in the universe may be in the form of so-called dark matter that is nonluminous and therefore difficult to detect. Such hypothetical dark matter could conceivably bring the universe's average density up to the level required by Zel'dovich's model.

The question of whether dark matter exists has been extremely important to cosmology for several reasons. According to simple mathematical models of the expanding universe, the cosmos will continue to expand indefinitely if its average density is less than the critical value of about 10^{-29} gram per cubic centimeter (the equivalent of about one hydrogen atom in every 3 cubic feet). On the other hand, if the total mass is greater than that amount, the expansion will eventually stop, and the universe will begin collapsing on itself and perhaps ultimately flash out of existence in what might be called the "Big Crunch." Since most astronomers estimate that the total amount of luminous matter observed to date is no more than about 15 percent of the critical value, additional dark matter could make the difference between an ever-expanding ("open") universe and one that will eventually collapse (a "closed" universe).

Dark matter, in any of a variety of forms, may either be spread uniformly throughout the universe or concentrated in particular locations. The outer regions of galaxies are one likely place where matter might exist. In the early 1970s, studies at Kitt Peak National Observatory showed that many spiral galaxies rotate at what appear to be impossibly high speeds. According to accepted theories, such galaxies should fly apart, unless some other force stabilizes them. Many astronomers believe that the gravitational influence of large amounts of matter at the periphery of these galaxies best explains their otherwise baffling stability.

Physicists have often suggested that elementary particles, possibly of a type as yet undiscovered, might account for substantial amounts of dark matter in the universe. For example, some very massive particles may contribute to the dark matter, although these particles have not been observed even in high-energy accelerator experiments. If these particles do occasionally reach Earth in the form of cosmic rays, they have not yet been detected.

A once-popular dark matter candidate — neutrinos — has fallen into disfavor. Some cosmologists have suggested that the immense number of neutrinos in the universe might collectively contribute a vast quantity of mass if each neutrino possesses a tiny nonzero mass. As of the late 1980s, experiments to test this possibility have had conflicting results. Furthermore, computer models of galaxy

Figure 7. Dark Matter in Spiral Galaxies. Galaxy NGC4594 in the Virgo Cluster, known as the "Sombrero" because of its appearance from Earth, is one of thousands of known spiral galaxies in the universe. The recent discovery that the outer regions of many spiral and disk-shaped galaxies rotate faster than initially expected has led astronomers to suspect that those outer regions contain large amounts of matter that cannot be seen. Matter that neither glows nor reflects light is generally invisible to astronomers and so is referred to as "dark matter." Some astronomers have even suggested that most of the mass in the universe may be dark matter, in such forms as unobserved particles, interstellar dust, and gas. *Courtesy: National Optical Astronomy Observatories, Kitt Peak.*

formation have shown that neutrinos with nonzero masses should have produced far more galaxy clustering than astronomers have actually observed.

Another candidate for dark matter is the so-called weakly interacting massive particles, or WIMPs. These particles interact only through the weak and gravitational forces, not through the strong or the electromagnetic forces. Other candidates include dim stars, planetary objects not associated with stars, interstellar dust, black holes, and hypothetical elementary particles such as the photinos, gluinos, and gravitinos predicted by the speculative theory of supersymmetry described in the article "Experimental Particle Physics."

Weakly Interacting Massive Particles

If dark matter does exist in the form of WIMPs, some physicists have suggested that these particles might explain a long-standing mystery of solar physics. Since the late 1960s, when the number of neutrinos reaching Earth from the Sun was first measured, physicists have not been able to explain why far fewer neutrinos have been observed than expected. Based on the current understanding of the nuclear reactions taking place inside the Sun, most astrophysical theories predict that three times as many neutrinos should reach Earth from the Sun as neutrino detectors have measured.

Astrophysicists believe that solar neutrinos are produced primarily in the Sun's core, where extremely high temperatures favor the occurrence of certain otherwise uncommon nuclear reactions. According to theory, the temperature in the Sun's core strongly influences the rate of these reactions. The unexpectedly low production rate of solar neutrinos could indicate that the Sun's core temperature is somewhat lower than had been assumed.

Astrophysicists John Faulkner at the University of California in Santa Cruz and Ron Gilliland at the National Center for Atmospheric Research in Boulder, Colorado, have suggested that the presence of a relatively small number of WIMPs could cool the Sun's core sufficiently to reduce its rate of neutrino production to the observed amount. In theory, WIMPs could accomplish this cooling by interacting frequently enough with ordinary matter to acquire energy from hot particles in the solar core, but not often enough to keep the WIMPs from escaping the core and carrying energy to the Sun's outer regions. If the WIMPs interacted very frequently with the hot particles in the solar core, they could not travel large distances within the Sun and therefore could not carry energy from one part of the Sun to another.

While Faulkner originally believed that neutrinos with a nonzero mass might be responsible for cooling the Sun's core, experimental evidence has shown that, even if neutrinos have mass, they probably do not exist in sufficiently large numbers to produce the predicted cooling of the Sun's core. Instead, scientists have come to favor weakly interacting particles with a mass in the range of 5 GeV.

According to calculations published by Faulkner and Gilliland in the 1985, WIMPs within the Sun's core could transport heat from the core without altering the Sun's brightness, which is in agreement with the brightness predicted by theory. Although WIMPs could move large distances within the Sun, the Sun's gravity would trap the particles and prevent them from escaping the Sun completely.

Also in 1985, William Press and David Spergel at the Harvard-Smithsonian Center for Astrophysics in Cambridge, Massachusetts, suggested that WIMPs could cool the Sun's core and thereby reduce its rate of neutrino production. Assuming that a specified number of WIMPs (or cosmions, as they were called by Press and Spergel) were created during the Big Bang, the two researchers calculated the number of WIMPs that would now be trapped within the Sun. This number — about one cosmion for every trillion particles in the Sun — is consistent with what would be expected for a star of the Sun's size and current brightness.

Many particle physicists believe that photinos are the most promising candidates for WIMPs. This possibility is not without problems, however. According to supersymmetry theory, photinos are their own antiparticles; this means that two photinos that collided inside the Sun would annihilate one another, and over a long period of time most of the Sun's photinos would be eliminated. Particle physicists still do not agree about how often such annihilations should occur. If high-energy accelerator experiments confirm the existence of photinos, then physicists will be anxious to determine whether they have the characteristics of the WIMPs proposed to exist in the Sun's core.

Inflationary Universe

Although the standard model of cosmology accurately describes many characteristics of the observed universe, it also leaves many features of the cosmos unexplained. For example, it does not explain the staggering degree of uniformity of the 3 K background radiation. The standard model simply accepts this uniformity, also called isotropy, as a given.

Scientists believe that the 3 K radiation was emitted by high-temperature matter about 100,000 years after the universe began. Long before then, however, the universe had expanded to so large a size that it should not have remained in overall thermal equilibrium, that is, maintained the same temperature throughout. The different parts of the universe would have maintained a uniform temperature only as long as they could exchange energy with one another and thus eliminate temperature differences.

According to the standard model, this thermal equilibrium should not have been possible after the universe was about 10^{-35} second old. From that time onward, any small temperature differences that arose between parts of the universe should have remained. Thus, the standard model predicts that the cosmic background radiation should be far less uniform than now observed.

In 1980, physicist Alan Guth, now at MIT, proposed a significant modification of the Big Bang model, a concept called the "inflationary universe," to explain the mysterious isotropy and other problems arising from the standard model. Guth's theory was later amended and expanded by the ideas of Andrei Linde in the U.S.S.R. and

Andreas Albrecht and Paul Steinhardt at the University of Pennsylvania. Inflation does not supersede the Big Bang theory; instead, the Big Bang simply takes over where inflation leaves off — after the first 10^{-30} second of cosmic time.

While the standard model assumes that the universe has expanded uniformly since the beginning of time, the inflationary model assumes that the universe initially expanded evenly and then underwent a brief period of tremendous expansion at about 10^{-35} second. In 10^{-30} second, the observable universe expanded by a factor of 10^{25} or perhaps by as much as $10^{100,000,000}$. At the end of this period of ultrarapid expansion, the universe continued to expand, but at the much reduced pace of the standard model.

The inflationary model suggests that this phenomenal expansion occurred after a pre-existing field became locked into a high-energy condition called a false vacuum. In this state, a *repulsive* gravitational field prevailed (rather than the usual attractive form of gravity), which drove the universe's expansion. The rapid inflation "supercooled" the universe — much as water can be cooled to below its freezing temperature — thus causing the four once-unified forces of nature to become distinct forces. The high-energy state of the false vacuum fell instantly, the temperature rose abruptly, and this led to the present uniform expansion of the Big Bang.

In the inflationary model, the universe remained small enough to maintain thermal equilibrium for a slightly longer time than in the standard model, and it also achieved a higher degree of thermal uniformity. The observed near-perfect uniformity of the 3 K background radiation reflects the more complete thermal equilibrium that the inflationary model predicts for the early moments of the universe.

The inflationary model also seems to solve some other cosmic mysteries. It suggests that all the matter in the universe may have been created in that first inflationary instant — from little or no pre-existing mass or energy. This means that the universe possesses zero net energy today: the positive energy locked up in mass exactly balances the negative energy of the universe's gravitational field. Inflation also explains away magnetic monopoles by providing a reason why they should be extremely rare. The huge size to which the universe has inflated has spread the monopoles out over such a vast volume that the chance of detecting one is extremely low.

Most remarkably, the inflationary model predicts that the universe should be precisely balanced between being open and closed. For all practical purposes, the universe would expand forever, continually slowing down but never quite enough to begin a collapse. The mass density of the universe may exactly equal the critical density, poising it permanently between an open and closed state. In this way, the inflationary model lends additional credence to the notion of dark matter, which would have to exist if the universe contains enough mass to reach the critical density. The discovery of just enough dark matter to bring the universe to the critical density would provide strong support for the inflationary model.

Cosmic Strings

Among the most remarkable objects astrophysicists have proposed in recent years are massive "cosmic strings" (not to be confused with the so-called superstrings of theoretical particle physics). Cosmic strings, if they exist, may be relics from the first inflationary moments of the universe. Some theorists suggest that extremely long cosmic strings or cosmic string loops may be fantastically thin tubes (10^{-30} centimeter in diameter) of trapped false vacuum with a number of exotic properties, some of which may help explain the formation of galaxies and their large-scale clustering.

Theory suggests that cosmic strings would be dense, more than 10^{22} grams per centimeter of length. Theoretically, cosmic strings may wiggle and vibrate and travel close to the speed of light. They may loop back upon themselves and over time form smaller cosmic strings, which would become massive nucleating sites for the formation of galaxies.

Cosmologists are now eagerly awaiting the results of a number of ongoing and planned searches for evidence of cosmic strings. One of these searches involves looking for tiny anomalies in the microwave cosmic background radiation, changes in intensity that would be only about one part in 100,000 or less. Yet other researchers have begun searching for evidence of so-called gravitational lensing by cosmic strings, a situation that would create a peculiar double image of a galaxy.

Great Attractor

Alan Dressler, an astronomer at the Carnegie Institution's Mount Wilson and Las Campanas observatories, and his colleagues announced in 1987 that the Milky Way Galaxy, as well as nearby superclusters of galaxies in a gigantic volume of space, seem to be systematically moving, or "streaming," in one direction. This led the scientists to suggest that the galaxies may be gravitationally affected by an as-yet-unidentified mass dubbed the "Great Attractor." The data suggest that the Great Attractor may be about 500 million light-years away and may have a mass 20 times greater than the local supercluster of galaxies.

The possibility of streaming flow and the existence of a Great Attractor has stirred great controversy among cosmologists. Most agree that it will take time to prove or disprove whether the apparent streaming flow is actually occurring.

Superconductivity

In 1986, researchers from the IBM Zurich Research Laboratory in Switzerland startled the scientific world when they announced the discovery of a class of ceramic materials that exhibit superconductivity at much higher temperatures than previously known. For 75 years, physicists had known that certain metals lost their resistance to the flow of electricity and became perfect electrical conductors when they were cooled below a critical temperature. Physicist H. Kamerlingh Onnes at the University of Leiden in the Netherlands first observed superconductivity in 1911, while studying the electrical conductivity of mercury at extremely low temperatures.

The 1986 discovery of high-temperature ceramic superconductors opened up a completely new area of research. Since then, scientists have developed a number of other ceramic compounds that become superconductors at temperatures higher than the boiling point of liquid nitrogen (about –321°F) — a fact of great significance, since users of superconductors can obtain liquid nitrogen much more easily and cheaply than liquid helium, the extremely cold liquid required to cool conventional metallic superconductors.

The leaders of the IBM team that made the initial discovery, K. Alex Müller and J. Georg Bednorz, received the 1987 Nobel Prize in Physics for their work. In a matter of months after Müller and Bednorz announced their breakthrough, the previously quiet field of superconductivity research was transformed into one of the most frenzied in science, with researchers worldwide striving to find materials with ever-higher superconducting temperatures and to develop a theoretical framework to explain the phenomenon. The excitement reached a peak during an historic, all-night superconductivity session at the meeting of the American Physical Society in March 1987. The jam-packed session generated so much excitement that it was nicknamed the "Woodstock of physics."

Earlier experiments had demonstrated that electricity would flow through a closed wire loop of a superconducting metal for days with no measurable loss. The critical temperature at which superconductivity appears differs for each material. Before 1986, no superconducting material had been found with a critical temperature higher than about 23 K (23 Kelvin above absolute zero, the lowest possible temperature. One Kelvin is equal to 1.8 degrees Fahrenheit. A temperature of 23 K is equal to –418°F.)

In even the best conventional conductors, such as aluminum and copper, electrical energy is lost as heat because of the material's resistance to the flow of electrons. Since superconductors have no electrical resistance, they eliminate such waste. Conventional metallic superconducting materials (those known before Müller and Bednorz's discovery) can carry electrical currents that would melt ordinary conductors. For these reasons, physicists have used conventional superconducting wire to fabricate extremely powerful, compact magnets and to develop superconducting electrical generating and storing systems.

Given the tremendous potential of superconductivity, industry and government were spending several hundred million dollars per year to develop and manufacture superconducting systems, even before the 1986 discovery. The chief drawback to using such systems has been that they require expensive and elaborate refrigeration equipment to maintain the superconducting materials at the extremely low temperatures needed for superconductivity.

Müller and Bednorz's initial ceramic material — a barium-lanthanum-copper oxide — became superconducting at about 30 K (about –400°F). In February 1987, C.W. (Paul) Chu of the University of Houston and his collaborators developed a ceramic material that became superconducting at a much higher temperature — about 90 K (about –300°F). Since then, other researchers have developed ceramics that become superconducting at even higher temperatures. The record critical temperature at the beginning of 1989 was 125 K (–235°F) for a thallium compound developed by the IBM Almaden Laboratories in California. This represented just one of several new high-temperature superconductors developed in 1988 that used bismuth or thallium. The compounds contain the elements bismuth, strontium, calcium, and copper (Bi, Sr, Ca, Cu) or thallium, barium, calcium, and copper (Tl, Ba, Ca, Cu) in addition to oxygen. In 1988, researchers at AT&T Bell Laboratories announced the development of a bismuth oxide ceramic superconductor that contains no copper atoms. This announcement was surprising, because copper atoms were believed to be essential for high-temperature superconductivity.

No one yet knows, however, whether any of these high-temperature materials will be suitable for many practical applications. A material's superconducting properties are influenced not only by its temperature but also by external magnetic fields and the intensity of current passing through it. The new materials have not yet demonstrated the ability to maintain their superconductivity under the normal operating conditions for some applications. Furthermore, although the new ceramic superconductors are in some ways easier to manufacture than conventional metallic ones, they are also brittle and inflexible — properties that would make it difficult, for example, to form them into wire.

Discovery of High-Temperature Superconductivity

Before early 1986, the highest critical temperature recorded for any superconductor was 23.3 K for an alloy of the metallic elements niobium and germanium (written in chemical symbols as Nb_3Ge), developed by Louis Testardi at Bell Laboratories in Murray Hill, New Jersey. Because the Nb_3Ge alloy is brittle and difficult to manufacture into useful objects, scientists generally use an easier-to-handle alloy of niobium and titanium (NbTi), with a critical temperature of only 10 K (–442°F), for most applications requiring superconductivity.

In January 1986, Müller and Bednorz observed the remarkable appearance of superconductivity at about 30 K in an oxide of lanthanum, barium, and copper. They had begun experimenting with such oxides because they believed the materials might be superconductive at higher temperatures than had previously been achieved.

In the summer of 1985, their research was influenced by a French paper describing the properties of barium-lanthanum-copper oxide, $BaLa_4Cu_5O_{13.4}$. This ceramic possesses a crystal structure like that of a mineral called perovskite. In fact, the most characteristic feature of all the ceramic superconductors seems to be the existence of flat structures — sheets or chains — of copper and oxygen atoms. Surprisingly, when scientists replace the rare earth and alkaline earth elements in these ceramics with others in the same chemical family, the materials still exhibit high-temperature superconductivity.

The ceramic oxides Müller and Bednorz chose to study were made by baking the component materials at temperatures under 1,000°C (about 1,800°F). From a detailed study of samples, Müller and Bednorz identified La_2CuO_4 as a key superconducting candidate. Their paper announcing the discovery of the high-temperature superconductor was published in *Zeitschrift für Physik B* during 1986. By the end of the year, their results were confirmed by researchers at the University of Tokyo, the University of Houston, and AT&T Bell Laboratories. In February 1987, Zhongxian Zhao and his collaborators at the Academia Sinica in Beijing, China, reported achieving a critical temperature of about 90 K with a yttrium-barium-copper oxide.

As these higher temperatures were achieved, many physicists began to doubt whether the standard "BCS" theory of superconductivity (named for the initials of University of Illinois physicists John Bardeen, Leon Cooper, and Robert Schrieffer, who developed the theory) could explain superconductivity in these ceramic materials. The adequacy of the theory had already been questioned after the discovery of two other types of superconductors — organic and "heavy-electron" superconductors — since 1979. Although these two types of materials are superconducting only at temperatures near absolute zero, they intrigue physicists because of their unusual conduction mechanisms.

The BCS model does seem to provide an adequate description of conventional metallic and heavy-electron superconductors. Physicists do not yet agree, however, about whether the BCS theory can also explain the mechanisms at work in organic and ceramic superconductors.

BCS Model of Superconductivity

Electrical conduction in ordinary solids involves the movement of charged particles (electrons or ions) through a fairly rigid lattice (structured network) of atoms. Metals such as aluminum and copper consist of a regular array of atoms whose outermost electrons are not tightly bound to any single atom but instead move freely through the lattice. For example, when a person attaches the ends of a metal wire to the oppositely charged poles of a battery, electrons pass through the wire to create an electric current. In general, scientists refer to current-carrying electrons as conduction electrons.

Electrons generally do not flow unimpeded through a lattice of atoms, however, since imperfections in the lattice and thermal vibrations of the atoms deflect them from their forward motion. When such interactions occur, the electrons transfer energy to the lattice, heating the conductor. While this conversion of electrical energy into heat is fine in a toaster, it wastes power in a transmission line and can damage materials carrying strong electric currents.

Metals are not perfect electrical conductors because their resistance causes them to heat up as a current passes through. Substances with tightly bound outer electrons and amorphous materials such as glass, whose atoms do not form regular arrays, tend to be poor conductors and may be used as insulators.

According to the BCS theory, lattice vibrations in a material below the critical temperature facilitate the formation of loosely bound pairs of conduction electrons called Cooper pairs, which can move without friction through the regular atomic lattice of a metal. Although the two electrons remain separated by a distance equal to the size of many atoms, they effectively behave like a single particle as they move through a metal.

The ability of Cooper pairs to move without friction through a lattice depends upon a property called quantum statistics. All subatomic particles may be divided into one of two classes — those with even quantum statistics (also called Bose-Einstein particles), and those with odd quantum statistics (also called Fermi-Dirac particles). Particles with even quantum statistics can enter into a special type of collective motion, in which a group of particles moving in one direction tends to pull any identical particles nearby into the same line of motion. In this state of collective motion, particles with even quantum statistics can move essentially without friction. By contrast, particles with odd quantum statistics behave in an opposite way, repelling one another.

Although individual isolated electrons have odd statistics, two electrons bound together in a Cooper pair behave as a single particle with even statistics and, as a result, move collectively through the lattice of a superconductor without friction. Before the BCS model, no theory had predicted that particles with odd statistics could ever behave like particles with even statistics.

Since the force binding the two electrons in a Cooper pair is weak, this pairing can be easily disrupted (and the superconductivity destroyed) by any of three factors: thermal vibrations; impurities in the metal, which can distort the lattice structure; and strong magnetic fields, which can affect the orientation of individual electrons. As a result, superconductivity occurs only in relatively pure samples and disappears when the temperature (the intensity of thermal vibrations) rises above a critical value. Magnetic fields stronger than another critical value specific for each material also stop superconductivity.

Theories of High-Temperature Superconductivity

While Cooper pairs exist in the new ceramic superconductors in much the same way as in conventional metallic superconductors, physicists have not agreed on a mechanism by which pairs can form and be maintained at relatively high temperatures. Another fundamental disagreement concerns whether a single theory can explain high-temperature superconductivity over the full range of temperatures at which it has been observed. Because of the many theoretical uncertainties, no one knows whether materials can be made with a critical temperature at or above room temperature. Some researchers, including William A. Goddard at Caltech, have developed theories predicting that the maximum critical temperature for copper oxide superconductors is in the range of about 225 K (–55°F). MIT physicist Keith Johnson has developed a chemical bonding theory of high-temperature superconductivity that predicts an upper limit of 235 K (–37°F) for any superconductor.

The common feature of the major classes of high-temperature ceramic superconductors seems to be crystal planes of copper oxide. Some physicists suspect that some kind of quantum-mechanical interaction between these planes causes the high-temperature phenomenon. Others believe that magnetism plays a critical role, and still others attribute the superconductivity to a variety of other mechanisms.

Suspension Effect in High-Temperature Superconductors

By chance, Palmer N. Peters of the National Aeronautics and Space Administration's Space Science Laboratory in Huntsville, Alabama, discovered an entirely new superconductivity phenomenon in 1987. When Peters was pulling a magnet away from a small sample of a specially treated, high-temperature ceramic superconductor (yttrium-barium-copper oxide), he found that the superconductor became suspended in midair beneath the magnet.

To explain the phenomenon, physicists theorized that current flow within the superconductor set up a magnetic field with a polarity opposite to that in the magnet. This field preserves a stable equilibrium between the downward pull of gravity and the upward pull of the magnet. Two ordinary magnets, one held above the other, would possess a balance point between gravity and their magnetic attraction, but this point would be unstable, a fact that makes the stable suspension effect observed by Peters all the more puzzling.

Superconducting Technology

Most existing superconducting systems have used the niobium-titanium alloy NbTi, cooled by refrigeration equipment to the temperature of liquid helium, 4 K (–452°F). In situations that require especially powerful magnetic fields, researchers have sometimes used the brittle, expensive niobium-tin alloy Nb_3Sn, because it remains superconducting in magnetic fields almost twice as strong as those that disrupt the superconducting state of NbTi. If the high-temperature ceramic superconductors can be made into durable components capable of withstanding high currents and magnetic fields, they could have a wide range of applications, ranging from extremely sensitive magnetic field detectors to large-scale transportation and power systems.

Josephson effect devices. In 1962, Brian Josephson proposed an application for Cooper pairs in what is now called the "Josephson junction," in which a thin insulating oxide layer separates two superconductors. In the DC (direct current) Josephson effect, Cooper pairs tunnel through the oxide barrier to produce a current that is extremely sensitive to external magnetic fields. Superconducting Quantum Interference Devices (SQUIDs), sensitive to magnetic fields as small as 10^{-10} gauss (about one ten-thousandth the magnetic field produced by the electric currents in a human heart), exploit the DC Josephson effect.

In the AC (alternating current) Josephson effect, Cooper pairs respond very sensitively to voltage changes as they move back and forth across a Josephson junction. This property could make the junctions suitable for use in very-high-speed switching devices in powerful computers and other electronic equipment. Most physicists believe that high-temperature superconductors will first see widespread use in SQUIDS, rather than in heavy industrial applications such as magnetically levitated trains and power transmission lines.

Superconducting generators and motors. Utilities may someday rely on large superconducting generators to produce electricity in commercial power plants. A conventional electrical generator runs at about 98.5 percent efficiency, meaning that it turns 98.5 percent of the mechanical energy supplied to it (by a steam-fed turbine, for example) into electrical energy; it loses the other 1.5 percent as waste heat. According to a 1988 report by Argonne National Laboratory, a generator that incorporated high-temperature superconductors could conceivably achieve an energy-conversion efficiency of 99.7 percent. What may seem like a small improvement in efficiency actually amounts to an 80 percent reduction in energy waste, an energy savings that could cut the life-cycle costs of an electrical generator by as much as 60 percent, according to the Argonne report.

Because current will circulate almost indefinitely in a closed superconducting loop, some scientists have suggested that huge superconducting coils, perhaps a mile or more in diameter, could be used to store excess electrical energy produced during times of low energy demand for later use at times when demand increases. This would allow utility companies to avoid much of the expense of installing enough generating equipment to meet the relatively few hours of peak demand each day. With the use of a storage coil to store excess production from a main power plant, a relatively small plant could handle the peak demand. The Argonne report suggested that the use of magnetic energy storage systems based on high-temperature superconductors could give the U.S. the equivalent of up to 15 percent additional generating capacity by allowing utilities to use their existing plants more efficiently.

John Rogers of the Los Alamos National Laboratory has already designed a small storage coil based on conventional superconductors, not to store power overnight, but instead to dampen troublesome oscillations in the power transmission lines between the cities of Los Angeles and Seattle. Rogers has found that the coil could significantly reduce these oscillations by storing electricity without loss and releasing it as needed.

If superconducting transmission lines became widely available, they could lead to sweeping changes in the power industry. To reduce the amount of power lost during transmission, power plants are now generally built relatively near the towns and cities where the power is used. In contrast, with superconducting transmission lines, power plants could be located many hundreds to thousands of miles from their users and transmit power without the high losses that now make this impractical. For example, if such systems were available, excess electricity from a power plant in one region where demand was low could be transmitted thousands of miles to meet high power demands in another area.

In 1983, the General Electric Co. completed tests of a 20-megawatt generator based on conventional superconductors, but neither General Electric nor Westinghouse,

Figure 8. Superconducting Generator. General Electric's prototype 20-megawatt superconducting generator can produce twice as much power as a conventional generator of the same size and wastes less energy. The superior efficiency of this prototype generator is the result of the superconducting wire in its 13-foot-long superconducting rotor, shown suspended in front of the machine. As in conventional generators, when the rotor is inserted inside the generator and turned within the generator's magnetic field, an electric current is induced in the rotor's coils. Unlike the coils of conventional rotors, however, those of superconducting rotors conduct electricity with zero loss. To operate as a superconductor, the rotor must be maintained at the frigid temperature of liquid helium (–452°F). *Courtesy: General Electric Research and Development Center.*

which developed a comparable unit, has found a market for their new products. Electric utility companies have not adopted superconducting generators because the technology is relatively new and largely untested. Generators built of reliable room-temperature superconductors would be very attractive, however, because they would not require the elaborate refrigeration systems needed for generators with parts made of conventional low-temperature superconductors.

Transportation applications. Superconductors may also be used in compact, efficient motors, where coils are made of conventional superconducting wire. In 1980, the Navy successfully tested a 400-horsepower motor made with conventional superconducting materials on a 60-foot ship in the Chesapeake Bay, and it hopes to develop a 40,000-horsepower unit for use in destroyers by the early 1990s.

Conventional superconducting magnets have also made possible an unusual type of train, called "maglev," which is short for magnetic levitation. A typical maglev train travels about 4 inches above a guideway (U-shaped in cross-section), suspended there by the repulsion between magnets in the guideway and superconducting magnets on the bottom of the train. At the same time, the magnets on the bottom of the train are attracted by a series of special propulsion magnets on the U-shaped guideway. The propulsion magnets are spaced a few feet apart along the length of the guideway and are activated automatically in sequence to pull the train forward.

Conventional trains waste a great deal of energy through friction, which increases with speed. This and other mechanical limitations prevent conventional trains from traveling at extremely high speeds. By eliminating wheel-rail contact, the maglev design dramatically reduces friction, increasing efficiency and allowing trains to travel faster. One experimental maglev train, built with conventional superconducting materials by the Japanese National Railways, has traveled at a speed of 400 miles per hour.

Uses in medicine and physics. Superconducting magnets are now used in magnetic resonance imaging (MRI), an advanced technique for diagnosing disease. This non-invasive imaging method, which doctors use to produce images of the body's interior, requires powerful and stable magnetic fields to achieve high resolution. Unlike conventional electromagnets, whose magnetic fields fluctuate in response to small changes in their power supplies, superconducting magnets — which do not depend on power supplies once they have been energized — are extremely stable. Consequently, powerful superconducting magnets are ideal for MRI. In fact, some analysts believe that MRI magnets will be the largest commercial application of conventional superconductors over the next few years.

Conventional superconducting magnets are also being used in other settings that require very powerful magnetic fields. The Mirror Fusion Test Facility reactor at Lawrence Livermore Laboratory incorporates twenty-six 75-ton conventional superconducting magnets to contain the hot, electrically charged particles that make up the plasma in the reactor.

The particle accelerators used in high-energy physics experiments likewise require powerful magnets to steer high-speed particles along curved paths. These magnets consume so much energy that the Fermilab accelerator exhausted its $8 million annual power budget after only 22 weeks of operation in 1983. Fermilab has completed the installation of about 1,000 superconducting magnets around the 4-mile circumference of its accelerator, reducing the facility's energy consumption by 75 percent.

In addition, superconducting magnets will serve as the key elements in the Superconducting Super Collider (SSC) to be built in Texas. This colossal accelerator will be 53 miles in circumference and will be capable of accelerating particles to unprecedented energies. Some project planners have wondered whether the SSC design should be revised to try to incorporate the new high-temperature superconductors developed during the past few years. However, to be effective, the new superconductors would have to carry very high currents without losing their superconductivity. They would also have to possess mechanical properties that would allow engineers to make them into working magnets. As of 1989, materials made of high-temperature superconductors were relatively brittle and unable to maintain the high current densities required for large superconducting magnets. Yet given the amazing progress that has occurred in superconductivity research during the late 1980s, many physicists are optimistic that these limitations can eventually be overcome.

Fusion

The ultimate goal of fusion research is to develop a powerful, relatively clean, and inexpensive energy source fueled by readily available materials, such as deuterium — an isotope of hydrogen plentiful in seawater. Long-term progress toward this goal continued through the 1980s, with several of the fusion projects in Europe, the U.S., and the U.S.S.R. expected to reach the long-sought "breakeven point" within the next few years. A reactor at breakeven, by definition, produces as much power as it consumes. Some analysts have predicted that commercial fusion power might be available early in the twenty-first century, while critics continue to question whether fusion will ever provide the almost unlimited, clean energy source that fusion advocates have predicted.

A successful fusion reactor would be able to sustain and control thermonuclear reactions akin to those responsible for the immense energy released by the Sun and hydrogen bombs. The heat generated by such reactions would be used to create steam to drive electricity-generating turbines. To duplicate in a fusion reactor the temperatures and pressures of stars has proved to be an extraordinarily difficult technological problem. The prospect of producing abundant, cheap energy has, however, sustained a long-term effort, begun in the 1950s, to develop a practical fusion reactor. In fiscal year 1989, the U.S. Department of Energy (DOE), the chief U.S. funding source for non-military fusion research, was expected to spend more than $350 million on projects in controlled fusion. Europe, Japan, and the U.S.S.R. are spending comparable amounts of money on nonmilitary fusion research projects.

Theory of Fusion

Fusion reactions release the energy that binds the nucleus together in a manner analogous to the release in ordinary fire of the energy contained in chemical bonds. In combustion — the burning of wood, for example — molecules in the substance being burned are combined with oxygen in the air to form water, carbon dioxide, and other compounds. Chemical bonds in the molecules are broken and new bonds are created, with the release of some of the initial binding energy as heat or light. Since this combustion reaction will take place only if the molecules are in a high-energy state — at a high temperature — a flame or a spark is needed to start the fire. But once begun, the fire releases more energy than it consumes.

Nuclear fusion is different from combustion in that it involves the release of nuclear binding energy — the energy that binds protons and neutrons together in the nucleus, rather than that which keeps electrons bound in atoms and molecules. At sufficiently high temperatures, two colliding atomic nuclei can fuse. In the process, the bonds holding the neutrons and protons together in each nucleus are disrupted and then re-established as the new, larger nucleus is formed.

Only fusion reactions in which the two reactant nuclei have less binding energy than the fused nucleus are energy-producing. The excess total energy is released in the form of high-speed neutrons, protons, and other subnuclear particles that fly out from the colliding nuclei. The binding energy carried off as kinetic energy by these particles can be harnessed for practical use. The particles can be used, for example, to bombard a solid target, and the resulting heat can, in turn, be used to convert water into steam to drive electricity-generating turbines.

The forces holding a nucleus together are about a million times more powerful than those binding electrons to the nucleus; accordingly, about a million times more energy for a given mass can be released in fusion reactions than in ordinary chemical reactions such as combustion. To accomplish nuclear fusion, however, temperatures of millions of degrees Celsius are required rather than the few hundred degrees to ignite an ordinary fire. At these enormous temperatures, the thermal motion is so high that individual nuclei come close enough to one another to react, despite the powerful electrostatic forces that normally keep them apart. Because extreme temperature is essential for fusion, the reactions are often called thermonuclear to distinguish them from the much less energetic fission reactions.

At multimillion-degree temperatures, found naturally only inside celestial bodies such as stars, ordinary solid matter is highly ionized — electrons are freed from atoms, creating a gas of nuclei and free electrons called a plasma. The aurorae that appear in the night sky in northern latitudes are examples of a natural plasma. In fact, most of the visible universe — stars and glowing nebulae — has matter in a plasma state.

To produce a thermonuclear fusion reaction on Earth, physicists must ionize a gas to create a plasma and heat it to temperatures of millions of degrees. Furthermore, the plasma must somehow be contained in a small volume at least long enough to sustain an energy-producing reaction. Since the enormous temperatures associated with a thermonuclear reaction would instantly vaporize an ordinary container (and heat transfer from the plasma to the walls of a material container would extinguish or "quench" the high-temperature plasma), scientists have concentrated on two general approaches to contain thermonuclear fusion reactions: magnetic confinement and inertial confinement. In the former, intense magnetic fields are used to contain a plasma more or less continuously; in the latter, multiple laser beams are used to compress tiny pellets of fuel and cause fusion in them.

Types of fusion reactions. One of the reactions conducted in fusion test reactors is the fusion of two deuterium nuclei. Deuterium — a relatively rare, heavy form of hydrogen — has a nucleus consisting of one neutron and one proton, while an ordinary hydrogen nucleus consists of a single proton. Deuterium occurs naturally in water at a concentration of one atom for every 6,700 ordinary hydrogen atoms. The energy yield of fusion reactions is, however, so much greater than the yield of chemical combustion reactions that the fusion of the deuterium nuclei in a gallon of seawater (1/250 ounce) would release as much energy as the burning of 300 gallons of gasoline. The deuterium in the top 10 feet of the world's oceans could meet all human energy needs for 50 million years. Another attraction is that extracting deuterium from water costs only pennies per gallon.

Fusion of ordinary hydrogen nuclei also yields energy, but requires more extreme conditions than the fusion of deuterium nuclei. Such proton-proton fusion does take place, however, in the Sun's extremely dense and hot core at temperatures of many millions of degrees. Another thermonuclear reaction, which releases more than four times as much energy as deuterium-deuterium fusion, is the fusion of deuterium with tritium — a form of hydrogen with a nucleus consisting of one proton and two neutrons.

Radioactive tritium does not occur naturally but can be produced at relatively low cost in conventional nuclear reactors or could ultimately be bred in fusion reactors.

Several experimental fusion reactors under construction or being planned will use deuterium-tritium reactions because of their higher energy yields, although these reactions also have disadvantages. The high-energy neutrons produced by deuterium-tritium fusion are captured by nuclei in the reactor materials, making atoms in the structure radioactive. As a result, remote-handling equipment must be used. Furthermore, key reactor parts would eventually become corroded and damaged by the intense radiation.

Lawson criterion. The behavior of hot plasma inside a fusion reactor is generally described by three crucial quantities: the plasma temperature, the particle density in the plasma, and the duration or lifetime of the fusion reaction. At sufficiently high temperatures, fusion reactions will readily take place. The fusion reaction time is, in practice, limited by the period during which the plasma can be contained in one place before the plasma particles fly apart; this is called the plasma confinement time. No prototype fusion reactor has yet proved capable of sustaining a fusion reaction for more than a fraction of a second. But even during a brief moment of sustained fusion, useful energy could be obtained from a reacting or "burning" plasma, provided that it was dense enough.

Physicist John Lawson of the U.K.'s Harwell Laboratory studied the combinations of temperature, density, and duration needed for energy-producing fusion reactions. In 1957, he formulated a widely used set of conditions for achieving breakeven, the condition in which a fusion reaction generates as much energy as it consumes. Since any fusion reactor must produce more energy than it consumes if it is to generate power, the breakeven point is a major goal for all fusion projects.

For deuterium-tritium reactions, the Lawson criterion is a temperature of approximately 100 million degrees Celsius, and any combination of plasma density and reaction time large enough to ensure that their product — the Lawson parameter — exceeds about 10^{14} nuclei-seconds per cubic centimeter. (This value applies only to fusion reactions in magnetic confinement reactors; a higher value applies to fusion in inertial confinement reactors.)

Although the breakeven condition has not yet been achieved, a succession of reactors designed over the past 20 years has steadily increased the plasma temperature, density, and confinement time to a level relatively close to this goal — within a factor of two, a relative small amount when compared with the progress since fusion research began. In particular, a magnetic confinement machine at Princeton University has come so close that investigators there are striving for "ignition" — a self-sustaining fusion reaction beyond the breakeven point — by 1996.

Magnetic Confinement Reactors

Electrically charged particles, such as the electrons and nuclei that make up a plasma, travel on curved trajectories when they move through a magnetic field. This principle is the basis of magnetic confinement reactors — fusion reactors that use powerful magnets to create a magnetic bottle to contain the hot plasma. Elaborately designed electromagnets, often made of superconducting wire, surround the reaction chamber and are shaped in such a way that plasma particles approaching the reactor walls are deflected back into the interior, thus preventing the cooling or quenching of the reaction plasma by heat transfer to the walls.

Because the amount of magnetic bending depends on a particle's mass and velocity, which are not the same for each type of particle, this magnetic confinement mechanism does not work perfectly, and some of the plasma particles continually leak out. As a result, reactor designers do not try to contain the plasma permanently; rather they attempt to contain it long enough to extract fusion energy and to replace lost particles by continuously injecting new ones into the reactor. As fusion researchers have gained experience with the behavior of plasmas under different conditions, however, they have developed new reactor designs that have made it possible to steadily increase containment times.

Tokamak reactors. The most common type of magnetic confinement reactor is the toroidal, or doughnut-shaped, reactor called a tokamak, after a series of machines of that name built at the Kurchatov Institute of Atomic Energy in Moscow, U.S.S.R., in the 1960s. One set of large wire coils surrounding the reactor chamber produces the magnetic confinement field. Another set of coils is used to heat the reactor's plasma to million-degree temperatures by causing an electric current to run through the plasma itself.

Like the wires in a toaster, the plasma has electrical resistance and so heats up when the electric current passes through it. This method of heating is generally called ohmic heating, after the nineteenth-century German physicist Georg Simon Ohm, who established the basic law

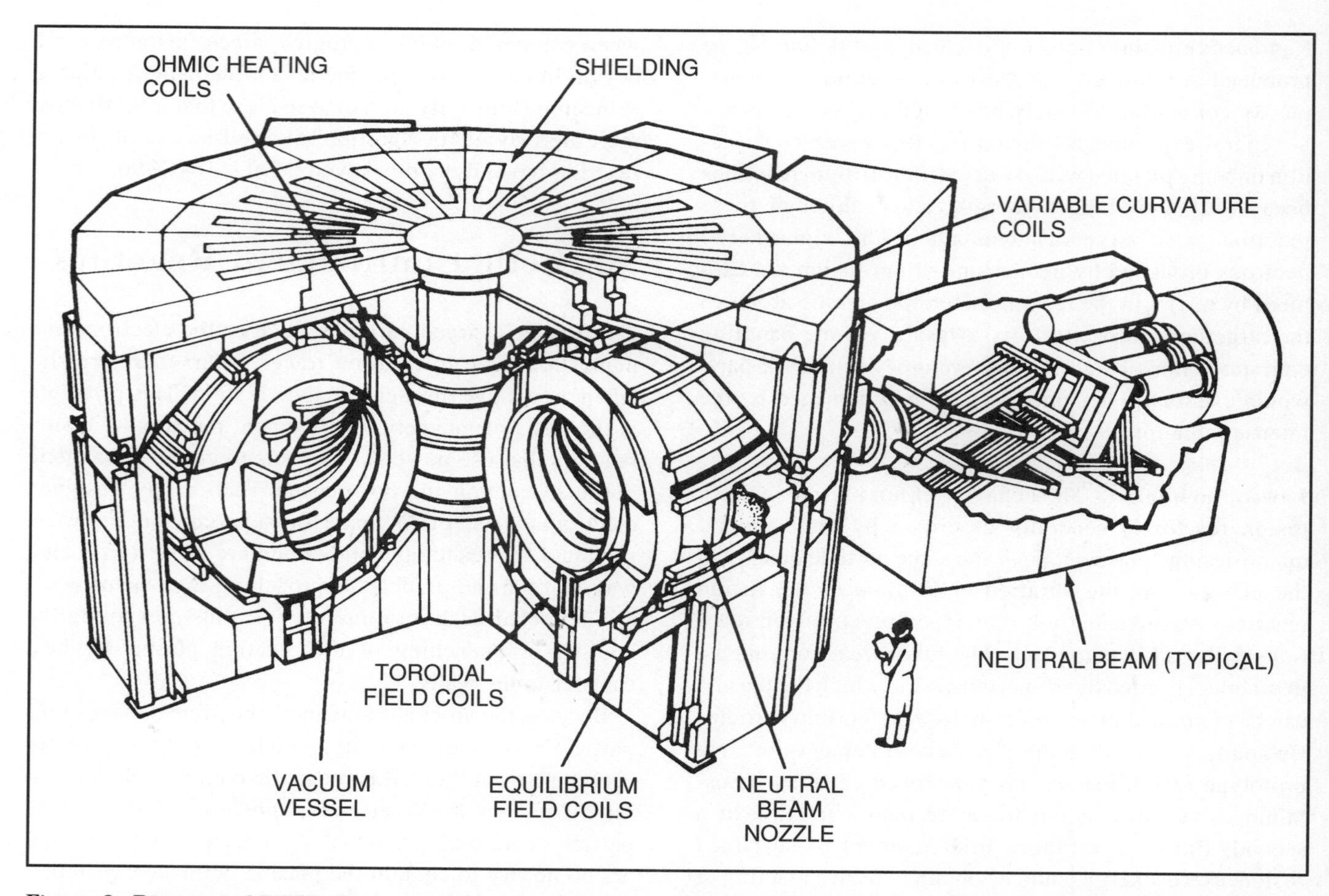

Figure 9. Diagram of TFTR. The powerful magnetic fields generated by TFTR's superconducting magnets — the toroidal and equilibrium field coils and the variable curvature coils — contain the reacting plasma of charged particles within the reactor's doughnut-shaped vacuum vessel. Heating coils supply the energy to increase the plasma's temperature, while a beam of neutral particles is continuously injected into the reaction chamber to replace particles leaking from the vessel. At sufficiently high temperature and density, the plasma will release fusion energy in the form of high-speed particles. *Courtesy: Princeton Plasma Physics Laboratory.*

describing the flow of currents through a medium with electrical resistance. In the plasma, the electric current causes a general motion of plasma particles in two directions; positively charged particles move in the direction of the current flow, while negatively charged particles move in the opposite direction. Because this type of flow, if sustained, would disrupt the plasma, ohmic heating can be applied only in brief pulses.

In tokamaks of a more advanced design, a different method is sometimes used to heat the plasma; a beam of high-energy, electrically neutral particles is injected into the plasma. Through random collisions, the injected, high-speed particles transfer their energy to the slower-moving plasma particles, thereby heating the plasma.

In practice, a tokamak might typically be operated for 1 second at a time, at 10-minute intervals. As researchers develop reactor materials better able to withstand the damaging heat and high-energy particles emitted in fusion reactions, longer reaction periods at closer intervals will become possible, increasing the total amount of energy that the tokamak can produce.

The largest tokamak in the U.S. is the Tokamak Fusion Test Reactor (TFTR) at Princeton University. TFTR was originally expected to reach plasma temperature, density, and containment-time levels that satisfy the Lawson criterion — that is, reach breakeven — by 1986, but cutbacks in the federal fusion budget have delayed the expected date to the early 1990s. TFTR was built at a cost of $314 million and began operating in December 1982. Its reactor vessel is a 7-foot-diameter tube wrapped around to form a doughnut-shape with a 24-foot outside diameter. When its surrounding magnets and support facilities are considered, the tokamak is as big as a medium-sized house.

In 1988, the TFTR reached a plasma temperature of more than 300 million degrees Celsius, surpassing the 140-million-degree record established by the smaller Princeton Large Torus Tokamak in 1978. To produce these high temperatures, TFTR's fusion reactions involve deuterium and tritium nuclei rather than two deuterium nuclei. Because the neutrons released in deuterium-tritium fusion rapidly make the reactor vessel walls radioactive, the tests to achieve breakeven will be restricted to six trials, each of 1-second duration.

Depending on funding, Princeton also hopes to achieve breakeven by 1996 with another machine, its proposed Compact Ignition Tokamak (CIT). CIT's goal is to produce and sustain a fusion reaction, that is, ignition, a major milestone in fusion development. The CIT design is being developed in a national collaborative effort involving major U.S. fusion laboratories — including Princeton, MIT, Oak Ridge, General Atomics, and Lawrence Livermore. Experience already gained with the Alcator C tokamak at MIT is helping in the design of CIT.

Another tokamak as large as TFTR is the Joint European Torus (JET), located at the Culham Laboratory near Oxford, U.K., and funded jointly by the nations of the European Economic Community. The reactor cost $500 million to build and started operating in June 1983. JET has a high-frequency heating system to permit the continuous transfer of energy into the plasma, in contrast to an ohmic heating system, which permits heating only in brief pulses. High-frequency heating, in which a powerful beam of electromagnetic waves is used to move the plasma particles back and forth at high speed, can be applied continuously. The electromagnetic beam heats the plasma by speeding up the random motion of the plasma particles, but does not displace particles over large distances and so does not disrupt the plasma.

Using deuterium-tritium reactions, researchers at JET hope to achieve breakeven in the next few years. Once tritium is used at JET, the reactor will have to be switched over to remote-control operation in order to protect the workers from radiation exposure from the radioactive isotope.

A third large tokamak, the JT-60 in Tokai-mura, Japan, produced its first plasma in 1985. An advanced neutral-beam injector has been installed at JT-60 to continuously replenish particles lost from the plasma without disrupting it. A high-frequency heating system has also been installed to introduce additional energy into the plasma. The Japanese researchers hope to achieve breakeven with JT-60 soon, even though the reactor uses the deuterium-deuterium reaction, rather than the higher-energy deuterium-tritium reaction. The U.S.S.R. has also built a large tokamak in Moscow. T-15, as the reactor is called, has a high-frequency heating system similar to that in the JET and uses superconducting magnets to contain the plasma.

Although smaller than the reactors discussed above, the Alcator C tokamak at MIT earned a place in history during 1983 by being the first reactor to exceed the minimum Lawson parameter for breakeven, with a mark of 6×10^{13} nuclei-seconds per cubic centimeter; however, the plasma temperature of 17 million Celsius was far short of the 100 million degrees required for breakeven. Alcator C was not designed to reach breakeven; instead, it was developed to test several reactor design elements on a small scale for eventual use in larger machines.

Mirror machines. While tokamaks have been the most successful magnetic confinement reactors to date, several other reactor designs have also been developed. One such design is the so-called mirror machine, in which a hot plasma is contained in a long, straight reactor chamber by powerful deflecting magnets at the chamber's ends. These magnets, which have wire coils in the shape of a baseball's seams, are called mirror magnets because they reflect charged particles that approach them. Some mirror reactors, including the Tandem Mirror Experiment (TMX) device at Lawrence Livermore Laboratory, use pairs of mirror magnets at each end.

The straight-line geometry of the mirror design allows the reaction chamber to be enlarged inexpensively, simply by lengthening the chamber. Replacing reactor walls damaged by neutrons from deuterium-tritium reactions is also relatively simple and inexpensive in a mirror machine. Neither mirror machines nor any of the other competing devices with unusual geometries have, however, performed as well as the older tokamak design.

The largest mirror machine is the Mirror Fusion Test Facility (MFTF), also at Livermore Laboratory. A $372-million project to transform this device into a larger and more powerful reactor with superconducting magnets — MFTF-B — was completed in the 1986. Although the MFTF-B performed according to specifications during start-up tests, it has not been used since, because its operating budget was cut amid fears that it would set back work on the more promising tokamaks by absorbing too large a portion of the U.S. fusion research budget.

International Thermonuclear Experimental Reactor

In the 1950s, the major fusion research projects in the U.S. and U.S.S.R. were largely kept secret for military reasons. Since then, however, much fusion research has been openly published and shared. INTER, the International Thermonuclear Experimental Reactor, is another step in international cooperation in fusion power research. This multinational project involving the U.S., Japan, the European Community, and the U.S.S.R., is sponsored by the International Atomic Energy Agency in Vienna, Austria, and was established to develop detailed designs for a magnetic confinement reactor. A principal goal of the INTER project is the development of a reactor capable of operating under conditions similar to those envisioned for a future commercial fusion plant. The target completion date for the reactor is the turn of the century.

Inertial Confinement Reactors

Inertial confinement devices are very different from magnetic confinement reactors. Instead of powerful containment magnets and heating coils, they use high-power lasers or particle beams to bombard tiny pellets of fuel in an attempt to initiate fusion reactions. Laser-driven devices are the most common type of inertial confinement reactor. When sufficiently intense beams of either light or high-energy particles strike a deuterium-containing fuel pellet 0.01 inch in diameter, the outer shell of the pellet vaporizes, producing a powerful inward-directed shock wave. Under suitable conditions, this shock wave can crush the remainder of the pellet inward — imploding it with tremendous force and simultaneously heating it.

In the brief instant of maximum compression, which may last as little as a trillionth of a second, electrons are stripped from the atoms in the pellet, creating a plasma of charged particles of enormous density and temperature. If the implosion is powerful enough, it may initiate a thermonuclear fusion reaction. Since no magnetic fields or other forces act continuously to contain the plasma, however, it almost immediately begins to rush outward, terminating the fusion reaction. In principle, however, a usable amount of energy can be released in the time before the reaction ends. Moments later, another pellet can be injected into the reaction chamber to continue the cycle.

While magnetic confinement uses large, elaborate magnets to contain a hot plasma, inertial confinement uses simple inertia to keep plasma particles inside an imploding pellet for a brief instant, but still long enough to allow a fusion reaction to occur. In fact, hydrogen bombs are based on roughly the same principle as inertial confinement reactors; however, a thermonuclear weapon uses a fission bomb initiator rather than laser beams to compress its fusion fuel.

The plasma conditions in inertial confinement reactors are much different from those inside a tokamak, where, in the future, reactions may proceed for seconds or minutes at a time, with typical plasma densities of 10^{14} particles per cubic centimeter. Although the fusion reactions inside an imploded pellet may last just 10^{-12} second, the vastly higher plasma density of 10^{26} particles per cubic centimeter would still allow a significant energy yield. The two approaches are simply far different ways of achieving comparable Lawson parameters. Inertial confinement tries to reach the Lawson criterion by achieving enormous densities for short intervals, while magnetic confinement attempts to contain a much less dense plasma for a much longer time.

To extract useful energy from an inertial confinement device, a layer of liquid lithium metal could be placed around the spherical reaction chamber. High-energy neutrons produced by the fusion reaction would be absorbed in this layer and heat it; this energy could then be used to generate electricity by conventional means — for example, by producing steam to drive a turbine.

Laser fusion devices. Laser fusion research is under way in several countries, including France, Japan, the U.K., and the U.S. Because some weapons experts believe that the technology of ultrahigh-power laser and particle beams will have important military applications, much research on inertial confinement fusion has been classified, and information about it is not generally available to the public.

The main goal of laser fusion research is, however, well known: to focus the maximum amount of power as symmetrically as possible onto a carefully designed fuel pellet. An asymmetrical implosion would simply propel a pellet to the side of its reaction chamber, rather than crush it. To produce a sufficiently symmetrical shock wave around a pellet, a laser fusion device typically focuses 10 or more laser beams at regularly spaced angles onto the pellet. For example, a device called OMEGA at the University of Rochester in New York uses 24 laser beams to achieve an energy intensity that varies by no more than 10 percent over the surface of a pellet.

Early in the 1980s, the world's most powerful laser fusion reactor was the Gekko XII reactor, with its 12-

beam laser, at Osaka University in Japan. The Gekko XII laser has been able to produce 50-trillion-watt pulses of infrared light, each lasting one ten-billionth of a second. Gekko XII has been used to develop a slightly different implosion technique called "indirect drive," in which laser light compresses a fuel pellet by setting off a contained shock wave. The fuel pellet is enclosed in an outer shell, and laser light entering the shell through two tiny holes vaporizes the outer layers of the pellet, creating a hot plasma within the shell. Pressure from this plasma crushes the pellet inward. Experiments at Lawrence Livermore Laboratory have used this technique to compress pellets to less than 1 percent of their original size.

In mid-1985, the Nova laser fusion reactor at Lawrence Livermore became the world's most powerful laser fusion reactor. Nova, which cost $200 million to build and began operating in April 1985, uses a single laser light source to create 10 separate beams. These beams concentrate intense bursts of energy on the tiny fuel pellet at the center of Nova's reaction chamber.

To power the laser, massive energy-storing devices accumulate energy from conventional electrical generators and release it in bursts of electrical energy roughly equivalent to the kinetic energy of a 1-pound ball moving at 1,000 miles per hour. This energy is used to produce powerful pulses of laser light, each of which lasts no longer than one ten-billionth of a second. During each pulse, as much as 100 trillion watts of power are focused on the pellet — 50 times greater than the total output of all the electrical power plants on Earth combined. As of early 1989, the huge Nova device was nearing breakeven. The Livermore Laboratory is now proposing to build an even more powerful laser fusion device called Athena, which will cost about $700 million.

The U.S. DOE, encouraged by recent progress in laser confinement fusion, is proposing to build a laboratory microfusion facility (LMF) capable of surpassing breakeven by the late 1990s. The classified Halite/Centurion program — a reportedly promising collaboration of weapons designers and inertial confinement researchers — has provided additional impetus to the inertial confinement effort.

Particle-beam fusion devices. Pulsed beams of high-energy subatomic particles can also be focused on a fuel pellet to produce an implosive shock wave. Because subatomic particles can penetrate into the fuel pellets and interact directly with atomic nuclei there, particle beams can transfer energy to their targets more efficiently than laser beams. Particle beams can also produce more bursts per minute, but they do have some disadvantages as well. They are more difficult to focus than light beams and cannot be tested in small-scale experiments, because such tests do not provide a reliable indication of how full-scale particle-beam devices would perform. Despite these drawbacks, particle-beam devices have emerged as a promising alternative to laser-fusion devices.

The particle beams used in fusion experiments are usually ion beams, which are intense streams of protons or heavier atomic nuclei accelerated to high energies by specialized accelerators. Japan, the U.K., the U.S., and the U.S.S.R. have all conducted particle-beam fusion experiments. Among the most advanced of these projects is the Particle Beam Fusion Accelerator (PBFA) at Sandia National Laboratories.

In PBFA, protons are accelerated by pulses of electrical energy released from a massive bank of storage devices called capacitors. Each pulse of electricity is fed into a series of electrodes spaced along an accelerator tube. The electrical charge on the electrodes accelerates protons injected into the tube until they reach high speeds. PBFA can deliver 1.5 trillion watts per square centimeter to its target in bursts of protons each lasting about ten billionths of a second.

A larger accelerator at Sandia called PBFA II accelerates lithium ions instead of protons, and these bombard fuel pellets of deuterium and tritium. PBFA II contains 36 particle accelerators, which are arranged like the spokes of a wheel and which focus their beams symmetrically on a fuel pellet one-eighth of an inch in diameter at the center of the device. PBFA II may deliver more than 100 trillion watts of power per square centimeter in pulses lasting about 50 billionths of a second. Sandia researchers believe that PBFA II may eventually achieve breakeven.

Prospects for Fusion

Forecasts for fusion power vary widely, from predictions that a successful fusion power industry will develop within a few decades to opinions that fusion power is unworkable. Some critics of fusion power, including MIT Professor of Nuclear Engineering Lawrence Lidsky, have suggested that even if the quest for fusion power does succeed, it may not be worth the great effort and expense involved. Lidsky has outlined a number of major technical obstacles that stand in the way of commercial fusion power, including radiation hazards and reactor material failures. Lidsky believes that even if all the technical problems could be solved, a fusion reactor would typically be able to produce only about 1,000 megawatts of electrical power, about the same output as a fission reactor of comparable physical size.

Although fusion reactors would have some advantages, such as readily available, cheap fuel and less dangerous waste products, those benefits may not justify the expense of developing a fusion power industry, according to Lidsky. In addition, he has noted that the rush to achieve power-producing reactors has led most fusion projects to adopt the deuterium-tritium reaction because it releases a great deal of energy. The high-energy neutrons produced in this reaction, however, will corrode the reactor components and make them radioactive. Thus, deuterium-tritium reactors would have the problem of nuclear waste that has troubled fission reactors. Physicist Herman Bondi, former chief scientist at the U.K. Department of Energy, has said: "To describe fusion as a clean way of producing energy is irresponsible."

On the other side, Harold Furth, director of fusion research at Princeton University, is one of many fusion advocates who maintain that the current problems of fusion power can be solved. Furth believes that the problem of neutron-induced reactor damage in magnetic-confinement reactors can be solved either by changing the arrangement of the magnets or by using more radiation-resistant construction materials. Furth has claimed that if the present rate of progress in fusion power research can be sustained, then controlled fusion should be able to supply a significant fraction of the world's energy needs by 2025.

Room-Temperature Fusion?

Two chemists shocked the scientific world in March 1989 by announcing that they had developed a simple method to produce sustained fusion in the laboratory. Unlike conventional fusion techniques, which require extremely high temperatures and expensive equipment, the new method reportedly required only a relatively simple electrolytic cell. The chemists who made the surprise announcement, Martin Fleischmann of the University of Southampton in the U.K. and B. Stanley Pons of the University of Utah, claimed that they had produced fusion at ordinary room temperature and that their method generated substantially more energy than it consumed.

In the technique, a vessel is filled with heavy water, a compound in which the hydrogen isotope deuterium replaces the ordinary hydrogen atoms in water. A positive and negative electrode are placed into the heavy water to make an electrolytic cell, and a current is run between the electrodes. As the current passes through the cell, it breaks the heavy water molecules into deuterium nuclei and oxygen ions. The positively charged deuterium nuclei are drawn to the negative electrode, which is a rod made of palladium, a soft, silvery metal often used as a catalyst. The palladium atoms in the rod are arranged in a tight network, or lattice. Deuterium nuclei are drawn into the rod's palladium lattice, where they are held close to other deuterium nuclei. Ordinarily, two deuterium nuclei tend to repel one another, because they are each positively charged.

When enough deuterium nuclei are drawn into the palladium rod, a phenomenon called quantum mechanical tunneling reportedly occurs, in which some of the deuterium nuclei overcome their mutual repulsion and fuse to create helium, tritium, and neutrons. According to Fleischmann and Pons, the reactions release four times as much energy (in the form of heat) as they consume. "This generation of heat continues over long periods, and is so large that it can only be attributed to a nuclear process," Fleischmann said. The water reportedly also emits gamma rays, evidence, they said, that neutrons produced by the reaction are being captured in a surrounding water container.

In spite of this explanation, most fusion experts remained highly skeptical about Fleischmann and Pons's claims, attributing their results to experimental error. Even researchers who believed that such a fusion reaction might be possible were sharply critical of the way in which Fleischmann and Pons reported their achievement. Instead of first publishing a paper summarizing their research in a scientific journal, they announced their findings at a press conference, a highly unorthodox procedure. At the time of the conference, Fleischmann and Pons also provided few details about their research.

As more information about the experiment became available, hundreds of research teams around the world hastily attempted to reproduce Fleischmann and Pons's findings. Within weeks, a few groups of scientists in California, Texas, Italy, and the U.S.S.R. reported ambiguous evidence that appeared to be consistent with cold fusion. The vast majority of the research teams, however, had found no evidence of cold fusion after several months of study. Several skeptical fusion theorists criticized the design of the experiments that seemed to support cold fusion, and they proposed alternative explanations for the results obtained.

The disappointing outcomes of most early studies led many teams to reduce or abandon their cold fusion research by the summer of 1989. In July, a U.S. government scientific panel said that it doubted whether the findings reported by Fleischmann, Pons, and others were actually signs of "a new nuclear process," and noted that, "The experiments reported to date do not present convincing evidence that useful sources of energy will result from the phenomena attributed to cold fusion." The panel concluded that the U.S. government should not allocate money to establish a new cold fusion research program. Nevertheless, it said that there were still enough unanswered questions about the phenomena to justify modest funding of cold fusion experiments.

Index

A

B

C

D

E

F

G

H

I

J

K

L

M

N

O

P

Q

R

S

T

U

V

W

X

Y

Z